MAGNETIC NANOPARTICLES
From Fabrication to Clinical Applications

MAGNETIC NANOPARTICLES
From Fabrication to Clinical Applications

THEORY TO THERAPY
CHEMISTRY TO CLINIC
BENCH TO BEDSIDE

EDITED BY **Nguyễn T. K. Thanh**

CRC Press
Taylor & Francis Group
Boca Raton   London   New York

CRC Press is an imprint of the
Taylor & Francis Group, an **informa** business

CRC Press
Taylor & Francis Group
6000 Broken Sound Parkway NW, Suite 300
Boca Raton, FL 33487-2742

© 2012 by Taylor & Francis Group, LLC
CRC Press is an imprint of Taylor & Francis Group, an Informa business

No claim to original U.S. Government works

International Standard Book Number: 978-1-4398-6932-1 (Hardback)
International Standard Book Number: 978-0-3675-0792-3 (Paperback)

Library of Congress Cataloging-in-Publication Data

Magnetic nanoparticles : from fabrication to clinical applications / editor, Nguyen Thi Kim Thanh.
    p. cm.
  "A CRC title."
  Includes bibliographical references and index.
  ISBN 978-1-4398-6932-1 (hardcover : alk. paper)
  1. Nanostructures--Magnetic properties. 2. Nanoparticles--Magnetic properties. 3. Medical instruments and apparatus. I. Thanh, Nguyen Thi Kim.

QC176.8.N35M3385 2012
610.28'4--dc23   2011043033

Visit the Taylor & Francis Web site at
http://www.taylorandfrancis.com

and the CRC Press Web site at
http://www.crcpress.com

This book is dedicated to my wonderful children,
Lan Thanh Nguyen Le and Nam Duc Nguyen Le;
and young researchers all over the world, who will harness
the potential of magnetic nanoparticles for mankind's better health.

Contents

PART I   Introduction

PART II   Fabrication and Characterisation of MNPs

PART III   Biofunctionalisation of MNPs for Biomedical Application

PART IV   Ex Vivo *Application of MNPs*

PART V   In Vivo *Application of MNPs*

Foreword

The field of nanotechnology is advancing at a rapid pace, so that yesterday's breakthroughs could be outdated today. It is very important for scientists to communicate effectively to avoid duplication of research. Translational research on tailored magnetic nanoparticles for biomedical applications spans a variety of disciplines, which makes the task of putting together the most significant advances into a practical format a truly challenging task.

This book, written by some of the most qualified experts in the field, not only fills a hole in the literature, but also bridges the gaps between all the different areas in this field successfully.

It provides a comprehensive review from diagnostics to clinical applications. It not only covers introductory fundamental sciences, basic research and the derived clinical methods, but also describes current developments. More importantly, it also looks ahead at the challenges and opportunities that await. This book is a toolbox of solutions and ideas for scientists in the field and for young researchers interested in magnetic nanoparticles.

The editor is a recognised and well-respected scientist in the field of nanoscience and has an excellent track record of publications in high-impact journals. She has chaired and organised conferences in the field and has delivered invited talks at numerous international meetings. Furthermore, in 2010, she edited a theme issue "Nanoparticles" for the prestigious journal *Philosophical Transactions of the Royal Society A*, dealing with the synthesis, characterisation and application of advanced nanoparticles. Her research is on fabrication and biofunctionalisation of the next generation of nanomaterials. This has allowed her to have a full understanding of the research topics covered in this book, which has impressive contributions from leading international scientists, engineers and clinicians in the field. This book will soon become a standard reference for researchers at different levels from graduate and postgraduate students to active researchers in chemistry, biochemistry, material science, biology, physics, engineering and the clinical fields as well as in the interdisciplinary areas of nanotechnology. Additionally, it will be in high demand for specialists working on innovations in medical diagnostics and therapeutics in the field of cancer and other diseases.

This book should also be of use to college teachers, funding agencies, policy makers, regulators and investors who are interested in nanotechnology.

Professor Mostafa A. El-Sayed
Julius Brown Chair and Regents Professor
Director, Laser Dynamics Lab
School of Chemistry and Biochemistry
Georgia Institute of Technology
Atlanta, Georgia

Preface

This book provides comprehensive and up-to-date information on the ever-expanding research area of nanoscale science and technology with a focus on magnetic nanoparticles (MNPs) and their applications in health care.

It covers the fundamental aspects of MNPs, their structure and magnetic properties relationships and techniques that are commonly used for their characterisation. It provides a detailed description of the fabrication of different types of MNPs, including next-generation MNPs using chemical and biological approaches. It also discusses the strategy for functionalisation of MNPs for specific biological targets and provides examples of functionalised MNPs with molecules designed for biomedical applications.

The book covers many *ex vivo* diagnostic and therapeutic applications such as immunoassays and removal of blood-borne toxin in the body. The development of biochemical assays and devices, lab-on-a-chip, new magnetic separation techniques for cell sorting, nanomagnetic gene transfection and targeting are also covered.

In vivo applications featured include manipulating magnetically labelled single cells for cell therapies and noninvasive delivery of drug t-PA for arterial thrombolysis. One of the most genial ways to deliver water-insoluble substances for tumour cells *in vivo* is the use of blood cells as drug carriers. Besides using biological and chemical approaches, engineers have developed magnet systems to guide nanoparticles towards targeted locations under the action of external magnetic fields. Magnetic fluid hyperthermia of novel nanocrystalline oxides for cancer therapy, magnetoliposomes for site-specific killing of malignant cells and magnetic microbubbles used for imaging, moving and delivering therapeutic agents are also included.

The most current development on *in vivo* imaging, magnetic particles imaging (MPI), is featured, and its applications in renal angiographic monitoring, cancer, inflammation imaging and stem cell tracking are described. Surgical magnetic systems and tracers for cancer staging, sentinel lymph node biopsy utilising *in vivo* MNPs and clinical results for breast cancer are highlighted. Finally, safety considerations and pharmacokinetics of MNPs are discussed in detail.

The chapters have been written by leading experts in a broad range of disciplines (physics, chemistry, biochemistry, biology, medicine, toxicology, engineering and entrepreneurship), who not only present the most cutting-edge research for active researchers in the field, but also provide fundamental knowledge in each field to encourage students and other young researchers to harness the true potential of MNPs for better health care.

Nguyễn Thị Kim Thanh
Department of Physics and Astronomy
University College London
and
The Davy Faraday Research Laboratory
The Royal Institution of Great Britain
London, United Kingdom
Email: ntk.thanh@ucl.ac.uk

Acknowledgements

I would first like to acknowledge my maternal grandfather, who was a university graduate engineer, for nurturing a scholarly tradition in science and engineering in the family and inspiring my mother to become a mechanical engineer. My father, who received his PhD in chemistry in Europe, also had a significant influence on my career track. I would also like to acknowledge Sofia Kovalevskaya (1850–1891), who has inspired me since my childhood. I still admire the passion and devotion she had for her work. Additionally, my husband's expertise in magnetism has had a profound effect on my recent research direction. I was working with noble metal nanoparticles in 2001 when he inadvertently got me interested in magnetic nanoparticles (MNPs). I have since become fascinated with MNPs as, unlike any other nanoparticle systems, they are complex. The chemical composition, crystal structure, size, shape, ligand attachment to nanoparticles' surface, etc., all influence their magnetic properties, expanding the possibilities for designing bespoke systems. In addition, the way in which external magnetic fields interact with MNPs can provide a long list of applications, as outlined in this book.

From my research on MNPs, it is critical that one appreciate other disciplines to make a real impact in clinical applications. My interest in disseminating and invigorating the research in this very exciting field prompted this project. I would like to thank Lance Wobus, Kathryn Younce, Jonathan Pennell, Randy Burling, and their colleagues at CRC Press/Taylor & Francis, and Suganthi Thirunavukarasu, project manager at SPi Global, for working closely with me to publish this timely book.

I would also like to thank Professor Donald Bethell for passing on his lifetime editorial experience and guiding me through at a very early stage of editing this book. Thanks also to Professor P. Andrew Evans for his advice on having detailed outlines of each chapter before writing commenced, which proved extremely valuable. I am indebted to his support through some difficult times during this project.

My sincere gratitude to all authors, who are leading experts in the field and herein have covered cutting-edge research while at the same time provided much-needed fundamental background for nonspecialist readers.

Thanks to Drs. Axel Rosengart and Michael Kaminski and their colleagues for writing not one but two interesting chapters on the clinical application of MNPs.

Thanks also to Dr. Daniel Ortega for some early discussions and his support throughout the project. I would also like to acknowledge Bettina Kozissnik, who was of immense assistance during an extremely busy period of editing. I am deeply indebted to reviewers for their time and careful reading of the chapters and for providing insightful comments in a timely manner. My thanks also to George Frodsham, Professor Geoff Ozin, Professor Hedi Mattoussi, Dr. Hugh Seton, Professor Ian K. Robinson, Dr. Ian Prior, Dr. Ian Robinson, Dr. Kris Page, Dr. Lanry Yung, Dr. Laura Parkes, Professor Mary Gulumian, Dr. Paul Southern, Professor Peter Dobson, Professor Salah-Eldin Abdelmoneim, Dr. Valadimir Kolesnichenko and Dr. Yuri Barnakov for useful discussions.

Finally, I would like to acknowledge my wonderful colleagues and friends from around the world for their kindness and their wholehearted support throughout this project.

Editor

Dr. Nguyễn Thị Kim Thanh, FRSC, CChem, CSci, MRI, is a UCL-RI Reader (Associate Professor) in nanotechnology and Royal Society University Research Fellow at the Department of Physics and Astronomy, University College London, and The Davy-Faraday Research Laboratory, The Royal Institution of Great Britain, United Kingdom.

In 1992, she graduated and received the award for top academic achievement in chemistry from Vietnam National University in Hanoi. She was then selected to study at the University of Amsterdam under a NUFFIC (the Netherlands organisation for international cooperation in higher education) programme, from where she embarked on a career in research and obtained her MSc in chemistry. Two years later, in 1994, she moved to London to undertake an EU-funded PhD in biochemistry. She then pursued postdoctoral work in medicinal chemistry at Aston University, Birmingham, United Kingdom, in 1999.

In 2001, she moved to the United States to take advantage of pioneering work in nanotechnology in the Department of Chemistry and Advanced Material Research Institute at the University of New Orleans. Two and a half years later, in 2003, she joined the Liverpool Centre for Nanoscale Science, United Kingdom, and it was not before long that she was awarded a prestigious Royal Society University Research Fellowship (2005–2014) and University of Liverpool Lectureship. She was based at the Department of Chemistry, which was ranked 7th in the United Kingdom in the 2008 research assessment exercise (RAE), and School of Biological Sciences.

In January 2009, she was appointed a UCL-RI Readership (Associate Professor) in nanotechnology based at the Department of Physics and Astronomy, University College London, and The Davy Faraday Research Laboratory, The Royal Institution of Great Britain, the oldest independent scientific research body in the world. There she leads a very dynamic research team focused on the design, synthesis and study of the physical properties of nanomaterials as well as their applications in biomedicine.

Dr. Thanh has been an invited speaker at over 60 institutes and scientific meetings. Furthermore, she has been a guest editor of *Philosophical Transactions A of the Royal Society* on the theme issue of Nanoparticles published in September 2010. She overcame fierce competition to be the lead exhibitor for Royal Society Science Summer Exhibition on a 'Nanoscale science: A giant leap for mankind' in London in July 2010 to celebrate the 350th anniversary of the Royal Society. *New Scientist* ranked the exhibition as one of the best.

Currently, she is a member of editorial board of *Advances in Natural Sciences: Nanoscience and Nanotechnology* as well as a committee member of the Royal Society of Chemistry Colloid & Interface Science Group and the Society of Chemical Industry Colloid & Surface Chemistry Group. She has organised many conferences and has been given the responsibility to organise the 'Functional Nanoparticles for Biomedical Applications' Symposium at 244th ACS meeting in Philadelphia, Pennsylvania, in 2012. She has also been selected to serve as the chair of the scientific committee of a future prestigious RSC Faraday discussion on the same theme in 2014.

Foreword Author

Professor Mostafa A. El-Sayed and his group have contributed to many physical and materials chemistry research areas. They have developed new techniques such as magnetophotoselection, picosecond Raman spectroscopy and phosphorescence microwave double resonance spectroscopy, using which they were able to answer fundamental questions regarding ultrafast dynamical processes involving molecules, solids and photobiological systems. Since they moved to Georgia Institute of Technology, El-Sayed and his group have focused on the study of the physical, chemical and photothermal processes of metallic and semiconductor nanostructures of different shapes. They have published over 560 peer-reviewed papers that have accumulated over 24,000 citations. The group has several patents on nanocatalysis and the use of gold nanoparticles in cancer diagnosis and selective phototherapy. Professor El-Sayed has given over 45 special named lectures and over 300 invited or plenary talks at national and international meetings. He has served on numerous international and national committees such as the advisory boards of the U.S. National Science Foundation, the Japanese Molecular Science Institute in Okazaki and Basic Energy Sciences of the U.S. Department of Energy. Please visit http://www.chemistry.gatech.edu/faculty/El-Sayed/ for more information.

Contributors

Mahmoud H. Abdel Kader
Photochemistry Department
German University in New Cairo
New Cairo, Egypt

Kristin Andreas
Berlin-Brandenburg Center for Regenerative
    Therapies
Charité-Universitätsmedizin Berlin
Berlin, Germany

Dhirendra Bahadur
Department of Metallurgical Engineering
    and Materials Sciences
Indian Institute of Technology Bombay
Mumbai, India

Rinti Banerjee
Department of Biosciences and Bioengineering
Indian Institute of Technology Bombay
Mumbai, India

Hans Bäumler
Center for Tumor Medicine
Institute of Transfusion Medicine
and
Berlin-Brandenburg Center for Regenerative
    Therapies
Charité-Universitätsmedizin Berlin
Berlin, Germany

Sylvie Begin-Colin
Unité Mixte de Recherche
National Centre for Scientific Research
European Engineering School in Chemistry,
    Polymers and Materials
Institute of Physics and Chemistry of Materials
University of Strasbourg
and
Ecole de Chimie Polymères et Matériaux
Strasbourg, France

Haitao Chen
Amgen, Inc.
Thousand Oaks, California

Kerry A. Chester
UCL Cancer Institute
University College London
London, United Kingdom

Steven M. Conolly
Department of Bioengineering
and
Department of Electrical Engineering and
    Computer Sciences
University of California, Berkeley
Berkeley, California

Marie-Hélène Delville
Centre National de la Recherche Scientifique
Institut de Chimie de la matière condensée de
    Bordeaux
University of Bordeaux
Pessac, France

Didier A. Depireux
Institute for Systems Research
University of Maryland
College Park, Maryland

Jon Dobson
Department of Biomedical Engineering
and
Department of Materials Science
    and Engineering
University of Florida
Gainesville, Florida

and

Institute for Science and Technology
    in Medicine
Keele University
Keele, United Kingdom

Kenneth J. Dormer
Department of Physiology
University of Oklahoma Health Sciences
    Center
Oklahoma City, Oklahoma

Michael Douek
Department of Reasearch Oncology
King's College London

and

Guy's & St. Thomas Hospitals
London, United Kingdom

Trevor Douglas
Department of Chemistry and Biochemistry
and
Center for Bio-Inspired Nanomaterials
Montana State University
Bozeman, Montana

Etienne Duguet
Centre National de la Recherche Scientifique
Institut de Chimie de la matière condensée de
    Bordeaux
University of Bordeaux
Pessac, France

Robert Eckersley
Department of Imaging Sciences
Imperial College London
London, United Kingdom

Mostafa A. El-Sayed
Laser Dynamics Laboratory
Georgia Institute of Technology
School of Chemistry and Biochemistry
Atlanta, Georgia

Delphine Felder-Flesch
Unité Mixte de Recherche
National Centre for Scientific Research
Institute of Physics and Chemistry of Materials
University of Strasbourg
Strasbourg, France

Angeliki Fouriki
Institute of Science and Technology
    in Medicine
Keele University
Keele, United Kingdom

Florence Gazeau
Unité Mixte de Recherche
Laboratoire Matière et Systèmes Complexes
National Centre for Scientific Research
and
Université Paris Diderot
Paris, France

Radostina Georgieva
Center for Tumor Medicine
Institute of Transfusion Medicine
Charité-Universitätsmedizin Berlin
Berlin, Germany

and

Department of Medical Physics, Biophysics
    and Radiology
Trakia University
Stara Zagora, Bulgaria

Manashjit Gogoi
Department of Biosciences and Bioengineering
Indian Institute of Technology Bombay
Mumbai, India

Patrick Goodwill
UC Berkeley Bioengineering
Berkeley, California

Luke A.W. Green
Department of Physics and Astronomy
University College London

and

The Davy Faraday Research Laboratory
The Royal Institution of Great Britain
London, United Kingdom

Peter Hawkins
Faculty of Health and Life Sciences
Institute of Bio-Sensing Technology
University of the West of England
Bristol, United Kingdom

Manish K. Jaiswal
Department of Metallurgical Engineering
    and Materials Sciences
Indian Institute of Technology, Bombay
Mumbai, India

Michael D. Kaminski
Chemical Sciences and Engineering Division
Argonne National Laboratory
Argonne, Illinois

Arash Komaee
Aerospace Engineering
University of Maryland
College Park, Maryland

Bettina Kozissnik
UCL Cancer Institute
University College London
London, United Kingdom

and

The Davy Faraday Research Laboratory
The Royal Institution of Great Britain
Great Britain, United Kingdom

Kannan M. Krishnan
Department of Materials Science
    and Engineering
University of Washington
Seattle, Washington

Roger Lee
Institute for Systems Research
University of Maryland
College Park, Maryland

Lars O. Liepold
Department of Chemistry and Biochemistry
and
Center for Bio-Inspired Nanomaterials
Montana State University
Bozeman, Montana

Xianqiao Liu
DuPont
Shanghai, China

Richard Luxton
Faculty of Health and Life Sciences
Institute of Bio-Sensing Technology
University of the West of England
Bristol, United Kingdom

Shinya Maenosono
Japan Advanced Institute of Science
    and Technology
Ishikawa, Japan

Eric Mayes
Endomagnetics Ltd
London, United Kingdom

Stéphane Mornet
Centre National de la Recherche Scientifique
Institut de Chimie de la matière condensée de
    Bordeaux
University of Bordeaux
Pessac, France

Helen Mulvana
Department of Imaging Sciences
Imperial College London
London, United Kingdom

Aleksandar Nacev
Fischell Department of Bioengineering
University of Maryland
College Park, Maryland

Daniel Ortega
Department of Physics and Astronomy
University College London

and

The Davy-Faraday Research Laboratory
The Royal Institution of Great Britain
London, United Kingdom

Nicole Pamme
Department of Chemistry
University of Hull
Hull, United Kingdom

Quentin Pankhurst
Department of Physics and Astronomy
University College London

and

The Davy-Faraday Research Laboratory
The Royal Institution of Great Britain
London, United Kingdom

Emil Pollert
Institute of Physics
Academy of Sciences of the Czech Republic
Praha, Czech Republic

Roland Probst
Fischell Department of Bioengineering
University of Maryland
College Park, Maryland

Dietmar Rempfer
Mechanical Materials and Aerospace
    Engineering Department
Illinois Institute of Technology
Chicago, Illinois

Axel J. Rosengart
Neuroscience Intensive Care Unit
Departments of Neurology, Neurosciences
    and Neurosurgery
New York Presbyterian Hospitals
Weill Cornell Medical College
New York, New York

Isaac Rutel
Department of Radiological Sciences
University of Oklahoma Health Sciences
    Center
Oklahoma City, Oklahoma

Ivo Safarik
Department of Nanobiotechnology
Institute of Nanobiology and Structural Biology
    of GCRC
Academy of Sciences
Ceske Budejovice, Czech Republic

and

Regional Centre of Advanced Technologies
    and Materials
Palacky University
Olomouc, Czech Republic

Mirka Safarikova
Department of Nanobiotechnology
Institute of Nanobiology and Structural Biology
    of GCRC
Academy of Sciences
Ceske Budejovice, Czech Republic

Taher A. Salah
Nanotechnology Centre
Regional Center for Food and Feed
Agricultural Research Centre
Cairo, Egypt

Hazem M. Saleh
Otolaryngology Unit
National Institute of Laser Enhanced Sciences
Cairo University
Giza, Egypt

Azeem Sarwar
Fischell Department of Bioengineering
University of Maryland
College Park, Maryland

Benjamin Shapiro
Fischell Department of Bioengineering
and
Institute for Systems Research
University of Maryland
College Park, Maryland

Patricia Caviness Stepp
Department of Orthopardic Surgery
University of Pittsburgh
Pittsburgh, Pennsylvania

Nadine Sternberg
Center for Tumor Medicine
Institute of Transfusion Medicine
Charité-Universitätsmedizin Berlin
Berlin, Germany

Eleanor Stride
Department of Mechanical Engineering
University College London
London, United Kingdom

Meng-Xing Tang
Department of Bioengineering
Imperial College London
London, United Kingdom

Nguyễn Thị Kim Thanh
Department of Physics and Astronomy
University College London

and

The Davy Faraday Research Laboratory
The Royal Institution of Great Britain
London, United Kingdom

Trinh Thang Thuy
School of Materials Science
Japan Advanced Institute of Science
    and Technology
Ishikawa, Japan

Masaki Uchida
Department of Chemistry and Biochemistry
and
Center for BioInspired Nanomaterials
Montana State University
Bozeman, Montana

Claire Wilhelm
Unité Mixte de Recherche
National Centre for Scientific Research
Laboratoire Matière et Systèmes Complexes
and
Université Paris Diderot
Paris, France

P. Stephen Williams
Department of Biomedical Engineering
Lerner Research Institute
Cleveland Clinic
Cleveland, Ohio

Yumei Xie
Pacific Northwest National Laboratory
Richland, Washington

Karel Závěta
Institute of Physics
Academy of Sciences of the Czech Republic
Praha, Czech Republic

Part I

Introduction

1 Structure and Magnetism in Magnetic Nanoparticles

Daniel Ortega

CONTENTS

Scale reduction in materials leads to profound changes in their inner structure, which in turn greatly modifies the intrinsic electronic and optical properties. Magnetic properties change as well. In the case of magnetic nanoparticles, these changes have meant the departure from some of the established laws governing the magnetic phenomena observed in bulk materials for the time being. The implications of these new phenomena on developing new technologies are manifold. On the one hand, the research done in magnetic nanoparticles over the last decades – more than 200,000 peer-reviewed publications – is now definitely moving from the theoretical grounds to the real applications arena, especially in medicine, as attested by the increasing number of both patents and companies formed around them[1,2].

On the other hand, and also in relation to medical applications, the effects that many nanostructures could have on health still remain unknown, challenging the *Food and Drug Agency* (FDA) in the United States or the *European Medicines Agency* (EMEA) to face continuous regulatory issues concerning the commercialisation of nanotechnology-based products[3,4] and their eventual consumption, certainly slowing down the rate at which they would become available to the final consumer.

The aim of this chapter is to present a concise overview of key concepts in magnetism applied to nanoparticles along with some of the instrumental techniques reasonably available in many laboratories to study the properties of magnetic nanoparticles. In accordance with the topic of this book, references to issues that could have a bearing on their potential uses in biomedicine and other related disciplines are also provided where appropriate. The contents are presented in a practical way in favour of a better understanding for the non-specialised reader, which will be able to subsequently resort to other fundamental[5–9] and more specific textbooks[10–12] dealing with the physics underlying the magnetic phenomena in a more rigorous fashion. Finally, a good deal of what is said here applies to many types of magnetic materials, but given their relevance in medicine compared to other fine-particle systems, the text is rather focused on ferrimagnetic iron oxides, namely *maghemite* (γ-Fe_2O_3) and *magnetite* (Fe_3O_4).

1.1 HISTORICAL BACKGROUND

Although the existence of nanoparticles dates back a long time ago, our knowledge of them has been only possible through the relentless development of more capable instrumental techniques that have allowed us to study the atomic structure of materials and measure new phenomena never detected before. In fact, the study of small magnetic particles have always been present in magnetism since its early times, and as this field has been evolving over time, the concept of 'small particle' has been also changing. Perhaps the basis of nanoparticle research, or even the present nanoscience, was established by Michael Faraday in the middle 1800s after his groundbreaking observations on the optical properties of gold colloids, as they were reported before the Royal Society in the Bakerian Lecture series[13]. This was actually a landmark in experimentally demonstrating how spatial confinement transforms the fundamental properties of matter. Two decades before this discovery, Faraday, along with other renowned researchers like Ampère, Oersted, Maxwell, Biot and Savart, contributed to the establishment of classical electromagnetism. Later, the validity of classical physics to explain the magnetic order in materials was severely questioned by Bohr and Van Leeuwen in their controversial theorem[5,14], stating that an ensemble of particles moving in a constant magnetic field shows zero magnetisation in a situation of thermal equilibrium; in other words, Maxwell's equations and classic statistical mechanics yield no magnetic order at all. This apparent contradiction uncovered the necessity of a revolutionary approach in the physics of magnetism, which came by the hand of quantum mechanics. It was not until the early 1900s when the seeds of quantum mechanics started to germinate in the area of magnetism, especially after the discovery of the electron and its spin[15,16], giving way to the golden Age of Understanding[17], during which most of the ideas underpinning the modern magnetism were put together. The work of Heisenberg, De Broglie, Pauli and other quantum physicists provided then a sound basis for Langevin's theory of paramagnetism[18] and Weiss's molecular field theory[19]. Despite the relevance of these theoretical developments elevated the science of magnetism to a new dimension, they exerted little influence in the practical applications of magnetic materials. This is not excessively surprising if, for instance, we recall that electricity had been used long before the electron was discovered. Regrettably, it was the military research during and after the World War II on permanent magnets and the interaction of magnetic materials with electromagnetic radiation – more specifically that within the radiofrequency and microwave ranges – that caused a vast technological progress in new applications of magnetism. By that time, Louis Néel was developing the concept of ferrimagnetism and anti-ferromagnetism[20], and soon after he contributed in a definitive manner to the understanding of magnetism in fine particles with the theory of superparamagnetism[21], a field also benefited from the work of William Fuller Brown[22].

The understanding of magnetic nanoparticles has been nourished from different areas within magnetism, but the rising interest in disordered spin systems during the 1970s and the knowledge accumulated in this topic over subsequent years undoubtedly constituted the most important influence[23]. The use of concepts like *frustration*[24] or the *Almeida–Thouless transition*[25] to explain the non-homogeneous magnetisation profile across nanoparticles confirms this sort of heritage from disordered systems like spin glasses[23].

Nowadays we are witnessing the great boom of magnetic nanoparticles as essential components in therapies and screening techniques progressively incorporated into several areas of the medical practice. A number of these cutting-edge technologies will be presented throughout the chapters of this book. Aside from the evident benefits of their small size to reach out places inside the human body otherwise difficult to be accessed, the interest in using magnetic nanoparticles in medicine chiefly relies on their ability to be easily oriented using external magnetic fields by virtue of their superparamagnetic behaviour (see Section 1.4). This would allow the particles to be magnetically actuated through external magnetic fields and no magnetic moment will remain upon field removal, avoiding the unwanted drawbacks derived from aggregation, like clogging of blood or lymphatic vessels.

1.2 PARADIGMATIC IRON OXIDES

The family of iron oxides is formed by 16 compounds, among which we can distinguish oxides, mixed oxides, hydroxides and oxyhydroxides. Yet in prehistory, some iron oxides – along with other minerals also contained in the original rocks – were used as pigments for paints, and later on others were being progressively incorporated to obtain new colours[26], somehow expanding their field of applications towards cosmetics. More modern uses came with the ability of extracting metallic iron from oxides, making available a myriad of new possibilities in construction and tool production. Since then, the continuous evolution of the iron oxides family has been quite remarkable and always linked to many cutting-edge technologies developed in different times. From abrasives to fertilisers and jewels, the list of applications of iron oxides did nothing but grow and grow through the centuries. More on the basic research side, written sources from the nineteenth century reported on the early synthesis of iron oxide by the alchemist Kunckel by heating metallic iron until red hot[27]. Subsequent studies contributed to a deeper knowledge of their structures and magnetic properties, although some of them were wrongly ascribed to the known behaviours so far (diamagnetic, paramagnetic and ferromagnetic); that was the case of maghemite, classified in principle as ferromagnetic until a formal model of ferrimagnetism was proposed by Néel in 1948. Precisely during the 1940s, the iron oxides started to gain a remarkable relevance in the field of magnetic recording, which was originally invented by Poulsen in 1900. Perhaps no other magnetic application of iron oxides has caused the same technological and economic impact in our lives as the magnetic tapes for audio and video recording did. In fact, it arguably constituted a pioneering use of maghemite, this time in the form of ellipsoidal particles suspended in a solvent and subsequently deposited and attached to a plastic thin film by a polymeric binder. The resulting tape topography is shown in Figure 1.1a. The information is stored by a recording head in the form of narrow bands (Figure 1.1b) that can be only imaged by magnetic techniques, such as magnetic force microscopy (MFM; Table 1.1). The search for a better recording performance lead to the development of other materials with smaller particle size and higher coercivity (this term will be defined in Section 1.3.2), like cobalt-substituted maghemites, iron cobalt or chromium oxide particles[28].

With such a track of record, one could expect that after being the core of many yet obsolete technologies, the ever changing practical uses of iron oxides could have come to an end due to the vast development in advanced alloys and other complex oxides with enhanced properties, but the discovery of new effects at the nanoscale and the advanced instrumentation to produce and characterise nanomaterials doubled the number of technological applications of the iron oxides. These have given rise to the development of, for example, magnetic resonance contrast agents[49], high-performance catalysts for the oxidation of organic compounds[50] or masks for nanolithography[51].

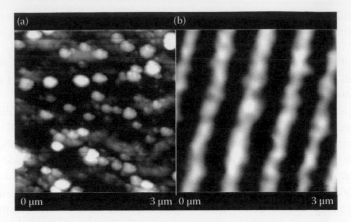

FIGURE 1.1 Topographical (a) and magnetic force gradient (b) images of a commercial video tape as obtained by magnetic force microscopy. The dark–bright contrasts in image (b) represent the regions where the tape has been magnetised in the plane of the recording surface.

TABLE 1.1
Some Techniques for Characterising Magnetic Materials

Category	Technique	Effect	References
Neutron	Small-angle neutron scattering (SANS)	Elastic scattering of neutrons caused by changes in refractive index at the nanoscale	[29,30]
	Extended x-ray absorption fine structure (EXAFS)	Difference in x-ray absorption caused by matching electron binding energies	[31]
	X-ray magnetic circular dichroism (XMCD)	Difference in the absorption of right and left circularly polarised x-rays	[32]
Electron microscopy	Photoelectron emission microscopy (PEEM)	Imaging of surface electrons emitted by the action of photons over a sample	[33,34]
	Electron holography	Phase change of electron waves that have travelled through a thin sample	[35,36]
	Lorentz microscopy	Interaction between the electron beam of the microscope and the local magnetic induction of a material	[37,38]
Scanning probe microscopy	Scanning tunnelling microscopy (STM) and spin-polarised STM	Tunnelling current between a fine conductive tip and a surface	[39]
	Magnetic force microscopy (MFM)	Magnetic force exerted on a magnetic tip scanned over a surface	[39,40]
Resonance, nuclear	Ferromagnetic resonance	Resonant absorption of radiofrequency radiation under the action of a magnetic field	[41–43]
	Muon spin rotation/resonance/ relaxation (μ-SR) spectroscopy	Change in spin orientation of implanted muons by structural and/or magnetic environments	[44,45]
	Mössbauer spectroscopy	Recoilless resonant absorption of gamma radiation	[46,47]
Magnetometry	Superconducting quantum interference device (SQUID) magnetometry, micro-SQUID (μ-SQUID)	Change of magnetic flux in a superconducting (Josephson) junction	[6,9]
	Vibrating sample magnetometry (VSM)	Change of magnetic flux by a sample oscillating at a fixed frequency	[6,9,48]

1.2.1 FERRIMAGNETIC OXIDES: MAGNETITE AND MAGHEMITE

The magnetism in materials stems from the magnetic moment of the electrons. The motion of electrons in atoms possesses both an orbital angular moment $\mathbf{m}_l$, associated to their angular momentum $\mathbf{l}$, and a spin moment $\mathbf{m}_S$, proportional to their spin $\mathbf{s}$:

$$\mathbf{m}_l = g_l \mathbf{l}$$
$$\mathbf{m}_S = -g_S \mu_B \mathbf{s},$$
(1.1)

where

g_l and g_S are the orbital and spin *g-factors*
$\mathbf{s}$ is the *spin* of the electron
μ_B is the *Bohr magneton* (9.274×10^{-24} A·m^2)

The coupling of $\mathbf{m}_l$ and $\mathbf{m}_S$ or, in other words, the interaction of electron spin with its motion gives rise to the *spin–orbit coupling*. Among others, the spin–orbit coupling constitutes the origin of the *magnetocrystalline anisotropy* (MCA), later defined and discussed in Section 1.4. Adding another level of complexity, within a material there will be interactions between atoms*, each with associated spin moments $\mathbf{S}$. The interactions between pairs of neighbouring atoms can be computed by extending to the whole lattice the Hamiltonian ($\mathcal{H}$) that Heisenberg originally formulated for pairs of atoms:

$$\mathcal{H} = -2 \sum_{i,j} J_{ij} \, \mathbf{S}_i \cdot \mathbf{S}_j,$$
(1.2)

where

i and j represent the positions of the atoms within the lattice
J is the exchange constant

If $J > 0$, there is a *ferromagnetic* interaction that couples the spins parallel ($\uparrow\uparrow$); if $J < 0$, then the interaction is *anti-ferromagnetic*, and now the spins are coupled anti-parallel ($\uparrow\downarrow$). The first type of interaction is typical in ferromagnets, whereas the second one is found in both ferrimagnetic and anti-ferrimagnetic materials (Figure 1.2). In summary, the *exchange interaction* refers to the way

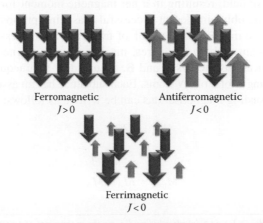

Ferromagnetic
$J > 0$

Antiferromagnetic
$J < 0$

Ferrimagnetic
$J < 0$

FIGURE 1.2 Types of magnetic order according to the sign of the exchange constant.

---

* For the sake of simplicity, the atoms can be considered as multi-electron entities. As such, they will have a net spin moment $\mathbf{S}$ containing the contribution of all the electron spin moments.

in which electrons are coupled, and this coupling is governed by Pauli's principle restrictions: two electrons in the same orbital sub-level cannot have the same spin quantum number.

At first sight, *ferrimagnetic** compounds could be undistinguishable from their ferromagnetic counterparts: both exhibit a remarkable magnetisation in absence of an applied field – or spontaneous magnetisation – and above a certain critical temperature, called *Curie temperature*, their magnetic order vanishes through a paramagnetic transition. Thus, it becomes clear that the differences between these two apparently similar behaviours have to be sought in a particular crystalline arrangement. A closer inspection of the maghemite and magnetite structures, the ferrimagnetic members of the iron oxides family, will serve to this purpose.

Magnetite is a mixed Fe^{2+} and Fe^{3+} oxide in a ratio of 1:2 with an inverse *spinel* structure[†] where oxygen atoms are forming a cubic close packing along the [111] direction. Denoting tetrahedral and octahedral sites as T_d and O_h, respectively, the structural formula can be written as follows:

$$[Fe^{3+}]_{T_d}[Fe^{3+}Fe^{2+}]_{O_h}O_4. \tag{1.3}$$

$Fe^{2+}/Fe^{3+} = 0.5$ is the stoichiometric ratio, but this can vary to a relatively large extent depending on the preparation method followed and the incorporation of other cations into the spinel structure[52].

Maghemite can be regarded as an oxidised magnetite, since the presence of Fe^{2+} cations in the latter makes it unstable against oxidation at ambient oxygen pressure if no specific means are used to prevent it, such as the use of an inert atmosphere (nitrogen, argon) during preparation or coating[‡] before exposition to open air. It has to be noted that the oxidation of magnetite to maghemite is largely favoured when it is forming small particles, but the kinetics of the process is dramatically slowed down in the case of bulk samples in such a way that no remarkable oxidation occurs. The cation distribution for maghemite structure, also an inverse spinel-type like magnetite, is shown in Equation 1.4. In maghemite, there are no Fe^{2+} ions and one-third of the O_h sites are vacant, which tend to compensate the charge imbalance generating a neutral structure:

$$\frac{3}{4}[Fe^{3+}]_{T_d}\left[Fe^{3+}_{\frac{5}{3}}V_{\frac{1}{3}}\right]_{O_h}O_4, \tag{1.4}$$

where V represents the vacancies. Given these crystallographic similarities, their magnetic properties share a common explanation, as it might be expected. Both structures are composed of two magnetic sublattices with anti-parallel magnetisations, but they do not cancel because each one 'feels' a different effective field, resulting in a net magnetic moment for the overall structure. In essence, this was the incredibly simple but successful reasoning proposed by Néel to tackle the problem of interpreting the magnetic behaviour of ferrites[20]. Regardless of the difference in the oxidation state of iron atoms in both structures, maghemite and magnetite can be considered as composed of two sublattices, denoted as A and B, occupied by an unequal fraction of iron atoms x and y, respectively, having a total of $2N$ atoms. Each sublattice has an associated magnetic moment μ_A and μ_B, so the corresponding magnetisations can be written as follows:

$$M_A = 2Nx\mu_A$$

$$M_B = -2Ny\mu_B \tag{1.5}$$

$$M_{total} = M_A + M_B.$$

---

* The term ferrimagnetism is due to Louis Néel after his studies on the magnetism of *ferrites*.

† Named after the archetypal magnesium and aluminium oxide ($MgAl_2O_4$).

‡ This is usually carried out by using long-chain molecules anchored to the surface of the particles by intramolecular, for example covalent bonding, or intermolecular forces. Examples are oleic acid, oleylamine, mercaptosuccinic acid, dextran or starch, to name but a few.

In order to continue this reasoning, another concept must be introduced. To explain ferromagnetism, Weiss[19] proposed the existence of a fictitious *molecular field* (H_m) proportional to the magnetisation of the system and responsible for keeping the spins aligned within the material:

$$H_m = \gamma M + H,\tag{1.6}$$

where γ is the Weiss coefficient. It is worth noting that the magnitude of this molecular field is much larger than that of the magnetic fields than can be generated in a laboratory. Considering the possible interactions between sublattices as positive when they are coupled parallel (A–A and B–B) and negative when antiparallel (A–B), Equation 1.6 gives the associated molecular fields H_A and H_B characterised by their respective Weiss coefficients, resulting in

$$H_A = \gamma_{AA}M_A - \gamma_{AB}M_B$$
$$H_B = \gamma_{AB}M_A - \gamma_{BB}M_B.\tag{1.7}$$

Expressing γ_{AA} and γ_{BB} with respect to γ_{AB},

$$\alpha = \frac{\gamma_{AA}}{\gamma_{AB}}; \quad \beta = \frac{\gamma_{BB}}{\gamma_{AB}}.\tag{1.8}$$

The molecular fields for each sublattice, due to their own atoms and the atoms in the other sublattice, are given by

$$H_A = \gamma_{AB}(\alpha x M_A - y M_B)$$
$$H_B = \gamma_{AB}(\beta y M_B - x M_A).\tag{1.9}$$

Sublattice magnetisations can subsequently be derived from Equations 1.9:[6]

$$M_A = 2N\mu_A L\left(\frac{\mu_A \gamma_{AB}(\alpha x M_A - y M_B)}{k_B T}\right)$$
$$M_B = 2N\mu_B L\left(\frac{\mu_B \gamma_{AB}(\beta y M_B - x M_A)}{k_B T}\right).\tag{1.10}$$

The sets of Equations 1.9 and 1.10 evidence the different magnetisations from each sublattice and satisfactorily explain the behaviour of ferrimagnets both below and above the Curie temperature (T_C).

1.2.2 Related Phase Transformations and Characterisation Issues

Among all the possible transformations involving any iron oxides, a briefly description of the most relevant ones involving magnetite, maghemite or both is given in this section. These transformations could take place at any stage during their synthesis (precipitation, thermal decomposition, etc.) or the subsequent post-processing (thermal treatment, additional reactions, etc.).

1.2.2.1 Lepidocrocite to Maghemite (γ-FeO(OH) to γ-Fe$_2$O$_3$)

This is a *topotactic* transformation, in which the final crystalline network shows one or several equivalent orientational relations from the crystallographic point of view with respect to the initial phase, or in much simpler words, although the transformation involved atomic rearrangements and gain or loss of mass, the initial and final phase will be coherent. Unlike the *reconstruction* – implying dissolution and precipitation processes to form new phases – the topotactic one is a solid state transformation. In one of the few works published dealing with it[53], a model is proposed to describe how the dehydration of the lepidocrocite and a partial migration of iron atoms from octahedral to tetrahedral sites lead to the eventual collapse of its structure giving rise to the oxide structure. From the magnetic point of view, the transformation is noticeable as far as lepidocrocite is paramagnetic at room temperature and maghemite is ferrimagnetic, as shown in the preceding section.

1.2.2.2 Maghemite to Hematite (γ-Fe$_2$O$_3$ to α-Fe$_2$O$_3$)

This transformation has to be borne in mind if a thermal treatment of the maghemite phase is required as a part of the preparation protocol. Its occurrence largely depends on the initial state of maghemite (bulk/particulate, coated/uncoated, etc.) as well as the heating atmosphere. This is exemplified by the high number of different transformation temperatures – also within a wide temperature range – that has been reported so far[26]. Haematite (α-Fe$_2$O$_3$) is a canted anti-ferromagnet (see Section 1.6.3.1), so extra precautions have to be taken to avoid its formation if obtaining the highest magnetic moment out of our sample is a priority. In the case of particulate samples, De Biasi and Portella[54] suggested that the nucleation process occurs preferentially at their surface. Given the high number of atoms at the surface, the first nuclei tend to readily extend all over the particle as long as the transformation proceeds. All this would also apply for magnetite; the only difference would be a previous oxidation step to maghemite.

1.2.2.3 Magnetite to Maghemite (Fe$_3$O$_4$ to γ-Fe$_2$O$_3$)

As it was pointed out before, the spontaneous or induced oxidation of magnetite yields maghemite as a product; nonetheless, aerial oxidation is not the only process taking part in this transformation[55]. The pH of the medium can favour the appearance of ionic or electronic transfer phenomena between species: in basic media, Fe^{3+} cations incorporate into the magnetite structure after an oxygen reduction initiated at the surface of the nanoparticles, while in acidic media, Fe^{2+} are desorbed as aquocomplexes, giving rise to the apparition of vacancies to keep the charge balance in the network.

Distinguishing between maghemite and magnetite is not a trivial issue, especially in samples with a mixture of both phases. Additionally, in particulate media the oxidation of magnetite is more pronounced and hence the fact that a variable fraction of maghemite is very likely to be present. Following are some remarks worth mentioning to draw the attention of the reader to this characterisation issue that, if not properly addressed, would induce the researcher to inaccurate conclusions.

A quite traditional means of characterising the different iron oxides is by x-ray diffraction (XRD) – fundamentals on this technique can be found in Refs. [56,57] – which provides a sort of 'fingerprint' for each phase formed by the set of characteristic positions of those peaks representing the different families of diffracting planes; nevertheless, in the case of maghemite and magnetite nanoparticles, the outcomes of XRD measurements could be not as accurate as it might be expected mainly due to the following two facts. The first one, of a more general character and applicable to any sample with dimensions fall under 0.5 μm, is that diffractograms obtained from small particles show homogeneously broadened peaks. Apart from the crystallite* size itself, this broadening can be due to instrumental effects and crystal strains. Even in the case of well-crystallised samples, when the size of nanoparticles is reduced enough, certain peaks tend to disappear and the structural information is therefore lost. The pattern simulations in Figure 1.3a illustrate how crystallite size changes the profile shape, affecting in a more marked way to the less intense peaks. As mentioned before, the degree of reticular strain also plays a significant role on the modification of the profile (Figure 1.3b). In addition to a remarkable peak broadening, poor crystalline or severely strained samples could also lead to a loss of intensity.

---

* In x-ray diffraction, the word *crystallite* is related to the minimum size of a coherently diffracting domain within the crystal. A particle could contain more than one crystallite, so both terms do not have to be equivalent depending on the sample under study.

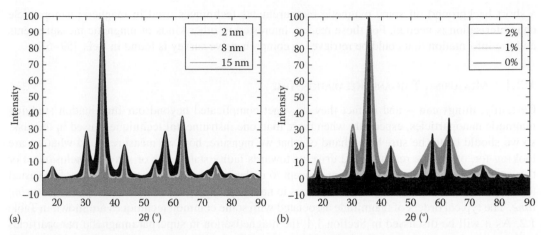

(a)

(b)

FIGURE 1.3 XRD pattern simulation of maghemite using (a) different crystallite sizes and (b) different reticular strain values.

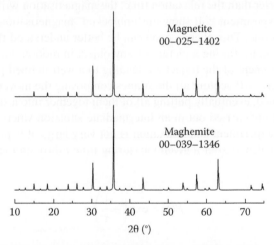

FIGURE 1.4 Simulated XRD patterns of magnetite and maghemite. Reference numbers of the used International Centre for Diffraction Data (ICDD) powder diffraction files are indicated.

The second fact is related to the isostructural character of maghemite and magnetite. Both are cubic, with a spinel structure and have similar unit cell parameters, indicating that the characteristic peaks of both structures are very difficult to distinguish even for particulate samples of several microns[58]. As suggested by the simulated patterns in Figure 1.4, there are only a few peaks that could help to differentiate which phase is which, but their low intensity will be easily blurred by the background noise or the crystallite size effect during the course of the experiment.

Thus, given the intricate identification of maghemite and magnetite by XRD, one has to resort to other techniques, for example, Mössbauer spectroscopy (see Section 1.3.4) or x-ray photoelectron spectroscopy (XPS), both sensitive to the oxidation state of the elements present in the sample. This is actually a major differentiation point between these structures, since no Fe^{2+} is present in maghemite.

1.3 MAGNETIC MEASUREMENTS

The list of instrumentation and methodologies available for magnetic measurements is quite extensive attending to all the types of signals that can be detected and the purposes served, as exemplified by Table 1.1. Before getting into the theoretical grounds of magnetism in magnetic nanoparticles, a short background on some magnetic measurements techniques used in magnetic nanoparticle characterisation is needed. For those readers interested in other kinds of magnetic measurements and the information that could be retrieved, a comprehensive survey is found in Refs. [59–61].

1.3.1 MEASURING TIME AND RELAXATION TIME

Certainly, things can – and in fact they do – get complicated beyond our imagination studying magnetic nanoparticles, especially when more than one instrumental technique is used in the task, so we should better be sure beforehand of what we measure, how we measure it and what we are looking for, otherwise results could drive us towards faulty statements or wrong conclusions. For this reason, maybe the most important concept to understand before getting into some of the usual techniques for measuring magnetic properties in nanoparticles systems is the concept of *measuring time*. The typical orders of magnitude associated with some common techniques are shown in Table 1.2. As it will be discussed in Section 1.4, the magnetisation in superparamagnetic nanoparticles undergoes continuous thermally driven fluctuations of certain amplitude, and each oscillation will take a characteristic time to be completed, known as *relaxation time* (τ). If the time frame defined by an experiment is shorter than the relaxation time, the magnetisation will appear as 'blocked'; in the opposite case, the experiment will show an 'unblocked' magnetisation typical of a particle in a superparamagnetic regime. This phenomenon can be better understood if compared to the effect of exposure time in cameras. Taking a picture of any object in motion with a short exposure time will capture a precise moment of the trajectory, leading to a well-defined photo. If we use a larger exposure time, the camera will acquire all the frames composing the movement of the object while the shutter remains opened, eventually putting all of them together into a single 'blurred' picture.

When measurements are carried out in an intermediate situation where the difference between characteristic times for experiment and relaxation is not very large, it is possible to know whether the magnetisation has switched after a given measuring time t through the probability function of the form:

$$P(t) = \exp\left(-\frac{1}{\tau}\right). \tag{1.11}$$

Wernsdorfer et al. assessed the validity of Equation 1.11 by means of μ-SQUID magnetometry experiments on single-domain cobalt nanoparticles[62]. There are several parameters that depend on the measuring time, such as the blocking temperature (Section 1.6.1.1), due to which they should not be used to uniquely characterise a system, although they can provide us with very valuable qualitative and semi-quantitative information.

TABLE 1.2

Characteristic Measurement Times Associated with Some Common Instrumental Techniques (Just the Order of Magnitude Is Presented)

Technique	Measurement Time τ_m(s)
DC susceptibility	60–100
AC susceptibility	10^2–10^4 (low-frequency experiments)
	10^{-1}–10^{-5} (classical experiments)
Mössbauer spectroscopy	10^{-7}–10^{-9}
Ferromagnetic resonance	10^{-9}
Neutron diffraction	10^{-8}–10^{-12}

1.3.2 DC Magnetic Measurements

In most of DC magnetic measurements, we determine the equilibrium, quasi-static magnetic properties. The usual procedure consists in measuring the magnetisation of a sample while increasing the applied field in a step-like manner, although this can also be done by continuously changing the field. Both methods should lead to the same results, but even if the field is changed very slowly, differences could arise due to the appearance of *after-effects* – once the field is removed, the magnetisation evolves to a certain extent with time – or *eddy currents* that oppose the induced change in magnetic flux.

For measuring nanoparticles, SQUID or VSM magnetometries are commonly chosen because of their versatility and high sensitivity for low mass or moment samples, detecting magnetic moments as low as 10^{-6}–10^{-8} emu in modern systems. In a SQUID, a homogeneous magnetic field is applied on the sample by a superconducting magnet while measuring the change in flux induced in a detection coil by gradually shifting the sample under study. In a VSM, the sample is oscillated at a fixed frequency (about 80 Hz) and the detection coil measures the induced voltage. Although the latter is able to deliver a great accuracy, sensitivity and, above all, speed in measurements, the induced eddy currents in the sample or the acoustic coupling between the motor drive and the detection system can affect the obtained results in various ways[63].

Regardless our choice of the measurement technique, keeping the different sources of artefacts to a minimum, along with a scrupulous sample preparation, is central for a successful characterisation. The high sensitivity of SQUID instruments makes them especially prone to be affected by a sizeable number of artefacts, which have been extensively reported in the related scientific literature[64–69].

1.3.2.1 Magnetisation Curves

In simple terms, when we measure the irreversible changes in the magnetisation of a sample as a result of applying cyclic positive and negative values of magnetic fields, we obtain a *hysteresis loop* (Figure 1.5). The path initially followed by the magnetisation as the field is increased is given by the so-called *virgin curve*, since the sample comes from an unmagnetised or virgin state; then the magnetic structure of the material will be irreversibly altered precluding this virgin curve to be retraced during the experiment. Among the different parameters that can be extracted from this peculiar graph, there are three that provide us with the basic information needed to describe the magnetic behaviour of a given material in a general fashion*. The *saturation magnetisation* (M_S) is essentially the limit value to which the curve tends within the high-field region, and is reached when all the magnetic moments in the material are aligned with the external field. Upon field decreasing, the sample does not recover its unmagnetised state, retaining a certain amount of magnetisation at zero field: the *remanence* or *remanent magnetisation* (M_r). The third parameter is also a consequence of the irreversible character of magnetic hysteresis; the *coercivity* or coercive field (H_c) represents the field we need to apply to completely demagnetise the sample. This parameter is very important, for example, in designing materials required to be in a non-magnetised state at a certain moment for a specific application or to be subjected to cycles of demagnetisation and remagnetisation, like magnetic recording media.

Sometimes we need a fast but reliable way to extract some important parameters from the hysteresis loops or magnetisation curves, like the saturation magnetisation. Within the high-field region, the hysteresis loops obtained from small particles can be simply fitted to the linear expression $M(H) = M_s + \chi_d H$, where χ_d is the high-field differential susceptibility accounting for the superficial spin disorder (Section 1.6.3.1). The magnetisation curves for the fitting procedure are obtained from

---

* Note that here just a simplified account of the magnetic hysteresis is given. For example, a single loop is usually studied in most of the cases, but the measured response could significantly differ from other spatial components of the applied field depending on the sample's nature[70]. A careful appraisal on the possible effects of this and other common assumptions should be done prior to any experiment.

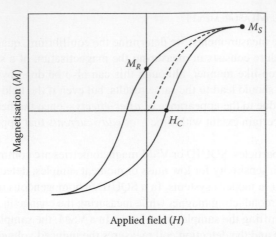

FIGURE 1.5 Main parameters of interest in a generic hysteresis loop. Virgin curve is indicated by the dashed line.

the average of the $H \rightarrow H_{max}$ branches of the loop. For anhysteretic loops, an empirical expression proposed by O. Fröhlich is applicable:[71]

$$M = \frac{\alpha H}{1 + \beta H}, \tag{1.12}$$

where the equality $\alpha/\beta = M_S$ is imposed by the condition that H must tend to infinite. Its simplicity makes it straightforward to use, producing reliable fittings when nothing but M_S is sought. There is an alternative though equivalent expression to the former equation developed by Kennelly:[72]

$$M = M_S \left[1 - c\frac{M_S}{H} + \left(c\frac{M_S}{H} \right)^2 + \cdots \right], \tag{1.13}$$

where c is just a dimensionless parameter. In the common practice, just the terms up to $1/H$ are taken into account.

Systems of superparamagnetic nanoparticles (Section 1.4) can be effectively approximated to an ensemble of paramagnetic ions that have a large magnetic moment. This approach allows for the use of *Langevin's paramagnetic equation*:

$$M_{SPM} = N_{SPM}\mu_{SPM} \left[\coth\left(\frac{\mu_0 \mu_{SPM} H}{k_B T} \right) - \frac{k_B T}{\mu_0 \mu_{SPM} H} \right], \tag{1.14}$$

where
   N_{SPM} is the number of superparamagnetic particles with average magnetic moment μ_{SPM}
   μ_0 is the magnetic permeability of the free space
   k_B is the Boltzmann constant

This function basically represents the classical limit of the quantum mechanical *Brillouin function*[6], and its use is justified by the fact that the magnetic moments in a set of superparamagnetic nanoparticles are larger than those of individual paramagnetic atoms, for which the Brillouin function must be employed. The Langevin equation provides a means to extract the average magnetic moment per particle from experimental isothermal magnetisation curves, but extreme care has to be taken even in the case of obtaining good fits, as the numerical outcomes can lead to physically meaningless results if the equation has been applied to highly anisotropic particles, or severe restrictions such as neglecting interparticle interactions in a system that actually has them, are assumed.

1.3.2.2 Zero-Field and Field-Cooled Curves

In a typical zero-field cooling/field-cooling (ZFC/FC) experiment (Figure 1.6), the sample is initially heated up to a temperature high enough as to ensure that it is in a superparamagnetic regime prior to the start of the measurement; in such a way, we ensure that the initial magnetic state is known*. Then the sample is subsequently cooled without magnetic field down to lowest achievable temperature in the magnetometer (T_{min}). From that point, the magnetisation (M_{ZFC}) is measured by heating up the sample under the action of an external field giving the ZFC branch of the experiment. The usual tendency of this curve is the same for most of the nanoparticle systems: at low temperatures, all the magnetic moments are blocked in random directions eventually leading to a (nearly) zero magnetisation. In practice, this will depend on the magnetic history of the material and the initial conditions of the experiments. As the temperature is further increased from T_{min}, M_{ZFC} rises to a maximum at the temperature T_{max} and then decreases again. T_{max} is related but not necessarily equal to the average *blocking temperature* (T_B), for which both the magnetic relaxation time of the nanoparticles and the measuring time coincide (see Section 1.6.1.1 for further considerations on this topic). The FC measurement is carried out in a similar fashion to the one already described for the ZFC experiments, although in this case the sample is cooled down under a constant magnetic field until T_{min} is reached again. Again as in the ZFC branch, the measurement, properly speaking, will begin when the system is heated up from T_{min} to room temperature or its surroundings. This process can be considered as reversible from a thermal point of view, since both the heating and cooling rates are the same.

In ZFC/FC experiments, the *irreversibility temperature* or just *irreversibility* (T_{irr}) – where the condition $M_{FC} - M_{ZFC} \neq 0$ is met – is that for which the separation between the two branches takes place, and represents the blocking temperature of those particles that must overcome a higher energy barrier (imposed by the anisotropy energy, as explained in Section 1.4) in their magnetic relaxation process.

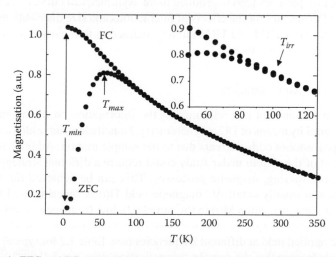

FIGURE 1.6 Generic ZFC and FC magnetisation curves for a system of nanoparticles.

---

* Obviously, if the sample present a more or less broad size distribution, it is possible that a fraction corresponding to the biggest particles is already blocked, that is in a non-superparamagnetic regime, even at the initial temperature.

It has to be borne in mind that the previously described procedure corresponds to a generic experiment and does not constitute any general protocol. Every sample demands particular experimental conditions with proper cooling rates and measuring fields, which have to be checked prior to the experiment using as much information as possible from other characterisation techniques.

1.3.2.3 Thermoremanent Magnetisation

If an assembly of superparamagnetic nanoparticles is cooled in a given field from a high enough temperature, a small bias is induced in the direction of magnetisation of the blocked particles in the temperature range below the highest T_B, that is, that of the biggest particles in the system; it is then said that the system has acquired a *thermoremanent magnetisation* (TRM). In a typical TRM experiment, once the sample has been slowly cooled down to the measuring temperature, the field is switched off, and after a certain waiting time, the magnetisation is measured. The procedure is then repeated for each temperature within a desired range, obtaining a remanence curve. The applied field and waiting time dependencies of TRM curves are of interest, since they provide information concerning the interparticle interactions and the size distribution present in the sample.

A similar process to that previously described happens in nature when a magnetic mineral is cooled down in the presence of a small field, usually Earth's magnetic field, from above its Curie temperature[73]. This process is of much interest in geophysical sciences to determine the direction and intensity of geomagnetic fields in ancient times.

1.3.2.4 Isothermal Remanent Magnetisation

In *isothermal remanent magnetisation* (IRM) experiments, after being at a high enough temperature, the sample is cooled in zero field and held at a low temperature. A field is then applied and eventually switched off, giving an IRM curve as a function of the applied field. The obtained maximum remanence is called *saturation remanence*, and materials with a higher coercivity should be more difficult to saturate in principle. Based on this observation, IRM curves could be useful to distinguish between some iron oxides, for example, hematite and magnetite, given the difference in coercivity usually found between both phases. Information related to interparticle interactions can be also extracted from IRM curves, whereas non-interacting single-domain particles give symmetrical curves, interacting particles tend to produce more asymmetrical curves[74]. A number of models have been developed to account for the effects of interactions and size distributions on the singularities experimentally found in IRM and TRM curves obtained in fine-particle systems. These can be found in Ref. [75] and references therein.

1.3.3 AC MAGNETIC MEASUREMENTS

We have seen so far the amount and relevance of the information on the magnetic properties of nanoparticles measured by means of DC magnetometry. Nonetheless, its results could be potentially affected by the appearance of eddy currents due to the sample motion. Additionally, the nature and properties of interest of the system under study could require a different strategy to gain access to dynamic, that is, time-varying, magnetic processes. This can be achieved through AC magnetic measurements, where a usually small AC magnetic field $\mathbf{H}(t)$ is superimposed to the DC field to induce a time-varying magnetisation $\mathbf{M}(t)$ in the sample. In low-frequency experiments, the result is just the magnetic *susceptibility* (χ), which is the slope of a magnetisation vs field curve ($\chi = d\mathbf{M}/d\mathbf{H}$) as a function of the applied field at different frequencies (see Table 1.2 for typical values). However, at higher measuring frequencies the sample magnetisation does not follow the $\mathbf{M}(H)$ curve anymore. The susceptibility can be written as a function of the driving frequency ω and the time t:

$$\chi(\omega,t) = \chi'(\omega,t) + i\chi''(\omega,t). \tag{1.15}$$

χ' is the *in-phase* or *real* susceptibility, and χ'' is the *out-of-phase* or *imaginary* component, which is proportional to the energy absorbed by the sample from the applied field. $\mathbf{M}(t)$ can be expressed as a Fourier series of both components:

$$\mathbf{M}(t) = \mathbf{H} \sum_{n=1}^{\infty} [\chi' \sin(n\omega t) + \chi'' \cos(n\omega t)], \qquad (1.16)$$

The result of the experiment will be the fundamental susceptibility if $n = 1$ in Equation 1.16, while bigger values of n will give higher harmonics of the susceptibility. Measurements implying higher harmonics are of interest when studying phase transitions or critical behaviours in spin systems; moreover, the second harmonic $n = 2$ is only observed if the system shows a spontaneous magnetisation, while odd harmonics are observed in spin glasses[76]. Both χ' and χ'' can be obtained from the measured magnetisation through the following relations:

$$\chi' = \frac{1}{\pi \mathbf{H}} \int_{0}^{2\pi} \mathbf{M}(t) \sin(n\omega t) d(\omega t)$$

$$\qquad (1.17)$$

$$\chi'' = \frac{1}{\pi \mathbf{H}} \int_{0}^{2\pi} \mathbf{M}(t) \cos(n\omega t) d(\omega t).$$

AC susceptibility offers some advantages over the DC measurements: (i) as it depends on the slope of the $M(H)$ instead of on its absolute value, it can detect small slope changes, (ii) both the real and imaginary components are sensitive to phase transitions, and (iii) the timescale can be easily changed in AC susceptometry by varying the measuring frequency, making it a valuable technique for studying relaxations times in fine-particle systems.

1.3.4 MÖSSBAUER SPECTROSCOPY

The beginning of this technique was set in 1957 after Rudolf Mössbauer's discovery of the recoilless resonant absorption of gamma radiation in a solid[46]. The ^{57}Fe Mössbauer spectroscopy is based on the use of the gamma photons emitted from the nuclei of a radioactive ^{57}Co source to study the interactions between ^{57}Fe nuclei and their environment within a sample[57]. Co decays to the second excited state of the ^{57}Fe nucleus with a spin of 5/2 and then to the first excited state with spin 3/2 (Figure 1.7).

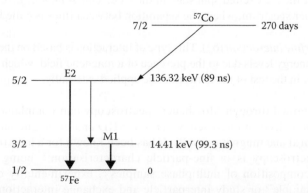

FIGURE 1.7 Radioactive disintegration of ^{57}Co. Due to the very low probability associated to the E2 transitions (0.0006%), the gamma radiation is mainly produced through M1 transitions.

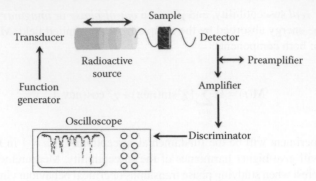

FIGURE 1.8 Scheme of the common experimental arrangement for Mössbauer spectroscopy.

The 14.41 keV transition between the excited 3/2 spin state and the fundamental 1/2 spin state produces the gamma photons used in ^{57}Fe Mössbauer experiments. Despite the fact that the natural abundance of ^{57}Fe is rather low (2.19%), it is still enough to produce usable gamma radiation.

The photons emitted by the source pass through the sample and those transmitted will reach the detector and subsequently will be counted (Figure 1.8). In order to study the nuclear energy levels associated with the different crystalline environments that the ^{57}Fe nuclei will interact with once they reach the sample, the energy of the gamma rays has to be sampled; this is done by constantly moving the source with respect to the absorber along both positive and negative values of the x axis, producing an energy shift in the gamma rays by Doppler effect that will allow for detecting the resonant energy levels in the sample of interest.

In short, Mössbauer experiments aim to study the *hyperfine interactions* resulting from the interaction of the nuclei and the electrical and magnetic fields associated with both the neighbouring electrons and the other nuclei. Accordingly, the relevant parameters of interest in a spectrum are as follows:

- *Isomer shift* (δ). This is related to the electrical distortion generated in the nuclear energy levels as a result of the interaction between the nucleus and the electrons occupying s-orbitals promoted by the change in size of the nuclei and excited states. δ yields information on the electronic state and chemical bonding of the iron atoms. The isomer shift in ^{57}Fe Mössbauer spectroscopy is given relative to a α-Fe foil standard sample.
- *Electric quadrupole interaction.* The unpaired valence electrons as well as the ionic charges will contribute to the generation of an electric gradient, which in turn removes the degeneration of the 3/2 excited spin state in the ^{57}Fe. This state will give rise to a splitting of the Mössbauer spectrum, where the separation between lines is called the *quadrupolar splitting* (δ).
- *Magnetic hyperfine interaction* (δ_m). This type of interaction is based on the Zeeman splitting of the nuclear energy levels due to the presence of a magnetic field, which can be originated inside the atom, in the rest of the network or be applied externally.

The information obtained through Mössbauer spectroscopy can complement or even improve that from transmission electron microscopy (TEM), XRD and/or magnetometry to cross-check and correlate structural and magnetic results. It has been known for a long time that how important Mössbauer spectroscopy is in fine-particle characterisation[77], being particularly useful to disentangle the composition of multiphase samples[78], experimentally determine the magnetic moment of a sample[79] or study interparticle and exchange interactions in many particles systems.[80]

1.4 MAGNETIC ORDER IN NANOPARTICLES AND SUPERPARAMAGNETIC REGIME

A bulk ferromagnet, such as iron, cobalt or nickel, tends to minimise its internal energy by spontaneously splitting into magnetic *domains*, which are regions containing magnetic moments coupled in the same direction. Each domain could be then represented by a single magnetisation vector accounting for all its magnetic moments per volume unit. They are separated from each other through *domain walls* that are likely to originate in the vicinity of defects present in the material, like dislocations or vacancies. It is worth noting that domain walls can be moved across the sample by applying small magnetic fields. The change in the magnetisation direction from one domain to its neighbour does not take place abruptly but in a gradual fashion, suggesting the existence of a *wall thickness* that depends on several energetic, crystallographic and geometric factors[6]. Besides other strictly energetic considerations, the bigger the sample size the higher the probability of magnetic domain formation, since the appearance of defects as nucleation sites for domain walls also increases. When sample dimensions are much reduced, like in a system of small particles, the energetic stability achieved through the formation of domains decreases considerably. For a given size or volume and no external magnetic field acting on the material, no domains will be present and the system spontaneously adopts a *single-domain* configuration in which the sample can be considered as uniformly magnetised throughout its volume and therefore represented by a single 'superspin' (Figure 1.9). Frenkel and Dorfman predicted the existence of single-domain particles[81] and pointed out the peculiarities of their magnetism, later revisited by Aharoni[82]. A great deal of experimental evidence regarding the magnetic structure of single-domain small particles have been contributed thereafter, mainly using MFM and cross-checking the outcomes against micromagnetic simulations. For example MFM images of 300 nm cobalt particles show uniformly magnetised single-domain states, and vortex configurations have been detected as transitional states when the magnetic moments were reversed by external fields[83]. Nonetheless, it seems that the multidomain–single-domain transition is not as well defined as it might be expected. The disagreement between theory and experimental data of remanent magnetisation measurements on small particles in geological samples lead geophysicists to the formulation of a new magnetic state, the *pseudo single-domain* (PSD), with a somewhat unclear nature despite the various proposed models to account for it[73].

As with other typical magnetic lengths scales, the *single-domain size* (R_{SD}) or the size for which a multidomain configuration is no longer stable, can be expressed as a function of the exchange length l_{ex}[17], which gives an idea of the competition between dipolar and exchange interactions (see Section 1.6.4):

$$R_{SD} = 36\kappa l_{ex} \tag{1.18}$$

with

$$l_{ex} = \sqrt{\frac{A}{\mu_0 M_S^2}} \tag{1.19}$$

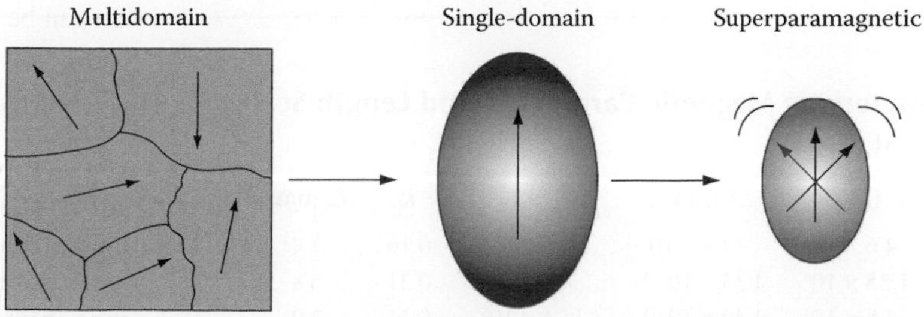

Multidomain   Single-domain   Superparamagnetic

FIGURE 1.9 Magnetic behaviours derived from the scale reduction in magnetic materials.

and

$$\kappa = \sqrt{\frac{K}{\mu_0 M_S^2}} \qquad (1.20)$$

where

K is the first anisotropy constant, which will be introduced later in this section
κ is the dimensionless hardness parameter
A is the exchange stiffness constant
μ_0 is the permeability of the free space (1.2566×10^6 J A^{-2} m^{-1})
M_S is the saturation magnetisation

κ accounts for the balance between dipolar interactions and anisotropy energy. Equation 1.18 might not be suitable for materials with either a large cubic anisotropy or low anisotropies and other expressions should be used instead[84]. It would be worth comparing R_{SD} with other parameters of interest in nanoparticles, such as the *superparamagnetic radius* (R_{SPM}). The latter represents the size from which a single-domain particle will begin to undergo thermal fluctuations:

$$R_{SPM} = \sqrt[3]{\frac{6k_B T}{K}}. \qquad (1.21)$$

Inserting the corresponding values for some oxides like maghemite and magnetite and some metals like iron or cobalt in Equations 1.18 through 1.20, the aforementioned parameters can be roughly estimated (Table 1.3). Maybe the most remarkable detail after a quick inspection of the results is the higher R_{SD} to R_{SPM} ratio for cobalt, owing to its high anisotropy constant. A more drastic size reduction is then needed in the case of cobalt to reach a superparamagnetic regime.

The possible orientation states that could be adopted by the magnetisation in a material are separated through an energy barrier imposed by the *anisotropy energy*. This energy tends to keep the magnetisation in a particular crystallographic direction, called *easy direction* or *easy axis*. More precisely, the easy direction dictates where the magnetisation will be spontaneously pointing at in the absence of an external field, and is mainly determined by a constant K intrinsic to the material. Thus, it is easy to understand that this kind of anisotropy is intimately related to the crystal system, that is, cubic, hexagonal, triclinic, etc., for which is named magnetocrystalline anisotropy. As pointed out later in Section 1.6, although MCA is the major contribution to the total anisotropy energy, other types of anisotropy should be added in accordance with the specific geometrical and structural factors of the sample. In hexagonal crystals, the MCA only depends on the angle formed

TABLE 1.3

Some Characteristic Magnetic Parameters and Length Scales for Different Metals and Oxides at 300 K

	K (J m^{-3})	A (J m^{-1})	M_S (A m^{-1})	κ	l_{ex} (nm)	R_{SD} (nm)	R_{SPM} (nm)	R_{SD}/R_{SPM}
Maghemite	4.6×10^3	$\sim 1 \times \sim 10^{-11}$	3.8×10^5	0.16	7.4	42.5	17.5	2.4
Magnetite	1.35×10^4	1.33×10^{-11}	4.8×10^5	0.21	6.8	52.7	12.2	4.3
Iron	4.8×10^4	1.49×10^{-11}	1.71×10^6	0.11	2.0	8.3	8.0	1.0
Cobalt	4.5×10^5	3.50×10^{-11}	1.42×10^6	0.42	3.7	56.4	3.8	14.8

between the *c* axis and the direction of the magnetisation; that is why it is often named as *uniaxial* anisotropy, and its associated energy can be expressed as follows:

$$E(\theta) = V(K_0 + K_1 \sin^2 \theta + K_2 \sin^4 \theta + \cdots), \tag{1.22}$$

where
K_n are the *n*th order anisotropy constants that can be either negative or positive
V is the sample volume
θ is the angle formed between the magnetisation and the anisotropy axis

As an approximation, the terms above the first-order one are often neglected in uniaxial magnetic nanoparticles, resulting in a much simpler expression than Equation 1.22:

$$E(\theta) = KV \sin^2 \theta \equiv -KV \cos^2 \theta. \tag{1.23}$$

K is positive if the magnetisation lies along the *c* axis (*easy axis configuration*) and negative if it lies somewhere within the plane defined by the other axes *a* and *b* (*hard axis configuration*)*. When a magnetic field is applied in the direction of the anisotropy axis, Equation 1.23 must be subsequently modified, yielding

$$E(\theta) = KV \sin^2 \theta - M_S H. \tag{1.24}$$

Note that the direction of the magnetisation is given by the interplay between the anisotropy axis and the applied field, in such a way that if the magnetic field is forming a certain angle φ with respect to the anisotropy axis (Figure 1.10), Equation 1.24 is then expressed as

$$E(\theta) = KV \sin^2 \theta - HM_S(\cos \theta \cos \varphi + \sin \theta \sin \varphi \cos \psi), \tag{1.25}$$

where Ψ is the angle between the projection of the magnetisation over the plane defined by the axes *a* and *b*. Unless the case of hexagonal crystals, fairly more complicated expressions are derived for

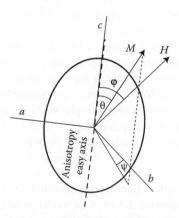

FIGURE 1.10 Usual axis system defined for a uniaxial nanoparticle under an external field.

---

* For simplicity, the first anisotropy constant K_1 will be denoted as just K in the following.

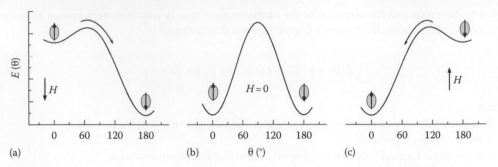

FIGURE 1.11 Switching of the magnetisation in a particle across the anisotropy energy barrier under the action of an external magnetic field H (a) pointing downwards, (b) no field and (c) pointing upwards.

other symmetries such as the cubic one[5,6]. Figure 1.11 shows the profile of the anisotropy energy barrier as calculated from Equation 1.24 and the switching of the magnetisation of a non-interacting uniaxial particle under the following three situations. After the application of an external field downwards (Figure 1.11a), the orientation state pointing in the field direction will become energetically favourable and the system will subsequently switch to that configuration. Removing the field (Figure 1.11b) will revert the system back to the equilibrium state dictated by Equation 1.24 and both orientation states will become equally probable. If the field is then applied upwards (Figure 1.11c), the energy barrier will favour a 'jump' towards the orientation parallel to the field, just the opposite of the first situation.

If we further reduce the size of a single-domain nanoparticle, in the absence of a magnetic field there will be a critical size above which the thermal energy will overcome the anisotropy barrier (Figure 1.11b) causing the magnetisation to rapidly fluctuate from one orientation state to the other*. Then it is said that the system has entered into a *superparamagnetic* regime (Figure 1.9). Bean and Livingston[85] were the first in introducing the term superparamagnetism to designate the thermal fluctuations of the magnetisation in fine particles at temperatures around the ambient one. Such fluctuations lead them to behave like a paramagnet[18], but having a magnetic moment several orders of magnitude higher. Initially introduced by Néel[21], the theory of superparamagnetism was subsequently developed by Brown[22] who, instead of considering the magnetisation vector pointing to a number of discrete orientations, proposed the existence of a distribution of magnetisation orientations under a random field. Despite these conceptual improvements, Brown eventually matched Néel's previous conclusions, although the Brownian model allows for the calculation of the relaxation time for any anisotropy. The set of these results is known as the Néel–Brown model of magnetic relaxation, and it is widely used in studies related to thermally activated magnetic processes both in single- and multi-particle systems. Under a practical point of view, and ignoring the fluctuations associated with the magnetic moments across and at the surface of the particles, it is commonly accepted that a system of nanoparticles is superparamagnetic when[8] (i) plotting the reduced magnetisation (M/M_S) curves versus H/T for a set of temperatures below the blocking temperature (Section 1.6.1.1) they tend to superimpose into a single one (Figure 1.12c and d) and (ii) there is neither coercivity nor remanence, that is, no hysteresis is observed. The H/T scaling behaviour holds as the H/T quotient appears in the Langevin equation's argument.

For the sake of simplicity, we have assumed in Equation 1.24 that the system is composed of non-interacting particles with uniaxial anisotropy though this is not necessarily the usual case that one may come across in real samples. Interactions (Section 1.6.4) are always present to a certain extent, and the second term of Equation 1.24 can be modified to accommodate an interaction field including exchange interactions between nanoparticles if needed.

---

* This process is also referred as *superparamagnetic relaxation*.

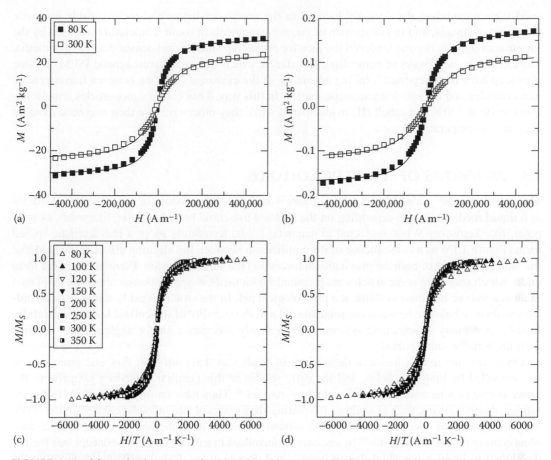

FIGURE 1.12 Magnetisation curves for (a) oleic acid–coated γ-Fe$_2$O$_3$ nanoparticles and (b) the same particles but embedded in a polystyrene matrix[86]. (c) and (d) show the reduced magnetisation versus H/T for the same samples, respectively.

There is a characteristic time for the thermal fluctuations of the magnetisation in a nanoparticle with uniaxial anisotropy to occur, the *relaxation time* τ, and the expression for its temperature dependence was proposed by Néel:[21]

$$\tau = \tau_0 \exp\left(\frac{E_B}{k_B T}\right), \tag{1.26}$$

where
   τ is the relaxation time
   E_B is the anisotropy energy barrier ($\sim KV$)
   k_B is the Boltzmann constant

The pre-exponential factor τ_0 is called *attempt time* and depends on several parameters such as temperature, saturation magnetisation or applied field, among others. For practical purposes, it is assumed to be constant with a value within the range 10^{-9}–10^{-13} s. Note that by changing the temperature or volume of the particles, τ can vary from 10^{-9} s to millions of years.

Given the time by when superparamagnetism was found in fine-particle systems, it constituted a serious drawback rather than an advantage for the technology to come. Researchers working in solid state magnetism were mainly focused in the search for better and more capable magnetic recording materials, and randomisation of magnetic moments at room temperature imposed by the superparamagnetic regime hindered the race for producing smaller and denser data storage media. More recently, new ways of controlling the size at which the superparamagnetic (SPM) regime shows up have been reported[87], taking advantage of the exchange coupling between ferromagnetic nanoparticles and an anti-ferromagnetic matrix. In this way, 4 nm cobalt nanoparticles remain ferromagnetic at 290 K in a cobalt (II) oxide matrix, while they otherwise lose their magnetic moment even at low temperatures.

1.5 PROPERTIES OF MAGNETIC COLLOIDS

Magnetic colloids* – also known as *ferrofluids* – are stable suspensions of magnetic nanoparticles in a liquid medium, which depending on the planned use could be either polar, like water, or non-polar, like kerosene. When subjected to magnetic fields, ferrofluids adopt a characteristic spiked form (Figure 1.13) as a consequence of the equilibrium between the aligning effect of the field and the flattening effect of both the gravitational forces and the surface tension. Ferrofluids are the form under which magnetic nanoparticles are prepared for biomedical applications, commonly dispersed in an aqueous or buffered medium at a physiological pH. In the case of weakly acidic or basic buffer solutions, a balance between the requisite pH and the stability of the colloid has to be carefully found, since many nanoparticle systems tend to slowly precipitate and/or agglomerate over time after an initially stable period.

Originally, the main interest in these kinds of fluids was that both their flow and viscosity can be controlled by magnetic fields, and the early studies of this peculiar behaviour gave rise to the development of a new area called *ferrohydrodynamics*[88]. They have an important number of technological applications, such as sealants of rotating shafts or coolants in loudspeakers[89], but a really burgeoning field of research has been built around their uses in medicine. An example is the idea of emulating muscle movements[90] by enclosing a ferrofluid in a membrane; this concept was further developed to build a ferrofluid-driven heart[91], and despite some design problems concerning the pumping capabilities and the optimal viscosity that questioned its viability, the idea was pushed forward[92]. Considering the generated number of scientific publications and the practical advancements already made, three major bio-applications of magnetic colloids can be highlighted: imaging contrasts, magnetic hyperthermia agents and drug delivery systems. Their fundamentals and recent developments will be presented in other chapters of this book.

FIGURE 1.13 Magnetite-based ferrofluid under the action of a permanent magnet showing the typical spikes produced as a result of the balance between gravitational and demagnetising energies.

* The *International Union of Pure and Applied Chemistry* (IUPAC) refers to colloidal systems as those where the size of the dispersed particles are roughly between 1 nm and 1 μm.

In order to produce a stable suspension, avoiding sedimentation or agglomeration due to interparticle interactions – with adverse effects for bio-applications – ferrofluids must fulfil several stability criteria[93]. The following are two examples of how they can be rationalised:

1. *Gravitational sedimentation.* The thermal energy of the particles ($E_T = k_B T$) has to be strong enough to keep the nanoparticles dispersed in the fluid overcoming the gravitational energy E_g:

$$E_g = \Delta \rho g V h, \tag{1.27}$$

where
  $\Delta \rho$ is the density difference between the particles and the solvent
  g is the gravity
  h is a typical dimension of the sample, for example, the longest dimension
  V is the volume

From the latter, we can easily deduce the diameter by assuming a spherical shape for the particles. Therefore the upper limit for the particle size to meet the stability criterion ($E_T \geq E_g$) is given by

$$d \leq \sqrt[3]{\frac{6k_B T}{\pi \Delta \rho g h}}. \tag{1.28}$$

For a magnetite ferrofluid in water, $\Delta \rho$ is about 4×10^3 kg·m^{-3}, and if we take $h = 0.1$ m, $g = 9.8$ m·s^{-1} and $T = 300$ K, it turns out from Equation 1.28 that particles under 12 nm will not settle down from the suspension.

2. *Sedimentation under the action of external fields.* If the particles are subjected to a magnetic field H, the thermal energy should allow the particles to freely move throughout the carrier overcoming the magnetic energy:

$$E_m = \mu_0 M_S V H, \tag{1.29}$$

where M_S is the spontaneous magnetisation. Thus, the upper limit for the particles to meet this new stability criterion ($E_T \geq E_m$) is of the form:

$$d \leq \sqrt[3]{\frac{6k_B T}{\pi \mu_0 M_S H}}. \tag{1.30}$$

μ_0 is the permeability of free space. For a magnetite ferrofluid with $M_S = 4.5 \times 10^5$ A·m^{-1} in a field of $H = 10^4$ A·m^{-1}, Equation 1.30 dictates that those particles with a diameter smaller than 10 nm will remain stable in the fluid. Similar reasoning will lead to other criteria related to agglomeration due to magnetic interactions and intermolecular attractions, where the role of the surfactant or coating is central. In both cases, a limiting diameter of ca. 10 nm is obtained, provided that a surfactant thickness of 2 nm is used.[89]

Another point of interest in designing applications based on ferrofluids lays upon the relaxation mechanisms by virtue of which the rotation of the constituent magnetic nanoparticles takes place in a liquid carrier as a response to a magnetic field. On the one hand, we saw in Section 1.4 that thermal fluctuations cause a rapid flipping of the magnetisation away from the anisotropy axis, entering in a superparamagnetic state. This relaxation mechanism – *Néel relaxation* – involves just a change in the orientation of all the spins represented by the magnetisation direction and not the physical rotation of the particle. On the other hand, the particles can rotate as a whole to a variable extent depending on the hydrodynamic parameters of both the particles and the medium; this is the so-called *Brown relaxation*:[22]

$$\tau_B = \frac{3V_h\eta}{k_BT},$$
(1.31)

where

V_h denotes the hydrodynamic volume (e.g. obtained from the hydrodynamic diameter measured through dynamic light scattering)

η is the viscosity of the liquid carrier

Figure 1.14 shows the evolution of both relaxation times with different particle diameters for cobalt- and magnetite-based ferrofluids as calculated from Equations 1.26 and 1.31. Bearing in mind that the fastest process will be the prevailing one, in smaller particles Néel relaxation dominates, but within a short range of particle sizes, relaxation times swiftly approach a crossover between both processes, and above a certain critical size, Brown relaxation mechanism takes over. Note how dissimilar this critical size can be for different particle compositions (the considered medium is water for both ferrofluids). In the surroundings of the crossover between the curves of Néel and Brown processes (d_{c1} and d_{c2} in Figure 1.14), where the weight of each contribution is blurred, the net relaxation time is simply given by their geometric mean[94].

The actual behaviour of magnetic fluids is much more complex than it has been described so far, and delivering a theoretical model accounting for the experimental observations has been proven a difficult task. For example, they tend to lag external magnetic fields, giving rise to a continuously changing situation certainly worsened by the internal rotation of the particles. Nonetheless, a set of basic equations for the magnetic relaxation has been put together considering dissipative processes from thermodynamic considerations[95].

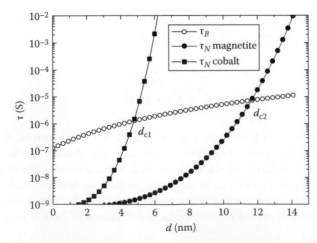

FIGURE 1.14 Theoretical curves for the Néel and Brownian relaxation times in ideal cobalt- and magnetite-based ferrofluids.

1.6 FACTORS INFLUENCING THE MAGNETISM OF NANOPARTICLES

After an introduction of some fundamental notions in nanoscale magnetism and their correlation with crystal structure, in the foregoing sections more specific aspects that further shape the magnetic behaviour of nanoparticle systems will be briefly presented. Apart from modifying the *intrinsic* magnetic properties of materials such as spontaneous magnetisation, anisotropy, Curie temperature or magnetostriction, they act on the different contributions to the anisotropy energy – at least the most significant ones – which in the absence of changes in temperature or external magnetic fields dictates to a significant extent the magnetism in the material. Foreseeing how these factors could be controlled during the preparation stage will allow us to design the magnetic properties for any desired application.

1.6.1 Size

It is well known that the more the size of an entity is decreased, the less the classical physics remain valid to describe the properties of the system, and magnetic particles are not an exception to this statement. There are several reasons for the distinct behaviour found in nanoparticles. The first one is a consequence of their finite size and has to do with the quantisation of energy states, often referred as *quantum confinement*, which relates to the fact that spacing between energy levels of the electron states in atoms increases inverse proportion to the square of the nanoparticle size. This effect gives rise to discrete energy levels instead of the denser set of the levels commonly found in bulk materials. The second reason is the increasing number of atoms located at interfaces and defects as a consequence of purely geometrical constrictions. In the case of nanoparticles, this occurs at their surface. Although the surface itself in bulk materials is very important and in many cases determines their technological interest, the fraction of surface atoms in that case is negligible in comparison with the inner ones. Considering nanoparticles as spheres, the fraction of atoms at the surface of a nanoparticle can be easily worked out from the surface to volume relationship $4\pi r^2 d/(4/3)\pi r^3 = 3d/r$, d being the atomic spacing and r the radius of the particle. Analogously, for a nanowire, we have $2\pi rd/\pi r^2 = 2d/r$. In the case of the cubic system, to which both magnetite and maghemite belong, the interplanar spacing can be estimated from the unit cell parameter (a) and the Miller indexes of adjacent planes using the following formula: $1/d^2 = (h^2 + k^2 + l^2)/a^2$. Maghemite has a unit cell parameter of $a = 0.8396$ nm, so in the case of the (311) set of planes, we have a resulting spacing of 0.253 nm. Inserting this value in the former equations for a set of radii (Figure 1.15),

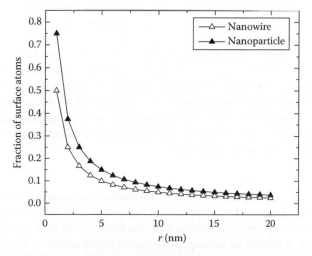

FIGURE 1.15 Fraction of superficial atoms as a function of the radius in nanoparticles. The same fraction for nanowires is also presented for comparison.

it can be seen that the number of surface atoms undergoes a steep increase in both types of nano-structures when the characteristic lengths fall below 5 nm, reaching the 75% for nanoparticles with a radius of 1 nm.

Yet in Section 1.4, a first size effect has been pointed out: the dimensions of nanoparticles are much smaller than the size of magnetic domains in the bulk material; therefore, a new magnetic structure is set up. Another important one is the onset of a superparamagnetic regime as a result of the magnetic moment randomisation imposed by the prominent role of the thermal energy over the magnetic one. In summary, it can be stated that finite-size effects imply the nanoparticle volume as a whole and no particular spatial localisation of the observed phenomena can be done; however, scale reduction brings other actors on stage that complete the real picture of the magnetism in nanomaterials.

Although many protocols have been developed in order to produce monodispersed nanoparticles with a specific size, most of the real systems present a statistical distribution of volumes, adding another degree of complexity that has to be taken into account when interpreting the outcomes of experimental results. A log-normal distribution is commonly found through sizing techniques such as microscopy:

$$f(y) = \frac{1}{\sqrt{2\pi}\sigma(y-\theta)} \exp\left(\frac{-\ln^2\left(\frac{(y-\theta)}{\mu}\right)}{2\sigma}\right), \quad (1.32)$$

where
  y stands for the particle diameter
  θ is the localisation parameter
  μ is the scale parameter
  σ is the shape parameter ($\theta = 0$ and $\mu = 1$ for a standard log-normal distribution)

Calculations made on systems with a volume distribution often include the statistical distribution function rather than the average volume; this is exemplified by the ZFC/FC models presented later in this section. For extracting some quantitative results in systems where the sizes are distributed following a bimodal-like distribution, it has been found very convenient to insert these bimodal functions experimentally obtained from measuring techniques in order to yield more physically meaningful fits to the Langevin equation[96]. If particle diameters were to be worked out from the fits to magnetisation curves acquired at different temperatures – in a non-rigorous nomenclature, the *magnetic diameter* – the obtained values can show unexpected dependencies. Figure 1.16 reflects this for a set of magnetic diameters from three nanoparticulate iron oxide samples[86], where an apparent reduction of the magnetic nanoparticle size is observed for decreasing temperatures. This behaviour has been interpreted in terms of the distribution of blocking temperatures with respect to the current measuring temperature, although the interparticle interactions that have not been taken into account could play a role in this departure.

1.6.1.1 Blocking Temperature

As we have seen before, the relaxation time of magnetic nanoparticles is affected by tempera-ture changes as a consequence of its exponential dependence on the thermal energy of the sys-tem. The characteristic frequency associated with the fluctuations of magnetic moments between anisotropy minima in a system of superparamagnetic nanoparticles undergoes a progressively damping or attenuation as the temperature is decreased in a controlled manner. For a particular

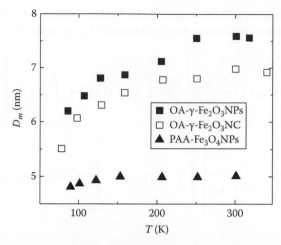

FIGURE 1.16 Magnetic diameter (D_m) calculated from magnetometry measurements at selected temperatures within the 80–300 K range for oleic acid–coated maghemite nanoparticles, oleic acid–coated maghemite nanocomposites and polyacrylic acid–coated magnetite nanoparticles[86].

temperature value, the relaxation time of the nanoparticles matches the characteristic measurement time; such temperature is regarded as the *blocking temperature*. There is a number of parameters that will affect how and when this moment is reached, such as the external field, interactions and particle size distribution, the latter being of a major influence as attested by the outcomes from several works[97–99].

In a significant number of publications, there is a certain tendency to use this parameter to characterise a particular system and compare it to others; moreover, it is also employed in calculating average particle sizes. In this regard, it has to be noted that the blocking temperature is not uniquely defined and depends on the instrumental technique and therefore on its measuring time, which in turn is not always precisely known. For example, in DC magnetometry, a τ_m of about 60–100 s is often considered for the sake of simplicity, but the uncertainty in this value will affect the estimated blocking temperature and so the particle volume. If $\tau_m = 100$ s, then Equation 1.26 leads to a simpler expression commonly used to roughly estimate T_B in a system of nanoparticles assumed to be uniaxial and non-interacting:

$$T_B = \frac{KV}{25k_B}. \tag{1.33}$$

Additionally, the particle diameter calculated from the volume sometimes implies certain assumptions to simplify calculations such as considering the particles as perfect spheres or neglecting the effects of dipolar or exchange interactions, as well as assuming an uniaxial anisotropy for the whole set of particles. This does not discard T_B and its evolution with temperature and the external field as a source of useful information to study the suitability of a certain nanoparticle system for a given application within the tested conditions. It has to be stressed once again that T_B must not be confused with the maximum temperature commonly observed in the ZFC branch of ZFC/FC experiments, denoted as T_{max}.

Hansen and Mørup derived a simple model – based on the Stoner–Wohlfart's one[100] – for estimating the blocking temperature of an ensemble of nanoparticles considering the system as a log-normal distribution of energy barriers[101]. Particles are assumed uniaxial and non-interacting. Calculations are made within two different temperatures ranges $T > T_B$ and $T_B \geq T$ to account for the unblocked and blocked state of the nanoparticles, respectively. In this approach, T_{irr} is taken as the temperature where M_{FC}–M_{ZFC} equals 10% of the magnetisation at T_{max} by convention:

$$M_{ZFC}(T) \propto \frac{M_S^2}{3KH}\left[\frac{E_{bm}}{k_B}\int_0^{\frac{T}{T_{Bm}}} T^{-1}y\,f(y)dy + \int_{\frac{T}{T_{Bm}}}^{\infty} f(y)dy\right] \qquad (1.34)$$

$$M_{FC}(T) \propto \frac{M_S^2}{3KH}\frac{E_{bm}}{k_B}\left[\int_0^{\frac{T}{T_{Bm}}} T^{-1}y\,f(y)dy + \int_{\frac{T}{T_{Bm}}}^{\infty} [T_B(yE_{bm})]^{-1}y\,f(y)dy\right], \qquad (1.35)$$

where

M_S accounts for the saturation magnetisation

K (=E_B/V, with V the particle volume) are the anisotropy constants

E_{bm} are the median energy barriers

y are the reduced energy barriers (E_B/E_{bm}), T_B (=$E_B/[k_B \ln(t_m/\tau_0)]$) and T_{Bm} (=$E_{bm}/[k_B \ln(t_m/\tau_0)]$) the blocking temperature and its median value

k_B is the Boltzmann constant

$f(y)$ is the statistical distribution function of energy barriers, a log-normal kind given by Equation 1.32

The first addends between square brackets in both Equations 1.34 and 1.35 stand for the fraction of blocked particles, while the second one represents the fraction of superparamagnetic particles. A measurement time $t_m = 100\,\mathrm{s}$ is assumed and, given the small influence of τ_0 value in the fits, a typical value of 10^{-10} s has been used. Despite its simplicity, this model can be used to extract some interesting conclusions on ZFC/FC experiments of nanoparticulate samples.

Focusing on the ZFC curves, which provide us with more information on the blocking processes, the results of some calculations using Equation 1.34 are shown in Figure 1.17a, illustrating the influence of the applied field during a ZFC experiment. The curves have been calculated using a particle size of 5 nm, the anisotropy constant of maghemite (4.6×10^3 J·m^{-3}) and neither interactions nor surface effects have been considered. On the one hand, the vertical dashed line across the curves indicates that the field

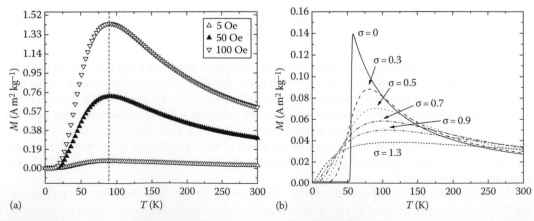

FIGURE 1.17 Applied magnetic field (a) and nanoparticle size distribution standard deviation (b) effects on the shape of theoretical ZFC magnetisation curves of maghemite nanoparticles as calculated from Equation 1.34. The vertical dashed line in (a) is merely a guide to the eye.

intensity shifts the position of T_{max} towards lower temperatures. That makes sense if we think about the system as a distribution of energy barriers: the magnetic field tends to decrease the amplitude of such barriers, favouring a specific orientation of the magnetisation (recall Figure 1.11) and also producing an overall increase in the total magnetisation of the system. On the other hand, the minimal influence of the field at the lowest temperatures side is in agreement with the low magnetisation state of the nanoparticles population as a consequence of the blocking process of the magnetic moments in random directions induced by the zero-field freezing during the experiment.

Figure 1.17b shows several simulations corresponding to different values of standard deviation in the size distribution function. As in the preceding case, a particle size of 5 nm has been considered, but now the measuring field has been set to 5 Oe. The particles have been assumed as uniaxial and non-interacting. In this occasion, there is a visible shift of T_{max} towards higher temperatures as the size dispersion of the system increases, since average T_{max} value is implicitly increasing. Also, there is a broadening and flattening of the ZFC curve shape around T_{max} as the size standard deviation increases.

Besides theoretical calculations, Equations 1.34 and 1.35 also fit experimental data quite well. Figure 1.18 shows the fit of the ZFC curve for a sample of maghemite nanoparticles grown in a silica matrix using two log-normal functions to reproduce the bimodal-like distribution experimentally found in the sample through TEM images[102].

Although the previously described approach gives us a rough idea of the basic interplay of various experimental parameters in ZFC/FC measurements, the model is basic on severe restrictions, which will affect the accuracy of the results. Very recently, another model has been proposed to better account for the evolution of the system magnetisation as a function of time for various initial conditions[104]. A dependence of T_B on the measuring field is detected and ascribed to the distribution of energy barriers and, as expected, a marked dependence of the T_B with the size distribution width is observed as well.

Another suitable technique for monitoring the evolution of the blocking temperature is Mössbauer spectroscopy. Figures 1.19 and 1.20 show the thermal dependence of the Mössbauer spectra obtained from the same maghemite nanoparticles sample analysed above. An initial broad sextet, accounting for the already blocked particles, along with a central singlet, containing the contribution of the superparamagnetic fraction of particles in the sample, is observed at higher temperatures. As the cooling proceeds, the sextet grows at the expense of the central singlet, depicting how the blocking process of the particles is taking place. The blocking temperature here in the

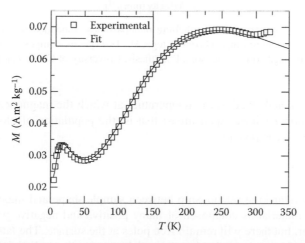

FIGURE 1.18 ZFC fit to Equation 1.34 for a sample of maghemite nanoparticles grown in an SiO_2 matrix. (From Ortega, D., Transparent composite materials with magnetic nanoparticles for current magneto-optic sensors, PhD thesis, University of Cádiz, Cádiz, Spain.)

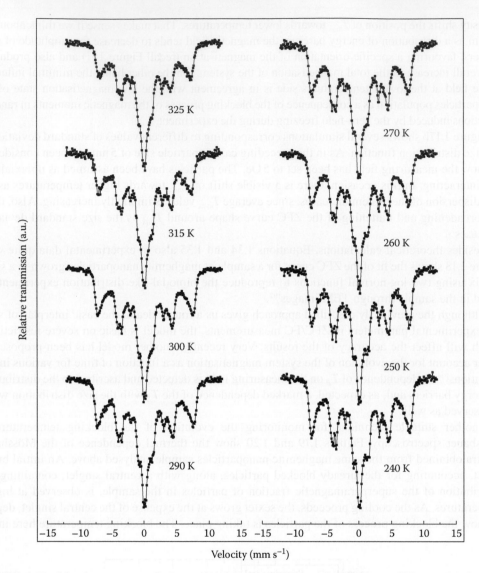

FIGURE 1.19 Mössbauer spectra from maghemite nanoparticles embedded in an SiO_2 matrix obtained within the 325–240 K temperature range. (From Ortega, D., Transparent composite materials with magnetic nanoparticles for current magneto-optic sensors, PhD thesis, University of Cádiz, Cádiz, Spain.)

Mössbauer spectra is then defined as the temperature at which the magnetically split (sextet) and unsplit (singlet) components have equal areas; that is, the population of blocked and unblocked particles is 50% for each component.

1.6.2 Morphology

The best way to understand the relationship between sample shape and magnetic properties is to consider a magnetic material as composed of many positive and negative poles; as a result they will cancel each other, but there will remain free poles at the surface. The latter then create then a magnetic field that tends to oppose the direction of the externally applied field: it is the so-called *demagnetising field* and it exclusively depends on the sample shape. As a major consequence, the

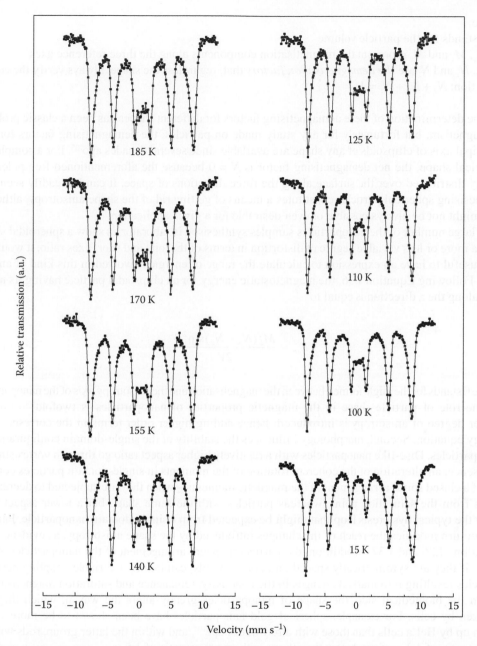

FIGURE 1.20 Mössbauer spectra from maghemite nanoparticles embedded in a silica matrix. (From Ortega, D., Transparent composite materials with magnetic nanoparticles for current magneto-optic sensors, PhD thesis, University of Cádiz, Cádiz, Spain.)

effective magnetic field acting on the sample will be different from the externally applied one. For a single-domain ferromagnet/ferrimagnet, there should be then a *shape anisotropy* due to the internal magnetostatic energy[24] of the form:

$$E_{sh} = \frac{1}{2V}(N_x M_x^2 + N_y M_y^2 + N_z M_z^2), \tag{1.36}$$

where

V stands for the particle volume

M_x, M_y and M_z represent the magnetisation components along the three reference axes

N_x, N_y and N_z are *the demagnetisation factors* that, regardless the shape, always verify the condition: $N_x + N_y + N_z = 1$

The determination of these demagnetising factors for different shapes has been a classic problem in magnetism, but fortunately for any study made on particles, the demagnetising factors for any principal axis of ellipsoids of any shape are available since several decades ago[105]. For a completely spherical shape, the net demagnetising factor is $N = 0$ because the aforementioned free poles are evenly distributed over the surface along the three directions of space. It can be readily seen that synthesising spherical particles constitutes a means of getting rid of the shape anisotropy, although this might not be always possible or even desirable for a given application.

A large number of the nanoparticles samples synthesised in laboratories show a spheroidal shape with a more or less variable degree of distortion in terms of the long and short axes ratio; it would be then useful to have an expression to calculate the range of energies involved in this kind of anisotropy. Following Equation 1.36, the magnetostatic energy for an ellipsoidal particle having its major axis along the z direction is equal to

$$E_{sh} = \frac{M_S^2(N_x - N_z)\sin^2\theta}{2V},\qquad(1.37)$$

where θ stands for the angle formed between the magnetisation and the anisotropy axis of the nanoparticle.

The role of particle shape in the magnetic properties of nanoparticles is twofold[106]. First, a higher degree of anisotropy is introduced, hence adding higher order terms in the corresponding energy equation. Second, morphology influences the stability of the single-domain configuration in nanoparticles. Disc-like nanoparticles with a relatively higher aspect ratio go through vortex states – in essence, an alteration of the coherent rotation of the moments in single-domain particles consisting of a closed spiral arrangement of the magnetic moments – when they are subjected to decreasing fields from the saturation point, whereas particles with the same shape but a lower aspect ratio show the typical hysteresis loop that might be expected from a single-domain nanoparticle. Particle size in turn modifies the reach of the changes introduced by the shape anisotropy, as evidenced by Equations 1.36 and 1.37. In addition, the particular spatial arrangement of the nanoparticles, especially if they are symmetrically spaced and not randomly, can induce a variable coupling between particles resulting into marked changes in the coercivity, remanence and saturation magnetisation.

On the biomedical side, the shape of nanoparticles greatly alters the way in which they are uptaken by cells. For example, spherical gold nanoparticles have been shown to be more easily taken up by HeLa cells than those with a rod-like shape[107], and within the latter group, rods with an aspect ratio of 1:3 perform better than those with an aspect ratio of 1:5.

1.6.3 Intrinsic and Extrinsic Surface Effects

It has been previously pointed out, in Section 1.6.1 how dramatically important the fraction of atoms at the surface of a nanoparticle becomes when its size falls below 10 nm. This fact reinforces the idea that ascribing deviations from the usual behaviour observed in small particles to surface effects is perfectly feasible. In magnetic nanoparticles, the study of this topic has attracted an enormous attention from both theoretical and experimental points of view as instrumental techniques began to reveal new phenomena and computing facilities exponentially increased their performance to cope with long and heavy calculations performed on manifold systems. A fantastic showcase of the most interesting findings in surface effects associated with magnetic nanoparticles has been compiled by Fiorani[24].

Depending on whether the superficial modifications are intrinsic to the material or are introduced by external agents, two types of surface effects can be distinguished.

1.6.3.1 Surface Spin Disorder

During the last decades, a common issue observed in magnetisation curves of magnetic oxides nanoparticles is that saturation is not reached even at high applied magnetic fields;[108] now it is well known that the frustration degree originated in the superficial spins is responsible for this effect. Following are some brief considerations on its structural origin.

Surface atoms show a lower coordination number than those located inside a nanoparticle, thus inducing an energy imbalance over the network that can be compensated either by creating new vacancies or undergoing a structural relaxation. Whichever the mechanism, the original long-range order cannot be recovered, and this atomic rearrangement will therefore modify the magnetic order resulting in frustrated interactions between the associated spins. The typical anti-ferromagnetic exchange of spins between sublattices in ferrimagnetic oxides (Section 1.2.1) will also be rearranged to find an equilibrium state, giving rise to the *spin canting* phenomenon. In this framework, the term 'canting' refers to a non-collinear coupling of spins at the surface of the nanoparticles, forming a disordered shell around an ordered core. This description is commonly referenced as a *core–shell* type model. An important remark to be done in this regard is that the superficial spin canting has not been observed in metallic nanoparticles[109]. This suggests that the magnetic frustration observed in oxide nanoparticles comes in fact from the anti-ferromagnetic arrangement adopted by the spins in these compounds (Section 1.2.1). To account for the modifications introduced by this effect in the anisotropy energy of the system, Bødker et al. proposed an expression for an effective anisotropy (K_{eff}) including the surface anisotropy contribution:[110]

$$K_{eff} = K_v + \frac{6}{D} K_s,$$ (1.38)

where

K_v represents the volumetric or bulk anisotropy of the nanoparticles

K_s is the surface anisotropy

Surface effects become more apparent at low temperatures, where a collective frozen state* is reached by the spins. As a consequence of the disordered arrangement of spins, this collective state cannot be ferromagnetic. Figure 1.21 displays the magnetisation curves obtained at different temperatures down to 10 K of 4.5 nm maghemite nanoparticles embedded in an SiO_2 matrix in a weight concentration of 3%. The particles are therefore fixed in random positions within the solid matrix. Whereas the curves in the range 70–300 K show a gradual increase in magnetisation at high fields, the 10 K curve experiences a marked change in slope as well as a rise of about 20% of the total magnetisation. In addition, the saturation state is not reached even at high applied fields. This behaviour can be explained in terms of the increase in the effective anisotropy introduced by the frustrated surface spins (Equation 1.38). According to numerical simulations of hysteresis loops of 10 nm ferrite-like nanoparticles with a random distribution of anisotropy axes at zero temperature[111], even in the absence of anisotropy, a non-negligible coercive field only due to dipole–dipole interactions appears for all the loops within a volume concentration range between 0.03% and 15%. In fact, Figure 1.21b to shows a noticeable increase in coercivity at 10 K (20,150 A·m^{-1}), but unlike the numerical simulations, the effective anisotropy contribution should be taken into account to explain the experimental results.

The consequences of surface spin disorder, in combination with finite-size effects, are also very remarkable in the case of nanoparticles of anti-ferromagnetic materials, like NiO, CoO or α-Fe$_2$O$_3$.

---

* The collective frozen state is a concept related to the study of spin glasses, and sometimes is employed in nanoparticles to describe some of the resemblances to the spin glasses.

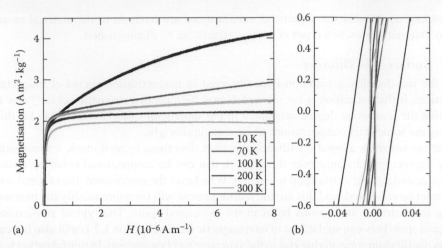

FIGURE 1.21 Magnetisation curves at different temperatures in the range 10–300 K for a sample of maghemite nanoparticles embedded in an SiO_2 matrix[103] (a) and detail showing the increase in coercivity at the lowest temperature (b). Note the change in the approach to saturation towards lower temperatures.

These were reviewed by Mørup et al.[112] One piece of experimental evidence found is the much higher magnetic moment measured in anti-ferromagnetic nanoparticles in comparison either with their bulk counterparts or even with many ferromagnetic or ferrimagnetic (oxide) nanoparticles, where the experimental magnetisation is lower than expected. The origin of this enhanced magnetisation has to be found in the uncompensated magnetic moment induced by (i) the different number of spins in each of the sublattices (Figure 1.2) due to finite-size effects and (ii) a thermoinduced magnetisation related to the fact that both sublattices are not completely anti-parallel as a result of thermal excitations in the uniform spin-precession mode[113]. In addition, the large coercivity measured in anti-ferromagnetic nanoparticles at low temperatures[114] supports the existence of a surface anisotropy produced by disordered spins, which allows for the description of the nanoparticles by the aforementioned core–shell structure. The anti-ferromagnetic core is then exchange coupled to the disordered shell, giving rise to the apparition of an *exchange bias* phenomenon[115], which demonstrates as a shift in the hysteresis loops after FC the sample, leading to an asymmetry in coercivity values.

1.6.3.2 Coatings

In many applications, nanoparticles require a surface treatment in order to protect them from reduction–oxidation processes, redissolution or uncontrolled growth, side reactions with other species that should remain intact in the medium or, on the contrary, increase their affinity for other molecules to further extend their functionalities. *Coating* is the word commonly chosen to designate the process by virtue of which a nanoparticle is covered with a certain reagent forming a stable layer of a variable thickness over its surface. The nature of the bond formed between nanoparticle and coating can range from a simple electrostatic interaction to a covalent bond, and this can substantially alter the genuine properties of the particle. Thiol molecules can surprisingly induce the onset of a permanent ferromagnetic order in gold, silver or copper nanoparticles that otherwise are diamagnetic[116,117]. Several measurements seem to indicate that the magnetic moment in these nanoparticles is located at the surface; a charge-transfer phenomenon involving a change in the d density of states would account for this effect. Also, room temperature ferromagnetism has been achieved in ZnO nanoparticles – typically diamagnetic – by attaching several long-chain thiol, amine and phosphine molecules to their surface[118], each one producing a different effect intensity. These two cases illustrate that a coating can literally turn magnetic what is not.

Important deviations from the Bloch's law have been reported for magnetite and maghemite nanoparticles with different coatings[86]. For bulk materials, it is well known that the variation of

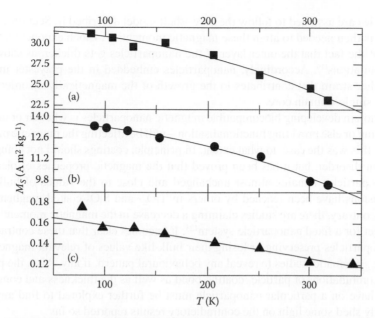

FIGURE 1.22 Comparison of the magnetisation thermal dependence measured in samples (a) oleic acid–coated maghemite, (b) polyacrylic acid–coated magnetite nanoparticles and (c) maghemite in a polyester-based nanocomposite[86]. The corresponding curve fittings to the Bloch's law are represented by solid lines.

saturation magnetisation with temperature $M_S(T)$ should follow the Bloch's law, which may be deduced from the spin wave theory:

$$M_S(T) = M_S(0)[1 - BT^n], \tag{1.39}$$

where

$M_S(0)$ is the spontaneous magnetisation at 0 K

B is a constant that depends on the spin wave stiffness and, thus, on the inverse of the exchange integral J[7]

The exponent n is equal to 3/2 for bulk materials. In the case of particles and clusters, the exponent n is higher than 3/2, and may reach 2 as a consequence of the reduction in the size of the particles[119]. This is the case of both oleic acid–coated maghemite and polyacrylic acid–coated magnetite nanoparticles, where the best fit of the experimental results is obtained when $n = 2$ (Figure 1.22a and b). However, when maghemite nanoparticles are embedded in a polyester resin instead of coated with oleic acid, the exponent from the modified Bloch's law is below 3/2, closer to 1. The best fit here was obtained for $n = 1.1$ (Table 1.4; Figure 1.22c).

TABLE 1.4

Fitting Parameters of Experimental Data to Bloch's Law

Sample	n	$M_S(0) (A\,m^{-2}\,kg^{-1})$	$B(K^{-n})$	R
OA-γ-Fe$_2$O$_3$ NPs	2	31.72	2.59×10^{-6}	0.991
PAA-Fe$_3$O$_4$ NPs	2	13.73	3.03×10^{-6}	0.995
OA-γ-Fe$_2$O$_3$ NC	1.1	0.17	4.00×10^{-4}	0.995

If the particles are assumed to follow the core–shell model described in Section 1.6.3.1, a higher magnetic field is then needed to align these magnetic moments. A Bloch exponent lower than 2 is a consequence of the fact that the outer layer of the nanoparticles gets thicker, as shown in previous theoretical simulations[119]. Accordingly, nanoparticles embedded in the polyester matrix are subjected to a higher strain that contributes to the growth of the magnetically disordered shell at the expense of the single-domain core.

A key question in developing biocompatible magnetic nanoparticles is whether or not the coating– just encapsulating or else providing functionalisation – will be affecting the magnetic properties of the product and, if this was the case, to what extent. In principle, coatings should not completely remove the surface spin disorder, but it has been proved that the magnetic properties of oleic acid–coated magnetite nanoparticles remains almost unchanged and close to the values of bulk magnetite[120]. Similar conclusions have been reached by others in TiO_2- and SiO_2-coated maghemite nanoparticles[121]. On the contrary, there are studies claiming a decrease in the magnetic moment[122], or a coating dependent effect for a fixed nanoparticle system[123]. It becomes clear that these contradicting reports on coated nanoparticles preserving or losing their bulk-like values of relevant magnetic parameters claim for more systematic studies to reveal any behavioural pattern, if any. Also, the possible effects that different combinations of particle–coating bond as well as the thickness and composition of the coating could have on a particular nanoparticle must be further explored to find any patterns that could eventually shed some light on the contradictory results reported so far.

1.6.4 INTERACTIONS

The concept of exchange interaction was previously described in Section 1.2.1. The coupling between atoms in magnetic materials may take place in different ways as a result of particular energetic considerations. These are summarised in Table 1.5 along with some examples.

In magnetic nanoparticles, the main type of interactions commonly found are dipolar and surface-mediated exchange among particles in close contact, the latter being much weaker than the intra-atomic exchange present in ferromagnetic materials. It has been previously pointed out in Section 1.4 that non-interacting nanoparticles tend to follow the Néel–Brown relaxation theory, but when interparticle interactions are non-negligible, the individual anisotropy energy barriers of nanoparticles are modified, and for strong interactions, the superparamagnetic relaxation may even be suppressed[124]. A typical means of preventing agglomeration and getting rid of interparticle inter-

TABLE 1.5

Main Types of Exchange Interactions in Magnetic Materials

Interactions	Origin	Examples
RKKY	Exchange coupling of magnetic moments over a relatively large distance through conduction electrons	MnPt, YDy
Direct exchange	Direct coupling of magnetic ions through overlapping magnetic orbitals	Metals
Super exchange	Coupling (typically anti-ferromagnetic) between two magnetic ions through the p-orbital of a non-magnetic anion	NiO
Double exchange	Coupling between magnetic ions in mixed valence configurations by hopping of the extra electron from one ion to the other through p-orbitals	Manganese-doped III–IV semiconductors
Anti-symmetric exchange	Perpendicular coupling with respect to the high-symmetry axis between already super exchange coupled spins	Hematite

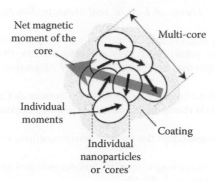

FIGURE 1.23 Schematic representation of a multi-core magnetic particle.

actions is to coat the nanoparticles with a surfactant molecule, but this is not necessarily enough to achieve a non-interacting system.

The study of interactions is of particular interest in nanoparticles with applications in medicine, as those aimed for drug targeting or magnetic hyperthermia are sometimes formulated under the form of coated *multi-cores* (Figure 1.23). In these preparations, the particles are in close contact, hence both dipole–dipole and exchange interactions are expected to act. In fact, recent Monte Carlo simulations on multi-core superparamagnetic magnetite nanoparticles[125] show that dipolar interactions contribute to a decrease in the overall magnetisation of multi-core aggregates. Given the relevance of the sort of applications these nanoparticles are aimed for, it would be desirable to have more experimental measurements for quantifying the actual reach of interactions in order to minimise them, or given the fact that the spatial arrangement of multi-cores gives little chances to do so, designing strategies to somehow take advantage of them.

ACKNOWLEDGEMENTS

The author acknowledges the support received from the European Commission's PEOPLE programme under the 7th Framework Programme. Also, the author would like to gratefully thank C. Blanco-Andújar, J. Owen, K. Page and P. Southern for their valuable support.

REFERENCES

1. Gupta, R., Davis, K., Ramsey, C., Meunier, A. and Kordzik, K. 2009. The metal oxide nanoparticle patent landscape. *Nanotechnology Law & Business*, 6, 354–373.
2. Bawa, R. 2008. Nanoparticle-based therapeutics in humans: A survey. *Nanotechnology Law & Business*, 5, 135–156.
3. Chowdhury, N. 2009. Regulation of nanomedicines in the EU: Distilling lessons from the pediatric and the advanced therapy medicinal products approaches. *Nanomedicine*, 5, 135–142.
4. Till, M. C., Simkin, M. M. and Maebius, S. 2005. Nanotech meets the FDA: A success story about the first nanoparticulate drugs approved by the FDA. *Nanotechnology Law & Business*, 2, 163–167.
5. Aharoni, A. 2000. *Introduction to the Theory of Ferromagnetism*, Oxford, U.K., Oxford University Press.
6. Chikazumi, S. 2009. *Physics of Ferromagnetism*, Oxford, MI, Oxford University Press.
7. Coey, J. M. D. 2010. *Magnetism and Magnetic Materials*, Cambridge, MA, Cambridge University Press.
8. Cullity, B. and Graham, C. 2008. *Introduction to Magnetic Materials*, Piscataway, NJ, Wiley-IEEE Press.
9. Jiles, D. 1998. *Introduction to Magnetism and Magnetic Materials*, New York, Chapman & Hall.
10. Skomski, R. 2008. *Simple Models of Magnetism*, Oxford, MI, Oxford University Press.
11. Weinberger, P. 2009. *Magnetic Anisotropies in Condensed Matter*, Boca Raton, FL, CRC Press.
12. Mayergoyz, I. D. 1990. *Mathematical Models of Hysteresis and Their Applications*, Berlin, Germany, Springer.
13. Faraday, M. 1857. The Bakerian lecture: Experimental relations of gold (and other metals) to light. *Philosophical Transactions of the Royal Society of London*, 147, 145–181.

14. Van Vleck, J. H. 1966. *The Theory of Electric and Magnetic Susceptibilities*, London, U.K., Oxford University Press.

15. Goudsmit, S. A. 1976. Fifty years of spin: It might as well be spin. *Physics Today*, 29, 40–43.

16. Uhlenbeck, G. E. 1976. Fifty years of spin: Personal reminiscences. *Physics Today*, 29, 43–48.

17. Coey, J. M. D. 2001. Magnetism in future. *Journal of Magnetism and Magnetic Materials*, 226, 2107–2112.

18. Langevin, P. 1905. Magnétisme et théeorie des électrons. *Annales de Chimie et de Physique*, 5, 70–127.

19. Weiss, P. 1907. L'hypothèse du champ moléculaire et la propriété ferromagnétique. *Journal de Physique*, 6, 661–690.

20. Néel, L. 1948. Propriétés magnétiques des ferrites: Ferrimagnétisme et antiferromagnétism. *Annales de Physique*, 3, 137–198.

21. Néel, L. 1949. Théorie du traînage magnétique des ferromagnétiques en grains fins avec applications aux terres cuites. *Annales Geophysicae*, 5, 99–136.

22. Brown, W. F. 1963. Thermal fluctuations of a single-domain particle. *Physical Review*, 130, 1677–1686.

23. Binder, K. and Young, A. P. 1986. Spin glasses: Experimental facts, theoretical concepts, and open questions. *Reviews of Modern Physics*, 58, 801–976.

24. Fiorani, D. 2005. *Surface Effects in Magnetic Nanoparticles*, New York, Springer Verlag.

25. Martínez, B., Obradors, X., Balcells, L., Rouanet, A. and Monty, C. 1998. Low temperature surface spin-glass transition in γ-Fe_2O_3 nanoparticles. *Physical Review Letters*, 80, 181–184.

26. Cornell, R. M. and Schwertmann, U. 2003. *The Iron Oxides: Structure, Properties, Reactions, Occurrences, and Uses*, Weinheim, Germany, Wiley-VCH.

27. Dronskowski, R. 2001. The little maghemite story: A classic functional material. *Advanced Functional Materials*, 11, 27–29.

28. Bate, G. (ed.) 1975. *Oxides for Magnetic Recording*, London, U.K., Wiley.

29. Michels, A. and Weissmüller, J. 2008. Magnetic-field-dependent small-angle neutron scattering on random anisotropy ferromagnets. *Reports on Progress in Physics*, 71, 066501.

30. Krycka, K. L., Booth, R., Borchers, J. A., Chen, W. C., Conlon, C., Gentile, T. R., Hogg, C., Ijiri, Y., Laver, M., Maranville, B. B., Majetich, S. A., Rhyne, J. J. and Watson, S. M. 2009. Resolving 3D magnetism in nanoparticles using polarization analyzed SANS. *Physica B: Condensed Matter*, 404, 2561–2564.

31. Dartyge, E., Baudelet, F., Brouder, C., Fontaine, A., Giorgetti, C., Kappler, J. P., Krill, G., Lopez, M. F. and Pizzini, S. 1995. Hard X-rays magnetic EXAFS. *Physica B: Condensed Matter*, 208–209, 751–754.

32. Srajer, G., Lewis, L. H., Bader, S. D., Epstein, A. J., Fadley, C. S., Fullerton, E. E., Hoffmann, A., Kortright, J. B., Krishnan, K. M., Majetich, S. A., Rahman, T. S., Ross, C. A., Salamon, M. B., Schuller, I. K., Schulthess, T. C., and Sun, J. Z. 2006. Advances in nanomagnetism via X-ray techniques. *Journal of Magnetism and Magnetic Materials*, 307, 1–31.

33. Kronmüller, H. and Parkin, S. S. P. 2007. *Handbook of Magnetism and Advanced Magnetic Materials*, New York, John Wiley & Sons, Ltd.

34. Brüche, E. 1933. Elektronenmikroskopische Abbildung mit lichtelektrischen Elektronen. *Zeitschrift für Physik*, 86, 448–450.

35. Dunin-Borkowski, R. E., Kasama, T., Wei, A., Tripp, S. L., Hytch, M. J., Snoeck, E., Harrison, R. J. and Putnis, A. 2004. Off-axis electron holography of magnetic nanowires and chains, rings, and planar arrays of magnetic nanoparticles. *Microscopy Research and Technique*, 64, 390–402.

36. Mccartney, M. R. and Smith, D. J. 2006. Electron holography of ferromagnetic materials. In: Kronmüller, H. and Parkin, S. (eds.) *Handbook of Magnetism and Advanced Magnetic Materials*, Vol. 3. New York: John Wiley & Sons, Ltd.

37. Kirk, K. J., Chapman, J. N. and Wilkinson, C. D. W. 1999. Lorentz microscopy of small magnetic structures. *Journal of Applied Physics*, 85, 5237–5242.

38. Zweck, J. and Uhlig, T. 2006. Lorentz microscopy of thin-film systems. In: Kronmüller, H. and Parkin, S. (eds.) *Handbook of Magnetism and Advanced Magnetic Materials*, Vol. 3. New York: Wiley-Blackwell.

39. Schwarz, A., Bode, M. and Wiesendanger, R. 2006. Scanning probe techniques: MFM and SP-STM. In: Kronmüller, H. and Parkin, S. (eds.) *Handbook of Magnetism*, Vol. 3. New York: Wiley-Blackwell.

40. Schwarz, A. and Wiesendanger, R. 2008. Magnetic sensitive force microscopy. *Nano Today*, 3, 28–39.

41. Barandiarán, J. M. and Schmool, D. S. 2000. Ferromagnetic resonance studies of multiphase ferromagnets. *Journal of Magnetism and Magnetic Materials*, 221, 178–186.

42. Schmool, D. S. and Schmalzl, M. 2007. Ferromagnetic resonance in magnetic nanoparticle assemblies. *Journal of Non-Crystalline Solids*, 353, 738–742.

43. Vonsovskii, S. V. 1964. *Ferromagnetic Resonance*, Oxford, U.K., Pergamon Press.

44. Dalmas De Réotier, P., Gubbens, P. C. M. and Yaouanc, A. 2004. Probing magnetic excitations, fluctuations and correlation lengths by muon spin relaxation and rotation techniques. *Journal of Physics: Condensed Matter*, 16, S4687–S4705.

45. Dalmas De Réotier, P. and Yaouanc, A. 1997. Muon spin rotation and relaxation in magnetic materials. *Journal of Physics: Condensed Matter*, 9, 9113–9166.

46. Greenwood, N. N. and Gibb, T. C. 1971. *Mössbauer Spectroscopy*, London, U.K., Chapman & Hall Ltd.

47. Murad, E. 1998. Clays and clay minerals: What can Mössbauer spectroscopy do to help understand them? *Hyperfine Interactions*, 117, 39–70.

48. Foner, S. 1956. Vibrating sample magnetometer. *Review of Scientific Instruments*, 27, 548.

49. Pouliquen, D., Le Jeune, J. J., Perdrisot, R., Ermias, A. and Jallet, P. 1991. Iron oxide nanoparticles for use as an MRI contrast agent: Pharmacokinetics and metabolism. *Magnetic Resonance Imaging*, 9, 275–283.

50. Garrido-Ramírez, E. G., Theng, B. K. G. and Mora, M. L. 2010. Clays and oxide minerals as catalysts and nanocatalysts in Fenton-like reactions – A review. *Applied Clay Science*, 47, 182–192.

51. Pileni, M. P. and Ngo, A. T. 2005. Mesoscopic structures of maghemite nanocrystals: Fabrication, magnetic properties, and uses. *ChemPhysChem*, 6, 1027–1034.

52. Schwertmann, U. and Cornell, R. 1991. *Iron Oxides in the Laboratory: Preparation and Characterization*, New York, Wiley-VCH.

53. Cudennec, Y. and Lecerf, A. 2005. Topotactic transformations of goethite and lepidocrocite into hematite and maghemite. *Solid State Sciences*, 7, 520–529.

54. De Biasi, R. S. and Portella, P. D. 1980. Magnetic resonance study of the transformation γ-Fe$_2$O$_3$ -> α-Fe$_2$O$_3$. *Physical Review B: Condensed Matter*, 22, 304–307.

55. Jolivet, J. P., Chaéac, C. and Tronc, E. 2004. Iron oxide chemistry. From molecular clusters to extended solid networks. *Chemical Communications*, 7, 481–487.

56. Cullity, B. D. and Stock, S. R. 1978. *Elements of X-ray Diffraction*, Reading, MA, Addison-Wesley.

57. Pecharsky, V. K. and Zavalij, P. Y. 2009. *Fundamentals of Powder Diffraction and Structural Characterization of Materials*, Berlin, Germany, Springer Verlag.

58. Rivers, J. M., Nyquist, J. E., Roh, Y., Terry, Jr., D. O. and Doll, W. E. 2004. Investigation into the origin of magnetic soils on the Oak Ridge Reservation, Tennessee. *Soil Science Society of America Journal*, 68, 1772–1779.

59. Czichos, H., Saito, T. and Smith, L. 2006. *Springer Handbook of Materials Measurement Methods*, Berlin, Germany, Springer Verlag.

60. Fiorillo, F. 2004. *Measurement and Characterization of Magnetic Materials*, San Diego, CA, Academic Press.

61. Tumanski, S. 2011. *Handbook of Magnetic Measurements*, Boca Raton, FL, CRC Press.

62. Wernsdorfer, W., Hasselbach, K., Mailly, D., Barbara, B., Benoit, A., Thomas, L. and Suran, G. 1995. DC-SQUID magnetization measurements of single magnetic particles. *Journal of Magnetism and Magnetic Materials*, 145, 33–39.

63. Clarke, J. and Braginski, A. I. 2004. *The SQUID Handbook: Vol. 1: Fundamentals and Technology of SQUIDs and SQUID Systems*, Berlin, Germany, Wiley-VCH.

64. Casán-Pastor, N., Gómez-Romero, P. and Baker, L. C. W. 2009. Magnetic measurements with a SQUID magnetometer: Possible artifacts induced by sample holder off centering. *Journal of Applied Physics*, 69, 5088–5090.

65. García, M. A., Fernández Pinel, E., De La Venta, J., Quesada, A., Bouzas, V., Fernández, J. F., Romero, J. J., Martín Gonzalez, M. S. and Costa-Kramer, J. L. 2009. Sources of experimental errors in the observation of nanoscale magnetism. *Journal of Applied Physics*, 105, 013925-013925-7.

66. Sawicki, M., Stefanowicz, W. and Ney, A. 2011. Sensitive SQUID magnetometry for studying nanomagnetism. *Semiconductor Science and Technology*, 26, 064006.

67. Ney, A. 2011. Magnetic properties of semiconductors and substrates beyond diamagnetism studied by superconducting quantum interference device magnetometry. *Semiconductor Science and Technology*, 26, 064010.

68. Stamenov, P. and Coey, J. M. D. 2006. Sample size, position, and structure effects on magnetization measurements using second-order gradiometer pickup coils. *Review of Scientific Instruments*, 77, 1–11.

69. Ney, A., Kammermeier, T., Ney, V., Ollefs, K. and Ye, S. 2008. Limitations of measuring small magnetic signals of samples deposited on a diamagnetic substrate. *Journal of Magnetism and Magnetic Materials*, 320, 3341–3346.

70. Bertotti, G. 1998. *Hysteresis in Magnetism: For Physicists, Materials Scientists and Engineers*, San Diego, CA, Academic Press.

71. Fröhlich, O. 1881. Investigations of dynamoelectric machines and electric power transmission and theoretical conclusions therefrom. *Electrotech Z*, 2, 134–141.
72. Kennelly, A. 1891. Magnetic reluctance. *Transactions of the American Institute of Electrical Engineers*, 8, 483–533.
73. Dunlop, D. J. and Özdemir, Ö. 2001. *Rock Magnetism: Fundamentals and Frontiers*, Cambridge, MA, Cambridge University Press.
74. Argyle, K. S. and Dunlop, D. J. 1990. Low-temperature and high-temperature hysteresis of small multidomain magnetites (215–540 nm). *Journal of Geophysical Research*, 95, 7069–7082.
75. Dormann, J., Fiorani, D. and Tronc, E. 1997. Magnetic relaxation in fine-particle systems. *Advances in Chemical Physics*, 98, 283–494.
76. Balanda, M. 2003. Dynamic susceptibility of magnetic systems. In: Haase, W. and Wróbel, S. (eds.) *Relaxation Phenomena*. Berlin, Germany: Springer.
77. Mørup, S. 1990. Mössbauer effect in small particles. *Hyperfine Interactions*, 60, 959–973.
78. Ortega, D., Garitaonandia, J., Barrera-Solano, C., Ramírez-Del-Solar, M., Blanco, E. and Domínguez, M. 2006. γ-Fe$_2$O$_3$/SiO$_2$ nanocomposites for magneto-optical applications: Nanostructural and magnetic properties. *Journal of Non-Crystalline Solids*, 352, 2801–2810.
79. Bahl, C. R. H., Hansen, M. F., Pedersen, T., Saadi, S., Nielsen, K. H., Lebech, B. and Mørup, S. 2006. The magnetic moment of NiO nanoparticles determined by Mössbauer spectroscopy. *Journal of Physics: Condensed Matter*, 18, 4161–4175.
80. Frandsen, C., Rasmussen, H. K. and Mørup, S. 2004. A Mössbauer study of the magnetization of γ-Fe$_2$O$_3$ nanoparticles in applied fields: The influence of interaction with CoO. *Journal of Physics: Condensed Matter*, 16, 6977–6981.
81. Frenkel, J. and Dorfman, J. 1930. Spontaneous and induced magnetisation in ferromagnetic bodies. *Nature*, 126, 274–275.
82. Aharoni, A. 1991. The concept of a single-domain particle. *IEEE Transactions on Magnetics*, 27, 4775–4777.
83. Fernandez, A., Gibbons, M. R., Wall, M. A. and Cerjan, C. J. 1998. Magnetic domain structure and magnetization reversal in submicron-scale Co dots. *Journal of Magnetism and Magnetic Materials*, 190, 71–80.
84. Nogués, J., Sort, J., Langlais, V., Skumryev, V., Suriñach, S., Muñoz, J. S. and Baró, M. D. 2005. Exchange bias in nanostructures. *Physics reports*, 422, 65–117.
85. Bean, C. P. and Livingston, J. D. 1959. Superparamagnetism. *Journal of Applied Physics*, 30, S120–S129.
86. Ortega, D., Vélez-Fort, E., García, D. A., García, R., Litrán, R., Barrera-Solano, C., Ramírez-Del-Solar, M. and Domínguez, M. 2010. Size and surface effects in the magnetic properties of maghemite and magnetite coated nanoparticles. *Philosophical Transactions of the Royal Society A: Mathematical, Physical and Engineering Sciences*, 368, 4407–4418.
87. Skumryev, V., Stoyanov, S., Zhang, Y., Hadjipanayis, G., Givord, D. and Nogués, J. 2003. Beating the superparamagnetic limit with exchange bias. *Nature*, 423, 850–853.
88. Neuringer, J. L. and Rosensweig, R. E. 1964. Ferrohydrodynamics. *Physics of Fluids*, 7, 1927–1937.
89. Odenbach, S. and Thurm, S. 2002. *Magnetoviscous Effects in Ferrofluids*, Berlin, Germany, Springer.
90. Rosensweig, R. 1985. *Ferrohydrodynamics*, Cambridge, MA, Cambridge University Press.
91. Nethe, A., Schoppe, T. and Stahlmann, H. D. 1999. Ferrofluid driven actuator for a left ventricular assist device. *Journal of Magnetism and Magnetic Materials*, 201, 423–426.
92. Mitamura, Y. and Mori, Y. 2007. Magnetic fluid-driven artificial hearts. In: Magjarevic, R. and Nagel, J. H. (eds.) *World Congress on Medical Physics and Biomedical Engineering 2006*. Berlin, Germany: Springer.
93. Odenbach, S. 2006. Ferrofluids. In: Buschow, K. H. J. (ed.) *Handbook of Magnetic Materials*, Vol. 16. Amsterdam, the Netherlands: Elsevier Science.
94. Martsenyuk, M. A., Raikher, Y. L. and Shliomis, M. I. 1974. On the kinetics of magnetization of ferromagnetic particle suspension. *Soviet Physics—JETP*, 38, 413–416.
95. Rosensweig, R. E. 2002. Basic equations for magnetic fluids with internal rotations. In: Odenbach, S. (ed.) *Ferrofluids*. Berlin, Germany: Springer Verlag.
96. Ortega, D., García, R., Marín, R., Barrera-Solano, C., Blanco, E., Domínguez, M. and Ramírez-Del-Solar, M. 2008. Maghemite–silica nanocomposites: Sol–gel processing enhancement of the magneto-optical response. *Nanotechnology*, 19, 475706.
97. Gittleman, J. I., Abeles, B. and Bozowski, S. 1974. Superparamagnetism and relaxation effects in granular Ni-SiO$_2$ and Ni-Al$_2$O$_3$ films. *Physical Review B*, 9, 3891–3897.
98. Madsen, D. E., Hansen, M. F. and Mørup, S. 2008. The correlation between superparamagnetic blocking temperatures and peak temperatures obtained from ac magnetization measurements. *Journal of Physics: Condensed Matter*, 20, 345209.1–345209.6.

99. Jiang, J. Z. and Mørup, S. 1997. Correlation between peak and median blocking temperatures by magnetization measurement on isolated ferromagnetic and antiferromagnetic particle systems. *Nanostructured Materials*, 9, 375–378.

100. Stoner, E. C. and Wohlfarth, E. P. 1948. A mechanism of magnetic hysteresis in heterogeneous alloys. *Philosophical Transactions of the Royal Society of London–Series A, Mathematical and Physical Sciences*, 240, 599–642.

101. Hansen, M. F. and Mørup, S. 1999. Estimation of blocking temperatures from ZFC/FC curves. *Journal of Magnetism and Magnetic Materials*, 203, 214–216.

102. Ortega, D., Domínguez, M., García, R. P., Garitaonandia, J. S., Ramírez-Del-Solar, M., Litrán, R., Barrera-Solano, C. and Blanco, E. 2008. Unpublished data.

103. Ortega, D. 2007. Transparent composite materials with magnetic nanoparticles for current magneto-optic sensors. PhD thesis, University of Cádiz, Cádiz Spain.

104. Usov, N. A. 2011. Numerical simulation of field-cooled and zero field-cooled processes for assembly of superparamagnetic nanoparticles with uniaxial anisotropy. *Journal of Applied Physics*, 109, 023913.

105. Osborn, J. 1945. Demagnetizing factors of the general ellipsoid. *Physical Review*, 67, 351–357.

106. Cowburn, R. P. 2000. Property variation with shape in magnetic nanoelements. *Journal of Physics D: Applied Physics*, 33, R1–R16.

107. Chithrani, B. D., Ghazani, A. A. and Chan, W. C. W. 2006. Determining the size and shape dependence of gold nanoparticle uptake into mammalian cells. *Nano Letters*, 6, 662–668.

108. Coey, J. M. D. and Khalafalla, D. 1972. Superparamagnetic γ-Fe$_2$O$_3$. *Physica Status Solidi A*, 11, 229–241.

109. Batlle, X. and Labarta, A. 2002. Finite-size effects in fine particles: Magnetic and transport properties. *Journal of Physics D: Applied Physics*, 35, R15.

110. Bødker, F., Mørup, S. and Linderoth, S. 1994. Surface effects in metallic iron nanoparticles. *Physical Review Letters*, 72, 282–285.

111. Ferré, R., Barbara, B., Fruchart, D. and Wolfers, P. 1995. Dipolar interacting small particles: Effects of concentration and anisotropy. *Journal of Magnetism and Magnetic Materials*, 140–144, 385–386.

112. Mørup, S., Madsen, D. E., Frandsen, C., Bahl, C. R. H. and Hansen, M. F. 2007. Experimental and theoretical studies of nanoparticles of antiferromagnetic materials. *Journal of Physics: Condensed Matter*, 19, 213202.

113. Mørup, S. and Frandsen, C. 2004. Thermoinduced magnetization in nanoparticles of antiferromagnetic materials. *Physical Review Letters*, 92, 217201.

114. Winkler, E., Zysler, R. D., Mansilla, M. V. and Fiorani, D. 2005. Surface anisotropy effects in NiO nanoparticles. *Physical Review B*, 72, 132409.

115. Iglesias, O., Batlle, X. and Labarta, A. 2007. Exchange bias and asymmetric hysteresis loops from a microscopic model of core/shell nanoparticles. *Journal of Magnetism and Magnetic Materials*, 316, 140–142.

116. Garitaonandia, J. S., Insausti, M., Goikolea, E., Suzuki, M., Cashion, J. D., Kawamura, N., Ohsawa, H., De Muro, I. G., Suzuki, K. and Plazaola, F. 2008. Chemically induced permanent magnetism in Au, Ag, and Cu nanoparticles: Localization of the magnetism by element selective techniques. *Nano Letters*, 8, 661–667.

117. Goikolea, E., Garitaonandia, J., Insausti, M., Gil De Muro, I., Suzuki, M., Uruga, T., Tanida, H., Suzuki, K., Ortega, D. and Plazaola, F. 2010. Magnetic and structural characterization of thiol capped ferromagnetic Ag nanoparticles. *Journal of Applied Physics*, 107, 09E317-09E317-3.

118. García, M., Merino, J., Pinel, E. F., Quesada, A., De La Venta, J., González, M. L. R., Castro, G., Crespo, P., Llopis, J. and González-Calbet, J. 2007. Magnetic properties of ZnO nanoparticles. *Nano Letters*, 7, 1489–1494.

119. Hendriksen, P. V., Linderoth, S. and Lindgård, P. A. 1993. Finite-size modifications of the magnetic properties of clusters. *Physical Review B*, 48, 7259–7273.

120. Guardia, P., Batlle-Brugal, B., Roca, A. G., Iglesias, O., Morales, M. P., Serna, C. J., Labarta, A. and Batlle, X. 2007. Surfactant effects in magnetite nanoparticles of controlled size. *Journal of Magnetism and Magnetic Materials*, 316, e756–e759.

121. Bittova, B., Poltierova-Vejpravova, J., Roca, A. G., Morales, M. P. and Tyrpekl, V. 2010. Effects of coating on magnetic properties in iron oxide nanoparticles. *Journal of Physics: Conference Series*, 200, 072012.

122. Köseoglu, Y. 2006. Effect of surfactant coating on magnetic properties of Fe$_3$O$_4$ nanoparticles: ESR study. *Journal of Magnetism and Magnetic Materials*, 300, e327–e330.

123. Mikhaylova, M., Jo, Y., Kim, D., Bobrysheva, N., Andersson, Y., Eriksson, T., Osmolowsky, M., Semenov, V. and Muhammed, M. 2004. The effect of biocompatible coating layers on magnetic properties of superparamagnetic iron oxide nanoparticles. *Hyperfine Interactions*, 156–157, 257–263.

124. Mørup, S. and Hansen, M. F. 2007. Superparamagnetic Particles. In: Kronmüller, H. and Parkin, S. (eds.) *Handbook of Magnetism and Advanced Magnetic Materials*, Vol. 4. New York: John Wiley & Sons, Ltd.
125. Schaller, V., Wahnström, G., Sanz-Velasco, A., Enoksson, P. and Johansson, C. 2009. Monte Carlo simulation of magnetic multi-core nanoparticles. *Journal of Magnetism and Magnetic Materials*, 321, 1400–1403.

Daniel Ortega obtained a chemistry degree with specialisation in materials science from the University of Cadiz, Spain. Daniel continued to do MSc and PhD in the area of condensed matter physics, more precisely in magnetic nanocomposites. He undertook his first postdoctoral position at the University of the Basque Country in Spain researching unusual magnetic properties in metallic and semiconductor nanoparticles, and his second postdoc in London at The Royal Institution of Great Britain as a European Commission's Marie Curie Fellow working in nanoparticles for magnetic hyperthermia. Currently, Daniel is a research fellow at UCL Physics, and his research is focused in bespoke magnetic nanoparticles with applications in medicine and developing new instrumental methods for their characterisation.

Part II

Fabrication and Characterisation of MNPs

2 Synthesis and Characterisation of Iron Oxide Ferrite Nanoparticles and Ferrite-Based Aqueous Fluids

Etienne Duguet, Marie-Hélène Delville, and Stéphane Mornet

CONTENTS

Applications of superparamagnetic magnetic fluids in biomedicine have been expanding over the last decade due to their tremendous potential for tagging and delivery strategies. Their applications in hyperthermia (HT), magnetic resonance imaging (MRI), gene delivery (GD), magnetic drug targeting (MDT), nanomedicine and regenerative medicine have been largely demonstrated. Key aspects of these magnetic fluids are that they exhibit magnetisation only in an applied magnetic field, they must form stable colloidal suspensions for *in vivo* biomedical applications and they can be directed to a desired site and serve for controlled targeting in clinical applications. Their successful application in healthcare is strongly dependent on their structural characteristics, specifically size and size distribution, shape, magnetic susceptibility, and surface chemistry. In this chapter, we review the various synthetic routes, and the surface modification providing fair colloidal stabilisation, including various coating materials as well as the usual characterisation techniques of these magnetic fluids.

2.1 FERRITE STRUCTURE AND CHEMICAL COMPOSITION

Ferrites can be divided into four different structural classes: hexagonal ferrites (magnetoplumbite structure, $MO \cdot 6Fe_2O_3$ with M = Pb, Ba or Sr), garnets ($3M_2O_3 \cdot 5Fe_2O_3$ with M = Y or another rare earth element), orthoferrites or perovskites (MFe_2O_3 with M is usually Y or another rare earth

47

element) and spinels. Among them, the latter draws special attention since it is the most widely used for biological applications. The $[A]_{Td}[B_2]_{Oh}O_4$ spinel structure (where Td and Oh make reference to tetrahedral and octahedral sites, respectively) is derived from the mineral spinel $MgO \cdot Al_2O_3$ or $MgAl_2O_4$ whose structure was elucidated by Bragg[1]. The spinel structure is called normal (or direct) or inverse depending on whether divalent ions occupy tetrahedral (A) or octahedral (B) sites of the face-centred cubic oxygen lattice, respectively. In the simplest structure, their analogous ferrites have the general formula $MO \cdot Fe_2O_3$ or MFe_2O_4, where M is the divalent metal ion. According to the prediction of the Néel theory of ferrimagnetism[2], ferrite spinels display ferrimagnetic properties resulting from the anti-ferromagnetic couplings between the magnetic moments of cations located in tetrahedral and octahedral sublattices.

In magnetic ferrites, the trivalent Al^{3+} is usually replaced by paramagnetic Fe^{3+} or by Fe^{3+} in combination with other trivalent ions while the divalent cation Mg^{2+} is replaced by a paramagnetic transition metal Mn^{2+}, Fe^{2+}, Co^{2+}, Ni^{2+} to provide unpaired electron spins and therefore part of the magnetic moment of a spinel. Others divalent ions such as non-paramagnetic ions, for example, Cu^{2+}, Zn^{2+}, Mg^{2+}, in combination with other paramagnetic divalent ions, but also combinations of monovalent Li^+ with an additional amount of $0.5\ Fe^{3+}$, can still be used to disproportionate the Fe^{3+} ions on the crystal lattice sites to provide or increase the magnetic moment. Finally, insertions of trivalent (Fe^{3+}, Mn^{3+}) or even tetravalent (Ti^{4+}, Sn^{4+}) ions are also possible provided that their ionic radii are compatible with the volume of available interstitial (A and B) sites, and the electroneutrality is preserved. The magnetic properties of ferrites depend not only on the metal ions located in the interstices but also on their distribution. By adjusting the ratio between the magnetic/non-magnetic ions in the different sites, the ferrites may display enhanced properties, such as higher magnetic permeability or electrical resistivity than those of simple ferrite such as magnetite (Fe_3O_4). A good example is found with zinc ferrites. While bulk $ZnFe_2O_4$ material displays a non-magnetic normal spinel structure, this property makes that the most efficient soft ferrite spinels are the Zn-based mixed ferrites of manganese–zinc $Mn_xZn_yFe_{2+z}O_4$ ($x + y + z = 1$), nickel–zinc $Ni_xZn_yFe_2O_4$ ($x + y = 1$) or nickel–zinc–copper $Ni_xZn_yCu_zFe_2O_4$ ($x + y + z = 1$).

In the case of NPs, the situation is often more complicated and confused because of the inversion of the cation distribution, size effects and non-stoichiometry. Considering again bulk $ZnFe_2O_4$ in its normal spinel form, the formula can be written $[A_\delta B_{1-\delta}]_{Td}[A_{1-\delta}B_\delta]_{Oh2}O_4$, where δ is the inversion parameter. When $\delta = 1$, the diamagnetic Zn^{2+} ions occupy only A sites and all the Fe^{3+} ions are in B sites coupled between each other *via* a superexchange pathway through A sites. The B–B interactions being very weak, the normal spinel $ZnFe_2O_4$ shows long-range anti-ferromagnetic ordering at the Néel temperature, $T_N = 9–11\ K^{3–5}$. If the non-inverted stoichiometric NPs have an anti-ferromagnetic ground state ($T_N = 13\ K$)[6] like the bulk material, the ground state of nanosized inverted zinc ferrites is magnetic with a large magnetisation that is generally explained by the distribution of the cation[7–9]. Indeed, in the partially inverted stoichiometric spinel, some Zn^{2+} ions occupy B sites while some Fe^{3+} ions are in the tetrahedral ones leading to a magnetically active A sublattice that strongly interacts with the B sublattice. Values of δ ranging from 0.03 to 0.6 are reported in the literature, depending strongly upon the preparation procedures. The magnetisation has been found to increase with grain size reduction. This feature is generally associated with the increase of the cation inversion[9–11]. The size, polydispersity, the mutual interaction between magnetic NPs and their surface states[12] generally complicate the interpretation of the magnetic properties of NPs-based powders.

In the context of biomedical applications, the most popular ferrites are the ferrous ferrite called magnetite Fe_3O_4 and its oxidised form gamma ferric oxide also called maghemite (γ-Fe_2O_3). This can be explained by at least two reasons: from the chemistry point of view, the control of the chemical and structural composition of mixed metal oxides (with more than two cations) can be difficult to obtain in nanoparticle form due to the different reactivities (thermodynamic and kinetic) of the metal elements and the possibility to find different oxidation degrees in different sites between two cations (case of Mn^{2+}/Mn^{3+} with Fe^{2+}/Fe^{3+}). Moreover in the biomedical field, it is also important to take into account the high toxicity of most transition metals.

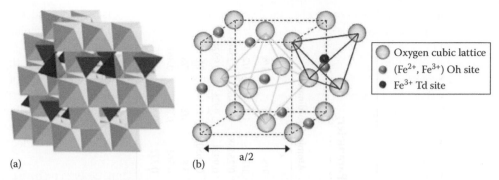

(a) (b) a/2

FIGURE 2.1 (a) Structure of magnetite Fe_3O_4 through a polyhedral model with alternating octahedra and tetrahedra layers. (From Jolivet, J.P., Chaneac, C., and Tronc, E., Iron oxide chemistry. From molecular clusters to extended solid networks, *Chem. Commun.*, 2004, 481–487, Copyright 2004. Reprinted in part by permission from Royal Society of Chemistry.) (b) Ball-and-stick model of unit cell and arrangement of octahedra and tetrahedra.

Magnetite is a black ferrimagnetic mineral containing both Fe[II] and Fe[III], while maghemite is a red brown ferrimagnetic mineral isostructural with magnetite but with cation deficient sites. In the cubic structure of both magnetite and maghemite, 1/3 of the interstices are in tetrahedral coordination with oxygen and 2/3 are in octahedral coordination (Figure 2.1).

In magnetite, all of these positions are filled with Fe^{2+}. Magnetite has a completely inverse spinel structure[14] according to $[Fe^{3+}]_{Td}[Fe^{3+}Fe^{2+}]_{Oh}O_4$ with a half metallic structure and exhibits semi-conductive properties resulting from a fast electron hopping between the iron ions Fe^{2+} and Fe^{3+} of the octahedral sublattice as evidenced by Mössbauer spectroscopy. This high electron mobility makes the magnetite NPs very sensitive to oxidation and easily transformable into non-stoichiometric spinel particles $[Fe^{3+}]_{Td}[Fe^{3+}_{1+2z/3}Fe^{2+}_{12z}V_{z/3}]_{Oh}O_4$, where V stands for a cationic vacancy. For a completely oxidised form, this gives the maghemite formula $[Fe^{3+}]_{Td}[Fe^{3+}_{5/3}V_{1/3}]_{Oh}O_4$. Synthetic maghemite often displays superstructure forms that arise as a result of cation and, therefore, vacancy ordering. The extent of vacancy ordering is related to the crystallite size and the nature of precursor and the amount of Fe[II] in the structure. Completely ordered maghemite has a primitive cubic or a tetragonal cell, and the x-ray pattern shows 'superstructure lines'; otherwise it is cubic[15]. Structural data of both magnetite and maghemite are summarised in Table 2.1. Such crystal structures result in a net spontaneous magnetisation of the spinel, characterised by their saturation magnetisation at 300 K (92 emu g^{-1} for magnetite and 78 emu g^{-1} for maghemite).

2.2 SYNTHESIS ROUTES OF SPINEL FERRITE NPs

The synthesis of iron oxide NPs has been an area of intense investigation by chemists and materials scientists for five decades[16]. Early investigations have developed a number of methods, such as physical grinding, gas phase vapour deposition and aqueous-solution-based routes including precipitation of ferrous and/or ferric salts and sol–gel process[17]. Breakthroughs in the synthesis of highly uniform semiconductor quantum dots in the mid-1990s (involving high-temperature organometallic routes)[18,19] opened up new efficient synthetic routes to produce monodisperse magnetic NPs.

For fulfilling the requirements of biomedical applications, the three most investigated synthetic routes of iron oxide NPs remain the alkaline precipitation in water and in water-in-oil microemulsions and the thermal decomposition of organometallic iron in organic liquids (Figure 2.2). Only these strategies are described and discussed here (Table 2.2). Less common routes, including recent ones such as sonochemistry, spray pyrolysis and laser pyrolysis, have been reviewed elsewhere[20–22]. Finally, the polyol route generally leads to metal NPs that may be converted into metal oxide NPs; it has also been recently reviewed elsewhere[23,24].

TABLE 2.1

Crystallographic Data for Magnetite and Cubic Maghemite (z Number of Formula per Unit Cell; a Lattice Parameter; α Lattice Angle)

Compound	System	Space Group	Stacking of Close Packed Anions	a (nm)	α (°)	Z	Density (g cm⁻³)		Atomic Coordinates	
								x	y	z
Magnetite Fe₃O₄	Cubic	$Fd\bar{3}m$	ABCABC [111]	0.8396	90	8	5.18			
								Fe₁ 1/8	1/8	1/8
								Fe₂ 1/2	1/2	1/2
								O 0.25468	0.25468	0.25468
Maghemite γ-Fe₂O₃	Cubic (or tetragonal)	P4₃32 (or P4₁2₁2)	ABCABC [111]	0.8347	90	8	4.87			
								Fe₁ 0.9921	0.9921	0.9921
								Fe₂ 0.8650	0.6150	7/8
								Fe₃ 3/8	1/8	7/8
								O₁ 0.861	0.861	0.861
								O₂ 0.372	0.377	0.876

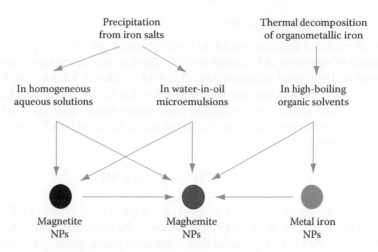

FIGURE 2.2 Main synthetic pathways of magnetite and maghemite NPs.

TABLE 2.2

Brief Description and Compared Advantages and Drawbacks of the Main Routes Investigated for the Synthesis of Iron Oxide NPs for Biomedical Purpose

Method/Brief Description	Advantages	Drawbacks
Alkaline precipitation in aqueous solutions from ferrous and/or ferric salt solutions	Most common Facile and reproducible Control of the composition Large amount of NPs in a single batch Tunable mean particle size Hydrophilic surface Giving access to a substantial literature on surface modification	Quite large size distribution
Alkaline precipitation in microemulsion droplets from ferrous and/or ferric salt aqueous solutions dispersed in oil	Good control of size distribution for sub-10nm NPs	Difficult scale-up due to the huge amount of oil required Surface contamination by surfactant
Thermal decomposition of organometallic iron in high-boiling organic liquids	Good control of size and shape Good crystallinity	Hydrophobic particles Cost of the metal precursors Toxicity of the organic residues (precursors, solvents, stabilising agents)

Among the requirements to fulfil, the mean size and size distribution, the crystallinity and the purity are the most critical because they are intimately related to the biocompatibility and performance of the magnetic properties. For instance, since the blocking temperature of superparamagnetic NPs depends on particle size, a wide particle size distribution will result in a wide range of blocking temperatures and therefore non-ideal magnetic behaviour for many applications. When the synthesis route does not lead to size-monodisperse NPs, size selection is a process where an electrolyte solution or a non-solvent is added to a stable colloid solution to disrupt it, causing larger particles to precipitate and leaving smaller and nearly monodisperse particles in the supernatant[23]. But such further stages are often the source of low yields and contamination of NPs.

2.2.1 Alkaline Precipitation in Water

Its long history, inexpensive reagents and facile protocol have made alkaline precipitation a common synthetic route of magnetite and maghemite NPs. The size, shape, and composition of the NPs very much depend on the type of salts used (e.g. chlorides, sulphates, nitrates), the Fe^{2+}/Fe^{3+} ratio, the reaction temperature, the pH value and ionic strength of the media[25]. Magnetite synthesis generally consists of the co-precipitation of ferrous and ferric hydroxides by addition of a base (typically NaOH or NH_4OH) to a solution of Fe(II) and Fe(III) salts at a 1:2 molar ratio[26]. The overall reaction may be written as follows:

$$Fe^{2+} + Fe^{3+} + 8OH^- \rightarrow Fe_3O_4 + 4H_2O$$

According to the thermodynamics of this reaction, a complete precipitation of Fe_3O_4 should be expected between pH 9 and 14. With this synthesis, once the synthetic conditions are fixed, the quality of the magnetite NPs is fully reproducible. Alternative routes use only ferrous[27,28] or ferric[29] precursors in oxidising or reducing conditions, respectively. The magnetic saturation values of magnetite NPs are experimentally determined to be in the range of 30–50 emu g^{-1}, which is lower than the bulk value, that is, 92 emu g^{-1}. In addition, it has been shown that by adjusting the pH and the ionic strength of the precipitation medium, it is possible to control the mean size of the particles over one order of magnitude (from 15 to 2 nm)[30]. The size decreases as the pH and the ionic strength in the medium increase. Both parameters affect the chemical composition of the surface and, consequently, the electrostatic surface charge of the particles.

Formation of iron oxides in aqueous systems involves nucleation and crystal growth[16]. Homogeneous nucleation occurs spontaneously in bulk solution when the supersaturation exceeds a certain critical value. The essential requirement for precipitation to take place is the formation of stable, embryonic clusters of molecules or ions, the so-called nuclei, in solution. These embryonic nuclei form as the result of collisions of ions or molecules in the bulk solution prior to nucleation. The embryos are continually breaking up and reforming. When the embryo exceeds a certain critical size, the nucleus grows by a decrease of free energy. Although a certain degree of supersaturation is needed for crystal growth, it is much less than that required for nucleation. Crystal growth involves a number of steps including diffusion of the growth units to the crystal surface and diffusion and adsorption processes at the surface itself. The overall rate of growth is determined by the slowest of these steps. Crystal growth is controlled by mass transport and by the surface equilibrium of addition and removal of the growth units. Hereby, the driving force for dissolution (removal) increases with decreasing particle size. Thus, within an ensemble of particles with slightly different sizes, the large particles will grow at the cost of the small ones. This mechanism is called Ostwald ripening and is generally believed to be the main path of crystal growth[31].

After (co-)precipitation, the ageing proceeds by heating the magnetite colloidal dispersion in the presence of a suitable surfactant (e.g. oleic acid) for conferring the required steric stabilisation to the resulting ferrofluid when the particles are dispersed in non-polar media[32]. Alternatively, it may be treated with concentrated acid or base solutions to obtain stable (through electrostatic stabilisation) ferrofluids in acid or alkaline aqueous media[33]. These ageing methods allow producing particles with a diameter of 3–20 nm, varying with the experimental conditions, for example, the $Fe^{3+}:Fe^{2+}$ molar ratio, nature and concentration of the alkali medium and temperature.

As already mentioned, magnetite is not thermodynamically stable, in particular, in the shape of NPs, and slowly oxidises into maghemite:

$$4Fe_3O_4 + O_2 \rightarrow 6Fe_2O_3$$

Therefore, magnetite NPs can be subjected to deliberate oxidation to convert them into maghemite NPs, by dispersing them in acidic medium and then oxidising by iron(III) nitrate.

The maghemite particles obtained are then chemically stable in alkaline and acidic medium and may be peptised in such media[33]. This multi-step route has become the most used for the preparation of aqueous ferrofluid of maghemite.

However, even if the magnetite particles are converted into maghemite after their initial formation, the experimental challenge in the synthesis of Fe_3O_4 by co-precipitation lies in control of the particle size and thus achieving a narrow particle size distribution[25]. Particles prepared by co-precipitation unfortunately tend to be rather polydisperse indicating that nucleation and crystal growth took place simultaneously over the bulk of the reaction. Since the classical model proposed by LaMer and Dinegar[34], many works have been devoted to describe the production of monodisperse (i.e. uniform in shape and size) inorganic colloids by precipitation from homogeneous solution. This method is based in separating the nucleation and crystal growth periods, because it is well known that a short burst of nucleation and subsequent slow controlled growth is crucial to produce monodisperse particles. This may be accomplished by dilution or addition of organic agents to modulate water activity. The effect of organic ions on the formation of metal oxides can be rationalised by two competing mechanisms. Chelation of the metal ions can prevent nucleation and lead to the formation of larger particles because the number of nuclei formed is small and the system is dominated by particle growth. On the other hand, the adsorption of additives on the nuclei and the growing crystals may inhibit the growth of the particles, which favours the formation of small units. For example, size-tunable maghemite NPs were prepared by initial formation of magnetite in the presence of the trisodium salt of citric acid, in an alkaline medium, and subsequent oxidation at 90°C for 30 min by iron(III) nitrate[35]. The particle size can be varied from 2 to 8 nm by adjusting the molar ratio of citrate ions and metal ions (Fe^{2+} and Fe^{3+}). The effects of several organic anions, such as carboxylate and hydroxy carboxylate ions, on the formation of iron oxides or oxyhydroxides have been studied extensively[36–38].

To produce particles with good size control, a modified method relies on the principle of particle formation in constrained domains such as macromolecular aqueous solutions[39]. Indeed, particle size can be limited by capturing the particles in a polymeric corona potentially allowing 'one-pot' synthesis of the polymer-coated iron oxide NPs[40,41]. These hydrophilic organic coronas have multiple utilities for reducing aggregation of NPs during storage (steric stabilisation), reducing degradation *in vivo*, increasing plasma half-life and influencing biodistribution[20]. They may also provide residues for subsequent functionalisation reactions, although their weak and uncontrolled interactions with NPs surface (essentially van der Waals interactions) may lead to depletion phenomena. The coronas are generally made of dextran and more rarely of albumin, carboxydextran, chitosan, starch or heparin[42].

2.2.2 Alkaline Precipitation in Microemulsion Systems

A microemulsion is defined as a thermodynamically stable, isotropic dispersion of two immiscible liquids, stabilised by a monolayer of surfactant at the interface of both liquids[43]. In water-in-oil microemulsion systems, small aqueous droplets (1–50 nm in size) are dispersed in the organic phase. When a metal salt is incorporated in the aqueous phase of the microemulsion, it will reside in the aqueous microdroplets surrounded by oil. These microdroplets will continuously collide, coalesce and break again. Conceptually, when iron salts and hydroxide anions are dissolved in water-in-oil microemulsions, they will form a magnetite precipitate on mixing. The growth of these particles in microemulsions can be conceptualised as a progress of inter-droplet exchange and nuclei aggregation. The finely dispersed precipitate so produced can be extracted from the surfactants. Compared to simple aqueous precipitation, microemulsion has certain advantages, due to the small size of the microemulsion droplets that act as nanoreactors, confining the reaction and preventing agglomeration. This method produces nanometre-sized iron oxide NPs with a quite good control of the size distribution, especially for sub-10 nm particles. In addition, the size of the microemulsion droplets can be adjusted by changing the ratio of water/surfactant/oil[44,45], but only a modest degree of control over particle size and shape can be achieved for largest NPs[17]. The literature often reports the use of sodium bis(2-ethylhexylsulphosuccinate) (AOT), cetyltrimethylammonium bromide (CTAB) or sodium dodecylsulphate (SDS) as ionic surfactants, which nevertheless seem lead to NPs suffering from poor crystallinity that can sometimes be remedied through annealing. The use of non-ionic surfactants, such as Triton X-100®, Igepal CO-520® and Brij-97®, avoids the complication of the presence of complexing functional species and offers great future potential. Finally, at room temperature, the yield is quite low, but it can be increased by performing high-temperature synthesis under reflux conditions[46]. Further drawbacks to the microemulsion method are scale-up that is difficult due to the huge amount of oil (organic phase) required and the surfactants that adhere to the particles and are difficult to remove[39].

This procedure may be easily extended to produce Mn-, Co-, Ni- and $ZnFe_2O_4$ NPs (see Chapter 4).

2.2.3 THERMAL DECOMPOSITION

The thermal decomposition method yields generally more crystalline and size-monodisperse NPs in the diameter range of 5–30 nm. It has been widely used for producing Fe or Fe-alloy NPs from the 1960s until nowadays. More recently, it was deliberately applied to iron oxides[47], and various approaches and modifications have since been developed.

Iron carbonyl, acetate, acetylacetonate, carboxylate, cupferronate, and chloride are some of the commonly used precursors for thermal decomposition in high-boiling organic liquids (e.g. o-dichlorobenzene, octyl ether)[48]. Fatty acids or amines are generally used as stabilising agents, leading to hydrophobic NPs that require additional surface-modification steps to make them hydrophilic. Nevertheless, water-dispersible NPs were recently produced using 2-pyrrolidone as both stabilising agent and liquid medium[49].

When the oxidation number of iron is zero as in the case of $Fe(CO)_5$, the thermal decomposition results in the formation of essentially amorphous iron NPs[25]. The NPs formed in this fashion could be readily oxidised by air or other oxidation reagents, but γ-Fe_2O_3 NPs can also be directly generated by introducing a mild oxidising agent (e.g. trimethylamine oxide) into the reaction solution[50]. With other precursors, where iron is 'pre-oxidised', iron oxide NPs are directly obtained[47]. The size and morphology of the NPs are influenced by the nature and relative amounts of iron precursors, solvents and stabilising agents, as well as reaction conditions, for example, temperature and time. Large particles may also be obtained using a seeded growth process.

The thermal decomposition process also allows preparing ferrite NPs. It necessitates adding organometallic precursors of another metal M to the reaction mixture, leading to the formation of FeM-alloy NPs, which may be oxidised simultaneously or subsequently[48].

Potential issues of the thermal decomposition process are the possible presence of stabilising agents that may hamper efficient subsequent surface modifications of the NPs and the use of toxic solvents, which may reduce the NPs biocompatibility.

2.3 MAGNETITE EXTRACTION FROM LIVING ORGANISMS

Magnetotactic bacteria (MTBs) are members of a unique group of mainly aquatic bacteria that biomineralise intracellularly membrane-bound nanometre-sized crystals of magnetite or iron sulphide greigite (Fe_3S_4)[51]. These organelles, called magnetosomes, are most often arranged as a chain within the cells (Figure 2.3) and help MTBs to orient along the magnetic field lines of earth's magnetic field. The magnetic crystals are usually homogeneous in shape and size (35–120 nm) and may be recovered after cell lysis and magnetic sorting. They are naturally stable in water, thanks to the lipid bilayer coating their surface. Therefore, MTBs are regarded as potent eco-friendly green nanofactories, and increasing efforts in recent years resulted in the establishment of robust and reliable techniques for their mass cultivation. Today, the productivity in magnetite can reach the value

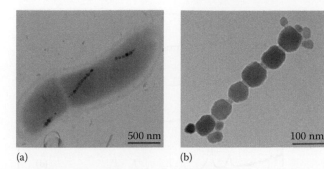

(a)   (b)

FIGURE 2.3 Transmission electron micrographs of (a) *Magnetospirillum magneticum* strain AMB-1 with chains of magnetosomes inside, (b) nanocrystalline magnetite chain harvested from lysed bacteria (magnetite nanocrystals are held together by a thin phospholipid membrane material after lysis). (Reprinted with permission Prozorov, T., Palo, P., Wang, L. et al., Cobalt ferrite nanocrystals: Out-performing magnetotactic bacteria, *ACS Nano*, 1, 228–233, 2007. Copyright 2007 American Chemical Society.)

of few milligram per litre per day[52]. According to a parallel strategy, specific proteins have been cloned from bacteria and covalently attached to self-assembling macromolecules. After addition of metal cation salts, magnetic NPs have been synthesised, including cobalt-ferrite crystals that are not produced naturally by bacteria[53]. To be exhaustive, algae and fungi have also been observed able to produce magnetite NPs[54].

2.4 AQUEOUS MAGNETIC FLUIDS

The stabilisation of magnetic NPs in aqueous media is a crucial key point when biological applications are concerned since aqueous magnetic fluids that are stable against aggregation in both a biological medium and a magnetic field are required. This stabilisation may be carried out either at the final stage of the functionalisation process of NPs or at an intermediate stage. It consists in the formation of physically/chemically stable colloidal suspensions where NPs do not aggregate, dissociate or chemically react to the solvent or any dissolved gas with time. The stability of a magnetic colloidal suspension results from the equilibrium between attractive and repulsive forces.

Theoretically, four kinds of forces can contribute to the inter-particle potential in the system. van der Waals forces induce strong attractive short-range isotropic attractions. In the case of two spheres of radius R separated by a much smaller distance d than R, the interaction potential can be written as

$$V_{VDW} = -\frac{AR}{12d} \tag{2.1}$$

where A is the effective Hamaker constant of the system, depending on the intrinsic Hamaker constants of water and solid. When magnetic suspensions are considered, magnetic dipolar forces between two particles must be added. This type of attractive magnetic interaction is effective over much longer range[23]. In the presence of an external magnetic field, further magnetisation of these clusters can occur increasing their aggregation[55].

On the other hand, common repulsive interactions include electrostatic repulsion of like-charged surfaces and steric and steric repulsion between surfaces coated with polymers (Figure 2.4)[56]. The electrostatic repulsive forces can be partially screened by adding salt to the suspension.

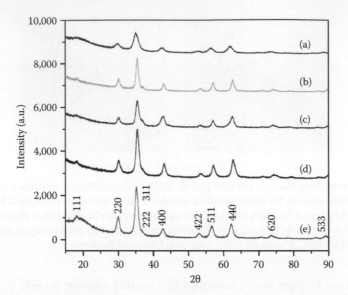

FIGURE 2.4 Powder XRD patterns of MFe$_2$O$_4$ NPs prepared in oleyl alcohol solutions: M = Mn (a), Fe (b), Co (c), Ni (d) and Zn (e). (Reprinted with permission from Adireddy, S., Lin, C., Palshin, V. et al., Size-controlled synthesis of quasi-monodisperse transition-metal ferrite nanocrystals in fatty alcohol solutions, *J. Phys. Chem. C*, 113, 20800–20811, 2009. Copyright 2009 American Chemical Society.)

The theoretical description of these two forces is known as the Derjaguin–Landau–Verwey–Overbeek (DLVO) theory[57,58]. Attraction between NPs can also be due to the presence of other solutes, such as high-molecular-weight molecules resulting in depletion flocculation.

2.4.1 ELECTROSTATIC STABILISATION

Electrostatic repulsion arises when electric charges are introduced at the NPs surface. These charges may come from the adsorption of charged ions at the surface, or they may be due to ionisation of groups such as hydroxyl groups at the surface of oxides. The structure of the solution in the immediate vicinity of the particle surface is in fact relatively complex. It is often described by a double-layer model: the inner layer, called the Stern layer, where the water molecules are highly structured by solvation of surface charges, and the diffuse layer that involves counter-ions, the latter being subjected simultaneously to electrostatic surface interactions and thermal agitation[30,60]. This kind of stabilisation is efficient when the ionic strength of the solution is not too high. When it is high, electrostatic repulsions are significantly reduced, causing irreversible aggregation of the NPs. The pH can also affect this form of stabilisation depending on the kind of ions adsorbed at the surface of the particle. The charge of certain ionic ligands depends on the solution pH. For example, molecules functionalised by carboxyl groups are negatively charged for pH values greater than 5–6 (COO$^-$) and neutral at lower pH values (COOH). Hence, particles stabilised by such ligands are only correctly dispersed at neutral or alkaline pH values. The use of electrostatic forces to counter-balance the van der Waals attraction between NPs in water is nevertheless limited to the case where the particles are able to sustain a surface charge, and it is very effective for iron oxide–based fluids.

In the specific case of iron oxides, the surface iron atoms act as Lewis acids and coordinate with molecules that donate lone-pair electrons. Therefore, in aqueous solutions, the Fe atoms coordinate with water, which dissociates readily to leave a surface functionalised with hydroxyl groups. These hydroxyl groups are amphoteric and react with either acids or bases[61]. The surface of the oxide can then be positive or negative with an isoelectric point (IEP) at pH 6.8 around which the surface charge density is too low to prevent the particles flocculation[15].

Therefore, small-charged organic species are generally introduced on the surface of NPs. The presence of these ionic groups on the surface of NPs allows conjugation with ionic ligands. With the molecule bound to the nanoparticle, the charge resides on the outside of the particle giving way to coulombic repulsion and thus dispersion of the NPs. The charge screening and the extent of coulombic repulsion can be controlled by the addition of salts, leading to precipitation of NPs or 'salting out'[62]. This aspect is crucial since in order to be effective in biological applications, NPs should remain colloidally stable when encountering physiological ionic strengths and the range of pH found in biological systems. Indeed, the physiological salt concentration is around 100 mM and therefore generally sufficient to cause precipitation of NPs lacking additional stabilisation. For this reason, unfortunately, ionically stabilised magnetic fluids are generally unsuitable for biomedical application.

The domain of colloidal stability of NPs can be shifted on the pH scale by coating with species such as citric acid/citrate[63], orthophosphoric acid/phosphate[64], surfactants such as oleic acid[65], lauric acid, alkane sulphonic acids and alkane phosphonic acids[66] and other species that may easily gain or lose protons. The potential at which coulombic repulsion is effective and NPs tend to aggregate is displaced at pH values around the pK_a of the surface functional group, making acidic anionic ligands (phosphate, citrate) suitable for stabilisation in basic to mild acidic conditions[67,68], whereas cationic ligands, such as alkylammoniums, offer stabilisation from acidic to mildly alkaline conditions[69]. The ionic stabilisation of NPs with charged species in biological media can have a significant effect on their fate. While positively charged NPs will be removed rapidly from the blood stream and suffer from liver and spleen clearance, the negatively charged NPs have a longer circulation time[42].

2.4.2 STERIC STABILISATION

An alternative to the ionic stabilisation is providing a physical steric barrier to prevent aggregation. Steric stabilisation can be achieved by coating NPs with a corona of hydrophilic macromolecular stabilisers[23]. Steric repulsion plays with the fact that the strength of van der Waals forces falls off very quickly with the distance d. This alternative consists in coating the surface with neutral and hydrophilic molecules that are long enough to maintain a certain distance between the NPs, thereby rendering the van der Waals attraction negligible. A variety of different macromolecules can be used, provided that they interact strongly with the particle surface and that they cannot form 'bridges' between different particles. Additionally, to be used *in vivo*, these steric stabilisers shall be biocompatible, biodegradable and protein-repellent in order to make the NPs long circulating. In some situations, the hydrophilic corona is used to load drug molecules for drug delivery purpose.

Among these steric stabilisers, poly(ethylene glycol) (PEG) and carbohydrates such as starch, dextran and chitosan are often used[42]. They may be adsorbed onto the NPs surface during the NPs synthesis (see Section 2.2.1) or grafted[70], anchored or adsorbed in a post-synthesis stage (see Chapters 5 and 6). When hydrophobic NPs are used, amphiphilic macromolecules may be advantageously employed[71].

Because the higher the molecular weight of the steric stabilisers, the higher the NPs hydrodynamic diameter, a compromise shall be found between the efficacy of the steric stabilisation and the efficacy of the targeting strategy. Indeed, some *in vivo* applications require a small hydrodynamic volume for efficient trans-membrane permeation and excretion[72].

The combination of both electrostatic and steric stabilisation *via* large molecular polyanionic coating on the surface of magnetic NPs provides a large enhancement of the stability of aqueous magnetic fluids[73]. The macromolecular humic acid, a notable fraction of the natural organic matter, contains mainly carboxylic groups that form surface complexes on the Fe–OH sites of iron oxides. It was demonstrated that the electrosteric stability of the magnetite NPs depends on the layer thickness of the macromolecular humic acid.

2.4.3 ENCAPSULATION IN A SILICA SHELL

Another widespread method to control the colloidal stability of iron oxide NPs in water is to encapsulate the magnetic NPs in a silica shell of controlled thickness. This silica coating allows not only increasing the distance d between the NPs and therefore to reduce the magnetic dipolar forces but also shifting the pH range of colloidal stability because the IEP of silica is at pH 2. Hence, the presence of a silica layer at the surface means that subsequent chemical modifications can be carried out at neutral pH, because under these conditions, the silica particles are stabilised by negative surface charges, whereas the naked iron oxide cores would flocculate. Moreover, the inert silica coating improves the chemical stability of NPs and therefore their biocompatibility[74] and provides a surface whose chemistry is well documented.

Essentially, two chemical routes are used for silica encapsulation. The first one derives from the Stöber's method using tetraethoxysilane (TEOS) in highly polar mixtures of ethanol and water alkalised by ammonia[75,76]. The second one is based on a water-in-oil microemulsion process using surfactants, TEOS, ammonia and positively charged iron oxide NPs in a non-polar liquid[77]. This strategy is very similar to the alkaline precipitation in microemulsion systems discussed in Section 2.2.2. and provides the same advantages and drawbacks (Table 2.2).

2.5 CHARACTERISATION TECHNIQUES

The most important technique for crystalline structure characterisation is x-ray diffraction (XRD) using both conventional and synchrotron radiation sources. The minimum size required for a crystal to diffract x-rays is on the order of a few unit cells (ca. 2–3 nm). Thus, this technique is well suited for characterisation of magnetic NPs used in the biomedical field for which size is rarely less than 4–5 nm. The XRD pattern of a powdered sample is a plot of the observed diffracted intensity of the x-rays against the Bragg angle, that is, the angle at which the x-rays strike the crystal and for which the maximum interference is observed. An XRD pattern of a crystalline phase consists of a number of reflections (peaks) of different intensities, which provide a fingerprint of the crystalline structure at the atomic level. Numerous data can be deduced from a diffraction pattern. The intensity can be used to quantify the proportion of iron oxide formed in a mixture by comparing experimental peak and reference peak intensities. From these patterns, the specific distances between the atomic layers (d_{hkl}-values) can be calculated using the Bragg equation (2.2), and from the set of d_{hkl}-values, the metal oxide can be identified (2.3):

$$d_{hkl} = \frac{\lambda}{2\sin\theta} \tag{2.2}$$

with

$$a = d_{hkl}\sqrt{h^2 + k^2 + l^2}, \quad \text{for a cubic lattice} \tag{2.3}$$

where
  λ is the wavelength
  a is the cubic lattice parameter
  hkl are the Miller's indexes

In addition to phase identification, XRD provides information about crystal size (and hence surface area) and crystal perfection, structural parameters (unit cell edge lengths) and degree of substitution of Fe^{3+} by other trivalent cations (isomorphous substitution). In the last case, structural incorporations can be deduced from a shift in the position of the XRD peaks (Figure 2.4). A shift only occurs if the replacing cation is sufficiently different in size from that of the Fe^{3+} cation. To measure the angular position, intensity and shape of the peaks, the step-counted pattern is usually computer-fitted using either empirical single-peak models based on a range of different peak profiles or the Rietveld model that produces a whole-pattern fit. The Rietveld model compares a calculated pattern based on the simultaneously refined model for crystal structure, diffraction optics, instrumental factors and other characteristics with the experimental pattern. Preferred orientation, asymmetric crystal development and peak broadening due to small crystals and structural disorder can also be accounted for. A full account of the theory and practice of XRD may be found elsewhere[78].

The average crystal size can also be calculated from the XRD line broadening using the Scherrer formula (2.4), which describes the corrected width of an XRD line at an angle θ as a function of the mean size of the coherently scattering domain D perpendicular the hkl plane:

$$D = \frac{K\lambda}{\beta\cos\theta} \tag{2.4}$$

with

$$\beta = \sqrt{B^2 + b^2} \tag{2.5}$$

where
 β is the true width of the x-ray peak at its half height
 K is a shape factor (equal to 0.89 for a sphere)

The true width can be corrected of the instrumental width by the formula (2.5), where B and b are the widths at half height of the more intense peak of the sample and of a well-crystallised sample reference. Generally, magnetic NPs used for biomedical applications are monocrystalline, and the size of the scattering domain D corresponds in first approximation to the NPs size. However, the Scherrer formula does not take into account the presence of a passivation layer or surface structural defects and surface strains due to the used synthesis pathway or to the post-synthesis oxidation processes. All these surface defects contribute to the broadening of the peaks, and therefore, the true crystal size is underestimated when this parameter is not taken into account. This restriction is overcome with the line profile analysis of Warren-Averbach[79,80] which enables the effects of strain to be separated from that of particle size and supplies more precise information about the particle size distribution.

The size and size distribution of magnetic NPs can be estimated from complementary techniques such as Mössbauer spectroscopy[81], extended x-ray absorption fine structure (EXAFS), infrared (IR) spectroscopy[82], transmission electron microscopy (TEM), photocorrelation spectroscopy (XPS) or also magnetometry. The mean diameter of the particles obtained from electron microscopy is often compared with the mean diameter from XRD and with the magnetic size obtained by the Langevin function treatment of the magnetisation curve. While there is a quite satisfying concordance between the dimensions obtained with TEM and traditional x-ray analysis, often the values obtained for the same sample between these two methods and magnetic data are fairly different. The magnetic diameter is generally smaller than the diameter obtained by TEM.

This feature is generally attributed to the presence of defects that contribute to the formation of a less-ordered layer on the particle surface. In some cases, authors have reported that even the stabilising agent could influence the surface structure of the particles and, thus, the magnetisation values[12,83,84]. Because NPs contain a limited number of cubic cell units, they can be considered to be a less-ordered system that is neither completely crystalline nor completely amorphous. This explains that the precise determination of the size of NPs often requires the combination of various analysis techniques. It should be noticed that NPs can be characterised as a liquid suspension mainly by small-angle x-ray scattering (SAXS) and small-angle neutron scattering (SANS)[85,86]. Energy dispersive x-ray diffraction (EDXRD), suitable for the systems with a low degree of crystallinity[87–89], is also relevant for the structural analysis of NPs even in suspension[90]. Other characterisation techniques such as AC and DC magnetometry, Mössbauer spectroscopy and Neutron techniques are covered in Chapter 1.

In most cases, XRD alone became insufficient to distinguish ferrites structural forms because the x-ray scattering power of the Fe^{3+} ions is almost the same as that of the other metal ions involved and can have similar detectable XRD patterns at the nanometre scale (Figure 2.4). This is almost true for both magnetite and maghemite, which possess the same inverse spinel structure. Moreover, the relative peak intensity for bulk samples may not be readily applied to NPs because of line broadening and the effect of the crystal shape. The structures of iron oxides can also be determined by neutron diffraction from which the interaction of the magnetic moment of the neutron with the spinel structure can make this distinction. IR[91] and Raman[92] spectroscopies, electron diffraction and high-resolution electron microscopy (HR-TEM) provide additional useful information. Finally, XPS can also be used to determine the oxidation state of iron in small particles because the core electron lines of ferrous and ferric ions can be both detected and distinguished on the spectra[93–95].

The concept of size of NPs is quite ambiguous. Indeed, it can include or not the different layers, that is, the crystalline part of the inorganic core, the amorphous layers of the inorganic core, the potential silica shells and/or (macro)molecular corona, and the hydrated layer. Additionally, in most cases, NPs are polydisperse, and the control of the size distribution is compulsory. This size heterogeneity leads to different values (even if characterising the same size) depending on the measurement technique.

The size of the particle core can be determined by TEM giving the global size of the particle core (crystalline and amorphous parts) and providing details on the size distribution and the shape of the particles as well as a number-weighted mean value, when performed on a statistically significant number of NPs. It should be stressed that sample preparations can affect the results and induce aggregation of the NPs. TEM measurements then may not reflect the size and the distribution observed in the aqueous dispersion. HR-TEM gives access to the atomic arrangement. It is then used to study lattice vacancies and defects, lattice fringe, glide plane and surface atomic arrangement of crystalline NPs[96,97]. SANS is also used to determine the size, polydispersity, shape (shape ratio) and structure of NPs[98].

Photon correlation spectroscopy (PCS), also called dynamic light scattering (DLS) or quasi-elastic light scattering (QELS), is a common technique to reach the average size of NPs dispersed in a liquid phase. It gives rise to the diffusion coefficient of NPs in solution and as a consequence to the hydrodynamic radius of a corresponding sphere as well as the size distribution[99].

Diffuse reflectance IR spectroscopy can be used to provide additional structural characterisations on maghemite by analysis of different vibrational modes. The extent of vacancy ordering inside γ-Fe_2O_3 nanoparticles can be easily observed by registering the IR spectra of different maghemite samples (Figure 2.5)[91]. Thus, considering samples prepared by co-precipitation, the one with the largest particle size (14 nm) shows the IR features of γ-Fe_2O_3 crystallites, which are at least partially ordered, as evidenced by the multiple lattice absorption bands between 800 and 200 cm^{-1}. The spectrum of common maghemite prepared by alkaline co-precipitation shows two

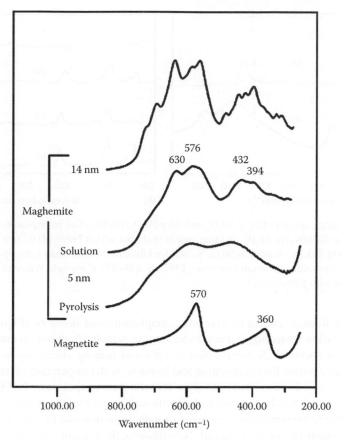

FIGURE 2.5 IR spectra for magnetite and maghemite nanoparticles prepared by different co-precipitation methods. (From Tartaj, P., Morales, M.P., Veintemillas-Verdaguer, S. et al.: The preparation of magnetic nanoparticles for applications in biomedicine. *J. Phys. D Appl. Phys.* 2003. 36. R182–R197 and references therein. Copyright Wiley-VCH Verlag GmbH & Co. KGaA. Reprinted with permission.)

wide vibrational Fe–O bands at 630 and 576 cm^{-1} typical of oxide nanoparticle surface whose spinel structure is disordered, while band in the 820–830 cm^{-1} range is characteristic of surface hydroxyl groups and corresponds to the stretching mode $\nu_{s(FeOH)}$[100]. Meanwhile, in the sample with the lowest particle size (5 nm) a significant reduction in the number of lattice absorption bands associated with increasing disorder is detected[101]. The difference in the IR spectra of two samples that have a similar particle size (5 nm) but have been prepared by two different techniques (solution and pyrolysis) can be noted. Particularly, the IR spectrum of the sample prepared by pyrolysis only displays two broad maxima at around 600 and 450 cm^{-1} indicating a random distribution of vacancies and therefore is expected to behave differently in the presence of an applied magnetic field. It is also noted from Figure 2.5 that IR spectroscopy is a simple tool to differentiate between maghemite and magnetite crystalline phases.

Raman spectroscopy may be also useful to get insights into the chemical composition and structure of magnetic fluids based on ferrites. Metal oxides and oxyhydroxides, the main metal corrosion products, strongly absorb IR radiation, but they are usually poor light scatterers. This is probably the reason for the larger number of studies on rust composition using IR spectrometry compared with Raman spectroscopy. Despite of this drawback, Raman spectroscopy presents at least two important advantages over IR spectroscopy: there is no need for sample preparation and the spectra of water–metal interfaces are easily and quickly obtained[102]. The investigation of the effect of the

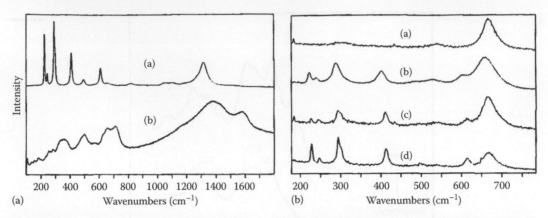

FIGURE 2.6 (a) Raman spectra of (a) α-Fe$_2$O$_3$ and (b) γ-Fe$_2$O$_3$; (b) Effect of laser power on the Raman spectrum of magnetite: (a) 0.7 mW; (b) 7 mW; (c) same spot as in (b) but with 0.7 mW; (d) 0.7 mW after exposure to a flame. (From De Faria, D.L.A., Venâncio Silva, S., and De Oliveira, M.T.: Raman microspectroscopy of some iron oxides and oxyhydroxides. *J. Raman Spectrosc.* 1997. 28. 873–878. Copyright Wiley-VCH Verlag GmbH & Co. KGaA. Redrawn with permission.)

laser power on the Raman spectra of magnetite, maghemite and hematite (Figure 2.6b) revealed that sample degradation frequently occurs under intense sample illumination and may lead to the misinterpretation of spectra[92]. A comparison of light and heating effects on magnetite was also made, and its transformation into maghemite and hematite with temperature increase was followed by Raman spectroscopy. It was possible to show that the Raman spectra of Fe$_3$O$_4$ and γ-Fe$_2$O$_3$ are clearly distinct (Figure 2.6a). In contrast to hematite and magnetite, the maghemite bands are not well defined, and the resolution seems to depend on sample preparation because it is directly related to the degree of crystallinity of the material. Nevertheless, the Raman spectrum of maghemite can be characterised by three broad structures at around 350, 500 and 700 cm^{-1}.

Colloidal fluids are generally characterised by zeta potential (ζ-potential) measurements. Zeta potential is defined as the electrical potential at the shear plane (also known as slipping plane). The shear plane is commonly considered smaller than the double layer, despite no exact relationship has been formulated yet. The zeta potential is a function of surface charge density, shear plane location and surface structure, and it is a very important parameter with respect to many features of the dispersed materials. In solution, the presence of a net charge on a metal oxide nanoparticle affects the distribution of surrounding ions, resulting in an increase in the concentration of counter-ions (ions of opposite charge to the particle) in the vicinity of the particle. This implies that the double layer is determined by the ionic strength of the solution[103]. When the electrolyte concentration is modified, the changes in the shear plane location may be caused either by changes in the double layer thickness and polarisation or by modification of the surface morphology. Thus, it is difficult to differentiate the shift in the shear plane from zeta potential measurements in different electrolyte concentration conditions. As a consequence, ζ-potential cannot be measured directly, but it has to be calculated from experimental techniques (streaming current or potential, electrophoretic mobility, electric conductivity for diluted systems[104] and ultrasounds for concentrated ones)[105]. Usually, the zeta potential of colloidal suspensions is measured from electrokinetic techniques. An electrokinetic phenomenon occurs when an external field acts on a colloidal suspension. When a particle moves in an arbitrary electrolyte solution, a thin liquid layer will move with the particle, too. The layer between the moving and stationary liquid defines the slipping plane, and the potential in that plane is the electrokinetic potential (also called ζ-potential). Inside the slip plane, the particles are considered to be solid spheres with no bulk conductivity. It is to notice that ζ-potential values obtained with different methodologies are hardly comparable and the standardisation of these measurements is far to be achieved. Indeed, the value of the ζ-potential for a determined charged interface should be independent of the electrokinetic technique used. However, there are a lot of reports where significant differences are found between the values of ζ-potential obtained for the same charged interface using different electrokinetic techniques or theoretical approaches[106–109]. Zeta potential values obtained from electric conductivity experiments are generally much higher than those obtained from electrophoretic mobility or streaming potential. ζ-potential values provided from electrophoretic mobility measurements are usually higher than those obtained from streaming potential[104]. Commonly used expressions in commercial instruments to convert the electrokinetic mobility into ζ-potential derive from approximations of the Henry equation by a separate analytical theory of the two following cases. The first case is represented by the most commonly used Smoluchowski equation[110], which is valid for relatively conducting aqueous solutions (polar solvents) where the double layer is usually much thinner than the particle radius a ($\kappa a \gg 1$) (where κa represents the ratio of the particle radius to electrical double layer thickness). The second opposite case ($\kappa a \ll 1$) is represented by the Huckel equation, which is valid for low-conducting liquids (non-polar solvent). Both, deriving from Henry equation, give appropriate zeta potentials only for rigid spheres at quite low mobility values (low values of zeta potential). Most of the colloidal nanoparticles used for medical purpose are coated with polymer or polyelectrolytes, and they cannot be really considered as rigid spheres[111]. To take into account the influence of the coating on the zeta potential values, a soft particle model has been proposed by combining the theory of rigid spherical colloids with the theory of completely permeable polyelectrolytes or polymers[112]. Nevertheless, even in this case, the definition of appropriate boundary conditions is needed to solve the complex algorithms.

However, the study of the evolution of zeta potential as a function of pH of surface-modified nanoparticles compared with pristine nanoparticles can be useful to compare their IEP and to investigate the stability of theses dispersion. The comparison of IEPs can help in the understanding of the nature of the surface site groups and on the quality of the surface modification. Thus, by working with similar apparatus and protocols, ζ-potential values can be very informative as shown in Figure 2.7,

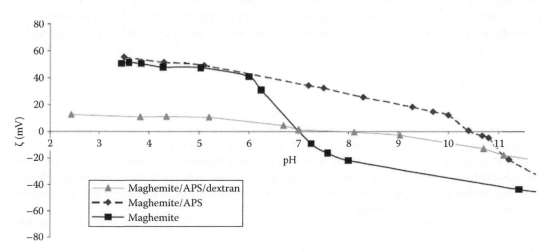

FIGURE 2.7 Evolution of zeta potential curves versus pH of pristine maghemite NPs, surface-aminated maghemite NPs (APS: 3-aminopropyltrimethoxysilane) and maghemite NPs encapsulated in a cross-linked dextran corona. (Reprinted from *J. Magn. Magn. Mater.*, 293, Mornet, S., Portier, J., and Duguet, E., A method for synthesis and functionalization of ultrasmall superparamagnetic covalent carriers based on maghemite and dextran, 127–134, Copyright 2005, with permission from Elsevier.)

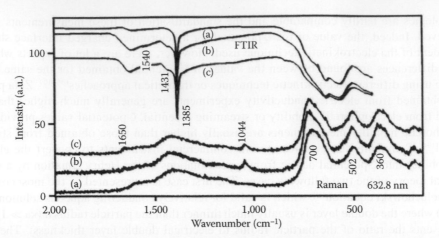

FIGURE 2.8 FTIR and Raman spectra of (a) maghemite, (b) aspartate-modified maghemite and (c) glutamate-modified maghemite. (Reprinted from *J. Magn. Magn. Mater.*, 225, Sousa, M.H., Rubim, J.C., Sobrinho, P.G., and Tourinho, F.A., Biocompatible magnetic fluid precursors based on aspartic and glutamic acid modified maghemite nanostructures, 67–72, Copyright 2001, with permission from Elsevier.)

where pristine maghemite NPs, surface-aminated maghemite NPs and maghemite NPs encapsulated in a cross-linked dextran corona were compared. Nevertheless, this technique has to be supported by complementary surface analysis methods such as spectroscopy analyses, for example, IR spectroscopy, Raman spectroscopy and XPS.

Finally, Raman spectroscopy appeared useful for surface characterisation purpose. The development of Raman instrumentation having high light throughput, equipped with high-performance charge-coupled device (CCD) detectors, the use of optical microscopes to focus the laser on the sample and collect the scattered radiation and the notch filter technology caused a tremendous increase in the use of Raman spectroscopy as an analytical tool. For instance, molecules adsorbed on surfaces at monolayer levels can be detected by Raman spectroscopy even if the system does not present surface-enhanced Raman characteristics[113,114]. Furthermore, if the system is in solution and strongly absorbing, as is the case of magnetic ferrofluids, the use of confocal spectroscopy makes possible the acquisition of good quality Raman spectra[115]. It has been shown that while Fourier transform infrared (FTIR) spectroscopy gives almost no information regarding to the surface species, Raman spectroscopy in and off resonance gives suitable information regarding to the adsorbate structure and bonding. Absorption of amino acid, such as glutamic acid and aspartic acid, was investigated by both spectroscopy techniques on maghemite nanoparticles surface (Figure 2.8). These techniques revealed that amino acids strongly absorb in the form of their respective salts, suggesting the formation of chemical bond of chelate type involving Fe(III) surface species and a carboxylate group. In this study, it has been observed that Raman spectroscopy was able to detect the adsorbates at a surface concentration in the range of nanomole per square centimetre, which is very interesting in the field of nanobiotechnology where the number of biological species adsorbed on the nanoparticle surface can be very low.

A great number of techniques are available for a complete characterisation of iron oxide NPs (Table 2.3) allowing to check the reproducibility of the synthesis protocols and to guarantee the physico-chemical features of the magnetic fluids based on iron oxide NPs.

TABLE 2.3

Main Characterisation Techniques Suited to the Full Characterisation of Iron Oxide NPs for Biomedical Purpose

NPs Part	Features	Conventional Techniques						Less Conventional Techniques
		XRD Patterns	TEM Images	PCS Spectra	ζ-Potential vs. pH Curves	Magnetisation Curves	IR Spectra	
Crystalline part of the iron oxide core	Chemical composition	×					×	XPS, Raman spectroscopy
	Average diameter	×						HR-TEM
Whole iron oxide core	Average diameter		×			×		Mossbauer spectroscopy, SAXS, SANS, EDXRD
	Size distribution		×					Mossbauer spectroscopy, SAXS, SANS, EDXRD, XPS
	Amorphous layer thickness							HR-TEM
Silica shell	Shell thickness		×					HR-TEM, XPS
(macro)molecular corona	Corona thickness at dehydrated state		×					
	Chemical composition				×			XPS
	Corona thickness at hydrated state			×				
Whole NPs	Hydrodynamic diameter (at hydrated state)			×				
	Surface charges density				×			

2.6 CONCLUSION

In this chapter, we have reviewed various synthetic methods to iron oxide NPs and compared the advantages and drawbacks of the main investigated routes for biomedical purposes. Different methods to stabilise colloidal systems for *in vivo* medium, including various coating materials, were described. An approach for the usual characterisation techniques of these magnetic fluids was also performed.

REFERENCES

1. Bragg, W.H. 1915. The structure of magnetite and the spinels. *Nature* 95, 561.
2. Néel, L. 1948. Propriétés magnétiques des ferrites. Ferrimagnétisme et antiferromagnétisme. *Annales de Physique* 3, 137–198.
3. Burghart, F.J., Potzel, W., Kalvius, G.M. et al. 2000. Magnetism of crystalline and nanostructured $ZnFe_2O_4$. *Physica B: Condensed Matter*, 289–290, 286–290.
4. Ho, J.C., Hamdeh, H.H., Chen, Y.Y. et al.1995. Low-temperature calorimetric properties of zinc ferrite nanoparticles. *Physical Review B* 52, 10122–10126.
5. Schiessl, W., Potzel, W., Karzel, H. et al. 1996. Magnetic properties of the $ZnFe_2O_4$ spinel. *Physical Review B* 53, 9143–9152.
6. Pannaparayil, T., Komarnen, S., Marande, R. and Zadarko, M. 1990. Subdomain zinc ferrite particles: Synthesis and characterization. *Journal of Applied Physics* 67, 5509–5511.
7. Goya, G.F., Rechenberg, H.R., Chen, M. and Yelon, W.B. 2000. Magnetic irreversibility in ultrafine $ZnFe_2O_4$ particles. *Journal of Applied Physics* 87, 8005–8007.
8. Yokoyama, M., Oku, T., Taniyama, T. et al. 1995. Intraparticle structure in ultra-fine $ZnFe_2O_4$ particles. *Physica B: Condensed Matter* 213–214, 251–253.
9. Hochepied, J.F., Bonville, P. and Pileni, M.P. 2000. Nonstoichiometric zinc ferrite nanocrystals: Syntheses and unusual magnetic properties. *Journal of Physical Chemistry B* 104, 905–912.
10. Chinnasamy, C.N., Narayanasamy, A., Ponpandian, N. et al. 2000. Magnetic properties of nanostructured ferrimagnetic zinc ferrite. *Journal of Physics: Condensed Matter* 12, 7795–7805.
11. Anantharaman, M., Jagathesan, S., Malini, K. et al. 1998. On the magnetic properties of ultra-fine zinc ferrites. *Journal of Magnetism and Magnetic Materials* 189, 83–88.
12. Grasset, F., Labhsetwar, N., Li, D. et al. 2002. Synthesis and magnetic characterization of zinc ferrite nanoparticles with different environments: Powder, colloidal solution, and zinc ferrite-silica core-shell nanoparticles. *Langmuir* 18, 8209–8216.
13. Jolivet, J.P., Chaneac, C. and Tronc, E. 2004. Iron oxide chemistry. From molecular clusters to extended solid networks. *Chemical Communications* 2004, 481–487.
14. Shull, C.G., Wollan, E.O. and Koehler, W.C. 1951. Neutron scattering and polarization by ferromagnetic materials. *Physical Review* 84, 912–921.
15. Cornell, R.M. and Schwertmann, U. 2003. *The Iron oxides. Structure, Properties, Reactions, Occurrences and Uses*. Weinheim, Germany: Wiley-VCH Verlag GmbH & Co. KGaA.
16. Schwertmann, U. and Cornell, R.M. 2000. *Iron Oxides in the Laboratory: Preparation and Characterization*. Weinheim, Germany: Wiley-VCH Verlag GmbH.
17. Dave, S.R. and Gao, X. 2009. Monodisperse magnetic nanoparticles for biodetection, imaging, and drug delivery: A versatile and evolving technology. *Nanomedicine and Nanobiotechnology* 1, 583–609 and references therein.
18. Murray, C.B., Norris, D.J. and Bawendi, M.G. 1993. Synthesis and characterization of nearly monodisperse CdE (E = sulfur, selenium, tellurium) semiconductor nanocrystallites. *Journal of the American Chemical Society* 115, 8706–8715.
19. Peng, X., Manna, L., Yang, W.D. et al. 2000. Shape control of CdSe nanocrystals. *Nature* 404, 59–61.
20. Tartaj, P., Morales, M.P., Veintemillas-Verdaguer, S. et al. 2003. The preparation of magnetic nanoparticles for applications in biomedicine. *Journal of Physics D: Applied Physics* 36, R182–R197 and references therein.
21. Durán, J.D.G., Arias, J.L., Gallardo, V. and Delgado, A.V. 2008. Magnetic colloids as drug vehicles. *Journal of Pharmaceutical Sciences* 97, 2948–2983 and references therein.
22. Mahmoudi, M., Sant, S., Wang, B. et al. 2011. Superparamagnetic iron oxide nanoparticles (SPIONs): Development, surface modification and applications in chemotherapy. *Advanced Drug Delivery Reviews* 63, 24–46 and references therein.

23. Laurent, S., Forge, D., Port, M. et al. 2008. Magnetic iron oxide nanoparticles: Synthesis, stabilisation, vectorization, physicochemical characterizations, and biological applications. *Chemical Reviews* 108, 2064–2110 and references therein.
24. Frimpong, R.A. and Hilt, J.Z. 2010. Magnetic nanoparticles in biomedicine: Synthesis, functionalization and applications. *Nanomedicine* 5, 1401–1414 and references therein.
25. Lu, A.H., Salabas, E.L. and Schüth, F. 2007. Magnetic nanoparticles: Synthesis, protection, functionalization, and applications. *Angewandte Chemie International Edition* 46, 1222–1244 and references therein.
26. Massart, R. and Cabuil, V. 1987. Effect of some parameters on the formation of colloidal magnetite in alkaline-medium: Yield and particle-size control. *Journal de Chimie Physique et de Physico-Chimie Biologique* 84, 967–973.
27. Sugimoto, T. and Matijević, E. 1980. Formation of uniform spherical magnetite particles by crystallization from ferrous hydroxide gels. *Journal of Colloid and Interface Science* 74, 227–243.
28. Shinkai, M., Honda, H. and Kobayashi, T. 1991. Preparation of fine magnetic particles and application for enzyme immobilization. *Biocatalysis* 5, 61–69.
29. Sun, Y.K., Ma, M., Zhang, Y. and Gu, N. 2004. Synthesis of nanometer-size maghemite particles from magnetite. *Colloids and Surfaces A: Physicochemical and Engineering Aspects* 245, 15–19.
30. Jolivet, J.P. 2000. *Metal Oxide Chemistry and Synthesis: From Solutions to Solid State*. New York: Wiley.
31. Boistelle, R. and Astier, J.P. 1988. Crystallization mechanisms in solution. *Journal of Crystal Growth* 90, 14–30.
32. López-López, M.T., Durán, J.D.G., Delgado, A.V. and González-Caballero, F. 2005. Stability and magnetic characterization of oleate-covered magnetite ferrofluids in different nonpolar carriers. *Journal of Colloid and Interface Science* 291,144–151.
33. Massart, R. 1981. Preparation of aqueous magnetic liquids in alkaline and acidic media. *IEEE Transactions on Magnetics* MAG-17, 1247–1248.
34. LaMer, V.K. and Dinegar, R.H. 1950. Theory, production and mechanism of formation of monodispersed hydrosols. *Journal of the American Chemical Society* 72, 4847–4854.
35. Bee, A., Massart, R. and Neveu, S. 1995. Synthesis of very fine maghemite particles. *Journal of Magnetism and Magnetic Materials* 149, 6–9.
36. Ishikawa, T., Kataoka, S. and Kandori, K. 1993. The influence of carboxylate ions on the growth of β-FeOOH particles. *Journal of Materials Science* 28, 2693–2698.
37. Ishikawa, T., Takeda, T. and Kandori, K. 1992. Effects of amines on the formation of β-ferric oxide hydroxide. *Journal of Materials Science* 27, 4531–4535.
38. Kandori, K., Kawashima, Y. and Ishikawa, T. 1992. Effects of citrate ions on the formation of monodispersed cubic hematite particles. *Journal of Colloid and Interface Science* 152, 284–288.
39. Lin, M.M., Kim, D.K., El Haj, A.J. and Dobson, J. 2008. Development of superparamagnetic iron oxide nanoparticles (SPIONS) for translation to clinical applications. *IEEE Transactions on Nanobiosciences* 7, 298–305 and references therein.
40. Molday, R.S. and Mackenzie, D. 1982. Immunospecific ferromagnetic iron-dextran reagents for the labeling and magnetic separation of cells. *Journal of Immunological Methods* 52, 353–357.
41. Palmacci, S. and Josephson, L. 1993, US Patent 5262176.
42. Mornet, S., Vasseur, S., Grasset, F. and Duguet, E. 2004. Magnetic nanoparticles design for medical diagnosis and therapy. *Journal of Materials Chemistry* 14, 2161–2175 and references therein.
43. Gupta, A.K. and Gupta, M. 2005. Synthesis and surface engineering of iron oxide nanoparticles for biomedical applications. *Biomaterials* 26, 3995–4021 and references therein.
44. Feltin, N. and Pileni, M.P. 1997. New technique for synthesizing iron ferrite magnetic nanosized particles. *Langmuir* 13, 3927–3933.
45. Gupta, A.K. and Wells, S. 2004. Surface-modified superparamagnetic nanoparticles for drug delivery: Preparation, characterization, and cytotoxicity studies. *IEEE Transactions on Nanobiosciences* 3, 66–73.
46. Lee, Y., Lee, J., Bae, C.J. et al. 2005. Large-scale synthesis of uniform and crystalline magnetite nanoparticles using reverse micelles as nanoreactors under reflux conditions. *Advanced Functional Materials* 15, 503–509.
47. Rockenberger, J., Scher, E.C. and Alivisatos, A.P. 1999. A new nonhydrolytic single-precursor approach to surfactant-capped nanocrystals of transition metal oxides. *Journal of the American Chemical Society* 121, 11595–11596.
48. Jeong, U., Teng, X., Wang, Y. et al. 2007. Superparamagnetic colloids: Controlled synthesis and niche applications. *Advanced Materials* 19, 33–60 and references therein.
49. Li, Z., Wei, L., Gao, M. and Lei, H. 2005. One-pot reaction to synthesize biocompatible magnetic nanoparticles. *Advanced Materials* 17, 1001–1005.

50. Griffiths, C.H., Ohoro, M.P. and Smith, T.W. 1979. Structure, magnetic characterization, and oxidation of colloidal iron dispersions. *Journal of Applied Physics* 50, 7108–7115.

51. Blakemore, R.P. 1975. Magnetotactic bacteria. *Science* 190, 377–379.

52. Faivre, D. and Schüler, D. 2008. Magnetotactic bacteria and magnetosomes. *Chemical Reviews* 108, 4875–4898.

53. Prozorov, T., Palo, P., Wang, L. et al. 2007. Cobalt ferrite nanocrystals: Out-performing magnetotactic bacteria. *ACS Nano* 1, 228–233.

54. Narayanan, K.B. and Sakthivel, N. 2010. Biological synthesis of metal nanoparticles by microbes. *Advances in Colloid and Interface Science* 156, 1–13.

55. Hamley, I.W. 2003. Nanotechnology with soft materials. *Angewandte Chemie International Edition* 42, 1692–1712.

56. Vincent, B., Edwards, J., Emment, S. and Jones, A. 1986. Depletion flocculation in dispersions of steri-cally-stabilised particles ("soft spheres"). *Colloids and Surfaces* 18, 261–281.

57. Derjaguin, B.V. and Landau, L. 1941. Theory of the stability of strongly charged lyophobic sols and of the adhesion of strongly charged particles in solutions of electrolytes. *Acta Physicochimica URSS* 14, 633–662.

58. Verwey, E.J.W. and Overbeek, J.T.G. 1948. *Theory of the Stability of Lyophobic Colloids*. Amsterdam, the Netherlands: Elsevier.

59. Adireddy, S., Lin, C., Palshin, V. et al. 2009. Size-controlled synthesis of quasi-monodisperse transition-metal ferrite nanocrystals in fatty alcohol solutions. *Journal of Physical Chemistry C* 113, 20800–20811.

60. Thanh, N.T.K. and Green, L.A.W. 2010. Functionalisation of nanoparticles for biomedical applications. *Nano Today* 5, 213–230 and references therein.

61. Lefebure, S., Dubois, E., Cabuil, V. et al. 1998. Monodisperse magnetic nanoparticles: Preparation and dispersion in water and oils. *Journal of Materials Research* 13, 2975–2981.

62. Atkins, P.W. and De Paula, J. 2006. *Atkins' Physical Chemistry*. Oxford/New York: Oxford University Press.

63. Turkevich, J., Stevenson, P.C. and Hillier, J. 1951. A study of the nucleation and growth processes in the synthesis of colloidal gold. *Discussions of the Faraday Society* 11, 55–75.

64. Daou, T.J., Begin-Colin, S., Grenèche, J.M. et al. 2007. Phosphate adsorption properties of magnetite-based nanoparticles. *Chemistry of Materials* 19, 4494–4505.

65. Hajdú, A., Tombácz, E., Illés, E. et al. 2008. Magnetite nanoparticles stabilized under physiological conditions for biomedical application. *Progress in Colloid and Polymer Science* 135, 29–37.

66. Sahoo, Y., Pizem, H., Fried, T. et al. 2001. Alkyl phosphonate/phosphate coating on magnetite nanoparticles: A comparison with fatty acids. *Langmuir* 17, 7907–7911.

67. Lévy, R., Thanh, N.T.K., Doty, R.C. et al. 2004. Rational and combinatorial design of peptide capping ligands for gold nanoparticles. *Journal of the American Chemical Society* 126, 10076–10084.

68. Racuciu, M., Creanga, D.E. and Airinei, A. 2006. Citric-acid-coated magnetite nanoparticles for biological applications. *European Physical Journal E: Soft Matter and Biological Physics* 21, 117–121.

69. Voisin, P., Ribot, E.J., Miraux, S. et al. 2007. Use of lanthanide-grafted inorganic nanoparticles as effective contrast agents for cellular uptake imaging. *Bioconjugate Chemistry* 18, 1053–1063.

70. Mornet, S., Portier, J. and Duguet, E. 2005. A method for synthesis and functionalization of ultrasmall superparamagnetic covalent carriers based on maghemite and dextran. *Journal of Magnetism and Magnetic Materials* 293, 127–134.

71. Pellegrino, T., Manna, L., Kudera, S. et al. 2004. Hydrophobic nanocrystals coated with amphiphilic polymer shell: A general route to water soluble nanocrystals. *Nano Letters* 4, 703–707.

72. Longmire, M., Choyke, P.L. and Kobayashi, H. 2008. Clearance properties of nano-sized particles and molecules as imaging agents: Considerations and caveats. *Nanomedicine* 3, 703–717.

73. Hajdú, A., Illés, E., Tombácz, E. and Borbáth, I. 2009. Surface charging, polyanionic coating and colloid stability of magnetite nanoparticles. *Colloids and Surfaces A: Physicochemical and Engineering Aspects* 347, 104–108.

74. Lesnikovich, A.E., Shunkevich, T.M., Naumenko, V.N. et al. 1990. Dispersity of magnetite in magnetic liquids and the interaction with a surfactant. *Journal of Magnetism and Magnetic Materials* 85, 14–16.

75. Stöber, W., Fink, A. and Bohn, E. 1968. Controlled growth of monodisperse silica spheres in the micron size range. *Journal of Colloid and Interface Science* 26, 62–69.

76. Pinho, S.L.C., Pereira, G.A., Voisin, P. et al. 2010. Fine tuning of the relaxometry of γ-Fe$_2$O$_3$@SiO$_2$ nanoparticles by tweaking the silica coating thickness. *ACS Nano* 4, 5339–5349.

77. Aubert, T., Grasset, F., Mornet, S. et al. 2010. Functional silica nanoparticles synthesized by water-in-oil microemulsion processes. *Journal of Colloid and Interface Science* 341, 201–208.

78. Bish, D.L. and Post, J.E. 1989. *Modern Powder Diffraction (Reviews in Mineralogy Volume 20).* Washington, DC: Mineralogical Society of America.
79. Warren, B.E. 1969. *X-Ray Diffraction.* Reading, MA: Addison-Wesley.
80. Ungár, T., Gubicza, J., Ribárik, G. and Borbély, A. 2001. Crystallite size distribution and dislocation structure determined by diffraction profile analysis: Principles and practical application to cubic and hexagonal crystals. *Journal of Applied Crystallography* 34, 298–310.
81. Serna, C.J., Bodker, F., Morup, S. et al. 2001. Spin frustration in maghemite nanoparticles. *Solid State Communications* 118, 437–440.
82. Morales, M.P., Serna, C.J., Bodker, F. and Morup, S. 1997. Spin canting due to structural disorder in maghemite. *Journal of Physics: Condensed Matter* 9, 5461–5467.
83. Yee, C., Kataby, G., Ulman, A. et al. 1999. Self-assembled monolayers of alkanesulfonic and -phosphonic acids on amorphous iron oxide nanoparticles. *Langmuir* 15, 7111–7115.
84. Tartaj, P., Morales, M.P., Veintemillas-Verdaguer, S. et al. 2006. Synthesis, properties and biomedical applications of magnetic nanoparticles. In *Handbook of Magnetic Materials,* Vol. 16, ed. K.H.J. Buschow, Elsevier, Amsterdam, pp. 403–482.
85. Shen, L., Stachowiak, A., Fateen, S.E.K. et al. 2001. Structure of alkanoic acid stabilized magnetic fluids. A small-angle neutron and light scattering analysis. *Langmuir* 17, 288–299.
86. Moeser, G.D., Green, W.H., Laibinis, P.E. et al. 2004. Structure of polymer stabilized magnetic fluids: Small-angle neutron scattering and mean-field lattice modeling. *Langmuir* 20, 5223–5234.
87. Caminiti, R., Carbone, M., Panero, S. et al. 1999. Conductivity and structure of poly(ethylene glycol) complexes using energy dispersive X-ray diffraction. *Journal of Physical Chemistry B* 103, 10348–10355.
88. Atzei, D., Ferri, T., Sadun, C. et al. 2001. Structural characterization of complexes between iminodiac-etate blocked on styrene-divinylbenzene matrix (Chelex 100 resin) and Fe(III), Cr(III), and Zn(II) in solid phase by energy-dispersive X-ray diffraction. *Journal of the American Chemical Society* 123, 2552–2558.
89. Sadun, C., Bucci, R. and Magri, A.L. 2002. Structural analysis of the solid amorphous binuclear complexes of iron(III) and aluminum(III) with chromium(III)-DTPA chelator using energy dispersive X-ray diffraction. *Journal of the American Chemical Society* 124, 3036–3041.
90. Di Marco, M., Port, M., Couvreur, P. et al. 2006. Structural characterization of ultrasmall superparamagnetic iron oxide (USPIO) particles in aqueous suspension by energy dispersive X-ray diffraction (EDXD). *Journal of the American Chemical Society* 128, 10054–10059.
91. Morales, M.P., Veintemillas-Verdaguer, S., Montero, M.I. et al. 1999. Surface and internal spin canting in γ-Fe_2O_3 nanoparticles. *Chemistry of Materials* 11, 3058–3064.
92. De Faria, D.L.A., Venâncio Silva, S. and De Oliveira, M.T. 1997. Raman microspectroscopy of some iron oxides and oxyhydroxides, *Journal of Raman Spectroscopy* 28, 873–878.
93. Graat, P. and Somers, M.A.J. 1998. Quantitative analysis of overlapping XPS peaks by spectrum reconstruction: Determination of the thickness and composition of thin iron oxide films. *Surface and Interface Analysis* 26, 773–792.
94. McIntyre, N.S. and Zetaruk, D.G. 1977. X-ray photoelectron spectroscopic studies of iron oxides. *Analytical Chemistry* 49, 1521–1529.
95. Fujii, T.F., De Groot, M.F., Sawatzky, G.A. et al. 1999. In situ XPS analysis of various iron oxide films grown by NO_2-assisted molecular-beam epitaxy. *Physical Review. B, Condensed Matter and Materials Physics* 59, 3195–3202.
96. Brice-Profeta, S., Arrio, M.A., Tronc, E. et al. 2005. Magnetic order in γ-Fe_2O_3 nanoparticles: A XMCD study. *Journal of Magnetism and Magnetic Materials* 288, 354–365.
97. Inouye, K., Endo, R., Otsuka, Y. et al. 1982. Oxygenation of ferrous ions in reversed micelle and reversed microemulsion. *Journal of Physical Chemistry* 86, 1465–1469.
98. Lindner, P. and Zemb, T. 2004. *Neutrons, X-Rays and Light: Scattering Methods Applied to Soft Condensed Matter.* Elsevier, Amsterdam.
99. De Jaeger, N., Demeyere, H., Finsy, R. et al. 1991. Particle sizing by photon correlation spectroscopy Part I: Monodisperse lattices: Influence of scattering angle and concentration of dispersed material. *Particle and Particle Systems Characterization* 8, 179–186.
100. Preudhomme, J. 1974. Correlation between infrared spectra and crystal chemistry of spinels. *Annales de Chimie,* 9, 31–41.
101. Morales, M.P., Pecharroman, C., Gonzalez-Carreño, T. and Serna, C.J. 1994. Structural characteristics of uniform γ-Fe_2O_3 particles with different axial (length/width) ratios. *Journal of Solid State Chemistry* 108, 158–163.
102. Reid, E.S., Cooney, R.P., Hendra, P.J. and Fleischmann, M. 1977. A Raman spectroscopic study of corrosion of lead electrodes in aqueous chloride media. *Journal of Electroanalytical Chemistry* 80, 405–408.

103. Hunter, R.J. 2001. *Foundations of Colloid Science.* 2nd edn. Oxford, U.K.: Clarendon Press.
104. El-Gholabzouri, O., Cabrerizo-Vílchez, M.A. and Hidalgo-Álvarez, R. 2006. Zeta potential of polystyrene latex determined using different electrokinetic techniques in binary liquid mixtures. *Colloids and Surfaces A: Physicochemical and Engineering Aspects* 291, 30–37.
105. Dukhin, A.S., Ohshima, H., Shilov, V.N. et al. 1999. Electroacoustics for concentrated dispersions. *Langmuir* 15, 3445–3451.
106. Russel, A.S., Scales, P.J., Mangelsdorf, C.S. et al. 1995. High-frequency dielectric response of highly charged sulfonate latices. *Langmuir* 11, 1553–1558.
107. Midmore, B.R., Pratt, G.V. and Herrington, T.M. 1996. Evidence for the validity of electrokinetic theory in the thin double layer region. *Journal of Colloid and Interface Science* 184, 170–174.
108. Gusev, I. and Horvath, C. 2002. Streaming potential in open and packed fused silica capillaries. *Journal of Chromatography A* 948, 203–223.
109. Di Marco, M., Guilbert, I., Port, M. et al. 2007. Colloidal stability of ultrasmall superparamagnetic iron oxide (USPIO) particles with different coatings. *International Journal of Pharmaceutics* 331, 197–203.
110. Thode, K., Müller, R.H. and Kresse, M. 2000. Two-time window and multiangle photon correlation spectroscopy size and zeta potential analysis – Highly sensitive rapid assay for dispersion stability. *Journal of Pharmaceutical Sciences* 89, 1317–1324.
111. Di Marco, M., Sadun, C., Port, M. et al. 2007. Physicochemical characterization of ultrasmall superparamagnetic iron oxide particles (USPIO) for biomedical application as MRI contrast agents. *International Journal of Nanomedicine* 2, 609–622.
112. Ohshima, H. 1995. Electrophoresis of soft particles. *Advances in Colloid and Interface Science* 62, 189–235.
113. Hugot-Le Goff, A., Joiret, S. and Falaras, P. 1999. Raman resonance effect in a monolayer of polypyridyl ruthenium(II) complex adsorbed on nanocrystalline TiO_2 via phosphonated terpyridyl ligands. *Journal of Physical Chemistry B* 103, 9569–9575.
114. Esser, N. 1999. Analysis of semiconductor surface phonons by Raman spectroscopy. *Applied Physics A: Materials Science & Processing* 69, 507–518.
115. Sousa, M.H., Tourinho, F.A. and Rubim, J.C. 2000. Use of Raman micro-spectroscopy in the characterization of $M^{II}Fe_2O_4$ (M = Fe, Zn) electric double layer ferrofluids. *Journal of Raman Spectroscopy* 31, 185–191.
116. Sousa, M.H., Rubim, J.C., Sobrinho, P.G. and Tourinho, F.A. 2001. Biocompatible magnetic fluid precursors based on aspartic and glutamic acid modified maghemite nanostructures. *Journal of Magnetism and Magnetic Materials* 225, 67–72.

Etienne Duguet received his physical-chemistry engineering diploma from the Graduate School of Chemistry and Physics of Bordeaux in 1989 (including a MSc in polymer chemistry and a MSc in solid state chemistry) and his PhD in polymer chemistry from the University of Bordeaux in 1992. He is currently full professor, and his research at the Institut de Chimie de la Matière Condensée de Bordeaux focuses on the synthesis of hybrid organic–inorganic materials based on inorganic particles derivatised through molecular surface modification, polymer encapsulation, dissymetrisation and/or functionalisation for optical or medical applications.

Marie-Hélène Delville is research director at French National Centre for Scientific Research CNRS. She received her PhD in organometallic chemistry under the guidance of Prof. D. Astruc at the University of Bordeaux 1, in 1988. Just after finishing her doctorate, she joined the CNRS in his team. She moved to the Institute of Condensed Matter of Bordeaux in 1996, where she is currently working. Her research interests are focused on the fundamental and practical aspects involved in the synthesis of organic–inorganic colloidal nanoobjects with special emphasis on the synthesis, control of shape, surface functionalisation of mineral oxide particles and sol–gel chemistry. Her research also includes their use in biomedical applications and their potential toxicity.

Stéphane Mornet received his PhD in physico-chemistry of condensed matter at the University of Bordeaux in 2002. After a 4 year post-doctoral fellowship at the European Institute of Chemistry and Biology of Bordeaux, he undertook a post-doctoral stay from 2006 to 2007 in Italy at the Institute for Health and Consumer Protection (Ispra), where he worked for the European Commission in the field of nanoparticles health risk. He is currently a researcher at the French National Centre for Scientific Research (CNRS) at the Institute of Condensed Matter Chemistry of Bordeaux, and his research, at the interface of chemistry and biology, focuses on the synthesis of magnetic, metallic and luminescent nanoparticles, their surface functionalisation and conjugation with biomolecules for imaging and therapy purposes.

3 Protein Cage Magnetic Nanoparticles
Inspiration, Synthesis and Biomedical Utility

Masaki Uchida, Lars O. Liepold, and Trevor Douglas

CONTENTS

3.1 INTRODUCTION

Magnetic materials of biological origin have been an inspiration to material scientists. The single-domain magnetic materials synthesised by magnetotactic bacteria[1], the magnetite (Fe_3O_4) in chiton teeth (and its apparent formation from an antiferromagnetic iron oxide phase)[2,3], and the uncompensated surface spins on the ferrihydrite mineral core of ferritin[4], are examples of biologically produced magnetic materials with unique and useful properties. Biomineralisation illustrates control over material composition, polymorph selection, orientation and morphology under mild reaction conditions[5,6]. A major goal in biomimetic nanomaterials chemistry is to utilise our understanding of the fundamental biomineralisation processes to synthesise target materials incorporating specific magnetic properties by design. Control over magnetic materials synthesis using protein–mineral interfaces to control properties has generated a truly biomimetic approach to the synthesis of materials well beyond the bounds of biology. In particular, the use of protein cage nanoparticles as templates for the directed growth of magnetic nanoparticles (MNPs) will be the particular focus of this chapter.

Protein cage architectures such as those of small virus capsids and ferritins are self-assembled from a limited and defined number of subunit building blocks[7,8]. These protein cage architectures typically range in size from tens to hundreds of nanometres and are spherical or rod shaped. By combining both chemical and genetic modifications of the subunits, one can impart novel functions to these protein cage architectures that are quite different than their native function in biology. There are three distinct surfaces of the assembled protein cage architecture that can be manipulated to impart function by design. These are the external surface, the interior surface and the surface that forms the interface between the subunits. Manipulation of the interactions at the subunit interface allows control over disassembly and reassembly[9,10]. The interior interface can be used for chemical attachment of small molecules and for encapsulation of nanoparticles sequestered inside the cage. The exterior surface can also be modified by small-molecule attachment or by presentation of cell-specific targeting ligands (peptides, antibodies and carbohydrates). A library of protein cages, which support a range of size, synthetic chemistries and genetic manipulations have been employed. These include the following: the 58 nm capsid of the *Salmonella typhimurium* bacteriophage P22[11–13], the 28 nm capsid of Cowpea chlorotic mottle virus (CCMV)[14], the 31 nm capsid of Cowpea mosaic virus (CPMV)[15], the 28 nm capsid of brome mosaic virus (BMV)[16], the 27 nm capsid of bacteriophage MS2[17], the 15 nm lumazine synthase (LS)[18], the 12 nm ferritin cage[19], the 12 nm small heat-shock protein (sHsp) from *Methanococcus jannaschii*[20], and the 9 nm Dps protein cage[21] (Figure 3.1)[22].

Many closed-shell protein cage architectures define an interior space that can be used to selectively direct synthesis or entrapment of a guest cargo. A significant advantage to the use of protein cages is the ability to produce mono-disperse nanoparticles with a precisely defined size and shape, and since the properties of nanophase materials are intimately related to their dimensions, any heterogeneity in size is reflected as heterogeneity in their physical properties[7,8]. The controlled formation of magnetic nanomaterials within protein cage architectures thus has broad applications in a variety of emerging technologies including medical diagnostics.

To understand the formation of MNPs in a controlled manner, ferritin is a great example from nature to exploit because its inherent biological function is the sequestration and storage of iron as a biomineral of ferrihydrite[23–25]. Lessons from ferritin have inspired material scientists to utilise this and other protein cages to fabricate nanoparticles of iron oxide and beyond. Using approaches that mimic the structure–function relationships in ferritin proteins, that is selective mineral nucleation and controlled particle growth, creates the potential for protein cage nanoparticles for imaging and other magnetic applications. Key to the success of this approach is an ability to produce and modify protein cages to impart novel functionality such as cell-targeting capability in addition to the size- and composition-dependent magnetic properties. In this chapter, we will first discuss exploitation

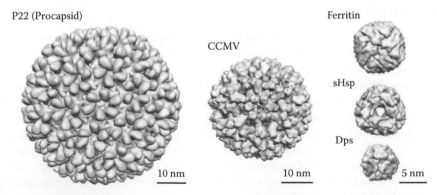

FIGURE 3.1   Space-filling images of protein cage architectures including viral capsids P22 pro-capsid form (58 nm diameter) and CCMV (28 nm); ferritin (12 nm), small heat-shock protein (sHsp, 12 nm) and DNA-binding protein from starved cells (Dps, 9 nm). The images were reproduced using the UCSF Chimera package (http://www.cgl.ucsf.edu/chimera/) from the Resource for Biocomputing, Visualization, and Informatics at the University of California (supported by NIH P41 RR-01081). (From Pettersen, E.F. et al., *J. Comput. Chem.*, 25, 1605, 2004.)

of protein cages as directing synthesis of MNPs, then introduce the application of protein cage-templated particles as biomedical imaging agents.

3.2 SYNTHESIS OF MNPs USING PROTEIN CAGE ARCHITECTURES

3.2.1 Biomineralisation in Ferritin as an Inspiration

Ferritins are a class of proteins that are almost ubiquitous across all domains of life[23–25]. While there are some differences in amino acid sequences between prokaryotic, eukaryotic and archaeal ferritins, they share a high degree of structural similarity. Ferritins are 24-subunit proteins that self-assemble into a cage-like architecture of 12 nm exterior diameter with an interior cavity of 8 nm in diameter[19]. A hydrated ferric oxide/phosphate is mineralised in the interior cavity[23–25].

The role of ferritin *in vivo* is to sequester iron as a hydrated form of iron oxide (or phosphate) predominantly as the mineral ferrihydrite $(FeOOH)$[4,23–25]. The magnetic properties of this mineral have been studied extensively[26–32]. Ferrihydrite is an antiferromagnetic oxyhydroxide, but each ferrihydrite particle encapsulated inside ferritin exhibits superparamagnetism due to its small size and large relative surface area with uncompensated spins[4,31,33]. The blocking temperature of this composite material is approximately 15 K, below which hysteresis from the uncompensated moments are observed. The cores of mammalian ferritins appear to be crystalline, and high-resolution electron microscopy reveals either single crystals or small crystalline domains interspersed with some iron oxide of low crystallinity[34].

Native mammalian ferritins are composed of a mixture of two different types of subunit, light (L) and heavy (H) chain[23]. A conserved dinuclear iron-binding site known as the ferroxidase centre, which catalyses the oxidation of Fe^{2+} to Fe^{3+}, exists in the H chain but is absent in the L chain. The ferroxidase site is located in the centre of the four-helix bundle in each subunit. *In vivo*, ferritin is responsible for sequestering and storing toxic iron as an innocuous mineral of iron oxide through an overall protein-mediated reaction represented in Reaction 3.1[35,36].

$$4Fe^{2+} + O_2 + 6H_2O \rightarrow FeOOH_{core} + 8H^+ \qquad (3.1)$$

The actual biological process of iron oxidation and encapsulation is a multi-step process involving Fe^{2+} entry, oxidation of Fe^{2+} to Fe^{3+}, hydrolysis, nucleation and crystal growth[25,36]. It is considerably more complex than Reaction 3.1 and some of the intimate steps are probably still unrevealed. Iron entry into eukaryotic ferritin (Fn) is believed to occur *via* a channel around threefold symmetry axis formed at the interface between subunits because electrostatic calculations on the recombinant human H-chain Fn reveal electrostatic gradients at the threefold axes that act as a guiding force directing cations through the channel towards the interior of the protein cage[37,38]. X-ray crystal structure analysis of frog M ferritin revealed the presence of multiple metal ions along the threefold channel that link exterior surface to the cage interior. Metal-binding sites also exist along the inner cavity surface between the exit of the channel and ferroxidase site[39]. These metals are believed to represent the path of sites transiently occupied by ferrous ions en route from the exterior of the cage to the ferroxidase sites. Fe^{2+} oxidation is enzymatically catalysed by a reaction at the ferroxidase centre resulting in the formation of diferric oxo/hydroxo mineral precursors[24,25]. The nucleation of an iron oxide material from this insoluble ferric ion is facilitated at the interior protein interface, and the particle grows from this nucleus but is limited by the size constraints of the cage interior. Although little was known about how and where the catalytic products migrate from the ferroxidase centre to the cavity, nuclear magnetic resonance (NMR) studies provided identification of the postcatalytic path of the mineral precursors from the active site to the cavity where the ferrihydrite biomineral is grown[40]. The precursors appear to proceed along a 20 Å long path inside the four-helix bundle of each subunit until they reach the terminus of the bundle around the fourfold axis of the protein. As the exit of the path is close enough to the other three sites around the fourfold axis, emerging mineral precursors would be facilitated in their transformation into a larger biomineral[40]. The mineralised iron particles are electron dense and are the approximate dimensions of the interior of the protein cage (5–7 nm diameter). When iron is allowed to undergo oxidative hydrolysis *in vitro* in the absence of Fn, an uncontrolled homogeneous nucleation results in mineralisation and precipitation of iron oxide. There are approximately 15 common polymorphs of iron oxide or iron oxyhydroxide[41], but under the narrow range of conditions compatible with biology and in the absence of macromolecular directing agents, the mineral phases of lepidocrocite (γ-FeO(OH)) or goethite (α-FeO(OH)) will be formed. It is interesting to note that in the presence of Fn, only a particular phase of iron oxide (ferrihydrite) is formed. Ferrihydrite is less crystalline than lepidocrocite or goethite and is characterised by electron or x-ray diffraction studies to commonly have either two or six diffraction lines but has recently been structurally characterised using pair distribution function (PDF) analysis[4,33]. This kinetically trapped phase of iron oxide is not usually a particularly stable phase but is stable when prepared inside Fn, indicating the ability of biomolecules to direct and selectively stabilise a particular polymorph.

The role of Fn as controlling size, shape and polymorph of biomineral formed inside of its cavity has inspired material scientists to exploit the protein cage as a template for inorganic nanoparticles beyond its native mineral, ferrihydrite. Mann et al. pioneered this field by demonstrating that iron oxide nanoparticles could be artificially synthesised within empty ferritin (apo-ferritin) cages under conditions of elevated temperature and pH[42–45]. The mineralised ferritin cages have iron oxide cores with homogeneous size distribution (7.3 ± 1.4 nm) that are almost indistinguishable from naturally formed iron oxide cores in holo-ferritin. This work has opened a new avenue for making nanoparticles using biomimetic processes, allowing iron oxides, as well as a variety of inorganic and metal nanoparticles, to be synthesised within apo-ferritin using similar methods. Recently, recombinant ferritin cages have been developed, which have demonstrated advantages for the encapsulation of maghemite (γ-Fe$_2$O$_3$) (or magnetite (Fe$_3$O$_4$))[46,47]. Using recombinant human H chain ferritin cages (HFn), which possess the ferroxidase centre, it has been shown that γ-Fe$_2$O$_3$ cores formed close to the theoretical core diameter, calculated from input Fe loading with narrow size distribution (Figure 3.2).

3.2.2 Ferritin as a Template for the Synthesis and Fabrication of Magnetic Nanoparticles

One of the advantages of protein cages such as Fn as a template of nanoparticle synthesis is its homogeneity in size and shape. As size and shape of an MNP affect its physical properties, control of these factors is significantly important to obtain designed MNPs. Protein cages, like ferritin, provide both a size- and shape-constrained reaction environment, which allows us to tailor the magnetic properties of the synthesised MNPs[47–50]. For example, size-dependent magnetic control has been demonstrated either by changing the size of template protein cages or by changing the loading factor of a metal oxide precursor[47,50]. The blocking temperature of γ-Fe$_2$O$_3$ particles encapsulated in protein cages increases with increasing particle size[47,50]. Anisotropy energy density of the particles increases with decreasing particle size (Figure 3.3a)[49,50]. In addition, a fit to the Neel–Arrhenius equation suggests that the protein shell works as an insulator and prevents magnetic interaction between particles (Figure 3.3b)[47,50].

Recent investigation of the nanoscale structure of protein cage-templated inorganic particles by PDF analysis from the total x-ray scattering revealed that, in the case of some materials such as γ-Fe$_2$O$_3$ and Mn$_3$O$_4$, domain sizes in the protein-templated minerals are significantly smaller than that in minerals synthesised under protein-free conditions[51]. This suggests that the presence of multiple nucleation sites in the confined geometry of protein cages leads to the formation of smaller crystalline domains[48,51]. Furthermore, the constrained reaction environment of the cavity allows the synthesis of nanoparticle composites with unique magnetic characteristics. For example, magnetic exchange bias behaviour, which is of considerable interest for technological applications such as

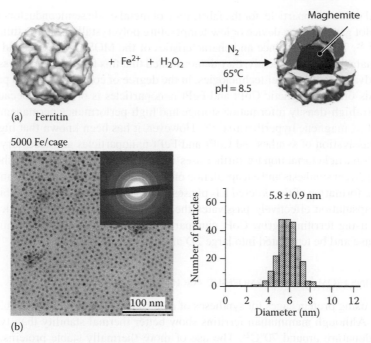

FIGURE 3.2 (a) Schematic representation of maghemite (or magnetite) nanoparticle synthesis in HFn. (b) TEM image of mineralised particles in HFn with 5000 Fe atoms/cage (left). Inset shows electron diffraction of the particles, which is consistent with maghemite (or magnetite). Size distribution of the particles analysed from the TEM image (right).

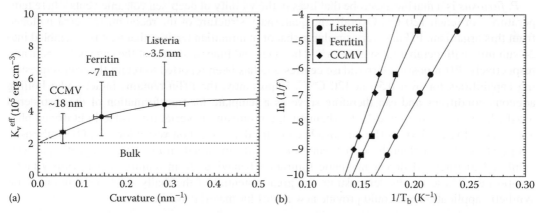

FIGURE 3.3 (a) Temperature-adjusted average anisotropy energy density (K_v^{eff}) plotted versus particle curvature. Large errors in the energy density estimates reflect the uncertainties in the diameter measurements. The dashed line represents the bulk value. (b) Neel–Arrhenius fits to the frequency dependence of the blocking temperature (T_b) of magnetite nanoparticles encapsulated in various protein cages. Here, the inverse blocking temperatures of the ferritin and CCMV samples have been scaled by 5 and 30, respectively, to more clearly display the linearity of all three data sets. The observed linear behaviour of the data is indicative of non-interacting particles. (Copyright 2005 The American Institute of Physics. Reproduced with permission[50].)

high-density recording and sensors, was observed if ferrimagnetic $Co_xFe_{3-x}O_4$ and antiferromagnetic Co_3O_4 nanoparticles were formed together inside of a single ferritin cage[52].

The size homogeneity of the protein cages makes them ideal building units for fabrication of nanodevices by bottom-up approaches of 2D and/or 3D assembly of each unit. Yamashita et al. have utilised ferritin and other protein cages as hierarchical building blocks, following the synthesis of an entrapped metal oxide nanoparticle, for the fabrication of metal oxide semiconductors (MOS) such as a floating nanodot gate memory device or low-temperature polycrystalline silicon thin film transistor flash memory[53–55]. The performance and characteristics of the MOS devices depend on size, shape and density of nanodot array. Recently, they demonstrated that this fabrication process, using protein cages, has an advantage over other technologies, in the degree of control over these parameters[56].

The synthesis of ferromagnetic CoPt and FePt nanoparticles is of interest because of its great potential for ultrahigh-density information storage and high-performance permanent magnet applications, as well as magnetic hyperthermia[57–61]. However, it has been known that high temperature and effective passivation of synthesised CoPt and FePt nanoparticles are necessary for stable ferromagnetic nanoparticle formation for further uses[57–61]. The ferritin cage has been shown to act as a template for the direct synthesis and encapsulation of ferrimagnetic CoPt and FePt nanoparticles[62–64]. While complete formation of the ordered $L1_0$ phases is not achieved with this mild biomimetic synthesis, the encapsulation effectively passivates the nanoparticles. This biomimetic approach has been exploited using ferritin to grow CoPt nanoparticles that were subsequently annealed into the ordered $L1_0$ phase and be templated into large 2D arrays[62–64].

3.2.3 Ferritins from Hyperthermophilic Archaeon

A limitation in using protein cages for syntheses of materials is the temperature stability of the protein templates. Although mammalian ferritins show better thermal stability than typical proteins, they begin to denature around $70°C$[23]. The use of more thermally stable proteins, isolated from hyperthermophilic bacteria and archaea, could overcome this limitation and expand the usefulness of protein cages as templates for materials synthesis. In particular, the ferritin isolated from the hyperthermophilic archaeal anaerobe *Pyrococcus furiosus* (PfFn) has been shown to act as a template for material syntheses at elevated temperature.

P. furiosus is a marine anaerobe that lives in the vicinity of deep-sea volcanic vents where temperature can reach $120°C$[65]. Overall, the quaternary structure of the recombinant ferritin cloned from this organism, PfFn, is very similar to that of mammalian ferritin, that is, it is assembled into 24 subunits with octahedral symmetry[66]. The outer and inner diameter of the cage is 12 and 8 nm, respectively. PfFn possess ferroxidase centres and has been reported to retain the iron sequestering capabilities for over 0.5 h at $120°C$[65,67]. Importantly, the PfFn remains intact under boiling aqueous conditions and can therefore serve as a template for transformation of the disordered ferrihydrite in its cavity to α-Fe_2O_3 (hematite), a promising material for visible light driven photocatalysis (Figure 3.4)[68]. PfFn has also been utilised as a reaction vessel for γ-Fe_2O_3 synthesis at temperatures higher than those that can be used with mammalian Fn[69]. The magnetic particles exhibited distinctly different magnetic saturation behaviour from mammalian ferritin γ-Fe_2O_3 composites. These results demonstrate the great potential of thermally stable protein cages for synthetic application and could provide new routes for material syntheses.

3.2.4 Dps Proteins

DNA-binding protein from nutrient starved cells (Dps) was originally isolated from *Escherichia coli*[70], and since its discovery structural and functional homologues have been isolated in many other bacteria[71–73], as well as archaea[74–76]. There is a clear relationship between Fn and the Dps proteins – they both are members of the ferritin superfamily. While there is some structural similarity between Fn and Dps proteins, the Dps protein cage is assembled from 12 subunits with tetrahedral (23) symmetry[72]. The cage has exterior diameter of 9 nm and interior diameter of 5 nm[21,72]. The Dps architecture contains two types of threefold symmetry channels. One of these channels is lined with hydrophilic amino acids that can provide access for cations from bulk solution to the interior surface similar in the threefold symmetry channel of Fn. It is proposed that once metal cations are inside the protein cage, interaction with the negatively charged interior surface can occur.

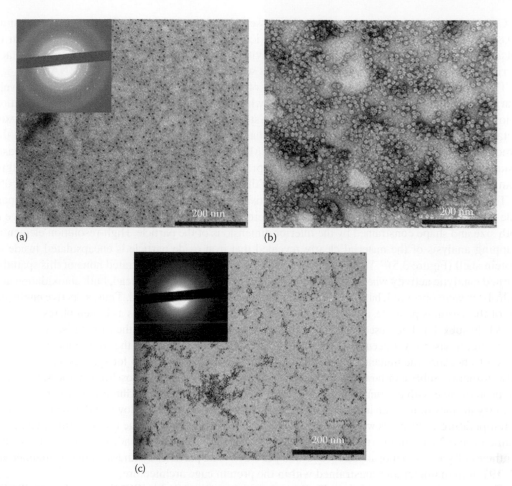

(a)

(b)

(c)

FIGURE 3.4 Transmission electron micrographs of (a) unstained, (b) stained refluxed PfFn and (c) unstained ferrihydrite in PfFn. The dark electron dense cores in (a) have an electron diffraction pattern (inset) consistent with hematite while no diffraction was observed in (c) consistent with disordered ferrihydrite. (Copyright 2010 The Royal Society of Chemistry. Reproduced with permission[68].)

This interaction facilitates a similar oxidative mineralisation reaction described for ferritin. The electrostatic surface of the interior of a Dps from the bacterium *Listeria innocua* (*Lis*Dps) is similar to the interior surface of Fn, with clusters of glutamic acid residues that can be involved in mineral core nucleation.

As is described in Section 3.2.4, Dps proteins possess a cage-like structure and negatively charged interior surface similar to ferritins. Therefore, constrained MNPs including γ-Fe$_2$O$_3$[77–79] and Co$_3$O$_4$[80] have been successfully synthesised in *Lis*Dps. However, the resulting nanoparticles are extremely small as Dps proteins have smaller interior cavity of 5 nm than that of 8 nm in ferritins. For example, it has been reported that the particle diameter of γ-Fe$_2$O$_3$ formed inside of *Lis*Dps was 4.1 ± 1.1 nm[77].

3.2.5 VIRUS CAPSIDS

Viral capsids possess conceptually cage-like architecture similar to Fn. They exist in a diverse range of sizes (18–500 nm), chemical and thermal stability and are potentially available as templates for variety of materials applications. Because the wide diversity of viral capsids will expand utility of biomolecules to serve for the syntheses of materials with interesting properties, viral capsids have increasingly drawn the attention of material scientists since late 1990s. We, and others, have exploited various virus capsids as size-constrained reaction containers for preparation of MNPs.

CCMV is an RNA-containing plant virus assembled from 180 identical coat protein subunits (19.8 kDa each). The quaternary structure is an icosahedral architecture, approximately 28 nm in outside diameter with 24 nm diameter interior cavity[14]. In the wild-type CCMV, the interior surface carries high positive charge density that interacts with the negatively charged viral RNA.

The crystallisation of guest molecules inside of CCMV can be achieved by utilising positively charged interior surface of the wild-type capsid, which facilitates the aggregation and crystallisation of ions. The virus has been shown to be a good system for the crystallisation and growth of polyoxo-metalate species (vanadate, molybdate and tungstate)[81]. To test the role of electrostatic effects in the directing mineralisation, a genetic mutant of CCMV was constructed (subE) in which all the basic residues (R and K) on the N-terminus of the coat protein were substituted for glutamic acids (E), thus dramatically altering the electrostatic character of the interior surface of the assembled protein cage[82].

The subE mutant was able to catalyse the oxidative hydrolysis of Fe^{2+} to form an iron oxide nanoparticle (lepidocrocite) within the interior of the protein cage[82]. The iron oxide particle within the cage is both size and shape constrained by the interior dimensions of the particle. High-resolution elemental mapping analysis of the material clearly revealed that iron oxide particle is encapsulated inside of protein shell (Figure 3.5)[82]. Wild-type CCMV, on the other hand, demonstrated none of this spatially defined catalytic activity when it was incubated in the presence of Fe^{2+}. Instead, bulk autoxidation and hydrolysis were observed, but no virion-encapsulated mineral was detected. Thus, selective engineering of the virion is possible, illustrating the plasticity of these architectures as biotemplates.

While subE has been used as a constrained reaction environment for the mineralisation of lepidocrocite, it has not yet been used for making ferrimagnetic particles of the iron oxide maghemite (γ-Fe_2O_3) because the mutant is not stable under the conditions necessary for synthesising the magnetic mineral. SubE can however be stabilised to these conditions by crosslinking the subunits of the protein cage with a bifunctional cross linker such as glutaraldehyde. Glutaraldehyde reacts with primary amines to form an inter-subunit link that locks the protein cage down and widens the range of temperature stability as well as pH stability. The result is a protein cage that is 30 nm in exterior diameter and 24 nm in internal diameter that is stable to the synthetic conditions necessary for the synthesis of γ-Fe_2O_3 nanoparticles (65°C and pH 8.5)[83]. The particles synthesised in this manner are 18–19 nm in diameter and constrained within the protein cage architecture.

The magnetic properties of the γ-Fe_2O_3 nanoparticles synthesised in CCMV are substantially different from γ-Fe_2O_3 nanoparticles synthesised in smaller protein cages that is, mammalian ferritin and *Lis*Dps. The blocking temperature for the 18.8 nm γ-Fe_2O_3 particles formed inside of CCMV was found to be 200 K whereas those for the γ-Fe_2O_3 formed in *Lis*Dps and ferritin were 5 and 30 K, respectively[50]. This result clearly shows advantage of large viral capsids to prepare MNPs, by which we can tune magnetic properties of the templated MNP. We have recently demonstrated that the *S. typhimurium* bacterio-phage P22[11–13,84], which has significantly larger capsid than CCMV, can template ferrihydrite particles[85].

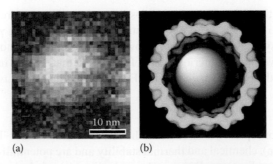

(a)   (b)

FIGURE 3.5   (a) Spatially resolved spectral image of an iron oxide nanoparticle formed within the genetic variant of CCMV, subE. Blue and yellow represent N and Fe, respectively. (b) Schematic representation of a guest material encapsulated within CCMV from cryo-TEM reconstruction. (Copyright 2007 Wiley-VCH. Reproduced with permission[82].)

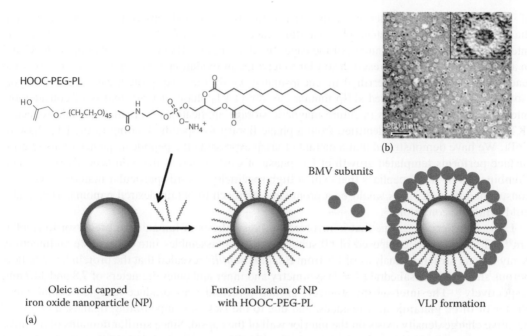

HOOC-PEG-PL

BMV subunits

Oleic acid capped
iron oxide nanoparticle (NP)

Functionalization of NP
with HOOC-PEG-PL

VLP formation

(a)

(b)

FIGURE 3.6 (a) Schematic illustration of virus-like particle (VLP) assembly around preformed iron oxide nanoparticle (NP). In the first step, a micelle of 1,2-distearoyl-*sn*-glycero-3-phosphoethanolamine-*N*-[carboxy(polyethylene glycol)-2000] (HOOC-PEG-PL) is formed around an oleic acid capped iron oxide NP. In the second step, BMV capsid subunits interact electrostatically with the functionalised NP and self-assemble around it forming a VLP. (b) TEM images of VLPs formed by self-assembling of BMV capsid proteins around iron oxide NPs coated with HOOC-PEG-PL. Inset shows a single VLP at a higher magnification.

P22 is approximately 58 nm in exterior diameter with a 50 nm diameter interior cavity. A ferrihydrite particle of 48 nm in diameter is formed inside of the capsid without disrupting its overall structure[85]. This result suggests great potential of P22 as a reaction vessel for synthesising particles with superior magnetic properties. CPMV, which possesses size and shape similar to CCMV, is another virus exploited for syntheses of MNPs[86,87]. For example, it has been demonstrated that a mutant of CPMV can form FePt particles, of approximately 30 nm in diameter inside of the capsid[86].

Alternative approaches to obtain protein cage-encapsulated MNPs have been demonstrated. Huang et al. demonstrated that capsid subunits of BMV could be directed to assemble around preformed MNPs (Figure 3.6)[88,89]. First, oleic acid capped iron oxide nanoparticles were mixed with a phospholipid (PL) 1,2-distearoyl-*sn*-glycero-3-phosphoethanolamine-*N*-[carboxy(polyethyleneglycol)-2000] (HOOC-PEG-PL). The PL tails of the molecules interact with the inorganic nanoparticle and a micelle is formed around the nanoparticles. Second, the iron oxide nanoparticles coated with the HOOC-PEG-PL molecules were mixed with monomers of the BMV capsid protein. The carboxyl group of the HOOC-PEG-PL molecule interacts electrostatically with the BMV capsid monomer and directs assembly of the monomer around the iron oxide nanoparticle. Interestingly, an assembled virus-like particle larger in size than native BMV capsid could be obtained[89]. This suggests the versatility of assembling viral subunits around a preformed core, since it could load larger cargo than a native capsid can accommodate.

3.2.6 OTHER CAGE-LIKE PROTEINS

The sHsp isolated from the hyperthermophilic archaeon *M. jannaschii* is another cage-like protein that has been utilised for syntheses of MNPs. The sHsp protein consists of 24 identical subunits with octahedral symmetry similar to Fn[20,90,91]. The assembled protein cage has an exterior diameter of 12 nm with a 6.5 nm interior cavity. sHsp differ from Fn and Dps in that it has large, 3 nm diameter, pores at the threefold axis that allow easy access even for bulky molecules between the interior and exterior environment of the cage. In addition, the sHsp cage is stable up to 70°C and in a pH range of 5–11. In a reaction similar to ferritin, air oxidation of Fe^{2+} in the presence of sHsp leads to the formation of ferrihydrite encapsulated within the sHsp protein cage[92]. More importantly, a genetically modified sHsp mutant can direct $L1_0$ phase of CoPt that is of considerable interest for addressable bits in future magnetic storage applications[93]. A small peptide sequence (KTHEIHSPLLHK) was identified from a phage library for specific binding to the $L1_0$ phase of CoPt. We have demonstrated that a mutant of sHsp expressing the peptide sequence on its interior surface performs templated growth of $L1_0$ phase of CoPt, whereas the wild-type sHsp does not. Combination of small peptides that exhibit high specificity towards particular materials and size-constrained protein cages would be a promising approach to obtain desired nanomaterials under mild synthetic conditions[95].

LS is a bacterial enzyme that is involved in the synthesis of lumazine, a precursor to riboflavin[94]. The enzyme is composed of 60 subunits and also assembles into a cage-like architecture. X-ray crystal structure analysis of LS from *Bacillus subtilis* revealed that the protein has a hollow porous shell with icosahedral (T = 1) symmetry and inner and outer diameters of 7.8 and 14.7 nm, respectively[18]. The inter-subunit contacts around threefold axes produce a region containing a cluster of three glutamic acid residues, and due to the lack of compensating ligands, a region of negative charge density exists on the interior wall of the capsid. Since similar domains of negative charge are present at the nucleation sites in ferritin, it was presumed that the negatively charged regions in LS could act as nucleation sites for iron oxide. Indeed reported the synthesis of γ-FeOOH (lepidocrocite), demonstrating the potential of LS as a nanoreactor for iron oxide mineralisation[95].

3.3 MAGNETIC PROTEIN CAGES AS MAGNETIC RESONANCE IMAGING AGENTS

3.3.1 Ferritin as an Endogenous T_2 Contrast Agent

One of the most promising avenues for MNPs templated in protein cages are their applications in biomedicine, especially their development as magnetic resonance imaging (MRI) agents[96,97]. Non-invasive and high-sensitivity imaging is an irreplaceable tool for diagnosis, prevention and treatment of numerous diseases[98]. Development of imaging agents such as superparamagnetic particles for MRI is an important element for the progression of imaging technology[99–103]. Nanoscale supramolecular systems have been explored as reaction templates and delivery vehicles for imaging agents because nanoparticles exhibit unique size-dependent properties and are able to pass biological barriers such as blood–brain barrier and blood vessel pores that are normally inaccessible to larger particles. Although a variety of proteins with cage-like architecture have been utilised as templates for MNPs, ferritin has been the most intensively studied protein cage for its application as a T_2 MRI agent. As described in Section 3.2.1, the inherent biological function of ferritin is sequestration and storage of iron in the form of ferrihydrite nanoparticle. While ferrihydrite is an antiferromagnetic oxide, the iron oxide core in ferritin is superparamagnetic because of the presence of cation vacancies in the crystal[31], uncompensated surface spins[4,33], and the possible natural incorporation of small amounts of other iron oxide polymorphs within the cage[104]. This superparamagnetic property of mineral core makes ferritin an endogenous T_2 contrast agent. In the presence of an applied magnetic field, superparamagnetic particles generate inhomogeneity in the local magnetic field which acts to shorten the length of time for the precession of surrounding water proton to lose coherence and become randomised, that is, T_2 relaxivity, after an excitation pulse.

Iron is an essential element for life but it is also toxic. Therefore, iron homeostasis is tightly regulated and abnormal iron metabolism leads to diverse pathological changes[105]. Ferritin plays a key role in controlling iron homeostasis. The expression of ferritin is tightly regulated through the

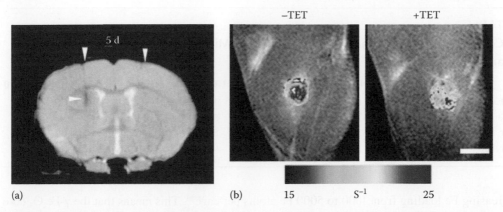

FIGURE 3.7 (a) *In vivo* result of MRI reporter expression in the mouse brain. Adenovirus containing the MRI reporter was inoculated into the striatum. T_2-weighted image 5 days after injection showing robust contrast of the inoculated sites (left arrows, MRI reporter; right arrow, AdV-lacZ control). (Reprinted by permission from Macmillan Publishers Ltd. *Nat. Med.*, Genove, G., Demarco, U., Xu, H., Goins, W., and Ahrens, E., A new transgene reporter for *in vivo* magnetic resonance imaging, 11, 450–454, Copyright 2005.) (b) *In vivo* MRI detection of tetracycline (TET)-regulated ferritin expressing tumours after inoculation of C6-TET-EGFP-HA-ferritin tumour cells in the hind limb of nude mice. Expression of ferritin is suppressed by administration of TET. TET was supplied in drinking water, starting 2 days before inoculation. (Reproduced from Cohen, B. et al., *Neoplasia*, 7, 109, 2005, Copyright 2005, Neoplasis Press Inc. With permission.)

interaction of iron regulatory RNA-binding proteins[106]. For this reason, endogenous ferritin *in vivo* has been considered as useful marker for MRI examination to assess the amount of iron in liver, spleen and brain[30,107,108]. Ferritin deposition in thrombi created in a rabbit model was also verified using a positive contrast magnetic resonance (MR) technique[109].

In addition, ferritin has recently been suggested as a candidate reporter gene to monitor transgene expression using MRI[110–113]. For example, Genove et al.[112] demonstrated that injection of adenovirus containing this MRI reporter gene into mouse striatum resulted in visible MRI contrast due to ferritin overexpression (Figure 3.7a). Cohen et al. demonstrated tetracycline (TET)-regulated overexpression of ferritin in tumour cells. Nude mice that were subjected to inoculation of the cells to create tumours showed induced expression of ferritin and increased r_2 relaxation on TET withdrawal (Figure 3.7b)[110]. These studies suggest ferritin could be used in a manner similar to green fluorescence protein and luciferase for *in vivo* optical imaging of gene expression.

One disadvantage of endogenous ferritin is the fact that it is only weakly superparamagnetic and is therefore a poor MRI contrast agent with 10- to 100-fold lower relaxivity than commercially available synthetic iron oxide MRI contrast agents (on a per iron atom basis)[30,31,114]. One approach to modulate this limitation is to aggregate ferritin into a larger particle[114]. It has recently been demonstrated that when multiple ferritins are chemically attached around a polystyrene nanoparticle of 20–1000 nm diameter, the material exhibits 25- to 30-fold greater r_2 relaxivity per iron concentration than endogenous ferritin alone[115].

3.3.2 MAGNETOFERRITIN AS A T_2 CONTRAST AGENT

Another way to overcome the inherent limitation of endogenous ferritin is utilisation of 'magnetoferritin' in which a superparamagnetic γ-Fe_2O_3 (or Fe_3O_4) nanoparticle has been synthesised inside the apo-ferritin cage[116,117]. Nanoparticles of γ-Fe_2O_3 (or Fe_3O_4) are strongly superparamagnetic and exhibit much higher r_2 compared with ferrihydrite. Apo-ferritin can be obtained *via* either demineralisation of endogenous ferritin or heterologous expression of recombinant ferritin. As is described in Section 3.2.1, recombinant human H ferritin, HFn, has been used as a template for γ-Fe_2O_3 particle formation[46,47]. The average size of the particles in HFn increases from 3.6 to 5.9 nm with

TABLE 3.1

Comparison of Size, r_1 and r_2 Relaxivity of the Mineralised HFn, Ferumoxides and Ferumoxtran-10

	HFn1000Fe	HFn3000Fe	HFn5000Fe	Ferumoxides	Ferumoxtran-10
Exterior diameter (nm)	12	12	12	58.5 ± 185.8	29.5 ± 23.1
Iron oxide diameter (nm)	3.6 ± 0.7	5.1 ± 0.9	5.8 ± 0.9	15.2 ± 8.9	4.9 ± 1.5
r_1 (mM^{-1} s^{-1})	2.3	4.2	8.4	7.8	12
r_2 (mM^{-1} s^{-1})	11	31	93	119	101

increasing Fe loading from 1000 to 5000 Fe atoms per cage[46]. This means that the γ-Fe$_2$O$_3$ core of the mineralised HFn and the exterior diameter are comparable with ultrasmall superparamagnetic iron oxide (USPIO) contrast agents. As expected, the r_1 and r_2 relaxivities of the γ-Fe$_2$O$_3$ mineralised HFn increases with increasing Fe loading per cage (i.e. the particle size), and the r_1 and r_2 of the HFn loaded with 5000 Fe atoms (HFn5000Fe) are comparable with those of commercially available small superparamagnetic iron oxide (SPIO) and USPIO materials (Table 3.1).

One of the unique features of the mineralised HFn with regards to MRI contrast is that it is readily taken up by macrophage cells and therefore provides a strong T$_2$* effect (Figure 3.8) [46]. Since macrophage cells play a significant role in the progression of inflammatory responses[118–120], imaging of macrophages involved in inflammatory diseases such as atherosclerosis is useful to assess and diagnose these diseases[121,122]. Recent *in vivo* studies using an atherosclerotic mouse model revealed that HFn injected (i.v.) into the mouse accumulates significantly more in macrophage-rich atherosclerotic plaque lesions formed in the left carotid arteries of the mouse than in the right carotid artery of the same mouse, which is kept in normal condition and serves as an internal control[123]. Immunofluorescence assessment, of fluorescently labelled HFn, revealed co-localisation of the injected ferritin with macrophage cells in these lesions. Furthermore, MRI assessment of the mouse demonstrated that 40%–50% loss of MR signal was observed from the macrophage-rich plaque lesion after 24 and 48 h of the mineralised HFn injection in comparison with measurements before injection. No significant signal loss was observed in the control carotid artery. This signal loss is likely due to T$_2$* effect of the mineralised HFn, which is preferentially accumulated in the plaque lesion. These results clearly indicate potential of the mineralised HFn as an *in vivo* MRI agent for early detection of atherosclerosis.

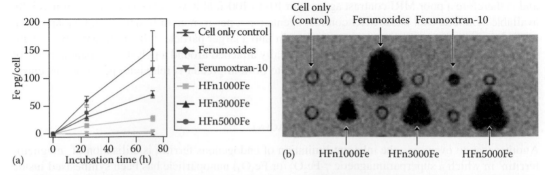

FIGURE 3.8 (a) Fe content taken up by macrophage cells after incubation with 165 μg mL^{-1} of various contrast agents. Amount of Fe taken up by macrophage cultured with HFn5000Fe was comparable with those cultured with ferumoxides. (b) *In vitro* MR images of macrophage cells incubated with 165 μg mL^{-1} of various contrast agents for 24 h. The MR imaging was performed at 1.5 T.

3.3.3 ENCAPSULATION OF T_1 CONTRAST AGENTS IN PROTEIN CAGES

The presence of paramagnetic ions effectively reduce the T_1 relaxation time of surrounding water protons through a process known as paramagnetic relaxation enhancement (PRE) by catalysing the relaxation of protons on water bound to these ions[124-127]. Gadolinium (Gd^{3+})-based contrast agents are most commonly used, as Gd^{3+} has seven unpaired electrons that results in efficient PRE[128,129]. One major advantage of T_1 over T_2 contrast agents is that they produce a bright signal, which can be clearly distinguished from background. However, the utility of T_1 contrast agents is currently limited by the inability to detect nanomolar concentrations of the contrast agents *in vivo*[130]. Designing contrast agents that are more sensitive to MR scanners will allow for the detection of targets that are smaller and have a lower abundance. More sensitive T_1 contrast agents will reduce the minimum contrast agent dosage required, which is important since adverse side effects, such as nephrogenic systemic fibrosis (NSF), occur in some patients using current dosages[131,132].

In the previous sections, we discussed templated synthesis of superparamagnetic nanoparticles inside of ferritin and other protein cages, which can serve as T_2 contrast agents. Protein cages play a role as platforms to direct the synthesis of materials, protect the material from exterior environment and impart cell-specific targeting moieties. Conceptually, the same advantages are conferred for preparation of T_1 contrast agents when these protein cages are utilised.

Several groups working with protein cage nanoparticles have made important contributions in the development of protein cages as platforms for addition of MR contrast functionality. To date, the range of these protein cage-based contrast agents can be divided into five major categories: (1) endogenous metal-binding sites[133], (2) genetic insertion of a metal-binding peptide[134], (3) entrapment of small molecule chelates in a cavity of a protein cage[135,136], (4) chemical attachment of small molecule chelates[134,137-140] and (5) protein cage-branched polymer hybrid particles[141,142] (Figure 3.9a). Of these five categories, the first two are less significant from a clinical standpoint due to their low Gd-binding affinities, since free Gd is toxic[131]. The majority of the protein cage-based MRI agents were synthesised using the strategy of chemically attaching clinically employed contrast agents, such as gadolinium-tetraazacyclo dodecanetetraacetic acid (DOTA-Gd) or gadolinium-diethylene-triamine-pentaacetic acid (DTPA-Gd) or the metal chelator hydroxypyridonate (HOPO)-Gd to protein cage. The ionic relaxivity (r_{1ionic}; relaxivity rate per Gd ion) data are relatively similar for all protein cage-based contrast agents that have been developed by covalently attaching DTPA-Gd, DOTA-Gd or HOPO-Gd[143] (categories 4 and 5).

Improvements in the Gd payload, particle relaxivity ($r_{1particle}$; relaxivity rate per protein cage) and relaxivity per unit volume and mass have been accomplished by developing a hybrid material that consists of a branched polymer, containing DTPA-Gd, grown on the interior of a protein cage nanoparticle. Both the sHsp (Hsp-BP-DTPA-Gd) and the P22 bacteriophage (P22-BP-DTPA-Gd) have been developed as platforms for this type of hybrid material[141,142,144]. Hsp-BP-DTPA-Gd fully utilises the interior volume of the sHsp cage, resulting in a high Gd payload and relaxivity density[141]. The particle relaxivity value of this construct is $4200\,mM^{-1}\,s^{-1}$, which is among the highest reported to date. The relaxivity per unit mass is higher than the clinically employed contrast agent Magnevist (Gd-DTPA). The larger structure of P22-BP-DTPA-Gd allowed for a payload of 4,400 Gd ions per capsid and an $r_{1particle} = 41,300\,mM^{-1}\,s^{-1}$ both of these values are the highest for a T_1 contrast agent based on a chemically attached Gd chelate on a protein cage[142].

For Gd-based contrast agent, much of the variation in r_{1ionic} can be attributed to differences in three important parameters that control the performance of the agent[134,139-142,145]. These parameters include the rotational correlation time of the Gd ion (τ_R), the exchange lifetime of the Gd-bound water (τ_M) and number of waters in direct contact with Gd (q) (Figure 3.9b). Long τ_R values are ideal for efficient relaxation resulting in more sensitive contrast in MRI. A balance must be struck between maximising the number of water molecules coordinated to the Gd ion (q) and the stability of the metal–chelator complex, since the two are, in general, inversely related. Finally, the water

	Endogenous Protein-Based Gd^{3+} Binding	Engineered Protein-Based Gd^{3+} Binding	Encapsulation of Gd^{3+} chelate	Chemical Attachment of Gd^{3+} Chelate				Chemical Attachment of a Branch Polymer Containing a Gd^{3+} Chelate		Gd-DTPA
r_{1ionic}	261	308	80	16.9	15.5	69	41.6	25.4	21.7	3.4
$r_{1particle}$	42,500	62,400	NA	7,200	4,150	4,200	3,900	4,039	41,300	3.4
MHz	35	35	20	64	64	21	31	31	28	43
Gd per particle	163	203	NA	426	268	61	94	159	1,903	1
r_{1ionic} per volume	3.7	5.4	NA	0.7	0.3	0.4	0.4	2.8	0.4	8.9
r_{1ionic} per mass	11.7	17.1	NA	2.9	1.1	1.2	1.6	10.2	2.5	6.2
Clinically relevant binding	No	No	Yes	Yes	Yes	Yes	Yes	Yes	Yes	Yes
Reference	[133]	[134]	[135,136]	[138]	[137]	[134]	[139]	[141]	[142]	[143]

(a)

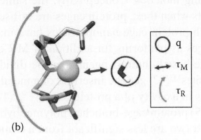

(b)

FIGURE 3.9 (a) PRE summary of five categories of protein cage-based Gd contrast agents. (b) Schematic illustrations of physical parameters that effect PRE. Parameters important to PRE. Shown here is Gd-DTPA, a clinically used contrast agent. The number of Gd-bound water molecules (q), the water exchange lifetime (τ_M) and the rotational correlation time (τ_R) all influence PRE.

exchange lifetime of the Gd-bound water, referred to as τ_M, must be optimised to achieve a maximal relaxation rate. These parameters can be determined by fitting r_1 versus field strength plots or *nuclear magnetic resonance dispersion* (NMRD) profiles to the Solomon–Bloembergen–Morgan (SBM) analytical model for relaxivity[124,127]. The results of fitting experiments for Hsp-BP-DTPA-Gd and P22-BP-DTPA-Gd suggest that while the water exchange lifetime (τ_M) is higher than optimal (water exchange is too slow), the rotational correlation time (τ_R) of the Gd ions have been optimised in both constructs. The optimal τ_R values in both these systems are a result of attaching the Gd chelate to a rigid-branched polymer backbone.

Another consideration for MR image collection is the r_1/r_2 ratio of the contrast agent. For contrast agents used in T_1-weighted MR images, higher r_1/r_2 ratios are preferred with common ratios of 0.5–0.9[146]. T_1-weighted images are a readout of the NMR signal at specific volume units within a sample. Both T_1 and T_2 relaxation processes occur for all protons contained in a sample. In T_1-weighted images, a large NMR signal corresponds to fast T_1 relaxation and appear as bright regions in the MR image. The T_2 relaxation process decreases the NMR signal. This is true since T_2 relaxation is a de-phasing event of the in-phase precession that is produced from the initial radio frequency (RF) pulse. This RF pulse is used to tilt the magnetisation of the sample from the longitudinal axis into the transfer axis. The decrease in the NMR signal from the T_2 relaxation process counteracts the signal gains that result from the enhancement of the T_1 relaxation rate due to the presence of Gd. So the most sensitive contrast agents used in T_1-weighted images have high T_1 relaxation enhancement and low T_2 relaxation enhancement, that is, high r_1/r_2 ratio. The r_1/r_2

ratio of P22-BP-DTPA-Gd for the field strength of 0.73 T is $r_1/r_2 = 0.80$. Above 0.73 T, the r_1 value decreases while the r_2 value remains relatively constant. This results in a reduction in the r_1/r_2 ratio to 0.53 for the field strength of 1.4 T.

A more detailed understanding of the complex nature of how r_1 and r_1/r_2 vary with field strength can be gained by imputing values of τ_M and τ_R that correspond to the sHsp and the P22 BP-DTPA-Gd constructs into the SBM analytical model for relaxivity[141,142]. These experiments suggest that optimal image contrast (high values of r_1 and r_1/r_2) should occur at field strengths less than or equal to 1 T. If higher field strengths (>1 T) are desired, optimal image contrast will occur when τ_R is decreased to 10^{-9} s and the τ_M is decreased to 10^{-8} s.

3.3.4 Cell Targeting of MNP within Protein Cages

Targeted delivery of MRI agents to desired tissues and cells is key to improving sensitivity and decreasing the required dose[99,100]. A widespread strategy for targeted delivery has exploited the display of small molecules, aptamers, peptides or antibodies on the surface of imaging agents, which specifically recognise tissues and cells of interest. Protein cages possess the versatility to be functionalised with these targeting moieties on their exterior surface using either chemical or genetic approaches[47,135,147–149]. As the protein cages are highly symmetrical architectures composed of subunit multimers, multivalent display of functional groups in a spatially defined manner can be achieved. For example, the amino acid sequence RGD-4C (CDCRGDCFC), which was identified by *in vivo* phage selection techniques and is known to bind integrins $\alpha_v\beta_3$ and $\alpha_v\beta_5$[150], has been genetically introduced as an N-terminal fusion protein into HFn[47,151]. As the N-terminus of the HFn is positioned on the exterior surface of the cage, it is reasonably expected that 24 copies of the RGD-4C peptide are displayed on the exterior surface of the HFn. The HFn mutant (RGD4C-Fn) exhibited the same cage-like structure indistinguishable from the wild-type HFn by either size or morphology[47]. In addition, there was no significant difference in average particle size and size distribution of γ-Fe$_2$O$_3$ formed between HFn and RGD4C-Fn under same iron loading per cage. These results suggest that the introduction of the RGD-4C peptide has minimal effect on the assembly of subunits into the cage architecture or on its ability to direct γ-Fe$_2$O$_3$ formation. Using fluorescently labelled RGD4C-Fn, and analysis using fluorescence-activated cell sorting, it was demonstrated that the RGD4C-Fn clearly exhibited enhanced targeting interaction with cells known to overexpress integrins $\alpha_v\beta_3$ (C32 amelanotic melanoma cells and THP-1 monocytes)[47,151]. More importantly, *in vivo* evaluation of the magnetite encapsulated RGD4C-Fn exhibits improved accumulation of the protein cage in the macrophage-rich plaque lesions, in atherosclerotic mice, and far better signal reduction than unmineralised HFn[152,153].

Another cell-targeting peptide that has been tested is LyP-1 (CGNKRTRGC), which targets tumour-associated lymphatic vessels and macrophages[154,155]. Using the sHsp as a platform, the LyP-1 peptide has been incorporated as a C-terminal fusion to the sHsp subunit (LyP–Hsp) as the C-terminal region of the protein is believed to be exposed on the exterior surface of the assembled cage[156]. Both *in vitro* and *in vivo* studies have demonstrated that LyP–Hsp shows high affinity to macrophages and accumulates significantly more in a macrophage-rich lesion of the atherosclerotic mouse model than sHsp without the LyP-1 peptide (Figure 3.10).

Chemical modification is another approach to impart cell-targeting ligands to exterior surfaces of protein cages[135,148,149]. For example, Destito et al. demonstrated conjugation of folic acid on the surface of CPMV by using the copper-catalysed azide–alkyne cycloaddition reaction. The obtained particles are effectively targeted to tumour cells *via* the folic-acid receptor, which is upregulated in these cells.

The results introduced earlier clearly indicate that incorporation of cell-targeting molecules such as peptides into protein cage platform is a promising approach to impart tissue- and cell-specific imaging capability to the encapsulated MRI agents.

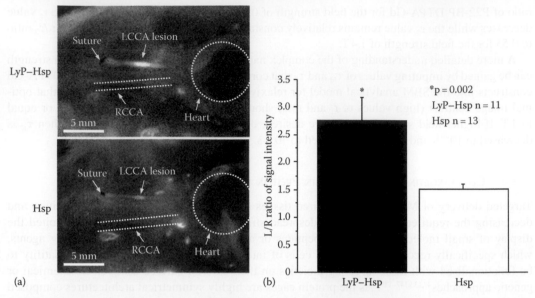

FIGURE 3.10   (a) *In situ* fluorescence imaging of atherosclerotic mice 48 h after Cy5.5 labelled Hsp or LyP–Hsp injection. A more intense fluorescence signal was observed in the left common carotid artery (LCCA), in which macrophage-rich plaque lesion had been formed in advance the injection, than the control right common carotid artery (RCCA). (b) Fluorescence signal intensity ratio of LCCA/RCCA obtained from *in situ* imaging. The LyP–Hsp-injected mice showed higher LCCA/RCCA than the Hsp-injected mice. (Copyright 2011 American Chemical Society. Reproduced with permission[156].)

3.4   FUTURE DIRECTIONS

Studies on mineralisation process of ferritin have revealed that the protein has evolved in a highly elaborated manner to actively uptake ferrous ions and promote mineral formation[23–25,38–40]. Better understanding of fundamental biomineralisation mechanisms involved in ferritin and other biomineral systems will provide insights by which we can engineer protein systems to develop more sophisticated and versatile platforms for creating functional nanomaterials. For example, introduction of an artificially designed catalytic active centre into a protein cage would be one of many promising approaches to expand utility of a protein cage[157]. For biomedical applications, development of multi-mode imaging agents and 'theranostic' materials is being actively explored.[158]. As protein cages could be highly adaptable platforms to encapsulate multi-cargos in a single cage, they are expected to be a useful platform for the development of multimodal nanoprobes. The field of biomimetic chemistry has evolved to a stage where ideas from molecular biology, inorganic chemistry, protein chemistry and materials science can be routinely combined in the pursuit of novel materials.

REFERENCES

1. Arakaki, A., Nakazawa, H., Nemoto, M., Mori, T., and Matsunaga, T. 2008. Formation of magnetite by bacteria and its application. *Journal of the Royal Society Interface,* 5, 977–999.
2. Weaver, J. C., Wang, Q. Q., Miserez, A., Tantuccio, A., Stromberg, R., Bozhilov, K. N., Maxwell, P., Nay, R., Heier, S. T., DiMasi, E., and Kisailus, D. 2010. Analysis of an ultra hard magnetic biomineral in chiton radular teeth. *Materials Today,* 13, 42–52.
3. Gordon, L. M., and Joester, D. 2011. Nanoscale chemical tomography of buried organic–inorganic interfaces in the chiton tooth. *Nature,* 469, 194–197.
4. Michel, F. M., Hosein, H.-A., Hausner, D. B., Debnath, S., Parise, J. B., and Strongin, D. R. 2010. Reactivity of ferritin and the structure of ferritin-derived ferrihydrite. *BBA – General Subjects,* 1800, 871–885.

5. Mann, S. 2002. *Biomineralization*. Oxford University Press, Weinheim, Germany.
6. Weiner, S. 2008. Biomineralization: A structural perspective. *Journal of Structural Biology*, 163, 229–234.
7. Douglas, T., and Young, M. 2006. Viruses: Making friends with old foes. *Science*, 312, 873–875.
8. Uchida, M., Klem, M. T., Allen, M., Flenniken, M. L., Gillitzer, E., Varpness, Z., Suci, P., Young, M. J., and Douglas, T. 2007. Protein cage architecture: Containers as templates for materials synthesis. *Advanced Materials*, 19, 1025–1042.
9. Kang, S., Oltrogge, L. M., Broomell, C. C., Liepold, L. O., Prevelige, P. E., Young, M., and Douglas, T. 2008. Controlled assembly of bifunctional chimeric protein cages and composition analysis using non-covalent mass spectrometry. *Journal of the American Chemical Society*, 130, 16527–16529.
10. Gillitzer, E., Suci, P., Young, M., and Douglas, T. 2006. Controlled ligand display on a symmetrical protein-cage architecture through mixed assembly. *Small*, 2, 962–966.
11. Chen, D. H., Baker, M. L., Hryc, C. F., DiMaio, F., Jakana, J., Wu, W., Dougherty, M., Haase-Pettingell, C., Schmid, M. F., Jiang, W., Baker, D., King, J. A., and Chiu, W. 2011. Structural basis for scaffolding-mediated assembly and maturation of a dsDNA virus. *Proceedings of the National Academy of Sciences of the United States of America*, 108, 1355–1360.
12. Parent, K. N., Khayat, R., Tu, L. H., Suhanovsky, M. M., Cortines, J. R., Teschke, C. M., Johnson, J. E., and Baker, T. S. 2010. P22 coat protein structures reveal a novel mechanism for capsid maturation: Stability without auxiliary proteins or chemical crosslinks. *Structure*, 18, 390–401.
13. Jiang, W., Li, Z., Zhang, Z., Baker, M. L., Prevelige, P. E., Jr., and Chiu, W. 2003. Coat protein fold and maturation transition of bacteriophage P22 seen at subnanometer resolutions. *Nature Structural Biology*, 10, 131–135.
14. Speir, J. A., Munshi, S., Wang, G. J., Baker, T. S., and Johnson, J. E. 1995. Structures of the native and swollen forms of cowpea chlorotic mottle virus determined by x-ray crystallography and cryoelectron microscopy. *Structure*, 3, 63–78.
15. Lin, T. W., Chen, Z. G., Usha, R., Stauffacher, C. V., Dai, J. B., Schmidt, T., and Johnson, J. E. 1999. The refined crystal structure of cowpea mosaic virus at 2.8 A resolution. *Virology*, 265, 20–34.
16. Lucas, R. W., Larson, S. B., and McPherson, A. 2002. The crystallographic structure of brome mosaic virus. *Journal of Molecular Biology*, 317, 95–108.
17. Golmohammadi, R., Valegard, K., Fridborg, K., and Liljas, L. 1993. The refined structure of bacteriophage MS2 at 2.8 A resolution. *Journal of Molecular Biology*, 234, 620–639.
18. Ladenstein, R., Schneider, M., Huber, R., Bartunik, H. D., Wilson, K., Schott, K., and Bacher, A. 1988. Heavy riboflavin synthase from *Bacillus subtilis*. Crystal structure analysis of the icosahedral beta 60 capsid at 3.3 A resolution. *Journal of Molecular Biology*, 203, 1045–1070.
19. Lawson, D. M., Artymiuk, P. J., Yewdall, S. J., Smith, J. M. A., Livingstone, J. C., Treffry, A., Luzzago, A., Levi, S., Arosio, P., Cesareni, G., Thomas, C. D., Shaw, W. V., and Harrison, P. M. 1991. Solving the structure of human H-ferritin by genetically engineering intermolecular crystal contacts. *Nature*, 349, 541–544.
20. Kim, K. K., Kim, R., and Kim, S. H. 1998. Crystal structure of a small heat-shock protein. *Nature*, 394, 595–599.
21. Grant, R. A., Filman, D. J., Finkel, S. E., Kolter, R., and Hogle, J. M. 1998. The crystal structure of Dps, a ferritin homolog that binds and protects DNA. *Nature Structural Biology*, 5, 294–303.
22. Pettersen, E. F., Goddard, T. D., Huang, C. C., Couch, G. S., Greenblatt, D. M., Meng, E. C., and Ferrin, T. E. 2004. UCSF Chimera – A visualization system for exploratory research and analysis. *Journal of Computational Chemistry*, 25, 1605–1612.
23. Harrison, P. M., and Arosio, P. 1996. Ferritins: Molecular properties, iron storage function and cellular regulation. *Biochimica Et Biophysica Acta - Bioenergetics*, 1275, 161–203.
24. Lewin, A., Moore, G. R., and Le Brun, N. E. 2005. Formation of protein-coated iron minerals. *Dalton Transactions*, 22, 3597–3610.
25. Liu, X., and Theil, E. C. 2005. Ferritins: Dynamic management of biological iron and oxygen chemistry. *Accounts of Chemical Research*, 38, 167–175.
26. Gider, S., Awschalom, D., Douglas, T., Mann, S., and Chaparala, M. 1995. Classical and quantum magnetic phenomena in natural and artificial ferritin proteins. *Science*, 268, 77–80.
27. Gilles, C., Bonville, P., Rakoto, H., Broto, J., Wong, K., and Mann, S. 2002. Magnetic hysteresis and superantiferromagnetism in ferritin nanoparticles. *Journal of Magnetism and Magnetic Materials*, 241, 430–440.
28. Gider, S., Awschalom, D., Douglas, T., Wong, K., Mann, S., and Cain, G. 1996. Classical and quantum magnetism in synthetic ferritin proteins. *Journal of Applied Physics*, 79, 5324–5326.

29. Kilcoyne, S., and Cywinski, R. 1995. Ferritin: A model superparamagnet. *Journal of Magnetism and Magnetic Materials,* 140, 1466–1467.

30. Gossuin, Y., Muller, R., Gillis, P., and Bartel, L. 2005. Relaxivities of human liver and spleen ferritin. *Magnetic Resonance Imaging,* 23, 1001–1004.

31. Gossuin, Y., Gillis, P., Hocq, A., Vuong, Q. L., and Roch, A. 2009. Magnetic resonance relaxation properties of superparamagnetic particles. *Wiley Interdisciplinary Reviews: Nanomedicine and Nanobiotechnology,* 1, 299–310.

32. Papaefthymiou, G. C. 2010. The Mössbauer and magnetic properties of ferritin cores. *BBA – General Subjects,* 1800, 886–897.

33. Michel, F. M., Ehm, L., Antao, S. M., Lee, P. L., Chupas, P. J., Liu, G., Strongin, D. R., Schoonen, M. A., Phillips, B. L., and Parise, J. B. 2007. The structure of ferrihydrite, a nanocrystalline material. *Science,* 316, 1726–1729.

34. Mann, S., Bannister, J. V., and Williams, R. J. 1986. Structure and composition of ferritin cores isolated from human spleen, limpet (*Patella vulgata*) hemolymph and bacterial (*Pseudomonas aeruginosa*) cells. *Journal of Molecular Biology,* 188, 225–232.

35. Mayer, D. E., Rohrer, J. S., Schoeller, D. A., and Harris, D. C. 1983. Fate of oxygen during ferritin iron incorporation. *Biochemistry,* 22, 876–880.

36. Bou-Abdallah, F. 2010. The iron redox and hydrolysis chemistry of the ferritins. *BBA – General Subjects,* 1800, 719–731.

37. Theil, E., Takagi, H., Small, G., He, L., Tipton, A., and Danger, D. 2000. The ferritin iron entry and exit problem. *Inorganica Chimica Acta,* 297, 242–251.

38. Douglas, T., and Ripoll, D. R. 1998. Calculated electrostatic gradients in recombinant human H-chain ferritin. *Protein Science,* 7, 1083–1091.

39. Tosha, T., Ng, H.-L., Bhattasali, O., Alber, T., and Theil, E. C. 2010. Moving metal ions through ferritin-protein nanocages from three-fold pores to catalytic sites. *Journal of the American Chemical Society,* 132, 14562–14569.

40. Turano, P., Lalli, D., Felli, I. C., Theil, E. C., and Bertini, I. 2010. NMR reveals pathway for ferric mineral precursors to the central cavity of ferritin. *Proceedings of the National Academy of Sciences of the United States of America,* 107, 545–550.

41. Cornell, R., and Schwertmann, U. 2003. *The Iron Oxides,* 2nd edn. Weinheim, Germany: Wiley-VCH.

42. Meldrum, F. C., Wade, V. J., Nimmo, D. L., Heywood, B. R., and Mann, S. 1991. Synthesis of inorganic nanophase materials in supramolecular protein cages. *Nature,* 349, 684–687.

43. Meldrum, F. C., Heywood, B. R., and Mann, S. 1992. Magnetoferritin: In vitro synthesis of a novel magnetic protein. *Science,* 257, 522–523.

44. Mann, S., Archibald, D. D., Didymus, J. M., Douglas, T., Heywood, B. R., Meldrum, F. C., and Reeves, N. J. 1993. Crystallization at inorganic–organic interfaces: Biominerals and biomimetic synthesis. *Science,* 261, 1286–1292.

45. Wong, K. K. W., Douglas, T., Gider, S., Awschalom, D. D., and Mann, S. 1998. Biomimetic synthesis and characterization of magnetic proteins (magnetoferritin). *Chemistry of Materials,* 10, 279–285.

46. Uchida, M., Terashima, M., Cunningham, C. H., Suzuki, Y., Willits, D. A., Willis, A. F., Yang, P. C., Tsao, P. S., McConnell, M. V., Young, M. J., and Douglas, T. 2008. A human ferritin iron oxide nano-composite magnetic resonance contrast agent. *Magnetic Resonance in Medicine,* 60, 1073–1081.

47. Uchida, M., Flenniken, M. L., Allen, M., Willits, D. A., Crowley, B. E., Brumfield, S., Willis, A. F., Jackiw, L., Jutila, M., Young, M. J., and Douglas, T. 2006. Targeting of cancer cells with ferrimagnetic ferritin cage nanoparticles. *Journal of the American Chemical Society,* 128, 16626–16633.

48. Martínez-Pérez, M. J., de Miguel, R., Carbonera, C., Martínez-Júlvez, M., Lostao, A., Piquer, C., Gómez-Moreno, C., Bartolomé, J., and Luis, F. 2010. Size-dependent properties of magnetoferritin. *Nanotechnology,* 21, 465707.

49. Li, H., Klem, M. T., Sebby, K. B., Singel, D. J., Young, M., Douglas, T., and Idzerda, Y. U. 2009. Determination of anisotropy constants of protein encapsulated iron oxide nanoparticles by electron magnetic resonance. *Journal of Magnetism and Magnetic Materials,* 321, 175–180.

50. Gilmore, K., Idzerda, Y. U., Klem, M. T., Allen, M., Douglas, T., and Young, M. 2005. Surface contribution to the anisotropy energy of spherical magnetite particles. *Journal of Applied Physics,* 97 (10), 10B301.

51. Jolley, C. C., Uchida, M., Reichhardt, C., Harrington, R., Kang, S., Klem, M. T., Parise, J. B., and Douglas, T. 2010. Size and crystallinity in protein-templated inorganic nanoparticles. *Chemistry of Materials,* 22, 4612–4618.

52. Klem, M. T., Resnick, D. A., Gilmore, K., Young, M., Idzerda, Y. U., and Douglas, T. 2007. Synthetic control over magnetic moment and exchange bias in all-oxide materials encapsulated within a spherical protein cage. *Journal of the American Chemical Society*, 129, 197–201.

53. Yamashita, I., Iwahori, K., and Kumagai, S. 2010. Ferritin in the field of nanodevices. *BBA – General Subjects*, 1800, 846–857.

54. Yamashita, I. 2008. Biosupramolecules for nano-devices: Biomineralization of nanoparticles and their applications. *Journal of Materials Chemistry*, 18, 3813–3820.

55. Miura, A., Hikono, T., Matsumura, T., Yano, H., Hatayama, T., Uraoka, Y., Fuyuki, T., Yoshii, S., and Yamashita, I. 2006. Floating nanodot gate memory devices based on biomineralized inorganic nanodot array as a storage node. *Japanese Journal of Applied Physics*, 45, L1–L3.

56. Yamada, K., Yoshii, S., Kumagai, S., Miura, A., Uraoka, Y., Fuyuki, T., and Yamashita, I. 2007. Effects of dot density and dot size on charge injection characteristics in nanodot array produced by protein supramolecules. *Japanese Journal of Applied Physics Part 1 - Regular Papers Brief Communications & Review Papers*, 46, 7549–7553.

57. Nakaya, M., Kanehara, M., and Teranishi, T. 2006. One-pot synthesis of large FePt nanoparticles from metal salts and their thermal stability. *Langmuir*, 22, 3485–3487.

58. Sun, S. H. 2006. Recent advances in chemical synthesis, self-assembly, and applications of FePt nanoparticles. *Advanced Materials*, 18, 393–403.

59. Maenosono, S., and Saita, S. 2006. Theoretical assessment of FePt nanoparticles as heating elements for magnetic hyperthermia. *IEEE Transactions on Magnetics*, 42, 1638–1642.

60. Maenosono, S., Suzuki, T., and Saita, S. 2008. Superparamagnetic FePt nanoparticles as excellent MRI contrast agents. *Journal of Magnetism and Magnetic Materials*, 320, L79–L83.

61. Kim, J. M., Rong, C. B., Liu, J. P., and Sun, S. H. 2009. Dispersible ferromagnetic FePt nanoparticles. *Advanced Materials*, 21, 906–909.

62. Warne, B., Kasyutich, O. I., Mayes, E. L., Wiggins, J. A. L., and Wong, K. K. W. 2000. Self assembled nanoparticulate Co: Pt for data storage applications. *IEEE Transactions on Magnetics*, 36, 3009–3011.

63. Mayes, E., Bewick, A., Gleeson, D., Hoinville, J., Jones, R., Kasyutich, O., Nartowski, A., Warne, B., Wiggins, J., and Wong, K. K. W. 2003. Biologically derived nanomagnets in self-organized patterned media. *IEEE Transactions on Magnetics*, 39, 624–627.

64. Mayes, E. L., and Mann, S. 2004. Mineralization in nanostructured biocompartments: Biomimetic ferritins for high-density data storage. In *Nanobiotechnology: Concepts, Applications and Perspectives*, edited by Niemeyer, C. M. and Mirkin, C. A. Weinheim, Germany: Wiley-VCH.

65. Tatur, J., Hagedoorn, P.-L., Overeijnder, M. L., and Hagen, W. R. 2006. A highly thermostable ferritin from the hyperthermophilic archaeal anaerobe *Pyrococcus furiosus*. *Extremophiles*, 10, 139–148.

66. Tatur, J., Hagen, W. R., and Matias, P. M. 2007. Crystal structure of the ferritin from the hyperthermophilic archaeal anaerobe *Pyrococcus furiosus*. *Journal of Biological Inorganic Chemistry*, 12, 615–630.

67. Honarmand Ebrahimi, K., Hagedoorn, P.-L., Jongejan, J. A., and Hagen, W. R. 2009. Catalysis of iron core formation in *Pyrococcus furiosus* ferritin. *Journal of Biological Inorganic Chemistry*, 14, 1265–1274.

68. Klem, M. T., Young, M., and Douglas, T. 2010. Biomimetic synthesis of photoactive α-Fe_2O_3 templated by the hyperthermophilic ferritin from *Pyrococcus furiosus*. *Journal of Materials Chemistry*, 20, 65–67.

69. Parker, M. J., Allen, M. A., Ramsay, B., Klem, M. T., Young, M., and Douglas, T. 2008. Expanding the temperature range of biomimetic synthesis using a ferritin from the hyperthermophile *Pyrococcus furiosus*. *Chemistry of Materials*, 20, 1541–1547.

70. Almiron, M., Link, A. J., Furlong, D., and Kolter, R. 1992. A novel DNA-binding protein with regulatory and protective roles in starved *Escherichia coli*. *Genes Development*, 6, 2646–2654.

71. Marjorette, M., Pena, O., and Bullerjahn, G. S. 1995. The DpsA protein of *Synechococcus* sp. strain PCC7942 is a DNA-binding hemoprotein. *The Journal of Biological Chemistry*, 270, 22478–22482.

72. Ilari, A., Savino, C., Stefanini, S., Chiancone, E., and Tsernoglou, D. 1999. Crystallization and preliminary X-ray crystallographic analysis of the unusual ferritin from *Listeria innocua*. *Acta Crystallographica Section D-Biological Crystallography*, 55, 552–553.

73. Ceci, P., Ilari, A., Falvo, E., and Chiancone, E. 2003. The Dps protein of *Agrobacterium tumefaciens* does not bind to DNA but protects it toward oxidative cleavage. *Journal of Biological Chemistry*, 278, 20319–20326.

74. Reindel, S., Schmidt, C. L., Anemuller, S., and Matzanke, B. F. 2002. Characterization of a non-heme ferritin of the Archaeon *Halobacterium salinarum*, homologous to Dps (starvation-induced DNA-binding protein). *Biochemical Society Transactions,* 30, 713–715.

75. Wiedenheft, B., Mosolf, J., Willits, D., Yeager, M., Dryden, K. A., Young, M., and Douglas, T. 2005. An archaeal antioxidant: Characterization of a Dps-like protein from *Sulfolobus solfataricus. Proceedings of the National Academy of Sciences of the United States of America,* 102, 10551–10556.

76. Ramsay, R., Wiedenheft, B., Allen, M., Gauss, G. H., Lawrence, C. M., Young, M., and Douglas, T. 2006. Dps-like protein from the hyperthermophilic archaeon *Pyrococcus furiosus. Journal of Inorganic Biochemistry,* 100, 1061–1068.

77. Allen, M., Willits, D., Mosolf, J., Young, M., and Douglas, T. 2002. Protein cage constrained synthesis of ferrimagnetic iron oxide nanoparticles. *Advanced Materials,* 14, 1562–1565.

78. Kang, S., Jolley, C., Liepold, L., Young, M., and Douglas, T. 2009. From metal binding to nanoparticle formation; monitoring biomimetic iron oxide synthesis within protein cages using mass spectrometry. *Angew Chemie International Edition,* 48, 4772–4776.

79. Ceci, P., Chiancone, E., Kasyutich, O., Bellapadrona, G., Castelli, L., Fittipaldi, M., Gatteschi, D., Innocenti, C., and Sangregorio, C. 2010. Synthesis of iron oxide nanoparticles in *Listeria innocua* Dps (DNA-binding protein from starved cells): A study with the wild-type protein and a catalytic centre mutant. *Chemistry: A European Journal,* 16, 709–717.

80. Allen, M., Willits, D., Young, M., and Douglas, T. 2003. Constrained synthesis of cobalt oxide nanomaterials in the 12-subunit protein cage from *Listeria innocua. Inorganic Chemistry,* 42, 6300–6305.

81. Douglas, T., and Young, M. 1998. Host–guest encapsulation of materials by assembled virus protein cages. *Nature,* 393, 152–155.

82. Douglas, T., Strable, E., Willits, D., Aitouchen, A., Libera, M., and Young, M. 2002. Protein engineering of a viral cage for constrained nanomaterials synthesis. *Advanced Materials,* 14, 415–418.

83. Allen, M., Prissel, M., Young, J. M., and Douglas, T. 2007. Constrained metal oxide mineralization: Lessons from ferritin applied to other protein cage architectures. In *Hand Book of Biomineralization: Biomimetic and Bioinspired Chemistry*, edited by Behrens, P. and Baeuerlein, E. Weinheim, Germany: Wiley-VCH.

84. Botstein, D., Waddell, C. H., and King, J. 1973. Mechanism of head assembly and DNA encapsulation in *Salmonella* phage p22. I. Genes, proteins, structures and DNA maturation. *Journal of Molecular Biology,* 80, 669–695.

85. Reichhardt, C., Uchida, M., O'Neil, A., Li, R., Prevelige, P. E., and Douglas, T. 2011. Accepted for publication. Templated assembly of organic–inorganic materials using the core shell structure of the P22 bacteriophage. *Chemical Communications*, 47, 6326–6328.

86. Shah, S. N., Steinmetz, N. F., Aljabali, A. A. A., Lomonossoff, G. P., and Evans, D. J. 2009. Environmentally benign synthesis of virus-templated, monodisperse, iron-platinum nanoparticles. *Dalton Transactions*, (40), 8479–8480.

87. Aljabali, A. A. A., Sainsbury, F., Lomonossoff, G. P., and Evans, D. J. 2010. Cowpea mosaic virus unmodified empty viruslike particles loaded with metal and metal oxide. *Small*, 6, 818–821.

88. Huang, X., Stein, B. D., Cheng, H., Malyutin, A., Tsvetkova, I. B., Baxter, D. V., Remmes, N. B., Verchot, J., Kao, C., Bronstein, L. M., and Dragnea, B. 2011. Magnetic virus-like nanoparticles in N. benthamiana plants: A new paradigm for environmental and agronomic biotechnological research. *ACS Nano,* 5, 4037–4045.

89. Huang, X., Bronstein, L. M., Retrum, J., Dufort, C., Tsvetkova, I., Aniagyei, S., Stein, B., Stucky, G., McKenna, B., Remmes, N., Baxter, D., Kao, C. C., and Dragnea, B. 2007. Self-assembled virus-like particles with magnetic cores. *Nano Letters,* 7, 2407–2416.

90. Kim, K. K., Yokota, H., Santoso, S., Lerner, D., Kim, R., and Kim, S. H. 1998. Purification, crystallization, and preliminary x-ray crystallographic data analysis of small heat shock protein homolog from *Methanococcus jannaschii*, a hyperthermophile. *Journal of Structural Biology,* 121, 76–80.

91. Kim, R., Kim, K. K., Yokota, H., and Kim, S. H. 1998. Small heat shock protein of *Methanococcus jannaschii*, a hyperthermophile. *Proceedings of the National Academy of Sciences of the United States of America,* 95, 9129–9133.

92. Flenniken, M. L., Willits, D. A., Brumfield, S., Young, M., and Douglas, T. 2003. The small heat shock protein cage from *Methanococcus jannaschii* is a versatile nanoscale platform for genetic and chemical modification. *Nano Letters,* 3, 1573–1576.

93. Klem, M. T., Willits, D., Solis, D. J., Belcher, A. M., Young, M., and Douglas, T. 2005. Bioinspired synthesis of protein-encapsulated CoPt nanoparticles. *Advanced Functional Materials,* 15, 1489–1494.

94. Schott, K., Ladenstein, R., König, A., and Bacher, A. 1990. The lumazine synthase–riboflavin synthase complex of *Bacillus subtilis*. Crystallization of reconstituted icosahedral beta-subunit capsids. *Journal of Biological Chemistry,* 265, 12686–12689.

95. Shenton, W., Mann, S., Cölfen, H., Bacher, A., and Fischer, M. 2001. Synthesis of nanophase iron oxide in lumazine synthase capsids. *Angewandte Chemie International Edition,* 40, 442–445.

96. Gupta, A. K., and Gupta, M. 2005. Synthesis and surface engineering of iron oxide nanoparticles for biomedical applications. *Biomaterials,* 26, 3995–4021.

97. Pankhurst, Q. A., Thanh, N. K. T., Jones, S. K., and Dobson, J. 2009. Progress in applications of magnetic nanoparticles in biomedicine. *Journal of Physics D: Applied Physics,* 42, 224001.

98. Massoud, T. F., and Gambhir, S. S. 2003. Molecular imaging in living subjects: Seeing fundamental biological processes in a new light. *Genes & Development,* 17, 545–580.

99. Jun, Y. W., Lee, J. H., and Cheon, J. 2008. Chemical design of nanoparticle probes for high-performance magnetic resonance imaging. *Angewandte Chemie International Edition,* 47, 5122–5135.

100. Sun, C., Lee, J. S. H., and Zhang, M. 2008. Magnetic nanoparticles in MR imaging and drug delivery. *Advanced Drug Delivery Reviews,* 60, 1252–1265.

101. Na, H. B., Song, I. C., and Hyeon, T. 2009. Inorganic nanoparticles for MRI contrast agents. *Advanced Materials,* 21, 2133–2148.

102. Cormode, D. P., Jarzyna, P. A., Mulder, W. J. M., and Fayad, Z. A. 2010. Modified natural nanoparticles as contrast agents for medical imaging. *Advanced Drug Delivery Reviews,* 62, 329–338.

103. Ito, A., Shinkai, M., Honda, H., and Kobayashi, T. 2005. Medical application of functionalized magnetic nanoparticles. *Journal of Bioscience and Bioengineering,* 100, 1–11.

104. Gálvez, N., Fernández, B., Sánchez, P., Cuesta, R., Ceolín, M., Clemente-León, M., Trasobares, S., López-Haro, M., Calvino, J. J., Stéphan, O., and Domínguez-Vera, J. M. 2008. Comparative structural and chemical studies of ferritin cores with gradual removal of their iron contents. *Journal of the American Chemical Society,* 130, 8062–8068.

105. Theil, E. C., and Goss, D. J. 2009. Living with iron (and oxygen): Questions and answers about iron homeostasis. *Chemical Reviews,* 109, 4568–4579.

106. Cohen, B., Ziv, K., Plaks, V., Harmelin, A., and Neeman, M. 2009. Ferritin nanoparticles as magnetic resonance reporter gene. *Wiley Interdisciplinary Review Nanomedicine and Nanobiotechnology,* 1, 181–188.

107. Stark, D. D., Bass, N. M., Moss, A. A., Bacon, B. R., McKerrow, J. H., Cann, C. E., Brito, A., and Goldberg, H. I. 1983. Nuclear magnetic resonance imaging of experimentally induced liver disease. *Radiology,* 148, 743–751.

108. Vymazal, J., Righini, A., Brooks, R. A., Canesi, M., Mariani, C., Leonardi, M., and Pezzoli, G. 1999. T1 and T2 in the brain of healthy subjects, patients with Parkinson disease, and patients with multiple system atrophy: Relation to iron content. *Radiology,* 211, 489–495.

109. Mani, V., Briley-Saebo, K. C., Hyafil, F., and Fayad, Z. A. 2006. Feasibility of in vivo identification of endogenous ferritin with positive contrast MRI in rabbit carotid crush injury using GRASP. *Magnetic Resonance in Medicine,* 56, 1096–1106.

110. Cohen, B., Dafni, H., Meir, G., Harmelin, A., and Neeman, M. 2005. Ferritin as an endogenous MRI reporter for noninvasive imaging of gene expression in C6 glioma tumors. *Neoplasia,* 7, 109–117.

111. Cohen, B., Ziv, K., Plaks, V., Israely, T., Kalchenko, V., Harmelin, A., Benjamin, L. E., and Neeman, M. 2007. MRI detection of transcriptional regulation of gene expression in transgenic mice. *Nature Medicine,* 13, 498–503.

112. Genove, G., Demarco, U., Xu, H., Goins, W., and Ahrens, E. 2005. A new transgene reporter for in vivo magnetic resonance imaging. *Nature Medicine,* 11, 450–454.

113. Deans, A. E., Wadghiri, Y. Z., Bernas, L. M., Yu, X., Rutt, B. K., and Turnbull, D. H. 2006. Cellular MRI contrast via coexpression of transferrin receptor and ferritin. *Magnetic Resonance in Medicine,* 56, 51–59.

114. Bennett, K. 2008. Controlled aggregation of ferritin to modulate MRI relaxivity. *Biophysical Journal,* 95, 342–351.

115. Sukerkar, P. A., Rezvi, U. G., Macrenaris, K. W., Patel, P. C., Wood, J. C., and Meade, T. J. 2011. Polystyrene microsphere–ferritin conjugates: A robust phantom for correlation of relaxivity and size distribution. *Magnetic Resonance in Medicine,* 65, 522–530.

116. Bulte, J. W. M., Douglas, T., Mann, S., Frankel, R. B., Moskowitz, B. M., Brooks, R. A., Baumgarner, C. D., Vymazal, J., Strub, M. P., and Frank, J. A. 1994. Magnetoferritin: Characterization of a novel superparamagnetic MR contrast agent. *JMRI–Journal of Magnetic Resonance Imaging,* 4, 497–505.

117. Bulte, J. W., Douglas, T., Mann, S., Vymazal, J., Laughlin, P. G., and Frank, J. A. 1995. Initial assessment of magnetoferritin biokinetics and proton relaxation enhancement in rats. *Academic Radiology,* 2, 871–878.

118. Libby, P. 2002. Inflammation in atherosclerosis. *Nature,* 420, 868–874.

119. Ross, R. 1999. Mechanisms of disease – Atherosclerosis – An inflammatory disease. *New England Journal of Medicine,* 340, 115–126.

120. Li, A. C., and Glass, C. K. 2002. The macrophage foam cell as a target for therapeutic intervention. *Nature Medicine,* 8, 1235–1242.

121. Jaffer, F. A., Libby, P., and Weissleder, R. 2006. Molecular and cellular imaging of atherosclerosis: Emerging applications. *Journal of the American College of Cardiology,* 47, 1328–1338.

122. Underhill, H. R., Hatsukami, T. S., Fayad, Z. A., Fuster, V., and Yuan, C. 2010. MRI of carotid atherosclerosis: Clinical implications and future directions. *Nature Reviews Cardiology,* 7, 165–173.

123. Terashima, M., Uchida, M., Kosuge, H., Tsao, P. S., Young, J. M., Conolly, S. M., Douglas, T., and McConnell, M. V. 2011. Human ferritin for macrophage imaging in atherosclerosis. *Biomaterials,* 32, 1430–1437.

124. Helm, L. 2006. Relaxivity in paramagnetic systems: Theory and mechanisms. *Progress in Nuclear Magnetic Resonance Spectroscopy,* 49, 45–64.

125. Caravan, P. 2006. Strategies for increasing the sensitivity of gadolinium based MRI contrast agents. *Chemical Society Reviews,* 35, 512–523.

126. Aime, S., Cabella, C., Colombatto, S., Geninatti Crich, S., Gianolio, E., and Maggioni, F. 2002. Insights into the use of paramagnetic Gd(III) complexes in MR-molecular imaging investigations. *Journal of Magnetic Resonance Imaging,* 16, 394–406.

127. Lauffer, R. B. 1987. Paramagnetic metal-complexes as water proton relaxation agents for NMR imaging: Theory and design. *Chemical Reviews,* 87, 901–927.

128. Caravan, P., Ellison, J., McMurry, T., and Lauffer, R. 1999. Gadolinium(III) chelates as MRI contrast agents: Structure, dynamics, and applications. *Chemical Reviews,* 99, 2293–2352.

129. Werner, E. J., Datta, A., Jocher, C. J., and Raymond, K. N. 2008. High-relaxivity MRI contrast agents: Where coordination chemistry meets medical imaging. *Angewandte Chemie International Edition,* 47, 8568–8580.

130. Ahrens, E. T., Rothbacher, U., Jacobs, R. E., and Fraser, S. E. 1998. A model for MRI contrast enhancement using T1 agents. *Proceedings of the National Academy of Sciences of the United States of America,* 95, 8443–8448.

131. Rocklage, S. M., Worah, D., and Kim, S. H. 1991. Metal-ion release from paramagnetic chelates: What is tolerable. *Magnetic Resonance in Medicine,* 22, 216–221.

132. Penfield, J. G., and Reilly, R. F., Jr. 2007. What nephrologists need to know about gadolinium. *Nature Clinical Practice Nephrology,* 3, 654–668.

133. Allen, M., Bulte, J. W. M., Liepold, L., Basu, G., Zywicke, H. A., Frank, J. A., Young, M., and Douglas, T. 2005. Paramagnetic viral nanoparticles as potential high-relaxivity magnetic resonance contrast agents. *Magnetic Resonance in Medicine,* 54, 807–812.

134. Liepold, L., Anderson, S., Willits, D., Oltrogge, L., Frank, J. A., Douglas, T., and Young, M. 2007. Viral capsids as MRI contrast agents. *Magnetic Resonance in Medicine,* 58, 871–879.

135. Geninatti Crich, S., Bussolati, B., Tei, L., Grange, C., Esposito, G., Lanzardo, S., Camussi, G., and Aime, S. 2006. Magnetic resonance visualization of tumor angiogenesis by targeting neural cell adhesion molecules with the highly sensitive gadolinium-loaded apoferritin probe. *Cancer Research,* 66, 9196–9201.

136. Aime, S., Frullano, L., and Crich, S. G. 2002. Compartmentalization of a gadolinium complex in the apoferritin cavity: A route to obtain high relaxivity contrast agents for magnetic resonance imaging. *Angewandte Chemie International Edition,* 41, 1017–1019.

137. Prasuhn, D. E., Jr., Yeh, R. M., Obenaus, A., Manchester, M., and Finn, M. G. 2007. Viral MRI contrast agents: Coordination of Gd by native virions and attachment of Gd complexes by azide–alkyne cycloaddition. *Chemical Communications,* (12), 1269–1271.

138. Anderson, E. A., Isaacman, S., Peabody, D. S., Wang, E. Y., Canary, J. W., and Kirshenbaum, K. 2006. Viral nanoparticles donning a paramagnetic coat: Conjugation of MRI contrast agents to the MS2 capsid. *Nano Letters,* 6, 1160–1164.

139. Hooker, J. M., Datta, A., Botta, M., Raymond, K. N., and Francis, M. B. 2007. Magnetic resonance contrast agents from viral capsid shells: A comparison of exterior and interior cargo strategies. *Nano Letters,* 7, 2207–2210.

140. Datta, A., Hooker, J. M., Botta, M., Francis, M. B., Aime, S., and Raymond, K. N. 2008. High relaxivity gadolinium hydroxypyridonate–viral capsid conjugates: Nanosized MRI contrast agents. *Journal of the American Chemical Society,* 130, 2546–2552.
141. Liepold, L. O., Abedin, M. J., Buckhouse, E. D., Frank, J. A., Young, M. J., and Douglas, T. 2009. Supramolecular protein cage composite MR contrast agents with extremely efficient relaxivity properties. *Nano Letters,* 9, 4520–4526.
142. Liepold, L., Abedin, M. J., Qazi, S., Johnson, B., Prevelige, P. E., Frank, J., and Douglas, T. Submitted for publication. The synthesis and NMRD characterization of a t1 MRI contrast agent derived from a Gd containing polymer grown inside a viral capsid.
143. Dunand, F. A., Borel, A., and Helm, L. 2002. Gd(III) based MRI contrast agents: Improved physical meaning in a combined analysis of EPR and NMR data? *Inorganic Chemistry Communications,* 5, 811–815.
144. Abedin, M. J., Liepold, L., Suci, P., Young, M., and Douglas, T. 2009. Synthesis of a cross-linked branched polymer network in the interior of a protein cage. *Journal of the American Chemical Society,* 131, 4346–4354.
145. Pierre, V. C., Botta, M., Aime, S., and Raymond, K. N. 2006. Substituent effects on Gd(III)-based MRI contrast agents: Optimizing the stability and selectivity of the complex and the number of coordinated water molecules. *Inorganic Chemistry,* 45, 8355–8364.
146. Mulder, W. J. M., Strijkers, G. J., van Tilborg, G. A. F., Griffioen, A. W., and Nicolay, K. 2006. Lipid-based nanoparticles for contrast-enhanced MRI and molecular imaging. *NMR in Biomedicine,* 19, 142–164.
147. Flenniken, M. L., Willits, D. A., Harmsen, A. L., Liepold, L. O., Harmsen, A. G., Young, M. J., and Douglas, T. 2006. Melanoma and lymphocyte cell-specific targeting incorporated into a heat shock protein cage architecture. *Chemistry & Biology,* 13, 161–170.
148. Destito, G., Yeh, R., Rae, C. S., Finn, M. G., and Manchester, M. 2007. Folic acid-mediated targeting of cowpea mosaic virus particles to tumor cells. *Chemistry & Biology,* 14, 1152–1162.
149. Huang, R. K., Steinmetz, N. F., Fu, C. Y., Manchester, M., and Johnson, J. E. 2011. Transferrin-mediated targeting of bacteriophage HK97 nanoparticles into tumor cells. *Nanomedicine,* 6, 55–68.
150. Arap, W., Pasqualini, R., and Ruoslahti, E. 1998. Cancer treatment by targeted drug delivery to tumor vasculature in a mouse model. *Science,* 279, 377–380.
151. Uchida, M., Willits, A. D., Muller, K., Willis, A. F., Jackiw, L., Jutila, M., Young, M., Porter, A. E., and Douglas, T. 2009. Intracellular distribution of macrophage targeting ferritin-iron oxide nanocomposite. *Advanced Materials,* 21, 458–462.
152. Kitagawa, T., Kosuge, H., Uchida, M., Dua, M. M., Iida, Y., Bogyo, M., Dalman, R. L., Douglas, T., and McConnell, M. V. In preparation.
153. Kitagawa, T., Kosuge, H., Uchida, M., Dua, M. M., Iida, Y., Bogyo, M., Dalman, R. L., Douglas, T., and McConnell, M. V. in press. RGD-conjugated human ferritin nanoparticles for imaging vascular inflammation and angiogenesis in experimental carotid and aortic disease. *Molecular Imaging and Biology.* DOI:10.1007/s11307-011-0495-1.
154. Laakkonen, P., Porkka, K., Hoffman, J. A., and Ruoslahti, E. 2002. A tumor-homing peptide with a targeting specificity related to lymphatic vessels. *Nature Medicine,* 8, 751–755.
155. Fogal, V., Zhang, L., Krajewski, S., and Ruoslahti, E. 2008. Mitochondrial/cell-surface protein p32/gC1qR as a molecular target in tumor cells and tumor stroma. *Cancer Research,* 68, 7210–7218.
156. Uchida, M., Kosuge, H., Terashima, M., Willits, D. A., Liepold, L. O., Young, M. J., McConnell, M. V., and Douglas, T. 2011. Protein cage nanoparticles bearing the LyP-1 peptide for enhanced imaging of macrophage-rich vascular lesions. *ACS Nano,* 5, 2493–2502.
157. Abe, S., Hirata, K., Ueno, T., Morino, K., Shimizu, N., Yamamoto, M., Takata, M., Yashima, E., and Watanabe, Y. 2009. Polymerization of phenylacetylene by rhodium complexes within a discrete space of apo-ferritin. *Journal of the American Chemical Society,* 131, 6958–6960.
158. Jaffer, F. A., Libby, P., and Weissleder, R. 2009. Optical and multimodality molecular imaging: Insights into atherosclerosis. *Arteriosclerosis, Thrombosis, and Vascular Biology,* 29, 1017–1024.

Professor Trevor Douglas is the letters and sciences distinguished professor at Montana State University, where he is also the director of the Centre for Bio-Inspired Nanomaterials. Dr. Douglas was born in Cape Town, South Africa, and earned his undergraduate degree (biochemistry) from University of California at San Diego (1986) and his PhD (inorganic chemistry) at Cornell University (1991). He completed post-doctoral work at Bath University in the United Kingdom, working in the area of biomineralisation. He has pioneered the use of viruses as synthetic materials for the templated synthesis of nanomaterials. Applications for these materials include the development of novel drug delivery, diagnostic imaging probes, magnetic storage and catalysis. At Montana State University, he established the Centre for Bio-Inspired Nanomaterials, a collaborative multidisciplinary centre with biologists, chemists, engineers and physicists to explore the use of biological macromolecules as templates for materials fabrication.

Dr. Masaki Uchida is an assistant research professor at Montana State University. He received his BA degree (industrial chemistry) in 1997 and PhD degree (material chemistry) in 2002 from

Kyoto University, Japan, under the supervision of Prof. Tadashi Kokubo. He conducted post-doctoral training at National Institute of Advanced Science and Technology, Japan, under the guidance of Dr. Atsuo Ito (2002–2006) and at Montana State University under the guidance of Prof. Trevor Douglas (2006–2010). His current research interests include biomineralisation, biomimetic material synthesis and biomedical materials.

Dr. Lars O. Liepold is a post-doctoral researcher at Montana State University, Bozeman, Montana. He received his BA at St. Johns University in chemistry (2000). In 2009, he received his PhD degree in chemistry from Montana State University under the Supervision of Prof. Trevor Douglas. His research interest includes MRI contrast agent development, native spray mass spectrometry, metabolomics and mass spectrometry methodology.

4 Next Generation Magnetic Nanoparticles for Biomedical Applications

Trinh Thang Thuy, Shinya Maenosono,
and Nguyễn Thị Kim Thanh

CONTENTS

4.1   INTRODUCTION

Superparamagnetic iron oxides (SPIOs) including γ-Fe_2O_3 (maghemite) and Fe_3O_4 (magnetite) nanoparticles (NPs) are biocompatible and relatively easy to synthesise; these properties make them the most used magnetic nanoparticles (MNPs) in biomedicine to date. They have been studied for several decades and have contributed to both diagnostics such as magnetic resonance imaging (MRI) contrast agents and therapeutics such as magnetic hyperthermia[1-4]. However, the relatively low saturation magnetisation (M_S) of SPIOs (~300–400 emu·cm^{-3}) limits their potential in these applications[4,5]. The response of SPIO NPs to an external magnetic field may be sufficient for imaging purposes to some extent, but their suitability in magnetically targeted drug delivery is doubted, that is, they could not effectively be directed within the human body by magnetic forces because the saturation magnetisation is too low[5]. Enhancement of the magnetic moment of MNPs is key for improvement of many applications in biomedicine. Considering the characteristic size of biological systems, that is, 10–100 μm for a cell, 20–450 nm for a virus, 5–50 nm for a protein and 2 nm in width and 10–100 nm in length for a gene, MNPs with smaller dimensions than normally used SPIO NPs are preferred as they would increase the spatial resolution. Using MNPs, which have higher saturation magnetisation and higher magnetocrystalline anisotropy energy than SPIOs, one can significantly improve efficiency in various biomedical applications. Moreover, these magnetically superior ultrasmall MNPs could lead to revolutionary and novel clinical applications.

Recently, mono- and bimetallic superparamagnetic MNPs have become readily available thanks to the development of a range of synthetic techniques. In general, the metallic MNPs exhibit higher magnetic properties than oxide MNPs. For example, elemental Fe and Co have M_S of about 1700 and 1400 emu·cm^{-3}, respectively[6]. Bimetallic FePt, CoPt, FeCo and $SmCo_5$ alloys each have M_S of about 1000, 800, 1900 and 900 emu·cm^{-3}, respectively[7-9]. With higher M_S, MNPs experience higher driving forces under a magnetic field, and thus, the efficacy of drug delivery or magnetic separation will be greatly improved[10]. In the case of hyperthermia therapy, the optimal size of MNPs as heating elements varies depending on the magnetocrystalline anisotropy energy[11]. Roughly, the optimal MNP size for hyperthermia is inversely proportional to the magnetocrystalline anisotropy constant[11]. On the other hand, the heating rate increases with increasing M_S[10,11]. For these reasons, the next generation magnetic nanoparticles (n-MNPs) increasingly attract attention in various biomedical fields. We consider SPIO NPs as the classical type of MNPs. In this chapter, we review techniques for chemical synthesis, magnetic properties and biomedical applications of various types of MNPs, which is not SPIO, including mono- and bimetallic alloys, Mn-based MNPs and heterostructured MNPs, and we class them as the next generation magnetic nanoparticles. n-MNPs are expected to be highly promising magnetic nanoprobes, nanotracers, nanocarriers or nanoheaters and will be very useful for many clinical applications introduced in this book such as removal of blood-borne toxin in the body (Chapter 7), magnetic separation of cells (Chapters 8, 10 and 11), magnetic immunoassay (Chapter 9), magnetic gene transfection (Chapter 12), magnetic cell targeting for cell therapies (Chapter 13), magnetic drug delivery (Chapters 14 through 16 and 19), hyperthermica cancer treatment (Chapters 17 and 18) and magnetic particle imaging (Chapter 20).

4.2   SYNTHETIC METHODS FOR NEXT GENERATION MNPs

Various synthetic methods have been intensively studied to obtain different types of nanomaterials such as semiconductor, metal and oxide NPs. These synthetic methods are applicable to the synthesis of n-MNPs[12,13]. The advantages and drawbacks of various synthetic methods for n-MNPs are briefly summarised in Table 4.1. To date, reduction and thermal decomposition are the most used synthetic methods.

TABLE 4.1

Summary Comparison of the Synthetic Methods

Method	Reaction Temperature	Solvent	Size Distribution	Shape Control	Yield
Thermolysis	100°C–320°C	Organic	Very narrow	Very good	High
Reduction	20°C–90°C	Water	Narrow	Not good	High
Reverse micelle	20°C–50°C	Organic	Narrow	Good	Low
Sonochemical	20°C–30°C	Water	Narrow	Not good	High
Hydrothermal	220°C	Polar	Very narrow	Very good	Medium

Source: Reproduced with kind permission from Springer Science + Business Media: *Angew. Chem. Int. Ed.*, Magnetic Nonoparticles: Synthesis, protection, functionalisation, and application, 46, 2007, 1222–1244, Lu, A.-H., Salabas, E.L., and Schüth, F., Copyright Wiley-VCH Verlag GmbH & Co. KGaA.

4.2.1 THERMAL DECOMPOSITION

Thermal decomposition or thermolysis of organometallic compounds in high boiling point organic solvents with surfactants is a common method to produce monodispersed MNPs. Organometallic compounds, such as metal acetylacetonates, $[M(acac)_n]$ (M = Fe, Mn, Co, Ni, Cr), cupferronates, $[M^{(x)}Cup_x]$ (M = metal ion; Cup = $C_6H_5N(NO)O^-$) or metal carbonyls, have been widely used as precursors in the literature[13]. In the case of alloy MNPs, single molecule bimetallic precursors, such as $[NEt_4]$ $[FeCo_3(CO)_{12}]$, $[NMe_3CH_2Ph]_2[Fe_3Pt_3(CO)_{15}]$, $[NMe_4]_2[FeNi_5(CO)_{13}]$ or $[NMe_3CH_2Ph]_2[Fe_4Pt(CO)_{16}]$, have occasionally been used to obtain uniformly composed MNPs[14]. Fatty acids, alkylamines or their combination are often used as surface capping ligands in this synthetic method. Metal precursors and organic capping ligands are put into a high boiling point solvent such as di-octylether or octadecene and then heated at high temperature (ca. 300°C). Then, the metal precursor thermally decomposes, leading to nucleation and growth of MNPs. Immediately after the nucleation, capping ligands cover the surfaces of the MNPs to protect them from aggregation and further crystal growth. In this process, conditions for supersaturation are easily met, so the size of nuclei tends to be relatively small. The ratios of organometallic precursors, surface capping ligands and solvent are generally the decisive parameters for controlling the size and morphology of resulting MNPs. The reaction temperature, reaction time and ageing period are also crucial factors for precise control of size, shape and composition. One drawback of this method is that the as-synthesised MNPs are hydrophobic in nature as they are synthesised under organic conditions and coated with hydrophobic stabilisers.

4.2.2 REDUCTION

4.2.2.1 Aqueous Reduction

This method employs metal salts as precursors. Metal ions are reduced in aqueous medium by reducing agents, such as hydrogen, sodium borohydride ($NaBH_4$) or hydrazines ($N_2H_4 \cdot H_2O$, $N_2H_4 \cdot HCl$); nucleation and growth follow under ambient conditions. Because borohydrides are strong reducing agents, the reduction of metal cations is rapid, and thus, the process is relatively difficult to control[15]. However, the presence of suitable surfactants can help to effectively separate nucleation and growth and prevent MNPs from aggregation. The method is well established and straightforward. As with thermal decomposition, size, shape and composition can be controlled by varying metal precursor and ligand concentrations, pH and reaction time. One advantage of this method is that the resulting MNPs are inherently water soluble. For example, water-soluble Co MNPs have been synthesised using $CoCl_2$ as a precursor, $NaBH_4$ as a reducing agent and alkyl thioether end-functionalised poly(methacrylic acid) (PMAA-DDT) polymer as a capping ligand[16]. On the other hand, a disadvantage is that resulting MNPs contain boron as an impurity, which affects the magnetic properties of the particles[17].

4.2.2.2 Reverse Micelle Methods

Reverse micelles are water-in-oil droplets stabilised by a surfactant, such as sodium-2-bis-(2-ethylhexyl)sulphosuccinate, Na(AOT). By solubilising metal precursors and reducing agents (i.e. $NaBH_4$, $N_2H_4 \cdot HCl$) in the water confined in the reverse micelles, one can easily create a plethora of discrete nanoscale reactors. By mixing two types of micelles (one contains metal ions and the other contains reducing agents), the reduction reaction of metal ions is promoted as a result of an exchange of micelle contents. Reverse micelle methods have been routinely used to produce a very wide range of MNPs[13]. By varying the water content ($w = [H_2O]/[S]$, S = surfactant), the size of the reverse micelles can be readily tuned, leading to size control. One advantage of this synthetic technique is that relatively uniform MNPs can be synthesised in terms of size and composition, which is relatively difficult to achieve *via* chemical reduction and thermolysis. For example, monodisperse FePt NPs with tunable sizes up to 20 nm and narrow composition distribution were synthesised using a reverse micelle method[18]. However, there are some drawbacks such as limitation of surfactants and co-surfactants, which can be used to form the desired reverse micelle microemulsions for the synthesis of n-MNPs. Additionally, large amounts of solvent are necessary to synthesise appreciable amounts of MNPs.

4.2.2.3 Polyol Reduction

In this process, metal precursors and stabilisers are dissolved in a polyol solvent, such as ethylene glycol, diethylene glycol, tetraethylene glycol or propanediol, and then, the reaction mixture is heated at high temperature. The reaction temperature is typically near the boiling point of the polyol solvent. Upon heating, the metal precursor forms intermediates, which are then reduced by the polyol reductant and by nucleation of the MNPs. The advantage of this process is the resulting MNPs usually have a well-defined morphology with narrow size distribution.

4.2.3 SONOCHEMISTRY

Sonochemistry arises from acoustic cavitation: the formation, growth and implosive collapse of bubbles in a liquid[19,20]. Cavitation-induced sonochemistry creates localised 'hot spots' inside the collapsing bubble with extreme conditions including temperatures of over 500 K, pressures of ~1000 bar and heating and cooling rates of $>10^{10}$ K·s^{-1} [19,20]. These extraordinary conditions permit access to a range of chemical reaction space normally not accessible, which allows for the synthesis of a wide variety of nanomaterials. To achieve high yields, the precursors should be highly volatile since the primary reaction site is the vapour inside the cavitating bubbles[21]. In addition, the solvent vapour pressure should be low at the sonication temperature, because significant solvent vapour inside the bubble reduces the bubble collapse efficiency[19]. The extremely rapid cooling rate strongly favours the formation of amorphous products. By changing the reaction medium, various nanomaterials have been synthesised[22]. The metallic precursors used in the sonochemical synthesis of n-MNPs are typically metal carbonyls, such as $Fe(CO)_5$, $Co_2(CO)_8$, $Co(CO)_3(NO)$ or $Fe_3Pt_3(CO)_{15}$[23,24]. A broad range of surface capping ligands, such as carboxylic acid, alkylamine, alkanethiols and polymers, can be used in the synthesis[22,24].

4.2.4 OTHER SYNTHETIC ROUTES

Other avenues for synthesis have also been explored; hydrothermal or microwave syntheses are two examples[12]. In a typical hydrothermal synthesis, the reaction mixture is heated beyond the boiling point of water under an elevated pressure to exploit the enhanced solubility of inorganic precursors in water under the extreme conditions[25]. Microwave heating has addressed the problem of heating inhomogeneity in reaction solutions and has been demonstrated to enhance reaction rates, selectivity and reaction yields. Several types of n-MNPs including Ni MNPs and Ni–Co core–shell

MNPs have been successfully synthesised using microwaves as a mode of heating, and the resulting MNPs are found to have small size, narrow size distribution and high purity[26,27].

4.2.5 SURFACE CAPPING

The chemical and colloidal stability of n-MNPs is an important issue because stability is crucial for bioapplication. The surface capping of MNPs is needed to protect the MNPs against oxidation or corrosion, and aggregation/flocculation. Moreover, surface capping molecules can provide functional groups, such as carboxylic acid, amine or thiol, through which MNPs can be conjugated to biological relevant molecules of interest *via* electrostatic interaction, (l-ethyl-3-dimethylaminopropyl) carbodiimide hydrochloride (EDC) coupling or maleimide methods (see Chapter 5). On other hand, the formation of the surface capping could modify the magnetic properties of MNPs (see Section 4.2.5.3 and Chapter 6).

4.2.5.1 Enhancement of Chemical Stability

Metallic MNPs, especially monometallic MNPs such as Fe, Co or Ni, oxidise easily and so it is necessary to protect them for practical applications. There are two types of protection strategies: first, capping by organic ligands;[28–30] second, capping by an inorganic shell, such as SiO_2[31], carbon[32], Ag[33], Au[34] or oxides[35,36]. Individually protected MNPs are readily dispersible in a solvent and become highly stable in ambient and even severe conditions. These protection strategies have been recently reviewed by Lu et al.[13] Although the formation of a protective shell around the MNPs can enhance the chemical stability of the magnetic cores, it is still challenging to maintain high colloidal stability of these systems[13].

4.2.5.2 Enhancement of Colloidal Stability

The aggregation/flocculation of MNPs stems from dipole interactions and van der Waals forces[37]. In order to enhance the colloidal stability of water-dispersible MNPs, the attractive forces must be balanced out by electrostatic, steric repulsive forces or combination of these forces[38]. There have been some reports attempting to prepare stable water-dispersible MNPs[39–44]. For example, FePt MNPs capped with oleic acid (OA) and oleylamine (OLA) become highly stable in water after ligand exchange of OA/OLA system with tetramethylammonium hydroxide (TMAOH)[39]. By replacing the surface capping ligands with TMAOH, the surface of FePt MNPs becomes negatively charged so electrostatic repulsion overcomes attractive interactions to stabilise the MNPs in water, as shown in Figure 4.1. Similarly, by replacing the capping ligands with 2-aminoethanethiol (AET), positively charged water-dispersible FePt NPs can be prepared[40]. Using 11-mercaptoundecanoic acid (MUA) as a capping ligand, both steric and electrostatic repulsions stabilise FePt MNPs in water[41].

Steric stabilisation of MNPs has typically been done *via* surface capping with water-soluble polymers, such as trifluoroethylester–polyethylene glycol–thiol[42–44]. MNPs coated with water-soluble polymer are dispersible in water with high colloidal stability. For example, CoPt MNPs capped with PMAA-pentaerythritol tetrakis (3-mercaptopropionate) (PMAA-PTMP) are highly stable in water across a wide range of pH, high electrolyte concentration and in cell culture medium[44].

4.2.5.3 Influence of Surface Capping on Magnetic Properties

The correlation between the surface capping and the magnetic properties of MNPs has been recently reported[45–50]. The noble metal shell coated on surfaces of MNPs has been found to influence the magnetic properties of MNPs[45,46]. For example, Au-coated Co MNPs have a lower magnetic anisotropy than uncoated Co MNPs, whereas Au coating Fe MNPs enhances the magnetic anisotropy[45]. The complicated interplay between the magnetic core and the metal shell determines the magnetic properties[46]. The effect of the coating is less clear and probably caused by the alloy formation between magnetic cores and the noble metal shells[45]. On the other hand, organic capping ligands also influence the magnetic properties of MNPs. They modify the spin magnetic moment and surface magnetic anisotropies of atoms at the surface of the MNPs[47]. In short, the magnetisation was reduced due to the formation of a nonmagnetic shell (surface dead layer) *via* binding of the polar end group of the capping ligands with the MNPs[48–50]. For example, the influence of the surface ligands

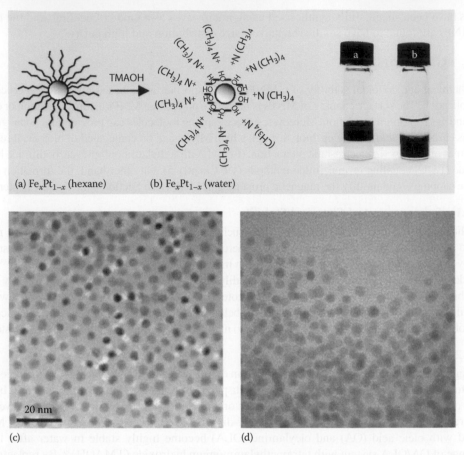

FIGURE 4.1 Top panel: dispersions of FePt NPs capped in OA and OLA in hexane (a) and FePt NPs coated with TMAOH in water (b). Bottom panel: TEM images (c, d) of FePt NPs in (a) and (b), respectively. (Reprinted with permission from Salgueiriño-Maceira, V., Liz-Marzán, L.M., and Farle, M., Water-based ferrofluids from Fe_xPt_{1-x} nanoparticles synthesized in organic media, *Langmuir*, 20, 6946–6950, 2004. Copyright 2004 American Chemical Society.)

on the magnetic properties of FePt clusters has been partially interpreted based on a density functional calculation[50]. It was found that the decrease in the magnetic moment of Fe atoms at adsorption sites originates from the interplay between the strong hybridisation of the majority *d* states of Fe atoms with majority *p* states of O, N or S atoms in the functional group of capping ligands. Importantly, whatever the case, the influence of surface capping on magnetic properties of MNPs is not desirable in most cases. Therefore, one needs to carefully design the surface capping considering the trade-off between stability and magnetic properties depending on the desired application.

4.3 MAGNETIC PROPERTIES AND BIOMEDICAL APPLICATIONS OF NEXT GENERATION MNPs

4.3.1 GENERAL ASPECTS OF MAGNETIC PROPERTIES OF MNPs

To compare the magnetic performance of n-MNPs to SPIOs, we recall some quantitative parameters including the critical single-domain radius, r_C; the critical superparamagnetic radius, r_{SP}; saturation magnetisation, M_S; magnetocrystalline anisotropy constant, K (uniaxial anisotropy constant, K_u); blocking temperature, T_B; and Curie temperature, T_C. Table 4.2 summarises the intrinsic magnetic properties of various type of n-MNPs together with iron oxides.

TABLE 4.2

Intrinsic Magnetic Properties of Various MNPs. M_S, K and T_C Values Are Estimated for Bulk Materials at Room Temperature

Materials	K_u (10^7 erg·cm^{-3})	M_S (emu·cm^{-3})	T_C (K)	r_C (nm)	r_{SP} (nm)
bcc-Fe	—	1745.9	1044	—	10
fcc-Co	0.45	1460.5	1388	30	10
hcp-Co	0.27	1435.9	1360	34	—
fcc-Ni	—	522.2	627	30	17
$L1_0$-MnAl	1.7	560	650	355	5.1
$L1_0$-FePt	6.6–10	1140	750	170	2.8–3.3
$L1_0$-FePd	1.8	1100	760	100	5
FeCo	—	1910	—	50	10
Fe$_3$Co	—	1993	—	—	10
$L1_2$-Co$_3$Pt	2.0	1100	—	105	4.8
$L1_0$-CoPt	4.9	800	840	305	3.6
SmCo$_5$	11–20	910	1000	355–480	2.2–2.7
γ-Fe$_2$O$_3$	—	380	—	30	20
Fe$_3$O$_4$	—	415	—	—	15

Sources: Wohlfarth, E.P., *Ferromagnetic Materials*, Vol. 1. North-Holland Publishing Company, 1980; Benenson, W. et al., *Handbook of Physics*, Springer-Verlag Inc., New York, 2002; Weller, D. et al., *Magnetic Nanoparticles*, Wiley-VCH Verlag GmbH & Co. KGaA, Weinheim, Germany, 2000; Mohn, P., *Magnetism in the Solid State*, Springer, Berlin, Germany, 2006; Krishnan, K.M., *IEEE Trans. Magn.*, 46, 2523, 2010; Gutfleisch, O. et al., *Avd. Eng. Mater.*, 7, 208, 2005.

4.3.2 Monometallic MNPs

4.3.2.1 Iron MNPs

Iron is a ferromagnetic element and has the highest saturation magnetisation at room temperature of all the elements. Fe MNPs exhibit superparamagnetic behaviour in the size range below ~20 nm. The syntheses of monodispersed Fe MNPs have been reviewed in 2005 by Huber[24]. A typical synthesis approach for Fe MNPs is the thermal decomposition of iron pentacarbonyl [Fe(CO)$_5$][54,55], which was firstly developed in the late 1970s[54]. In the synthesis, Fe MNPs with diameter of about 1.5–20 nm were synthesised in the presence of functional polymers as shown in Figure 4.2.

In general, Fe MNPs are extremely reactive and subject to facile oxidation, often resulting in the formation of an iron oxide shell[54]. Because pure Fe MNPs are unstable in ambient conditions, it is important to protect them from oxidation by coating them with another chemically stable biocompatible material such as an Au shell[56,57]. The Au shell has strong affinity to thiols, which often present in biological molecules, make it very versatile surface for biofunctionalisation. On the other hand, it has been reported that the Au shell also enhances the magnetic anisotropy of the Fe core presumably due to the formation of an inhomogeneous Fe–Au alloy[45]. Recently, carbon-coated Fe (Fe@C) MNPs were investigated in preclinical trials as possible vehicles for magnetically targeted chemotherapy[58,59]. Fe@C MNPs of 20 nm diameter were synthesised using arc discharge methods (Figure 4.3a). The porous carbon shell permits rapid adsorption and slow release of therapeutic agents[5]. Preclinical studies using the chemotherapy agent doxorubicin tagged to Fe@C MNPs have been conducted. Histopathological analysis confirmed that the magnetic carriers could be selectively drawn to a tumour in each animal's left kidney, close to an implanted magnet (Figure 4.3b and c)[58].

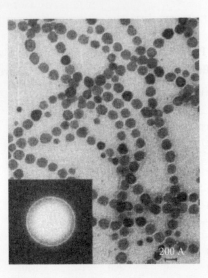

FIGURE 4.2 TEM image of Fe MNPs synthesised *via* the decomposition of Fe(CO)$_5$ in poly(4-vinylpyridin-estyrene)/dichlorobenzene. The Fe MNPs have formed chain-like structures due to dipolar interactions that are moderated by adsorbed polymer chains. (Reprinted with permission from Griffiths, C., Ohoro, M., and Smith, T., The structure, magnetic characterization, and oxidation of colloidal iron dispersions, *J. Appl. Phys.*, 50, 7108–7115, 1979. Copyright 1979 American Institute of Physics.)

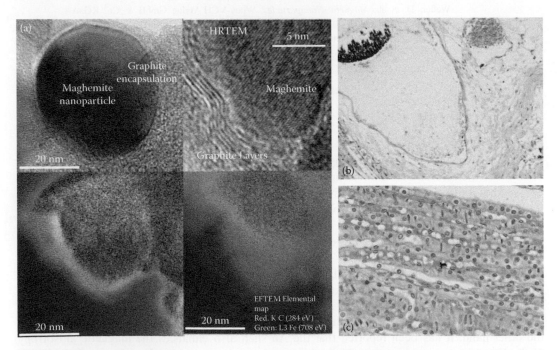

FIGURE 4.3 (a) High-resolution and energy-filtering TEM images of carbon-coated Fe MNPs. (b, c) Histopathology analysis of (b) the left kidney shows MNPs (stained with haematoxylin–eosin) aligned along the magnetic field lines of an implanted permanent magnet, and (c) practically no MNPs are observed in the right kidney, where no magnet was implanted. (Reprinted from *Nanotoday*, 1, Gould, P., Nanomagnetism shows *in vivo* potential, 34–39; *J. Magn. Magn. Mater.*, 311, Fernández-Pacheco, R., Marquina, C., Valdivia, J.G., Gutiérrez, M., Romeroc, M.S., Cornudella, R., Laborda, A., Viloria, A., Higuera, T., García, A., de Jalón, J.A.G., and Ibarra, M.R., Magnetic nanoparticles for local drug delivery using magnetic implants, 318–322, Copyright 2006, 2007, with permission from Elsevier.)

4.3.2.2 Cobalt MNPs

Cobalt is a well-known ferromagnetic element along with iron and is commonly used as an alloying element in permanent magnets. Bulk Co has hexagonal close-packed (*hcp*) and face-centred cubic (*fcc*) phases. In the case of small particles, however, the *fcc* phase preferably exists[60]. At the nanoscale, a new so-called ε phase was found by Dinega and Bawendi in 1999[61]. *hcp*-Co is magnetically harder than *fcc*-Co due to its high magnetocrystalline anisotropy and coercivity as compared to the low magnetocrystalline anisotropy and coercivity of *fcc*-Co[62,63]. ε-Co has the magnetocrystalline anisotropy lies between those of *hcp* and *fcc* phase of Co, and it is also considered as a soft magnetic material similar to *fcc*-Co[64]. The magnetic moment per atom in ε-Co estimated as $1.70\,\mu_B$ (Borh magneton) is comparable to those of the bulk *fcc*-Co ($1.75\,\mu_B$) and *hcp*-Co ($1.72\,\mu_B$)[61].

A large number of approaches for the chemical synthesis of Co MNPs have been reported, including thermal decomposition of $Co_2(CO)_8$[61,65,66], sonochemical synthesis[67], reverse micelle[68–70], chemical reduction[15,16,71] and polyol reduction[72]. The two most popular approaches for the Co MNP synthesis are reverse micelle (Figure 4.4) and thermal decomposition methods. The spontaneous formation of an oxide shell on the surfaces of Co MNPs drastically changes the magnetic behaviour of the MNPs. For example, an enhanced magnetoresistance due to the strong exchange coupling between the ferromagnetic Co core and the antiferromagnetic CoO shell has been observed[73]. For surface capping of Co MNPs, several stabilisers including Au[45], peptides[74] and thermoresponsive polymers[43,75] have been proposed. Thermoresponsive polymer exhibits a non-linear response to temperature; capping MNPs with such a material allows them to be used as controlled drug carriers as well as smart heaters for hyperthermia[76]. The capability of polymer-coated water-soluble Co MNPs as MRI contrast agents has been evaluated; the longitudinal (r_1) and transverse (r_2) relaxivities of the Co MNPs (3.9 nm diameter) were found to be 7.4 and $88\,s^{-1}\cdot mM^{-1}$, respectively[77]. These values are similar to those reported for iron oxide with larger size, indicating the potential of Co MNPs as negative-contrast agents.

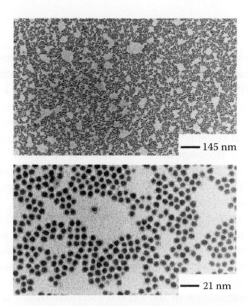

FIGURE 4.4 TEM images at different magnifications of 5.8 nm *fcc*-Co MNPs synthesised *via* the reverse micelle method. (Reproduced with kind permission from Springer Science + Business Media: *Adv. Mater.*, Self-organization of magnetic nanosized cobalt particles, 10, 1998, 259–261, Petit, C., Taleb, A., and Pileni, M.-P., Copyright Wiley-VCH Verlag GmbH & Co. KGaA.)

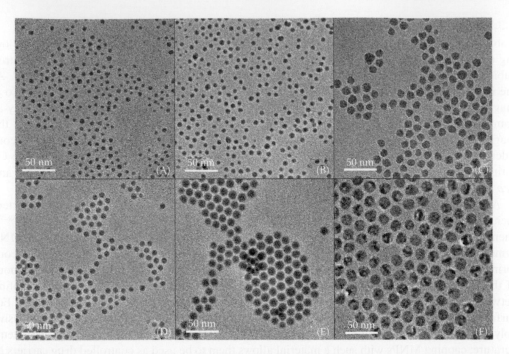

FIGURE 4.5 TEM images of Ni MNPs synthesised *via* thermal decomposition of Ni salts in the presence of OA, *n*-trioctylamine (TOA) and *n*-trioctylphosphine (TOP). The particle size can be tuned in the range of 4–16 nm (A–F), corresponding to different tailored reaction conditions. (Reprinted with permission from Winnischofer, H., Rocha, T.C.R., Nunes, W.C., Socolovsky, L.M., Knobel, M., and Zanchet, D., Chemical synthesis and structural characterization of highly disordered Ni colloidal nanoparticles, *ACS Nano*, 2, 1313–1319, 2008. Copyright 2008 American Chemical Society.)

4.3.2.3 Nickel MNPs

Nickel MNPs are a highly interesting material because they have both superparamagnetic and optical properties[78,79]. The optical properties of Ni MNPs include both surface plasmon resonance (SPR) and fluorescence, and the SPR peak of Ni MNPs (2.5 nm diameter) was located at 417 nm, as shown in Figure 4.6d (inset)[79]. The origin of the SPR band of Ni MNPs is thought to be the creation of NiO clusters and Ni^{2+} ions[78]. A combination of fluorescent and magnetic properties would lead to new applications in biological systems[80]. Several synthetic approaches towards Ni MNPs have been developed, including thermal decomposition[81–83], chemical reduction[15,17,84], reverse micelle[85] and polyol reduction[86]. The thermal decomposition method typically results in monodispersed Ni MNPs as shown in Figure 4.5[81]. Chemical reduction using unimolecular templates such as dendrimers is also capable of preparing well-defined small Ni MNPs[84].

In terms of biomedical applications, a good attempt at direct conjugation of Ni MNPs with an enzyme was published recently, and the enzyme used was bovine pancreatic α-chymotrypsin (CHT)[79]. The Ni–CHT nanobioconjugates of diameter of 2.5 nm exhibited fluorescence in the visible region (peak at 500 nm), as shown in Figure 4.6a and c. The fluorescence of the NPs arises due to their molecule-like electronic structure. The origin of the emission is not very clear; however, it could be due to the recombination of the excited electrons from excited states in the *sp* band with the holes in the low-lying *d* band (interband transition). It should be noted that the excitation spectra of the Ni–CHT nanobioconjugates show peaks centred at 320 and 360 nm (Figure 4.6c). Thus, the SPR peak centred at 417 nm (Figure 4.6d, inset) is not responsible for the luminescence. Furthermore, Ni–CHT nanobioconjugates exhibited superparamagnetic properties at

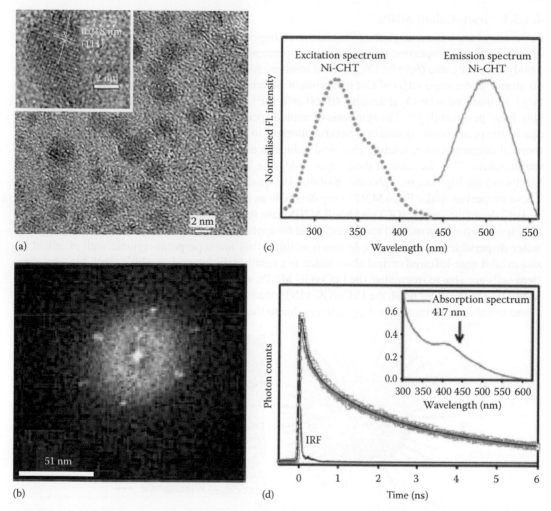

FIGURE 4.6 (a) TEM image of dialysed Ni–CHT nanobioconjugates synthesised *via* solution chemical reduction using NaBH$_4$ as reductant. (b) Fast Fourier transform (FFT) pattern of the Ni–CHT nanobioconjugate. (c) Excitation and emission spectra of Ni–CHT nanobioconjugates. (d) Fluorescence decay of Ni–CHT nanobioconjugate (λ_{ex} = 375 Å) monitored at 500 nm. Inset: the absorbance spectrum of Ni–CHT nanobioconjugates. IRF stands for instrument response function. (From Verma, P., Giri, A., Thanh, N.T.K., Tung, L.D., Mondal, O., Pal, M., and Pal, S. K., Superparamagnetic fluorescent nickel-enzyme nanobioconjugates: Synthesis and characterization of a novel multifunctional biological probe, *J. Mater. Chem.*, 20, 3722–3728, 2010. Reproduced by permission of The Royal Society of Chemistry.)

room temperature and have a magnetocrystalline anisotropy of 6.96×10^5 erg·cm^{-3}. Such a multifunctional superparamagnetic, fluorescent and biologically active nanobioconjugate may be of relevance in the MNP-based diagnostic and therapeutic applications.

4.3.3 BIMETALLIC ALLOY MNPs

Bimetallic alloy MNPs are more chemically stable than monometallic MNPs. Furthermore, the magnetocrystalline anisotropy energy of alloy MNPs is enhanced due to spin–orbit coupling and hybridisation between *d* states of incorporated metals[87]. In alloy MNPs, the magnetic properties are not only dependent on size but also strongly dependent on the composition and crystal structure. Therefore, requirements for alloy MNPs are uniformity in size, shape and composition.

4.3.3.1   Iron–Cobalt MNPs

FeCo alloys are known as permendur and are an important soft magnetic material because of their unique magnetic properties; they exhibit large permeability and very high M_S. Fe and Co can form a body-centred-cubic (*bcc*) Fe_xCo_{100-x} solid solution. Although FeCo MNPs reach a maximum M_S at 40 atomic percentage (at%) of Co, a maximum coercivity (H_C) at 80 at% of Co and H_C is twice as large as those of γ-Fe_2O_3 at around 40–60 at% Co[88]. Equiatomic compositions offer a considerably large permeability[89]. The synthesis of monodispersed FeCo MNPs remains a challenging task due to the poor chemical stability. Several attempts to synthesise FeCo MNPs have been reported; thermal decomposition, solution chemical reduction and polyol reduction methods are reported in the literature[14,90–93]. Because of their superior M_S (see Table 4.2) compared with other MNPs, FeCo MNPs exhibit high magnetophoretic mobility in liquid medium under an external magnetic field[94]. These properties make FeCo MNPs very desirable as nanotracers or nanocarriers.

FeCo/graphite carbon (GC) core/shell MNPs have recently been synthesised by a scalable chemical vapour deposition method and investigated for applications in MRI[95–97]. The FeCo/GC MNPs are water dispersible and have multiple functionalities: they are superparamagnetic with excellent M_S and exhibit near-infrared optical absorbance as a result of the layered graphitic shell. Mesenchymal stem cells are able to internalise the FeCo/GC MNPs, showing high negative-contrast enhancement in MRI[95]. On the other hand, the FeCo/GC MNPs loaded with doxorubicin (DOX) (Figure 4.7) were found to enhance intracellular drug delivery due to their high M_S[97].

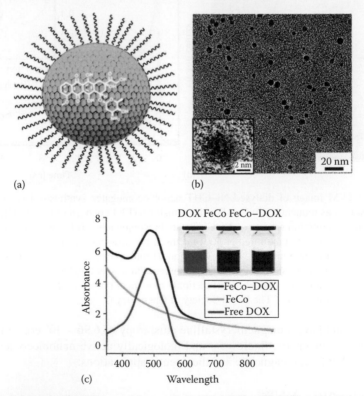

FIGURE 4.7   (a) Schematic of DOX π-stacking on FeCo/GC MNPs. The FeCo core (shown in green) is surrounded by a single layer of graphite. DOX (shown in white) loads noncovalently on the graphite surface of FeCo/GC MNPs. (b) TEM images of 4 nm FeCo/GC–DOX conjugates. Inset: high resolution transmission electron microscopy (HRTEM) image of a single FeCo/GC–DOX MNP. (c) UV-vis absorbance spectra of free DOX, FeCo/GC–DOX (FeCo–DOX) or FeCo/GC (FeCo). Both free DOX and FeCo/GC–DOX show a characteristic peak around 490 nm. Inset: suspensions of DOX, FeCo/GC–DOX and FeCo/GC. (Reprinted with permission from Sherlock, S.P., Tabakman, S.M., Xie, L., and Dai, H., Photothermally enhanced drug delivery by ultrasmall multifunctional FeCo/graphitic shell nanocrystals, *ACS Nano*, 5, 1505–1512, 2011. Copyright 2011 American Chemical Society.)

4.3.3.2 Nickel–Cobalt MNPs

The NiCo alloy system is important because of its magnetostriction and zero magnetocrystalline anisotropy at a composition of $Ni_{80}Co_{20}$[98]. The NiCo MNPs are of particular interest as MRI contrast agents. Recently, a simple synthesis route for single-walled flux-closure Ni_7Co_3 nanorings coated with poly(*N*-vinyl-2-pyrrolidone) (PVP) has been reported[99]. Ni_7Co_3 alloy–Au composites (Figure 4.8a and b) were then synthesised using the Ni_7Co_3 nanorings as both reducing agent and sacrificial template[100]. The Ni_7Co_3 alloy–Au composites exhibit an enhancement of contrast in the T_2-weighted MRI (Figure 4.8e).

4.3.3.3 Iron–Platinum MNPs

FePt MNPs have received much attention in recent years due to their potential applications in many fields including high-density magnetic storage[101], catalysis[102] and biomedicine[29,103–112]. The chemically synthesised FePt MNPs have a chemically disordered *fcc* structure in which Fe and Pt atoms are randomly arranged and are superparamagnetic. Post-synthesis thermal annealing, typically at a temperature over 580 °C, transforms the crystalline structure from *fcc* to chemically ordered face-centred tetragonal (*fct*) ($L1_0$) phase. $L1_0$-phase FePt MNPs have a large magnetocrystalline anisotropy (see Table 4.2) and, thus, exhibit large coercivity at room temperature, even when their size is as small as several nanometres. For biomedical applications, such as magnetic separation[29,103–106], MRI[107–109] and magnetic hyperthermia[110,111], the superparamagnetic *fcc* phase FePt MNP is also an attractive material. The typical chemical methods for the synthesis of FePt MNPs are thermal decomposition[101,113,114], reverse micelle[18], solution chemical reduction[115] and polyol reduction[116,117]. Thermal decomposition of iron pentacarbonyl [$Fe(CO)_5$] and reduction of platinum acetylacetonate [$Pt(acac)_2$] in the presence of 1,2-hexadecanediol is a commonly used synthetic method to obtain monodisperse FePt MNPs[101]. To synthesise more uniform FePt MNPs (Figure 4.9) in terms of both size and composition distributions, other Fe precursors such as Fe^{3+} precursors[113] and iron(III) ethoxide [$Fe(OEt)_3$][114] have been utilised. However, the synthesis of FePt MNPs with larger size (larger than 20 nm) keeping an equiatomic composition still remains challenging.

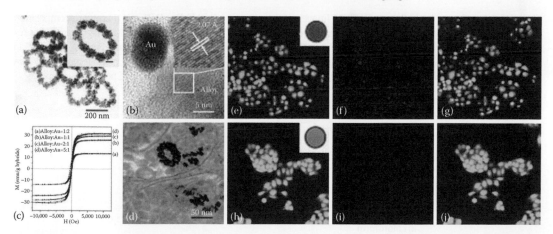

FIGURE 4.8 (a, b) TEM and HRTEM images of Ni_7Co_3 alloy–Au nanoring (5:1). (c) Hysteresis loops of Ni_7Co_3 alloy–Au nanoring formed by loading Ni_7Co_3 with various concentration of Au. (d) TEM image indicating the dispersion of the magnetic Ni_7Co_3–Au hybrid NPs inside HeLa cells transfected with green fluorescent protein (GFP)-expressing vector were cultured with Ni_7Co_3–Au (1:1); (e) GFP image; (f) two-photon fluorescence (TPF) and (g) overlaid images. Controls without any nanomaterials: (h) GFP image; (i) TPF and (j) overlaid images. The insets in (e) and (h) are the corresponding MRI images. (Reproduced with kind permission from Springer Science + Business Media: *Adv. Funct. Mater.*, Magnetic alloy nanorings loaded with gold nanoparticles: synthesis and applications as multimodal imaging contrast agents, 20, 2010, 3701–3706, Lu, Y., Shi, C., Hu, M.-J., Xu, Y.-J., Lu, L., Wen, L.-P., Zhao, Y., Xu, W.-P., and Yu, S.-H., Copyright Wiley-VCH Verlag GmbH & Co. KGaA.)

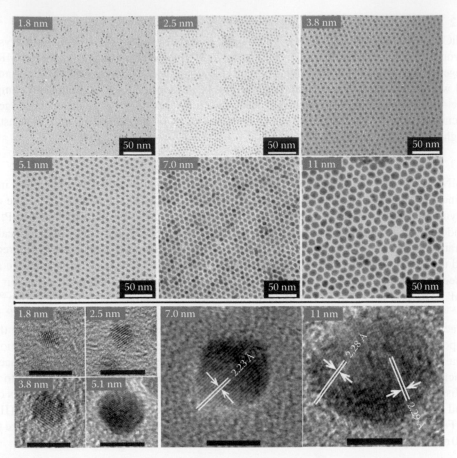

FIGURE 4.9 Top and middle panels: TEM images of the as-synthesised FePt MNPs *via* thermal decomposition of stable Fe^{3+} salts in the presence of fatty acids. Bottom panel: the corresponding HRTEM images (scale bar for HRTEM: 5 nm). (Reprinted with permission from Zhao, F., Rutherford, M., Grisham, S.Y., and Peng, X., Formation of monodisperse FePt alloy nanocrystals using air-stable precursors: Fatty acids as alloying mediator and reductant for Fe^{3+} precursors, *J. Am. Chem. Soc.*, 131, 5350–5358, 2009. Copyright 2009 American Chemical Society.)

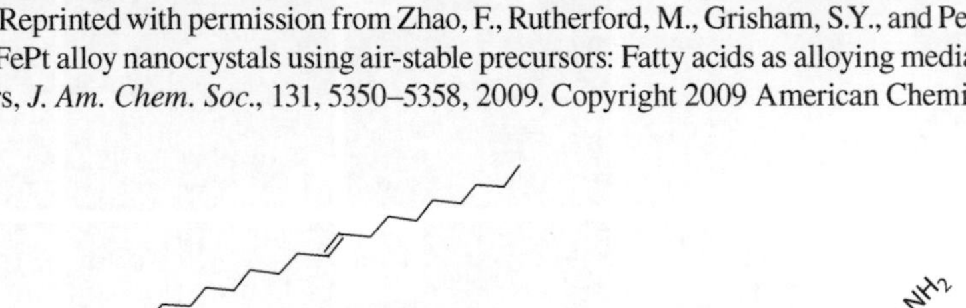

FIGURE 4.10 Schematic illustration of the ligand exchange from OA to AET, cysteamine. (Reprinted from *J. Magn. Magn. Mater.*, 320, Tanaka, Y. and Maenosono, S., Amine-terminated water-dispersible FePt nanoparticles, L121–L124, Copyright 2008, with permission from Elsevier.)

The surfaces of FePt MNPs can be modified with various kinds of surface capping agents including organic ligands and inorganic shells. For example, cysteamine-capped water-dispersible FePt MNPs are synthesised, as shown in Figure 4.10[40]. In this system, amino groups exposed to the outside environment enable MNPs to be stable in acidic conditions and are responsible for bioconjugation. On the other hand, vancomycin (Van) is covalently linked to FePt MNPs *via* cysteamide and can bind to the terminal peptide, D-Ala-D-Ala, on the cell wall of a Gram-positive bacterium *via* hydrogen bonds (Figure 4.11)[105]. The FePt–Van conjugate exhibits high sensitivity to bacteria whose cell walls express D-Ala-D-Ala as the terminal peptides and capture those bacteria at a concentration as low as ~4 cfu mL^{-1} (cfu: colony forming units). FePt MNPs have also been functionalised with mercaptoalkanoic acids and used as a general agent to bind histidine-tagged proteins forming a conjugate that is promising for instant and sensitive detection of pathogens at ultralow concentrations[106]. Recent studies have demonstrated that FePt MNPs could serve as a dual modality contrast agent for computed tomography (CT)/MRI molecular imaging[108]. The cysteamine-coated and cysteamine–silica-coated FePt MNPs have been confirmed to be stronger T_2 contrast agents than

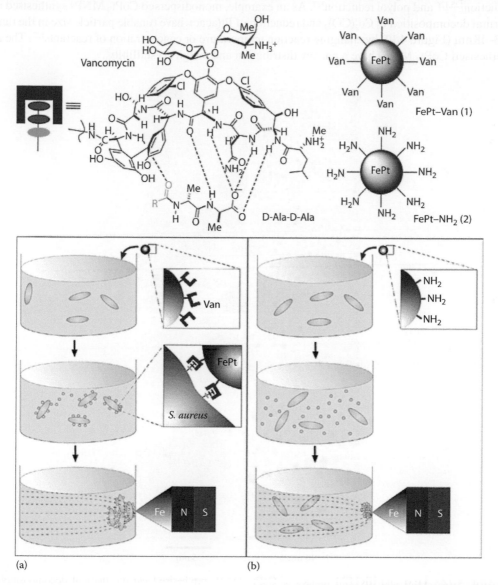

FIGURE 4.11 Top panel: the structures of Van, the FePt–Van MNPs (1) and the control, FePt–NH$_2$ MNPs (2). Bottom panel: illustration of the capture of bacteria by Van-conjugated MNPs *via* a plausible multivalent interaction (a) and the corresponding control experiment (b). (Reprinted with permission from Gu, H., Ho, P.-L., Tsang, K.W.T., Wang, L., and Xu, B., Using biofunctional magnetic nanoparticles to capture vancomycin-resistant *Enterococci* and other gram-positive bacteria at ultralow concentration, *J. Am. Chem. Soc.*, 125, 15702–15703, 2003. Copyright 2003 American Chemical Society.)

commercial SPIO-based contrast agents[109]. In association with the potential for radiofrequency-induced hyperthermia, FePt MNPs also hold great potential as a future theranostic platform.

4.3.3.4 Cobalt–Platinum MNPs

CoPt alloys exhibit various crystal phases including fcc, $L1_0$ (CoPt), $L1_1$(Co$_3$Pt) and $L1_2$ (CoPt$_3$)[118]. Among these phases, the $L1_0$ structure with equiatomic composition has the highest magnetocrystalline anisotropy ($K \sim 4.9 \times 10^7$ erg·m^{-3}). A general method applied for the synthesis of CoPt MNPs involves the thermal decomposition of Co(II) or Co(0) precursors and the reduction of Pt(acac)$_2$ in the presence of capping ligands[119–124]. Alternative ways to synthesise CoPt MNPs are chemical reduction[125,126] and polyol reduction[127]. As an example, monodispersed CoPt$_3$ MNPs synthesised via thermal decomposition of Co$_2$(CO)$_8$ and reduction of Pt(acac)$_2$ have tunable particle size in the range of 3–18 nm (Figure 4.12) by changing reaction temperature or concentration of reactants[122]. The as-synthesised CoPt$_3$ MNPs have a narrow distribution and a high crystallinity.

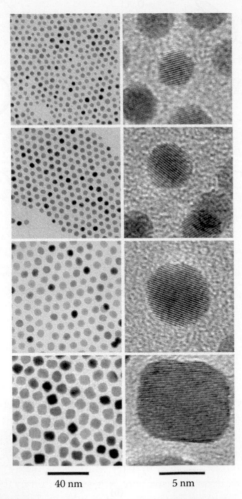

40 nm   5 nm

FIGURE 4.12 TEM and HRTEM images of CoPt$_3$ MNPs synthesised via the thermal decomposition of Co$_2$(CO)$_8$ and reduction of Pt(acac)$_2$ in the presence of 1-adamantanecarboxylic acid and hexadecylamine as stabilising agents. The mean sizes are 3.7, 4.9, 6.3 and 9.3 nm (top to bottom), corresponding reaction temperatures 220°C, 200°C, 170°C and 145°C, respectively. (Reprinted with permission from Shevchenko, E.V., Talapin, D.V., Schnablegger, H., Kornowski, A., Festin, Ö., Svedlindh, P., Haase, M., and Weller, H., Study of nucleation and growth in the organometallic synthesis of magnetic alloy nanocrystals: The role of nucleation rate in size control of CoPt$_3$ nanocrystals, $J. Am. Chem. Soc.$, 125, 9090–9101, 2003. Copyright 2003 American Chemical Society.)

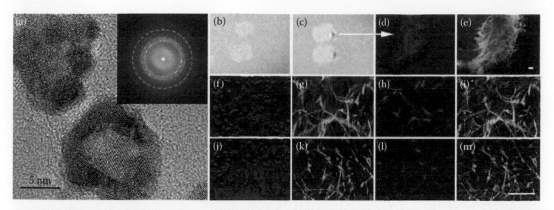

FIGURE 4.13 (a) HRTEM image of CoPt hollow MNPs synthesised in the presence of 0.12 mM cysteine–cysteine–alanine–leucine–asparagine–asparagine (CCALNN) peptide and 0.24 mM *O*-[2-(3-mercaptopropionyl-amino)ethyl]-*O'*-methylpolyethylene glycol (PEG-SH) polymer (M_w = 5000 g · mol^{-1}) (inset: the corresponding FFT image). (b–m) CoPt-labelled NSCs were detected by MRI and maintained multipotency after transplantation into spinal cord slices: two spinal cord slices were recorded simultaneously under magnetic resonance (MR) scanner as two bright elliptic regions adjacent to each other, carrying transplanted NSCs either with or without CoPt NPs labelling (c or b, respectively); MRI of spinal cord slices showed regions of hypointense signal attributable to CoPt-labelled NSCs (c), arrowheads indicate grafted CoPt-labelled NSCs, which were represented by two dark spots in spinal cord slices (c); no hypointense regions were detected by MRI when similar numbers of unlabelled cells were implanted (b); corresponding Hoechst 33342 labelling (d) and anti-nestin immuohistochemical staining (e) were carried out immediately after the MRI procedure on exactly the same spinal cord slices; CoPt-labelled NSCs still maintained multipotency after transplantation, which differentiated into astrocytes (f: Hoechst 33342; g: anti-nestin; h: anti-glial fibrillary acidic protein (GFAP); i: merged) and neurons (j: Hoechst 33342; k: anti-Tuj1; l: anti-GFAP; m: merged) in the spinal cord slice. Bars = 10 μm. (From Lu, L.T., Tung, L.D., Long, J., Fernig, D.G., and Thanh, N.T.K., Facile and green synthesis of stable, water-soluble magnetic CoPt hollow nanostructures assisted by multi-thiol ligands, *J. Mater. Chem.*, 19, 6023–6028, 2009; Meng, X., Seton, H.C., Lu, L.T., Prior, I.A., Thanh, N.T.K., and Song, B., Magnetic CoPt nanoparticles as MRI contrast agent for transplanted neural stem cells detection, *Nanoscale*, 3, 977–984, 2011. Reproduced by permission of The Royal Society of Chemistry.)

Recently, water-dispersible hollow CoPt MNPs were successfully synthesised *via* a simple reduction method in aqueous solution by the Thanh group[44]. In this approach, the presence of the multi-thiol functional group of the ligands is essential for the formation of the hollow MNPs (Figure 4.13a) and allows tuning of the particle size. CoPt hollow NPs are stable in a wide range of pH = 1.5–10 and up to 2 M NaCl for many months. Neural stem cells (NSCs) are a promising treatment for repairing spinal cord injuries as they have the ability to generate tissue, but there is no effective way of monitoring the cells for long periods of time after transplantation. The team labelled NSCs with hollow CoPt NPs, injected them into spinal cord slices and took images of their progress over time[128]. As results, the hollow CoPt MNPs are found to have a significant T_2-shortening ability, and MRI of grafted NSCs labelled with CoPt NPs was demonstrated as a useful tool to evaluate organotypic spinal cord slice models and has potential applications in other biological systems (Figure 4.13b–m).

4.3.4 MANGANESE-BASED MNPS

Manganese was one of the earliest reported examples of paramagnetic MRI contrast agents because of its efficient positive contrast enhancement[129]. The Mn-based contrast agents have been extensively reviewed by Pan et al.[130] Despite the promising earlier studies into the biological application of Mn, Mn-based MNPs have since been less studied. Mn-based MNPs can be classified into three categories: Mn-doped iron oxide (ferrite), manganese oxide and lanthanum manganite (i.e. $La_{1-x}Ag_xMnO_{3+\delta}$, $La_{1-x}Sr_xMnO_3$). Very recently, manganese oxide and lanthanum manganite MNPs have been investigated as novel MNPs for both MRI contrast agents and hyperthermia, these are explained in the following sections.

4.3.4.1   Manganese Oxide MNPs

There are two interesting forms of manganese oxide MNPs: MnO and Mn_3O_4. The Mn_xO_y MNPs have been developed mainly as MRI contrast agents. Biocompatible MnO MNPs conjugated with a tumour-specific antibody were shown to behave as selectively imaging agents for breast cancer cells in the metastatic brain tumour[131]. Hollow Mn_3O_4 MNPs were prepared by further oxidation of MnO MNPs to form $MnO@Mn_3O_4$ core/shell MNPs followed by removal of the MnO core[132]. The hollow Mn_3O_4 MNPs exhibited potential as a dual contrast agent for both T_1- and T_2-weighted MRI. The hollow Mn_3O_4 MNPs were functionalised using 3,4-dihydroxy-L-phenylalanine as an adhesive moiety for targeted cancer detection by MRI and simultaneous delivery of therapeutic siRNA[133]. Silica-coated Mn_3O_4 ($Mn_3O_4@SiO_2$) MNPs were also synthesised and then aminated through silanisation, which enabled further covalent conjugation of fluorescent dye onto the surface[134]. These biocompatible magneto-fluorescent MNPs exhibited potential as positive MRI contrast agents as well as fluorescent labels of cancer cells.

4.3.4.2   Lanthanum Manganite MNPs

Lanthanum manganite MNPs, such as $La_{0.7}Sr_{0.3}MnO_3$ (LSMO) or $La_{1-x}Ag_xMnO_{3+\delta}$, (LAMO), have been developed and are interesting due to their high Curie temperature (T_C). There are few reports available for the synthesis of the lanthanum manganite MNPs in the literature[135–139]. LSMO has a T_C value of 380 K and a large magnetic moment at room temperature[140]. On the other hand, the T_C of LAMO MNPs can be tuned in the range of 314–317 K by adjusting the Ag doping level[139], and this is an important feature for magnetic hyperthermia. In addition, LAMO MNPs are highly stable and their magnetic properties are hardly affected by surrounding media[139]. These features make them promising candidates for MRI contrast agents and programmable nanoheaters for magnetic hyperthermia[138].

4.3.5   Heterostructured MNPs

MNPs composed of different components with different magnetic phases in heterostructures are highly functional systems due to their unique chemical and magnetic properties and have high potential for many applications. This combination of phases results in MNPs with much more chemical stability and changes in magnetic response due to spin exchanged coupling between the phases. For example, the combinations between magnetically hard materials (FePt) and magnetically soft (Fe_3O_4) or antiferromagnetic (MnO) materials in core@shell structures enhance magnetic properties of $FePt@Fe_3O_4$ and FePt@MnO MNPs[141,142]. Particularly, FePt@MnO MNPs with an FePt core size of 3.5 nm in diameter and a tunable MnO shell exhibit a superparamagnetic behaviour. The M_S of the FePt@MnO MNPs with core size of 3.5 nm and shell thickness of 6 nm was 978 emu·cm^{-3}, which is higher than that of uncoated FePt MNPs (553 emu·cm^{-3})[142]. The enhanced M_S is a result of the interaction between magnetic spin of FePt and MnO, which could significantly reduce the thickness of the magnetic dead layer or canted spin layer due to broken symmetry at the surface of the FePt MNPs[143]. Systems with metallic MNP cores and a controllable oxide shell in core@shell nanostructure have attracted much attention for biological application. In addition to the magnetic enhancement, the formation of oxide shells provides much higher chemical stability and a better biocompatibility when compared to uncoated MNPs.

In general, a method for the synthesis of core@shell MNPs with metallic core and corresponding oxide shell is control of the oxidation process of the cores by changing the reaction conditions. The metal-based MNPs such as Fe, Co or Ni are readily oxidised if exposed to air, water or other mild oxidising agents. Co@CoO, $FePt@Fe_3O_4$, $Fe@Fe_3O_4$ and Ni@NiO were synthesised by the controlled oxidation method by using different conditions[73,141,144,145]. Ni@NiO MNPs are known to have good affinity for polyhistidine that is used in the separation and purification of proteins[146]. FePt–Fe$_x$O$_y$ heterodimer MNPs exhibited tunable single-phase-like magnetic behaviour, distinct from that of their individual components[147]. The MNPs have also proven the possibility to reach the proton nuclear relaxivity values comparable or even higher than those achievable with the commercial Endorem contrasting agent, which is an aqueous colloid of SPIO associated with dextran for intravenous administration as a MRI contrast medium[147].

On the basis of yolk–shell nanostructures, which are a special type of core–shell structure with interstitial hollow spaces between the core and shell sections, reported by Alivisatos and co-workers[148], another new designed system of FePt@CoS$_2$ yolk–shell MNPs has been synthesised *via* the Kirkendall effect using FePt MNPs as seeds[149]. These MNPs have been tested for cytotoxicity by using HeLa cells. Figure 4.14 shows the mechanism of cytotoxicity of FePt@CoS$_2$ yolk–shell MNPs against the HeLa cells: the FePt MNPs with size of about 2 nm in diameter have no surface protection group after the formation of CoS$_2$ shells, and thus, they are highly reactive and cytotoxic. The results showed a high anticancer activity of FePt@CoS$_2$ yolk–shell NPs. The high activity is a result of the CoS$_2$ porous shell formed through the Kirkendall effect ensuring the slow diffusion of platinum ions out of the shells and may lead to a new class of candidates for use as anticancer drugs.

Other heterostructured MNPs are composed of a magnetic component and a nonmagnetic component, and these are called magnetic metal MNPs (i.e. FePt–Au heterodimer or Co@Au core–shell nanostructures)[150,151] or magnetic semiconductor MNPs (FePt@CdX, CoPt@CdX or NiPt@CdX

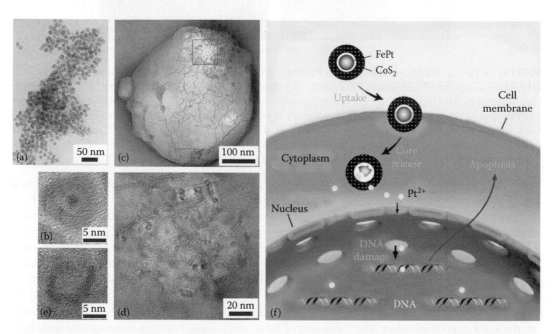

FIGURE 4.14 Transformation of FePt@CoS$_2$ yolk–shell MNPs before and after incubation with HeLa cells. (a) TEM and (b) HRTEM images of the MNPs dispersed in deionised water for more than 3 days but not incubated with HeLa cells. (c) Representative TEM image of mitochondria from HeLa cells incubated with $4.0 \, \mu g \cdot mL^{-1}$ of the MNPs for 3 days. (d) Magnification of the TEM image in (c). (e) HRTEM image of the MNPs in mitochondria of HeLa cells: a few hollow CoS$_2$ nanocrystals without FePt cores are randomly dispersed in the organelles (mitochondria). (f) Illustration of a possible mechanism accounting for the MNPs killing HeLa cells. After cellular uptake, FePt MNPs were oxidised to give Fe^{3+} (omitted for clarity) and Pt^{2+} ions (yellow). The Pt^{2+} ions enter into the nucleus (and mitochondria), bind to DNA and lead to apoptosis of the HeLa cell. (Reprinted with permission from Gao, J., Liang, G., Zhang, B., Kuang, Y., Zhang, X., and Xu, B., FePt@CoS$_2$ yolk-shell nanocrystals as a potent agent to kill HeLa cells, *J. Am. Chem. Soc.*, 129, 1428–1433, 2007. Copyright 2007 American Chemical Society.)

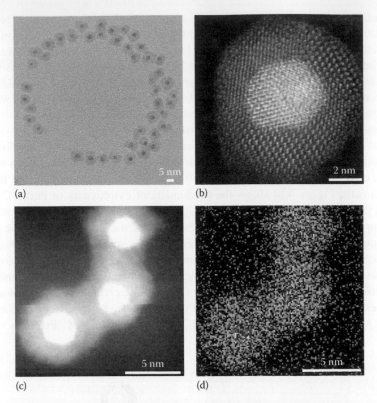

(a)   (b)

(c)   (d)

FIGURE 4.15 (a) TEM image of the FePt@CdSe core–shell NPs. (b,c) high-angle annular dark-field (HAADF)-scanning STEM images of a single and three single FePt@CdSe NPs, respectfully, at different magnifications. (d) Overlay of 2D energy-dispersive x-ray spectroscopy (EDS) elemental mapping of three FePt@CdSe MNPs: Fe K edge is shown in white, Pt M edge is shown in red, Se L edge is shown in green and Cd L edge is shown in blue. (From Trinh, T.T., Mott, D., Thanh, N.T.K., and Maenosono, S., One-pot synthesis and characterization of well defined core@shell structure of FePt@CdSe nanoparticles, *RSC Adv.*, 1, 100, 2011. Reproduced by permission of The Royal Society of Chemistry.)

core–shell MNPs, X: Se, S)[152–155]. Such materials can exhibit superparamagnetic properties as a result of the magnetic cores at room temperature and SPR as a result of an Au or otherwise fluorescent semiconducting shell. As an example, FePt@CdSe MNPs with a well-defined core@shell structure (Figure 4.15) were synthesised *via* one-spot synthesis[155]. The FePt@CdSe MNPs with a core size of about 4.3 nm and shell thickness of about 2.5 nm displayed a fluorescence emission around 600 nm. They exhibited superparamagnetic behaviour at room temperature, and the blocking temperature was about 55 K, which was almost the same as uncoated FePt NPs, while the saturation magnetisation increased from 19 emu · g⁻¹ for the FePt NPs to 23 emu · g⁻¹. The enhanced M_S of FePt@CdSe MNPs can be a result of the passivation of the surface of FePt NPs by the CdSe shell (or the CdO interfacial layer). The formation of these shells could reduce a nonmagnetic shell (surface dead layer), which is formed by the interaction of organic ligands to the surface of FePt NPs, or a canted spin layer due to broken symmetry at the surface. In addition, the exchange coupling may contribute to the increase in M_S. These multifunctional MNPs simultaneously offer magnetic manipulation and optical detection; they are promising and open up new windows in biomedicine[80].

4.4 CONCLUSION AND FUTURE OUTLOOK

Recent advancements in the chemical synthesis and biological applications of n-MNPs have been reviewed. n-MNPs exhibit superior magnetic properties compared to conventional SPIOs and, thus, are expected to be highly utilised in various biomedical applications. In this chapter, we classified the n-MNPs into four main types: monometallic, bimetallic alloy, Mn-based MNPs and heterostructured MNPs. Accompanying the recent advance of nanobiotechnology, various materials have been revisited as desirable candidates of MNPs for biomedical applications, because they exhibit superior magnetic properties including superparamagnetism and much higher saturation magnetisation compared to conventional SPIOs. Despite being extremely reactive and thus readily oxidised, monometallic MNPs demonstrate a high-spin magnetic moment. Bimetallic alloy MNPs are chemically more stable and exhibit higher magnetocrystalline anisotropy energy due to the exchange interaction between their components. Manganese-based oxide MNPs are expected to be a new class of MNPs that are useful as both MRI contrast agents and heating elements for hyperthermia. Combination of magnetic components with other optically active components in a nanoheterostructure leads to multifunctionality opening a new window for biomedical applications. By replacing SPIOs with n-MNPs, one can push forward the boundaries of this exciting field of magnetic medicine and create entirely new biomedical applications.

ACKNOWLEDGEMENTS

We thank Derrick Mott, Luke Green, Valadimir L. Kolesnichenko, Yuri Barnakov, Marie-Hélène Delville and Etienne Duguet for useful discussion. Nguyễn Thị Kim Thanh thanks the Royal Society for her University Research Fellowship.

REFERENCES

1. Gupta, A. K. and Gupta, G. 2005. Synthesis and surface engineering of iron oxide nanoparticles for biomedical applications. *Biomaterials*, 26, 3995–4021.
2. Laurent, S., Forge, D., Port, M., Roch, A., Robic, C., Elst, L. V. and Muller, R. N. 2008. Magnetic iron oxide nanoparticles: Synthesis, stabilization, vectorization, physicochemical characterizations, and biological applications. *Chem. Rev.*, 108, 2064–2110.
3. Veiseh, O., Gunn, J. W. and Zhang, M. 2010. Design and fabrication of magnetic nanoparticles for targeted drug delivery and imaging. *Adv. Drug Deliv. Rev.*, 62, 284–304.
4. Sun, C., Lee, J. S. H. and Zhang, M. 2008. Magnetic nanoparticles in MR imaging and drug delivery. *Adv. Drug Deliv. Rev.*, 60, 1252–1265.
5. Gould, P. 2006. Nanomagnetism shows in vivo potential. *Nanotoday*, 1, 34–39.
6. Wohlfarth, E. P. 1980. *Ferromagnetic Materials*, Vol. 1. North-Holland Publishing Company, Amsterdam, the Netherlands.
7. Benenson, W., Harris, J. W., Stocker, H. and Lutz, H. 2002. *Handbook of Physics*. Springer-Verlag, New York.
8. Weller, D., Moser, A., Folks, L., Best, M. E, Lee, W., Toney, M. F., Schwickert, M., Thiele, J.-U. and Doerner, M. F. 2000. High K_u materials approach to 100 Gbits/in^2. *IEEE Trans. Magn.*, 36, 10–15.
9. Gubin, S. P. 2009. *Magnetic Nanoparticles*. Wiley-VCH Verlag GmbH and Co. KGaA, Weinheim, Germany.
10. Pankhurst, Q. A., Connolly, J., Jones, S. K. and Dobson, J. 2003. Applications of magnetic nanoparticles in biomedicine. *J. Phys. D Appl. Phys.*, 36, R167–R181.
11. Maenosono, S. and Saita, S. 2006. Theoretical assessment of FePt nanoparticles as heating elements for magnetic hyperthermia. *IEEE Trans. Magn.*, 42, 1638–1642.
12. Cushing, B. L., Kolesnichenko, V. L. and O'Connor, C. J. 2004. Recent advances in the liquid-phase syntheses of inorganic nanoparticles. *Chem. Rev.*, 104, 3893–3946.
13. Lu, A.-H., Salabas, E. L. and Schüth, F. 2007. Magnetic nanoparticles: synthesis, protection, functionalization, and application. *Angew. Chem. Int. Ed.*, 46, 1222–1244.
14. Robinson, I., Zacchini, S., Tung, L. D., Maenosono, S. and Thanh, N. T. K. 2009. Synthesis and characterization of magnetic nanoalloys from bimetallic carbonyl clusters. *Chem. Mater.*, 13, 3021–3026.
15. Li, Y. D., Li, L. Q., Liao, H. W. and Wang, H. R. 1999. Preparation of pure nickel, cobalt, nickel–cobalt and nickel–copper alloys by hydrothermal reduction. *J. Mater. Chem.*, 9, 2675–2677.
16. Lu, L. T., Robinson, I., Tung, L. D., Tan, B., Long, J., Cooper, A. I., Fernig, D. J. and Thanh, N. T. K. 2008. Size and shape control for water-soluble magnetic cobalt nanoparticles using polymer ligands. *J. Mater. Chem.*, 18, 2453–2458.
17. Glavee, G. N., Klabunde, K. J., Sorensen, C. M. and Hadjipanayis, G. C. 1994. Borohydride reduction of nickel and copper ions in aqueous and nonaqueous media. Controllable chemistry leading to nanoscale metal and metal boride particles. *Langmuir*, 10, 4726–4730.

18. Yan, Q., Purkayastha, A., Kim, T., Kröger, R., Bose, A. and Ramanath, G. 2006. Synthesis and assembly of monodisperse high-coercivity silica capped FePt nanomagnets of tunable size, composition, and thermal stability from microemulsions. *Adv. Mater.*, 18, 2569–2573.

19. Suslick, K. S. 1990. Sonochemistry. *Science,* 247, 1439–1445.

20. Flint, E. B. and Suslick, K. S. 1991. The temperature of cavitation. *Science*, 253, 1397–1399.

21. Suslick, K. S., Cline, Jr. R. E. and Hammerton, D. A. 1986. The sonochemical hot spot. *J. Am. Chem. Soc.*, 108, 5641–5642.

22. Suslick, K. S., Hyeon, T. and Fang, M. 1986. Nanostructured materials generated by high-intensity ultrasound: Sonochemical synthesis and catalytic studies. *Chem. Mater.*, 8, 2172–2179.

23. Bang, J. H. and Suslick, K. S. 2010. Applications of ultrasound to the synthesis of nanostructured materials. *Adv. Mater.*, 22, 1039–1059.

24. Huber, D. L. 2005. Synthesis, properties, and applications of iron nanoparticles. *Small*, 1, 482–501.

25. Wang, X., Zhuang, J., Peng, Q. and Li, Y. 2005. A general strategy for nanocrystal synthesis. *Nature*, 437, 121–124.

26. Tsuji, M., Hashimoto, M. and Tsuji, T. 2002. Fast preparation of nano-sized nickel particles under microwave irradiation without using catalyst for nucleation. *Chem. Lett.*, 31, 1232–1233.

27. Yamauchi, T., Tsukahara, Y., Yamada, K., Sakata, T. and Wada, Y. 2011. Nucleation and growth of magnetic Ni–Co (core–shell) nanoparticles in a one-pot reaction under microwave irradiation. *Chem. Mater.*, 23, 75–84.

28. Liu, X., Guan, Y., Ma, Z. and Liu, H. 2004. Surface modification and characterization of magnetic polymer nanospheres prepared by miniemulsion polymerization. *Langmuir*, 20, 10278–10282.

29. Hong, R., Fischer, N. O., Emrick, T. and Rotello, V. M. 2005. Surface PEGylation and ligand exchange chemistry of FePt nanoparticles for biological applications. *Chem. Mater.*, 17, 4617–4621.

30. Sahoo, Y., Pizem, H., Fried, T., Golodnitsky, D., Burstein, L., Sukenik, C. N. and Markovich, G. 2001. Alkyl phosphonate/phosphate coating on magnetite nanoparticles: A comparison with fatty acids. *Langmuir*, 17, 7907–7911.

31. Kobayashi, Y., Horie, M., Konno, M., Rodriguez-Gonzalez, B. and Liz-Marzan, L. M. 2003. Preparation and properties of silica-coated cobalt nanoparticles. *J. Phys. Chem. B*, 107, 7420–7425.

32. Lu, A.-H., Li, W., Matoussevitch, N., Spliethoff, B., Bönnemann, H. and Schüth, F. 2005. Highly stable carbon-protected cobalt nanoparticles and graphite shells. *Chem. Commun.*, 41, 98–100.

33. Sobal, N. S., Hilgendorff, M., Moehwald, H., Giersig, M., Spasova, M., Radetic, T. and Farle, M. 2002. Synthesis and structure of colloidal bimetallic nanocrystals: The non-alloying system Ag/Co. *Nano Lett.*, 41, 2, 621–624.

34. Lin, J., Zhou, W., Kumbhar, A., Wiemann, J., Fang, J., Carpenter, E. E. and O'Connor, C. J. 2001. Gold-coated iron (Fe@Au) nanoparticles: Synthesis, characterization, and magnetic field-induced self-assembly. *J. Solid State Chem.*, 159, 26–31.

35. Boyen, H.-G., Kästle, G., Zürn, K., Herzog, T., Weigl, F., Ziemann, P., Mayer, O., Jerome, C., Moller, M., Spatz, J. P., Garnier, M. G. and Oelhafen, P. 2003. A micellar route to ordered arrays of magnetic nanoparticles: From size-selected pure cobalt dots to cobalt–cobalt oxide core–shell systems. *Adv. Funct. Mater.*, 13, 359–364.

36. Carpenter, E. E., Calvin, S., Stroud, R. M. and Harris, V. G. 2003. Passivated iron as core–shell nanoparticles. *Chem. Mater.*, 15, 3245–3246.

37. Xu, C. J. and Sun, S. 2007. Monodisperse magnetic nanoparticles for biomedical applications. *Polym. Int.*, 56, 821–826.

38. Qin, Y. and Fichthorn, K. A. 2003. Molecular-dynamics simulation of forces between nanoparticles in a Lennard-Jones liquid. *J. Chem. Phys.* 119, 9745–9754.

39. Salgueiriño-Maceira, V., Liz-Marzán, L. M. and Farle, M. 2004. Water-based ferrofluids from Fe_xPt_{1-x} nanoparticles synthesized in organic media. *Langmuir*, 20, 6946–6950.

40. Tanaka, Y. and Maenosono, S. 2008. Amine-terminated water-dispersible FePt nanoparticles. *J. Magn. Magn. Mater.*, 320, L121–L124.

41. Bagaria, H. G., Ada, E. T., Shamsuzzoha, M., Nikles, D. E. and Johnson, D. T. 2006. Understanding mercapto ligand exchange on the surface of FePt nanoparticles. *Langmuir*, 22, 7732–7737.

42. Latham, A. H. and Williams, M. E. 2006. Versatile routes toward functional, water-soluble nanoparticles via trifluoroethylester–PEG–thiol ligands. *Langmuir*, 22, 4319–4326.

43. Robinson, I., Alexander, C., Tung, L. D., Fernig, D. G. and Thanh, N. T. K. 2009. Fabrication of water-soluble magnetic nanoparticles using thermo-responsive polymers. *J. Magn. Magn. Mater.*, 321, 1421–1423.

44. Lu, L. T., Tung, L. D., Long, J., Fernig, D. G. and Thanh, N. T. K. 2009. Facile and green synthesis of stable, water-soluble magnetic CoPt hollow nanostructures assisted by multi-thiol ligands. *J. Mater. Chem.*, 19, 6023–6028.

45. Paulus, P. M., Bönnemann, H., van der Kraan, A. M., Luis, F., Sinzig, J. and de Jongh, L. J. 1999. Magnetic properties of nanosized transition metal colloids: The influence of noble metal coating. *Eur. Phys. J. D*, 9, 501–504.

46. Hormes, J., Modrow, H., Bönnemann, H. and Kumar, C. S. S. R. 2005. The influence of various coatings on the electronic, magnetic, and geometric properties of cobalt nanoparticles. *J. Appl. Phys.*, 97, 10R102(1–6).

47. van Leeuwen, D. A., van Ruitenbeek, J. M., de Jongh, L. J., Ceriotti, A., Pacchioni, G., Häberlen, O. D. and Rösch, N. 1994. Quenching of magnetic moments by ligand–metal interactions in nanosized magnetic metal clusters. *Phys. Rev. Lett.*, 73, 1432–1435.

48. Wu, X. W., Liu, C., Li, L., Jones, P., Chantrell, R. W. and Weller, D. 2004. Nonmagnetic shell in surfactant-coated FePt nanoparticles. *J. App. Phys.*, 95, 6810–6812.

49. Tanaka, Y., Saita, S. and Maenosono, S. 2008. Influence of surface ligands on saturation magnetization of FePt nanoparticles. *Appl. Phys. Lett.*, 92, 093117(1–3).

50. Trinh, T. T., Ozaki, T. and Maenosono, S. 2011. Influence of surface ligands on the electronic structure of Fe-Pt clusters: A density functional theory study. *Phys. Rev. B*, 83, 104413(1–10).

51. Mohn, P. 2006. *Magnetism in the Solid State*. Springer, Berlin, Germany.

52. Krishnan, K. M. 2010. Biomedical nanomagnetics: A spin through possibilities in imaging, diagnostics, and therapy. *IEEE Trans. Magn.*, 46, 2523–2558.

53. Gutfleisch, O., Lyubina, J., Müller, K.-H. and Schultz, L. 2005. FePt hard magnets. *Avd. Eng. Mater.*, 7, 208–212.

54. Griffiths, C., Ohoro, M. and Smith, T. 1979. The structure, magnetic characterization, and oxidation of colloidal iron dispersions. *J. Appl. Phys.* 50, 7108–7115.

55. Pei, W., Kakibe, S., Ohta, I. and Takahashi, M. 2005. Controlled monodisperse Fe nanoparticles synthesized by chemical method. *IEEE Trans. Magn.*, 41, 3391–3393.

56. Carpenter, E. E. 2001. Iron nanoparticles as potential magnetic carriers. *J. Magn. Magn. Mater.*, 225, 17–20.

57. Chen, M., Yamamuro, S., Farrell, D. and Majetich, S. A. 2003. Gold-coated iron nanoparticles for biomedical applications. *J. Appl. Phys.* 93, 7551–7553.

58. Fernández-Pacheco, R., Marquina, C., Valdivia, J. G., Gutiérrez, M., Romeroc, M. S., Cornudella, R., Laborda, A., Viloria, A., Higuera, T., García, A., de Jalón, J. A. G. and Ibarra, M. R. 2007. Magnetic nanoparticles for local drug delivery using magnetic implants. *J. Magn. Magn. Mater.*, 311, 318–322.

59. Fernández-Pacheco, R., Ibarra, M. R., Valdivia, J. G., Marquina, C., Serrate, D., Romero, M. S., Gutiérrez, M. and Arbiol, J. 2005. Carbon coated magnetic nanoparticles for local drug delivery using magnetic implants. *NanoBiotechnology*, 1, 300–303.

60. Kitakami, O., Satao, H., Shimada, Y., Sato, F. and Tanaka, M. 1997. Size effect on the crystal phase of cobalt fine particles. *Phys. Rev. B*, 56, 13849–13854.

61. Dinega, D. P. and Bawendi, M. G. 1999. A solution-phase chemical approach to a new crystal structure of cobalt. *Angew. Chem. Int. Ed.*, 38, 1788–1791.

62. Yang, H. T., Shen, C. M., Su, Y. K., Yang, T. Z., Gao, H. J. and Wang, Y. G. 2003. Self-assembly and magnetic properties of cobalt nanoparticles. *Appl. Phys. Lett.*, 82, 4729–4731.

63. Gambardella, P., Rusponi, S., Veronese, M., Dhesi, S. S., Grazioli, C., Dallmeyer, A., Cabria, I., Zeller, R., Dederichs, P. H., Kern, K., Carbone, C. and Brune, H. 2003. Giant magnetic anisotropy of single cobalt atoms and nanoparticles. *Science*, 300, 1130–1133.

64. Puntes, V. F. and Krishnan, K. M. 2001. Synthesis, structural order and magnetic behavior of self-assembled ε-Co nanocrystal arrays. *IEEE Trans. Mag.*, 37, 2210–2212.

65. Puntes, V. F., Krishan, K. M. and Alivisatos, A. P. 2001. Colloidal nanocrystal shape and size control: The case of cobalt. *Science*, 291, 2115–2117.

66. Puntes, V. F., Zanchet, D., Erdonmez, C. K. and Alivisatos, A. P. 2002. Synthesis of hcp-Co nanodisks. *J. Am. Chem. Soc.*, 124, 12874–12880.

67. Gibson, C. P. and Putzer, K. J. 1995. Synthesis and characterization of anisometric cobalt nanoclusters. *Science*, 267, 1338–1340.

68. Petit, C., Taleb, A. and Pileni, M.-P. 1998. Self-organization of magnetic nanosized cobalt particles. *Adv. Mater.*, 10, 259–261.

69. Legrand, J., Ngo, A.-T., Petit, C. and Pileni, M.-P. 2001. Domain shapes and superlattices made of cobalt nanocrystals. *Adv. Mater.*, 13, 58–62.

70. Vaucher, S., Li, J. M., Dujardin, E. and Mann, S. 2002. Molecule-based magnetic nanoparticles: Synthesis of cobalt hexacyanoferrate, cobalt pentacyanonitrosylferrate, and chromium hexacyanochromate coordination polymers in water-in-oil microemulsions. *Nano Lett.*, 2, 225–229.

71. Glavee, G. N., Klabunde, K. J., Sorensen, C. M. and Hadjipanayis, G. C. 1993. Borohydride reduction of cobalt ions in water. Chemistry leading to nanoscale metal, boride, or borate particles. *Langmuir*, 9, 162–169.

72. Poul, L., Jouini, N. and Fiévet, F. 2000. Layered hydroxide metal acetates (metal = zinc, cobalt, and nickel): Elaboration via hydrolysis in polyol medium and comparative study. *Chem. Mater.*, 12, 3123–3132.

73. Peng, D. L., Sumiyama, K., Konno, T. J., Hihara, T. and Yamamuro, S. 1999. Characteristic transport properties of CoO-coated monodispersive Co cluster assemblies. *Phys. Rev. B*, 60, 2093–2100.

74. Thanh, N. T. K., Puntes, V. F., Tung, L. D. and Fernig, D. G. 2005. Peptides as capping ligands for in situ synthesis of water soluble Co nanoparticles for bioapplications. *J. Phys. Conf. Ser.*, 17, 70–76.

75. Robinson, I., Alexander, C., Lu, L. T., Tung, L. D., Fernig, D. G. and Thanh, N. T. K. 2007. One-step synthesis of monodisperse water-soluble "dual-responsive" magnetic nanoparticles. *Chem. Commun.*, 44, 4602–4604.

76. Thanh, N. T. K., Robinson, I. and Tung, L. D. 2007. Magnetic nanoparticles: Synthesis, characterisation and biomedical applications. *Dekker Encycl. Nanosci. Nanotechnol.*, 1, 1–10.

77. Parkes, L. M., Hodgson, R., Tung, L. D., Lu, L. T., Robinson, I., Fernig, D. G. and Thanh, N. T. K. 2008. Cobalt nanoparticles as a novel magnetic resonance contrast agent-relaxivities at 1.5 and 3 Tesla. *Contrast Media Mol. Imaging*, 3, 150–156.

78. Yeshchenko, O. A., Dmitruk, I. M., Alexeenko, A. A. and Dmytruk, A. M. 2008. Optical properties of sol–gel fabricated Ni/SiO$_2$ glass nanocomposites. *J. Phys. Chem. Solids*, 69, 1615–1622.

79. Verma, P., Giri, A., Thanh, N. T. K., Tung, L. D., Mondal, O., Pal, M. and Pal, S. K. 2010. Superparamagnetic fluorescent nickel-enzyme nanobioconjugates: Synthesis and characterization of a novel multifunctional biological probe. *J. Mater. Chem.*, 20, 3722–3728.

80. Gao, J., Zhang, W., Huang, P., Zhang, B., Zhang, X. and Xu, B. 2008. Intracellular spatial control of fluorescent magnetic nanoparticles. *J. Am. Chem. Soc.*, 130, 3710–3711.

81. Winnischofer, H., Rocha, T. C. R., Nunes, W. C., Socolovsky, L. M., Knobel, M. and Zanchet, D. 2008. Chemical synthesis and structural characterization of highly disordered Ni colloidal nanoparticles. *ACS Nano*, 2, 1313–1319.

82. Han, M., Liu, Q., He, J. H., Song, Y., Xu, Z., Zhu, J. 2007. Controllable synthesis and magnetic properties of cubic and hexagonal phase nickel nanocrystals. *Adv. Mater.*, 19, 1096–1100.

83. Cordente, N., Respaud, M., Senocq, F., Casanove, M.-J., Amiens, C. and Chaudret, B. 2001. Synthesis and magnetic properties of nickel nanorods. *Nano Lett.*, 1, 565–568.

84. Mitran, E., Dellinger, B. and McCarley, R. L. 2010. Highly size-controlled, low-size-dispersity nickel nanoparticles from poly(propylene imine) dendrimer–Ni(II) complexes. *Chem. Mater.*, 22, 6555–6563.

85. Chen, D. H. and Wu, S. H. 2000. Synthesis of nickel nanoparticles in water-in-oil microemulsions. *Chem. Mater.*, 12, 1354–1360.

86. Tzitzios, V., Basina, G., Gjoka, M., Alexandrakis, V., Georgakilas, V., Niarchos, D., Boukos, N. and Petridis, D. 2006. Chemical synthesis and characterization of hcp Ni nanoparticles. *Nanotechnology*, 17, 3750–3755.

87. Burkert, T., Eriksson, O., Simak, S. I., Ruban, A. V., Sanyal, B., Nordström, L. and Wills, J. M. 2005. Magnetic anisotropy of L1$_0$ FePt and Fe$_{1-x}$Mn$_x$Pt. *Phys. Rev. B*, 71, 134411(1–8).

88. Li, X. G., Murai, T., Saito, T. and Takahashi, S. 1998. Thermal stability, oxidation behavior and magnetic properties of Fe–Co ultrafine particles prepared by hydrogen plasma–metal reaction. *J. Magn. Magn. Mater.*, 190, 277–288.

89. Sourmail, T. 2005. Near equiatomic FeCo alloys: Constitution, mechanical and magnetic properties. *Prog. Mater. Sci.*, 50, 816–880.

90. Chaubey, G. S., Barcena, C., Poudyal, N., Rong, C., Gao, J., Sun, S. and Liu, J. P. 2007. Synthesis and stabilization of FeCo nanoparticles. *J. Am. Chem. Soc.*, 129, 7214–7215.

91. van Wonterghem, J., Mørup, S., Koch, C. J. W., Charles, S. W. and Wells, S. 1986. Formation of ultra-fine amorphous alloy particles by reduction in aqueous solution. *Nature*, 322, 622–623.

92. Desvaux, C., Amiens, C., Fejes, P., Renaud, P., Respaud, M., Lecante, P., Snoeck, E. and Chaudret, B. 2005. Multimillimetre-large superlattices of air-stable iron–cobalt nanoparticles. *Nat. Mater.*, 4, 750–753.

93. Bönnemann, H., Brand, A. A., Brijoux, W., Hofstadt, H. W., Frerichs, M., Kempter, V., Maus-Kodama, D., Shinoda, K., Sato, K., Konno, Y., Joseyphus, R. J., Motomiya, K., Takahashi, H., Matsumoto, T., Sato, Y., Tohji, K. and Jayadevan, B. 2006. Chemical synthesis of sub-micrometer- to nanometer-sized magnetic FeCo dice. *Adv. Mater.*, 18, 3154–3159.

94. Hütten, A., Sudfeld, D., Ennen, I., Reiss, G., Hachmann, W., Heinzmann, U., Wojczykowski, K., Jutzi, P., Saikaly, W. and Thomas, G. 2004. New magnetic nanoparticles for biotechnology. *J. Biotech.*, 112, 47–63.

95. Seo, W. S., Lee, J. H., Sun, Z., Suzuki, Y., Mann, D., Liu, Z., Terashima, M., Yang, P., Mcconnell, M. V., Nishimura, D. G. and Dai, H. 2006. FeCo/graphitic-shell nanocrystals as advanced magnetic-resonance-imaging and near-infrared agents. *Nat. Mater.*, 5, 971–976.

96. Lee, J. H., Sherlock, S. P., Terashima, M., Kosuge, H., Suzuki, Y., Goodwin, A., Robinson, J., Seo, W. S., Liu, Z. and Luong, R. 2009. High-contrast in vivo visualization of microvessels using novel FeCo/GC magnetic nanocrystals. *Magn. Reson. Med.*, 62, 1497–1509.

97. Sherlock, S. P., Tabakman, S. M., Xie, L. and Dai, H. 2011. Photothermally enhanced drug delivery by ultrasmall multifunctional FeCo/graphitic shell nanocrystals. *ACS Nano*, 5, 1505–1512.

98. Clark, C. A. 1956. The dynamic magnetostriction of nickel–cobalt alloys. *Br. J. Appl. Phys.*, 7, 355–360.

99. Hu, M.-J., Lu, Y., Zhang, S., Guo, S.-R., Lin, B., Zhang, M. and Yu, S.-H. 2008. High yield synthesis of bracelet-like hydrophilic Ni–Co magnetic alloy flux-closure nanorings. *J. Am. Chem. Soc.*, 130, 11606–11607.

100. Lu, Y., Shi, C., Hu, M.-J., Xu, Y.-J., Lu, L., Wen, L.-P., Zhao, Y., Xu, W.-P. and Yu, S.-H. 2010. Magnetic alloy nanorings loaded with gold nanoparticles: Synthesis and applications as multimodal imaging contrast agents. *Adv. Funct. Mater.*, 20, 3701–3706.

101. Sun, S., Murray, C. B., Weller, D., Folks, L. and Moser, A. 2000. Monodisperse FePt nanoparticles and ferromagnetic FePt nanocrystal superlattices. *Science*, 287, 1989–1992.

102. Kim, J., Lee, Y. and Sun, S. 2010. Structurally ordered FePt nanoparticles and their enhanced catalysis for oxygen reduction reaction. *J. Am. Chem. Soc.*, 132, 4996–4997.

103. Gu, H., Ho, P.-L., Tsang, K. W. T., Yu, C.-W. and Xu, B. 2003. Using biofunctional magnetic nanoparticles to capture gram-negative bacteria at an ultra-low concentration. *Chem. Commun.*, 1966–1967.

104. Gu, H., Xu, K., Xu, C. and Xu, B. 2006. Biofunctional magnetic nanoparticles for protein separation and pathogen detection. *Chem. Commun.*, 941–949.

105. Gu, H., Ho, P.-L., Tsang, K. W. T., Wang, L. and Xu, B. 2003. Using biofunctional magnetic nanoparticles to capture vancomycin-resistant *Enterococci* and other gram-positive bacteria at ultralow concentration. *J. Am. Chem. Soc.*, 125, 15702–15703.

106. Xu, C., Xu, K., Gu, H., Zhong, X., Guo, Z., Zheng, R., Zhang, X. and Xu, B. 2004. Nitrilotriacetic acid-modified magnetic nanoparticles as a general agent to bind histidine-tagged proteins. *J. Am. Chem. Soc.*, 126, 3392–3393.

107. Maenosono, S., Suzuki, T. and Saita, S. 2008. Superparamagnetic FePt nanoparticles as excellent MRI contrast agents. *J. Magn. Magn. Mater.*, 320, L79–L83.

108. Chou, S.-W., Shau, Y.-H., Wu, P.-C., Yang, Y.-S., Shieh, D.-B. and Chen, C.-C. 2010. In vitro and in vivo studies of FePt nanoparticles for dual modal CT/MRI molecular imaging. *J. Am. Chem. Soc.*, 132, 13270–13278.

109. Chen, S., Wang, L., Duce, S. L., Brown, S., Lee, S., Melzer, A., Sir Cuschieri, A. and Andre, P. 2010. Engineered biocompatible nanoparticles for in vivo imaging applications. *J. Am. Chem. Soc.*, 132, 15022–15029.

110. Seehra, M. S., Singh, V., Dutta, P., Neeleshwar, S., Chen, Y. Y., Chen, C. L., Chou, S. W. and Chen, C. C. 2010. Size-dependent magnetic parameters of fcc FePt nanoparticles: Applications to magnetic hyperthermia. *J. Phys. D Appl. Phys.*, 43, 145001–145007.

111. Kitamoto, Y. and He, J.-S. 2009. Chemical synthesis of FePt nanoparticles with high alternate current magnetic susceptibility for biomedical applications. *Electrochim. Acta*, 54, 5969–5972.

112. Xu, C., Yuan, Z., Kohler, N., Kim, J., Chung, M. A. and Sun, S. 2009. FePt nanoparticles as an Fe reservoir for controlled Fe release and tumor inhibition. *J. Am. Chem. Soc.*, 131, 15346–15351.

113. Zhao, F., Rutherford, M., Grisham, S. Y. and Peng, X. 2009. Formation of monodisperse FePt alloy nanocrystals using air-stable precursors: Fatty acids as alloying mediator and reductant for Fe^{3+} precursors. *J. Am. Chem. Soc.*, 131, 5350–5358.

114. Saita, S. and Maenosono, S. 2005. Formation mechanism of FePt nanoparticles synthesized via pyrolysis of iron(III) ethoxide and platinum(II) acetylacetonate. *Chem. Mater.*, 17, 6624–6634.

115. Elkins, K. E., Vedantam, T. S., Liu, J. P., Zeng, H., Sun, S., Ding, Y. and Wang, Z. L. 2003. Ultrafine FePt nanoparticles prepared by the chemical reduction method. *Nano Lett.*, 3, 1647–1649.

116. Liu, C., Wu, X., Klemmer, T., Shukla, N., Yang, X., Weller, D., Roy, A. G., Tanase, M. and Laughlin, D. 2004. Polyol process synthesis of monodispersed FePt nanoparticles. *J. Phys. Chem. B*, 108, 6121–6123.

117. Jeyadevan, B., Hobo, A., Urakawa, K., Chinnasamy, C. N., Shinoda, K. and Tohji, K. 2003. Towards direct synthesis of fct-FePt nanoparticles by chemical route. *J. Appl. Phys.*, 93, 7574–7576.

118. Hansen, M. 1958. *Constitution of Binary Alloys. Metallurgy and Metallurgical Engineering Series.* McGraw-Hill, New York.

119. Ely, T. O., Pan, C., Amiens, C., Chaudret, B., Dassenoy, F., Lecante, P., Casanove, M. J., Mosset, A., Respaud, M. and Broto, J. M. 2000. Nanoscale bimetallic Co_xPt_{1-x} particles dispersed in poly(vinylpyrrolidone): Synthesis from organometallic precursors and characterization. *J. Phys. Chem. B*, 104, 695–702.

120. Shevchenko, E. V., Talapin, D. V., Rogach, A. L., Kornowski, A., Haase, M. and Weller, H. J. 2002. Colloidal synthesis and self-assembly of CoPt(3) nanocrystals. *J. Am. Chem. Soc.*, 124, 11480–11485.

121. Chen, M. and Nikles, D. E. 2001. Synthesis of spherical FePd and CoPt nanoparticles. *J. Appl. Phys.*, 91, 8477–8479.

122. Shevchenko, E. V., Talapin, D. V., Schnablegger, H., Kornowski, A., Festin, Ö., Svedlindh, P., Haase, M. and Weller, H. 2003. Study of nucleation and growth in the organometallic synthesis of magnetic alloy nanocrystals: The role of nucleation rate in size control of $CoPt_3$ nanocrystals. *J. Am. Chem. Soc.*, 125, 9090–9101.

123. Wang, Y. and Yang, Y. 2005. Synthesis of CoPt nanorods in ionic liquids. *J. Am. Chem. Soc.*, 127, 5316–5317.

124. Tzitzios, V., Niarchos, D., Gjoka, M., Boukos, N. and Petridis, D. 2005. Synthesis and characterization of 3D CoPt nanostructures. *J. Am. Chem. Soc.*, 127, 13756–13757.

125. Vasquez, Y., Sra, A. K. and Schaak, R. E. 2005. One-pot synthesis of hollow superparamagnetic CoPt nanospheres. *J. Am. Chem. Soc.*, 127, 12504–12505.

126. Jolley, C. C., Uchida, M., Reichhardt, C., Harrington, R., Kang, S., Klem, M. T., Parise, J. B. and Douglas, T. 2010. Size and crystallinity in protein-templated inorganic nanoparticles. *Chem. Mater.*, 22, 4612–4618.

127. Tzitzios, V., Niarchos, D., Gjioka, M., Fidler, J. and Petridis, D. 2005. Synthesis of CoPt nanoparticles by a modified polyol method: Characterization and magnetic properties. *Nanotechnology*, 16, 287–291.

128. Meng, X., Seton, H. C., Lu, L. T., Prior, I. A., Thanh, N. T. K. and Song, B. 2011. Magnetic CoPt nanoparticles as MRI contrast agent for transplanted neural stem cells detection. *Nanoscale*, 3, 977–984.

129. Lauterbur, P. C., Mendonça-Dias, H. M. and Rudin, A. M. 1978. Augmentation of tissue water proton spin-lattice relaxation by in vivo addition of paramagnetic ions. In *Frontiers of Biological Energetics*, eds. P. L. Dutton, J. S. Leigh, and A. Scarpa, Vol. 1, pp. 752–759. New York: Academic Press.

130. Pan, D., Caruthers, S. D., Senpan, A., Schmieder, A. H., Wickline, S. A. and Lanza, G. M. 2011. Revisiting an old friend: Manganese-based MRI contrast agents. *WIREs Nanomed. Nanobiotechnol.*, 3, 162–173.

131. Na, H. B., Lee, J. H., An, K., Park, Y. I., Park, M., Lee, I. S., Nam, D.-H., Kim, S. T., Kim, S.-H., Kim, S.-W., Lim, K.-H., Kim, K.-S., Kim, S.-O. and Hyeon, T. 2007. Development of a T_1 contrast agent for magnetic resonance imaging using MnO nanoparticles. *Angew. Chem. Int. Ed.*, 46, 5397–5401.

132. Shin, J., Anisur, R. M., Ko, M. K., Im, G. H., Lee, J. H. and Lee, I. S. 2009. Hollow manganese oxide nanoparticles as multifunctional agents for magnetic resonance imaging and drug delivery. *Angew. Chem. Int. Ed.*, 48, 321–324.

133. Bae, K. H., Lee, K., Kim, C. and Park, T. G. 2011. Surface functionalized hollow manganese oxide nanoparticles for cancer targeted siRNA delivery and magnetic resonance imaging. *Biomaterials*, 32, 176–184.

134. Yang, H., Zhuang, Y., Hu, H., Du, X., Zhang, C., Shi, X., Wu, H. and Yang, S. 2010. Silica-coated manganese oxide nanoparticles as a platform for targeted magnetic resonance and fluorescence imaging of cancer cells. *Adv. Funct. Mater.*, 20, 1733–1741.

135. Urban, J. J., Ouyang, L., Jo, M.-H., Wang, D. S. and Park, H. 2004. Synthesis of single-crystalline $La_{1-x}Ba_xMnO_3$ nanocubes with adjustable doping levels. *Nano Lett.*, 4, 1547–1550.

136. Uskoković, V., Košak, A. and Drofenik, M. 2006. Preparation of silica-coated lanthanum–strontium manganite particles with designable Curie point, for application in hyperthermia treatments. *Int. J. Appl. Ceram. Technol.*, 3, 134–143.

137. Rajagopal, R., Mona, J., Kale, S. N., Bala, T., Pasricha, R., Poddar, P., Sastry, M., Prasad, B. L. V., Kundaliya, D. C. and Ogale, S. B. 2006. $La_{0.7}Sr_{0.3}MnO_3$ nanoparticles coated with fatty amine. *Appl. Phys. Lett.*, 89, 023107(1–3).

138. Melnikov, O. V., Gorbenko, O. Y., Mărkelova, M. N., Kaul, A. R., Atsarkin, V. A., Demidov, V. V., Soto, C., Roy, E. J. and Odintsov, B. M. 2009. Ag-doped manganite nanoparticles: New materials for temperature-controlled medical hyperthermia. *J. Biomed. Mater. Res. A*, 91, 1048–1055.

139. Pradhan, A. K., Bah, R., Konda, R. B., Mundle, R., Mustafa, H., Bamiduro, O., Rakhimov, R. R., Wei, X. and Sellmyer, D. J. 2008. Synthesis and magnetic characterizations of manganite-based composite nanoparticles for biomedical applications. *J. Appl. Phys.*, 103, 07F704(1–3).

140. Park, J.-H., Vescovo, E., Kim, H.-J., Kwon, C., Ramesh, R. and Venkatesan, T. 1998. Direct evidence for a half-metallic ferromagnet. *Nature*, 392, 794–796.

141. Zeng, H., Li, J., Wang, Z. L., Liu, J. P. and Sun, S. 2004. Bimagnetic core/shell FePt/Fe$_3$O$_4$ nanoparticles. *Nano Lett.*, 4, 187–190.

142. Kang, S., Miao, G. X., Shi, S., Jia, Z., Nikles, D. E. and Harrell, J. W. 2006. Enhanced magnetic properties of self-assembled FePt nanoparticles with MnO shell. *J. Am. Chem. Soc.*, 128, 1042–1043.

143. Thomson, T., Toney, M. F., Raoux, S., Lee, S. L., Sun, S., Murray, C. B. and Terris, B. D. J. 2004. Structural and magnetic model of self-assembled FePt nanoparticle arrays. *J. Appl. Phys.*, 96, 1197–1201.

144. Peng, S., Wang, C., Xie, J. and Sun, S. 2006. Synthesis and stabilization of monodisperse Fe nanoparticles. *J. Am. Chem. Soc.*, 128, 10676–10677.

145. Johnston-Peck, A. C., Wang, J., Tracy, J. B. 2009. Synthesis and structural and magnetic characterization of Ni(core)/NiO(shell) nanoparticles. *ACS Nano*, 3, 1077–1084.

146. Lee, I. S., Lee, N., Park, J., Kim, B. H., Yi, Y.-W., Kim, T., Kim, T. K., Lee, I. H., Paik, S. R. and Hyeon, T. 2006. Ni/NiO core/shell nanoparticles for selective binding and magnetic separation of histidine-tagged proteins. *J. Am. Chem. Soc.*, 128, 10658–10659.

147. Figuerola, A., Fiore, A., Corato, R. D., Falqui, A., Giannini, C., Micotti, E., Lascialfari, A., Corti, M., Cingolani, R., Pellegrino, T., Cozzoli, P. D. and Manna, L. 2008. One-pot synthesis and characterization of size-controlled bimagnetic FePt–iron oxide heterodimer nanocrystals. *J. Am. Chem. Soc.*, 130, 1477–1487.

148. Yin, Y. D., Rioux, R. M., Erdonmez, C. K., Hughes, S., Somorjai, G. A. and Alivisatos, A. P. 2004. Formation of hollow nanocrystals through the nanoscale Kirkendall effect. *Science*, 304, 711–714.

149. Gao, J., Liang, G., Zhang, B., Kuang, Y., Zhang, X. and Xu, B. 2007. FePt@CoS$_2$ yolk–shell nanocrystals as a potent agent to kill HeLa cells. *J. Am. Chem. Soc.*, 129, 1428–1433.

150. Choi, J.-S., Jun, Y.-W., Yeon, S.-I., Kim, H. C., Shin, J.-S. and Cheon, J. 2006. Biocompatible heterostructured nanoparticles for multimodal biological detection. *J. Am. Chem. Soc.*, 128, 15982–15983.

151. Robinson, I., Tung, L. D., Maenosono, S., Wältid, C. and Thanh, N. T. K. 2010. Synthesis of core–shell gold coated magnetic nanoparticles and their interaction with thiolated DNA. *Nanoscale*, 2, 2624–2630.

152. Gao, J., Zhang, B., Gao, Y., Pan, Y., Zhang, X. and Xu, B. 2007. Fluorescent magnetic nanocrystals by sequential addition of reagents in a one-pot reaction: A simple preparation for multifunctional nanostructures. *J. Am. Chem. Soc.*, 129, 11928–11935.

153. Tian, Z.-Q., Zhang, Z.-L., Jiang, P., Zhang, M.-X., Xie, H.-Y. and Pang, D.-W. 2009. Core/shell structured noble metal (alloy)/cadmium selenide nanocrystals. *Chem. Mater.*, 21, 3039–3041.

154. Kim, H., Achermann, M., Balet, L. P., Hollingsworth, J. A. and Klimov, V. I. J. 2005. Synthesis and characterization of Co/CdSe core/shell nanocomposites: Bifunctional magnetic-optical nanocrystals. *J. Am. Chem. Soc.*, 127, 544–546.

155. Trinh, T. T., Mott, D., Thanh, N. T. K. and Maenosono, S. 2011. One-pot synthesis and characterization of well defined core@shell structure of FePt@CdSe nanoparticles. *RSC Advances.*, 1, 100–108.

Trinh Thang Thuy obtained an MS degree from Japan Advanced Institute of Science and Technology (JAIST), Japan in 2009. He is now in the third year of study for a PhD under the guidance of Dr. Shinya Maenosono at JAIST. His focus is on the synthesis and characterisation of magnetic nanoparticles

Associate Professor Shinya Maenosono leads his research group at JAIST (http://www.jaist. ac.jp/~shinya/english/index.html). His research in JAIST has focussed on two main areas of interest in the field of materials chemistry and nanotechnology. The first area involved wet chemical synthesis of semiconductor nanoparticles with controlled size, shape and composition for energy conversion device applications. The second area has focused on the synthesis and bioapplication development of monometallic and alloyed multimetallic nanoparticles.

Dr. Nguyễn Thị Kim Thanh is the editor of this book. For more information, please see the editor's biography or visit her website: http://www.ntk-thanh.co.uk

Part III

Biofunctionalisation of MNPs
for Biomedical Application

5 Strategies for Functionalisation of Magnetic Nanoparticles for Biological Targets

Bettina Kozissnik, Luke A.W. Green, Kerry A. Chester, and Nguyễn Thị Kim Thanh

CONTENTS

Functionalisation of magnetic nanoparticles (MNPs) plays an important role in achieving their full potential in biomedical applications. The coupling strategy used for linking the functionalising agent to the particle is a key component for success and will depend both on the coating of the MNP and the available functional groups on the targeting moiety. This chapter discusses the most common attachment methods and gives an overview of the essential biological molecules used for functionalising MNPs. We focus mainly on MNPs, but for comprehensive-ness we refer to non-MNPs where applicable.

5.1 STRATEGIES FOR CONJUGATION

5.1.1 CARBODIIMIDE COUPLING

Carbodiimide coupling comes from the name of the chemical commonly used to mediate the forma-tion of amide or phosphoramidate covalent bonds; there are a variety of available carbodiimides: 1-ethyl-3-(dimethylaminopropyl)carbodiimide (EDC or EDAC), 1-cyclohexyl-3-(2-morpholinoethyl) carbodiimide (CMC), dicyclohexyl carbodiimide (DCC) and diisopropyl carbodiimide (DIC)[1].

For conjugation of nanoparticles (NPs) with biomolecules, the 'zero length' covalent linkages afforded by this method are most commonly mediated with EDC[2]. The advantages of this coupling agent over the others are that it involves no lengthy linker species, allowing the hydrodynamic radius of the NP to be minimised and both the reagent and the by-product (isourea) are water soluble and can be removed *via* dialysis. For optimum yield, EDC works best in the pH range 4.5–7.5. Care should be taken with molecules with both carboxylic and amine groups as cross-linking can occur; this is not necessarily detrimental to the end product, but something that the reader should be aware of. Tyrosine and chemicals exhibiting sulfhydryl groups can result in undesirable side reactions[1]. The reaction mechanism forming this 'zero length' bond is outlined in Figure 5.1.

While it is possible to form amide bonds directly between a terminal carboxylic acid group on the surface of NPs and free amines on biological species like proteins, reaction efficiency can be improved through the use of additives such as N-hydroxysuccinimide (NHS)[1,2,4–6]. The efficiency of the coupling reaction is increased by stabilising the O-acylisourea intermediate by formation of the succinimide ester. This is achieved by addition of NHS or sulfo-NHS (Figure 5.1). NHS derivatives can also be used to conjugate two amines *via* a linker. Lin et al.[7] employed bifunctional suberic acid bis-N-hydroxysuccinimide (DSS) to link aminosilane-coated MNP with anti-serum amyloid P in a two-step process.

With the aim of targeting and treating thrombosis, recombinant tissue plasminogen activator (rtPA) has been conjugated with EDC to form polyacrylic acid-coated magnetite *via* an amide bond; enzyme activity of conjugated rtPA was maintained at up to 87% ± 1% of free rtPA. These results indicate that lower treatment doses could be used[8].

Depending on the enzyme (dispase, chymotrypsin and streptokinase), Koneracka et al.[9,10] showed that retention of up to 50%–80% of enzymatic activity is achievable; optimum conditions for such immobilisations were later investigated by the same group. More recently, nattokinase and lum-brukinase were attached to NPs[11].

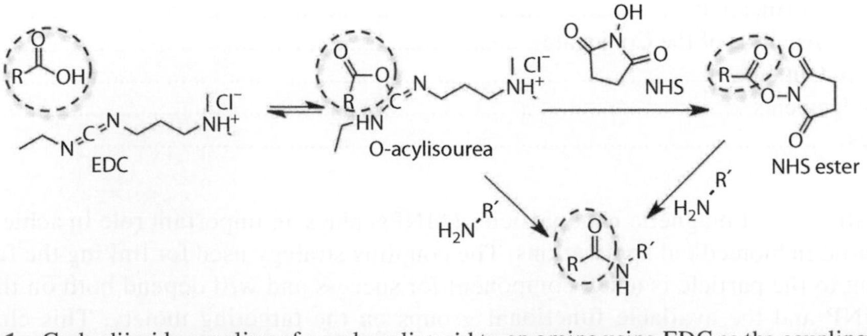

FIGURE 5.1 Carbodiimide coupling of a carboxylic acid to an amine using EDC as the coupling agent. An N-hydroxysuccinimide (NHS) ester intermediate may be formed to increase reaction efficiency. (From *Nano Today*, 5, Thanh, N.T.K. and Green, L.A.W., Functionalisation of nanoparticles for biomedical applications, 213–230, Copyright 2010, Elsevier.)

Carboxyls are not the only functional groups that can be coupled to amines *via* EDC coupling; single-stranded DNA (ssDNA) with a phosphate group at the 5′ end was covalently linked to the amino group of magnetite NPs by forming a phosphoramidate bond in the presence of EDC[12]. Insulin, immunoglobulin G (IgG), vancomycin and streptavidin have all been directly function-alised on to magnetic iron oxide NPs; the respective biological species were first conjugated to EDC and then covalently bound to the NPs *via net* dehydration reactions, leading to ester linkages, rather than amides[13–16].

In addition, chitosan and linoleic acid have been connected *via* EDC to generate self-organised chitosan–linoleic acid particles in which 12 nm superparamagnetic iron oxide nanoparticles (SPIONs) were incorporated for magnetic resonance imaging (MRI) signal enhancement in the liver and reduced cytotoxicity[17].

5.1.2 Cyanogen Bromide Conjugation

Cyanogen bromide allows the transformation of available hydroxide groups on the surface of the MNP into a reactive cyanate ester, which reacts with free amines and forms an isourea bond. The transformation is performed in two steps, as shown in Figure 5.2; the first step follows a nucleophilic elimination mechanism, where the nucleophilic hydroxyl group on the alcohol attacks the electron-deficient carbon atom and eliminates the bromine to give hydrogen bromide. The second step then consists of a nucleophilic addition of the amine to the electron-deficient cyanide group.

Liu et al.[18] used cyanogen bromide conjugation to functionalise SPIONs with neutravidin (NA). This then allowed for the attachment of biotin-linked fluorescein isothiocyanate (FITC)-labelled phosphorothioate-modified oligodeoxy nucleotides (sODNs) NPs to monitor the gene transcription in the mouse brain. Wu et al.[19] applied this conjugation method to attach the fluorescent arginine–glycine–aspartic acid (RGD) peptide to silica NPs for *in vitro* targeting and imaging of $\alpha_v\beta_3$ of human breast cancer cells. Using cyanogen bromide is a straightforward process; however, its high toxicity makes it an undesirable method for conjugation aimed at *in vivo* application. To overcome this obstacle, Liu et al.[18] filtered and dialysed their conjugate, SPION-NA, against 20 times the volume of sodium citrate buffer.

Another way to conjugate MNPs with hydroxyl groups on their surface is the Josephson method, which uses fewer toxic chemicals than cyanogen bromide[20]. MNPs are cross-linked with a mixture

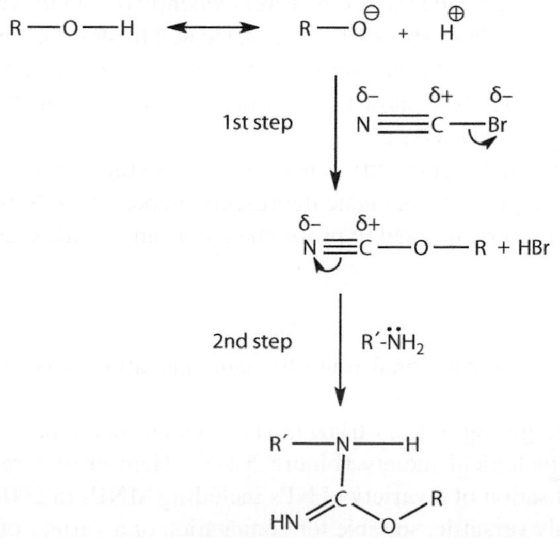

FIGURE 5.2 Schematic representation of cyanogen bromide coupling of an amine to a hydroxyl MNP using CNBr to first form a cyanate ester and consequently an isourea bond with the ligand.

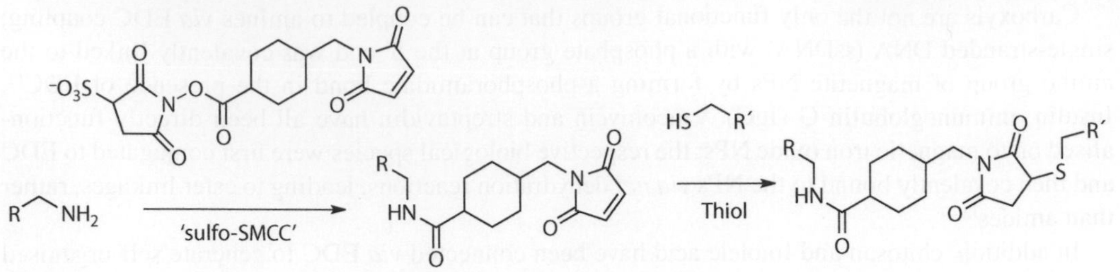

FIGURE 5.3 Maleimide coupling of amine and thiol using sulfo-SMCC as linker. (From *Nano Today*, 5, Thanh, N.T.K. and Green, L.A.W., Functionalisation of nanoparticles for biomedical applications, 213–230, Copyright 2010, Elsevier.)

of poly(ethylene glycol) diglycidyl ether and epichlorohydrine followed by ammonia to functionalise the particle with amino groups. The amines on the surface of the particle then allow the application of a variety of other conjugation chemistries, such as maleimide coupling[21].

5.1.3 MALEIMIDE COUPLING

A maleimide may be used to conjugate primary amines to thiols as illustrated in Figure 5.3[1,3]. When using this kind of coupling it should be noted that the pH is absolutely crucial. In the range of pH 6.5–7.5, the maleimide end is more specific to sulphydryls than amines and proceeds faster by a factor of 10^3. In order to increase the preference of conjugation to amines, more alkaline conditions are required. The most commonly used maleimide-derived coupling reagent is sulfosuccinimidyl-4-(maleimidomethyl)cyclohexane-1-carboxylate (sulfo-SMCC).

Maleimide coupling has found wide application; Lee et al., for example, have used this approach to link iron oxide and dye-doped silica NPs, creating MRI- and fluorescence-active 'core satellite' imaging agents[22]. In another example, poly(ethylene glycol) (PEG)-NPs were functionalised with maleimide in order to allow linkage with cysteine units of single chain Fv antibody fragments. The resultant radio-immunonanoparticles were used in conjunction with an alternating magnetic field (AMF) for directed treatment of cancer[23].

Conjugation of molecules prior to ligation with NPs is not uncommon; before conjugation to γ-Fe$_2$O$_3$, poly(D,L-lactide-*co*-glycolide) (PLGA) was covalently bound to Arg peptide *via* maleimide coupling, these particles were later successfully translocated to stem cell nuclei with no cytotoxicity[24]. In another case, biotin was bound similarly to poly(*N*-isopropylacrylamide) (N-iPAm) and exchanged onto iron oxide MNPs to show temperature-sensitive binding to streptavidin, as a result of chain collapse of the polymeric chain[25].

Employment of maleimide functionality is not restricted to the conjugation of biological entities to MNPs and can also be applied to conjugate fluorescent markers to MNPs. These provide an additional imaging mode and can be utilised to prove chemical functionality, as shown by Liu et al.[26]

5.1.4 CLICK CHEMISTRY

Another common, but less conventional route for bioconjugation is the Cu(I)-catalysed alkyne–azide cycloaddition reaction, also known as 'CuAAC', or 'click chemistry' involving the coupling of an alkyne to an azide giving a 1,2,3-triazole ring, which then can serve as a strong covalent bond between NP and biological moiety (Figure 5.4)[3,27]. Hein et al.[28] reviewed the use of click chemistry for functionalisation of a variety of NPs including MNPs in 2008. This process has been demonstrated to be highly versatile, suitable for conjugation of a variety of species including small molecules[29,30]. Specifically, chemicals conjugated *via* click chemistry are shown to exhibit stability *in vivo* for more than 5 hrs, demonstrating the good biocompatibility of the 1,2,3-triazole ring[31].

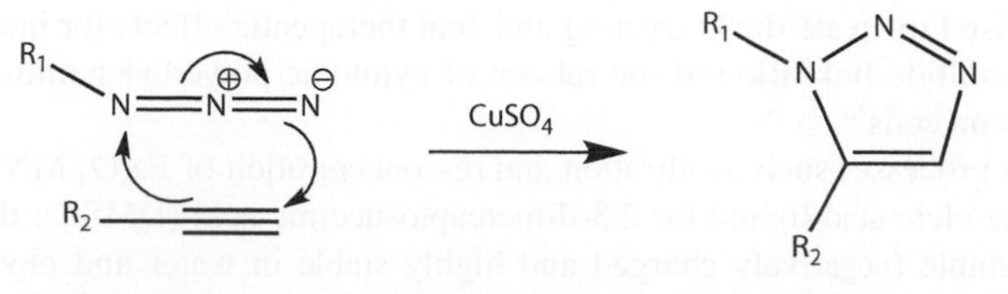

FIGURE 5.4 Cu-mediated alkyne–azide cycloaddition ('click reaction'). The reaction proceeds *via* Cu(I) intermediate to $CuSO_4$. (From *Nano Today*, 5, Thanh, N.T.K. and Green, L.A.W., Functionalisation of nanoparticles for biomedical applications, 213–230, Copyright 2010, Elsevier.)

Versatility is provided through the fact that either the alkyne or the azide could be expressed on the biological moiety, with the complementary functional group expressed on the NP surface. Combined with the variety of ligand head groups available for NP–ligand bond formation, this procedure has much potential[1,3].

The coupling of two functional groups of use is desirable for many applications not only in the laboratory but also in the clinic. In one case, the specificity of folic acid for breast cancer has been utilised to deliver a temperature-responsive drug 'vehicle' to the area of interest. Here, coupling of β-cyclodextrin (β-CD) (previously bound to SPIONs) to folic acid was performed *via* the click reaction, then application of an AMF induced a temperature increase leading to a release of hydrophobic drugs from the β-CD inclusion cavity[32]. Click chemistry has also been demonstrated as an effective method (80% yield) for conjugating tetradecapeptide to dye-functionalised SPIONs for target-specific imaging of prostate cancer[33]. This has potentially wide clinical application, as prostate cancer is the second most common form of cancer in men in North America and current diagnostic techniques are questionable and limited[34].

Though click chemistry can be applied in most cases, conjugation of oligonucleotides with an alkyne group is rather problematic, as Cu(I) is shown to degrade DNA-oligonucleotides[35]. To circumvent this obstacle, tris-hydroxypropyl triazolylamine (THPTA) was used to aid click coating of SPIONs with a dense monolayer of oligonucleotides; these particles were successfully introduced into HeLa (cervical cancer) cells without the use of transfection agents[36].

The clinical benefits of creating multimodal NPs are immediately apparent. F^{19}, commonly used in positron emission tomography, has successfully been conjugated to azide-PEG iron oxide NPs, previously functionalised with a near-IR fluorochrome (NIRF)[37]. The toxicity of the Cu-catalysed reaction has potential to be overcome by replacing this Cu catalyst with a fluorine-substituted cyclooctyne reagent; this reagent has been shown to dynamically introduce chemicals to live cells with no apparent toxicity; however, this reagent has not yet been used in conjunction with MNPs[38]. As well as problems of toxicity, Cu has been shown to present problems with luminescence quenching when used to conjugate chemicals to quantum dots (QDs); a copper-free buffer has been developed to overcome this problem in QD chemistry[39]. In 2010, a different 'click' chemical reaction was successfully used for the first time to couple a thiol with an alkene. This new chemical methodology employed UV light to promote the thiol–ether formation[40].

5.1.5 DISULFIDE BRIDGES

Disulfide bridges are single sulfur–sulfur bonds, which in biology act as a type of intraprotein cross-linker and have been used for the reversible chemical coupling of MNPs[41]. Tian et al.[42] showed that the oxidation and reduction of disulfide bridges between silica NPs and Fe_3O_4 NPs were facilitated by glutathione disulfide, an oxidative bond formation, and dithiothreitol (DTT), a reductive bond cleavage. While current work appears to be focused on the formation of hybrid nanostructures, this approach could also be adapted for drug delivery, because cleaving agents, such as glutathione (GSH), are present at considerable amounts *in vivo*[43]. Disulfide bonds can be exploited for therapeutic advantage. For example, microenvironments inside cancer cells can exhibit chemical conditions, which can be utilised to create dual targeting and dual therapeutic effects for imaging and therapy. A pH-sensitive disulfide link allowed the release of cytotoxic polyethyleneimine and therapeutic siRNA in C6 glioma cells[44].

Post-treatment processes such as filtration and re-concentration of Fe_3O_4 MNPs can be avoided by exchanging the oleic acid ligand for 2,3-dimercaptosuccinic acid (DMSA); these new particles are particularly stable (negatively charged and highly stable in water and physiological assays). Raman spectroscopy confirmed that intermolecular disulfide bridges formed between the DMSA molecules[45].

MRI gives information about the location of MNPs, but does not provide information about the chemical environment around them. However, by conjugating NIRF-activated arginyl peptides to MNPs *via* disulfide and thioether linkages, selective detection of DTT and trypsin was possible[46].

5.1.6 HISTIDINE-TAGGED PROTEINS

In physiology, histidine is an essential amino acid needed for the production of histamine, for growth and for general tissue repair. Histidine is also claimed to have vasodilatory properties, and, due to its ability to act as an electron donor, it can neutralise both oxygen and hydroxyl radicals[47].

Histidine has been successfully covalently bound to iron oxide MNPs in a one-pot synthesis; however, the application was directed at catalysis as opposed to biomedical use[47]. The addition of six terminal histidine residues (6 His tag) on a protein allows the protein to act as a chelating agent, coordinating to a metal cation held to the NP surface by nitriloacetic acid (NTA) groups. This is utilised in the magnetic-assisted purification of proteins extracted directly from cell lysate[48-50]. Furthermore, this process is capable of detecting 6 His-tagged proteins at concentrations as low as 0.5 pM without non-specific binding to undesired proteins occurring[50]. In addition to assisted purification of proteins from cell lysate; employment of the '6-His' coupling strategy allowed the use of 'magnetomicelles' for recapture of T7 polymerase in biocatalysis[51]. Silicatein, a His-tagged recombinant protein was isolated and purified through biomagnetic separation of marine sponge[52].

Iron oxide is not the only choice of magnetic material for biomagnetic separation. Ni/NiO MNPs efficiently separated His-tagged green fluorescent protein from a supernatant; this method has proven useful for separation of His-FcBD protein from *Escherichia coli* cell lysate[53].

5.1.7 IONIC COUPLING

An obvious application of ionic coupling is the conjugation of charged biological species such as DNA to NPs. Magnetic cationic liposomes can be ionically conjugated to DNA as the positively charged surface can bind to the negatively charged groups on DNA[54].

Due to the nature of ionic coupling, a variation in pH can encourage or discourage binding of cationic, anionic or even neutral species to oppositely charged NPs. Gadolinium has been indirectly conjugated to MNPs *via* a silica matrix to generate a multimodal MRI contrast agent; [III]In has also been tried, leading the way for guided delivery of therapeutic radiotherapy[55,56]. *N*-methylimidazolium chloride-modified iron oxide NPs have been synthesised by means of a silane linker; *N*-alkylimidazolium salts are attracting much interest as ionic liquids, the counter ions were also successfully exchanged for DNA[57].

5.2 BIOLOGICAL MOLECULES FOR CONJUGATION

5.2.1 OLIGONUCLEOTIDES AND NUCLEIC ACIDS

Oligonucleotides and other nucleic acids, such as DNA, are polymers composed of the four bases: adenine, cytosine, guanine and thymine, connected by a phosphate backbone. To form a double strand, the oligonucleotides have to be complementary, adenine interacts with thymine *via* two hydrogen bonds, while cytosine interacts with guanine *via* three hydrogen bonds. In the case of RNA, thymine is replaced by uracil.

Depending on the intended use, different configurations of DNA can be attached to MNPs. In most cases, DNA can easily be thiolated to gold NPs, functionalised *via* straightforward adsorption of thiols onto the gold surface. Furthermore, mostly ssDNA is used, as in the experiments by Mirkin et al.[58] This group coated gold NPs with thiol-functionalised ssDNA through coordinate (dative covalent) bonds to form specific colloids either in a two-strand or three-strand system. In the two-strand system, one batch of NPs was coated with one kind of ssDNA, while the second batch was coated with partially complementary ssDNA. The three-strand system NPs were coated with only one kind of ssDNA and a linker ssDNA was introduced to form self-assembling colloids. Alivisatos et al.[59] developed the approach further and successfully tested the feasibility of arranging DNA–Au NPs in head-to-head (antiparallel) dimer, head-to-tail (parallel) dimer and parallel trimer configurations.

Following on from these studies, much research has gone into optimising the oligonucleotide surface coverage by adjusting the salt and temperature conditions[60,61], by introduction of oligonucleotide spacer segments[61] and by co-adsorption of diluent strands[61]. Further methods include enzymatic manipulation[62] to control the number of oligonucleotides on each particle, functionalisation of particles with oligonucleotides through polyvalent oligonucleotides using 'click' chemistry[36] or even asymmetric functionalisation[63].

Additionally, the aggregation of gold NPs leads to the formation of a new absorption band at longer wavelengths as a result of electric dipole–dipole interaction and coupling between the plasmons of neighbouring particles in the formed aggregates[64]. In the future, this principle will not only allow the assembly of more complex nanostructured materials, but also the development of analytical tools[65].

Yung et al. developed a simple colorimetric quantitative DNA assay to identify single nucleotide mutations and single nucleotide polymorphisms for the diagnosis of genetic diseases such as Duchenne muscular dystrophy[66]. The assay consists of a mixture of two sets of gold NP DNA probes (nAu-18b and nAu-70b) to which a 43-base synthetic target DNA, as shown in Figure 5.5, is added.

The nAu-70b probe was designed to facilitate target DNA binding and therefore contains a matching 30-base sequence of the target DNA. In contrast, the nAu-18b probe facilitates mismatch discrimination; therefore, a complex between the two probes will only form if the remaining bases of the target DNA match perfectly; if there is a single mutation in this part of the target DNA, there is no dimer formation. They demonstrated that their assay works with both single- and double-stranded DNA (dsDNA)[66]. Since clinical samples are amplified through the polymerase chain reaction (PCR) resulting in dsDNA, this assay could serve as a low-cost alternative to current clinical diagnostics.

5.2.2 Peptides

Peptides are small amino acid polymers with less than 20 amino acids. They can be used for a number of different applications like protein purification, antibody recognition through tags like the hexahistidine tag (6 His tag) or the myc tag, a 10-amino acid peptide (EQKLISEEDL).

Target DNA

nAu-70b   nAu-18b

FIGURE 5.5 Schematic representation of gold nanoparticle DNA probes nAu-70b and nAu-18b interacting with targeting DNA. (Reproduced and adapted from *Biosens. Bioelectron.*, 25, Qin, W.J., Yim, O.S., Lai, P.S., and Yung, L.Y.L., Dimeric gold nanoparticle assembly for detection and discrimination of single nucleotide mutation in Duchenne muscular dystrophy, 2021–2025, Copyright 2010, with permission from Elsevier.)

Peptides have been used extensively for coating gold NPs; Levy et al.[67] designed 58 different peptides and found that pentapeptide CALNN stabilises citrate-capped gold NPs. The thiol group of the cysteine binds to the NPs surface, which leads to the self-assembly of the peptides into a densely packed layer, probably forming parallel beta strands, providing the NPs with stability in a wide pH range (4–10) and high salt concentrations (up to 1.5 M NaCl). Peptide capping ligands, such as CALNN, provide a novel and unique way of producing monodisperse magnetic metal Co NPs[68] and magnetic hollow bimetallic alloy CoPt NPs[69]. These NPs are stable in a wide range of pH (1.5–10) and up to 2 M NaCl for many months. The carboxyl groups of the peptides on the NP surface will be used for conjugation with biomolecules of interest for further biomedical applications.

Another way of using peptides is to help enhance cellular uptake, as described by Liu[70]. Liu used so-called cell-penetrating peptides (CPPs) or protein transduction domains (PTD)[71], which are small peptides containing a high percentage of basic amino acids, allowing them to facilitate delivery of large macromolecules across the cell membrane. Such macromolecules can include 100 kDa proteins[72] and NPs up to 40 nm in diameter[20]. Liu[70] coated carboxylated quantum dots with the modified CPPs, HR9 and PR9 and demonstrated an increase in transduction efficiency by direct membrane translocation. For further reference and information about other CPPs such as TAT, Gupta et al.[73] have published an excellent review on the topic. Though peptides are relatively costly to make and prone to biodegradation, Okuyama et al.[74] introduced small-molecule carriers (SMoCs) as an alternative way to enhance cellular uptake. SMoCs were specifically designed to mimic the helical conformation found in well-known CPPs like HIV-1 TAT and the antennapedia homeodomain. This was achieved by using unsubstituted guanidine groups in different confirmations. Using their SMoCs they achieved transportation of geminin, a 23.5 kDa protein DNA replication licensing repressor protein, into different cells[74].

5.2.3 PROTEINS

Proteins are polypeptides or long amino acid polymers, which form 3D structures. Antibodies are one of the most important members of this class of molecule when selective targeting of the MNP is required. There are a variety of different antibody formats available for targeting, and some of the most commonly used IgG and its derived fragments are shown in Figure 5.6. IgG consists of two heavy and two light peptide chains, together forming two antigen-binding arms with the Fc region for effector function[75]. The variable region (Fv) contains the complementary determining regions (CDRs),

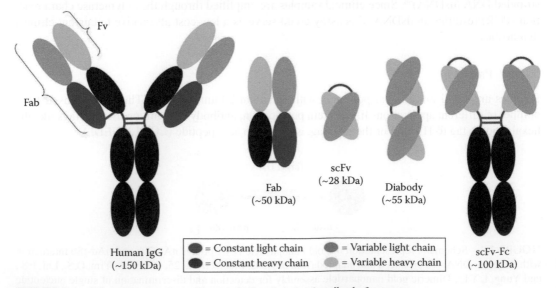

= Constant light chain	= Variable light chain	
= Constant heavy chain	= Variable heavy chain	

Fv

Fab

scFv
(~28 kDa)

Fab
(~50 kDa)

Diabody
(~55 kDa)

Human IgG
(~150 kDa)

scFv-Fc
(~100 kDa)

FIGURE 5.6 Schematic structure of human IgG and derived antibody fragments.

which determine the binding specificity of the antibody. These are described in more detail in Section 5.3. Advances in antibody engineering over the last two decades have allowed the development of a staggering array of stable antibody formats, with different targets, molecular weights and sites for conjugation[76-78]. These are all important parameters to consider when assessing use in biofunctionalisation of MNPs. Depending on the antibody format chosen for conjugation, it can increase the hydrodynamic diameter significantly and this can affect cellular uptake *in vitro* or biodistribution and blood half-life *in vivo*.

The clearance of the conjugate is also an important factor when exploiting the enhanced permeability and retention effect (EPR)[79] to deliver NPs to the tumour tissue. NPs slip through the leaky vasculature of the tumour and increase in concentration over time.

Vigor et al.[80] demonstrated that it is feasible to increase this enrichment of NPs for MRI by functionalising SPION with single-chain Fv antibody fragments (scFv) specific to the carcinoembryonic antigen (CEA), an oncofoetal target overexpressed by a range of adenocarcinomas, including colon, pancreatic and oesophageal cancer[81].

Antibody functionalised NPs can also be used for a whole variety of other applications, such as diagnosis of diseases like tuberculosis. For example, Qin et al.[82] have developed a rapid test for *Mycobacterium tuberculosis*, in which protein A-labelled RuBpy-doped silica NPs are used for immunofluorescence microscopy. The fluorescent dye-doped NPs not only enhance the signal but also show higher photostability than regular fluorescent dyes.

Another important protein group used for functionalisation of MNPs are enzymes. A prominent member of this group is trypsin, whose activity is specific to cleaving the peptide bonds between L-lysyl and L-argininyl. Li et al. prepared trypsin-coated NPs with different surface functionality, to compare how the different conjugation methods affected its activity and to further see if the immobilised trypsin was comparable to the free trypsin. Interestingly, MNPs with –COOH and –SH functionality showed higher activity and stability than free trypsin. This indicates that MNPs can be important tools in proteomics, leading the way to other potential biomedical applications[4].

5.2.4 PHOSPHOLIPIDS

Phospholipids consist of a hydrophilic head and two hydrophobic fatty acid tails and are mostly found in cell membranes as bilayers. They have been employed as micelles around quantum dots enhancing their stability and biocompatibility for *in vivo* imaging and have also been developed as a drug-delivery system[83] (see Chapter 18). In order to develop a treatment for rheumatoid arthritis, Zykova[84] incorporated methotrexate (MT) into 50 nm phospholipid-coated NPs. This resulted in an increased MT blood half-life and a higher activity of the drug compared to its free form.

Based on their previous work on 'magnetofection', Plank et al.[85,86] further developed magnetic and acoustically active lipospheres for gene therapy[87,88]. To achieve this, lipospheres were coated with MNPs and nucleic acids, and the resultant functionalised MNPs were administered to mice. The subsequent application of a magnetic field gradient allowed an improved deposition of the lipospheres in lungs, an effect also obtained by the application of ultrasound alone. A combination of magnetic field and ultrasound had no synergistic effect regarding the enrichment of the lipospheres; however, Plank et al.[87] observed a superior site-specific plasmid deposition.

5.2.5 CARBOHYDRATES

Carbohydrates are polymers often referred to as sugars or saccharides containing hydroxyl functionality, they range from single-unit monosaccharides to multi-unit polysaccharides. Within the body they fulfil a number of different functions: polysaccharides serve as energy storage; ribose and desoxyribose are part of the backbone for RNA and DNA; ribose is also important for energy transfer as a part of coenzyme adenosine triphosphate (ATP). Furthermore, carbohydrates also play important roles in the immune system.

Carbohydrates, like dextran or starch, remain the coating material of choice for NPs in biomedical applications due to their good biocompatibility and low cytotoxicity. Prominent examples include MRI contrast agents, such as Resovist® (Bayer-Schering), Endorem® (Geurbet) or Feridex I.V.® (AMAG Pharmaceuticals) and anaemia drugs like Feraheme® (AMAG Pharmaceuticals). Carbohydrate coatings are an ideal platform for surface modification as they allow the use of a number of different chemistries, such as the previously mentioned cyanogen bromide conjugation for dextran and carbodiimide coupling for chitosan (containing free amines) coated particles. However, there are challenges with using carbohydrates as NP coating. These include adsorption of plasma proteins, a short half-life[89], storage stability, difficulty in controlling particle size distribution, morphology and stability of magnetic preparations in different media[90]. Sometimes these issues can be advantageous, like in the case of the dextran-coated Resovist®; this MRI agent is easily taken up by the liver and therefore improves the imaging of liver metastases.

Because of the previously mentioned challenges, another non-carbohydrate coating is frequently used: PEG. It is an amphiphilic and a chemically inert polymer that minimises the adsorption of plasma proteins and therefore helps the particle evade immediate natural elimination mechanisms, such as the reticuloendothelial system (RES). This results in 'stealth' particles, which also minimise non-specific uptake into cells and show improved blood half-life[91,92].

Perrault et al. compared pegylated gold NPs with hydrodynamic diameters between 20 and 100 nm and found that half-life of these NPs could be enhanced by decreasing the hydrodynamic diameter and increasing the molecular weight of the PEG coating. Their data suggest that the particles coated with large PEG chains (5–10 kDa) resulting in 60–100 nm diameter would be ideal for pharmacokinetics and accumulation in the tumours[93].

5.2.6 HORMONES

Hormones are molecules secreted by glands into the bloodstream or directly into ducts to regulate physiological functions such as growth, the immune system and metabolism. Examples of hormones include insulin, histamine and dopamine (DA). Even though DA is a neurotransmitter, this property is less relevant for MNPs where it is used in a completely different manner; in 2004, while other groups still looked at thiolated organic compounds to functionalise NPs, Xu et al. looked at ways to stabilise iron oxide NPs with much smaller molecules. They found that DA serves as a robust high-affinity anchor to attach NTA to iron oxide NPs. As previously mentioned in Section 5.1.6, NTA forms chelates with 6-His tags and is therefore often used in protein purification. Xu et al. investigated the specificity of NTA-DA-MNPs for His-tagged proteins and discovered that they were more efficient than the commercially available HiTrap affinity column, which uses microbeads as its stationary phase. After applying a cell lysate to both materials under the same conditions, the NTA-DA-MNP only specifically captured His-tagged proteins, while the HiTrap column, also captured a number of other proteins[49].

Since it was claimed that the adsorption of DA onto iron oxide would lead to the degradation of the NPs[94], Amstad et al. studied the effect on iron oxide NP stability of eight different anchor catechol derivatives linked *via* 5 kDa PEG. They found that conventional DA analogues failed to stabilise iron oxide NPs at elevated temperatures and desorb upon dilution under physiological conditions; however, their newly designed anchor groups like nitrodopamine or nitro-DOPA resulted in ultrastable iron oxide NP suspensions[95].

5.3 BIOLOGICAL INTERACTIONS

5.3.1 ANTIBODY–ANTIGEN

One of the most important biological interactions in biomedicine is the one between antibody and antigen and this specificity is determined by the CDR loops, presented on the variable parts of light and heavy chains. These are illustrated in Figure 5.7 showing a cartoon of human IgG with the

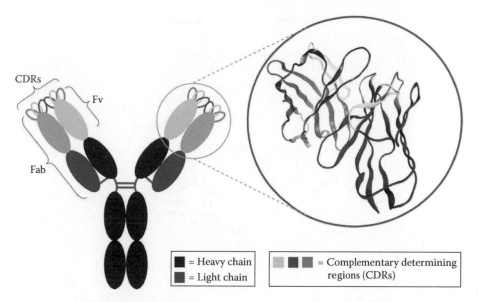

FIGURE 5.7   Human IgG with indicated CDRs on the example of the scFv MFE-23.

CDRs depicted as coloured loops. The same colours are used to highlight the CDR loops on the example of a crystal structure of MFE-23, a murine scFv. Antibodies tend to bind to their targets with K_d in the nanomolar range. Considering the size of human IgG, it is quite remarkable how little is actually dedicated to the antigen recognition and binding. Using recombinant protein expression and specific antibody engineering techniques, it is possible to reduce the antibodies to their essential parts and thus design different antibody formats, as described earlier.

5.3.2   AVIDIN–BIOTIN

The interaction between avidin and biotin is the strongest non-covalent biological interaction known today with a dissociation constant K_d in the order of 4×10^{-14} M[96].

Since its discovery by Snell et al.[97,98], the binding energies causing this incredibly strong bond were investigated. It was soon clear that it must be the result of multiple interactions along the two molecules[99,100]. Such interactions include van der Waals forces between tryptophan contacts[101,102] and multiple hydrogen bonds deep within the binding pocket. To specifically investigate the role of the hydrogen bonds in this interaction, Hyre et al.[103] used single mutants, S45A and D128A, previously described[104,105] in which the sites of the two strongest hydrogen bonds are mutated. Additionally, a new mutant S45A/D128A carrying both mutations was designed. Thermodynamic analysis revealed that the double mutation lowered the dissociation constant K_d by $\sim 2 \times 10^{-7}$ M, 11-fold lower than expected from the binding losses of the single mutations.

Holmberg et al.[106] showed that by functionalising MNPs with streptavidin, elevated temperatures caused this strong bond to be reversibly broken in aqueous solution, which could allow the reuse of streptavidin surfaces on assays or biosensor chips.

5.3.3   DNA–DNA

As stated in Section 5.2.1, DNA, the building block of life, consists of two complementary polymer strands, made of a series of four different bases (adenine, cytosine, guanine and thymine) connected by a sugar-phosphate backbone allowing the storage of genetic information. By matching hydrogen bonds, adenine interacts with thymine, while cytosine preferably interacts with guanine as shown in Figure 5.8.

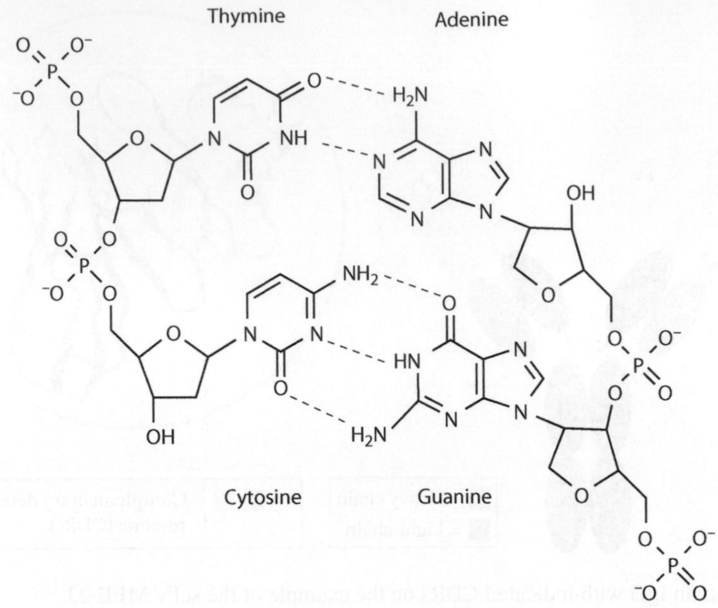

FIGURE 5.8 Schematic representation of DNA showing the four bases adenine, cytosine, guanine and thymine interacting *via* corresponding hydrogen bonds along a phosphate backbone.

This interaction has been used to attach DNA to NPs and is exploited in a number of different applications, such as the previously mentioned dimer and trimer formation[58,59], enzymatic self-assembly systems[107] and micron-sized colloids[108]. More literature on DNA-functionalised colloids can be found in the excellent review by Nienke Geerts and Erika Eiser[109]. However, such interactions between DNA strands can also be used for analytical purposes, such as the detection of single nucleotide mutations for diagnosis of genetic diseases[62,65,66].

5.3.4 LIGAND–RECEPTORS

One of the most important interactions in cell signalling is the one between a ligand and its receptor. Many ligands act as extracellular signalling molecules. They can be as small as a single ion that binds to a specific receptor, either on the cell surface or within the cell, inducing a cellular response, such as proliferation, differentiation or apoptosis. Examples include hormones, neurotransmitters and growth factors[110].

The topic of ligand–receptor interactions is rather complex and beyond the scope of this chapter; for further reading on this topic, the authors would strongly recommend *Cellular Signal Processing* by Marks et al.[111]

5.4 QUALITY CONTROL

Targeting MNPs is an exciting concept, but in recent years, attention has been mainly paid to the feasibility of targeting NPs *in vitro* and *in vivo*. In these experiments, SPIONs have been functionalised, by attaching targeting molecules like antibodies, peptide ligands and nucleic acids, as previously discussed. Attachment has been achieved by a number of different chemistries such as carbodiimide coupling. However, these chemistries are not site specific, and it is essential to be able to demonstrate target binding and product reproducibility. Here, we present methods to address this need.

5.4.1 EXPERIMENTAL DESIGN

For biofunctionalisation, quality control starts with the design of the experiment. It is important to establish the surface chemistry of the MNPs; with this information, the most effective conjugation technique can be formulated. This can be done in two ways: either the chemistry is adapted to the molecule that is going to be attached, or the molecule is modified according to the chemistry.

In the first example, the functional groups of the MNPs and the biological moiety are considered, and the chemistry is tailored to suit. So cyanogen bromide is chosen if an antibody containing available lysines, and therefore free amines, is to be conjugated to MNPs carrying hydroxyl groups on their surface. In the latter case, the molecule or particle is tailored to the chemistry, that is one or both may require further adaptation to suit the reaction constraints. If we consider the earlier example of the hydroxyl MNP and the available lysines on the antibody, in the antibody there may be lysines within the CDRs. If this antibody is attached to the MNP using cyanogen bromide coupling, the antibody will be bound to the MNP in a random orientation, as shown in Figure 5.9. This could lead to a lack of functionality as the conjugation bond could also occur in the CDRs using lysines responsible for binding to the antigen. In order to avoid this, the CDRs can be modified in a way that removes the lysines without affecting the activity of the antibody or by attaching a tag to the antibody that allows the use of different conjugation chemistry. Therefore, the researcher specifically determines the orientation of the antibody, and the first step towards a successful conjugation is laid.

Up to the time of writing this chapter, we have not found any work carried out on site-specific conjugation of MNPs to antibodies and the control targeting efficiency in the literature; we are currently working on this very specific issue.

5.4.2 PURIFICATION

Characterisation is essential (see Chapter 2), especially after chemically modifying the surface of the NPs. To achieve this, most analytical techniques require purification of the conjugate. A common way to purify particle suspensions is dialysis, where molecules are separated by diffusion, driven by the different concentrations of molecules on either side of the membrane. When choosing this technique to purify conjugate suspensions, the size of the molecule being removed must be appropriate to the pore size of the dialysis membrane.

Hence, there are a few drawbacks, as dialysis membranes or cassettes are expensive; limited number of sizes and molecular weight ranges are available. Good purification is not often achieved, as unwanted substances are diluted out, but not completely eliminated.

FIGURE 5.9   Antibody orientation using non-specific versus specific attachment methods.

Another method used for purification is size exclusion chromatography, where molecules travel different distances through a matrix depending on their size. Knowledge about the size of molecule chosen for functionalisation is essential, so that an appropriate column length and matrix can be selected. Furthermore, to allow the particles to move through the matrix, the coating of the MNP and the surface charge (ζ-potential) should also be considered.

Size exclusion chromatography allows a proper separation of conjugate and unbound targeting molecule; nevertheless, depending on the volume of the column, it will dilute the sample and concentration might be necessary.

5.4.3 Analysis of the Conjugate

When working with biological moieties, besides conventional NP characterisation methods such as dynamic light scattering or ζ-potential, techniques from biochemistry or molecular biology can also be sometimes adapted to analyse conjugates. For example, if the conjugate is functionalised with an antibody, an enzyme-linked immunosorbent assay (ELISA) can be performed as described by Vigor et al.[80] This assay specifically tests for the presence of an antibody–antigen interaction and therefore can give important information on the success of conjugation. However, the ELISA technique may not be able to distinguish between bound and unbound antibody and therefore the conjugate must be purified prior to the test. Furthermore, controls are absolutely essential using this method, and therefore besides the antigen, pure or on cells, an antibody detecting the MNP is also recommended (e.g. use of anti-dextran antibody for a dextran-coated MNP).

Another technique from molecular biology is quartz crystal microbalance (QCM) technology, which is used to determine protein–protein interactions. It is based on the piezoelectric effect; a crystal oscillates upon application of an alternating current, resulting in a standing shear wave. When the crystal, situated on a chip (Figure 5.10), becomes coated with a substrate, a shift in frequency can be detected (the baseline). If the analyte flows over the chip, it will bind to the substrate, causing another shift in frequency. Depending on the strength of the interaction, the analyte will come off resulting in a drop in frequency, giving a release profile.

Depending on the surface of the crystal, it can be functionalised with the same chemistries used for attaching biological molecules to MNP.

This technique allows a closer look on how the conjugation actually affected the binding affinity of the attached targeting moiety. Our first results indicate great potential of this method. However, as the system is not able to distinguish between conjugated and free molecule, purification is essential.

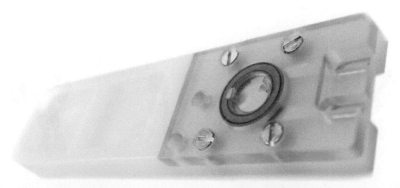

FIGURE 5.10 QCM sensor chip, as developed by Attana AB. (Reproduced with kind permission from Attana AB.)

5.5 FUTURE OUTLOOK

As many exciting potential therapeutic applications of MNP begin to be realised, it becomes progressively more important to fully understand the *in vitro* and *in vivo* behaviour of NPs, allowing them to be tuned for the different tasks. In order to do so, characterisation, biofunctionalisation, purification and re-characterisation are essential. It is timely that regulatory bodies such as the European Medicines Agency (EMA) and the Food and Drug Administration (FDA) are becoming increasingly interested in MNPs and their application in the clinic. Compliance with the requirements of these regulatory bodies will doubtless reinforce the importance of full characterisation.

ACKNOWLEDGEMENTS

Bettina Kozissnik and Kerry Chester thank Paul J. Gane from University College London for generating an image of MFE-23 based on the crystal structure and Attana AB, especially Liselotte Kaiser and Staffan Grenklo, for their help with analysis of MNP conjugates. They also thank Cancer Research UK, The UCL Cancer Institute Research Trust, The Royal Institution of Great Britain and The Engineering and Physical Sciences Research Council for support. The authors thank Lin-Yue Lanry Yung and Hedi Mattoussi for their useful discussion. Nguyễn Thị Kim Thanh thanks the Royal Society for her University Research Fellowship.

REFERENCES

1. Hermanson, G. T. 2008. *Bioconjugate Techniques*, Academic Press, London, U.K.
2. Wang, T. H. and Lee, W. C. 2003. Immobilization of proteins on magnetic nanoparticles. *Biotechnology and Bioprocess Engineering*, 8, 263–267.
3. Thanh, N. T. K. and Green, L. A. W. 2010. Functionalisation of nanoparticles for biomedical applications. *Nano Today*, 5, 213–230.
4. Li, D., Teoh, W. Y., Gooding, J. J., Selomulya, C. and Amal, R. 2010. Functionalization strategies for protease immobilization on magnetic nanoparticles. *Advanced Functional Materials*, 20, 1767–1777.
5. Wang, F. H., Yoshitake, T., Kim, D. K., Muhammed, M., Bjelke, O. and Kehr, J. 2003. Determination of conjugation efficiency of antibodies and proteins to the superparamagnetic iron oxide nanoparticles by capillary electrophoresis with laser-induced fluorescence detection. *Journal of Nanoparticle Research*, 5, 137–146.
6. Lee, C. H., Chen, C. B., Chung, T. H., Lin, Y. S. and Lee, W. C. 2010. Cellular uptake of protein-bound magnetic nanoparticles in pulsed magnetic field. *Journal of Nanoscience and Nanotechnology*, 10, 7965–7970.
7. Lin, P. C., Chou, P. H., Chen, S. H., Liao, H. K., Wang, K. Y., Chen, Y. J. and Lin, C. C. 2006. Ethylene glycol-protected magnetic nanoparticles for a multiplexed immunoassay in human plasma. *Small*, 2, 485–489.
8. Ma, Y. H., Wu, S. Y., Wu, T., Chang, Y. J., Hua, M. Y. and Chen, J. P. 2009. Magnetically targeted thrombolysis with recombinant tissue plasminogen activator bound to polyacrylic acid-coated nanoparticles. *Biomaterials*, 30, 3343–3351.
9. Koneracka, M., Kopcansky, P., Antalik, M., Timko, M., Ramchand, C. N., Lobo, D., Mehta, R. V. and Upadhyay, R. V. 1999. Immobilization of proteins and enzymes to fine magnetic particles. *Journal of Magnetism and Magnetic Materials*, 201, 427–430.
10. Koneracka, M., Kopcansky, P., Timko, M. and Ramchand, C. N. 2002. Direct binding procedure of proteins and enzymes to fine magnetic particles. *Journal of Magnetism and Magnetic Materials*, 252, 409–411.
11. Ren, L. L., Wang, X. M., Wu, H., Shang, B. B. and Wang, J. Y. 2010. Conjugation of nattokinase and lumbrukinase with magnetic nanoparticles for the assay of their thrombolytic activities. *Journal of Molecular Catalysis B: Enzymatic*, 62, 190–196.
12. Zhu, N. N., Zhang, A. P., He, P. G. and Fang, Y. Z. 2004. DNA hybridization at magnetic nanoparticles with electrochemical stripping detection. *Electroanalysis*, 16, 1925–1930.
13. Gupta, A. K., Berry, C., Gupta, M. and Curtis, A. 2003. Receptor-mediated targeting of magnetic nanoparticles using insulin as a surface ligand to prevent endocytosis. *IEEE Transactions on Nanobioscience*, 2, 255–261.
14. Ho, K. C., Tsai, P. J., Lin, Y. S. and Chen, Y. C. 2004. Using biofunctionalized nanoparticles to probe pathogenic bacteria. *Analytical Chemistry*, 76, 7162–7168.

15. Lee, C. W., Huang, K. T., Wei, P. K. and Yao, Y. D. 2006. Conjugation of gamma-Fe$_2$O$_3$ nanoparticles with single strand oligonucleotides. *Journal of Magnetism and Magnetic Materials*, 304, E412–E414.

16. Lin, Y. S., Tsai, P. J., Weng, M. F. and Chen, Y. C. 2005. Affinity capture using vancomycin-bound magnetic nanoparticles for the MALDI-MS analysis of bacteria. *Analytical Chemistry*, 77, 1753–1760.

17. Lee, C. M., Jeong, H. J., Kim, S. L., Kim, E. M., Kim, D. W., Lim, S. T., Jang, K. Y., Jeong, Y. Y., Nah, J. W. and Sohn, M. H. 2009. SPION-loaded chitosan–linoleic acid nanoparticles to target hepatocytes. *International Journal of Pharmaceutics*, 371, 163–169.

18. Liu, C. H., Kim, Y. R., Ren, J. Q., Eichler, F., Rosen, B. R. and Liu, P. K. 2007. Imaging cerebral gene transcripts in live animals. *Journal of Neuroscience*, 27, 713–722.

19. Wu, P., He, X. X., Wang, K. M., Tan, W. H., Ma, D., Yang, W. H. and He, C. M. 2008. Imaging breast cancer cells and tissues using peptide-labeled fluorescent silica nanoparticles. *Journal of Nanoscience and Nanotechnology*, 8, 2483–2487.

20. Josephson, L., Tung, C. H., Moore, A. and Weissleder, R. 1999. High-efficiency intracellular magnetic labeling with novel superparamagnetic-tat peptide conjugates. *Bioconjugate Chemistry*, 10, 186–191.

21. Gruttner, C., Muller, K., Teller, J., Westphal, F., Foreman, A. and Ivkov, R. 2007. Synthesis and antibody conjugation of magnetic nanoparticles with improved specific power absorption rates for alternating magnetic field cancer therapy. *Journal of Magnetism and Magnetic Materials*, 311, 181–186.

22. Lee, J. H., Jun, Y. W., Yeon, S. I., Shin, J. S. and Cheon, J. 2006. Dual-mode nanoparticle probes for high-performance magnetic resonance and fluorescence imaging of neuroblastoma. *Angewandte Chemie International Edition*, 45, 8160–8162.

23. Natarajan, A., Xiong, C. Y., Gruettner, C., Denardo, G. L. and Denardo, S. J. 2008. Development of multivalent radioimmunonanoparticles for cancer imaging and therapy. *Cancer Biotherapy and Radiopharmaceuticals*, 23, 82–91.

24. Lee, S. J., Jeong, J. R., Shin, S. C., Huh, Y. M., Song, H. T., Suh, J. S., Chang, Y. H., Jeon, B. S. and Kim, J. D. 2005. Intracellular translocation of superparamagnetic iron oxide nanoparticles encapsulated with peptide-conjugated poly(D,L lactide-*co*-glycolide). *Journal of Applied Physics*, 97, 10Q913.

25. Narain, R., Gonzales, M., Hoffman, A. S., Stayton, P. S. and Krishnan, K. M. 2007. Synthesis of monodisperse biotinylated p(NIPAAm)-coated iron oxide magnetic nanoparticles and their bioconjugation to streptavidin. *Langmuir*, 23, 6299–6304.

26. Liu, X. Q., Novosad, V., Rozhkova, E. A., Chen, H. T., Yefremenko, V., Pearson, J., Torno, M., Bader, S. D. and Rosengart, A. J. 2007. Surface functionalized biocompatible magnetic nanospheres for cancer hyperthermia. *IEEE Transactions on Magnetics*, 43, 2462–2464.

27. Kolb, H. C., Finn, M. G. and Sharpless, K. B. 2001. Click chemistry: Diverse chemical function from a few good reactions. *Angewandte Chemie International Edition*, 40, 2004–2021.

28. Hein, C. D., Liu, X. M. and Wang, D. 2008. Click chemistry, a powerful tool for pharmaceutical sciences. *Pharmaceutical Research*, 25, 2216–2230.

29. Lu, A. H., Salabas, E. L. and Schuth, F. 2007. Magnetic nanoparticles: Synthesis, protection, functionalization, and application. *Angewandte Chemie International Edition*, 46, 1222–1244.

30. White, M. A., Johnson, J. A., Koberstein, J. T. and Turro, N. J. 2006. Toward the syntheses of universal ligands for metal oxide surfaces: Controlling surface functionality through click chemistry. *Journal of the American Chemical Society*, 128, 11356–11357.

31. Von Maltzahn, G., Ren, Y., Park, J. H., Min, D. H., Kotamraju, V. R., Jayakumar, J., Fogal, V., Sailor, M. J., Ruoslahti, E. and Bhatia, S. N. 2008. In vivo tumor cell targeting with "click" nanoparticles. *Bioconjugate Chemistry*, 19, 1570–1578.

32. Hayashi, K., Ono, K., Suzuki, H., Sawada, M., Moriya, M., Sakamoto, W. and Yogo, T. 2010. High-frequency, magnetic-field-responsive drug release from magnetic nanoparticle/organic hybrid based on hyperthermic effect. *ACS Applied Materials & Interfaces*, 2, 1903–1911.

33. Martin, A. L., Hickey, J. L., Ablack, A. L., Lewis, J. D., Luyt, L. G. and Gillies, E. R. 2010. Synthesis of bombesin-functionalized iron oxide nanoparticles and their specific uptake in prostate cancer cells. *Journal of Nanoparticle Research*, 12, 1599–1608.

34. CDC. 2011. Cancer among men [Online]. Centre of Disease Control and Prevention. Available: http://www.cdc.gov/cancer/dcpc/data/men.htm (accessed April 2011).

35. Kanan, M. W., Rozenman, M. M., Sakurai, K., Snyder, T. M. and Liu, D. R. 2004. Reaction discovery enabled by DNA-templated synthesis and in vitro selection. *Nature*, 431, 545–549.

36. Cutler, J. I., Zheng, D., Xu, X. Y., Giljohann, D. A. and Mirkin, C. A. 2010. Polyvalent oligonucleotide iron oxide nanoparticle "click" conjugates. *Nano Letters*, 10, 1477–1480.

37. Devaraj, N. K., Keliher, E. J., Thurber, G. M., Nahrendorf, M. and Weissleder, R. 2009. [18]F labeled nanoparticles for *in vivo* PET-CT imaging. *Bioconjugate Chemistry*, 20, 397–401.

38. Baskin, J. M., Prescher, J. A., Laughlin, S. T., Agard, N. J., Chang, P. V., Miller, I. A., Lo, A., Codelli, J. A. and Bertozzi, C. R. 2007. Copper-free click chemistry for dynamic in vivo imaging. *Proceedings of the National Academy of Sciences of the United States of America*, 104, 16793–16797.

39. Bernardin, A., Cazet, A., Guyon, L., Delannoy, P., Vinet, F., Bonnaffe, D. and Texier, I. 2010. Copper-free click chemistry for highly luminescent quantum dot conjugates: Application to in vivo metabolic imaging. *Bioconjugate Chemistry*, 21, 583–588.

40. Rutledge, R. D., Warner, C. L., Pittman, J. W., Addleman, R. S., Engelhard, M., Chouyyok, W. and Warner, M. G. 2010. Thiol–ene induced diphosphonic acid functionalization of superparamagnetic iron oxide nanoparticles. *Langmuir*, 26, 12285–12292.

41. Clayden, J. 2001. *Organic Chemistry*. Oxford University Press, Oxford.

42. Tian, L., Shi, C. and Zhu, J. 2007. Reversible formation of hybrid nanostructures via an organic linkage. *Chemical Communications*, (37), 3850–3852.

43. El-Sayed, M. E. H., Hoffman, A. S. and Stayton, P. S. 2005. Rational design of composition and activity correlations for pH-sensitive and glutathione-reactive polymer therapeutics, *Journal of Controlled Release*, 104, 417–427.

44. Mok, H., Veiseh, O., Fang, C., Kievit, F. M., Wang, F. Y., Park, J. O. and Zhang, M. Q. 2010. pH-sensitive siRNA nanovector for targeted gene silencing and cytotoxic effect in cancer cells. *Molecular Pharmaceutics*, 7, 1930–1939.

45. Chen, Z. P., Zhang, Y., Zhang, S., Xia, J. G., Liu, J. W., Xu, K. and Gu, N. 2008. Preparation and characterization of water-soluble monodisperse magnetic iron oxide nanoparticles via surface double-exchange with DMSA. *Colloids and Surfaces A: Physicochemical and Engineering Aspects*, 316, 210–216.

46. Josephson, L., Kircher, M. F., Mahmood, U., Tang, Y. and Weissleder, R. 2002. Near-infrared fluorescent nanoparticles as combined MR/optical imaging probes. *Bioconjugate Chemistry*, 13, 554–560.

47. Unal, B., Durmus, Z., Baykal, A., Sozeri, H., Toprak, M. S. and Alpsoy, L. 2010. L-histidine coated iron oxide nanoparticles: Synthesis, structural and conductivity characterization. *Journal of Alloys and Compounds*, 505, 172–178.

48. Kim, J. S., Valencia, C. A., Liu, R. H. and Lin, W. B. 2007. Highly-efficient purification of native polyhistidine-tagged proteins by multivalent NTA-modified magnetic nanoparticles. *Bioconjugate Chemistry*, 18, 333–341.

49. Xu, C. J., Xu, K. M., Gu, H. W., Zheng, R. K., Liu, H., Zhang, X. X., Guo, Z. H. and Xu, B. 2004. Dopamine as a robust anchor to immobilize functional molecules on the iron oxide shell of magnetic nanoparticles. *Journal of the American Chemical Society*, 126, 9938–9939.

50. Xu, C. J., Xu, K. M., Gu, H. W., Zhong, X. F., Guo, Z. H., Zheng, R. K., Zhang, X. X. and Xu, B. 2004. Nitrilotriacetic acid-modified magnetic nanoparticles as a general agent to bind histidine-tagged proteins. *Journal of the American Chemical Society*, 126, 3392–3393.

51. Herdt, A. R., Kim, B. S. and Taton, T. A. 2007. Encapsulated magnetic nanoparticles as supports for proteins and recyclable biocatalysts. *Bioconjugate Chemistry*, 18, 183–189.

52. Shukoor, M. I., Natalio, F., Tahir, M. N., Divekar, M., Metz, N., Therese, H. A., Theato, P., Ksenofontov, V., Schroder, H. C., Muller, W. E. G. and Tremel, W. 2008. Multifunctional polymer-derivatized gamma-Fe$_2$O$_3$ nanocrystals as a methodology for the biomagnetic separation of recombinant His-tagged proteins. *Journal of Magnetism and Magnetic Materials*, 320, 2339–2344.

53. Lee, I. S., Lee, N., Park, J., Kim, B. H., Yi, Y. W., Kim, T., Kim, T. K., Lee, I. H., Paik, S. R. and Hyeon, T. 2006. Ni/NiO core/shell nanoparticles for selective binding and magnetic separation of histidine-tagged proteins. *Journal of the American Chemical Society*, 128, 10658–10659.

54. Shinkai, M. 2002. Functional magnetic particles for medical application. *Journal of Bioscience and Bioengineering*, 94, 606–613.

55. Lewin, M., Carlesso, N., Tung, C. H., Tang, X. W., Cory, D., Scadden, D. T. and Weissleder, R. 2000. Tat peptide-derivatized magnetic nanoparticles allow in vivo tracking and recovery of progenitor cells. *Nature Biotechnology*, 18, 410–414.

56. Rieter, W. J., Kim, J. S., Taylor, K. M. L., An, H. Y., Lin, W. L., Tarrant, T. and Lin, W. B. 2007. Hybrid silica nanoparticles for multimodal imaging. *Angewandte Chemie International Edition*, 46, 3680–3682.

57. Naka, K., Narita, A., Tanaka, H., Chujo, Y., Morita, M., Inubushi, T., Nishimura, I., Hiruta, J., Shibayama, H., Koga, M., Ishibashi, S., Seki, J., Kizaka-Kondoh, S. and Hiraoka, M. 2008. Biomedical applications of imidazolium cation-modified iron oxide nanoparticles. *Polymers for Advanced Technologies*, 19, 1421–1429.

58. Mirkin, C. A., Letsinger, R. L., Mucic, R. C. and Storhoff, J. J. 1996. A DNA-based method for rationally assembling nanoparticles into macroscopic materials. *Nature*, 382, 607–609.

59. Alivisatos, A. P., Johnsson, K. P., Peng, X. G., Wilson, T. E., Loweth, C. J., Bruchez, M. P. and Schultz, P. G. 1996. Organization of "nanocrystal molecules" using DNA. *Nature*, 382, 609–611.

60. Mahtab, R., Harden, H. H. and Murphy, C. J. 2000. Temperature- and salt-dependent binding of long DNA to protein-sized quantum dots: Thermodynamics of "inorganic protein"–DNA interactions. *Journal of the American Chemical Society*, 122, 14–17.

61. Demers, L. M., Mirkin, C. A., Mucic, R. C., Reynolds, R. A., Letsinger, R. L., Elghanian, R. and Viswanadham, G. 2000. A fluorescence-based method for determining the surface coverage and hybridization efficiency of thiol-capped oligonucleotides bound to gold thin films and nanoparticles. *Analytical Chemistry*, 72, 5535–5541.

62. Qin, W. J. and Yung, L. Y. L. 2005. Nanoparticle–DNA conjugates bearing a specific number of short DNA strands by enzymatic manipulation of nanoparticle-bound DNA. *Langmuir*, 21, 11330–11334.

63. Xu, X. Y., Rosi, N. L., Wang, Y. H., Huo, F. W. and Mirkin, C. A. 2006. Asymmetric functionalization of gold nanoparticles with oligonucleotides. *Journal of the American Chemical Society*, 128, 9286–9287.

64. Thanh, N. T. K. and Rosenzweig, Z. 2002. Development of an aggregation-based immunoassay for anti-protein A using gold nanoparticles. *Analytical Chemistry*, 74, 1624–1628.

65. Qin, W. J. and Yung, L. Y. L. 2009. Nanoparticle carrying a single probe for target DNA detection and single nucleotide discrimination. *Biosensors and Bioelectronics*, 25, 313–319.

66. Qin, W. J., Yim, O. S., Lai, P. S. and Yung, L. Y. L. 2010. Dimeric gold nanoparticle assembly for detection and discrimination of single nucleotide mutation in Duchenne muscular dystrophy. *Biosensors and Bioelectronics*, 25, 2021–2025.

67. Levy, R., Thanh, N. T. K., Doty, R. C., Hussain, I., Nichols, R. J., Schiffrin, D. J., Brust, M. and Fernig, D. G. 2004. Rational and combinatorial design of peptide capping ligands for gold nanoparticles. *Journal of the American Chemical Society*, 126, 10076–10084.

68. Thanh, N. T. K., Puntes, V. F., Tung, L. D. and Fernig, D. G. 2005. Peptides as capping ligands for in situ synthesis of water soluble Co nanoparticles for bioapplications. *Fifth International Conference on Fine Particle Magnetism*, 17, 70–76.

69. Lu, L. T., Tung, L. D., Fernig, D. G. and Thanh, N. T. K. 2009. Facile and green synthesis of stable, water-soluble magnetic CoPt hollow nanostructures assisted by multi-thiol ligands. *Journal of Materials Chemistry*, 19, 6023–6028.

70. Liu, B. R. 2011. Intracellular delivery of quantum dots mediated by a histidine- and arginine-rich HR9 cell-penetrating peptide through the direct membrane translocation mechanism. *Biomaterials*, 32, 3520–3537.

71. Wadia, J. S. and Dowdy, S. F. 2002. Protein transduction technology. *Current Opinion in Biotechnology*, 13, 52–56.

72. Schwarze, S. R., Ho, A., Vocero-Akbani, A. and Dowdy, S. F. 1999. In vivo protein transduction: Delivery of a biologically active protein into the mouse. *Science*, 285, 1569–1572.

73. Gupta, B., Levchenko, T. S. and Torchilin, V. P. 2005. Intracellular delivery of large molecules and small particles by cell-penetrating proteins and peptides. *Advanced Drug Delivery Reviews*, 57, 637–651.

74. Okuyama, M., Laman, H., Kingsbury, S. R., Visintin, C., Leo, E., Eward, K. L., Stoeber, K., Boshoff, C., Williams, G. H. and Selwood, D. L. 2007. Small-molecule mimics of an alpha-helix for efficient transport of proteins into cells. *Nature Methods*, 4, 153–159.

75. Edelman, G. M. 1973. Antibody structure and molecular immunology. *Science*, 180, 830–840.

76. Bradbury, A. R. M., Sidhu, S., Dubel, S. and Mccafferty, J. 2011. Beyond natural antibodies: The power of in vitro display technologies. *Nature Biotechnology*, 29, 245–254.

77. Weisser, N. E. and Hall, J. C. 2009. Applications of single-chain variable fragment antibodies in therapeutics and diagnostics. *Biotechnology Advances*, 27, 502–520.

78. Holliger, P. and Hudson, P. J. 2005. Engineered antibody fragments and the rise of single domains. *Nature Biotechnology*, 23, 1126–1136.

79. Brannon-Peppas, L. and Blanchette, J. O. 2004. Nanoparticle and targeted systems for cancer therapy. *Advanced Drug Delivery Reviews*, 56, 1649–1659.

80. Vigor, K. L., Kyrtatos, P. G., Minogue, S., Al-Jamal, K. T., Kogelberg, H., Tolner, B., Kostarelos, K., Begent, R. H., Pankhurst, Q. A., Lythgoe, M. F. and Chester, K. A. 2009. Nanoparticles functionalised with recombinant single chain Fv antibody fragments (scFv) for the magnetic resonance imaging of cancer cells. *Biomaterials*, 31, 1307–1315.

81. Chester, K. A., Mayer, A., Bhatia, J., Robson, L., Spencer, D. I. R., Cooke, S. P., Flynn, A. A., Sharma, S. K., Boxer, G., Pedley, R. B. and Begent, R. H. J. 2000. Recombinant anti-carcinoembryonic antigen antibodies for targeting cancer. *Cancer Chemotherapy and Pharmacology*, 46, S8–S12.

82. Qin, D. L., He, X. X., Wang, K. M., Zhao, X. J. J., Tan, W. H. and Chen, J. Y. 2007. Fluorescent nanoparticle-based indirect immunofluorescence microscopy for detection of *Mycobacterium tuberculosis*. *Journal of Biomedicine and Biotechnology*, 2007, 89364.

83. Dubertret, B., Skourides, P., Norris, D. J., Noireaux, V., Brivanlou, A. H. and Libchaber, A. 2002. In vivo imaging of quantum dots encapsulated in phospholipid micelles. *Science*, 298, 1759–1762.

84. Zykova, M. G. 2008. Antirheumatic activity of methotrexate in phospholipid nanoparticles (phosphogliv). *Biochemistry (Moscow) Supplemental Series B: Biomedical Chemistry*, 2, 71–74.

85. Xenariou, S., Griesenbach, U., Ferrari, S., Dean, P., Scheule, R., Cheng, S., Geddes, D., Plank, C. and Alton, E. 2004. Magnetofection to enhance airway gene transfer. *Molecular Therapy*, 9, S180–S180.

86. Plank, C., Scherer, F., Schillinger, U., Bergemann, C. and Anton, M. 2003. Magnetofection: Enhancing and targeting gene delivery with superparamagnetic nanoparticles and magnetic fields. *Journal of Liposome Research*, 13, 29–32.

87. Vlaskou, D., Mykhaylyk, O., Krotz, F., Hellwig, N., Renner, R., Schillinger, U., Gleich, B., Heidsieck, A., Schmitz, G., Hensel, K. and Plank, C. 2010. Magnetic and acoustically active lipospheres for magnetically targeted nucleic acid delivery. *Advanced Functional Materials*, 20, 3881–3894.

88. Holzbach, T., Vlaskou, D., Neshkova, I., Konerding, M. A., Wortler, K., Mykhaylyk, O., Gansbacher, B., Machens, H. G., Plank, C. and Giunta, R. E. 2010. Non-viral VEGF(165) gene therapy—Magnetofection of acoustically active magnetic lipospheres ("magnetobubbles") increases tissue survival in an oversized skin flap model. *Journal of Cellular and Molecular Medicine*, 14, 587–599.

89. Alexis, F., Pridgen, E., Molnar, L. K. and Farokhzad, O. C. 2008. Factors affecting the clearance and biodistribution of polymeric nanoparticles. *Molecular Pharmaceutics*, 5, 505–515.

90. Dias, A. M. G. C., Hussain, A., Marcos, A. S. and Roque, A. C. A. 2011. A biotechnological perspective on the application of iron oxide magnetic colloids modified with polysaccharides. *Biotechnology Advances*, 29, 142–155.

91. Gref, R., Luck, M., Quellec, P., Marchand, M., Dellacherie, E., Harnisch, S., Blunk, T. and Muller, R. H. 2000. "Stealth" corona-core nanoparticles surface modified by polyethylene glycol (PEG): Influences of the corona (PEG chain length and surface density) and of the core composition on phagocytic uptake and plasma protein adsorption. *Colloids and Surfaces B: Biointerfaces*, 18, 301–313.

92. Hamidi, M., Azadi, A. and Rafiei, P. 2006. Pharmacokinetic consequences of pegylation. *Drug Delivery*, 13, 399–409.

93. Perrault, S. D., Walkey, C., Jennings, T., Fischer, H. C. and Chan, W. C. W. 2009. Mediating tumor targeting efficiency of nanoparticles through design. *Nano Letters*, 9, 1909–1915.

94. Shultz, M. D., Reveles, J. U., Khanna, S. N. and Carpenter, E. E. 2007. Reactive nature of dopamine as a surface functionalization agent in iron oxide nanoparticles. *Journal of the American Chemical Society*, 129, 2482–2487.

95. Reimhult, E., Amstad, E., Gillich, T., Bilecka, I. and Textor, M. 2009. Ultrastable iron oxide nanoparticle colloidal suspensions using dispersants with catechol-derived anchor groups. *Nano Letters*, 9, 4042–4048.

96. Green, N. M. 1990. Avidin and streptavidin. *Methods in Enzymology*, 184, 51–67.

97. Eakin, R. E., Snell, E. E. and Williams, R. J. 1940. A constituent of raw egg white capable of inactivating biotin in vitro. *Journal of Biological Chemistry*, 136, 801–802.

98. Eakin, R. E., Snell, E. E. and Williams, R. J. 1941. The concentration and assay of avidin, the injury-producing protein in raw egg white. *Journal of Biological Chemistry*, 140, 535–543.

99. Hendrickson, W. A., Pahler, A., Smith, J. L., Satow, Y., Merritt, E. A. and Phizackerley, R. P. 1989. Crystal-structure of core streptavidin determined from multiwavelength anomalous diffraction of synchrotron radiation. *Proceedings of the National Academy of Sciences of the United States of America*, 86, 2190–2194.

100. Weber, P. C., Wendoloski, J. J., Pantoliano, M. W. and Salemme, F. R. 1992. Crystallographic and thermodynamic comparison of natural and synthetic ligands bound to streptavidin. *Journal of the American Chemical Society*, 114, 3197–3200.

101. Sano, T. and Cantor, C. R. 1995. Intersubunit contacts made by tryptophan-120 with biotin are essential for both strong biotin binding and biotin-induced tighter subunit association of streptavidin. *Proceedings of the National Academy of Sciences of the United States of America*, 92, 3180–3184.

102. Chilkoti, A., Tan, P. H. and Stayton, P. S. 1995. Site-directed mutagenesis studies of the high-affinity streptavidin–biotin complex—Contributions of tryptophan residue-79, residue-108, and residue-120. *Proceedings of the National Academy of Sciences of the United States of America*, 92, 1754–1758.

103. Hyre, D. E., Le Trong, I., Merritt, E. A., Eccleston, J. F., Green, N. M., Stenkamp, R. E. and Stayton, P. S. 2006. Cooperative hydrogen bond interactions in the streptavidin–biotin system. *Protein Science*, 15, 459–467.

104. Hyre, D. E., Le Trong, I., Freitag, S., Stenkamp, R. E. and Stayton, P. S. 2000. Ser45 plays an important role in managing both the equilibrium and transition state energetics of the streptavidin–biotin system. *Protein Science*, 9, 878–885.

105. Freitag, S., Chu, V., Penzotti, J. E., Klumb, L. A., To, R., Hyre, D., Le Trong, I., Lybrand, T. P., Stenkamp, R. E. and Stayton, P. S. 1999. A structural snapshot of an intermediate on the streptavidin–biotin dissociation pathway. *Proceedings of the National Academy of Sciences of the United States of America*, 96, 8384–8389.

106. Holmberg, A., Blomstergren, A., Nord, O., Lukacs, M., Lundeberg, J. and Uhlen, M. 2005. The biotin–streptavidin interaction can be reversibly broken using water at elevated temperatures. *Electrophoresis*, 26, 501–510.

107. Kanaras, A. G., Wang, Z. X., Bates, A. D., Cosstick, R. and Brust, M. 2003. Towards multistep nanostructure synthesis: Programmed enzymatic self-assembly of DNA/gold systems. *Angewandte Chemie International Edition*, 42, 191–194.

108. Milam, V. T., Hiddessen, A. L., Crocker, J. C., Graves, D. J. and Hammer, D. A. 2003. DNA-driven assembly of bidisperse, micron-sized colloids. *Langmuir*, 19, 10317–10323.

109. Geerts, N. and Eiser, E. 2010. DNA-functionalized colloids: Physical properties and applications. *Soft Matter*, 6, 4647–4660.

110. Alberts, B., Johnson, A., Julian, L., Raff, M., Roberts, K. and Walter, P. 2007. *Molecular Biology of the Cell*. Garland Science, London.

111. Marks, F., Klingmüller, U. and Müller-Decker, K. *Cellular Signal Processing*. Garland Science, New York.

Bettina Kozissnik obtained her MSc in Austria. During that time, she joined Dr. Urs Häfeli's team at the University of British Columbia in Canada for her practical training semester. At the moment, she is finishing her PhD under the guidance of Prof. Kerry Chester and Prof. Quentin Pankhurst at the University College London (UCL) in cooperation with The Davy-Faraday Research Laboratory at The Royal Institution of Great Britain. Her current research interests are targeted magnetic nanoparticle hyperthermia for cancer therapy.

Luke A.W. Green obtained his MChem with a Year in Europe degree from the University of York in 2008. After a short spell working at a hospital biochemistry department, he is now in the 2nd year of study for a PhD under the guidance of Dr. Nguyễn Thị Kim Thanh at UCL and is based at The Davy-Faraday Research Laboratory, The Royal Institution of Great Britain. His focus is on the synthesis of magnetic NPs for biomedical application.

Professor Kerry A. Chester leads the Recombinant Antibody Therapeutics Group at the UCL Cancer Institute (http://www.ucl.ac.uk/cancer/). Her main research interests are design and construction of antibody-based therapeutics and the interaction of these molecules with specific cancer targets. The work is largely translational and aimed at developing new agents for clinical use. To facilitate this, her group has established an academic good manufacturing practice (GMP) facility to make clinical grade recombinant proteins for use in Phase I/II cancer trials. Kerry's current research includes developing antibodies as cancer imaging agents, as therapeutic fusion proteins and to target nanomedicines.

Dr. Nguyễn Thị Kim Thanh is the editor of this book. For more information see the editor's biography or visit her website: http://www.ntk-thanh.co.uk

6 Functionalisation of Magnetic Iron Oxide Nanoparticles

Sylvie Begin-Colin and Delphine Felder-Flesch

CONTENTS

The continuous growth of nanotechnology has brought challenging innovations in medicine, revolutionising the field of diagnosis and therapy. Indeed, in the field of the synthesis and functionalisation of inorganic nanoparticles (NPs) for biomedical applications, most researches aim at developing multifunctional theranostic (i.e. including therapeutic and diagnostic functions) NPs, which can both identify disease states and simultaneously deliver therapy. To ensure such multifunctional activity, the key points are the design of the organic coating and its grafting at the surface of NPs while preserving properties of both NPs and molecules. Indeed the coating design is challenging: the molecules anchored at the surface of NPs should bring different functions such as dyes to combine optical imaging, targeting ligands to reach target tissue or cells or therapeutic agents. There must be also functions preventing NPs from agglomeration in a physiological environment and favouring biodistribution (Figure 6.1). Most molecules have to be designed to fulfil all these requirements.

To be used *in vivo*, functionalised iron oxide (IO) NPs have to be stable in biocompatible solutions at pH close to the physiological blood pH (7.4) and close to the plasmatic iso-osmolarity (320 mmol L^{-1}). Indeed, the iso-osmolarity condition is essential to obtain for the *in vitro* cellular

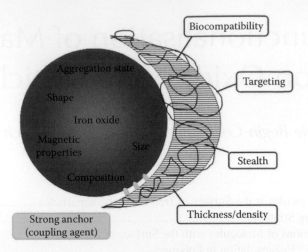

FIGURE 6.1 Schematic representation of a multifunctional nanoprobe.

marking. Furthermore, the absence of aggregates is mandatory for intravenous injection (risk of pulmonary embolism in case of macroaggregates or coagulation disorders in case of microaggregates, resulting in animal death). Upon entering the blood circulation, NPs are subjected to opsonisation, the non-specific fouling of plasma protein on the surface of NPs, and subsequent uptake by reticuloendothelial system (RES). For diagnostic or therapeutic applications, this non-specific uptake by macrophages can be used in some cases such as hepatic or ganglionic tumour imaging. With the exception of this specific case, biodistribution must be favourable, with little non-specific tissular uptake (mainly hepatic), and a quasi-complete elimination of the non-uptaken nano-objects, knowing that the most effective ways of elimination are the urinary and the hepatobiliary pathways. To prevent opsonisation of NPs and to increase the ability to evade the RES, the organic coating and its anchoring at the surface of NPs have to be tailored, and the particle size distribution of functionalised NPs must be optimised. Long blood circulation time would maximise the possibility to reach target tissue: suspensions of particles with average hydrodynamic (HD) sizes of 10–100 nm are optimal for *in vivo* delivery, as smaller ones (<10 nm) are rapidly removed by renal clearance, while bigger ones (>200 nm) are quickly sequestered by the RES.

The design of molecules for biomedical applications is therefore challenging. They have to bring several bioactive functions ensuring biocompatibility, stealth, targeting and therapeutic care. Furthermore, the molecules have to be strongly anchored at the surface of NPs to avoid their desorption under *in vivo* conditions, and the grafting way has to be optimised so that the particle size distribution stays below 100 nm. This chapter will aim at presenting the different organic coating currently developed for biomedical applications and the different strategies often used to anchor the molecules at the surface of IO NPs.

6.1 ANCHORING MOLECULES AT THE SURFACE OF IRON OXIDE NANOPARTICLES

6.1.1 GRAFTING STRATEGIES

The surface of a metal oxide is generally assumed to be covered with hydroxyl groups: M–OH (where M is the metal, in this case Fe)[1,2]. Therefore, as with all metal oxides, ferrites are amphoteric solids developing surface charges in water due to the protonation and deprotonation reactions of Fe–OH sites (see Chapter 2). An important point is thus the pH_{PZC} (pH at the point of zero charge, PZC), determined generally by acid/base potentiometric titration method and in the range of 6–8 for IO NPs[3–6]. Below pH_{PZC}, protonation reaction leads to the formation of $\equiv Fe–OH_2^+$ groups, while deprotonation occurring above pH$_{PZC}$ gives rise to $\equiv$Fe–O$^-$ groups. However, all Fe–OH surface sites are not equivalent, and less than 20% of the surface sites are generally considered to be ionisable[6]. One may notice that the term PZC is defined as a unique pH, where the net surface charge is zero for amphoteric oxides bearing only pH-dependent charges. When NPs are suspended in water, the colloidal stability generated by electrostatic interactions is determined by electrokinetics measurements, which give the evolution of the zeta potential (ζ) (the surface charge balance referred to the diffuse layer) as a function of pH. The pH at which the zeta potential value is zero is named isoelectric point (IEP). It is generally considered that at pH below and above this IEP value, the colloidal stability by electrostatic interaction is favoured when the absolute value of the zeta potential is higher than 30 mV. The pH$_{PZC}$ is equal to IEP when adsorbing ions are only OH$^-$ and H$_3$O$^+$. Therefore, zeta potential measurements as a function of pH are usually performed to investigate the colloidal stability of functionalised NPs and the influence of the organic coating on this stability.

Conjugation of molecules occurs mainly by interactions of molecules with these surface hydroxide groups. In water, electrostatic interactions, between the molecules and the surface of NPs, may be induced depending on the pH and on the pKa of functional groups carried on the molecules. Indeed, the IEP of naked IO is around 6.8; their surface is thus positively charged below this pH value and may interact electrostatically with molecules bearing negatively charged functions below this pH, that is, with pKa values well below 6.8. Stronger binding may be generated by interactions of the surface hydroxide groups with coupling agents associated with the molecules. These last interactions may occur either in water or organic solvents.

However, depending on the synthesis method, IO NPs may be either dispersed in water without molecules at their surfaces as after hydrolytic synthesis methods (i.e. co-precipitation technique) or *in situ* coated with surfactants during the synthesis step and mostly dispersed in organic solvent as after non-hydrolytic synthetic routes (i.e. thermal decomposition) (see Chapter 2; Figure 6.2). Two major approaches are therefore developed to functionalise the NPs: *in situ* coating (during the 'one-pot' synthesis) and post-synthetic coating[7]. During the *in situ* coating process, precursors of magnetic cores and coating materials are dissolved in the same reaction solution, and the nucleation of magnetic cores occurs within the coating materials. Therefore, the magnetic cores and the coatings of NPs form at the same time. *In situ* coatings are mainly observed with NPs synthesised by

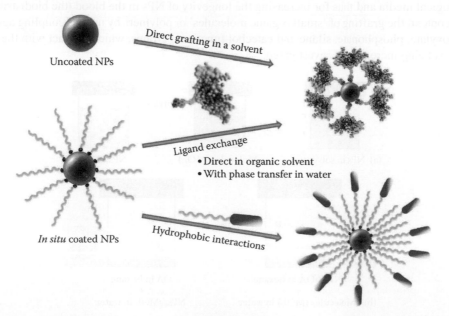

FIGURE 6.2 Grafting strategies in the post-synthetic coating process as a function of the NP synthesis methods leading either to uncoated or to *in situ* ligand-coated NPs.

co-precipitation or by synthesis method leading to surface free NPs in water. In general, biopolymers such as carbohydrates (dextran, chitosan, alginate, arabinogalactan), proteins, etc., synthetic polymers such as polyethylene glycol (PEG), poly(vinyl alcohol) (PVA), poly(acrylic acid) (PAA), poly(methylacrylic acid) (PMAA), poly(lactic acid) (PLA), polyvinylpyrrolidone (PVP), polyethyleneimine (PEI) and AB and ABC-type block copolymers containing the aforementioned polymers as segments or small molecules such as citric acid, tartaric acid, gluconic acid, dimercaptosuccinic acid (DMSA) and phosphoryl choline are often used as precipitating agents.

In the post-synthetic coating process, first, the magnetic cores are synthesised, and depending on the synthesis method, their surface is 'free' or capped by surfactants protecting against agglomeration (Figure 6.2). Then, the coating materials are introduced to the surface of the magnetic cores mainly by direct grafting, ligand exchange or hydrophobic interactions (Figure 6.2)[7–12].

Direct grafting is developed with uncoated NPs by introducing molecules in the NPs suspensions, and the solvent may be either water or organic (Figure 6.3)[13,14]. Coating of NPs by ligand exchange and hydrophobic interactions are performed mainly with NPs *in situ* coated by surfactants. In the case of the ligand exchange process, it may occur by introducing directly the molecules in the suspension of NPs in organic solvent or by a phase transfer process in water (Figure 6.3b). The original ligands on the surface can be displaced without changing the intrinsic properties of the iron core. Compared with surface encapsulation, the NPs resulting from this approach are more stable because stronger coordinate bonds formed during the surfactant exchange reaction overcome the inherent shortcomings of the surface-encapsulated NPs[15].

NPs grafted with molecules may be schematically described either as NPs capped with functional polymers, either through encapsulation or grafting, or as NPs coated by organic functional molecules, grafted through coupling agent or introduced through hydrophobic interactions (Figures 6.2 and 6.4). There are also examples of NPs encapsulated in micelles or liposomes, but they will not be detailed in this chapter.

Another important point is the nature of the coating interactions with the surface of NPs as this coating may encapsulate the NPs or may be physisorbed (electrostatic interactions, hydrophobic interactions) or chemisorbed at the surface of NPs (through anchoring groups). Current studies suggest that a strong binding of molecules at the surface of NPs is important for the colloidal stability in physiological media and thus for increasing the longevity of NPs in the blood (the biodistribution)[7]. In that context, the grafting of 'small organic molecules' or polymers by using a coupling agent such as carboxylate, phosphonate, silane and catechol functional groups, which interact with the surface of NPs, is being increasingly investigated.

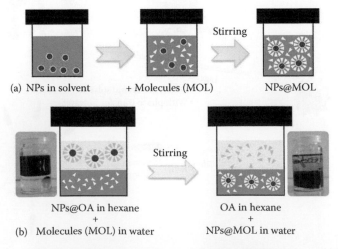

(a) NPs in solvent   + Molecules (MOL)   NPs@MOL

NPs@OA in hexane   OA in hexane
+   +
(b)   Molecules (MOL) in water   NPs@MOL in water

FIGURE 6.3 (a) Direct grafting of uncoated NPs by molecules (MOL) (white triangle) and (b) ligand exchange and phase transfer process in water of NPs coated by oleic acid (OA) (green triangle) (NPs@OA).

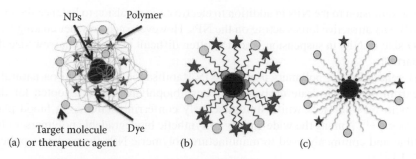

FIGURE 6.4 Schematic representation of NPs: (a) coated by polymers, (b) with molecules grafted at their surface, and (c) coated with functional molecules through hydrophobic interactions.

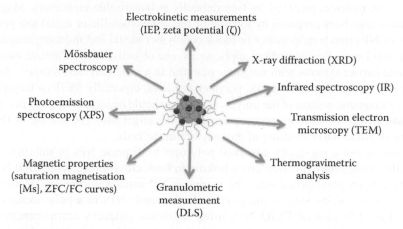

FIGURE 6.5 Main characterisation techniques of functionalised NPs (IEP, DLS and ZFC/FC = zero field cool/field cool curves).

Thus, different grafting strategies have been developed according to the nature of the molecules, the NPs synthesis methods, the surface of NPs and the type of solvents[8,12,16]. These strategies involve grafting (i.e. strong anchoring) of organic molecules or coating (i.e. encapsulation) with polymers or an inorganic layer, such as silica or metal or non-metal. Practically, it is desirable that in many cases the protecting shells not only stabilise the magnetic IO NPs, but can also be used for further biofunctionalisation. Most of the strategies aim at performing this grafting step by ensuring a good colloidal stability in water and/or in physiological media as well as controlling the grafting rate (the number of grafted molecules) and at preserving the properties of both molecules and NPs. The coating quality on the NPs can be monitored by measuring the HD size, surface charge (IEP) and other physicochemical parameters (Figure 6.5). The HD size of NPs can be obtained by dynamic light scattering (DLS) and is often studied under different solution conditions, such as various pH, ionic strength and temperature[17,18]. The surface charge of NPs is examined by measuring the zeta potential as a function of pH to determine the pH range stability of suspensions. Once the NPs are well characterised after the synthesis and coating stages, further derivation and evaluation towards biomedical applications can be performed.

6.1.2 Interactions of Molecules with the Surface of Nanoparticles

6.1.2.1 Encapsulation in Polymers

Encapsulation of NPs in a polymeric matrix may proceed through three main ways: the polymerisation is induced in a suspension of NPs, or the NPs synthesis occurs in a polymeric media or polymer molecules are grafted at the surface of NPs. Polymers are interesting as stabilising agents as they provide steric repulsion to the NPs in addition to electrostatic repulsion to balance the magnetic and the *van der Waals* attractive forces acting on the NPs. However, such polymer coating leads to large average HD size of NPs in suspension, and it is often difficult to obtain a narrow size distribution of coated particles.

The first generations of polymers used for NPs stabilisation were based on natural polymers, such as dextran or carbohydrate derivatives. These biopolymers were adopted for their ability to interact with NPs surfaces while simultaneously conferring stability in blood plasma (low-fouling surfaces)[12,19]. Among the wide variety of synthetic biodegradable polymers, only a few are biocompatible and commonly used to manufacture polymeric NPs for medical applications (see Section 6.2.1.1)[12,20].

Emulsion and dispersed media polymerisations have been utilised to encapsulate IO NPs into polymeric micro- and nanospheres[12,19,21]. These materials typically kinetically trap numerous IO NPs in polymeric matrices prepared by free radically polymerisable monomers. Magnetic nano-composites have also been prepared by loading of the organometallic or metal salt precursors for the synthesis of NPs into homopolymer or block copolymer media and inducing magnetic NP formation in the solid state. For biomedical applications, one objective is to formulate multifunctional nanoparticulate carrier systems with magnetic material in a solid polymer matrix[22]. An important challenge in the preparation of magnetic polymer particles, especially for drug targeting application, is that the magnetic content of the polymer particles should be large enough to permit magnetic guidance and delivery of these magnetic particles to the target site. The release of the drug will depend on the degradation or swelling of the polymer in the body.

A limitation of using physically adsorbed polymers to disperse NPs is stability. Attempts to address stability issues have involved cross-linking to form cross-linked IO (CLIO) NPs. Dextran molecules have been physisorbed onto the magnetic NP surface during the synthesis process, followed by a cross-linking step to encapsulate the magnetic NPs in a polysaccharide matrix[23]. The chemical modification of CLIO NPs using ammonia produces amine-functional dextran suitable for attaching biomolecules such as proteins and peptides or hydrophobic drugs such as doxorubicin[24].

In the case of the grafting of polymers, the anchoring of the polymer to the NPs surface can be achieved using two alternative approaches:[12,25] the functional polymer is directly grafted onto the NPs[26] or an initiator is fixed at the surface of the NPs, and the polymer is grown from the surface through, for example, an atom transfer radical polymerisation (ATRP)[27] or by initiator ring-opening polymerisation (ROP)[10]. Each of these approaches has advantages and disadvantages. The ROP yields a higher grafting density than the grafting of preformed polymers, but this last strategy allows better control of polymer architecture and functionality. Indeed, the ATRP method can offer polymeric shells with low polydispersity. With this method, it is easy to control the molecular weight, thereby, the thickness of the polymeric shell. When direct grafting is involved, the attachment of polymer chains to NPs can also be achieved *via* multiple interactions between the polymer chains and the particle surface. Several coupling agents can be introduced on the polymer chains and represent multiple ways of binding the NPs surface, but this process may reduce the packing density. In the following sections, we will discuss the nature of the interaction between the coupling agents and the NP surfaces.

6.1.2.2　Coating through Hydrophobic Interactions

A special case of the surface-encapsulation strategy is the formation of coatings *via* hydrophobic interactions, named also bilayer functionalisation[7,8,28–31]. Magnetic NPs that are synthesised by thermal decomposition method are often capped by hydrophobic surfactants. Those hydrophobic magnetic cores can be coated with amphiphilic molecules, forming micelle-like structures. An amphiphilic molecule consists of both hydrophilic and hydrophobic regions. The hydrophilic regions can interact with water by either ionic interactions or hydrogen bonding, stabilising NP-containing micelles in aqueous solutions. The hydrophobic regions are typically long-chain hydrocarbons that can interact with hydrophobic surfactants on the NPs surface such as alkylphosphonate surfactants and other surfactants such as ethoxylated fatty alcohols or phospholipids or 11-aminoundecano-ate[7,30,32,33]. Cetyl trimethylammonium bromide (CTAB) is a common surfactant, which can be self-assembled with oleylamine or oleic acid on the surface of hydrophobic NPs and make them hydrophilic. Various amphiphilic copolymers derived from PEG[7,24,29,34–37] or polysaccharides[38] or pluronic PF127 copolymers[39] have been successfully used to transfer hydrophobic NPs from organic solvents[8,29,40,41].

Moreover, the amphiphilic polymers also offer different chemical functional groups, including carboxylic acid, amine, thiol, carbonyl, hydrazide and biotin and facilitate subsequent immobilisation of various biological molecules *via* bioconjugation chemistry.

The approach has the advantage for large-scale preparation of hydrophilic NPs. However, there are several bottlenecks for this approach[15]. First, the obtained NPs cannot be stable under physiological conditions because of the weak interaction between hydrophobic double-layer structures. Second, the size of the amphiphilic copolymer-encapsulated NPs often exceeds 30 nm, and this would result in shorter blood circulation time *in vivo*. Furthermore, it is very difficult to manipulate the number of terminal functionalities for controllable bioconjugation due to the uncertainty in the number of functional groups in the polymer.

In order to form a stable surface coating, various surface-binding groups have been explored such as carboxylate (COO^-)[42,43], phosphate (PO_3^{3-})[13] and alcoholate (O^-)[32].

6.1.2.3 Interactions Using Coupling Agents

Covalently binding is an alternative to the physisorption as physisorbed molecules are easily desorbed when NPs are introduced to physiological media. Indeed, NPs coated with adsorbed molecules tend to aggregate in phosphate-buffered saline (PBS) due to displacement of carboxylic acid groups from the NP surface by phosphate salts[44–48]. The first experiments involving strong binding between the molecule and the NPs have been conducted with a well-known polymer, that is, dextran. The adopted strategy was based on the incorporation of aldehyde groups into the dextran structure, which can be conjugated to amine $(-NH_2)$ groups that have been previously grafted onto the magnetic NPs *via* a Schiff's base reaction[49].

The chemistry consisting in grafting covalently, at the surface of the NPs, a molecule bearing specific functions used afterwards to couple bioactive molecules (such as dye or fluorescent molecules, therapeutic agents or targeted molecules) through click or sulfhydryl, carboxyl, anhydride or imine chemistry, is described in Chapter 5[45,50]. We will focus here on the chemisorption of molecules at the surface of NPs and especially on the coupling groups, which are usually used (Figure 6.6). The most common and initially investigated functional group for anchoring on an IO core is the carboxylate group[10,45,51,52]. There are few reports that have investigated other anchoring agents such as sulphonate, phosphate and phosphonate, despite of their strong affinity towards the IO core[53,54]. Moreover, very recently, other high-affinity surface-capping agents have been explored to improve the coating stability, such as hydroxamic acid, phosphine oxide and catechol-based ligands.

6.1.2.3.1 *Silane Group*

The silanisation process is based on the covalent binding of silane-based molecules, such as (3-aminopropyl) trimethoxysilane (APTMS), to the surface of NPs, which bring functional groups on which bioactive molecules may be further coupled[12,45,50,55–57]. The silane coupling agent, usually an organosilane, has the following structure: $F_xSiR_{(4-x)}$. Silicon is located at the centre of the molecule and contains functional groups, F (vinyl, amine, chloro, etc.), for further coupling of bioactive molecules and other functional groups, R (methoxy, ethoxy, etc.). The organic groups R–Si of the molecule hydrolyze to silanol and form a metal hydroxide or siloxane bond with the inorganic materials (Figure 6.7). The functional group F can be connected covalently with organic biomaterials. Alkoxysilanes have been studied extensively since they can form highly stable polysiloxane shells on the surface of metal oxide NPs[12,58–64].

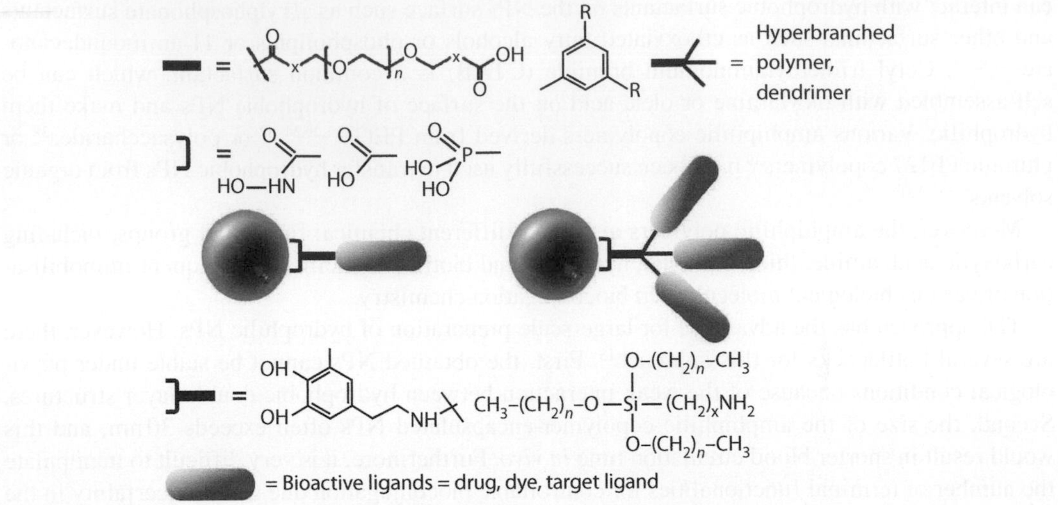

FIGURE 6.6 Schematic representation of the different coupling agents used to induce a strong binding between molecules and NPs.

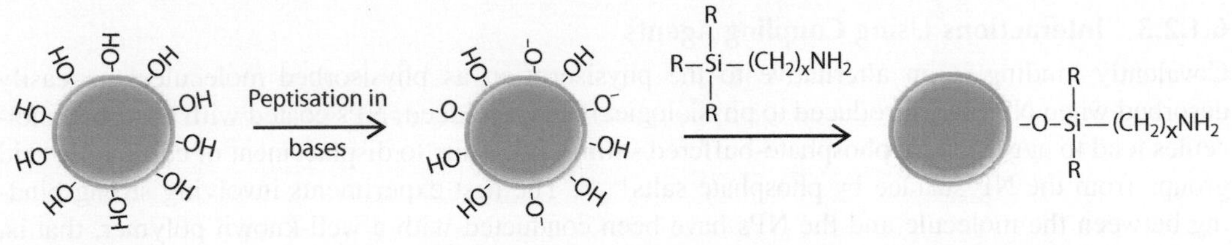

FIGURE 6.7 Schematic view of surface modification of magnetic NPs by a silanisation process (R = methoxy, ethoxy, etc.), and here the functional group (F) brings an amine function for further coupling with bioactive molecules.

The silane agent is often considered as a good candidate for modifying the surface of IO NPs directly. Silanes were found to render the IO NPs highly stable and water dispersible, and it was also found to form a protective layer against mild acid and alkaline environments[12]. The main advantages are the biocompatibility as well as a high density of surface functional end groups, such as alcohol, amine and thiol, which are useful for further biofunctionalisation using small bio-compounds[64] and carbohydrates[65]. For example, (3-aminopropyl)trimethoxysilane (APTMS) coated NPs, suspended in deionised water at pH = 5, could not be suspended in any buffers where they are aggregated and eventually precipitated[45]. APTMS-coated NPs on which carboxymethyl dextran was covalently grafted through the ammonium group were found stable for up to 48 h in all buffers. The colloidal stability is attributed to the steric repulsion offered by the covalently grafted dextran preventing agglomeration of NPs. Functional PEG–silane has been grafted to NPs using a ligand-exchange strategy and have improved their water solubility and biocompatibility[8,60].

6.1.2.3.2 Carboxylate

Another important and widely employed functionality for the modification of NPs surfaces is the carboxylic acid group, which can interact with the surface of NPs by a coordination process. The –COOH group has been employed for the NPs synthesis in organic solvents, with oleic acid used as a surfactant.

Molecules integrating all bioactive functions and also bearing a carboxylate function for the anchoring to the NPs surface, or small intermediary molecules, such as lipoamino acid (LAA) that can ensure the availability of a complementary attachment site for functional biomolecules, may be grafted[56]. 2,3-DMSA was widely used to prepare DMSA-coated Fe_3O_4 NPs, which are dispersible in PBS[7,29,66,67]. DMSA is an organic molecule composed of two carboxylate (pK_{COOH} = 2.7:3.4) and two thiol functions (pK_{SH} = 9.65). DMSA can form a stable coating on the IO surface *via* carboxylic chelate bonds and stabilise the shell of IO NPs by forming intermolecular disulphide cross-linkages.

Direct preparation of water-soluble NPs by thermal decomposition has also been reported. Gao et al. synthesised methoxy-polyethyleneglycol (MPEG)–COOH-coated or HOOC–PEG–COOH-coated Fe_3O_4 NPs by the thermal decomposition of $Fe(acac)_3$ in the presence of MPEG–COOH or HOOC–PEG–COOH[51]. The resultant nanocrystals are very soluble in aqueous solution and physiological saline and possess a long blood circulation time.

Concerning the stability of NPs in buffers, Goff et al.[68] reported that the strong adsorption affinity of phosphate ions onto IO surfaces affects the anchoring of functional groups, such as ammonium or carboxylate, which results in the desorption of these functional molecules. Indeed, the –COOH/NPs coordination bond has been observed to be sometimes labile and to break easily by increasing temperature or by exchanging with another carboxylic acid compound or stronger binding compounds. It should depend on the nature of the surface complex: monodentate, bridging bidentate or chelating bidentate, which certainly depend on the grafting conditions.

6.1.2.3.3   Carboxylate and OH Groups

Carboxylic groups have been reported to be more active or more important than the hydroxyl group. However, when the hydroxyl group is ortho-positioned with respect to the carboxyl groups, the hydroxyl group and carboxyl group will simultaneously coordinate to the iron atom, forming a chelating complex structure[7,69]. This surface complexation was observed with molecules such as PAA, citric, tartaric and gluconic acids used during one-pot synthesis processes.

6.1.2.3.4   Catechol

The ortho-dihydroxyphenyl (catechol) functional group strongly coordinates to various inorganic–organic surfaces[57,70–72]. Catechol unit in dopamine or dopamine–PEG derivative can coordinate to the surface of IO as a result of improved orbital overlap of the five-membered ring and a reduced steric effect on the iron complex, resulting in strong binding between dopamine moiety and the surface of the IO NPs[73]. A range of biologically important molecules have been bound to dopamine through its amine function: a carboxyl-ended DPA–PEG-based ligand, hyaluronic acid, conjugate peptide, amino acids, dendrimers, etc.[15,29,57,74] However, problems with the stability of this bond in water and biological fluids have been reported after long exposure periods[75].

6.1.2.3.5   Hydroxamic Acid

Hydroxamic acid derivatives are also developed to anchor molecules at the surface of NPs (Figure 6.6)[76–78]. They were mainly used in a ligand exchange process to coat NPs *in situ* coated with fatty acids during their synthesis step. Such hydroxamic acids anchoring groups enable an exchange even against strongly bound ligands. This is attributed to the formation of a five-member chelating ring instead of a four-member ring or monodentate binding, which occurs in the case of carboxylic acid ligands[78].

6.1.2.3.6   Phosphonate Groups

Recently, the phosphonate moiety was introduced as an effective anchoring agent because of the high ability of $-PO(OH)_2$ groups to complex metal ions and to form complexes, which are stable even at elevated temperature. Indeed, phosphonates have a strong tendency to adsorb onto a variety of metal oxide surfaces, making them attractive for use in a variety of applications[57,79].

One study aiming at comparing the grafting rate induced by a phosphonate or a carboxylate group has demonstrated that the phosphonate coupling agent leads to a higher grafting rate and a stronger molecule anchoring at the surface of NPs than the carboxylate group[14,79,80]. These bonds have been found to be more stable than the ones formed through the carboxylic acid function and have shown stability for several weeks at neutral pH[79]. Moreover, the magnetic properties are better preserved[14]. The P–C bond system in phosphonic acids has low toxicity, appreciable thermostability and is highly resistant to enzymatic cleavage[81].

The use of bifunctional phosphonic acid-based coupling agents with polar end groups (–COOH, –OH, –NH$_2$) not only renders the NPs hydrophilic and stable with respect to aggregation, but also imparts functionality on the surface to provide facile access to bioconjugates[43,57,82–85]. NPs have been coated by amine-derivatised N-phosphonomethyl iminodiacetic acid (PMIDA) and then bioactive molecules (a fluorescent dye: rhodamine B isothiocyanate, a cancer-targeting folic acid [FA] and the folate analogue methotrexate [MTX] through a pH labile ester linkage) have been coupled on PMIDA through amine groups[86]. The NPs have been synthesised by the well-known co-precipitation methods in presence of PMIDA and then further modified.

However, some problems may appear when bifunctional phosphonic acid molecules are grafted after the synthesis of NPs. Three hydrophilic oligoethyleneglycol-based dendrons displaying a phosphonic acid at the focal point and three biocompatible chains (oligoethyleneglycol) bearing functional groups such as COOH or NH$_2$ were designed (Figure 6.8a). The direct grafting of uncharged dendrons (only PEG chains, Figure 6.8a), by introducing the molecules to an uncoated co-precipitated aqueous NP suspensions, was demonstrated to occur at pH 5 by favouring interactions of negatively charged phosphonate groups with hydroxyl and positively charged groups at the IO surface[6,13]. Indeed, the phosphonate group displays two pKa values, 3.1 and 5.4, and is thus negatively charged in a large pH range. The zeta potential measurements as a function of pH support a positively charged surface below pH 6.8 and moreover, the lower the pH, the higher the surface charge density (Figure 6.8b). Therefore, taking into account that, after the grafting step, the zeta potential curve is shifted towards acidic pH[12] and that the IEP of dendronised NPs is around 4, a grafting step achieved at pH 5 was shown not only to favour electrostatic interaction between the NP surface and the phosphonate anchor, but also to ensure a given suspension stability of the decorated NPs.

In the case of charged dendrons, such grafting process at pH 5 was unsuccessful. Indeed, in the case of carboxylated dendrons, phosphonate as well as carboxylate functions are both negatively charged at this pH and thus interact competitively with the NPs surface. In the case of aminated dendrons, the amine function is positively charged at pH 5 and can then either interact with the negatively charged phosphonate or with the NPs surface, leading to lower grafting rate. Elaboration of new grafting strategies based on the different pKa values of these three functions (phosphonic acid, carboxylic acid and primary amine) was therefore mandatory to allow efficient grafting on the NPs (Figure 6.8b). As the pKa of the carboxylic acid/carboxylate couple is about 4 and the NPs are positively charged at pH lower than 6.8, the grafting process has been conducted at pH 3.5. With aminated dendrons, grafting experiments have been performed at pH higher than 11. However, at this pH value, the NP surface is negatively charged and the affinity of phosphonate for iron is not sufficient to induce

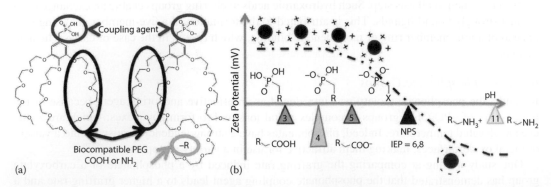

FIGURE 6.8 (a) Phosphonated dendrons: PEG dendron, carboxylated dendrons and aminated dendrons; (b) zeta potential curve of IO NPs together with pKa values of phosphonic and carboxylic acids and primary amine.

bonding. Finally, the grafting process has been investigated at pH 6.4, slightly lower than the IEP, so the surface is slightly positively charged and the repulsion and attraction strengths are minimised.

The amount of grafted dendrons deduced from chemical analyses of iron and phosphorus and by considering that the surface covered by one molecule is equal to $0.75\,nm^2$ (a value deduced from molecular modelling experiments) was 1.35 molecules nm^{-2} for PEG dendron, 1 molecule nm^{-2} for carboxylated dendron and 0.8 molecule nm^{-2} for aminated dendron. The theoretical amount of grafted molecules is calculated to be 1.4 molecules nm^{-2} and is almost reached with uncharged dendrons. The lower grafting rates observed with charged dendrons may be related either to steric hindrance induced by the longer oligoethyleneglycol chain bearing the charged group, or to the charge itself, which induces repulsive or attractive interactions limiting grafting. NPs coated with carboxylated dendrons were found to be very stable in iso-osmolar media.

Hence, surface bioengineering of nanomaterials involving phosphonic acid coupling agents appears to be highly promising.

6.1.2.3.7 Inorganic Coatings

Surface functionalisation can also be achieved by coating the IO particle with silica[57,87–90], Au[91] and other metals or metal oxides[7,12].

A silica shell is the most common approach for protection and modification of magnetic cores. It can prevent any direct contact of the magnetic core with additional reagents to avoid unwanted interactions, such as acidic corrosion. Furthermore, silica coatings have many advantages such as better biocompatibility, better stability under aqueous conditions (hydrophilicity), developed and facile surface modification approaches and easy control of interparticle interactions[12,92].

Different approaches have been explored to generate silica coatings on the surfaces of IO particles. The Stöber method and sol–gel processes are the most common choices for coating magnetic NPs with silica. The coating thickness can be tuned by varying the concentration of ammonium and the ratio of tetraethoxysilane (TEOS) to H_2O. Another method is based on microemulsion synthesis[93], in which micelles or inverse micelles were used to confine and control the coating of silica on core NPs. This last method might require much effort to separate the core–shell NPs from the large amount of surfactants associated with the microemulsion system, but controlled silica shell thickness may be obtained. Aerosol pyrolysis has also been used for silica coating, but the structure of composite NPs commonly is the mosaic type, such as hollow silica spheres with IO shells[94]. Dye molecules or other compounds can also be co-encapsulated into the silica shell[57].

In this way prepared, NPs are highly stable in aqueous dispersions. Though, various recent publications have shown that silica or silica-coated particles have a significant cytotoxicity and may induce oxidative stress[78]. In magnetic resonance imaging (MRI) applications, the shell thickness of $\gamma\text{-}Fe_2O_3@SiO_2$ NPs has a significant impact on their relaxivities by compromising the efficiency of NPs in relaxing the surrounding water molecules[92].

From the surface functionalisation point of view, the silica coating will offer versatile choices for coupling the particles with targeting molecules. The use of siloxane is the prevailing approach for the further modification of the surface of the silica coating[92,95].

In general, there are two ways for preparing metal functionalised IO NPs[12]. One is direct reduction of the single-metal ions at the surface of IO NPs. The most common route was by reduction of the single-metal ion on the surface of small molecules, polymer or SiO_2 functionalised IO NPs. Several authors have reported magnetic IO NPs coated with gold. Gold coatings provide not only stability to the NPs in solution, but also help in binding biological molecules through their –SH group for various biomedical applications.

6.1.3 CHARACTERISATION OF THE COATING

One important fact is to verify that the NPs are coated by organic molecules and that a bond has been formed between the coupling agent and the surface of NPs. The anchoring groups have different binding affinities, which should affect significantly the final performance of the NPs. Strong anchoring groups favour a higher density of molecules at the surface of NPs and thus afford higher stability, anti-biofouling ability and higher densities of bioactive molecule. For example, it has been shown both experimentally and theoretically that when the density of PEG layer reaches a given value, the unspecific adsorption of protein to the NPs may be completely prevented[96,97].

6.1.3.1　Identification of the Surface Complexes

The main characterisation techniques confirming grafting of molecules are infrared (IR) spectroscopy, thermogravimetric analysis (TGA) showing weight losses and sometimes photoemission spectroscopy.

X-ray photoelectron spectroscopy (XPS) analysis is used to validate the successful coating on the NPs surface by considering C1s, O1s, N1s, P2p and Fe2p spectra.

The C1s spectrum may display typical peaks at $\sim$285, 286.5, 288.8 and 289.5 $\pm$ 0.3 eV, attributed to C–C/C–H, C–O (ester bonds), O–C=O and C=O (carbonyl)[14,26,64,85,98]. The C1s peak attributed to C–C/C–H bonds is generally used as a reference and is fixed at 285 eV. Carbonates at the surface of NPs lead to a band at $\sim$290 eV. The appearance of C–N and N–C=O peaks at 285.8 $\pm$ 0.3 eV and 287.3 $\pm$ 0.3 eV, respectively, is due to the amine groups present on molecules as well as to the amide linkages formed to couple bioactive molecules.

The consideration of O1s peaks may help to discriminate between the types of complex formed at the surface of NPs. The oxygen 1s peak of uncoated IO NPs synthesised by co-precipitation is deconvoluted into four spectral bands at 530.1, 531.0, 532.1 and 533.7 $\pm$ 0.3 eV attributed respectively to the lattice oxygen (O_2-) (Fe–O–Fe) in the metal oxide, to hydroxides (Fe–OH), to water (Fe–OH_2) and to carbonates at the surface of IO[6]. Carbonates lead to an O1s band at 533.2 eV in correlation with the C1s band at 290 eV.

After coating with molecules, the O1s band is more complicated to analyse as there are the contributions of several other oxygen-based bonds. In particular, the carboxylate anchoring is difficult to analyse by this technique. Apart from the presence of Si in the XPS spectrum, the silane coating leads to the appearance of two pronounced peaks in the O1s spectrum at $\sim$531.5 and $\sim$532.5 eV. These peaks are located at approximately +2 eV with respect to the principal Fe–O peak ($\sim$530 eV) and can be mainly assigned to Fe–O–Si and Si–O bonds, respectively[64,99]. After grafting of phosphonate-based molecules, beside the bands due to Fe–O bonds, a peak located at a binding energy of 531.5 eV is observed and corresponds to contributions of P–O bonds (P=O, P–OH, Fe–O–P) in agreement with bibliographic data, which give the binding energies of M–O–P (M – metal) and P=O in the range 531.3–532.1 eV[6].

The P2p spectrum of phosphonate-based molecules exhibited two peaks at $\sim$133.8 and 134.7 eV, corresponding to P2p$_{3/2}$ and P2p$_{1/2}$, respectively[6,86,100]. When the molecules are grafted, a shift of the peaks towards lower binding energies is observed (peak located at 132.5–133.4 eV), which is characteristic of the formation of P–O–Fe bonds[100]. Such significant shift (1.3 eV) suggests that the electron density around P atoms is lower in phosphonated NPs and that the surface complex is at least binuclear.

In the N1s spectrum, peaks may appear at 400 and 406.08 eV attributed to tertiary nitrogen and amine bonds, respectively[26].

The Fe2p doublet with binding energy values of 710 and 725 eV implied the presence of Fe–O bonds, typical for IO. The Fe^{3+} and Fe^{2+} ions are distinguishable by XPS. Indeed, when Fe^{2+} ions are present at the surface, the satellite of the 2p$_{3/2}$ peak around 719 eV characteristic of the Fe^{3+} ions in γ-Fe_2O_3 becomes less resolved due to the main 2p$_{3/2}$ and 2p$_{1/2}$ peaks broadening and to rising intensity at about 716 eV of the satellite for the Fe^{2+} ions. The presence of Fe^{2+} at the probed surface of IO is shown by the quiet absence of the satellite around 719 eV between the main Fe2p peaks. Moreover, the Fe2p$_{3/2}$ peak position is around 710.6 eV for FeO (Fe^{2+}) and is at 711 eV for magnetite. The consideration of the Fe2p spectrum after the grafting step allows controlling if surface modification has or has not modified the iron composition of NPs[6,85].

In IR spectroscopy, characteristic bands of the molecules on the IR spectra of coated NPs may be highlighted[64,101]. Initially, the IR bands of alkyl chain (the asymmetric $-CH_2$ stretchings, $-CH_2$ deformation and $-CH$ deformation at around 2928 and 2850, 1486 and 1328 cm[-1], respectively)[102] are identified and afterwards other bands such as ether bonds ($C-O-C$ in PEG chains: $\sim$1050 cm[-1]), benzene groups, carboxylate or amine end groups. The presence of $-CO-NH-$ bands is characterised by two prominent bands at 1650 and 1547 cm[-1]. A characteristic $C-N$ peak was observed at about 1130 cm[-1].

The characteristic bands of end functions, mainly carboxylic acid and amine groups can also clearly be identified. For carboxylate: the carbonyl $C=O$ bond is often identified around 1730 ± 20 cm[-1]. After coupling of bioactive molecules, the $C=O$ stretching band is shifted or disappears. For amine functions, the vibrational modes at 3400, 1627, 1556 and 873 cm[-1] are attributed to the $N-H$ stretching vibration, $-NH_2$ bending mode of free amino groups, $C-N$ stretching mode and NH_2 wagging mode, respectively[102,103]. A band at 1095 cm[-1] confirms the presence of tertiary amine. Imine formation may be easily recognisable from the stretch of the $C=N$ bond group at ca. 1645 cm[-1][101].

Besides, the clear identification of the type of complex formed at the surface of IO appears often difficult. For silane-based molecules, a strong band is observed with peaks at 1108 and 1000 cm[-1], which are assigned to the oligomerisation of siloxane groups, leading to the formation of $Si-O-Si$ (Figure 6.9)[64]. These groups can condense around the magnetic NPs forming a highly cross-linked polysiloxane film, which entraps the NPs[49,103]. The interaction between iron atoms on the NP surface and siloxane ($Si-O$) groups from the APTMS molecules has been reported at about 584 cm[-1][103].

The binding of carboxylic acid to iron is apparent from IR spectra where the $C=O$ band of the carboxylic group at about 1687 cm[-1] disappeared and asymmetric $\nu_{as}(COO-)$ and symmetric $\nu_s(COO-)$ bands appeared[78]. These asymmetric $\nu_{as}(COO-)$ and symmetric $\nu_s(COO-)$ bands are usually observed in the range (1650–1500 cm[-1]) and (1340–1440 cm[-1]), respectively. In addition, information on the binding configuration of the carboxylate group to iron can be obtained from a comparison of the splitting of the asymmetric and symmetric carbonyl stretching modes. The largest difference (200–300 cm[-1]) would correspond to the monodentate interaction, the separation

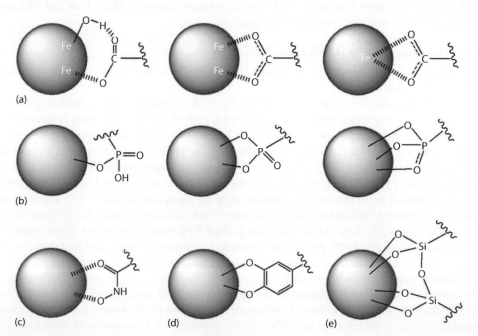

FIGURE 6.9 Schematic presentation of possible surface complexes as a function of the coupling agent: (a) carboxylate, (b) phosphonate, (c) hydroxamic, (d) catechol and (e) silane.

in the range 140–190 cm^{-1} to the bridging bidentate and the smallest separation (<110 cm^{-1}) to the chelate bidentate. Most reported surface complexes are either bridging bidentate or chelate bidentate (Figure 6.9)[78]. Nevertheless, this analysis is often complicated by a possible overlap with the bands of other functional groups of molecules.

The characterisation of the bonding with a catechol coupling agent is difficult by IR spectroscopy: no clear bands allow evidencing the binding[15]. Similarly, the complexation of hydroxamic group is not easy to observe: a peak around 1590 ± 5 cm^{-1} is assigned to the carbonyl vibration upon metal complexation[78,104].

Borggaard et al.[105] observed a strong adsorption affinity of phosphate ions onto well-crystallised oxides such as goethite (α-FeOOH), which was related to formation of a binuclear phosphate complex on the goethite surface by replacing singly coordinated OH groups. P–OH and P=O bonds in phosphonate-based molecules are characterised by the presence of bands at 900–1000 cm^{-1} and around 1200–1250 cm^{-1}, respectively[6]. However, the clear identification of P–O bands is mostly difficult as the bands overlap with other bands such as bands from phenyl groups (C=C, C–C, C–H bands). A fine IR characterisation of the uncharged PEG dendron, depicted in Figure 6.8, at different synthetic steps allowed identification (unambiguously, despite the high number of IR bands) of the phosphonate bands in the 800–1250 cm^{-1} region: the band around 1200 cm^{-1} was ascribed to the P=O bond, and the P–OH bands were located at 1020 and 999 cm^{-1}. After the grafting step, the phosphonate bands evolved drastically: the intensity of the P=O band strongly decreased, the P–OH bands disappeared, and a new band attributed to the formation of Fe–O–P bonds appeared at 975 cm^{-1}. The usually observed shoulder of the second ν(P–O–Fe) band at 1015 cm^{-1} attributed to the presence of a bidentate mode of coordination of phosphonate to the surface of NPs is not well visible[106]. With the disappearance of the P–OH bands, the phosphonate surface complex is thus suggested to be at least a binuclear bridging complex (Figure 6.9).

In the case of a silica coating, the bound Si–OH groups are characterised by the very broad IR absorption band in the 2800–3700 cm^{-1} region, whereas the so-called free Si–OH groups provide a narrow IR absorption band at 3630 cm^{-1}. The stretching band at 1635 cm^{-1} shows the presence of residual physisorbed water molecules, while the large bands centred at 1864, 1108 and 796 cm^{-1} are assigned to the Si–O and Si–O–Si stretching modes[92].

6.1.3.2 Determination of the Grafting Rate

The amount of grafted molecules is mainly determined by TGA, ultraviolet (UV) spectroscopy and from chemical analysis[17,80,85,96,107].

The amount of molecules at the surface of NPs by TGA is determined by considering the weight loss. However, one may notice that the decomposition path of grafted molecules is sometimes different from that of molecules alone[15,17]. The weight losses of molecules were usually observed at one stage within the temperature range of 150°C–450°C. Some phosphonated molecules were observed to decompose into two stages: the first higher weight loss in the same temperature range than before is attributed to the pyrolysis of molecules, and the second that is lower and occurs at higher temperature is attributed to the thermal decomposition of the phosphonate part[6,13,14,80]. The decomposition process of grafted molecules proceeds in more than one stage within the temperature range of 35°C–1000°C: mainly three stages when the grafted NPs were in water and in two stages when TGA is performed on dried NPs previously suspended in organic solvent. The first stage for water-suspended NPs is due to the evaporation of physically adsorbed water: it occurs below 200°C and the decrease in weight loss is below 10%[15].

The other two stages are attributed to the decomposition of grafted molecules, and it is suggested that the first stage is attributed to the thermal pyrolysis of the molecules and the second to the breaking of the bond between the NPs and the molecule[15,17]. However, with NPs grafted with phosphonate-based molecules, the last step corresponding to the phosphonate part is not observed when all molecules are grafted[108].

Few studies on biofunctionalisation of NPs deal with the calculation of the amount of grafted molecules. However, it may appear as a good way to discriminate between the current coupling agents as the grafting density will depend on their binding ability. Furthermore, one will need in the future to know the number of bioactive functions at the surface of NPs. Some grafting rates are reported in Table 6.1 and some conclusions may be extracted. The average HD size in suspension is smaller when the NPs are *in situ* coated by fatty acid during their synthesis and functionalised by a ligand exchange process, in comparison with NPs synthesised by the co-precipitation method, which tend to agglomerate. Large- and long-chain molecules lead to lower grafting rate than small molecules.

6.1.3.3 Effect of the Coating on Properties

The morphology of the coated particles can sometimes change from sphere-like to cube-like, indicating slight Fe_3O_4 surface corrosion during the exchange[15].

The surface coating may affect the magnetic properties of IO NPs. Few studies have been devoted to the specific effect of the organic coating and especially to the surface modification it generates[16,109,110]. Indeed, in most studies conducted on IO NPs, a decrease in the value of the saturation magnetisation (Ms) has been observed, which is attributed to finite size effects, structural disorder in the whole particle and surface effects, the latter being assigned to a large fraction of surface subcoordinated atoms and/or a disordered (spin-glass) or magnetically different surface layer, inducing spin canting[8,111]. However, it is difficult because no clear trends may, in fact, be extracted as, depending on the synthesis methods, the magnetic properties differ (with the presence of a surface canted layer), and the surface complexes are not clearly identified and in particular the strength of the surface bonding. However, the nature of the ligand, the linker between NPs and molecules, can affect the organisation of magnetic spins in the NPs, resulting in modified magnetic properties[16,25,112]. Phosphonate coupling agents have been demonstrated to preserve the magnetic properties after grafting[13–15,80,107,108]. Moreover, when *in situ* oleic acid-coated NPs synthesised by thermal decomposition are functionalised by ligand exchange, the Ms is maintained, whereas NPs synthesised by co-precipitation display higher Ms than naked NPs after the grafting step.

Phosphonate and carboxylate coupling agents have been compared, and the Ms of the magnetite NPs is lowered after functionalisation with the carboxylate molecules, whereas it is not affected by coupling with phosphonate agents. The magnetic structure of the oxide surface layer, on which functional molecules have been grafted, has been demonstrated to depend on the nature of the coupling agent through which the molecules are linked to the surface[14]. The carboxylate coupling agent induces a spin canting in the surface oxidised layer reducing the Ms. On the contrary, there is no canted surface structure when phosphonated molecules are linked covalently onto the IO NPs. The phosphonate surface complex is suggested to induce a surface magnetic ordering through superexchange magnetic interactions. Such studies are of particular importance to understand the change in magnetic properties when NPs are coated with an organic layer.

The Ms of silica-coated NPs is also observed to decrease as well as the blocking temperature (T_B) and the coercivity (H_c)[85]. The decrease of T_B and of H_c is related to lower dipolar interactions induced by the silica coating.

The surface coating has also been reported to affect the relaxivity properties of NPs. The chain length of polymers has an effect on the relaxivity properties. Indeed, the magnitude of MRI relaxivity is dependent on the number of water molecules disturbed by the magnetic field generated by the NPs, which may be influenced by the thickness and nature of the polymer layer. LaConte et al.[113] reported a decrease in R_2 with increasing chain length (or molecular weight) of the polymer coating. Duan et al.[114] examined the effect of polymer hydrophobicity on the magnetic properties (relaxivities) of NPs and found that the hydrophobic polymers diminished the relaxivity (R_2) behaviour. In contrast, NPs with a core size of approximately 10 nm coated with PEI, a hydrophilic-charged polymer, presented significantly enhanced relaxivity. A similar result was also observed for NPs with a core size greater than 30 nm, although the difference was less significant[25].

TABLE 6.1
Grafting Rate, Average HD Size and Ms of Some Functionalised NPs as the Function of Their Size, Their Synthesis and Grafting Ways

NPs	Molecules	Grafting Method	Grafting Amount (Determination Method)	Mean HD (nm)	Ms (emu g⁻¹)	References
5 nm by co-precipitation	Amine-derivatised PMIDA	Phosphonate CA introduced during the synthesis step	80.68 NH_2 per NP (UV visible absorbance)	30 ± 3 in water		[86]
8 nm one-step flame spray pyrolysis	Two homopolymers	Phosphonate CA	0.14–0.22 chain nm⁻²	90 in water	45	[86]
	One diblock polymer	After the synthesis of NPs, in water (pH = 8 and 9)	(TGA) 0.04–0.06 chain nm⁻²	150 in water		
8.8 nm thermal decomposition	Different generations of L-lysine-based dendritic ligands		G1(Lys) 984 ligand/particle G2(Lys) 644 G3(Lys) 496 (TGA)	12.3 ± 3.8 15.5 ± 4.3 25.0 ± 8.5	53.4 50.3 41.6	[15]
	Different generations of L-glutamic acid dendritic ligands	Ligand exchange	G1(Glu) 1083 ligand/particle G2(Glu) 569 G3(Glu) 367	13.4 ± 4.3 16.1 ± 4.0 26.2 ± 9.2	55.0 50.1 42.9	
10 nm thermal decomposition	PAA + NH_2–PEG–NH_2	Ligand exchange with PAA then linkage of NH_2–PEG–NH_2	1.5 PEG nm⁻² Kaiser test (amino groups) and TGA	53 in water and PBS	77	[96]
9 nm thermal decomposition	PEG-hydroxydopamine		1.3 PEG nm⁻²			[72]
39 nm co-precipitation	Linear oligo(phenylene vinylenes) (OPV) p, referred thereafter to as G0OPVp and pro-dendritic oligomers with a tapered shape (referred thereafter to as G1OPVp)	G0OPV2 G1OPV2 G0OPV2 G1OPV3 Direct grafting in THF	1.5 (1.4) mol nm⁻² 0.46 (0.35) 1.3 (1.5) 0.53 (0.35) TGA (UV)		67	[17]
12 nm co-precipitation	Phosphonate dendron	Direct grafting	1.3 molecules (TGA + chemical analyses)	50	70	[108]
10 nm co-precipitation (NPcop)	Phosphonate dendron	Direct grafting for NPcop	1.2 molecules nm⁻² (chemical analyses)	50	70	[107]
Thermal decomposition NPtd		Ligand exchange for NPtd	1.3 molecules nm⁻²	30	55	

In conclusion, the interest of multifunctional NPs for biomedical applications through grafting of multifunctional molecules has been widely studied. However, some fundamental studies are necessary to better master the anchoring of molecules at the surface of NPs and its effect on magnetic and colloidal properties.

6.2 BIOACTIVE MOLECULES GRAFTED AT THE SURFACE OF IRON OXIDE NANOPARTICLES AND THEIR EFFECTIVENESS FOR BIOMEDICAL APPLICATIONS

The structure, synthesis and biocompatibility of magnetic IO NPs have been well studied, and they have been widely used in the biosciences for MRI[11,115,116], cell labelling[117], bioseparation[118], targeted drug delivery[119,120], hyperthermia[11,121] and triggerable drug delivery systems.

In the following part, both engineering for biocompatibility and biodistribution and the syntheses of various conjugates for multimodal systems will be reviewed.

6.2.1 ENGINEERING FOR BIOCOMPATIBILITY, COLLOIDAL STABILITY AND BIODISTRIBUTION, AGAINST TOXICITY

Surface properties can be improved by chemical, physical and biological means to increase their biocompatibility: grafting of enzymes, drugs, proteins and antibodies to the polymer surface for targeting to organs and cells[122].

6.2.1.1 Polymers

Biodegradable polymers used for particle fabrication include poly(orthoesters), poly(anhydrides), poly(amides), poly(phosphazenes) and poly(phosphoesters)[123]. The use of copolymers to synthesise particles with desired physical properties is also possible: these include copolymers of aforementioned polymers with hydrophilic polymers such as PEG, poly(ethyleneoxide) (PEO) or poly(propylene oxide) (PPO) to impart stealth properties. Polymer properties such as molecular weight, charge, branching and stability have a great impact on the particle functions, including the mode of degradation, encapsulation efficiency, release rates and cellular internalisation[124].

Synthetic Polymers		Natural Polymers
Biodegradable	**Non-Biodegradable**	**Natural Polymers**
Polyesters: PLA, poly(glycolic acid), poly(hydroxyl butyrate), poly(ε-caprolactone), poly(β-malic acid) and poly(dioxanones)	*Cellulose derivatives*: carboxymethyl cellulose, ethyl cellulose, cellulose acetate, cellulose acetate propionate and hydroxypropyl methyl cellulose	*Polysaccharides*: agarose, alginate, carrageenan, hyaluronic acid, dextran, chitosan and cyclodextrins
Polyethers: polyethylene oxide and polyoxypropylene	*Silicones*: polydimethylsiloxane and colloidal silica	*Protein-based polymers*: collagen, albumin and gelatin
Polyanhydrides: poly(sebacic acid), poly(adipic acid) and poly(terphthalic acid)	*Acrylic polymers*: polymethacrylates, poly(methyl methacrylate) and polyhydro(ethyl-methacrylate)	
Polyamides: poly(imino carbonates) and polyamino acids	*Others*: polyvinylpyrrolidone, ethyl vinyl acetate, poloxamers and poloxamines	
Phosphorus-based: polyphosphates, polyphosphonates and polyphosphazenes		
Others: poly(cyano acrylates), polyurethanes, polyortho esters, polydihydropyrans and polyacetals		

Surface chemistry plays a key role in enhanced systemic circulation of NPs. One of the major breakthroughs in this area was the finding that NPs coated with hydrophilic polymer molecules, such as PEG, can resist serum protein adsorption and prolong the particle's systemic circulation[125]. Since then, numerous variations of PEG and other hydrophilic polymers have been tested for improved circulation[126]. Indeed, surfaces covered with PEG are biocompatible, that is, non-immunogenic, non-antigenic and protein resistant[127]. This is because PEG has uncharged hydrophilic residues and very high surface mobility leading to high steric exclusion[128]. Therefore, covalently immobilising PEG on the surfaces of superparamagnetic magnetic NPs is expected to effectively improve the biocompatibility of the NPs. In addition, it has been demonstrated that particles with PEG-modified surfaces can cross cell membranes, in non-specific cellular uptake[129]. Poloxamers and poloxamines, block copolymers of PEO and PPO have been extensively studied in this context[130].

Advantages of different polymers for NP coating to enhance colloidal stability are as follows:

Synthetic polymers	PEG improves the biocompatibility, blood circulation time and internalisation efficiency of the NPs
	PVP enhances blood circulation time and stabilises colloidal suspensions
	PVA prevents particles' coagulation, giving rise to monodisperse NPs
	PAA increases NPs stability and biocompatibility and helps in bioadhesion
	Poly(N-isopropylacryl-amide) allows thermosensitive drug delivery
Natural polymers	Dextran enhances blood circulation lifetime and stabilises colloidal suspensions
	Gelatin is used as a gelling agent, hydrophilic and biocompatible emulsifier
	Chitosan is used as non-viral gene delivery system, biocompatible and hydrophilic

Factors Influencing Polymers Biodegradation

Chemical structure and composition
Physical factors (shape, size and defects)
Morphology
Physicochemical factors (ionic strength, ion exchange and pH)
Molecular weight distribution
Annealing
Administration route and site of action
Degradation mechanism (enzymatic, hydrolysis and microbial)

6.2.1.2   Dendrimers[131]

The use of dendrimers or dendritic compounds for biomedical applications is a flourishing area of research, mainly because of their precisely defined structure and composition and also high tunable surface chemistry[132]. For biological applications, dendrimers and dendron discrete building blocks are indeed promising as the diversity of functionalisation brought by the arborescent structure simultaneously solves the problems of biocompatibility, low toxicity, large *in vivo* stability and specificity. Moreover, in addition to a controlled multifunctionalisation, dendrimers and dendron units allow a versatility of size (according to the generation) and of physicochemical properties (hydrophilic, lipophilic) that can be precisely tuned. The resulting effects on stability (dendrimer effect), pharmacokinetics and biodistribution can then clearly be identified. There is no doubt that dendrimer-based organic–inorganic hybrids represent highly advanced pharmaceutical tools, able to target a specific type of cell or organ, be tracked while doing it and deliver a specific drug *in situ*.

6.2.1.2.1   *Dendrimer-Stabilised IO NPs*

Dendrimer-stabilised NPs are referred to as a nanostructure, where the NP is surrounded with multiple dendrimers. In general, the formation process of these objects (larger than 5 nm) under appropriate conditions is *in situ*. In the presence of dendrimers, IO NPs can be synthesised and simultaneously stabilised: indeed, Strable et al.[133] synthesised stabilised ferromagnetic IO NPs in the presence of carboxylated G4.5 poly(amidoamine) (PAMAM) dendrimers. In their work, the electrostatic interaction of negatively charged carboxylated PAMAM dendrimers with positively charged IO NPs was considered to play an important role for the NPs stabilisation. The carboxylated dendrimer provides both a good nucleation surface and strongly adsorbed passivating layer on the oxide surface. PAMAM dendrimers with other functionalities (amine, alcohol) cannot stabilise IO NPs, thus indicating the role of electrostatic interaction for the NP stabilisation. Using these synthesised 30 nm magnetodendrimers, Bulte et al.[134] labelled mammalian cells through a non-specific membrane adsorption process with subsequent intracellular localisation in endosomes. Incubated magnetodendrimer doses as low as 1 mg of iron mL^{-1} allowed sufficient magnetic resonance (MR) cell contrast without compromising the cell viability and differentiation. The labelled neural stem cell could be readily detected *in vivo* by MRI at least 6 weeks after transplantation. Using the same magnetodendrimers, Bulte et al. were able to track the olfactory unsheathing glia grafted into the rat spinal cord *in vivo*[135], to detect the murine and human skin stem/progenitor cells[136] and to monitor stem cell therapy *in vivo* by MR imaging[137]. However, the so-elaborated magnetodendrimers did not have any specific surface modifications, not enabling the specific interaction between the particles and the target cells.

6.2.1.2.2　*Dendrimer-Assembled IO NPs*

These entities are formed on the basis of preformed NPs as a result of driving forces such as electrostatic interactions[138], covalent bonding[139] and the combination of different weak forces. For targeted imaging of cancer cells *in vitro* and *in vivo*, it would be ideal to assemble targeting ligand-modified dendrimers onto the NPs (e.g. IO NPs). In 2007, Shi et al.[140] synthesised and characterised a group of FA-modified carboxyl-functionalised G3 PAMAM dendrimers that were used to coat IO NPs through electrostatic interactions in order to achieve specific targeting to KB cells that overexpress folic acid receptor (FAR). It appeared that carboxyl-terminated PAMAM dendrimer assembled IO NPs could be uptaken by KB cells regardless of the repelling force between the negatively charged cells and the negatively charged particles. In the presence of a large amount of carboxyl terminal groups on the dendrimer surface, the receptor-mediated endocytosis of IO NPs assembled by FA-modified dendrimers was not facilitated: indeed, the surface charge of dendrimer-stabilised magnetic IO NPs in a biological medium is an important factor influencing their biological performance. Earlier studies showed that targeted dendrimers with a neutral surface (e.g. acetamide terminal groups) can specifically target cancer cells through ligand–receptor interaction[141]. This led to an idea to synthesise neutralised IO NPs for specific MRI of tumours[142]. In a recent study[143], IO NPs were assembled with multilayers of poly(glutamic acid) (PGA) and poly(L-lysine) (PLL), followed by assembly with G5NH2-FI (fluorescein isothiocyanate)–FA dendrimers. The interlayers were then cross-linked through 1-ethyl-3-(3-dimethylaminopropyl)carbodiimide hydrochloride (EDC) chemistry to covalently link the hydroxy groups of IO, the carboxy groups of PGA, the amino groups of PLL and the dendrimers (Figure 6.10). The remaining amino groups of the dendrimers were finally acetylated in order to neutralise the surface charge. Following the injection of either targeted IO–FA or non-targeted IO non-FA NPs, the *in vivo* MR imaging showed that the tumour MR signal intensity of mice injected with IO–FA NPs gradually decreases as a function of time (over 24 h), whereas, in sharp contrast, the tumour MR signal intensity of mice treated with IO non-FA NPs does not decrease significantly with time post-injection. This approach to the functionalisation of magnetic NPs can be applied to other small targeting molecules (peptides and growth factors), thereby providing a general cost-effective approach for *in vivo* MR detection of various biological systems.

6.2.1.3　Dendronised Ultrasmall Superparamagnetic IOs (USPIOs)

6.2.1.3.1　*Divergent 'Grafting from' Method*

The dendrimer part in this type of hybrid NP is synthesised in a step-by-step manner, similar to the conventional divergent approach used to synthesise dendrimers[144]. In an earlier study, Pan et al.[145] developed an approach to growing PAMAM dendrimers onto the surface of IO NPs to allow

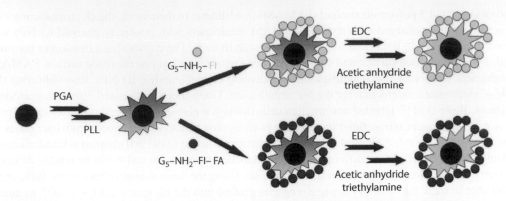

FIGURE 6.10 Schematic representation of the procedure for fabricating multifunctional shell-cross-linked IO NPs.

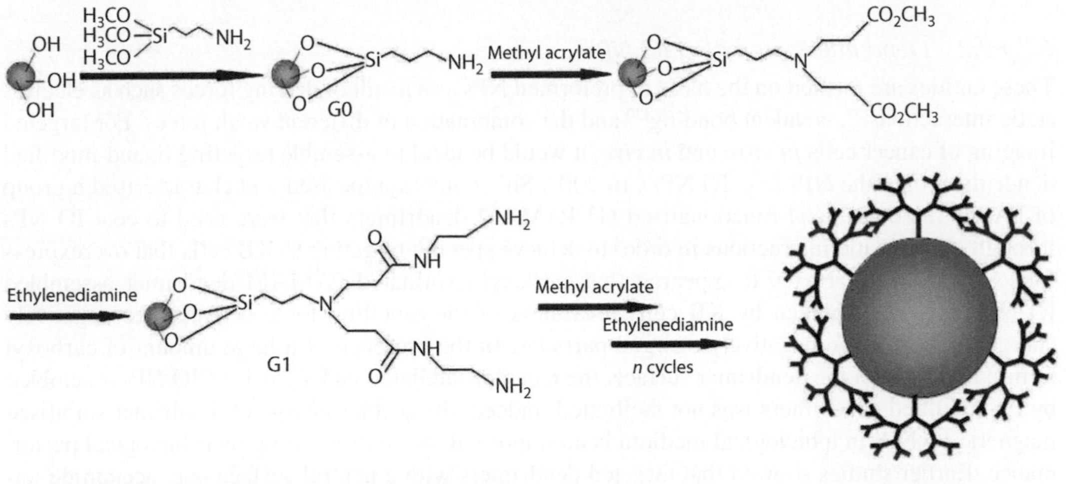

FIGURE 6.11 Magnetite NP modified with PAMAM dendrimers *via* a stepwise divergent approach.

enhanced immobilisation of bovine serum albumin (BSA). In their approach, the NPs were first modified by (3-aminopropyl)trimethoxysilane (APTMS) *via* silanisation, covering the NP surface with abundant primary amine groups. The APTMS-modified NPs were further alternately reacted with methyl acrylate and ethylene diamine, allowing divergent growth of PAMAM dendrimers onto the NP surface (Figure 6.11): the dendrimer-modified IO NPs were water dispersible and showed increased BSA binding with the generation number, due to the increased number of surface amine groups. Using these dendronised NPs, Pan et al.[146] were able to transfect anti-sense surviving oligodeoxynucleotide (asODN) into human tumour cell lines such as human breast cancer and liver cancer: specifically, the asODN–dendrimer–IO composites could enter into tumour cells within 15 min, causing marked down regulation of the surviving gene, protein and cell growth inhibition in both dose- and time-dependent manner. These results showed that PAMAM-modified magnetic NPs may represent a good gene delivery system.

6.2.1.3.2 Direct Grafting

In 2010, Begin-Colin, Felder-Flesch and co-workers demonstrated the great interest of dendronised IO NPs obtained through direct grafting of small-sized hydrophilic dendrons, allowing a phosphonate anchoring on the NP, for the development of efficient MRI contrast agents[107,108]. Indeed, such nano-objects were shown to display very good colloidal properties and higher relaxivity values than commercial polymer-functionalised NPs. The choice of phosphonic acid as coupling agent

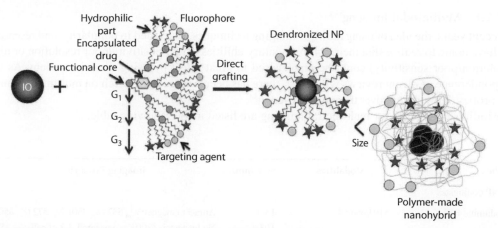

FIGURE 6.12 Schematic synthesis of dendronised NPs of tunable size.

was justified by previous studies that have evidenced a significantly higher grafting rate and a stronger binding than those obtained with carboxylate anchors[6,14,80,147]. A clear input was brought by the small-sized dendrons as they are discrete and monodisperse entities, and relevant characteristics like their size, hydrophilicity, molecular weight and biocompatibility can easily be tuned as a function of their generation[148]. Furthermore, a dendritic shell allows versatile and reproducible polyfunctionalisation at its periphery, which could lead to multimodal imaging probes through dye or fluorophore grafting, target-specific therapeutics and diagnostics through specific drug anchoring. Therefore, small-sized dendrons have an impressive future in the functionalisation of magnetic NPs[149], thanks to their highly controlled molecular structure and high tuneability leading to biocompatible, polyfunctional and water-soluble systems (Figure 6.12).

6.2.2 Conjugates for Multimodal Systems, for Theranostics

Synthesis of multifunctional IO NPs is a currently active research area[11,150], as they show great promise in biomedical imaging as multimodal imaging probes (Figure 6.13). Their polyfunctional surfaces also allow rational grafting of biological effectors or drugs, thus producing beneficial 'theranostics' allowing a double strike to fight cancer[151,152].

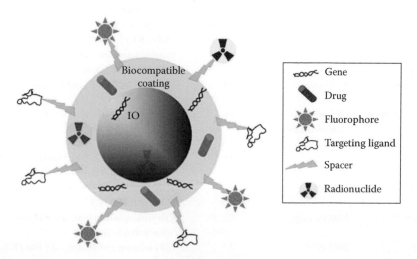

FIGURE 6.13 Schematic representation of a multifunctional nanoprobe.

6.2.2.1 Multimodal Imaging[153]

In recent years, the idea of using multiple imaging techniques has gained in popularity, and researchers have come to realise that their complementary abilities (high sensitivity/poor resolution or high resolution/poor sensitivity) could be emphasised to great effect by using them in tandem. As the preponderance of recent reviews will attest, there has been a surge in research on multimodal imaging probes development over the past few years[154].

Modifications of IO for multimodal imaging are listed in the following table.

Probes	Modalities	Size (nm)	Imaging Properties
IO NP conjugates			
Rhodamine B[155]	MRI/optical	13.3	Attract to magnet; λ_{em} 532 (λ_{ex} 350), λ_{em} 572 (λ_{ex} 850)
		108	No hysteresis (300 K), measured ΔB of cells; λ_{em} 577 (λ_{ex} 555)
AlexaFluor[156]	MRI/optical	80	Magnetic separation; microscopy, fluorescence activated cell storing (FACS)
		64.8	MRI cells; microscopy and FACS
Danzyl[157]	MRI/optical	39	$r1$ 41.2 mM^{-1} s^{-1}, $r2$ 110.6 mM^{-1} s^{-1} (3 T); λ_{em} 355 (λ_{ex} 530)
Indolequinone[158]	Magnetic/optical	9.5	Magnetic NP, used as quencher; λ_{em} 515 (λ_{ex} 495)
Diarylethene[159]	MRI/switchable optical	7	Superparamagnetic (300 K); λ_{em} 460/490 (λ_{ex} 312, 410)
Oligothiophene[160]	Magnetic/optical	17.9–26.9	Attract to magnet; $\lambda_{em} \approx$ 510, 620 (λ_{ex} 365)
Cy5.5, no target[161]	MRI/optical	31	$r1$ 23 mM^{-1} s^{-1}, $r2$ 59 mM^{-1} s^{-1} (0.47 T); cell and rat glioma imaging,
Cy5.5, target vascular cell adhesion molecule (VCAM)-1[162]	MRI/optical	31	VCAM-2 targeting, microscopy, MRI
Cy5.5 and annexin V[163]	MRI/optical	31	Annexin V conjugates, cell FACS, cell pellet MRI (4.7 T), 9.4 T imaging, coronary occlusion mice and microscopy
Cy5.5, target uMUC-1[164]	MRI/optical	35.8	EPPT peptide conjugation, $r1$ 26.43 mM^{-1} s^{-1}, $r2$ 53.44 mM^{-1} s^{-1} (0.47 T)
Cy5.5 and RGD[165]	MRI/optical	36	$r2$ 118 mM^{-1} s^{-1}
Cy5.5 and siRNA[166]	MRI/optical/therapy	—	Cell imaging
Enzyme-activity Cy5.5[167]	MRI/optical	62–68	$r1$ 27.8–29.9 mM^{-1} s^{-1}, $r2$ 91.2–92.5 mM^{-1} s^{-1}, 1.2–1.8 Cy5.5/particle
Cy5.5, target uPAR[168]	MRI/optical	10–15	Imaging pancreatic cancer mouse model, 3 T MRI and in vivo optical imaging (λ_{ex} 625, λ_{em} 700)
Cy5.5 and chlorotoxin (CTX)[58]	MRI/optical	10	MRI cells (4.7 T) and microscopy
64Cu-tetraazacyclododecanetetraacetic acid (DOTA)[169]	MRI/positron emission tomography (PET)	45	Superparamagnetic, no coercivity; MRI 3 T, PET
		32	$r1$ 14.46 mM^{-1} s^{-1}, $r2$ 72.55 mM^{-1} s^{-1} (1.5 T, 37°C); up to 22% radiolabelling yield
64Cu-DTPA[170]	MRI/PET	20	$r1$ 29 mM^{-1} s^{-1}, $r2$ 60 mM^{-1} s^{-1} (0.47 T, 39°C); MRI 7 T, PET–CT
111In-DOTA[171]	MRI/therapy	20, 30, 100	Cell studies, whole body autorad and pharmacokinetics
124I-CLIO[172]	MRI/PET	32	MRI solution and animal 1.5 T and PET
18F-PEG, Vivotag[173]	PET/CT	30	PET–CT phantom and mouse

(continued)

Probes	Modalities	Size (nm)	Imaging Properties
Core–shell IO NPs			
Fluorescein isothiocyanate (FITC), silica shell[174]	MRI/optical	150	Imaging only, cells, eye vein inject mice, 4.7 T MRI and microscopy
		50	Imaging only, cells, 1.5 T MRI and microscopy
Rhodamine, silica shell[175]	Magnetic/optical	13.7	Paramagnetic (ZFC), λ_{em} 552 (λ_{ex} 500)
		100–150	Superparamagnetic, no rt remanence, λ_{em} 560 (λ_{ex} 520)
		1.3–34	Superparamagnetic, no rt remanence, microscopy (rhodamine)
		30–80	Hysteresis (°C not given), microscopy cells (rhodamine isothiocyanate (RITC), FITC)
Terbium, silica shell[176]	MRI/optical	52	Superparamagnetic, no remanence, λ_{em} 544 (λ_{ex} 265)
(tris-bipyridine) ruthenium II (RuBPY), silica shell[177]	MRI/optical	20	Superparamagnetic, no rt remanence, $r2$ 30.4 mM^{-1} s^{-1} (FePt core), $r2$ 26.1 mM^{-1} s^{-1}, (Fe$_2$O$_3$ core), $\lambda_{em} \approx$ 610, abs 458, 522, 782
Pyrene, silica shell	Magnetic/optical	330	Superparamagnetic, no rt hysteresis; λ_{em} 375, 385, 394 (λ_{ex} 388)
Quant. dot, silica shell[178]	Magnetic/optical	170–200	Superparamagnetic, no rt remanence; λ_{em} 537 (quantum dot (QD), Au abs 518
DiR, PAA shell[179]	MRI/optical	88–100	$r2$ 202–208 mM^{-1} s^{-1} (0.47 T, 37°C); λ_{em} 595, abs 555
Pyrene, (3-mercaptopropyl) trimethoxysilane (MPTMS) shell[180]	Magnetic/optical	547	Superparamagnetic, no remanence; λ_{em} 375, 385, 394 (λ_{ex} 338)
Quant. dot, PAA shell[181]	MRI/optical/therapy	250	No rt remanence, attract to magnet; λ_{em} 593 (pH 7.4) (582, pH 2)
CdSe shell[182]	MRI/optical	>4	Superparamagnetic (not shown), blue, green and orange
		12–15	Paramagnetic, no remanence 298 K; λ_{em} 550, quantum yield (QY) 18%–20%
CdSe/ZnS shell[183]	Magnetic separation/ optical	30	Attract to magnet; λ_{em} 550
		25	Moderate magnetisation (2.28 emu g^{-1}); λ_{em} 565, abs 530
Eu:GdO3 shell[184]	Magnetic/optical	400	Linear (magnetisation vs. magnetic field); λ_{em} 615 (λ_{ex} 260)
LaF3:CeTb shell[185]	Magnetic/optical	30	Attract to magnet; λ_{em} 319/490/543/584/619 (λ_{ex} 270), abs 543
Y:Er:NaF4 shell[186]	Magnetic/optical	68	Superparamagnetic, zero coercivity; λ_{em} 539/658 (λ_{ex} 980)
Y2O2:Eu shell[187]	Magnetic/optical	100	Superparamagnetic, zero rt coercivity, 4.3 emu g^{-1}; λ_{em} 610 (λ_{ex} 260)
Gold shell[188]	MRI/optical/therapy	18/250	$r1$ 6.87 mM^{-1} s^{-1}, $r2$ 28.15 mM^{-1} s^{-1} (7 T, 25°C) imaging only 4.7 T; abs ($\approx$ 540, 700 nm for therapy)
Graphite shell (FeCo core)[189]	MRI/optical	30	Superparamagnetic rt, $r1$ 70 mM^{-1} s^{-1}, $r2$ 644 mM^{-1} s^{-1} (7 nm core), $r1$ 31 mM^{-1} s^{-1}, $r2$ 185 mM^{-1} s^{-1} (1.5 T); abs UV, 808 nm for therapy
Texas red-1,2-Bis (diphenylphosphino) ethane (DPPE)[190]	MRI/optical	10	Iron content only; microscopy
Doped IO NPs			
Terbium doped[191]	MRI/optical	13	Superparamagnetic, zero rt coercivity; λ_{em} 490/545/587/612 (λ_{ex} 235)

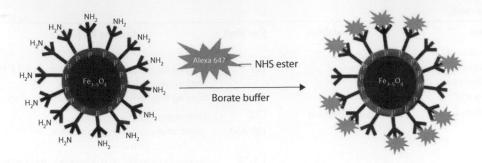

FIGURE 6.14 Post-functionalisation of dendronised IO NPs allowing multimodal imaging.

In 2011, Begin-Colin, Felder-Flesch and co-workers[192] investigated the synthesis of small-size dendrons and their grafting at the surface of IO NPs under, optimal conditions according to the nature of the peripheral functional groups. The double objective of the study was to obtain a good colloidal stability in water and osmolar media (with a mean HD diameter smaller than 100 nm) and to ensure the possibility of tuning the organic coating characteristics (e.g. morphology, functionalities, physicochemical properties, grafting of fluorescent or targeting molecules; Figure 6.14). *In vitro* and *in vivo* MRI and optical imaging were then demonstrated to be simultaneously possible using such versatile SPIOs covered by an optimised dendritic shell displaying either carboxylate or ammonium groups at their periphery.

The very high enhancement contrast ratio (EHC) values obtained for these dendronised NPs confirmed their high contrast power even at high magnetic field (7 T). For instance, on MR T_{2w} image at 7 T, EHC values were 15%–75% higher than those obtained for Endorem™. Both positive and negative EHC values were observed after injection and could be explained by local differences of concentration: indeed, at lower concentration (in the kidneys and initially in the bladder), the relative hypersignal can be explained by a dominating $T1$ effect, while at higher concentrations (liver, bladder and kidneys later in time), the $T2$ effect becomes very important and compensates widely for the $T1$ effect at the origin of a decreased signal (Figure 6.15). There was, however, a $T2$ saturation effect at very high concentrations, for example, in the liver, especially at the initial time. No significant liver (fluorescent intensity being equivalent to control), skin and spleen uptake was observed (no RES uptake). A very fast urinary fluorescence indicating urinary elimination was observed (no kidney signal). In contrast, a very fast (no persistent liver uptake at 20 min) and very high fluorescent signal was detected in intestinal tract, indicating hepatobiliary elimination (main elimination pathway; Figure 6.16). Observation of fluorescence in faeces confirmed this hypothesis.

6.2.2.2 Oncology/Drug Delivery

The selection of polymers for carriers in controlled-delivery devices requires the consideration of characteristics such as molecular weight, adhesion and solubility, depending on the type of system to be prepared, its action and the target site in the body. In some cases, polymeric materials for drug delivery must satisfy additional requirements, such as environmental responsiveness (e.g. pH- or temperature-dependent phase or volume transformations).

6.2.2.2.1 SLN Targeting[193]

In the last decade, methods for the precise localisation of sentinel lymph node (SLN) have drawn the tremendous attention of cancer surgeons and researchers in the field of medical diagnosis. Indeed, the accurate identification and characterisation of lymph nodes by imaging has important therapeutic and prognostic significance in patients with newly diagnosed cancers: the SLN is the first lymph node that receives lymphatic drainage from the site of a primary tumour. The sentinel node

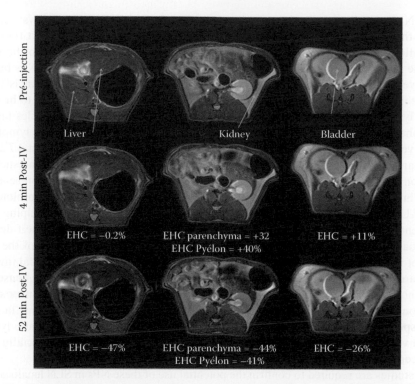

FIGURE 6.15 About 4 min spin echo $T2$-weighted axial images (RT/ET = 3000/9.5 ms, 2 mm slice thickness) were sequentially acquired before and immediately after IV injection of 500 μL of dendronised NPs@ COOH at 236.4 μg mL^{-1} iron concentration at 7 T with a small animal dedicated MRI (Biospin–Brücker®) during 60 min.

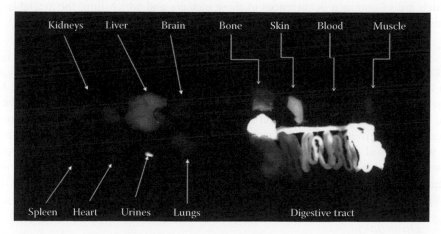

FIGURE 6.16 *Ex vivo* optical imaging of organs removed from a rat 20 min after IV injection of Alexa-labelled NPs@NH$_2$.

is much more likely to contain metastatic tumour cells than other lymph nodes in the same region. Among the various exploited methods for SLN diagnosis, nanocarriers[194] have received increasing attention as lymph node delivery agents. USPIO, in conjunction with enhanced MRI, are now being used as potential biomarkers for the diagnosis of lymph node metastases: they are transported into the interstitial space and reach the SLN *via* the lymphatic circulation, acting as $T2$-imaging agents, which can potentially identify the metastases independent of the lymph node size.

USPIO NPs are disseminated to the lymph nodes by two discrete pathways: (a) they can move into the medullar sinuses of the lymph node through venules *via* direct transcapillary passage, followed by phagocytosis by macrophages, and (b) they can move into the interstitial space in the body through endothelial transcytosis, followed by uptake of the NPs by draining lymphatic vessels and transport to the lymph nodes *via* afferent lymphatic channels[195]. USPIOs with minor macrophage uptake and prolonged blood half-life have been shown to be useful for metastatic lymph node imaging. Various studies have shown that USPIO particles taken up by the macrophages are transported to the interstitial spaces and subsequently to the lymph node *via* the lymph vessels. Thus, healthy lymph node tissues produce a dark signal with a *T*2 sequence since they are rich in macrophages, which phagocytose the USPIOs. However, no such contrast modification was observed in the metastatic lymph node tissue, which lacks the presence of macrophages, thus facilitating the effective imaging of this tissue. Additionally, the prolonged blood half-life of USPIOs allowed their progressive access to the lymph nodes thus helping in further enhancement of the image quality, which can be procured[196]. Nodal disease is a self-determining adverse prognostic aspect in many types of cancers. By and far, measurement of the node size by means of imaging is the only widely established method for assessing this nodal involvement. In this regards, the superparamagnetic agents (Ferumoxtran-10) have raised immense attention as normal nodes possess high affinity towards these agents following intravenous or subcutaneous injection: numerous studies on Ferumoxtran-10 have demonstrated its efficacy (a sensitivity of 78%, a specificity of 96% and a negative predictive value of 97%) in metastatic lymph node imaging in various types of cancer with significant results being reported, especially in case of breast cancer[197].

Further studies are required to confirm the potential use of these NPs in SLN localisation. Apart from the earlier discussed nanoparticulate systems, a variety of studies have been reported for SLN localisation with various other types of magnetic NPs. In one of the studies, SLN mapping of the stomach cancer has been reported using fluorescent magnetic NPs: here, the use of these NPs was found to overcome problems such as unpredictable lymphatic drainage patterns and skip metastases, which are of common occurrences in these cancers and which limit the available imaging techniques for their precise diagnosis[198].

6.2.2.2.2   Cisplatin

In some cases, it may be useful to target magnetic NPs to a tumour site (e.g. by magnetic accumulation[199]) then combine local drug release with hyperthermic treatment. This technique has been demonstrated with the chemotherapeutic agent cisplatin[200], which was adsorbed onto the surface of coated IO NPs, then released at hyperthermic (42°C) or ablative (60°C) temperatures. These complexes were relatively stable at physiological temperature for up to 14 days, while 50%–75% cisplatin was released from the surface within 180 min at hyperthermic or ablative temperatures, and all was released within 20 min at higher temperatures (121°C)[201]. While hyperthermia or cisplatin treatment alone reduced BP6 rat sarcoma cells viability by 20% *in vitro*, combined treatment reduced viability by nearly 80%.

6.2.2.2.3   Doxorubicin

In 2008, Hyeon and co-workers reported on a multifunctional polymer nanomedical platform for simultaneous cancer-targeted magnetic resonance or optical imaging and magnetically guided drug delivery (Figure 6.17)[202]. This platform was composed of four components: first, biodegradable poly(D,L-lactic-*co*-glycolic acid) (PLGA) NPs were used as a matrix for loading and subsequent controlled release of hydrophobic therapeutic agents into cells[203]. Second, two kinds of inorganic nanocrystals were incorporated into the PLGA matrix: superparamagnetic magnetite nanocrystals were employed for both magnetically guided delivery and *T*2 MRI contrast agent[42] and semiconductor NPs (quantum dots) were incorporated into the polymer particles for optical imaging[204]. Third, doxorubicin was used as a cancer therapeutic agent[205]. Finally, cancer-targeting

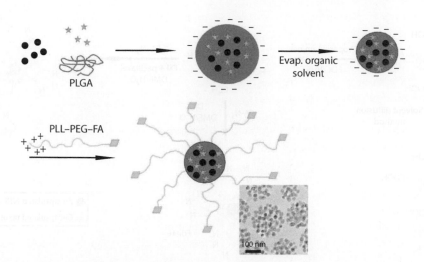

FIGURE 6.17 Synthetic procedure for the elaboration of multifunctional polymer NPs.

folate was conjugated onto the PLGA NPs by PEG groups to target KB cancer cells that have overexpressed folate receptors on their cell surface[206]. The cancer cells targeted with the multifunctional polymer NPs were detectable through MRI or confocal microscopy. Furthermore, magnetic guiding enhanced the cancer-targeting efficiency of the multifunctional polymer nanomedicine.

6.2.2.2.4 *Folic Acid*

Low-molecular-weight targeting agents such as FA have been demonstrated to preferentially target cancer cells, because the folate receptor (FAR) is frequently overexpressed on their surface (Figure 6.18)[207]. FAR is not only a tumour marker, but also has been shown to efficiently internalise molecules coupled to folate[208]. Furthermore, FA itself is stable and generally poorly immunogenic.

6.2.2.2.5 *Methotrexate*

In 2009, Das, Pramanik and co-workers[152] proposed a USPIO nanocore combining cancer-targeted magnetic resonance/optical imaging and pH-sensitive drug release. The multifunctional system was modified with a hydrophilic, biocompatible and biodegradable coating of PMIDA, which allowed further grafting of functional molecules such as rhodamine B isothiocyanate, FA and MTX. The theranostic agent could selectively target, detect and kill cancer cells overexpressing FAR, while allowing real-time monitoring of tumour response to drug treatment through dual-modal fluorescence and MRI.

6.2.2.2.6 *Bleomycin[209]*

Bleomycin (BLM) is an anti-neoplastic antibiotic used to treat several types of cancer (cervical, uterine, head and neck, testicular, penile cancer and certain types of lymphoma). In 2010, Cabuil et al.[210] proposed a nanoparticle-mediated delivery of BLM based on the use of multifunctional core–shell NPs made of γ-Fe$_2$O$_3$@SiO$_2$–PEG–NH$_2$ (Figure 6.19). Compared with a nanometric system incorporating drugs, the covalent grafting of the anti-neoplastic agent at the surface allowed a drastic reduction in the uncontrolled release of the agent and thus reduced unwanted side effects in healthy tissues. The authors succeeded in designing a biocompatible therapeutic hybrid able to convey BLM, while maintaining its cytotoxic activity and to allow interactions between BLM and both nucleus and mitochondria, thus resulting in an inhibition of cell growth.

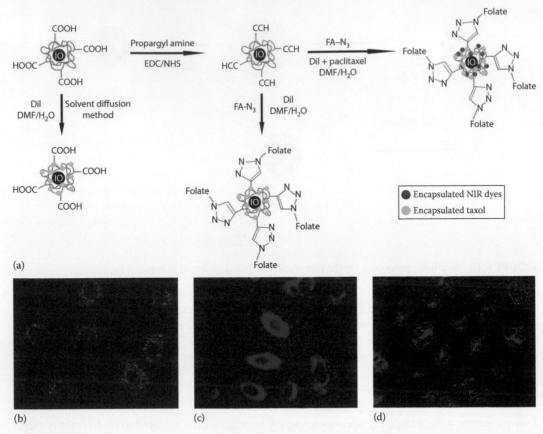

FIGURE 6.18 Multimodal imaging probe[179]. (a) Schematic representation of the synthesis of ther-anostics and multifunctional IO NPs. (b–d) Assessment of IO NP cellular uptake *via* confocal laser-scanning microscopy using lung carcinoma A549 cells. (b) No internalisation was observed in cells treated with carboxylated IO NPs, as no lipophilic fluorescent dye (DiI) fluorescence was observed in the cytoplasm. (c) Enhanced internalisation was observed upon incubation with the folate-immobilised IO NPs. (d) Cells incubated with paclitaxel (Taxol) and DiI co-encapsulating folate-functionalised IO NPs induced cell death.

6.2.2.2.7 Peptides

Peptides have been used to coat NPs as they allow the targeting of specific tissues *in vivo*. IO NPs targeting and imaging integrins that are expressed in tumours cells have been developed[165]: RGD peptide analogues have been chemically conjugated to the primary amines of amino CLIO NPs using disuccinimidyl suberimidate. The decorated NPs demonstrated integrin-mediated cellular uptake in BT-20 tumour cells *in vitro* and localised in BT-20 tumours *in vivo*. In another example, c(RGDyK) peptide was conjugated to 4-methylcatechol (MC)-coated ultrasmall IO NPs (4.5 nm)[211]: the resulting hybrids had an overall diameter of 8.4 nm, were stable under physiological conditions and showed fivefold higher cellular uptake in integrin-positive U87MG cells compared to integrin-negative cells. They also accumulated preferentially in tumours when intravenously administered in an U87MG tumour-bearing mice model. CTX, a 36-amino acid peptide that has a strong affinity for tumours of neuroectodermal origin[212], has also been introduced onto NPs. The NPs comprise IO NP coated with PEG or PEG-grafted chitosan copolymer, to which CTX was conjugated[213]. The probes showed higher internalisation and localisation properties in mice bearing xenograft tumours and in transgenic ND2: SmoA1 mice, respectively.

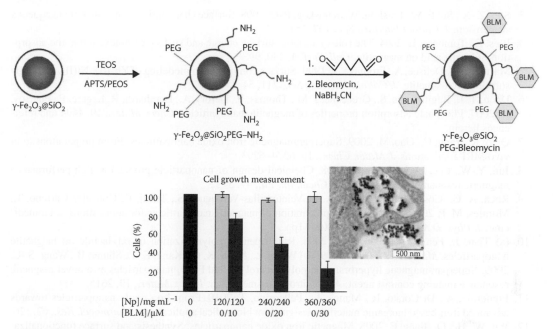

FIGURE 6.19 Top: NPs synthesis and BLM anchoring. Bottom: Percentage of living cells after 48 h of incubation at 37°C with NPs (red bars) or Cu(II)-BLM@NPs (blue bars). Inset: Representative TEM image of an HT1080 cell treated with Cu(II)–BLM@NPs after 30 min of incubation.

6.3 CONCLUSION

In the past 30 years, the explosive growth of nanotechnology has burst into challenging innovations in medicine, which is in the process of revolutionising the field of therapy, in particular, the local treatments of diseases. The main input of today's nanotechnology in biology is that it allows real progress to achieve temporal and spatial site-specific drug delivery, local therapy and imaging. This breakthrough was made possible by the development of multifunctional biocompatible nanosystems resulting from cutting-edge researches based on multidisciplinary approaches. IO NPs functionalised by bioactive molecules could be integrated in the new area of nanopharmaceuticals, which include either imaging systems or biologically active drug delivery products. Challenges remain to build new multifunctional platforms aimed at (i) carrying components/agents for diagnosis and therapy, specifically targeting the desired organ or containing functionalities to allow transport across biological barriers, (ii) improving the biodistribution and decreasing the toxicity and (iii) imaging and studying the interactions between the cell components.

To answer these challenges, the theranostic approach is being more and more developed and may bring necessary breakthroughs to *in vivo* applications. Nevertheless, fundamental studies are still needed, for example, the selection of coupling agents allowing strong binding and high grafting rate by preserving both molecules and NPs properties. Moreover, quantitative evaluation of the functions at the surface of NPs will be among the mandatory studies.

REFERENCES

1. Cornell, R. M., Schwertmann, U. 2006. *The Iron Oxide. Structure, Properties, Reactions, Occurrences and Uses*, Wiley-VCH, Weinheim, Germany.
2. Jolivet, J.-P., Froidefond, C., Pottier, A., Chanéac, C., Cassaignon, S., Tronc, E., Euzen, P. 2004. Size tailoring of oxide nanoparticles by precipitation in aqueous medium. A semi-quantitative modelling. *J. Mater. Chem.*, 14, 3281.

3. Sun, Z.-X., Su, F.-W., Forsling, W., Samskog, P.-O. 1998. Surface characteristics of magnetite in aqueous suspension. *J. Colloid Interface Sci.*, 197, 151.
4. Illés, E., Tombacz, E. 2004. The role of variable surface charge and surface complexation in the adsorption of humic acid on magnetite. *Coll. Surf. A*, 230, 99.
5. Marmier, N., Delisée, A., Fromage, F. 1999. Surface complexation modeling of Yb(III), Ni(II), and Cs(I) sorption on magnetite. *J. Colloid Interface Sci.,* 211, 54.
6. Daou, T. J., Bégin-Colin, S., Grenèche, J. M., Thomas, F., Derory, A., Bernhardt, P., Legaré, P., Pourroy, G. 2007. Phosphate adsorption properties of magnetite nanoparticles. *Chem. Mater.*, 19, 4494 and references herein.
7. Qiao, R., Yang, C., Gao, M. 2009. Superparamagnetic iron oxide nanoparticles: From preparations to in vivo MRI applications. *J. Mater. Chem.*, 19, 6274–6293.
8. Jun, Y.-W., Lee, J. H., Cheon, J. 2008. Chemical design of nanoparticle probes for high-performance magnetic resonance imaging. *Angew. Chem. Int. Ed.*, 47, 5122.
9. Roca, A. G., Costo, R., Rebolledo, A. F., Veintemillas-Verdaguer, S., Tartaj, P., Gonzalez-Carreno, T., Morales, M. P. 2009. Progress in the preparation of magnetic nanoparticles for applications in biomedicine. *J. Phys. D Appl. Phys.*, 42, 224002, 11pp.
10. (a) Tian, J., Feng, Y. K., Xu, Y. S. 2006. Ring opening polymerization of D,L-lactide on magnetite nanoparticles. *Macromol. Res.*, 14, 209; (b) Wang, L., Neoh, K. G., Kang, E. T., Shuter, B., Wang, S.-C. 2009. Superparamagnetic hyperbranched polyglycerol-grafted Fe_3O_4 nanoparticles as a novel magnetic resonance imaging contrast agent: An in vitro assessment. *Adv. Funct. Mater.*, 19, 2615.
11. Figuerola, A., Di Corato, R., Manna, L., Pellegrino, T. 2010. From iron oxide nanoparticles towards advanced iron-based inorganic materials designed for biomedical applications. *Pharmacol. Res.*, 62, 126.
12. Wu, W., He, Q., Jiang, C. 2008. Magnetic iron oxide nanoparticles: Synthesis and surface functionalization strategies. *Nanoscale Res. Lett.*, 3, 397.
13. Daou, T. J., Pourroy, G., Grenèche, J. M., Bertin, A., Felder-Flesch D., Begin-Colin, S. 2009. Water soluble dendronized iron oxide nanoparticles. *Dalton Trans.*, 38, 4442.
14. Daou, T. J., Grenèche, J. M., Pourroy, G., Buathong, S., Derory, A., Ulhaq-Bouillet, C., Donnio, B., Guillon D., Begin-Colin, S. 2008. Coupling agent effect on magnetic properties of functionalized magnetite-based nanoparticles. *Chem. Mater.*, 20, 5869.
15. Zhu, R., Jiang, W., Pu, Y., Luo, K., Wu, Y., He, B., Gu, Z. 2011. Functionalization of magnetic nanoparticles with peptide dendrimers. *J. Mater. Chem.*, 21, 5464.
16. Gupta, A. K., Gupta, M. 2005. Synthesis and surface engineering of iron oxide nanoparticles for biomedical applications. *Biomaterials*, 26, 3995.
17. Buathong, S., Ung, D., Daou, T. J., Ulhaq-Bouillet, C., Pourroy, G., Guillon, D., Ivanova, L., Bernhardt, I., Begin-Colin, S., Donnio, B. 2009. Thermal, magnetic and luminescent properties of dendronized ferrite nanoparticles. *J. Phys. Chem. C*, 113, 12201.
18. Suh, W. H., Suslick, K. S., Stucky, G. D., Suh, Y. H. 2009. Nanotechnology, nanotoxicology, and neuroscience. *Prog. Neurobiol.*, 87, 133.
19. Douziech-Eyrolles, L., Marchais, H., Hervé, K., Munnier, E., Soucé, M., Linassier, C., Dubois, P., Chourpa, I. 2007. Nanovectors of anticancer agents based on superparamagnetic iron oxide nanoparticles. *Int. J. Nanomedicine*, 2(4), 541.
20. Barratt, G. M. 2000. Therapeutic applications of colloidal drug carriers. *Pharm. Sci. Technol. Today*, 3, 163.
21. Pyun, J. 2007. Nanocomposite materials from functional polymers and magnetic colloids. *Polymer Rev.*, 47, 231.
22. Häfeli, U. O. 2004. Magnetically modulated therapeutic systems. *Int. J. Pharm.*, 277, 19.
23. Palmacci, S., Josephson, L. 1993. U.S. Patent 5262176.
24. Park, J., Yu, M. K., Jeong, Y. Y., Kim, J. W., Lee, K., Phan, V. N., Jon, S. 2009. Antibiofouling amphiphilic polymer-coated superparamagnetic iron oxide nanoparticles: Synthesis, characterization, and use in cancer imaging *in vivo*. *J. Mater. Chem.*, 19, 6412.
25. Boyer, C., Whittaker, M. R., Bulmus, V., Liu J., Davis, T. P. 2010. The design and utility of polymer-stabilized iron-oxide nanoparticles for nanomedicine applications. *NPG Asia Mater.*, 2(1), 23.
26. Boyer, C., Priyanto, P., Davis, T. P., Pissuwan, D., Bulmus, V., Kavallaris, M., Teoh, W. Y., Amal, R., Carroll, M., Woodward, R., St Pierre, T. 2010. Anti-fouling magnetic nanoparticles for siRNA delivery. *J. Mater. Chem.*, 20, 255.
27. Lattuada, M., Hatton, T. A. 2007. Functionalization of monodisperse magnetic nanoparticles. *Langmuir*, 23, 2158–2168.
28. Fang, C., Zhang, M. 2009. Multifunctional magnetic nanoparticles for medical imaging applications. *J. Mater. Chem.*, 19, 6258.

29. Cheng, K., Sun, S. 2010. Recent advances in syntheses and therapeutic applications of multifunctional porous hollow nanoparticles. *Nano Today*, 5, 183.

30. Sun, S., Zeng, H., Robinson, D. B., Raoux, S., Rice, P. M., Wang, S. X., Li, G. 2004. Monodisperse MFe_2O_4 (M = Fe, Co, Mn) nanoparticles. *J. Am. Chem. Soc.*, 126, 273.

31. Frey, N. A., Peng, S., Cheng, K., Sun, S. 2009. Magnetic nanoparticles: Synthesis, functionalization, and applications in bioimaging and magnetic energy storage. *Chem. Soc. Rev.*, 38, 2532.

32. Robinson, D. B., Persson, H. H. J., Zeng, H., Li, G. X., Pourmand, N., Sun, S. H., Wang, S. X. 2005. DNA-functionalized MFe_2O_4 (M = Fe, Co, or Mn) nanoparticles and their hybridization to DNA-functionalized surfaces. *Langmuir*, 21, 3096.

33. Hultman, K. L., Raffo, A. J., Grzenda, A. L., Harris, P. E., Brown T. R., O'Brien, S. 2008. Magnetic resonance imaging of major histocompatibility class II expression in the renal medulla using immunotargeted superparamagnetic iron oxide nanoparticles. *ACS Nano*, 2, 477.

34. Na, H. B., Lee, J., An, K., Park, Y. I., Park, M., Lee, I. S., Nam, D., Kim, S. T., Kim, S., Kim, S., Lim, K., Kim, K., Kim S., Hyeon, T. 2007. Development of a T1 contrast agent for magnetic resonance imaging Using MnO nanoparticles. *Angew. Chem. Int. Ed.*, 46, 5397.

35. Choi, S. H., Na, H. B., Park, Y. I., An, K., Kwon, S. G., Jang, Y., Park, M. H., Moon, J., Son, J. S., Song, I. C., Moon W. K., Hyeon, T. 2008. Simple and generalized synthesis of oxide-metal heterostructured nanoparticles and their applications in multimodal biomedical probes. *J. Am. Chem. Soc.*, 130, 15573.

36. Barcena, C., Sra, A., Chaubey, G., Khemtong, C., Liu, J., Gao, J. 2008. Zinc ferrite nanoparticles as MRI contrast agents. *Chem. Commun.*, 44, 2224.

37. Shin, J. S., Anisur, R., Ko, M., Im, G., Lee L., Lee, D. 2009. Hollow manganese oxide nanoparticles as multifunctional agents for magnetic resonance imaging and drug delivery. *Angew. Chem. Int. Ed.*, 48, 321.

38. Goodwin, A. P., Tabakman, S. M., Welsher, K., Sherlock, S. P., Prencipe G., Dai, H. 2009. Phospholipid-dextran with a single coupling point: A useful amphiphile for functionalization of nanomaterials. *J. Am. Chem. Soc.*, 131, 289.

39. Qin, J., Laurent, S., Jo, Y. S., Roch, A., Mikhaylova, M., Bhujwalla, Z. M., Muller, R. N., Muhammed M. 2007. A high-performance magnetic resonance imaging T2 contrast agent. *Adv. Mater.*, 19, 1874.

40. Yu, W. W., Chang, E., Falkner, J. C., Zhang, J. Y., Al-Somali, A. M., Sayes, C. M., Johns, J., Drezek, R., Colvin, V. L. 2007. Forming biocompatible and nonaggregated nanocrystals in water using amphiphilic polymers. *J. Am. Chem. Soc.*, 129, 2871.

41. Lin, C. A. J., Sperling, R. A., Li, J. K., Yang, T. Y., Li, P. Y., Zanella, M., Chang W. H., Parak, W. G. J. 2008. Design of an amphiphilic polymer for nanoparticle coating and functionalization. *Small*, 4, 334.

42. Huh, Y. M., Jun, Y. W., Song, H. T., Kim, S., Choi, J. S., Lee, J. H., Yoon, S., Kim, K. S., Shin, J. S., Suh J. S., Cheon, J. 2005. In vivo magnetic resonance detection of cancer by using multifunctional magnetic nanocrystals. *J. Am. Chem. Soc.*, 127, 12387.

43. White, M. A., Johnson, J. A., Koberstein J. T., Turro, N. J. 2006. Toward the syntheses of universal ligands for metal oxide surfaces: Controlling surface functionality through click chemistry. *J. Am. Chem. Soc.*, 128, 11356.

44. Miles, W. C., Goff, J. D., Huffstetler, P. P., Reinholz, C. M., Pthhayes, N., Caba, B. L., Davis R. M., Riffle, J. S. 2009. Synthesis and colloidal properties of polyether–magnetite complexes in water and phosphate-buffered saline. *Langmuir*, 25, 803.

45. Creixell, M., Herrera, A. P., Latorre-Esteves, M., Ayala, V., Torres-Lugo, M., Rinaldi, C. 2010. The effect of grafting method on the colloidal stability and in vitro cytotoxicity of carboxymethyl dextran coated magnetic nanoparticles. *J. Mater. Chem.*, 20, 8539.

46. Bajaj, A., Samanta, B., Yan, H., Jerry D. J., Rotello, V. M. 2009. Stability, toxicity and differential cellular uptake of protein passivated-Fe_3O_4 nanoparticles. *J. Mater. Chem.*, 19, 6328.

47. Wotschadlo, J., Liebert, T., Heinze, T., Wagner, K., Schnabelrauch, M., Dutz, S., Muller, R., Steiniger, F., Schwalbe, M., Kroll, T. C., Hoffken, K., Buske, N., Clement, J. H. 2009. Magnetic nanoparticles coated with carboxymethylated polysaccharide shells – Interaction with human cells. *J. Magn. Magn. Mater.*, 321, 1469.

48. Thomas, L. A., Dekker, L., Kallumadil, M., Southern, P., Wilson, M., Nair, S. P., Pankhurst Q. A., Parkin, I. P. 2009. Carboxylic acid-stabilised iron oxide nanoparticles for use in magnetic hyperthermia. *J. Mater. Chem.*, 19, 6529.

49. Mornet, S., Vasseur, S., Grasset, F., Duguet, E. 2004. Magnetic nanoparticle design for medical diagnosis and therapy. *J. Mater. Chem.*, 14, 2161.

50. Herrera, A. P., Barrera, C., Rinaldi, C. 2008. Synthesis and functionalization of magnetite nanoparticles with aminopropylsilane and carboxymethyldextran. *J. Mater. Chem.*, 18, 3650.

51. Li, Z., Wei, L., Gao, M. Y., Lei, H. 2005. One-pot reaction to synthesize biocompatible magnetite nanoparticles. *Adv. Mater.*, 17, 1001.

52. Hu, F. Q., Wei, L., Zhou, Z., Ran, Y. L., Li, Z., Gao, M. Y. 2006. Preparation of biocompatible magnetite nanocrystals for in vivo magnetic resonance detection of cancer. *Adv. Mater.*, 18, 2553.
53. Adden, N., Gamble, L. J., Castner, D. G., Hoffmann, A., Gross, G., Menzel, H. 2006. Phosphonic acid monolayers for binding of bioactive molecules to titanium surfaces. *Langmuir*, 22, 8197.
54. Yee, C., Kataby, G., Ulman, A., Prozorov, T., White, H., King, A., Rafailovich, M., Sokolov, J., Gedanken, A. 1999. Self-assembled monolayers of alkanesulfonic and -phosphonic acids on amorphous iron oxide nanoparticles. *Langmuir*, 15, 7111.
55. Barrera, C., Herrera, A. P., Zayas, Y., Rinaldi, C. 2009. Surface modification of magnetite nanoparticles for biomedical applications. *J. Magn. Magn. Mater.*, 321, 1397.
56. Kim, D. K., Dobson, J. 2009. Nanomedicine for targeted drug delivery. *J. Mater. Chem.*, 19, 6294.
57. Zhou, L., Yuan, J., Wei, Y. 2011. Core–shell structural iron oxide hybrid nanoparticles: From controlled synthesis to biomedical applications. *J. Mater. Chem.*, 21, 2823.
58. Veiseh, O., Sun, C., Gunn, J., Kohler, N., Gabikian, P., Lee, D., Bhattarai, N., Ellenbogen, R., Sze, R., Hallahan, A., Olson, J., Zhang, M. Q. 2005. Optical and MRI multifunctional nanoprobe for targeting gliomas. *Nano Lett.*, 5, 1003.
59. Bridot, J. L., Faure, A. C., Laurent, S., Riviere, C., Billotey, C., Hiba, B., Janier, M., Josserand, V., Coll, J. L., Elst, L. V., Muller, R., Roux, S., Perriat P., Tillement, O. 2007. Hybrid gadolinium oxide nanoparticles: Multimodal contrast agents for in vivo imaging. *J. Am. Chem. Soc.*, 129, 5076.
60. Zhang, Y., Kohler, N., Zhang, M. Q. 2002. Surface modification of superparamagnetic magnetite nanoparticles and their intracellular uptake. *Biomaterials*, 23, 1553.
61. Zhang, Y., Sun, C., Kohler, N., Zhang, M. Q. 2004. Self-assembled coatings on individual monodisperse magnetite nanoparticles for efficient intracellular uptake. *Biomed. Microdevices*, 6, 33.
62. Kohler, N., Fryxell, G. E., Zhang, M. Q. 2004. A bifunctional poly(ethylene glycol) silane immobilized on metallic oxide-based nanoparticles for conjugation with cell targeting agents. *J. Am. Chem. Soc.*, 126, 7206.
63. Jana, N. R., Earhart, C., Ying, J. Y. 2007. Synthesis of water-soluble and functionalized nanoparticles by silica coating. *Chem. Mater.*, 19, 5074.
64. De Palma, R., Peeters, S., Van Bael, M. J., Van den Rul, H., Bonroy, K., Laureyn, W., Mullens, J., Borghs, G., Maes, G. 2007. Silane ligand exchange to make hydrophobic superparamagnetic nanoparticles water-dispersible. *Chem. Mater.*, 19, 1821.
65. El-Boubbou, K., Gruden, C., Huang, X. 2007. Magnetic glyco-nanoparticles: A unique tool for rapid pathogen detection, decontamination, and strain differentiation. *J. Am. Chem. Soc.*, 129, 13392.
66. Maurizi, L., Bisht, H., Bouyer, F., Millot, N. 2009. Easy route to functionalize iron oxide nanoparticles via long-term stable thiol groups. *Langmuir*, 25, 8857.
67. Mejias, R., Perez-Yague, S., Gutierrez, L., Cabrera, L. I., Spada, R., Acedo, P., Serna, C. J., Lazaro, F. J., Villanueva, A., Morales, M. D., Barber, D. F. 2011. Dimercaptosuccinic acid-coated magnetite nanoparticles for magnetically guided in vivo delivery of interferon gamma for cancer immunotherapy. *Biomaterials*, 32, 293.
68. Goff, J. D., Huffstetler, P. P., Miles, W. C., Pothayee, N., Reinholz, C. M., Ball, S., Davis, R. M., Riffle, J. S. 2009. Novel phosphonate-functional poly(ethylene oxide)-magnetite nanoparticles form stable colloidal dispersions in phosphate-buffered saline. *Chem. Mater.*, 21, 4784.
69. (a) Gu, B. H., Schmitt, J., Chen, Z. H., Liang, L. Y., Mccarthy, J. F. 1994. Adsorption and desorption of natural organic-matter on iron-oxide – Mechanisms and models. *Environ. Sci. Technol.*, 28, 38; (b) Gu, B. H., Schmitt, J., Chen, Z. H., Liang, L. Y., Mccarthy, J. F. 1995. Adsorption and desorption of different organic matter fractions on iron oxide. *Geochim. Cosmochim. Acta*, 59, 219.
70. Xu, C., Xu, K., Gu, H., Zheng, R., Liu, H., Zhang, X. X., Guo Z., Xu, B. 2004. Dopamine as a robust anchor to immobilize functional molecules on the iron oxide shell of magnetic nanoparticles. *J. Am. Chem. Soc.*, 126, 9938.
71. Young, K. L., Xu, C., Sun, S. 2009. Conjugating methotrexate to magnetite (Fe_3O_4) nanoparticles via trichloro-s-triazine. *J. Mater. Chem.*, 19, 6400.
72. Amstad, E., Zücher, S., Mashaghi, A., Wong, J. Y., Textor, M., Reimhult, E. 2009. Surface functionalization of single superparamagnetic iron oxide nanoparticles for targeted magnetic resonance imaging. *Small*, 5, 1334.
73. Wang, L., Yang, Z. M., Gao, J. H., Xu, K. M., Gu, H. W., Zhang, B., Zhang, X. X., Xu, B. 2006. A biocompatible method of decorporation: Bisphosphonate modified magnetite nanoparticles to remove uranyl ions from blood. *J. Am. Chem. Soc.*, 128, 13358.
74. Cheng, K., Peng, S., Xu, C., Sun, S. 2009. Porous hollow Fe(3)O(4) nanoparticles for targeted delivery and controlled release of cisplatin. *J. Am. Chem. Soc.*, 131, 10637.
75. Shultz, M. D., Reveles, J. U., Khanna, S. N., Carpenter, E. E. 2007. Reactive nature of dopamine as a surface functionalization agent in iron oxide nanoparticles. *J. Am. Chem. Soc.*, 129, 2482.

76. Baldi, G., Bonacchi, D., Franchini, M. C., Gentili, D., Lorenzi, G., Ricci A., Ravagli, C. 2007. Synthesis and coating of cobalt ferrite nanoparticles: A first step toward the obtainment of new magnetic nanocarriers, *Langmuir*, 23, 4026.

77. Kim, M., Chen, Y., Liu Y., Peng, X. 2005. Super-stable, high quality Fe_3O_4 dendron-nanocrystals dispersible in both organic and aqueous solutions. *Adv. Mater.*, 17, 1429.

78. Hofmann, A., Thierbach, S., Semisch, A., Hartwig, A., Taupitz, M., Rühl, E., Graf, C. 2010. Highly monodisperse water-dispersable iron oxide nanoparticles for biomedical applications. *J. Mater. Chem.*, 20, 7842.

79. Sahoo, Y., Pizem, H., Fried, T., Golodnitsky, D., Burstein, L., Sukenik, C. N., Markovich, G. 2001. Alkyl phosphonate/phosphate coating on magnetite nanoparticles: A comparison with fatty acids; *Langmuir*, 17, 7907.

80. Daou, T. J., Buathong, S., Ung, D., Donnio, B., Pourroy, G., Guillon, D., Bégin, S. 2007. Investigation of the grafting rate of organic molecules on the surface of magnetic nanoparticles as a function of the coupling agent. *Sensor Actuator B*, 126, 159.

81. Lalatonne, Y., Paris, C., Serfaty, J. M., Weinmann, P., Lecouvey, M., Motte, L. 2008. Bis-phosphonates-ultra small superparamagnetic iron oxide nanoparticles: A platform towards diagnosis and therapy. *Chem. Commun.*, 44, 2553.

82. Burgos-Asperilla, L., Darder, M., Aranda, P., Vazquez, L., Vazquez, M., Ruiz-Hitzky, E. 2007. Novel magnetic organic–inorganic nanostructured materials. *J. Mater. Chem.*, 17, 4233.

83. Kobayashi, M., Matsuno, R., Otsuka H., Takahara, A. 2006. Precise surface structure control of inorganic solid and metal oxide nanoparticles through surface-initiated radical polymerization. *Sci. Technol. Adv. Mater.*, 7, 617.

84. Babu, K., Dhamodharan, R. 2008. Grafting of poly(methyl methacrylate) brushes from magnetite nanoparticles using a phosphonic acid based initiator by ambient temperature atom transfer radical polymerization (ATATRP). *Nanoscale Res. Lett.*, 3, 109.

85. Das, M., Mishra, D., Maiti, T. K., Basak, A., Pramanik, P. 2008. Bio-functionalization of magnetite nanoparticles using an aminophosphonic acid coupling agent: New, ultradispersed, iron-oxide folate nanoconjugates for cancer-specific targeting. *Nanotechnology*, 19, 415101.

86. Manasmita, M., Mishra, F., Dhak, P., Gupta, S., Kumar Maiti, T., Basak, A., Pramanik, P. 2009. Biofunctionalized, phosphonate-grafted, ultrasmall iron oxide nanoparticles for combined targeted cancer therapy and multimodal imaging. *Small*, 5, 2883.

87. Lai, C. W., Wang, Y. H., Lai, C. H., Yang, M. J., Chen, C. Y., Chou, P. T., Chan, C. S., Chi, Y., Chen Y. C., Hsiao, J. K. 2008. Iridium-complex-functionalized Fe_3O_4/SiO_2 core/shell nanoparticles: A facile three-in-one system in magnetic resonance imaging, luminescence imaging, and photodynamic therapy. *Small*, 4, 218.

88. Selvan, S. T., Patra, P. K., Ang, C. Y., Ying, J. Y. 2007. Synthesis of silica-coated semiconductor and magnetic quantum dots and their use in the imaging of live cells. *Angew. Chem. Int. Ed.*, 119, 2500.

89. Yi, D. K., Selvan, S. T., Lee, S. S., Papaefthymiou, G. C., Kundaliya, D., Ying, J. Y. 2005. Silica-coated nanocomposites of magnetic nanoparticles and quantum dots. *J. Am. Chem. Soc.*, 127, 4990.

90. Julian-Lopez, B., Boissière, C., Chanéac, C., Grosso, D., Vasseur, S., Miraux, S., Duguet, E., Sanchez, C. 2007. Mesoporous maghemite–organosilica microspheres: A promising route towards multifunctional platforms for smart diagnosis and therapy. *J. Mater. Chem.*, 17, 1563.

91. Cho, S. J., Jarrett, B. R., Louie, A. Y., Kauzlarich, S. M. 2006. Gold-coated iron nanoparticles: A novel magnetic resonance agent for T1 and T2 weighted imaging. *Nanotechnology*, 17, 640.

92. Pinho, S. L., Pereira, G. A., Voisin, P., Kassem, J., Bouchaud, V., Etienne, L., Peters, J. A., Carlos, L., Mornet, S., Geraldes, C. F. G., Rocha, J., Delville, M. H. 2010. Fine tuning of the relaxometry of Fe_2O_3@SiO_2 nanoparticles by tweaking the silica coating thickness. *ACS Nano*, 4, 5339.

93. (a) Santra, S., Tapec, R., Theodoropoulou, N., Dobson, J., Hebard, A., Tan, W. 2001. Synthesis and characterization of silica-coated iron oxide nanoparticles in microemulsion: The effect of nonionic surfactants. *Langmuir*, 17, 2900; (b) Chang, S.-Y., Liu, L., Asher, S. A. 1994. Creation of templated complex topological morphologies in colloidal silica. *J. Am. Chem. Soc.*, 116, 6745.

94. Tartaj, P., Gonzalez-Carreno, T., Serna, C. J. 2001. Single-step nanoengineering of silica coated maghemite hollow spheres with tunable magnetic properties. *Adv. Mater.*, 13, 1620; Tartaj, P., Gonzalez-Carreno, T., Serna, C. J. 2004. From hollow to dense spheres: Control of dipolar interactions by tailoring the architecture in colloidal aggregates of superparamagnetic iron oxide nanocrystals. *Adv. Mater.*, 16, 529.

95. Ashtari, P., He, X., Wang, K., Gong, P. 2005. An efficient method for recovery of target ssDNA based on amino-modified silica-coated magnetic nanoparticles. *Talanta*, 67, 548.

96. Liu, D., Wu, W., Ling, J., Wen, S., Gu, N., Zhang, X. 2011. Effective PEGylation of iron oxide nanoparticles for high performance in vivo cancer imaging. *Adv. Funct. Mater.*, 21, 1498.

97. (a) Szleifer, I. 1997. Protein adsorption on surfaces with grafted polymers: A theoretical approach. *Biophys. J.*, 72, 595; (b) Sofia, S. J., Premnath, V., Merrill, E. W. 1998. Poly(ethylene oxide) grafted to silicon surfaces: Grafting density and protein adsorption. *Macromolecules*, 31, 5059.

98. Huang, C., Neoh, K. G., Wang, L., Kang, E.-T., Shuter, B. 2010. Magnetic nanoparticles for magnetic resonance imaging: Modulation of macrophage uptake by controlled PEGylation of the surface coating. *J. Mater. Chem.*, 20, 8512–8520.

99. Turner, M. R., Duguet, E., Labrugere, C. 1997. Characterization of silane-modified ZrO_2 powder surfaces. *Surf. Interface Anal.* 25, 917.

100. Nooney, M. G., Murrell, T. S, Corneille, J. S., Rusert, E. I., Rossner, L. R., Goodman, D. W. 1996. A spectroscopic investigation of phosphate adsorption onto iron oxides. *J. Vac. Sci. Technol. A*, 14, 1357.

101. Panella, B., Vargas, A., Ferri, D., Baiker A. 2009. Chemical availability and reactivity of functional groups grafted to magnetic nanoparticles monitored in situ by ATR-IR spectroscopy. *Chem. Mater.*, 21, 4316.

102. Lin-Vien, D., Colthup, N., Fateley W., Graselli, J. 1991. *The Handbook of Infrared and Raman Characteristic Frequencies of Organic Molecules*, Academic Press, San Diego, CA.

103. Yamaura, M., Camilo, R. L., Sampaio, L. C., Macedo, M. A., Nakamura, M., Toma, H. E. 2004. Preparation and characterization of (3-aminopropyl) triethoxysilane-coated magnetite nanoparticles. *J. Magn. Magn. Mater.*, 279, 210.

104. Bruening, M., Moons, E., Yaron-Marcovich, D., Cahen, D., Libman, J., Shanzer, A. 1994. Polar ligand adsorption controls semiconductor surface potentials. *J. Am. Chem. Soc.*, 116, 2972.

105. Borggaard, O. K., Raben-Lange, B., Gimsing A. L., Strobel, B. W. 2005. Influence of humic substances on phosphate adsorption by aluminium and iron oxides. *Geoderma*, 127, 270.

106. Lartigue, L., Oumzil, K., Guari, Y., Larionova, J., Guerin, C., Montero, J.-L., Barragan-Montero, V., Sangregorio, C., Caneschi, A., Innocenti, C., Kalaivani, T., Arosio, P., Lascialfari, A. 2009. Water-soluble rhamnose-coated Fe_3O_4 nanoparticles. *Org. Lett.*, 11, 2992.

107. Basly, B., Felder-Flesch, D., Perriat, P., Pourroy, G., Begin-Colin, S. 2011. Properties and suspension stability of dendronised iron oxide nanoparticles for MRI applications. *Contrast Media Mol. Imag.*, DOI:10.1002/cmmi.

108. Basly, B., Felder-Flesch, D., Perriat, P., Billotey, C., Taleb, J., Pourroy, G., Begin-Colin, S. 2010. Dendronized iron oxide nanoparticles as contrast agent for MRI. *Chem. Commun.*, 46, 985.

109. (a) Vestal, C. R., Zhang, Z. J. 2003. Effects of surface coordination chemistry on the magnetic properties of $MnFe_2O_4$ spinel ferrite nanoparticles. *J. Am. Chem. Soc.*, 125, 9828; (b) Brice-Profeta, S., Arrio, M. A., Tronc, E., Menguy, N., Letard, I., Cartier dit Moulin, C., Nogués, M., Chanéac, C., Jolivet, J. P., Sainctavit, Ph. 2005. Magnetic order in γ-Fe_2O_3 nanoparticles: A XMCD study. *J. Magn. Magn. Mater.*, 288, 254; (c) Tronc, E., Ezzir, A., Cherkaoui, R., Chanéac, C., Nogués, M., Kachkachi, H., Fiorani, D., Testa, A. M., Grenèche, J. M., Jolivet, J. P. 2000. Surface-related properties of γ-Fe_2O_3 nanoparticles. *J. Magn. Magn. Mater.*, 221, 63; (d) Yee, C., Kataby, G., Ulman, A., Prozorov, T., White, H., King, A., Rafailovich, M., Sokolov, J., Gedanken, A. 1999. Self-assembled monolayers of alkanesulfonic and -phosphonic acids on amorphous iron oxide nanoparticles. *Langmuir*, 15, 7111.

110. Bittova, B., Poltierova-Vejpravova, J., Roca, A. G., Morales, M. P., Tyrpekl, V. 2010. Effects of coating on magnetic properties in iron oxide nanoparticles. *J. Phys. Conf. Ser.*, 200, 072012.

111. Morales, M. P., Veintemillas-Verdaguer, S., Montero, C. J., Serna, C. J. 1999. Surface and internal spin canting in γ-Fe_2O_3 nanoparticles. *Chem. Mater.*, 11, 3058.

112. Vestal, C. R., Zhang, Z. J. 2003. Effects of surface coordination chemistry on the magnetic properties of $MnFe_2O_4$ spinel ferrite nanoparticles. *J. Am. Chem. Soc.*, 125, 9828.

113. LaConte, L. E. W., Nitin, N., Zurkiya, O., Caruntu, D., O'Connor, C. J., Hu, X., Bao, G. 2007. Coating thickness of magnetic iron oxide nanoparticles affects R2 relaxivity. *J. Magn. Res. Imag.*, 26, 1634.

114. Duan, H., Kuang, M., Wang, X., Andrew Wang, Y., Mao, H., Nie, S. 2008. Reexamining the effects of particle size and surface chemistry on the magnetic properties of iron oxide nanocrystals: New insights into spin disorder and proton relaxivity. *J. Phys. Chem. C*, 112, 8127.

115. Laurent, S., Forge, D., Port, M., Roch, A., Robic, C., Vander Elst, L., Muller, R. N. 2008. Magnetic iron oxide nanoparticles: Synthesis, stabilization, vectorization, physicochemical characterizations, and biological applications. *Chem. Rev.*, 108, 2064.

116. (a) Na, H. B., Song, I. C., Hyeon, T. 2009. Inorganic nanoparticles for MRI contrast agents. *Adv. Mater.*, 21, 2133; (b) Qiao, R. R., Yang, C. H., Gao, M. Y. 2009. Superparamagnetic iron oxide nanoparticles: From preparations to in vivo MRI applications. *J. Mater. Chem.*, 19, 6274; (c) Hogemann, D., Josephson, L., Weissleder, R., Basilion, J. P. 2000. Improvement of MRI probes to allow efficient detection of gene expression. *Bioconjugate Chem.*, 11, 941; (d) Corot, C., Robert, P., Idee, J. M., Port, M. 2006. Recent advances in iron oxide nanocrystal technology for medical imaging. *Adv. Drug Deliv. Rev.*, 58, 1471.

117. Thorek, D. L. J., Chen, A., Czupryna, J., Tsourkas, A. 2006. Superparamagnetic iron oxide nanoparticle probes for molecular imaging. *Ann. Biomed. Eng.*, 34, 23.

118. Zhang, L., Qiao, S. Z., Jin, Y. G., Yang, H. G., Budihartono, S., Stahr, F., Yan, Z. F., Wang, X. L., Hao, Z. P., Lu, G. Q. 2008. Fabrication and size-selective bioseparation of magnetic silica nanospheres with highly ordered periodic mesostructure. *Adv. Funct. Mater.*, 18, 3203.

119. Mahmoudi, M., Sant, S., Wang, B., Laurent, S., Tapas, S. 2011. Superparamagnetic iron oxide nanoparticles (SPIONs): Development, surface modification and applications in chemotherapy. *Adv. Drug. Deliv. Rev.*, 63, 24.

120. Lee, J. W., Foote, R. S. 2009. *Micro and Nano Technologies in Bioanalysis: Methods and Protocols*, Vol. 544, Humana Press, New York.

121. Jordan, A., Scholz, R., Maier-Hauff, K., Johannsen, M., Wust, P., Nadobny, J., Schirra, H., Schmidt, H., Deger, S., Loening, S., Lanksch, W., Felix, R. 2001. Presentation of a new magnetic field therapy system for the treatment of human solid tumors with magnetic fluid hyperthermia. *J. Magn. Magn. Mater.*, 225, 118.

122. Angelova, N., Hunkeler, D. 1999. Rationalizing the design of polymeric biomaterials. *Trends Biotechnol.*, 17, 409.

123. Panyam, J., Labhasetwar, V. 2003. Biodegradable nanoparticles for drug and gene delivery to cells and tissue. *Adv. Drug Delivery Rev.*, 55, 329.

124. Pillai, O., Panchagnula, R. 2001. Polymers in drug delivery. *Curr. Opin. Chem. Biol.*, 5, 447.

125. Moghimi, S. M., Hunter, A. C., Murray, J. C. 2001. Long-circulating and target-specific nanoparticles: Theory to practice. *Pharmacol. Rev.*, 53, 283.

126. Alexis, F., Pridgen, E., Molnar, L. K., Farokhzad, O. C. 2008. Factors affecting the clearance and biodistribution of polymeric nanoparticles. *Mol. Pharm.*, 5, 505.

127. Gölander, C.-G., Herron, J. N., Lim, K., Claesson, P., Stenius, P., Andrade, J. D. 1992. Properties of immobilizated PEG films and the interaction with proteins. In: J. M. Harris, editor. *Poly(Ethylene Glycol) Chemistry, Biotechnical and Biomedical Applications*, Plenum Press, New York, p. 221.

128. (a) Zhang, M., Desai, T., Ferrari, M. 1998. Proteins and cells on PEG immobilized silicon surfaces. *Biomaterials*, 19, 953; (b) Zhang, M. Q., Ferrari, M. 1998. Enhanced blood compatibility of silicon coated with a self-assembled poly(ethyleneglycol) and monomethoxypoly(ethyleneglycol), micro- and nanofabricated structures and devices for biomedical environmental applications, in Progress in Biomedical Optics, P. L. Gourley and A. Katzir eds., Vol. 3258, pp. 15–19.

129. Caliceti, P., Schiavon, O., Mocali, A., Veronese, F. M. 1989. Evaluation of monomethoxypolyethyleneglycol derivatized superoxide dismutase in blood and evidence for its binding to blood cells. *Farmaco*, 44, 711.

130. Cheng, J., Teply, B. A., Sherifi, I., Sung, J., Luther, G., Gu, F. X., Levy- Nissenbaum, E., Radovic-Moreno, A. F., Langer, R., Farokhzad, O. C. 2007. Formulation of functionalized PLGA–PEG nanoparticles for in vivo targeted drug delivery. *Biomaterials*, 28, 869.

131. Shen, M., Shi, X. 2010. Dendrimer-based organic/inorganic hybrid nanoparticles in biomedical applications. *Nanoscale*, 2, 1596.

132. (a) Stiriba, S. E., Frey, H., Haag, R. 2002. Dendritic polymers in biomedical applications: From potential to clinical use in diagnostics and therapy. *Angew. Chem. Int. Ed.*, 41, 1329; (b) Cloninger, M. J. 2002. Biological applications of dendrimers. *Curr. Opin. Chem. Biol.*, 6, 742; (c) Rolland, O., Turrin, C.-O., Caminade, A.-M., Majoral, J.-P. 2009. Dendrimers and nanomedicine: Multivalency in action. *New J. Chem.*, 33, 1809.

133. Strable, E., Bulte, J. W. M., Moskowitz, B., Vivekanandan, K., Allen M., Douglas, T. 2001. Synthesis and characterization of soluble iron oxide–dendrimer composites. *Chem. Mater.*, 13, 2201.

134. Bulte, J. W. M., Douglas, T., Witwer, B., Zhang, S.-C., Strable, E., Lewis, B. K., Zywicke, H., Miller, B., van Gelderen, P., Moskowitz, B. M., Duncan L. D., Frank, J. A. 2001. Magnetodendrimers allow endosomal magnetic labeling and in vivo tracking of stem cells. *Nat. Biotechnol.*, 19, 1141.

135. Lee, I. H., Bulte, J. W. M., Schweinhardt, P., Douglas, T., Trifunovski, A., Hofstetter, C., Olson, L., Spenger, C. 2004. In vivo magnetic resonance tracking of olfactory ensheathing glia grafted into the rat spinal cord. *Exp. Neurol.*, 187, 509.

136. Tunici, P., Bulte, J. W. M., Bruzzone, M. G., Poliani, P. L., Cajola, L., Grisoli, M., Douglas, T., Finocchiaro, G. 2006. Brain engraftment and therapeutic potential of stem/progenitor cells derived from mouse skin. *J. Gene Med.*, 8, 506.

137. Bulte, J. W. M., Douglas, T., Witwer, B., Zhang, S.-C., Lewis, B. K., van Gelderen, P., Zywicke, H., Duncan I. D., Frank, J. A. 2002. Monitoring stem cell therapy in vivo using magnetodendrimers as a new class of cellular MR contrast agents. *Acad. Radiol.*, 9, S332.

138. Frankamp, B. L., Boal, A. K., Tuominen, M. T., Rotello, V. M. 2005. Direct control of the magnetic interaction between iron oxide nanoparticles through dendrimer-mediated self-assembly. *J. Am. Chem. Soc.*, 127, 9731.

139. (a) Li, Z., Huang, P., Zhang, X., Lin, J., Yang, S., Liu, B., Gao, F., Xi, P., Ren, Q., Cui, D. 2010. RGD-conjugated dendrimer-modified gold nanorods for *in vivo* tumor targeting and photothermal therapy. *Mol. Pharm.*, 7, 94.

140. Shi, X., Thomas, T. P., Myc, L. A., Kotlyar, A., Baker Jr., J. R. 2007. Synthesis, characterization, and intracellular uptake of carboxyl-terminated poly(amidoamine) dendrimer-stabilized iron oxide nanoparticles. *Phys. Chem. Chem. Phys.*, 9, 5712.

141. (a) Quintana, A., Raczka, E., Piehler, L., Lee, I., Myc, A., Majoros, I., Patri, A. K., Thomas, T. P., Mule, J., Baker Jr., J. R. 2002. Design and function of a dendrimer-based therapeutic nanodevice targeted to tumor cells through the folate receptor. *Pharm. Res.*, 19, 1310; (b) Thomas, T. P., Majoros, I. J., Kotlyar, A., Kukowska-Latallo, J. F., Bielinska, A., Myc, A., Baker Jr., J. R. 2005. Targeting and inhibition of cell growth by an engineered dendritic nanodevice. *J. Med. Chem.*, 48, 3729; (c) Shi, X., Wang, S., Meshinchi, S., Van Antwerp, M. E., Bi, X., Lee, I., Baker Jr., J. R. 2007. Dendrimer-entrapped gold nanoparticles as a platform for cancer-cell targeting and imaging. *Small*, 3, 1245.

142. Wang, S., Shi, X., Van Antwerp, M., Cao, Z., Swanson, S. D., Bi, X., Baker Jr., J. R. 2007. Dendrimer-functionalized iron oxide nanoparticles for specific targeting and imaging of cancer cells. *Adv. Funct. Mater.*, 17, 3043.

143. Shi, X., Wang, S. H., Swanson, S. D., Ge, S., Cao, Z., Van Antwerp, M. E., Landmark, K. J., Baker Jr., J. R. 2008. Dendrimer-functionalized shell-crosslinked iron oxide nanoparticles for in-vivo magnetic resonance imaging of tumors. *Adv. Mater.*, 20, 1671.

144. (a) Tomalia, D. A., Baker, H., Dewald, J. R., Hall, M., Kallos, G., Martin, S., Roeck, J., Ryder, J., Smith, P. 1986. Dendritic macromolecules: Synthesis of starburst dendrimers. *Macromolecules*, 19, 2466; (b) Tomalia, D. A., Naylor, A. M., Goddard III, W. A. 1990. Starburst dendrimers: Molecular-level control of size, shape, surface chemistry, topology, and flexibility from atoms to macroscopic matter. *Angew. Chem. Int. Ed. Engl.*, 29, 138.

145. Pan, B., Gao, F., Ao, L. 2005. Investigation of interactions between dendrimer-coated magnetite nanoparticles and bovine serum albumin. *J. Magn. Magn. Mater.*, 293, 252; b) Pan, B., Gao, F., Gu, H. 2005. Dendrimer modified magnetite nanoparticles for protein immobilization. *J. Colloid Interface Sci.*, 284, 1.

146. Pan, B., Cui, D., Sheng, Y., Ozkan, C., Gao, F., He, R., Li, Q., Xu, P., Huang, T. 2007. Dendrimer-modified magnetic nanoparticles enhance efficiency of gene delivery system. *Cancer Res.*, 67, 8156.

147. Boyer, C., Bulmus, V., Priyanto, P., Teoh, W. Y., Amal, R., Davis, T. P. 2009. The stabilization and biofunctionalization of iron oxide nanoparticles using heterotelechelic polymers. *J. Mater. Chem.*, 19, 111.

148. (a) Bertin, A., Steibel, J., Michou-Gallani, A.-I., Gallani, J.-L., Felder-Flesch, D. 2009. Development of a dendritic manganese-enhanced magnetic resonance imaging (MEMRI) contrast agent: Synthesis, toxicity (in vitro) and relaxivity (in vitro, in vivo) studies. *Bioconjugate Chem.*, 20, 760; (b) Bertin, A., Michou-Gallani, A.-I., Steibel, J., Gallani, J.-L., Felder-Flesch, D. 2010. Synthesis and characterization of a highly stable dendritic catechol-tripod bearing technetium-99m. *New J. Chem.*, 34, 267; (c) Bertin, A., Michou-Gallani, A.-I., Gallani, J.-L., Felder-Flesch. D. 2010. In vitro neurotoxicity of magnetic resonance imaging (MRI) contrast agents: Influence of the molecular structure and paramagnetic ion. *Toxicol. Vitro*, 24, 1386; (d) Villaraza, A. J. L., Bumb, A., Brechbiel, M. W. 2010. Macromolecules, dendrimers, and nanomaterials in magnetic resonance imaging: The interplay between size, function, and pharmacokinetics. *Chem. Rev.*, 110, 2921.

149. (a) Chandra, S., Mehta, S., Nigam, S., Bahadur, D. 2010. Dendritic magnetite nanocarriers for drug delivery applications. *New J. Chem.*, 34, 648; (b) Zhu, R., Jiang, W., Pu, Y., Luo, K., Wu, Y., He, B., Gu, Z. 2011. Functionalization of magnetic nanoparticles with peptide dendrimers. *J. Mater. Chem.*, 21, 5464.

150. (a) Hao, R., Xing, R., Xu, Z., Hou, Y., Gao, S., Sun, S. 2010. Synthesis, functionalization, and biomedical applications of multifunctional magnetic nanoparticles. *Adv. Mater.*, 22, 2729; (b) Fang, C., Zhang, M. 2009. Synthesis, functionalization, and biomedical applications of multifunctional magnetic nanoparticles. *J. Mater. Chem.*, 19, 6258.

151. Lammers, T., Subr, V., Ulbrich, K., Hennink, W. E., Storm, G., Kiessling, F. 2010. Polymeric nanomedicines for image-guided drug delivery and tumor-targeted combination therapy. *Nano Today*, 5, 197.

152. Das, M., Mishra, D., Dhak, P., Gupta, S., Kumar Maiti, T., Basak, A., Pramanik, P. 2009. Biofunctionalized, phosphonate-grafted, ultrasmall iron oxide nanoparticles for combined targeted cancer therapy and multimodal imaging. *Small*, 5(24), 2883.

153. Louie, A. 2010. Multimodality imaging probes: Design and challenges. *Chem. Rev.*, 110, 3146.

154. (a) McCarthy, J. R., Patel, P., Botnaru, I., Haghayeghi, P., Weissleder, R., Jaffer, F. A. 2009. Multimodal nanoagents for the detection of intravascular thrombi. *Bioconjugate Chem.*, 20, 1251; (b) Mulder, W. J. M., Strijkers, G. J., van Tilborg, G. A. F., Griffioen, A. W., Nicolay, K. 2006. Lipid-based nanoparticles for contrast-enhanced MRI and molecular imaging. *NMR Biomed.*, 19, 142; (c) Jennings, L. E., Long, N. J. 2009. 'Two is better than one' – Probes for dual-modality molecular imaging. *Chem. Commun.*, 3511; (d) Frullano, L., Meade, T. J. 2007. Multimodal MRI contrast agents. *J. Biol. Inorg. Chem.*, 12,

939; (e) Kim, J., Piao, Y., Hyeon, T. 2009. Multifunctional nanostructured materials for multimodal imaging, and simultaneous imaging and therapy. *Chem. Soc. Rev.*, 38, 372; (f) Cheon, J., Lee, J. H. 2008. Synergistically integrated nanoparticles as multimodal probes for nanobiotechnology. *Acc. Chem. Res.*, 41, 1630; (g) Park, K., Lee, S., Kang, E., Kim, K., Choi, K., Kwon, I. C. 2009. New generation of multifunctional nanoparticles for cancer imaging and therapy. *Adv. Funct. Mater.*, 19, 1553.

155. (a) Banerjee, S. S., Chen, D. H. 2009. A multifunctional magnetic nanocarrier bearing fluorescent dye for targeted drug delivery by enhanced two-photon triggered release. *Nanotechnology*, 20, 185103; (b) Gallagher, J. J., Tekoriute, R., O'Reilly, J. A., Kerskens, C., Gun'ko, Y. K., Lynch, M. 2009. Bimodal magnetic-fluorescent nanostructures for biomedical applications. *J. Mater. Chem.*, 19, 4081.

156. Gunn, J., Wallen, H., Veiseh, O., Sun, C., Fang, C., Cao, J. H., Yee, C., Zhang, M. Q. 2008. A multimodal targeting nanoparticle for selectively labeling T cells. *Small*, 4, 712.

157. Cheng, C. M., Chu, P. Y., Chuang, K. H., Roffler, S. R., Kao, C. H., Tseng, W. L., Shiea, J., Chang, W. D., Su, Y. C., Chen, M., Wang, Y. M., Cheng, T. L. 2009. Hapten-derivatized nanoparticle targeting and imaging of gene expression by multimodality imaging systems. *Cancer Gene Ther.*, 16, 83.

158. Hirata, N., Tanabe, K., Narita, A., Tanaka, K., Naka, K., Chujo, Y., Nishimoto, S. 2009. Preparation and fluorescence properties of fluorophore-labeled avidin–biotin system immobilized on Fe_3O_4 nanoparticles through functional indolequinone linker. *Biorg. Med. Chem.*, 17, 3775.

159. Yeo, K. M., Gao, C. J., Ahn, K. H., Lee, I. S. 2008. Superparamagnetic iron oxide nanoparticles with photoswitchable fluorescence. *Chem. Commun.*, 44, 4622.

160. Quarta, A., Di Corato, R., Manna, L., Argentiere, S., Cingolani, R., Barbarella, G., Pellegrino, T. 2008. Multifunctional nanostructures based on inorganic nanoparticles and oligothiophenes and their exploitation for cellular studies. *J. Am. Chem. Soc.*, 130, 10545.

161. (a) Pittet, M. J., Swirski, F. K., Reynolds, F., Josephson, L., Weissleder, R. 2006. Labeling of immune cells for in vivo imaging using magnetofluorescent nanoparticles. *Nat. Protoc.*, 1, 73; (b) Trehin, R., Figueiredo, J. L., Pittet, M. J., Weissleder, R., Josephson, L., Mahmood, U. 2006. Fluorescent nanoparticle uptake for brain tumor visualization. *Neoplasia*, 8, 302.

162. Kelly, K. A., Allport, J. R., Tsourkas, A., Shinde-Patil, V. R., Josephson, L., Weissleder, R. 2005. Detection of vascular adhesion molecule-1 expression using a novel multimodal nanoparticle. *Circ. Res.*, 96, 327.

163. (a) Schellenberger, E. A., Sosnovik, D., Weissleder, R., Josephson, L. 2004. Magneto/optical annexin V, a multimodal protein. *Bioconjugate Chem.*, 15, 1062; (b) Sosnovik, D. E., Schellenberger, E. A., Nahrendorf, M., Novikov, M. S., Matsui, T., Dai, G., Reynolds, F., Grazette, L., Rosenzweig, A., Weissleder, R., Josephson, L. 2005. Magnetic resonance imaging of cardiomyocyte apoptosis with a novel magneto-optical nanoparticle. *Magn. Reson. Med.*, 54, 718.

164. Moore, A., Medarova, Z., Potthast, A., Dai, G. P. 2004. In vivo targeting of underglycosylated MUC-1 tumor antigen using a multimodal imaging probe. *Cancer Res.*, 64, 1821.

165. Montet, X., Montet-Abou, K., Reynolds, F., Weissleder, R., Josephson, L. 2006. Nanoparticle imaging of integrins on tumor cells. *Neoplasia*, 8, 214.

166. Medarova, Z., Kumar, M., Ng, S. W., Yang, J. Z., Barteneva, N., Evgenov, N. V., Petkova, V., Moore, A. 2008. Multifunctional magnetic nanocarriers for image-tagged SiRNA delivery to intact pancreatic islets. *Transplantation*, 86, 1170.

167. Josephson, L., Kircher, M. F., Mahmood, U., Tang, Y., Weissleder, R. 2002. Near-infrared fluorescent nanoparticles as combined MR/optical imaging probes. *Bioconjugate Chem.*, 13, 554.

168. Yang, L., Mao, H., Cao, Z. H., Wang, Y. A., Peng, X. H., Wang, X. X., Sajja, H. K., Wang, L. Y., Duan, H. W., Ni, C. C., Staley, C. A., Wood, W. C., Gao, X. H., Nie, S. M. 2009. Molecular imaging of pancreatic cancer in an animal model using targeted multifunctional nanoparticles. *Gastroenterology*, 136, 1514.

169. Lee, H. Y., Li, Z., Chen, K., Hsu, A. R., Xu, C., Xie, J., Sun, S. H., Chen, X. Y. 2008. PET/MRI dual-modality tumor imaging using arginine–glycine–aspartic (RGD)-conjugated radiolabeled iron oxide nanoparticles. *J. Nucl. Med.*, 49, 1371.

170. Nahrendorf, M., Zhang, H. W., Hembrador, S., Panizzi, P., Sosnovik, D. E., Aikawa, E., Libby, P., Swirski, F. K., Weissleder, R. 2008. Nanoparticle PET–CT imaging of macrophages in inflammatory atherosclerosis. *Circulation*, 117, 379.

171. Natarajan, A., Gruettner, C., Ivkov, R., DeNardo, G. L., Mirick, G., Yuan, A., Foreman, A., DeNardo, S. J. 2008. Nanoferrite particle based radioimmunonanoparticles: Binding affinity and in vivo pharmacokinetics. *Bioconjugate Chem.*, 19, 1211.

172. Choi, J. S., Park, J. C., Nah, H., Woo, S., Oh, J., Kim, K. M., Cheon, G. J., Chang, Y., Yoo, J., Cheon, J. 2008. A hybrid nanoparticle probe for dual-modality positron emission tomography and magnetic resonance imaging. *Angew. Chem. Int. Ed.*, 47, 6259.

173. Devaraj, N. K., Keliher, E. J., Thurber, G. M., Nahrendorf, M., Weissleder, R. 2009.[18]F labeled nanoparticles for *in vivo* PET–CT imaging. *Bioconjugate Chem.*, 20, 397.

174. (a) Lu, C. W., Hung, Y., Hsiao, J. K., Yao, M., Chung, T. H., Lin, Y. S., Wu, S. H., Hsu, S. C., Liu, H. M., Mou, C. Y., Yang, C. S., Huang, D. M., Chen, Y. C. 2007. Bifunctional magnetic silica nanoparticles for highly efficient human stem cell labeling. *Nano Lett.*, 7, 149; (b) Wu, S. H., Lin, Y. S., Hung, Y., Chou, Y. H., Hsu, Y. H., Chang, C., Mou, C. Y. 2008. Multifunctional mesoporous silica nanoparticles for intracellular labeling and animal magnetic resonance imaging studies. *ChemBioChem*, 9, 53.

175. (a) Makovec, D., Campelj, S., Bele, M., Maver, U., Zorko, M., Drofenik, M., Jamnik, J., Gaberscek, M. 2009. Nanocomposites containing embedded superparamagnetic iron oxide nanoparticles and rhodamine 6G. *Colloids Surf. A*, 334, 74; (b) Lee, K., Moon, H. Y., Park, C., Kim, O. R., Ahn, E., Lee, S. Y., Park, H. E., Ihm, S. H., Seung, K. B., Chang, K., Yoon, T. J., Lee, C., Cheong, C., Hong, K. S. 2009. Magnetic resonance imaging of macrophage activity in atherosclerotic plaques of apolipoprotein E-deficient mice by using polyethylene glycolated magnetic fluorescent silica-coated nanoparticles. *Curr. Appl. Phys.*, 9, S15; (c) Yoon, T. J., Kim, J. S., Kim, B. G., Yu, K. N., Cho, M. H., Lee, J. K. 2005. Multifunctional nanoparticles possessing a 'magnetic motor effect' for drug or gene delivery. *Angew. Chem. Int. Ed.*, 44, 1068.

176. Ma, Z. Y., Dosev, D., Nichkova, M., Dumas, R. K., Gee, S. J., Hammock, B. D., Liu, K., Kennedy, I. M. 2009. Synthesis and characterization of multifunctional silica core–shell nanocomposites with magnetic and fluorescent functionalities. *J. Magn. Magn. Mater.*, 321, 1368.

177. Heitsch, A. T., Smith, D. K., Patel, R. N., Ress, D., Korgel, B. A. 2008. Multifunctional particles: Magnetic nanocrystals and gold nanorods coated with fluorescent dye-doped silica shells. *J. Solid State Chem.*, 181, 1590.

178. Salgueirino-Maceira, V., Correa-Duarte, M. A., Lopez-Quintela, M. A., Rivas, J. 2009. Advanced hybrid nanoparticles. *J. Nanosci. Nanotechnol.*, 9, 3684.

179. Santra, S., Kaittanis, C., Grimm, J., Perez, J. M. 2009. Drug/dye-loaded, multifunctional iron oxide nanoparticles for combined targeted cancer therapy and dual optical/magnetic resonance imaging. *Small*, 5, 1862.

180. Nagao, D., Yokoyama, M., Yamauchi, N., Matsumoto, H., Kobayashi, Y., Konno, M. 2008. Synthesis of highly monodisperse particles composed of a magnetic core and fluorescent shell. *Langmuir*, 24, 9804.

181. Li, L. L., Li, H. B., Chen, D., Liu, H. Y., Tang, F. Q., Zhang, Y. Q., Ren, J., Li, Y. 2009. Preparation and characterization of quantum dots coated magnetic hollow spheres for magnetic fluorescent multimodal imaging and drug delivery. *J. Nanosci. Nanotechnol.*, 9, 2540.

182. (a) Du, G. H., Liu, Z. L., Wang, D., Xia, X., Jia, L. H., Yao, K. L., Chu, Q., Zhang, S. M. 2009. Characterization of magnetic fluorescence Fe_3O_4/CdSe nanocomposites. *J. Nanosci. Nanotechnol.*, 9, 1304; (b) Selvan, S. T., Patra, P. K., Ang, C. Y., Ying, J. Y. 2007. Synthesis of silica-coated semiconductor and magnetic quantum dots and their use in the imaging of live cells. *Angew. Chem. Int. Ed.*, 46, 2448.

183. Wang, D. S., He, J. B., Rosenzweig, N., Rosenzweig, Z. 2004. Superparamagnetic Fe_2O_3 beads–CdSe/ZnS quantum dots core–shell nanocomposite particles for cell separation. *Nano Lett.*, 4, 409.

184. Dosev, D., Nichkova, M., Dumas, R. K., Gee, S. J., Hammock, B. D., Liu, K., Kennedy, I. M. 2007. Magnetic/luminescent core/shell particles synthesized by spray pyrolysis and their application in immunoassays with internal standard. *Nanotechnology*, 18, 055102.

185. He, H., Xie, M. Y., Ding, Y., Yu, X. F. 2009. Synthesis of Fe_3O_4@LaF_3:Ce,Tb nanocomposites with bright fluorescence and strong magnetism. *Appl. Surf. Sci.*, 255, 4623.

186. Lu, H. C., Yi, G. S., Zhao, S. Y., Chen, D. P., Guo, L. H., Cheng, J. 2004. Synthesis and characterization of multi-functional nanoparticles possessing magnetic, up-conversion fluorescence and bio-affinity properties. *J. Mater. Chem.*, 14, 1336.

187. Ma, Z. Y., Dosev, D., Nichkova, M., Gee, S. J., Hammock, B. D., Kennedy, I. M. 2009. Synthesis and biofunctionalization of multifunctional magnetic Fe_3O_4@Y_2O_3:Eu nanocomposites. *J. Mater. Chem.*, 19, 4695.

188. (a) Cho, S. J., Jarrett, B. R., Louie, A. Y., Kauzlarich, S. M. 2006. Gold-coated iron nanoparticles: A novel magnetic resonance agent for T_1 and T_2 weighted imaging. *Nanotechnology*, 17, 640; (b) Larson, T. A., Bankson, J., Aaron, J., Sokolov, K. 2007. Hybrid plasmonic magnetic nanoparticles as molecular specific agents for MRI/optical imaging and photothermal therapy of cancer cells. *Nanotechnology*, 18, 325101.

189. Seo, W. S., Lee, J. H., Sun, X. M., Suzuki, Y., Mann, D., Liu, Z., Terashima, M., Yang, P. C., McConnell, M. V., Nishimura, D. G., Dai, H. J. 2006. FeCo/graphitic-shell nanocrystals as advanced magnetic-resonance-imaging and near-infrared agents. *Nat. Mater.*, 5, 971.

190. Soenen, S. J. H., Vercauteren, D., Braeckmans, K., Noppe, W., De Smedt, S., De Cuyper, M. 2009. Stable long-term intracellular labelling with fluorescently tagged cationic magnetoliposomes. *ChemBioChem*, 10, 257.

191. Zhang, Y. X., Das, G. K., Xu, R., Tan, T. T. Y. 2009. Tb-doped iron oxide: Bifunctional fluorescent and magnetic nanocrystals. *J. Mater. Chem.*, 19, 3696.

192. Lamanna, G., Kueny-Stotz, M., Mamlouk, H., Basly, B., Ghobril, C., Berniard, A., Billotey, C., Pourroy, G., Begin-Colin, S., Felder-Flesch, D. 2011. Versatile dendronized iron oxides as smart nano-objects for multi-modal imaging. *Biomaterials*, DOI:10.1016/f.biomaterials.2011.07.026.

193. Jain, R., Dandekar, P., Patravale, V. 2009. Diagnostic nanocarriers for sentinel lymph node imaging. *J. Control. Release*, 138, 90.

194. Kueny-Stotz, M., Mamlouk-Chaouachi, H., Felder-Flesch, D. 2011. Synthesis of Patent Blue derivatized hydrophilic dendrons dedicated to sentinel node detection in breast cancer. *Tet. Lett.*, DOI: 10.1016/j.tetlet.2011.03.144.

195. Pankhurst, Q. A., Connolly, J., Jones, S. K., Dobson, J. 2003. Applications of magnetic nanoparticles in biomedicine. *J. Phys. D Appl. Phys.*, 36, R167.

196. (a) Dobson, J. 2006. Magnetic nanoparticles for drug delivery. *Drug Dev. Res.*, 67, 55; (b) Harisinghani, M. G., Barentsz, J., Hahn, P. F., Deserno, W. M., Tabatabaei, S., Hulsbergen van de Kaa, C., De la Rosette, J., Weissleder, R. 2003. Noninvasive detection of clinically occult lymph-node metastases in prostate cancer. *N. Engl. J. Med.*, 348, 2491.

197. Rogers, J. M., Lewis, J., Josephson, L. 1994. Visualization of superior mesenteric lymph nodes by the combined oral and intravenous administration of the ultrasmall superparamagnetic iron oxide, AMI-227. *Magn. Reson. Imaging*, 12, 1161.

198. (a) Chambon, C., Clément, O., Le Blanche, A., Schouman-Claeys, E., Frija, G. 1993. Superparamagnetic iron oxides as positive MR contrast agents: In vitro and in vivo evidence. *Magn. Reson. Imaging*, 11, 509; (b) Canet, E., Revel, D., Forrat, R., Baldy-Porcher, C., de Lorgeril, M., Sebbag, L., Vallee, J. P., Didier, D., Amiel, M. 1993. Superparamagnetic iron oxide particles and positive enhancement for myocardial perfusion studies assessed by subsecond T_1-weighted MRI. *Magn. Reson. Imaging*, 11, 1139.

199. Alexiou, C., Jurgons, R., Schmid, R., Hilpert, A., Bergemann, C., Parak, F., Iro, H. 2005. In vitro and in vivo investigations of targeted chemotherapy with magnetic nanoparticles. *J. Magn. Magn. Mater.*, 293, 389.

200. Babincova, M., Altanerova, V., Altaner, C., Bergemann, C., Babinec, P. 2008. In vitro analysis of cisplatin functionalized magnetic nanoparticles in combined cancer chemotherapy and electromagnetic hyperthermia. *IEEE Trans. Nanobiosci.*, 7, 15.

201. Kettering, M., Zorn, H., Bremer-Streck, S., Oehring, H., Zeisberger, M., Bergemann, C. Hergt, R., Halbhuber, K. J., Kaiser, W. A., Hilger, I. 2009. Characterization of iron oxide nanoparticles adsorbed with cisplatin for biomedical applications. *Phys. Med. Biol.*, 54, 5109.

202. Kim, J., Lee, J. E., Lee, S. H., Yu, J. H., Lee, J. H., Park, T. G., Hyeon, T. 2008. Designed fabrication of a multifunctional polymer nanomedical platform for simultaneous cancer-targeted imaging and magnetically guided drug delivery. *Adv. Mater.*, 20, 478–483.

203. (a) Panyam, J., Labhasetwar, V. 2003. Biodegradable nanoparticles for drug and gene delivery to cells and tissue. *Adv. Drug Delivery Rev.*, 55, 329; (b) Jeong, B., Bae, Y. H., Lee, D. S., Kim, S. W. 1997. Biodegradable block copolymers as injectable drug-delivery systems. *Nature*, 388, 860; (c) Yoo, H. S., Choi, H. K., Park, T. G. 2001. Protein-fatty acid complex for enhanced loading and stability within biodegradable nanoparticles. *J. Pharm. Sci.*, 90, 194; (d) Kakizawa, Y., Furukawa, S., Kataoka, K. 2004. Block copolymer-coated calcium phosphate nanoparticles sensing intracellular environment for oligodeoxynucleotide and siRNA delivery. *J. Contr. Release*, 97, 345; (e) Diwan, M., Park, T. G. 2003. Stabilization of recombinant interferon-alpha by pegylation for encapsulation in PLGA microspheres. *Int. J. Pharm.*, 252, 111.

204. (a) Klostranec, J. M., Chan, W. C. W. 2006. Quantum dots in biological and biomedical research: Recent progress and present challenges. *Adv. Mater.*, 18, 1953; (b) Brunchez, M. P., Moronne, M., Gin, P., Weiss, S., Alivisatos, A. P. 1998. Semiconductor nanocrystals as fluorescent biological labels. *Science*, 281, 2013; (c) Chan, W. C., Nie, S. 1998. Quantum dot bioconjugates for ultrasensitive nonisotopic detection. *Science*, 281, 2016; (d) Alivisatos, A. P. 2004. The use of nanocrystals in biological detection. *Nat. Biotechnol.*, 22, 47; (e) Dubertret, B., Skourides, P., Norris, D. J., Noireaux, V., Brivanlou, A. H., Libchaber, A. 2002. In vivo imaging of quantum dots encapsulated in phospholipid micelles. *Science*, 298, 1759; (f) Michalet, X., Pinaud, F. F., Bentolila, L. A., Tsay, J. M., Doose, S., Li, J. J., Sundaresan, G., Wu, A. M., Gambhir, S. S., Weiss, S. 2005. Quantum dots for live cells, in vivo imaging, and diagnostics. *Science*, 307, 538; (g) Medintz, I. L., Clapp, A. R., Brunel, F. M., Tiefenbrunn, T., Uyeda, H. T., Chang, E. L., Deschamps, J. R., Dawson, P. E., Mattoussi, H. 2006. Proteolytic activity monitored by fluorescence resonance energy transfer through quantum-dot-peptide conjugates. *Nat. Mater.*, 5, 581; (h) Medintz, I. L., Clapp, A. R., Melinger, J. S., Deschamps, J. R., Mattoussi, H. 2005. A reagentless biosensing assembly based on quantum dot-donor Förster resonance energy transfer. *Adv. Mater.*, 17, 2450; (i) Kim, S., Lim, Y. T., Soltesz, E. G., De Grand, A. M., Lee, J., Nakayama, A., Parker, J. A., Mihaljevic, T., Laurence, R. G., Dor, D. M., Cohn, L. H., Bawendi, M. G., Frangioni, J. V. 2004. Near-infrared fluorescent type II

quantum dots for sentinel lymph node mapping. *Nat. Biotechnol.*, 22, 93; (j) Voura, E. B., Jaiswal, J. K., Mattoussi, H., Simon, S. M. 2004. Tracking metastatic tumor cell extravasation with quantum dot nanocrystals and fluorescence emission-scanning microscopy. *Nat. Med.*, 2, 993.

205. Zunino, F., Capranico, G. 1990. DNA topoisomerase II as the primary target of anti-tumor anthracyclines. *Anti-Cancer Drug Des.*, 5, 307.
206. Lua, Y., Low, P. S. 2003. Immunotherapy of folate receptor-expressing tumors: Review of recent advances and future prospects. *J. Contr. Release*, 91, 17.
207. Weitman, S. D., Lark, R. H., Coney, L. R., Fort, D. W., Frasca, V., Zurawski, V. R., Kamen, B. A. 1992. Distribution of the folate receptor GP38 in normal and malignant cell lines and tissues. *Cancer Res.*, 52, 3396.
208. (a) Coney, L. R., Tomassetti, A., Carayannopoulos, L., Frasca, V., Kamen, B. A., Colnaghi, M. I., Zurawski, W. R. 1991. Cloning of a tumor-associated antigen: MOv18 and MOv19 antibodies recognize a folate-binding protein. *Cancer Res.*, 51, 6125; (b) Antony, A. C. 1992. The biological chemistry of folate receptors. *Blood*, 79, 2807.
209. (a) Uchegbu, I. F., Schatzlein, A. G., Tetley, L., Gray, A. I., Sludden, J., Siddique, S., Mosha, E. 1998. Polymeric chitosan-based vesicles for drug deliver. *J. Pharm. Pharmacol.*, 50, 453; (b) Lau, K. G., Hattori, S., Chopra, S., O'Toole, E. A., Storey, A., Nagai, T., Maitani, Y. 2005. Ultra-deformable liposomes containing bleomycin: In vitro stability and toxicity on human cutaneous keratinocyte cell lines. *Int. J. Pharm.*, 300, 4.
210. Georgelin, T., Bombard, S., Siaugue, J.-M., Cabuil, V. 2010. Nanoparticle-mediated delivery of bleomycin. *Angew. Chem. Int. Ed.*, 49, 8897.
211. Xie, J., Chen, K., Lee, H. Y., Xu, C., Hsu, A. R., Peng, S., Chen, X., Sun, S. 2008. Ultrasmall c(RGDyK)-coated Fe_3O_4 nanoparticles and their specific targeting to integrin $\alpha_v\beta_3$-rich tumor cells. *J. Am. Chem. Soc.*, 130, 7542.
212. (a) Lyons, S. A., O'Neal, J., Sontheimer, H. 2002. Chlorotoxin, a scorpion-derived peptide, specifically binds to gliomas and tumors of neuroectodermal origin. *Glia*, 39, 162; (b) Veiseh, M., Gabikian, P., Bahrami, S. B., Veiseh, O., Zhang, M., Hackman, R. C., Ravanpay, A. C., Stroud, M. R., Kusuma, Y., Hansen, S. J., Kwok, D., Munoz, N. M., Sze, R. W., Grady, W. M., Greenberg, N. M., Ellenbogen, R. G., Olson, J. M. 2007. Tumor paint: A chlorotoxin:Cy5.5 bioconjugate for intraoperative visualization of cancer foci. *Cancer Res.*, 67, 6882.
213. Sun, C., Veiseh, O., Gunn, J., Fang, C., Hansen, S., Lee, D., Sze, R., Ellenbogen, R. G., Olson, J., Zhang, M. 2008. In Vivo MRI Detection of gliomas by chlorotoxin-conjugated superparamagnetic nanoprobes. *Small*, 4, 372.

Sylvie Begin-Colin completed her PhD degree in material chemistry at University of Nancy (France) in 1992. She has then integrated with the CNRS as a researcher at the Laboratory of Science and Engineering of Material and of Metallurgy at the Mining Engineering School of Nancy and has developed research on physicochemical modifications induced by ball milling in oxides. She was then appointed professor in September 2003 at the European Engineering School in Chemistry, Material and Polymer (ECPM) of the University of Strasbourg and is currently Deputy Director of ECPM. Her research interests at the Institute of Physic and Chemistry of Materials (IPCMS)

of Strasbourg focus on the synthesis, functionalisation and organisation of oxide nanoparticles for biomedical, catalysis and spintronic applications.

Delphine Felder-Flesch was born in 1973, Colmar, France. In 2001, she obtained her PhD in supra-molecular organic chemistry (fullerene-based materials) at the University of Strasbourg, France, under the supervision of Dr. J.-F. Nierengarten and Prof. J.-F. Nicoud. After an ERASMUS training period during her PhD in the Laboratory of Organometallic Chemistry of Pr. S. B. Duckett, University of York, England, she started post-doctoral research at the ETH-Zürich, Switzerland, in the group of Prof. Dr. François Diederich working on porphyrin chemistry. Since October 2003, she is a CNRS research scientist at the Department of Organic Materials, Institute of Physics and Chemistry of Materials, Strasbourg. Currently, her research interests include dendritic nanosystems such as contrast agents for (manganese enhanced) MRI, dendritic radiopharmaceuticals (SPECT) for brain imaging, dendronised nanoparticles for MRI, but also carbonaceous nano-objects and more precisely the synthesis of [60]Fullerene hexakisadducts for the development of high-added value materials (thin films, liquid-crystals).

Part IV

Ex Vivo *Application of MNPs*

Part IV

Ex Vivo Application of MNPs

7 Removal of Blood-Borne Toxin in the Body Using Magnetic Nanospheres*

Michael D. Kaminski, Haitao Chen, Xianqiao Liu, Dietmar Rempfer, and Axel J. Rosengart

CONTENTS

7.1 INTRODUCTION AND PROBLEM DESCRIPTION

Blood-borne toxins, that is, substances circulating within the bloodstream causing adverse side effects, can originate from various sources. For example, ingestion of biohazards can occur with the intake of contaminated food, while other toxins penetrate through the skin or are readily taken up by the lungs *via* inhalation of particulates. There are also less frequent but more extreme potential cases of toxin releases such as from biological and chemical weapons of mass destruction. Of importance, the recent events at the Fukushima reactor in Japan highlight[1] the potential for mass release and exposure of civilians to radiological toxins. Similarly, the weaponised release of radiological toxins *via* a radiological dispersal device (so-called dirty bomb) or detonation of an improvised nuclear weapon could present an insurmountable triage setting with hospitals lacking sufficient methods of treating internalised radiological contaminations.

After such disaster scenarios, current treatment strategies for internal decontamination, that is, removing toxins from the body after their absorption into the bloodstream, are limited to a combination of chelator drug therapies or through preventative inoculation. The chelator drug therapies generally achieve only a modest reduction in the toxins' half-life and are only effective for a few selected radionuclides. Preventative inoculation, although appropriate for certain high-risk populations such as specialised military units or civilians (e.g. doctors, first responder teams), is not recommended to the general public due to proven or suspected acute and long-term risks.

* Argonne National Laboratory, operated by UChicago Argonne, LLC, for the U.S. Department of Energy under Contract No. DE-AC02-06CH11357.

Therefore, most treatments are limited to providing supportive medical care and reducing secondary symptoms caused by exposure.

The overall aim is to develop a rapid and selective treatment method to effectively and selectively reduce circulating toxins in exposed subjects. Towards this aim, we have described a new methodology that combines the development of functionalised, magnetic, biodegradable micro- and nanocarriers with a novel method for selective removal of nanocarrier-toxin complexes from the bloodstream. In this chapter, we will introduce a conceptual approach and outline our initial results.

Toxins, such as certain heavy metals, may be taken in from food, water or air. After toxin absorption and circulation throughout the body *via* the bloodstream, deposition into organs and tissue occurs. Heavy metal compounds interfere with the cellular functions of various organ systems such as the central nervous system (CNS), digestive tract, haematopoietic system (bone marrow), liver, kidneys, etc.

The treatment for heavy metal poisoning currently available is chelation therapy. In general, chelation refers to the formation of two or more bonds between a single, large organic compound (ligand) and the target metal atom. A chelating agent is given either directly into the vein (intravenously), by mouth (orally) or, for extended release kinetics, into the muscular tissues (intramuscularly) to bind to a target metal that facilitates its removal from circulation. Examples of clinically used chelators include penicillamine for copper toxicity, dimercaptosuccinic acid (DMSA) for arsenic or mercury poisoning, ethylenediamine tetraacetic acid (calcium disodium versenate) ($CaNa_2$–EDTA) for lead poisoning and the use of deferoxamine and deferasirox for acute iron poisoning. The chelator circulates within the bloodstream where it binds to circulating toxins to form a complex to reduce the amount of toxins available for organ deposition. However, in most scenarios, each toxin has various binding kinetics, and hence, ability to form complexes is difficult to predict. Therefore, the removal of these complexes from the body, that is, renal filtration, is less predictable and impossible in patients with advanced kidney disease.

Chelation therapy for radiological toxins was a field of vigorous research during the Cold War. More than 25 years ago, the Department of Energy led an effort to develop injectable chelators that would reduce the biological half-life of radioactivity and, thus, the incurred radiation dose. From that effort, only diethylenetriamine pentaacetic acid (DTPA) and Prussian Blue ($Fe_4[Fe(CN_6)]_3$) received approval in the United States for treatment of intake for actinides elements (americium, plutonium, uranium, curium) and caesium, respectively. These programmes were terminated because chelators, overall, did not significantly improve body *clearance* despite appropriate formation of chelator–toxin complexes[2,3].

There is a need for removing toxins from the blood of acutely exposed victims as most blood-borne toxins and biohazards cannot be effectively treated. For example, intake of overdosed pain killer (analgesics), anti-depressants, sedatives, anti-psychotics, stimulants and cardiovascular medicines, accounts for up to 40% of intoxications in population causing a significant number of deaths[4]. Drug overdose places an enormous medico-legal burden on society. Furthermore, only small number of antidotes is available for medications, and urgent treatments for intoxications still include supportive measures. These include whole bowel irrigation (to reduce toxin load), correction of electrolyte and fluid disturbances (for stabilisation), and only for some selective toxins, extracorporeal cleansing procedure such as haemodialysis is feasible.

Taken together, the clinically relevant detoxification methods are limited to employing plasmapheresis, haemodialysis and haemofiltration, extracorporeal immunoabsorption and direct injection of chelators and antibodies, which can only treat a minority of known toxins to human. Although these technologies have had an important impact on the treatment of specific diseases, the majority of radioactive and biohazard exposures, medication intoxications and self-generated (auto-immune) toxins cannot be adequately treated or removed from the human body. Overwhelmingly, supportive medical measures are the only treatment option available in many intoxication syndromes. There is an important need for improvement of contemporary detoxification methods. A novel and innovative treatment modality is needed to improve the efficiency and speed of internal detoxification of exposed humans. Removal should preferably take place directly from the bloodstream, which is the unifying conduit for all intake-to-organ deposition toxin pathways.

What are the requirements for an ideal detoxification platform technology? The system must (a) be minimally invasive, (b) address a wide variety of specific toxins, (c) allow for rapid detoxification of the victim, (d) utilise biocompatible and non-toxic substrates, (e) be useable in the field as well as in a hospital environment and (f) be suitable for acute or chronic treatment of large numbers of victims with different exposure scenarios.

One strategy for blood-borne toxin removal involves injecting functionalised magnetic nanospheres to sequestrate toxins. After injection and toxin binding, the magnetic nanospheres could be removed from the circulation *via* an efficient separator device. There are several advantages to utilising magnetic nanospheres as a novel building block in detoxification technology. First, nanospheres have large surface areas to attach specific toxin-binding ligands. Second, sphere composition and surface properties are adjustable, allowing manipulations to optimise toxin uptake and circulation time. Therefore, a nanosphere-based detoxification system inherently has vast design flexibility and is potentially suitable as a treatment for exposure to the vast arrays of different toxins endangering humans. Third, toxin-nanosphere complexes could not only be the building block for toxin removal, but also for toxin sensing and quantification.

The envisioned detoxification technology using functionalised magnetic nanospheres can be summarised as follows (Figure 7.1):

1. Functionalised magnetic nanospheres are synthesised. The nanosphere surface is treated to exhibit highly specific toxin binding and to permit a circulation time of several hours within the bloodstream, avoiding rapid clearance, that is, by spleen or liver.
2. Systemic (i.e. intravenous) injection of these functionalised nanospheres will lead to *in vivo* toxin sequestration, that is, the binding of circulating toxins and formation of functionalised-nanosphere complexes within the bloodstream.
3. Subsequent efficient removal of the toxin-nanosphere complexes from the bloodstream can be achieved by a minimally invasive method (i.e. access to a small-sized artery and vein of an extremity) using extracorporeal circulation of blood through a magnetic separator module. While blood passes through this module, a local magnetic field allows efficient retention of the toxin-nanosphere complexes within the device.

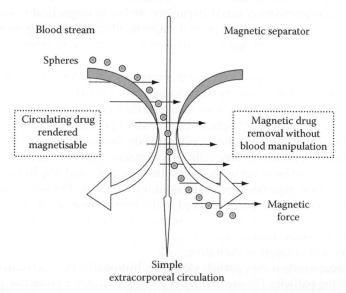

FIGURE 7.1 The schematic of a blood detoxification system using functionalised nanospheres and magnetic separation technique.

7.2   FUNCTIONALISED MAGNETIC NANOSPHERES

The main building block for the detoxification technology is a magnetisable nanosphere consisting of magnetic nanoparticles coated with polymers or encapsulated within a polymeric host matrix that is suitable for functionalisation. The magnetic nanoparticles can be any of several types that have been developed or are being developed, such as iron oxide (magnetite and maghemite, etc.), nickel, cobalt or iron. The matrix itself may consist of biocompatible and biodegradable polymers or proteins such as poly(lactic acid) (PLA), poly(lactic-co-glycolic acid) (PLGA), poly(ethylene glycol) (PEG), dextran, albumin, chitosan, etc. These matrix materials not only host the magnetic nanomaterials, but also impart non-immunogenic properties yielding extended vascular circulation. By conjugating ligands onto the terminus of the polymeric chains, selective functionality can be achieved. For small toxins that can diffuse into the porous polymeric matrix, such as dissolved radiological complexes, the functionality may be imbedded within the polymeric matrix and not necessarily present on the surface.

Magnetic nanospheres are increasingly investigated in biology and medicine for use in detection systems[5–8], biomolecule and cell separations[9–17], cell sorting[18,19], hyperthermia treatment[20–24] and controlled drug delivery[25–28]. Here, we restrict our discussion to the design of functionalised, magnetic nanospheres that are biocompatible and surface activated to be useful in human detoxification technologies.

7.2.1   Magnetic Nanosphere Design Considerations

For biomedical applications, circulating magnetic nanospheres should have an extended plasma half-life in order to maintain a reasonable *in vivo* activity, that is, successful blood toxin sequestration. The survival of the spheres within the bloodstream for prolonged time periods is largely dependent on both the size distribution and the surface characteristics of the spheres. For example, particle diameters greater than 5 µm allow the particles to readily cause capillary occlusion leading to multi-organ failure and death[29,30]. In contrast, particles with diameters less than 100 nm are much more likely to be absorbed[31]. These size boundaries are important restrictors for *in vivo* use. Furthermore, when employing individual spheres in the very low nanometre range, it becomes more difficult to sustain functional magnetite content and hence sustain the specific magnetisation. The reduction in specific magnetisation would impair the ability to magnetically steer or concentrate the nanospheres in the body or remove them in a magnetic filter. Additionally, monodispersity, that is, a rather narrow and well-controlled size distribution, is preferred in order to facilitate uniform magnetic trapping and blood clearance.

Nanospheres must have appropriate surface characteristics in order to resist immediate removal (within minutes) from the bloodstream by the reticuloendothelial (RES) system (liver, spleen, etc.).[32] Success in liposome and nanosphere systems identified the significance of hydrophilic PEG in prolonging intra-vascular circulation[33,34]. Spheres coated with PEG chains offer both steric and charge stability (they are near neutral). They also reduce antibody formation, protein adsorption (opsonisation) and subsequent engulfment by large cells such as macrophages (phagocytosis)[34]. Additionally, total blood circulation time increases as the molecular weight of covalently linked PEG increases[34]. Five hours after systemic injection, less than one-third of 20 kDa PEG-conjugated PLGA nanospheres (140 nm mean diameter) were measured in the liver compared with 65% of uncoated particles[34]. Similar prolongation approaches were summarised by Allen[35] and described by Li[36] and Dunn[37]. However, the exact mechanisms for macrophage avoidance are not known, and debate persists as to the best PEG length or derivative.

Therefore, to successfully reduce particle clearance, hydrophilic PEG derivatives are conjugated to the surfaces of the particles (Figure 7.2). This coating provides a protective physico-chemical boundary layer between the hydrophilic blood environment and the hydrophobic matrix, offering steric stability and near-neutral charge. One favourable approach is the covalent attachment of the

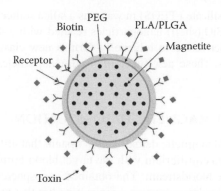

FIGURE 7.2 Example of a functionalised magnetic nanosphere that could be used in the proposed technology.

individual PEG chains directly to and individual PLA/PLGA polymer (co-block polymerisation) prior to nanosphere preparation to achieve maximal PEG coating[21].

Next, the nanosphere is functionalised with specific receptors (i.e. antibodies, chelators or ligands) for binding and sequestration of the target toxins within the bloodstream (Figure 7.2). The free terminus of the PEG chains (opposite to the end tethered to the surface) can be activated with either biotin or maleimide by synthesising PLA/PLGA-PEG-biotin or PLA/PLGA-PEG-maleimide tri-blocks. Both biotin and maleimide bridges provide binding sites for the attachment of avidin-bond receptors, or thiolated antibodies or antibody fragments, chemical ligands, or other toxin-binding structures. Other methods may use reactive amine or carbonyl groups.

Another option for functionalisation is to imbed functional groups internally to the carrier. For large molecular targets such as proteins and antigens, internal functionalisation would be futile since these large molecular structures would be sterically prevented from entering the porosity of the polymeric matrix of the carrier. However, for small targets such as free ions or weak ion complexes, diffusion through the PEG outerlayer and into the porous network of the polymer may prove feasible. Besides covalently attaching functional groups to the polymeric backbone of the carrier, one could imbed functional materials such as solid inorganic extractants such as molecular sieves or liquid organic extractants such as tributyl phosphate for uranium and plutonium sequestration. This hypothesis is yet to be tested but is intriguing.

The absolute quantity needed to detoxify 4–6 L blood volume depends, among other parameters, on toxin amounts and availability within the bloodstream, surface ligand density, affinity to toxin, sphere diameter and circulation half-lives. An absolute number is difficult to arrive at with our current knowledge since the potential values can vary by orders of magnitude (e.g. tens or thousands of ligands on the surface). *In vivo* detoxification experiments are needed to establish some baseline values.

7.2.2 Toxin Sensing and Nanosphere Decorporation

Stamopoulos et al.[38] proposed biocompatible ferromagnetic nanoparticle-targeted binding substance conjugates for use with haemodialysis. They employed Fe_3O_4 and bovine serum albumin (BSA) as the nanoparticle and targeted-binding substance, respectively. BSA was used to make the Fe_3O_4 more biocompatible and to exhibit high affinity for target blood-borne toxic substances. The Fe_3O_4 were prepared using a modification of a published procedure[38]. After the preparation, the conjugates exhibited a size range of 50–150 nm, and *in vitro* experiments demonstrated that Fe_3O_4–albumin conjugate extensively absorbed two model toxins, homocysteine and p-cresol, both in aqueous solutions with rapid kinetics.

As a next step, magnetic nanospheres can be functionalised to be used not only for toxin decorporation but also for toxin sensing. For example, Lee et al[39]. successfully attached a fluorescence receptor to nickel–silica nanoparticles for the removal of Pb^{2+} from blood. These nickel nanoparticles were prepared using a modified polyol process and were coated with silica shells using the Stöber method with tetraethyl orthosilicate (TEOS) in water as a silica source to yield $Ni(core)/SiO_2(shell)$ nanoparticles. The $Ni(core)/SiO_2(shell)$ nanoparticles reacted with a 4,4-difluoro-4-bora-3a,4-adiaza-s-indacene (BODIPY) derivative in toluene to form a new class of BODIPY-functionalised magnetic silica nanoparticles. These new functionalised nanoparticles were successfully used for Pb^{2+} detection and adsorption.

7.3 EXTRACORPOREAL MAGNETIC DETOXIFICATION

We propose an extracorporeal magnetic detoxification system that utilises functionalised, biocompatible nanospheres that attain equilibrium with the target blood-borne toxin after direct injection of these nanospheres into the bloodstream. The obtained nanosphere-toxin complexes are subsequently separated from the bloodstream using a magnetic filter that receives blood *via* extracorporeal circulation.

7.3.1 MAGNETIC SEPARATOR DESIGN CONSIDERATIONS

Alongside the known industrial applications of magnetic separation, a variety of magnetic separators for biological sample concentration and analysis are commercially available. Basically, there are two types of separators: batch and flow-through.

The simplest designs simply place strong rare-earth permanent magnets against the wall of a batch cell containing the magnetic particle suspension. Examples of commercial batch magnetic separators include Dynal MPC® series from Invitrogen Corporation, BioMag® Flask Separator from Polysciences, Inc., MCB 1200 processing system from Sigris Research, Inc., and others. Low separation volume and slow accumulation rates are the main limitations of those batch separators.

Flow-through magnetic separators are characterised by the flow of both liquid and suspended magnetic components through the separation system. These systems are usually more expensive, more complicated in design and separate faster than batch separators. The separator efficiency is increased by producing regions of high magnetic field and field gradient to capture flowing magnetic particles from the medium. Typically, columns with fine magnetic grade stainless steel wire wool or beads are placed between the poles of strong permanent magnets or electromagnets strongly distorting the magnetic fields local to the wire or bead surfaces to create large field gradients. Fluid containing magnetic components is pumped through the column, and the magnetic components are retained onto the magnetised wool or beads. Once the magnetic field is removed, the magnetic components can be retrieved by washing and gentle vibration of the column. So-called high-gradient magnetic separators have been used on both industrial and laboratory scale[40–44].

In contrast, a magnetic separator suitable for real-time clearing of magnetic nanospheres from circulating blood in humans has so far not been established. In fact, a magnetic separator for medical applications requires specific separator design characteristics as the device should permit high volumetric throughput together with high capture efficiency of nanosized blood-borne magnetic particles. At the same time, extracorporeal circulation of larger blood volumes through the device should avoid dangerous adverse effects such as blood clotting (coagulopathy), blood contaminations and mechanical cell destruction (cell lysis).

Based on the performances of current separator technologies and published prototypes[45], the engineering of a biomedical separator seems to pose specific challenges in addition to those outlined earlier. Conventional high-gradient magnetic separation (HGMS) systems utilise designs where the ferromagnetic elements are exposed directly to the fluid. However, positioning ferromagnetic elements within the bloodstream is not suitable for prospective biomedical applications due to the high risk of blood clotting, both locally within the device and systemically throughout the bloodstream, easily induced by interactions between blood components and the packing materials. On the other hand, nanosphere separation decreases as the ferromagnetic elements are removed from the fluid flow due

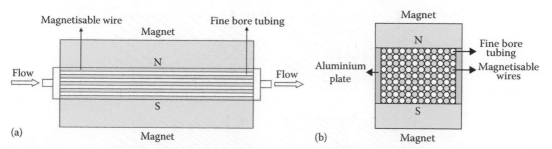

FIGURE 7.3 Proposed high-gradient magnetic separator: (a) axial section view and (b) transverse section view. (Reprinted from *Med. Hypotheses*, 68, Chen, H., Kaminski, M.D., Liu, X. et al., A novel human detoxification system based on nanoscale bioengineering and magnetic separation techniques, 1071–1079, 2007, Copyright 2008, with permission from Elsevier B.V.)

to the relatively short range perturbations of the magnetic field (field gradient) from those elements. Hence, the HGMS elements must remain in close proximity to, but not in direct contact with, the circulating blood. Several design options for the proposed 'filtration' system seem plausible. Certainly, different design strategies are needed for different modes of operation; for example, a zero-power, light-weight, strap-on separator device for emergent in-field detoxification versus a stationary hospital-based unit. An ideal unit would maximise the collection of magnetic nanospheres in a single pass at the highest volumetric flow rates tolerable to limit the treatment time required to clean the total volume of 4–6 L. Since blood is continuously mixed within the body, significantly more than 4–6 L will be required to pass through the unit to eliminate the circulating carriers. The exact volume will depend on the single pass magnetic filter efficiency, venous/arterial access point and the dynamics of toxin sorption/desorption with tissue.

With the aforementioned prior art and knowledge, a basic device was designed for emergency, in-field usage, that is, needed after mass contamination by a specific toxin. The magnetic separator utilises a vascular (arterial or venous) dual-lumen access catheter at one extremity to provide extracorporeal blood flow through a short, closed-loop, pre-sterilised and heparinised tube or catheter segment. A simple syphon hand-pump may ensure additional flow rate control, if needed. The blood receiving part of the catheter is diverted into a segment of multiple small tubes (in analogy to a capillary network), and these tubes are immersed in a magnetic field created by the ferromagnetic wire matrix between two parallel NdFeB permanent magnets. With this design, the magnetic nanospheres deflect towards and collect at the individual tube wall (Figure 7.3)[45].

This design combines both the principles of HGMS mentioned earlier, with many applications in the treatment of biological[45–48] and industrial fluids[41,43,49,50], and extracorporeal blood circulation. Extracorporeal blood circulation is characterised by temporarily removing blood from the body and then returning it as in haemodialysis[51–53]. The precise geometry of the device (quantity and dimensions of wires and tubing; size and position of the magnets, and vascular access route, etc.) and induced changes on blood coagulation (i.e. inhibiting thrombosis) and cells (i.e. avoiding destruction) are defined research areas and in part proprietary knowledge.

7.3.2 THEORETICAL MODELLING

The magnetic separator consists of a matrix of parallel biocompatible capillary tubing and fine magnetisable wires, which are immersed in an externally applied magnetic field. The finite element package COMSOL Multiphysics 3.3 was used in order to numerically solve the 3D partial differential equations that constitute the model and to predict the trajectories of the magnetic spheres as they travel with the blood from left to right through the tubing while under the influence of both hydrodynamic and magnetic forces. In order to reduce the complexity of the model, only one cell of each of the spatially periodic configurations (see Figure 7.4) was considered[54].

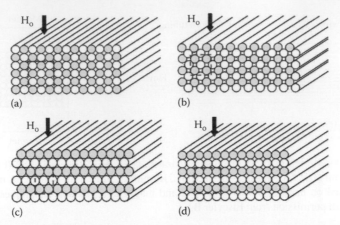

FIGURE 7.4 The possible configurations of the magnetisable wires (dark-coloured) and tubing (light-coloured) within the separator. Four basic configuration units (in dashed lines) were named from the designs: (a) 4-wire (0), (b) 4-wire (45), (c) 4-wire (Hexa) and (d) 6-wire. (Reprinted with permission from Chen, H., Bockenfeld, D., Rempfer, D., Kaminski, M.D., and Rosengart, A.J., Three-dimensional modeling of a portable medical device for magnetic separation of particles from biological fluids, *Phys. Med. Biol.*, 52, 5205–5218, 2007, Copyright 2007, IOP Publishing.)

Figure 7.4a shows a central tube with wires in each of the cross-flow directions: a 4-wire (0) configuration, where '0' indicates the angle (in degrees) between the applied magnetic field and one plane passing through the axes of tubing and the wires. The basic unit in Figure 7.4b also consists of four wires and one tube and is named 4-wire (45) configuration because the angle between the applied magnetic field and the tubing-wire plane is 45°. The basic unit in Figure 7.4c is named a 4-wire (Hexa) configuration in order to differentiate it from the others. Figure 7.4d shows the central tubing with a row of three wires above and below it, a 6-wire configuration.

A representative 3D control volume (CV) is provided (Figure 7.5) for reference[54]. The methodology utilised here is to determine the capture efficiency (CE_{num}). Numerically, this is determined by counting the fraction of magnetic sphere trajectories that enter the tubing and terminate at the tubing wall. The simplicity of the model was achieved by considering only the hydrodynamic (due to the blood flow) and the magnetic forces (due to the effect of the external magnetic field and magnetised wires). Inertial and lift forces, as well as the magnetic inter-particle forces that might lead to magnetic sphere agglomeration, were neglected. The magnetic sphere trajectory would be determined by the software, provided that the particle velocity $\mathbf{u}_p$ was given. Numerical capture efficiency, CE_{num}, is given by

$$CE_{num}(\%) = \left(1 - \frac{n_{p,out}}{n_{p,in}}\right) 100, \qquad (7.1)$$

where
$n_{p,in}$ is the chosen number of starting points for the magnetic sphere trajectories
$n_{p,out}$ is the number of trajectories that exit the tubing

To evaluate Equation 7.1, however, the variables that help determine $\mathbf{u}_p$ must be evaluated first. This can be done by resolving the force balance that includes hydrodynamic and magnetic forces upon a single magnetic sphere everywhere inside the tubing. These two types of forces were independently evaluated as follows. We first found the blood velocity $\mathbf{u}$ and the blood pressure P within the tubing by solving the continuity and 3D Navier–Stokes equations for blood, which was assumed to be an incompressible fluid of density ρ_B and viscosity η_B,

$$\nabla \cdot \mathbf{u} = 0, \qquad (7.2)$$

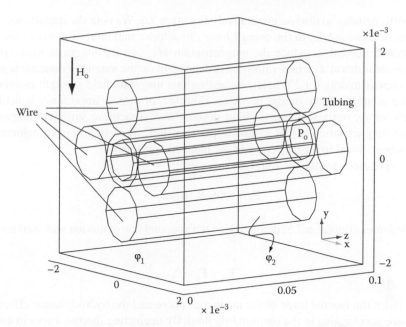

FIGURE 7.5 An example of the schematic of magnetic separator unit (3D). (Reprinted with permission from Chen, H., Bockenfeld, D., Rempfer, D., Kaminski, M.D., and Rosengart, A.J., Three-dimensional modeling of a portable medical device for magnetic separation of particles from biological fluids, *Phys. Med. Biol.*, 52, 5205–5218, 2007, Copyright 2007, IOP Publishing.)

$$\rho_B \left[\frac{\partial \mathbf{u}}{\partial t} + (\nabla \mathbf{u}) \cdot \mathbf{u} \right] = -\nabla P + \eta_B \nabla^2 \mathbf{u}. \tag{7.3}$$

The blood was assumed to enter the tubing at $z = 0$ in a direction parallel to the tubing walls and with parabolic profile with an average velocity U_o. At the wall of the tubing, the velocity in all directions is identically zero, and at the outlet of the tubing, the blood pressure was specified as $P_o = 1$ atm.

Next, the magnetic force upon the magnetic sphere was determined by evaluating the magnetic field $\mathbf{H}$ within the CV. This was done using the Maxwell equation for conservative fields within the CV:

$$\nabla^2 \phi = 0, \tag{7.4}$$

where ϕ is the scalar magnetic potential and is related to the field $\mathbf{H}$ according to

$$\mathbf{H} = -\nabla \phi. \tag{7.5}$$

The following expression was then used to evaluate the magnetic force ($\mathbf{F}_m$) on a magnetic nanosphere[55,56]:

$$\mathbf{F}_m = \frac{1}{2} \omega_{fm,p} V_p \mu_o \frac{M_{fm,p}}{H_1} \nabla(\mathbf{H}_1 \cdot \mathbf{H}_1), \tag{7.6}$$

where
  H_1 is the magnetic field strength at the non-magnetisable space not occupied by the wires
  μ_o is the magnetic permeability of vacuum
  V_p is the volume of the magnetic spheres
  $\omega_{fm,p}$ and $M_{fm,p}$ are the volumetric fraction and magnetisation of the ferromagnetic material in the magnetic nanospheres, respectively

Here, it is worth pausing to discuss qualitatively Equation 7.6. We note the importance of the magnetic gradient in Equation 7.6 to the overall force. To achieve maximal filtration in the device, the magnetic force is needed to induce the magnetisation $M_{fm,p}$ on the magnetic material. Since the magnetisation saturates at a certain value that is dependent on the magnetic material (e.g. magnetite saturates in approximately a 1.2 T field), exceeding this magnetic field strength is unnecessary to improving the magnetic force on the particles. However, of utmost importance is achieving high gradients. Therefore, optimal magnetic filter designs seek to achieve sufficient magnetic field to maximise the magnetisation of the magnetic particles but, more importantly, achieve very high magnetic gradients local to the trajectory of the magnetic particles.

With the expressions for the hydrodynamic forces[54,55]

$$\mathbf{F}_d = 6\pi\eta_B R_p(\mathbf{u} - \mathbf{u}_p),\tag{7.7}$$

and magnetic forces (Equation 7.6) defined, Newton's second law of motion was used for a magnetic nanosphere,

$$\mathbf{F}_m + \mathbf{F}_d = 0,\tag{7.8}$$

where $\mathbf{F}_i$ includes the inertial force of the magnetic sphere and the hydrodynamic effect associated with the sphere accelerating in the surrounding fluid. By neglecting inertial forces in an essentially aqueous solution, $\mathbf{F}_i$ can be set to zero.

Numerous simulations were completed that varied mean flow velocity (1–20 cm/s), specific magnetisation (for which we varied the type of magnetic material and magnetic mass content in the sphere), sphere radius (50–1000 nm), wire radius (0.25–1.0 mm) and composition (nickel, 430-type stainless steel, wairauite), fluid viscosity (deionised water to glycol solutions), wire configuration (see the following), strength of magnetic field (0.1–2.0 T) and tube radius (0.2–1.0 mm) and thickness (0.125–0.25 mm). These values were chosen based on availability of material (for wire, magnetic particles and magnets) and the desire to achieve a compact device (size of wires and tubes).

To verify the model predictions, flow experiments were employed. The prototype magnetic separators consisted of a piece of capillary tubing (0.5 mm in outer radius, 0.375 mm in inner radius and 15 cm in length) and multiple pieces of straight stainless steel wires (0.5 mm in radius and 10 cm in length with saturation magnetisation 175 emu/g). A relatively homogenous external magnetic field was created by two parallel rectangular NdFeB magnets (8.8 cm × 8.8 cm × 2.75 cm). The tubing-wire set-up was sandwiched between the magnets, producing relatively high magnetic gradients about the magnetised wires. The experimental set-up is shown in Figure 7.6[57]. The 1.7 μm polystyrene (PS) magnetic spheres (saturation magnetisation = 12.54 kA/m) were used as the test spheres in the experiment. The syringe pump drove the sample suspension through the separator where the magnetic spheres were collected against the tubing wall and a fraction drained into the receiving container.

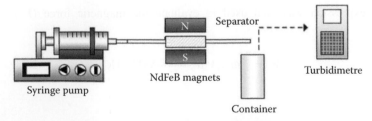

FIGURE 7.6 The magnetic filtration test set-up for comparing the experimental capture efficiency with that derived from the computational model. The magnetic assembly (flow tubules sandwiched between the two magnets) was fist-size. (Reprinted with permission from Chen, H., Kaminski, M.D., and Rosengart, A.J., 2009, Characterization of a prototype compact high gradient magnetic separator device for blood detoxification, *Separ. Sci. Tech.*, 44, 1954–1969, Copyright 2009 Taylor & Francis.)

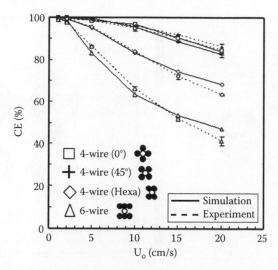

FIGURE 7.7 Comparison of numerical and experimental results at different mean flow velocities ($\mu_o H_o$ = 0.44 T) for the capture of 1.7 μm magnetic spheres. (Reprinted with permission from Chen, H., Bockenfeld, D., Rempfer, D., Kaminski, M.D., and Rosengart, A.J., Three-dimensional modeling of a portable medical device for magnetic separation of particles from biological fluids, *Phys. Med. Biol.*, 52, 5205–5218, 2007, Copyright 2007, IOP Publishing.)

The results are shown in Figure 7.7[57]. Clearly, the 4-wire (0) and 4-wire (45) designs produced superior CE, being >95% for flow rates up to 10 cm/s, even though the magnetic field set-up of these two designs differed dramatically (Figure 7.8)[54]. In the 4-wire (0) design, the wires created a strong magnetic field gradient in the *y*-direction, while in the 4-wire (45) design, the wires created strong fields in the *x*-direction (Figure 7.8) and there was strong coupling between the magnetic fields of the wires (visualised as the continuous red regions in the *y*-direction and the continuous blue in the *x*-direction). Therefore, with regard to capture efficiency, the direction of the applied magnetic field is not an important factor in the 4-wire configuration, provided the applied magnetic field is perpendicular to the flow. The important factor is the configuration of the wires and the tubes.

By adjoining the wires in the 4-wire (Hexa) and 6-wire designs, the magnetic field lines became smoother (Figure 7.8), thereby reducing the field gradients and the CE compared with the other 4-wire configurations. The reduced CE occurred despite a significant increase in the strength of the magnetic field in the 4-wire (Hexa) and 6-wire configurations (noted by the yellow in the flow tube region). These data highlight the importance of the magnetic field gradient in particle capture. In the 4-wire (0) and (45) configurations, the strength of the magnetic field can be low, but the field gradients must be high in order to achieve the large CE that was calculated (see Equation 7.6). In the 4-wire (Hexa) and 6-wire designs, the strong magnetic field could not offset the decrease in the magnetic field gradients caused by the adjoining wires.

7.3.3 EXTRACORPOREAL MAGNETIC DETOXIFICATION

A more complete magnetic separator device was built after theoretical analyses and initial experimental verification. The specific dimensions of the device are given in Figure 7.9[56] and are a result of computer simulations and experiments designed for use in the rat model. In order to obtain the volumetric throughput and separation efficiency >95%, the device consisted of 50 straight SS430 wires (10 cm in length and 0.25 mm in radius) and 50 polyetheretherketone (PEEK) fine bore tubes (0.25 mm inner radius and 0.30 mm outer radius). The wires and tubes were in the 4-wire (0) configuration, as shown in Figure 7.4a and were sandwiched between two NdFeB magnetic plates creating a static field of approximately 0.44 T. Human whole blood for testing was donated by healthy volunteers according to the university guidelines. The blood was drawn into blood collection

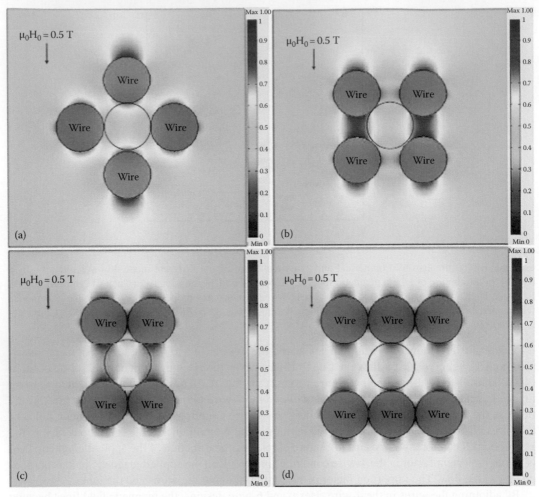

FIGURE 7.8 Magnetic field densities in the configurations (from COMSOL Multiphysics 3.3). (a) through (d) show four different wire-tube configurations that were modelled. (Reprinted with permission from Chen, H., Bockenfeld, D., Rempfer, D., Kaminski, M.D., and Rosengart, A.J., Three-dimensional modeling of a portable medical device for magnetic separation of particles from biological fluids, *Phys. Med. Biol.*, 52, 5205–5218, 2007, Copyright 2007, IOP Publishing.)

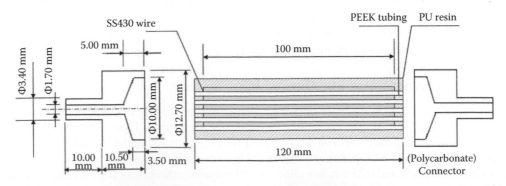

FIGURE 7.9 The dimensions of the separator device. An NdFeB magnet was placed above and below to create the magnetic field. (Reprinted with permission from Chen, H., Kaminski, M.D., Caviness, P.L. et al., 2007, Magnetic separation of micro-spheres from viscous biological fluids, *Phys. Med. Biol.*, 52, 1185–1196, Copyright 2007 IOP Publishing.)

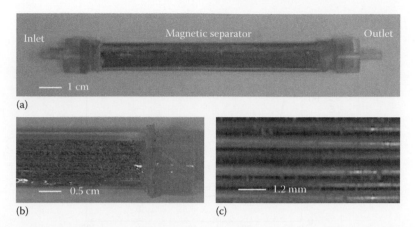

(a)

(b)   (c)

FIGURE 7.10 The compact magnetic separator device used in the experiment. (a) shows the entire unit, (b) shows a closer view of outlet manifold, and (c) shows close up of wires and tubes. (Reprinted with permission from Chen, H., Kaminski, M.D., Caviness, P.L. et al., 2007, Magnetic separation of micro-spheres from viscous biological fluids, *Phys. Med. Biol.*, 52, 1185–1196, Copyright 2007, Copyright 2007 IOP Publishing.)

tubes containing EDTA. The 1.7 μm PS magnetic spheres were radioactivated for gamma counter-quantification and used in the experiment.

The whole tube-wire matrix was built layer by layer, and each layer contained five tubes and five wires alternative to each other. The fluid inlet and outlet connectors were made of polycarbonate (Figure 7.10). The device was then primed by infusing heparin solution (5.0 mL).

The experimental set-up was similar to the one used earlier except that a 7521-55 *Masterflex®* peristaltic pump was used to create a circulating flow system. The fluid used in the initial experiments was 44% (v/v) ethylene glycol–water solution (viscosity of 3.5 cP at 20°C) to simulate the viscosity of human blood at 37°C. The change of sphere concentration in the sample was indicated by the gamma detection of radioactive Fe-59 in the magnetite, and the filter efficiency was calculated from concentration changes[57]. This design continuously circulated the suspension through the separator, so that the entire fluid volume passed many times through the separator.

Figure 7.11 shows the per cent remaining in suspension of magnetic spheres at various flow rates [analogous to (1-CE)]. The per cent remaining was measured from the relative sphere concentration,

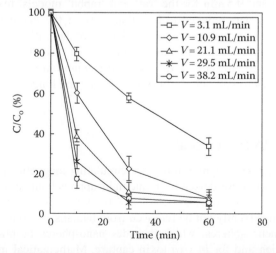

FIGURE 7.11 The effect of flow rate on per cent of 1.7 μm magnetic spheres at a dose of 0.04 mg/mL remaining in suspension. (Reprinted with permission from Chen, H., Kaminski, M.D., Caviness, P.L. et al., 2007, Magnetic separation of micro-spheres from viscous biological fluids, *Phys. Med. Biol.*, 52, 1185–1196, Copyright 2007 IOP Publishing.)

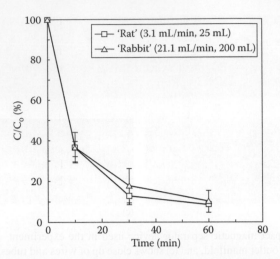

FIGURE 7.12   Magnetic separation of 1.7 μm magnetic spheres from blood simulant for simulated 'rat' and 'rabbit' models. Total of 5 mg and 10 mg spheres were injected for the 'rat' and 'rabbit' model flow experiments, respectively. (Reprinted with permission from Chen, H., Kaminski, M.D., Caviness, P.L. et al., 2007, Magnetic separation of micro-spheres from viscous biological fluids, *Phys. Med. Biol.*, 52, 1185–1196, Copyright 2007 IOP Publishing.)

that is, C/C_o, in the container with C_o as the initial concentration (lower C/C_o indicates higher recovery). What these data show is that, although we can achieve higher CE at lower flow velocity (see Figure 7.10), much higher flow velocities force the suspension to pass many more times through the separator. Even though the per-pass efficiency is lower (lower CE), the result of passing an effective flow volume of suspension many times through the separator proves to be more important for this test case.

Figure 7.12 shows magnetic separation of spheres using a device scaled to the 'rat' and 'rabbit' model with the blood volume and blood flow rates specifically chosen to simulate the circulatory system in rats and rabbits. The entire blood volume was passed many times through the separator. As expected, the results show similar sphere recoveries with similar turnover time for the two models (8.1 min and 9.5 min for the 'rat' and 'rabbit' models, respectively). The sphere concentration in the blood was reduced to less than 10% of the original concentration within 60 min, equalling seven and six passes through the device for the 'rat' and 'rabbit' models, respectively. Presumably, an even lower flow rate might provide the system with a higher one-pass CE. However, the advantage of the lower flow rate is usually compromised by the higher viscosity of the fluid due to the non-Newtonian behaviour of the blood[56] and longer turnover time as indicated from Figure 7.11.

7.4   CONCLUSION AND OUTLOOK

A blood detoxification system designed to remove circulating magnetic spheres based on a novel extracorporeal magnetic separator device is unique in its approach and aims to serve as a platform for the removal of various radioactive-, chemo- and bio-toxins from acutely exposed humans – many of these toxins can currently not be removed from humans. We outlined a general approach to functionalising magnetic spheres, which provides nanospheres or microspheres suitable for extended blood circulation and for *in vivo* toxin capture. Mathematical modelling and benchtop experiments were completed, and the results identify feasible, portable magnetic separator designs that can effectively capture magnetic spheres.

However, there are still unsettled points that warrant explanation. First, research is required in developing and testing functionalised magnetic nanoparticles for specific targets toxins, whether they are biological or radiological. The outline for functionalising nanospheres that we provided early in this chapter hides many details of which there is some uncertainty. First, long-circulation times are required to allow for magnetic filtration in the extracorporeal device. Long-circulation behaviour is achieved by neutralisation of the surface and the absence of charge groups (i.e. functional groups). Studies on the effect of attaching immunoglobulin residues onto PEGylated liposomes[58,59] showed the importance of species-specific antibodies and further evidence that antibody fragments and not full-length antibodies with constant and variable chains were better suited for long-circulation applications. The circulation half-lives for PEGylated liposomes were progressively reduced with increasing number of immunoglobulin (IgG) residues attached to the surface (up to 40 IgG per liposome)[59]. But they noted differences in this effect depending on the antibody fragment and its species of origin (e.g. human, mouse, rat).

For radiological toxins, prior programmes developed ligands based on strict criteria for selectivity, toxicity, solubility in blood, affinity, biological clearance, etc. since the injected chelators must circulate, bind and then be excreted. Therefore, it is not surprising that only DTPA and Prussian Blue resulted from the research. However, with the nanoparticles platform, one can revisit chelators that were rejected in the past. Solubility in blood is replaced by long-circulation requirements. Toxicity may not be as important as the chelators remains tethered from the nanoparticles surface. Biological clearance is not a requirement as long as the nanoparticles conjugate remains in circulation, so that it may be cleared in the magnetic filter.

Presumably, any surface functional groups would need to be carefully chosen to preserve the long-circulation half-lives of the spheres for detoxification applications, but mechanistic understanding of how best to choose/design these functional groups is still lacking. Certainly, this is an area in need of study that includes animal toxin models.

Next, the magnetic filter studies are promising, but more experiments are needed to validate the computational models for smaller magnetic spheres. We highlighted data in this chapter of the filtration of magnetic microspheres because these uniform, magnetic spheres were immediately available for experimental testing and comparison with the model. The breadth of parametric theoretical studies[54,60] is ahead of the experiments. However, the agreement between experiment and computations described in Figure 7.7 suggest that we can extrapolate to smaller nanospheres and specific magnetisations with some confidence in the calculations. In those calculations[54,60], the smaller magnetic nanospheres (<1 µm) will require slower flow velocities, much smaller tube diameters, or higher saturation magnetisations to achieve similar capture efficiency to those reported in Figure 1.7. Slower velocity and smaller tube diameters have the effect of reducing the volumetric throughput of the device, which is highly undesirable. A larger device can be engineered to increase the throughput, if needed, but at least several blood volumes would need to be passed through the filter for effective removal of the nanospheres. Combined with novel methods of synthesising uniform magnetic nanospheres <500 nm in diameter, one can move towards validating the capture efficiency models for nanospheres.

The potential advantages of functional, magnetic nanospheres for detoxification of the blood are exciting, but there is much uncertainty in determining whether the technology can meet the rigorous needs of long circulation, selective and high affinity binding and efficient separations. We look forward to the exciting research that will test hypotheses of this type of novel treatment modality.

ACKNOWLEDGEMENT

This work was supported by the Defense Advanced Research Program Agency: Defense Sciences Office and the Department of Energy under contract W-31-109-Eng-38.

REFERENCES

1. Tompkins B.Ed. 2011. Nuclear news special report: Fukushima Daiichi after earthquake and tsunami, *Nucl. News*, 54(4), 17.
2. 1999. *Medical Management of Radiological Casualties Handbook*. Bethesda, MD: Armed Forces Radiobiology Research Institute.
3. Fritsch P., Ramounet B., Burgada R. et al. 1998. Experimental approaches to improve the available chelator treatments for Np decorporation. *J Alloy Comp*, 271, 89–92.
4. Leroux J. C. 2007. Injectable nanocarriers for biodetoxification. *Nat Nanotechnol*, 2, 679–684.
5. Guven B., Basaran-Akgul N., Temur E., Tamer U. and Boyaci I. H. 2010. SERS-based sandwich immunoassay using antibody coated magnetic nanoparticles for *Escherichia coli* enumeration. *Analyst*, 136, 740–748.
6. Kim K. S. and Park J. K. 2005. Magnetic force-based multiplexed immunoassay using superparamagnetic nanoparticles in microfluidic channel. *Lab Chip*, 5, 657–664.
7. Li J., Gao H., Chen Z., Wei X. and Yang C. F. 2010. An electrochemical immunosensor for carcinoembryonic antigen enhanced by self-assembled nanogold coatings on magnetic particles. *Anal Chim Acta*, 665, 98–104.
8. Bi S., Zhou H. and Zhang S. S. 2010. A novel synergistic enhanced chemiluminescence achieved by a multiplex nanoprobe for biological applications combined with dual-amplification of magnetic nanoparticles. *Chem Sci*, 1, 681–687.
9. Hobley T. J., Ferre H., and Gomes C. S. G. et al. 2005. Advances in high-gradient magnetic fishing for downstream and bioprocessing. *J Biotechnol*, 118, S56–S56.
10. Gupta A. K. and Gupta M. 2005. Synthesis and surface engineering of iron oxide nanoparticles for biomedical applications. *Biomaterials*, 26, 3995–4021.
11. Heebøll-Nielsen A., Justesen S. F. L., Hobley T. J. and Thomas O. R. T. 2004. Superparamagnetic cation-exchange adsorbents for bioproduct recovery from crude process liquors by high-gradient magnetic fishing. *Separ Sci Tech*, 39, 2891–2914.
12. Shukoor M. I. N., Natalio F., Tahir M. N. et al. 2008. Multifunctional polymer-derivatized gamma-Fe_2O_3 nanocrystals as a methodology for the biomagnetic separation of recombinant His-tagged proteins. *J Magn Magn Mater*, 320, 2339–2344.
13. Che Y. H., Li Y., Slavik M. and Paul D. 2000. Rapid detection of *Salmonella typhimurium* in chicken carcass wash water using an immunoelectrochemical method. *J Food Protect*, 63, 1043–1048.
14. Kell A. J., Stewart G., Ryan S. et al. 2008. Vancomycin-modified nanoparticles for efficient targeting and preconcentration of Gram-positive and Gram-negative bacteria. *ACS Nano*, 2, 1777–1788.
15. Song E. Q., Hu J., Wen C. Y. et al. 2011. Fluorescent-magnetic-biotargeting multifunctional nanobioprobes for detecting and isolating multiple types of tumor cells. *ACS Nano*, 5, 761–770.
16. Shan Z., Wu Q., Wang X. et al. 2009. Bacteria capture, lysate clearance, and plasmid DNA extraction using pH-sensitive multifunctional magnetic nanoparticles. *Anal Biochem*, 398, 120–122.
17. Chen G. D., Alberts C. J., Rodriguez W. and Toner M. 2009. Concentration and purification of human immunodeficiency virus type 1 virions by microfluidic separation of superparamagnetic nanoparticles. *Anal Chem*, 82, 723–728.
18. Das M., Mishra D., Maiti T. K., Basak, A. and Pramanik, P. 2008. Bio-functionalization of magnetite nanoparticles using an aminophosphonic acid coupling agent: New, ultradispersed, iron-oxide folate nanoconjugates for cancer-specific targeting. *Nanotechnology*, 19, 415101.
19. Groman E. V., Bouchard J. C., Reinhardt C. P. and Vaccaro D. E. 2007. Ultrasmall mixed ferrite colloids as multidimensional magnetic resonance imaging, cell labeling, and cell sorting agents. *Bioconjugate Chem*, 18, 1763–1771.
20. Ito A., Kuga Y., Honda H. et al. 2004. Magnetite nanoparticle-loaded anti-HER2 immunoliposomes for combination of antibody therapy with hyperthermia. *Cancer Lett*, 212, 167–175.
21. Liu X., Novosad V. and Rozhkova E. et al. 2007. Surface functionalized biocompatible magnetic nanospheres for cancer hyperthermia. *IEEE Trans Magn*, 43, 2462–2464.
22. Chen B., Wu W., Wang X. 2010. Magnetic iron oxide nanoparticles for tumor-targeted therapy. *Curr Cancer Drug Targets*, 11, 184–189.
23. Li Z. X., Kawashita M., Araki N. et al. 2010. Magnetite nanoparticles with high heating efficiencies for application in the hyperthermia of cancer. *Mater Sci Eng C Mater Biol Appl*, 30, 990–996.
24. Sonvico F., Dubernet C., Colombo P. and Couvreur P. 2005. Metallic colloid nanotechnology, applications in diagnosis and therapeutics. *Curr Pharm Des*, 11, 2095–2105.
25. Li J. B., Ren J., Cao Y. and Yuan, W. 2009. Preparation and characterization of thermosensitive and biodegradable PNDH-g-PLLA nanoparticles for drug delivery. *React Funct Polymer*, 69, 870–876.

26. He Q., Zhang J., Chen F. et al. 2010. An anti-ROS/hepatic fibrosis drug delivery system based on salvi-anolic acid B loaded mesoporous silica nanoparticles. *Biomaterials*, 31, 7785–7796.

27. Chen J. P., Yang P. C., Ma Y. H. and Wu, T. 2011. Characterization of chitosan magnetic nanoparticles for in situ delivery of tissue plasminogen activator. *Carbohydr Polymer*, 84, 364–372.

28. Ying, X. Y., Du, Y. Z., Hong, L. H., Yuan, H. and Hu, F. Q. 2011. Magnetic lipid nanoparticles load-ing doxorubicin for intracellular delivery: Preparation and characteristics. *J Magn Magn Mater*, 323, 1088–1093.

29. Slack J. D., Kanke M., Simmons G. H. and DeLuca P. P. 1981. Acute hemodynamic effects and blood pool kinetics of polystyrene microspheres following intravenous administration. *J Pharm Sci*, 70, 660–664.

30. Chen H., Kaminski M. D., Pytel P., Macdonald L. and Rosengart A. J. 2008. Capture of magnetic carriers within large arteries using external magnetic fields. *J Drug Target*, 16, 262–268.

31. Moghimi S. M., Hunter A. C. and Murray J. C. 2001. Long-circulating and target-specific nanoparticles: Theory to practice. *Pharmacol Rev*, 53, 283–318.

32. Ishiwata K., Ido T., Monma M. et al. 1991. Potential radiopharmaceuticals labeled with Ti-45. *Appl Radiat Isot*, 42, 707–712.

33. Symon Z., Peyser A., Tzemach D. et al. 1999. Selective delivery of doxorubicin to patients with breast carcinoma metastases by stealth liposomes. *Cancer*, 86, 72–78.

34. Gref R., Minamitake Y., Peracchia M. T. et al. 1994. Biodegradable long-circulating polymeric nano-spheres. *Science*, 263, 1600–1603.

35. Allen T. M. and Moase E. H. 1996. Therapeutic opportunities for targeted liposomal drug delivery. *Adv Drug Deliv Rev*, 21(2), 117–133.

36. Li Y. P., Pei Y. Y. and Zhang X. Y. 2001. PEGylated PLGA nanoparticles as protein carriers: synthesis, preparation and biodistribution in rats, *J. Control. Release*, 71, 203–211.

37. Dunn S. E., Brindley A., Davis S. S., Davies M. C. and Illum L. 1994. Polystyrene-poly (ethylene glycol) (PS-PEG2000) particles as model systems for site specific drug delivery. 2. The effect of PEG surface density on the in vitro cell interaction and in vivo biodistribution. *Pharm Res*, 11, 1016–1022.

38. Stamopoulos D., Benaki D., Bouziotis P. and Zirogiannis P. N. 2007. In vitro utilization of ferromagnetic nanoparticles in hemodialysis therapy. *Nanotechnology*, 18, 495102.

39. Lee H. Y., Bae D. R., Park J. C. et al. 2009. A selective fluoroionophore based on BODIPY functionalized magnetic silica nanoparticles: Removal of Pb2+ from human blood. *Angew Chem Int Ed Engl*, 48, 1239–1243.

40. Meyer A., Hansen D. B., Gomes C. S. et al. 2005. Demonstration of a strategy for product purification by high-gradient magnetic fishing: Recovery of superoxide dismutase from unconditioned whey. *Biotechnol Prog*, 21, 244–254.

41. Karapinar N. 2003. Magnetic separation of ferrihydrite from wastewater by magnetic seeding and high-gradient magnetic separation. *Int J Miner Process*, 71, 45–54.

42. Hubbuch J. J. and Thomas O. R. 2002. High-gradient magnetic affinity separation of trypsin from porcine pancreatin. *Biotechnol Bioeng*, 79, 301–313.

43. Newns A. and Pascoe R. D. 2002. Influence of path length and slurry velocity on the removal of iron from kaolin using a high gradient magnetic separator. *Miner Eng*, 15, 465–467.

44. Richards A. J., Thomas T. E., Roath O. S. et al. 1993. Improved high gradient magnetic separation for the positive selection of human blood mononuclear cells using ordered wire filters. *J Magn Magn Mater*, 122, 364–366.

45. Chen H., Kaminski M. D., Liu X. et al. 2007. A novel human detoxification system based on nanoscale bioengineering and magnetic separation techniques. *Med Hypotheses*, 68, 1071–1079.

46. Soh N., Nishiyama H., Asano Y. et al. 2004. Chemiluminescence sequential injection immunoassay for vitellogenin using magnetic microbeads. *Talanta*, 64, 1160–1168.

47. Chosy E. J., Nakamura M., Melnik K. et al. 2003. Characterization of antibody binding to three cancer-related antigens using flow cytometry and cell tracking velocimetry. *Biotechnol Bioeng*, 82, 340–351.

48. Cabioglu N., Igci A., Yildirim E. O. et al. 2002. An ultrasensitive tumor enriched flow-cytometric assay for detection of isolated tumor cells in bone marrow of patients with breast cancer. *Am J Surg*, 184, 414–417.

49. Augusto P. A., Augusto P. and Castelo-Grande T. 2004. Magnetic shielding: Application to a new mag-netic separator and classifier. *J Magn Magn Mater*, 272, 2296–2298.

50. Nedelcu S. and Watson J. H. P. 2002. Magnetic separator with transversally magnetised disk permanent magnets. *Miner Eng*, 15, 355–359.

51. Grano V., Diano N., Portaccio M. et al. 2002. The alpha1-antitrypsin/elastase complex as an experimental model for hemodialysis in acute catabolic renal failure, extracorporeal blood circulation and cardiocircu-latory bypass. *Int J Artif Organs*, 25, 297–305.

52. Yang M. C. and Lin C. C. 2000. In vitro characterization of the occurrence of hemolysis during extracorporeal blood circulation using a mini hemodialyzer. *ASAIO J*, 46, 293–297.

53. Swartz R. D., Somermeyer M. G. and Hsu C. H. 1982. Preservation of plasma volume during hemodialysis depends on dialysate osmolality. *Am J Nephrol*, 2, 189–194.

54. Chen H., Bockenfeld D., Rempfer D., Kaminski M. D. and Rosengart A. J. 2007. Three-dimensional modeling of a portable medical device for magnetic separation of particles from biological fluids. *Phys Med Biol*, 52, 5205–5218.

55. Chen H., Ebner A. D., Rosengart A. J., Kaminski M. D. and Ritter J. A. 2005 Analysis of magnetic drug carrier particle capture by a magnetizable intravascular stent: 2. Parametric study with multiple-wire two-dimensional model. *J Magn Magn Mater*, 293, 616–632.

56. Chen H., Kaminski M. D., Caviness P. L. et al. 2007. Magnetic separation of micro-spheres from viscous biological fluids. *Phys Med Biol*, 52, 1185–1196.

57. Chen H., Kaminski M. D. and Rosengart A. J. 2009. Characterization of a prototype compact high gradient magnetic separator device for blood detoxification. *Separ Sci Tech*, 44, 1954–1969.

58. Harding J. A., Engbers C. M., Newman M. S., Goldstein N. I. and Zalipsky S. 1997. Immunogenicity and pharmacokinetic attributes of poly(ethylene glycol)-grafted immunoliposomes. *Biochim Biophys Acta*, 1327, 181–192.

59. Maruyama K., Takizawa T., Takahashi N. et al. 1997. Targeting efficiency of PEG-immunoliposome-conjugated antibodies at PEG terminals. *Adv Drug Deliv Rev*, 24, 235–242.

60. Bockenfeld D., Chen H., Kaminski M. D., Rosengart A. J. and Rempfer, D. 2010. Parametric study of a portable magnetic separator for separation of nanospheres from circulatory system. *Separ Sci Tech*, 45, 355–363.

Dr. Michael D. Kaminski leads a team of scientists, engineers, post-doctoral students and students in developing processes for various functionalised materials. He has been developing functionalised magnetic polymeric microspheres and nanospheres for 19 years when he started by developing them for selective radionuclide and metals separations. This included evaluating magnetic filtration for recovery of particles from treated waste stream, characterisation of the performance of these solvent-coated particles in differing solution including acids, salts, suspensions and scale-up designs for demonstration unit. He studied the potential of high-magnetic field gradients to separate magnetic components from diamagnetic feeds. He is currently the principal investigator in the development of a functionalised polymer gel for the selective removal of radioactive elements from urban surfaces such as concrete. He has studied functionalisation and sorption chemistry of nano-porous silica glass. He has spent 10 years developing magnetic and non-magnetic microparticles and nanoparticles as a platform for drug delivery/targeting, and the detoxification of blood-borne toxins. This work included synthesis and characterisation of these microspheres and nanospheres but also the design and modelling of an extracorporeal magnetic filter for their subsequent removal from vascular circulation.

Axel J. Rosengart, MD, PhD, is a clinician scientist with expertise in acute brain and spinal cord injuries. He received his PhD and training in Neuroscience in Germany and is currently the Director of Neurocritical Care at Weill Cornell Medical College, New York Presbyterian Hospitals. For the past 11 years, his research efforts has concentrated on magnetically guided drug delivery and detoxification technologies based on biocompatible, superparamagnetic, drug-loaded and sur-face functionalised nanocarriers. Much of this effort was supported by the U.S. Department of Defense (DARPA). Together with friend and colleague Dr. Michael Kaminski of Argonne National Laboratory in Illinois, their research pursues a new way to non-invasively deliver the clot buster tPA to improve the outcome of stroke patients. Furthermore, he is a co-inventor of an innova-tive detoxification technology focusing on magnetically based, selective removal of functionalised nanospheres for circulating blood. When he is away from his patients and the research laboratory, he loves to spend time with his wife, Seng, two children, Tristan and Jevin, and dog Mundo.

8 Magnetic Nanoparticles for *In Vitro* Biological and Medical Applications
An Overview

Ivo Safarik and Mirka Safarikova

CONTENTS

8.1 INTRODUCTION

Advances in information storage technologies since the 1950s have resulted in the preparation of different types of magnetic nanoparticles with defined properties. Since then, the applications of magnetic nanoparticles in different areas of biosciences and biotechnology have been studied intensively. Magnetic nano- and microparticles have been used for *in vitro* cells and biomolecules separation, immobilisation of biologically active compounds, for *in vivo* magnetic drug targeting, magnetic resonance imaging (MRI), magnetic hyperthermia tumour therapy and wastewater treatment. Due to rapid advances in nanotechnology, novel synthetic routes for fabricating magnetic nanoparticles with the defined properties—for example, coating, crystallinity and size uniformity—have been reported. The development of these new types of magnetic nanomaterials, together with their surface functionalisation and stabilisation, has led for their expanded biology, biotechnology and medical applications[1].

The broad family of magnetic field controllable materials, including both nano- and microparticles, high aspect ratio structures (nanotubes, nanowires), thin films, etc., is of general interest for bioapplications. Ferrofluids (magnetic fluids), magnetorheological fluids, magnetic synthetic polymers and biopolymers, magnetic inorganic materials, magnetically modified biological structures, magnetic particles with bound biomolecules and affinity ligands, etc. can serve as typical examples. In many cases, magnetically responsive composite materials have been prepared and used consisting of small magnetic particles (most often formed by magnetite, maghemite or various ferrites), usually in the nanometre to micrometer range, dispersed in a polymer, biopolymer or inorganic matrix; alternatively, magnetic particles can be adsorbed on the outer surface of diamagnetic particles, including both living and dead cells[2,3].

Different types of responses of such materials to external magnetic field enable their applications in numerous areas, namely[4,5]:

- Selective separation (removal) of magnetically responsive nano- and microparticles and other relevant materials from complex samples using an external magnetic field (e.g. using an appropriate magnetic separator, permanent magnet or electromagnet). This process is very important for bioapplications due to the fact that the absolute majority of biological materials has diamagnetic properties that enable efficient selective separation of magnetic materials. Using relatively simple technical approaches, magnetic separations can be efficiently performed both in small and large scales, and both in batch and flow-through arrangements[6,7].
- Targeting and localisation of magnetic particles to the desired place using an external magnetic field. In the bioengineering area, this property has been used in the course of rotary blood pump construction using magnetic fluid seals, which enabled mechanical contact-free rotation of the shaft[8]. Magnetic drug targeting allows the concentration of drugs at a defined target site generally and, importantly, away from the reticuloendothelial system (RES) with the aid of a magnetic field[9].
- Heat generation caused by magnetic particles subjected to high-frequency alternating magnetic field. This phenomenon is employed especially during magnetic fluid hyperthermia (MFH), for example, for cancer treatment. MFH is an example of localised therapy, which involves direct injection of magnetic nanoparticles into the tumour; when placed in an alternating magnetic field, the nanoparticles dissipate heat and destroy the tumour cells due to the elevating the temperature up to $42°C–46°C$[10].
- Increase of a negative T_2 contrast by magnetic iron oxides nanoparticles during MRI. Two types of iron oxide contrast agents are generally used: superparamagnetic iron oxides (SPIO) and ultrasmall superparamagnetic iron oxides (USPIO). For example, when superparamagnetic nanoparticles are delivered to the liver, healthy liver cells can uptake the particles, whereas diseased cells cannot; consequently, the healthy cells are darkened and the diseased regions remain bright[11].
- Great increase of apparent viscosity of magnetorheological fluids when subjected to a magnetic field. Typical magnetorheological fluids are the suspensions of micron sized, magnetisable particles (mainly iron) suspended in an appropriate carrier liquid, such as mineral oil, synthetic oil, water or ethylene glycol. A variety of additives (stabilisers and surfactants) are used to prevent gravitational settling and promote stable particles suspension[12].
- Magnetic nano- and microparticles can be used for magnetic modification of diamagnetic biological materials (e.g. prokaryotic and eukaryotic cells or plant-derived materials), biopolymers, organic polymers and inorganic materials, and for magnetic labelling of biologically active compounds and affinity ligands (e.g. antibodies [Abs], enzymes and aptamers)[2].

Biogenic (naturally occurring) magnetic nanoparticles and related structures, both present within the cells and organs or produced extracellularly, are important due to the following facts:

- Magnetite (and in some cases also greigite [Fe_3S_4]) nanoparticles have been found in many organisms, including magnetotactic bacteria, protozoa, insect, fish, amphibians, birds and mammals[11]. At least in some cases (e.g. magnetotactic bacteria or homing pigeons [*Columba livia*]), it has been proved that magnetic nanoparticles are involved in magnetoreception[13,14].
- Magnetic materials can serve as mechanical protective layers; in chitons (marine mollusks of the class Polyplacophora), the magnetite serves to harden the tooth caps, enabling the

chitons to eat endolithic algae[11,15]. A recently discovered gastropod has its foot covered by scales of conchiolin (a complex protein), mineralised with pyrite (FeS_2) and ferrimagnetic greigite (Fe_3S_4)[16].

- Dissimilatory Fe(III)-reducers, such as *Geobacter metallireducens* and *Shewanella putrefaciens*, as well as a vast number of other Fe(III)-reducing bacteria, including thermophilic species, produce crystals of magnetite as a by-product of their metabolism when grown in culture[17].

In most cases, synthetic (laboratory-produced) magnetically responsive nano- and microparticles and related structures have been prepared and used for *in vitro* experiments; however, biologically produced magnetic nanoparticles (e.g. magnetosomes produced by magnetotactic bacteria, see Figure 8.1) have also been successfully employed for selected bioapplications[18]. This review chapter shows typical examples of *in vitro* biological and medical applications of biocompatible magnetic nanomaterials.

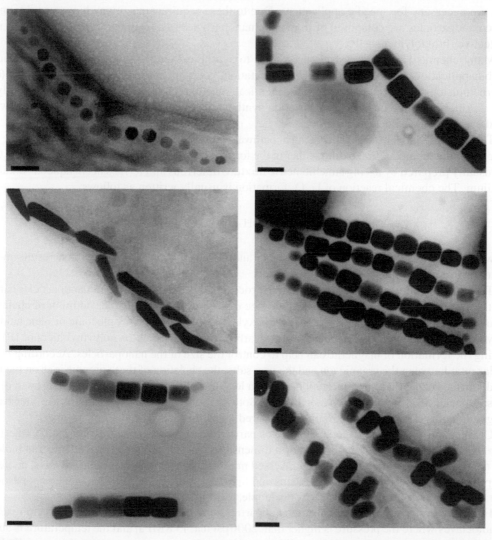

FIGURE 8.1 Examples of magnetosomes produced by different magnetotactic bacteria. The photographs were taken by Prof D. Schuler, Germany. (Reproduced from Safarik, I. and Safarikova, M., *Monatsh. Chem.* 133, 737–759, 2002. With permission.)

8.2   SYNTHESIS OF BIOCOMPATIBLE MAGNETIC NANOPARTICLES

Several types of magnetic materials have close connection to biological systems, processes and bioapplications. Materials found in the biological systems or produced in the laboratory are mainly composed of magnetite (Fe_3O_4), maghemite (γ-Fe_2O_3), greigite (Fe_3S_4), various types of ferrites ($MeO \cdot Fe_2O_3$, where Me = Ni, Co, Mg, Zn, Mn, ...), iron, nickel, etc[11].

Magnetic (nano)particles intended for biological and medical applications should be composed of non-toxic and biocompatible materials, exhibiting good response to an applied static/dynamic magnetic field tailored for the specific application, with negligible remanence to ensure minimal or no magnetic interactions and agglomeration when the external field is switched off, having well-defined size and monodispersity and having structural and chemical stability under different conditions. To enable broader application, the magnetic nanoparticles should be produced in an easy way at minimal cost. In addition, the surface of the particles should enable simple modification and stabilisation of the colloidal dispersions, both to ensure biocompatibility, to prevent non-specific interactions with the medium and body fluids, and to facilitate the attachment of functional groups that are necessary for specific applications. Due to these requirements, many biocompatible magnetic nanoparticles are composed of iron oxide nanoparticles as a magnetic core encapsulated in a protective (bio)polymer shell[19].

Many chemical procedures have been used to synthesise magnetic nano- and microparticles for bioapplications, such as classical co-precipitation, reactions in constrained environments (e.g. microemulsions), sol–gel syntheses, sonochemical reactions, hydrothermal reactions, hydrolysis and thermolysis of precursors, flow injection syntheses (FISs), electrospray syntheses, microwave synthesis and mechanochemical processes[19–25].

The simplest and most efficient chemical pathway to obtain magnetic particles is probably the co-precipitation technique. Iron oxides, either in the form of magnetite (Fe_3O_4) or maghemite (γ-Fe_2O_3), are usually prepared by ageing stoichiometric mixture of ferrous and ferric salts in aqueous alkaline medium. The chemical reaction of Fe_3O_4 formation is usually written as follows:

$$Fe^{2+} + 2Fe^{3+} + 8OH^- \rightarrow Fe_3O_4 + 4H_2O$$

However, magnetite (Fe_3O_4) is not very stable and is sensitive to oxidation, which results in the formation of maghemite (γ-Fe_2O_3).

The main advantage of the co-precipitation process is that a large amount of nanoparticles can be synthesised. However, the control of particle size distribution is limited. The addition of chelating organic anions (carboxylate or α-hydroxy carboxylate ions, such as citric, gluconic or oleic acid) or polymer surface complexing agents (dextran, carboxydextran, starch or polyvinyl alcohol [PVA]) during the formation of magnetite can help to control the size of the nanoparticles. According to the molar ratio between the organic ion and the iron salts, the chelation of these organic ions on the iron oxide surface can either limit nucleation and then lead to larger particles or inhibit the growth of the crystal nuclei, leading to small nanoparticles. Both sterically and ionically stabilised water-based magnetic fluids (ferrofluids) are routinely prepared using this approach[20,26,27].

Classical co-precipitation method generates particles with a broad size distribution. Synthesis of iron oxide nanoparticles with more uniform dimensions can be performed in synthetic and biological nanoreactors, such as water-swollen reversed micellar structures in non-polar solvents, apoferritin protein cages, dendrimers, cyclodextrins and liposomes[20]. Alternatively, magnetic nanoparticles can be prepared in (bio)polymer gels, for example, in iron(II) cross-linked alginate[28].

Hydrothermal syntheses of magnetite nanoparticles are performed in aqueous media in reactors or autoclaves where the pressure is higher than 2000 psi (ca. 13.8 MPa) and the temperature is above 200°C. In this process, the reaction conditions, such as solvent, temperature and time, usually have important effects on the products. The particle size of magnetite powders increased with a prolonged reaction time, and higher water content resulted in the precipitation of larger magnetite particles[20].

The sol–gel process is a suitable wet route to the synthesis of nanostructured metal oxides. This process is based on the hydroxylation and condensation of molecular precursors in solution, originating a 'sol' of nanometric particles. Further condensation and inorganic polymerisation leads to a 3D metal oxide network denominated wet gel. Further heat treatment is necessary to acquire the final crystalline state. The main parameters that influence the kinetics, growth reactions, hydrolysis, condensation reactions, and consequently, the structure and properties of the gel are solvent, temperature, nature and concentration of the salt precursors employed, pH and agitation[20].

The polyol process is a versatile chemical approach for the synthesis of nano- and microparticles with well-defined shapes and controlled sizes. Selected polyols (e.g. polyethylene glycol [PEG]) used as solvents exhibit high dielectric constants and can dissolve inorganic compounds. In addition, due to their relatively high boiling points, they offer a wide operating-temperature range for preparing inorganic compounds. Polyols also serve as reducing agents as well as stabilisers to control particle growth and prevent interparticle aggregation[20].

Recently, a novel synthesis of magnetite nanoparticles based on a flow injection synthesis (FIS) technique has been developed. The technique consists of continuous or segmented mixing of reagents under laminar flow regime in a capillary reactor. The FIS technique has shown some advantages, such as a high reproducibility because of the plug flow and laminar conditions, a high mixing homogeneity and an opportunity for a precise external control of the process. The obtained magnetite nanoparticles had a narrow size distribution in the range of 2–7 nm[20,29].

Spray and laser pyrolysis, typical representatives of aerosol technologies, are continuous chemical processes allowing for high rate production of nanoparticles. In spray pyrolysis, a solution of ferric salts and a reducing agent in organic solvent is sprayed into a series of reactors, where the aerosol solute condenses and the solvent evaporates. The resulting dried residue consisted of particles whose size depends upon the initial size of the original droplets. Maghemite particles with size ranging from 5 to 60 nm with different shapes have been obtained using different iron precursor salts in alcoholic solution[20].

Nearly monodispersed superparamagnetic maghemite nanoparticles (15–20 nm) were prepared by a one-step thermal decomposition of iron(II) acetate in air at 400°C. This synthetic route is simple, cost-effective and allows to prepare the high-quality superparamagnetic particles in a large scale[30].

A simple and quick microwave method to prepare high-performance magnetite nanoparticles directly from Fe^{2+} salt has been developed; monodispersed magnetite nanoparticles (ca. 80 nm in diameter) were prepared. This technique has been easily modified to prepare composite magnetite/silver nanoparticles[25].

Nanosized magnetite powders can also be synthesised *via* a mechanochemical reaction. Ball milling of ferrous and ferric chlorides with sodium hydroxide led to a mixture of magnetite and sodium chloride. To avoid agglomeration, the excess of sodium chloride is usually added to the precursor before ball milling. To prepare different size of particles, the as-milled powders were annealed at temperatures ranging from 100°C to 800°C for 1 h in appropriate atmosphere[21].

8.3 FUNCTIONALISATION OF MAGNETIC NANOPARTICLES

To obtain biocompatible magnetically responsive materials, it is usually necessary to stabilise the prepared iron oxide nanoparticles by appropriate modification of their surface or by their incorporation into appropriate biocompatible matrix. The modified magnetic nanoparticles should be stable against aggregation in both a biological medium and a magnetic field[1].

Several compounds with carboxylic, phosphate and sulphate functional groups are known to bind to the surface of magnetic particles and stabilise them. Citric acid can be successfully used to stabilise water-based magnetic fluids (ferrofluids) by coordinating *via* one or two of the carboxyl residues; this leaves at least one carboxylic acid group exposed to the solvent, which should be responsible for making the surface negatively charged and hydrophilic. Other ferrofluids

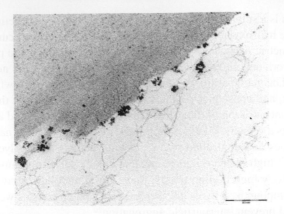

FIGURE 8.2 Ultrathin sections of magnetic sawdust particles observed in transmission electron microscopy. The bar corresponds to 200 nm. (Reproduced from Safarik, I., Safarikova, M., Weyda, F., Mosiniewicz-Szablewska, E., and Slawska-Waniewska, A., *J. Magn. Magn. Mater.*, 293, 371–376, 2005. With permission.)

can be stabilised by ionic interactions, using for example, perchloric acid or tetramethylammonium hydroxide[20,26].

In most cases, biocompatible (bio)polymers are used for magnetic particles stabilisation and modification. Ideal natural or synthetic polymeric materials used for particles stabilisation should have several important properties; they should be biocompatible and for many applications also biodegradable, non-toxic, non-thrombogenic, non-immunogenic and inexpensive. The 'ideal' magnetically responsive (bio)polymer biocompatible composite nanoparticles should have the following typical properties: particles diameter below 100 nm, stability in blood, no activation of neutrophils, no platelet aggregation, avoidance of the RES, non-inflammatory behaviour, prolonged circulation time, possible immobilisation of appropriate biologically active compounds (e.g. Abs) and scalable and cost-effective production[31].

Dextran, a polysaccharide polymer composed exclusively of α-D-glucopyranosyl units with varying degrees of chain length and branching has often been used as a polymer coating mostly because of its excellent biocompatibility. The formation of magnetite in the presence of dextran 40,000 was reported for the first time in 1980s[32]. The same procedure has been used for the preparation of approved magnetic resonance contrast agent Ferumoxtran-10 (Combidex, Sinerem). Other common biopolymer coatings are formed, for example, by carboxymethylated dextran, carboxydextran, starch, chitosan, alginate, arabinogalactan or glycosaminoglycan, while PEG and PVA represent biocompatible synthetic polymers[20].

Magnetic nanoparticles often form a magnetic part of magnetically responsive composite microparticles and high aspect materials formed from various synthetic polymers, biopolymers, carbon nanotubes, inorganic materials, plant materials such as sawdust (Figure 8.2) or microbial cells[2,33]. Superparamagnetic monodisperse microparticles composed of polystyrene matrix with entrapped maghemite nanoparticles (ca. 8 nm in diameter[34]), known as Dynabeads (Invitrogen), have been used in many bioapplications, especially in molecular biology, cell biology, microbiology and protein separation. Also other types of magnetic nanocomposite (bio)polymer microparticles have been prepared and used[23,24].

8.4 *IN VITRO* APPLICATION OF MAGNETIC NANOPARTICLES

Magnetically responsive nanoparticles and nanocomposites have been used in many *in vivo* and *in vitro* experiments in different branches of biology and medicine; in the following part, several typical examples of *in vitro* applications will be presented.

8.4.1 IMMOBILISATION OF BIOLOGICALLY ACTIVE COMPOUNDS AND AFFINITY LIGANDS

Immobilisation of enzymes, enzyme inhibitors, Abs, lectins, avidin/streptavidin, oligonucleotides and other biologically active compounds on different types of carriers is a very important technique used in various areas of biosciences and biotechnology[36]. Immobilisation of biologically active compounds on magnetic carriers enables their simple recovery from the assay or reaction systems by an external magnetic field. Alternatively, immobilised compounds can be targeted to the desired place.

Standard covalent conjugation strategies developed in enzymology, immunology and affinity chromatography, using amine, hydroxyl, carboxyl, aldehyde or thiol groups exposed on the surface of magnetic particles, can be employed for compounds immobilisation. Immobilised compounds can be used to express their activities in a desired process (e.g. immobilised enzymes) or as affinity ligands to capture or to modify the target molecules or cells (e.g. Abs or their fragments, avidin, streptavidin, lectins, oligonucleotides, aptamers, oligo- and polysaccharides, metal chelate binding groups, etc.).

Magnetic particles with immobilised Abs form the basis of immunomagnetic procedures, especially used in medical, food and water microbiology and in cell biology for selective capture of target prokaryotic and eukaryotic cells[37,38], and also for separations and bioassays of important biologically active compounds and xenobiotics[6,39]. Ideally, Abs should be oriented with its Fc part towards the magnetic particle, so that the Fab region is pointing outwards from the particle. Several procedures have been employed for direct binding of Abs to magnetic carriers, namely[38]:

- Covalent binding of Abs on activated magnetic (nano)particles (e.g. tosylactivated Dynabeads) or on magnetic particles carrying appropriate functional groups (e.g. amino, carboxy, hydroxy) using standard immobilisation procedures.
- Secondary Abs (i.e. Abs against primary Abs) are immobilised first on magnetic (nano) particles and then primary Abs are bound. The secondary Abs may, in this case, function as a spacer and lead to a favourable orientation of primary Abs.
- Streptavidin or avidin immobilised on magnetic carriers binds biotinylated Abs.
- Protein A and protein G immobilised on magnetic (nano)particles bind the Fc region of IgG of most mammalian species leaving antigen-specific sites free.
- Magnetic carriers with immobilised boronic acid derivative reversibly bind Abs through the interaction with the carbohydrate units on the Fc part of Abs, thus keeping the favourable Abs orientation.
- Primary Abs tagged with oligo dA are coated on magnetic (nano)particles with immobilised oligo dT utilising the hybridisation between oligo dA and oligo dT homopolymers.
- Magnetic carriers bearing hydrazide groups can be used for oriented immobilisation of Abs *via* their carbohydrate moiety.
- Adsorption of Abs on hydrophobic magnetic particles, especially those made of polystyrene (e.g. Dynabeads).

In addition to Abs, aptamers (artificial nucleic acid ligands with affinity against various low- and high-molecular-weight compounds), immobilised on magnetic (nano)carriers can be used in the similar way[40,41].

One of the most popular methods for non-covalent conjugation is employing the natural strong binding of two proteins (avidin or streptavidin) with low-molecular-weight compound biotin (vitamin H), exhibiting a dissociation constant of about 10^{-15} M. Avidin is a glycoprotein found in egg whites that contains four identical subunits of 16.4 kDa each, giving an intact molecular weight of approximately 66 kDa. Each subunit contains one binding site for biotin. Avidin is highly basic, having a pI of about 10. The high value of pI and the presence of carbohydrates in the avidin molecule can lead to non-specific interactions causing elevated background signals in some assays. That is why streptavidin isolated from *Streptomyces avidinii* is currently preferentially used.

Streptavidin (molecular mass of about 60 kDa) contains four subunits, each with a single biotin-binding site. Streptavidin is not a glycoprotein and has a much lower isoelectric point (5–6); these factors lead to better signal-to-noise ratios in assays using streptavidin–biotin interactions[42].

Magnetic nanoparticles and nanocomposites with immobilised streptavidin and avidin have been regularly used for efficient binding of biotinylated compounds; several types are available commercially as shown in Table 8.1.

Many different types of enzymes have been immobilised on magnetic nanoparticles and nanocomposites and used in both small- and large-scale applications. Immobilised enzymes can be used repeatedly, they are usually more stable during storage, and they can be easily manipulated using external magnetic field. Enzyme reactions are usually performed in mixed reactors with subsequent magnetic separation or in magnetically stabilised fluidised beds. In both cases, it is possible to perform enzyme reactions in mixtures containing solid impurities[43]. Trypsin immobilised on magnetic nanocarriers has been successfully applied in proteomics as an efficient agent for protein digestion. The covalent conjugation usually results in increase in trypsin thermostability and elimination

---

TABLE 8.1
Examples of Commercially Available Magnetic Nanoparticles and Nanocomposites with Immobilised Avidin and Streptavidin

Name	Diameter (nm)	Nanoparticle/Nanocomposite Characterisation	Manufacturer/Supplier
BcMag™ Monomer Avidin Magnetic Beads	1000	Nanometre scale SPIO core completely encapsulated by a high-purity silica shell with high density of avidin subunit monomer on the surface	Bioclone Inc, USA
BcMag™ Streptavidin Magnetic Beads	1000	Nanometre scale SPIO core completely encapsulated by a high-purity silica shell with high-density streptavidin on the surface	Bioclone Inc, USA
Bio-Adembeads Streptavidin	100–500	Superparamagnetic nanoparticles	Ademtech, France
Dynabeads® M-280 Streptavidin	2800	Dynabeads (polystyrene)	Invitrogen, USA
Dynabeads® M-270 Streptavidin	2800		
Dynabeads® MyOne™ Streptavidin C1	1000		
Dynabeads® MyOne™ Streptavidin T1	1000		
fluidMAG–Streptavidin	100, 200	Aqueous dispersion of starch-coated magnetite nanoparticles	Chemicell GmbH, Berlin, Germany
Magnetic nanoparticles, avidin labelled	25–30	SPIO nanoparticles coated with biocompatible polymer	Nanocs, Inc., New York, USA
Sera-Mag* Magnetic Streptavidin Particles	1000	Superparamagnetic microparticles prepared by a core–shell process	Seradyn, USA
SiMAG–Streptavidin	1000	Aqueous dispersion of silica particles containing maghemite nanoparticles with bound streptavidin	Chemicell GmbH, Berlin, Germany
Streptavidin MicroBeads	20–150	Superparamagnetic nanoparticles made from an iron oxide core and a dextran coating	Miltenyi Biotec, Germany

of its autolysis. Consequently, the immobilisation of trypsin allows fast in-solution digestion of proteins and their identification by MALDI-TOF mass spectrometry[30,44].

Isolated membrane-bound enzymes can be efficiently immobilised in magnetoliposomes; in a typical example, lipid-depleted, beef-heart cytochrome c-oxidase demonstrated a 15-fold enhancement of enzyme activity after its incorporation into magnetoliposomes by sonication[45].

In addition to isolated enzymes, also magnetically responsive whole-cell biocatalysts, such as magnetic fluid treated baker's yeast cells (*Saccharomyces cerevisiae*) have been prepared and used for hydrogen peroxide decomposition and sucrose conversion[46]. Typical examples of enzymes immobilised on magnetic nanoparticles and nanocomposites are given in Table 8.2.

TABLE 8.2
Examples of Enzymes Immobilised on Magnetic Nanoparticles or Nanocomposites

Enzyme	Source	Magnetic Carrier	Activation/ Immobilisation Procedure	Application	Reference
Alcohol dehydrogenase	Baker's yeast	Magnetite nanoparticles (10.6 nm)	Carbodiimide	Reduction of 2-butanone	[47]
α-Amylase	—	Cellulose-coated magnetite nanoparticles (2.5–22.5 nm)	Periodate oxidation	Degradation of starch	[48]
Cellulose	—	PVA/Fe₂O₃ magnetic nanoparticles (270 nm)	Immobilisation in microemulsion system	Degradation of microcrystalline cellulose	[49]
Chitosanase	*Bacillus pumilus*	Amylose-coated magnetic nanoparticles (200 nm)	Glycidol	Pentamers and hexamers of chitosan oligosaccharides production	[50]
Chymotrypsin	Bovine pancreas	Dynabeads M-280 tosylactivated	Tosylactivated beads used	Proteolysis of specific proteins	[51]
β-D-Galactosidase	*Aspergillus oryzae*	Magnetite-chitosan nanoparticles (30 nm)	Glutaraldehyde	Galacto-oligosaccharide synthesis	[52]
Glucose oxidase	*Aspergillus niger*	Magnetite nanoparticles (20 nm)	(3-Aminopropyl) triethoxysilane/ glutaraldehyde	Nanometric glucose sensors	[53]
Glycolate oxidase	*Medicago falcata*	Hydrothermally synthesised magnetic nanoparticles	Physical adsorption	Biosynthesis of glyoxylic acid	[54]
Lipase	*Pseudomonas cepacia*	Magnetite nanoparticles (16 nm)	Carbodiimide	Biodiesel production	[55]
Lipase	*Thermomyces lanuginose*	Magnetite nanoparticles (11.2 nm)	(3-Aminopropyl) triethoxysilane/ glutaraldehyde	Biodiesel production	[56]
Phosphatase, alkaline	—	Magnetite nanoparticles (40 nm)	Carbodiimide	Dephosphorylation of plasmid DNA	[57]
Trypsin (TPCK-treated)	Bovine pancreas	Amine-functionalised magnetic nanoparticles (50 nm)	Glutaraldehyde	Protein digestion	[58]

The formulas above are written as PVA/Fe_2O_3.

TABLE 8.3

Examples of Commercially Available Magnetic Nanoparticles and Nanocomposites with Immobilised Oligodeoxythimidine

Name	Diameter (nm)	Nanoparticle/Nanocomposite Characterisation	Manufacturer/Supplier
BcMag mRNA	1000	Nanometre scale SPIO core completely encapsulated by a high-purity silica shell grafted by oligo dT(25) on the surface	Bioclone Inc., USA
Dynabeads® oligo (dT)25	1000, 2800	Dynabeads (polystyrene)	Invitrogen, USA
Nucleo-Adembeads	100–500	Superparamagnetic nanoparticles	Ademtech, France
Oligo(dT) MicroBeads	20–150	Superparamagnetic nanoparticles made from an iron oxide core and a dextran coating	Miltenyi Biotec, Germany
Sera-Mag oligo (dT)30	1000	Superparamagnetic microparticles prepared by a core–shell process	Seradyn, USA

Molecular biology research can be often simplified by the application of magnetic nanoparticles and nanocomposites with immobilised nucleic acids or oligonucleotides. Several companies offer oligodeoxythymidine immobilised on magnetic particles (see Table 8.3), which can be used effectively for the rapid isolation of highly purified mRNA from eukaryotic cell cultures or total RNA preparations. These procedures are based on the hybridisation of the oligonucleotide dT sequence with the stable polyadenylated 3-termini of the eukaryotic mRNA. The length of the complementary sequence differs between 20 and 30 oligonucleotides. This sequence is usually directly bound covalently to the particle surface or indirectly by biotinylated oligonucleotides to the avidin- or streptavidin-coated magnetic particles[59].

As shown previously, selected polysaccharides (especially dextran) can be used to stabilise magnetic nanoparticles. Anyway, other polysaccharides can be used for coating magnetic nanoparticles, such as agarose, alginate, carrageenan, chitosan, gum arabic, heparin, pullulan and starch[60]. Covalent immobilisation of saccharides and their derivatives on magnetic carriers is not so common; immobilisation of 4-aminophenyl-β-D-thioglucopyranoside on silanised magnetic particles, subsequently used for affinity separation of β-galactosidase[61] and immobilisation of mannose on magnetic nanoparticles used for stem cells labelling[62], can serve as examples.

Several low-molecular-weight biologically active compounds have been immobilised on magnetic particles. Magnetic polymer nanospheres with immobilised Mn(III) porphyrin were synthesised and catalytic activity for cyclohexane hydroxylation was investigated; it was shown that these magnetic nanospheres are highly efficient and are recyclable catalysts[63]. Silanised magnetic nanoparticles were used to immobilise norvancomycin hydrochloride; by a separation assay of *Staphylococcus aureus* and *Escherichia coli*, it was proved that this type magnetic particle can selectively capture *S. aureus*[64]. Poly(L-lysine)-modified iron oxide nanoparticles were used for stem cell labelling[65]. Biotin immobilised on magnetite nanoparticles has been employed for the selective binding of avidin and its derivatives[66]. 3-Aminophenylboronic acid specifically interacts with compounds containing vicinal hydroxyl groups; after its immobilisation to magnetic (nano)carriers, it is possible to separate various types of saccharides and glycoproteins, including clinically important glycated haemoglobin[67].

Chemical modification of enzymes with the amphipathic macromolecule, PEG, leads to conjugates, which are soluble and active in organic solvents. The PEG–enzyme conjugates can be also prepared in magnetic forms, which stably disperse in both organic solvents and aqueous solutions. Magnetically modified lipase was used to catalyse ester synthesis in organic solvents and could be easily recovered by magnetic force without loss of enzyme activity[68,69].

For specific applications, commercially available magnetic nanoparticles, nanocomposites and microparticles with covalently immobilised biologically active compounds and affinity ligands can be used. The immobilised molecules comprehend, for example, streptavidin, avidin, annexin V, different types of primary and secondary Abs, protein A, protein G, pepsin, papain, oligonucleotides, biotin, nitrilotriacetic acid, affinity ligands for the isolation of recombinant fusion proteins, Reactive Blue 2, Reactive Red 120, etc. A wide variety of activated magnetic particles enabling a one-step immobilisation process is also commercially available.

Magnetic particles present within magnetic carriers are mainly used for magnetic manipulation. However, experiments were performed to show that the activity of enzymes immobilised on magnetic carries can be influenced due to the heat generated in alternating magnetic field, such as in case of invertase immobilised together with γ-Fe_2O_3 in a co-polymer gel of *N*-isopropylacrylamide and acrylamide[70]. Heat generation due to eddy currents and hysteresis loses induced by an alternating magnetic field was utilised for the enhancement of ethanol formation catalysed by yeast immobilised together with iron powder or Ba-ferrite in alginate beads. The ethanol concentration increased by 12%–14% with immobilised yeast; these effects corresponded to a 4 K rise in temperature inside the gel[71].

8.4.2 MAGNETIC SEPARATION OF IMPORTANT BIOLOGICALLY ACTIVE COMPOUNDS AND XENOBIOTICS

Isolation, separation and purification of various types of biologically active compounds, such as proteins and peptides, nucleic acids and oligonucleotides, carbohydrates, xenobiotics, as well as of other specific molecules, are used in almost all branches of biosciences and biotechnologies. In the area of biosciences and biotechnology, the isolation of biologically active compounds is usually performed using variety of chromatography, electrophoretic, ultrafiltration, precipitation and other procedures, affinity chromatography being one of the most important techniques. Affinity ligand techniques represent currently the most powerful tool available to the downstream processing both in terms of their selectivity and recovery. However, the disadvantage of all standard column liquid chromatography procedures is the impossibility of the standard column systems to cope with the samples containing particulate material so they are not suitable for work in early stages of the isolation/purification process, where suspended solid and fouling components are present in the sample. In this case, magnetic affinity, ion exchange, hydrophobic or adsorption batch separation processes, applications of magnetically stabilised fluidised beds or magnetically modified two-phase systems have shown their usefulness[6,72].

Magnetic adsorbents possess a uniquely attractive property that permits their rapid and highly selective removal from almost any type of bio-feedstock. The basic principle of batch magnetic separation is very simple. In most cases, magnetic carriers bearing an immobilised affinity, pseudo affinity, hydrophobic or mixed mode ligand or ion-exchange groups, or magnetic biopolymer particles having affinity to the isolated structure are mixed with a sample containing target compound(s). Samples may be crude cell lysates, whole blood, plasma, ascites fluid, milk, whey, urine, cultivation media, wastes from food and fermentation industry and many others; many of the treated materials contain suspended solid diamagnetic impurities. Following an incubation period, when the target compounds bind to the magnetic particles, the whole magnetic complex is easily and rapidly removed from the sample using an appropriate magnetic separator. After washing out the contaminants, the isolated target compound(s) can be eluted and used for further work (Figure 8.3).

Magnetic separation techniques have several advantages in comparison with standard separation procedures. This process is usually very simple, with only a few handling steps. All the steps of the purification procedure can take place in one single test tube or another vessel. There is no need for expensive equipment; a large amount of commercially available magnetic separators can be used (Figure 8.4). In some cases (e.g. isolation of intracellular proteins), it is even possible

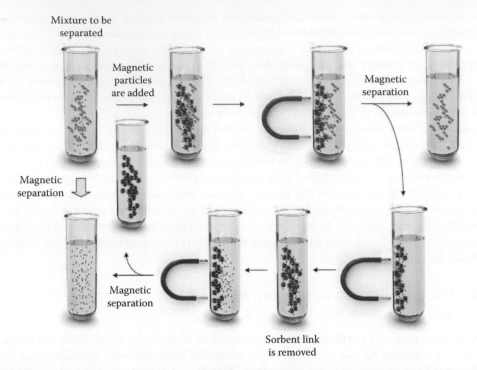

FIGURE 8.3 A simplified description of a typical batch magnetic separation. Red particles represent magnetic (nano)particles with affinity for the target compounds or cells. Large grey particles represent unwanted (contaminating) species and small grey particles represent the target species. (Reproduced from Yavuz, C.T., Prakash, A., Mayo, J.T., and Colvin, V.L., *Chem. Eng. Sci.*, 64, 2510–2521, 2009. With permission.)

to integrate the disintegration and separation steps and thus shorten the total separation time[74]. Moreover, the power and efficiency of magnetic separation procedures are especially useful at large-scale operations. The magnetic separation techniques are also the basis of various automated procedures, especially magnetic particle–based immunoassay systems for the determination of a variety of analytes, among them proteins and peptides. Several automated systems for the separation of proteins or nucleic acids have become available recently[6,75,76]. The detailed overview covering magnetic separation of proteins and peptides is available online free of charge (http://biomagres.com/content/2/1/7)[6].

Recombinant DNA techniques enable production of large quantities of various recombinant proteins. To enable simple separation of the target protein from the complex biological matrices, recombinant fusion proteins are usually produced. The fusion proteins contain an appropriate affinity tag structure, enabling one-step adsorption purification. Affinity tags (e.g. polyhistidine-tag, glutathione S-transferase (GST)-tag, Strep-tag II, polyarginine-tag, chitin-binding domain, maltose-binding protein, FLAG-tag, SNAP-tag, c-myc-tag, green fluorescent protein-tag) have been immobilised on various magnetic nano- and microparticles and the developed magnetic affinity adsorbents used for one-step separation of recombinant fusion proteins[77]. The examples of commercially available magnetic nanoparticles and nanocomposites with immobilised affinity tags are given in Table 8.4; examples of commercial magnetic microparticles for fusion proteins separation can be found in a review article[77].

The isolation of DNA or RNA is an important step before many biochemical and diagnostic processes. Many downstream applications such as detection, cloning, sequencing, amplification, hybridisation, cDNA synthesis, etc. cannot be carried out with the crude sample material. The presence of large amounts of cellular or other contaminating materials, for example, proteins or carbohydrates, in such complex mixtures often impedes many of the subsequent reactions and techniques.

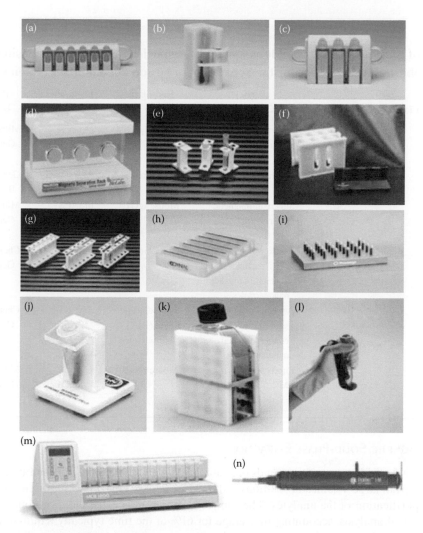

FIGURE 8.4 Examples of batch magnetic separators applicable for magnetic separation of target compounds and cells. (a) Dynal MPC – S for six microtubes (Dynal, Norway; now Invitrogen, USA); (b) Dynal MPC – 1 for one test tube (Dynal, Norway; now Invitrogen, USA); (c) Dynal MPC – L for six test tubes (Dynal, Norway; now Invitrogen, USA); (d) magnetic separator for six Eppendorf tubes (New England BioLabs, USA); (e) MagneSphere Technology Magnetic Separation Stand, two position (Promega, USA); (f): MagnaBot Large Volume Magnetic Separation Device (Promega, USA); (g) MagneSphere Technology Magnetic Separation Stand, twelve-position (Promega, USA); (h) Dynal MPC – 96 S for 96-well microtitre plates (Dynal, Norway; now Invitrogen, USA); (i) MagnaBot 96 Magnetic Separation Device for 96-well microtitre plates (Promega, USA); (j) BioMag Solo-Sep Microcentrifuge Tube Separator (Polysciences, USA); (k) BioMag Flask Separator (Polysciences, USA); (l) MagneSil Magnetic Separation Unit (Promega, USA); (m) MCB 1200 processing system for 12 microtubes based on MixSep process (Sigris Research, USA); (n) PickPen magnetic tool (Bio-Nobile, Finland). (Reproduced from Safarik, I. and Safarikova, M., *Biomagn. Res. Technol.,* 2, Article No. 7, 2004. With permission.)

In addition, DNA may contaminate RNA preparations and vice versa. DNA preparations can be efficiently isolated using magnetic silica particles[78,79], while mRNA is routinely isolated using magnetic (nano)particles with immobilised deoxythymidine[59]. It is also possible to isolate components of the cell lysate, which inhibit, for example, the DNA polymerase of a following polymerase chain reaction (PCR) like polysaccharides, phenolic compounds or humic substances using specific magnetic particles[59].

TABLE 8.4

Examples of Commercially Available Magnetic Nanoparticles and Nanocomposites with Immobilised Affinity Tags for the Separation of Fusion Recombinant Proteins

Recombinant Tag	Diameter (nm)	Particles Composition	Immobilised Affinity Ligand	Commercial Product	Producer/ Distributor
c-myc-tag	50	Superparamagnetic MicroBeads	Mouse IgG1	μMACS Anti-c-myc MicroBeads	Miltenyi Biotec, Germany
FLAG	250	Dextran–magnetite	Monoclonal Ab	Anti FLAG M1	Gentaur, Belgium
GST	50	Superparamagnetic MicroBeads	Mouse IgG1	μMACS Anti-GST MicroBeads	Miltenyi Biotec, Germany
	1000	Aqueous dispersion of silica particles containing maghemite nanoparticles	Glutathione	SiMAG– Glutathione	Chemicell, Germany
His (6xHis)	1000	Aqueous dispersion of silica particles containing maghemite nanoparticles	Iminodiacetic acid (IDA), Ni^{2+} charged	SiMAG– IDA/Nickel	Chemicell, Germany
	50	Superparamagnetic MicroBeads	Mouse IgG2b	μMACS Anti-His MicroBeads	Miltenyi Biotec, Germany

8.4.3 MAGNETIC SOLID-PHASE EXTRACTION

Analysis of biologically active compounds or xenobiotics often requires pre-concentration of the target analytes from large volumes of solutions and/or suspensions. This process is often accompanied by partial purification of the analytes. The sample preparation is often the most time-consuming step in chemical analysis, accounting in average for 61% of the time typically required to perform analytical tasks[80]. The sample preparation is also the source of much of the imprecision and inaccuracy of the overall analysis[81].

At present solid-phase extraction (SPE) is regularly used to isolate and pre-concentrate desired components from a sample matrix, especially on a small scale. A new procedure for SPE, based on the use of magnetic or magnetisable adsorbents called 'magnetic solid-phase extraction' (MSPE) has been developed for large-scale processes. In this procedure, magnetic adsorbent is added to a solution or suspension containing the target analyte. The analyte is adsorbed on to the magnetic adsorbent, and then the adsorbent with adsorbed analyte is recovered from the suspension using an appropriate magnetic separator. The analyte is consequently eluted from the recovered adsorbent and analysed[82] (Figure 8.5).

Different types of magnetic adsorbents, such as magnetic charcoal[82,84], reactive copper phthalocyanine dye immobilised to the silanised magnetite particles (blue magnetite)[82,85], magnetic derivatives of Chromosorb, Tenax TA, Tenax GR, Porapak, Chezacarb, polyamide and polyphenyleneoxide[84,86,87] and some others have been used for the pre-concentration of various organic xenobiotics, including dyes, tensides, nonylphenol derivatives, alkylphenols, etc.

MSPE has been efficiently used for the detection of compounds illegally used as fish drugs in aquacultures. Recently, malachite green (Basic Green 4, CI 42000) has found extensive use all over the world in the fish farming industry as a fungicide, ectoparasiticide and disinfectant; a similar situation is valid for crystal violet (Basic Violet 3, CI 42555). Both dyes are linked to an increased

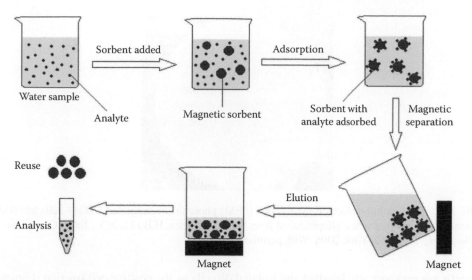

FIGURE 8.5 Magnetic SPE procedure. (Reproduced from Chen, L., Wang, T., and Tong, J., *TrAC Trend. Anal. Chem.*, 30, 1095–1108, 2011. With permission.)

risk of cancer, and that is why the use of malachite green has been banned in aquacultures intended to produce fish for human consumption[85].

Malachite green and crystal violet in the chromatic form specifically interact with copper phthalocyanine dye immobilised on magnetic particles. This feature enables specific separation and concentration of the target dyes (and other specifically binding compounds, such as polyaromatic hydrocarbons) not only from solutions, but also from suspension systems. After MSPE, the target dyes have been eluted with small amount of methanol, and spectra have been recorded. Concentrations of both dyes in the range 0.5–1.0 μg L^{-1} of water could be reproducibly detected. The dyes can be detected not only in potable water, but also in river water[85].

This procedure can find interesting applications also in medical analysis, such as for pre-concentration of drugs and drug metabolites from urine and other body fluids.

8.4.4 Magnetic Separation and Labelling of Cells

Two types of magnetic separations can be distinguished when working with cells. In the first type, cells to be separated demonstrate sufficient intrinsic magnetic moment so that magnetic separations can be performed without any modification. There are only a few types of such cells in nature, especially red blood cells (erythrocytes) containing high concentrations of paramagnetic haemoglobin and magnetotactic bacteria containing small magnetic particles within their cells. In the second type, one or more non-magnetic (diamagnetic) components of a mixture have to be tagged by a magnetic label to achieve the required contrast in magnetic susceptibility between the cell and the medium. The attachment of magnetic labels (mainly magnetic nanoparticles and microparticles) is usually mediated by affinity ligands of various nature, which can interact with target structures on the cell surface. Most often Abs against specific cell surface epitopes are used, but other specific ligands such as lectins can be employed, too. The newly formed complexes have magnetic properties and can be manipulated using an appropriate magnetic separator[38]. Many other possibilities are available for magnetic modification of microbial cells, as shown in a recent review article[88].

Target diamagnetic cells can be separated using two basic strategies, namely positive selection or depletion. The optimal separation strategy depends on the frequency of target cells in the cell sample, their phenotype compared with the other cells in the sample, the availability of reagents and a full consideration of how the target cells are to be used. Positive selection means that the desired

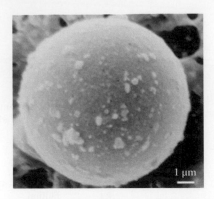

FIGURE 8.6 The scanning electron microscope (SEM) photo of the separated CD34+ cells labelled with immunomagnetic nanoparticles. (Reproduced from Chen, W., Shen, H.B., Li, X.Y., Jia, N.Q., and Xu, J.M., *Appl. Surf. Sci.*, 253, 1762–1769, 2006. With permission.)

target cells are magnetically labelled and isolated directly as the positive cell fraction (Figure 8.6). It is the most direct and specific way to isolate the target cells from a heterogeneous cell suspension. Positive selection is particularly well suited for the isolation of rare cells. Both fractions – labelled and unlabelled – can be recovered and used. Depletion means that the unwanted cells are magnetically labelled and eliminated from the cell mixture, and the non-magnetic, untouched fraction contains the cells of interest. Potential effects on the functional status of cells are minimised[38,89].

Target cells can be labelled in a different way. During direct labelling, magnetic particles with immobilised specific Abs or lectins are used to capture the target cells; this is the most rapid procedure. Direct labelling minimises the number of washing steps and thereby prevents cell loss. Indirect labelling is usually performed when no magnetic particles for direct magnetic labelling are available. In this case, target cells are labelled with a primary Ab that is unconjugated, biotinylated or fluorochrome conjugated. In a second step, appropriate magnetic particles are added to the suspension, enabling one of the following three strategies, namely[38,89]:

1. Binding of unlabelled primary Ab to magnetic particles conjugated to secondary Abs against the primary Ab used
2. Binding of biotinylated primary Ab to magnetic particles bearing immobilised streptavidin, avidin or anti-biotin Ab
3. Binding of fluorochrome-labelled primary Ab to magnetic particles conjugated to anti-fluorochrome Ab

8.4.4.1 Magnetic Separation and Labelling of Eukaryotic Cells

Magnetic sorting of eukaryotic cells has become a standard method for cell separation in many different fields, both in small and large scale. The isolation of almost any cell type is possible from complex cell mixtures, such as peripheral blood, haematopoietic tissue (spleen, lymph nodes, thymus, bone marrow, etc.), non-haematopoietic tissue (solid tumours, epidermis, dermis, liver, thyroid gland, muscle, connective tissue, etc.) or cultured cells[89].

Two basic variants for magnetic cells separation exist, which differ in two main features, namely the composition and size of the magnetic particles used for cell labelling (nanoparticles or microparticles) and the mode of magnetic separation (high-gradient magnetic separation or batch magnetic separation).

The most popular magnetic separation system employing magnetic nanoparticles is the magnetic-activated cell sorting (MACS) system (Miltenyi Biotec, Germany), which is characterised by the use of nano-sized superparamagnetic particles made from an iron oxide core and a dextran coating, ranging from 20 to 150 nm in diameter, and forming stable colloidal solutions.

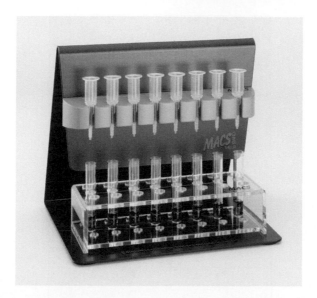

FIGURE 8.7 A typical example of laboratory-scale high-gradient magnetic separators. OctoMACS Separator (Miltenyi Biotec, Germany) can be used for simultaneous isolation of magnetically labelled cells or mRNA. (Reproduced from Safarik, I. and Safarikova, M., *Encyclopedia of Separation Science* (Wilson, I.D., Adlard, R.R., Poole, C.F., and Cook, M.R., Eds.). Academic Press, London, U.K., pp. 2163–2170, 2000. With permission.)

Magnetic separation is performed in a separation column filled with a matrix of ferromagnetic steel wool or iron spheres, which is placed inside the high-gradient magnetic separation system (Figure 8.7). The separator contains a strong permanent magnet creating a high-gradient magnetic field on the magnetisable column matrix; high magnetic gradients up to approximately 10^4 T m^{-1} are generated in the vicinity of the ferromagnetic matrix. The magnetic force is then sufficient to retain the target cells labelled with a very small number of magnetic nanoparticles. Columns of different size are commercially available[89,91].

Magnetic cell separation using the MACS system is performed in three basic steps as follows[89]:

1. Magnetic labelling of target cells in a cell suspension is performed by immunomagnetic nanoparticles (MicroBeads), which typically are directly covalently conjugated to a monoclonal Ab or other ligand specific for a certain cell type.
2. Magnetic separation of magnetically labelled cells; the cell suspension is passed through the separation column that contains a ferromagnetic matrix and is placed in a MACS separator. Labelled target cells are retained in the column *via* magnetic forces, whereas unlabelled cells flow through. By simply rinsing the column with buffer, the entire untouched cell fraction is obtained.
3. Elution of the labelled cell fraction; after removing the column from the magnetic field of the MACS separator, the retained labelled cells can easily be eluted with buffer.

The entire procedure can be performed in less than 30 min, and both cell fractions – magnetically labelled and untouched cells – are ready for further use, such as flow cytometry, molecular biology, cell culture, transfer into animals or clinical cellular therapy[89].

The large-scale magnetic separation of target cells using magnetic nanoparticles can be performed in the automated CliniMACS device (Miltenyi Biotec), which enables magnetic cell selection in a closed and sterile system (Figure 8.8). The use of clinical-grade isolation or depletion of cells is now a standard technique established in many cellular therapy centres.

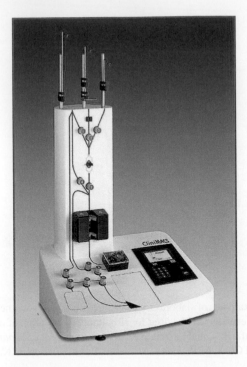

FIGURE 8.8 The automated system for clinical isolation of human cell subsets (CliniMACS, Miltenyi Biotec, Germany). (Reproduced from Safarik, I. and Safarikova, M., *Monatsh. Chem.* 133, 737–759, 2002. With permission.)

An alternative procedure is based on the use of magnetically responsive microparticles, such as Dynabeads, bearing immobilised primary or secondary Abs or (strept)avidin. In small scale, labelled cells (Figure 8.9) can be easily separated using standard magnetic separators. The Isolex® 300 System (Baxter, USA) is a semi-automated magnetic cell separation system designed to select and isolate CD34+ cells, *ex vivo*, from mobilised peripheral blood using anti-CD34 monoclonal Ab and superparamagnetic Dynabeads microspheres.

The positively selected cells, in many cases, may not show any interference from the larger magnetic particles and may also be analysed or used with the particles attached on them. In some

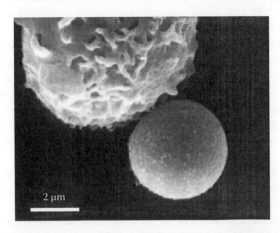

FIGURE 8.9 A scanning electron micrograph showing a human neutrophil adhering to an anti-CD45-coated Dynabeads M-450. (Reproduced from Shao, J.-Y. et al., *Proc. Natl. Acad. Sci. USA,* 95, 6797, 1998. With permission.)

cases, however, it is necessary to remove larger immunomagnetic particles from the cells after their isolation. The detachment process can be performed in several ways, namely[38]:

- Incubating captured cells overnight in cell culture medium and subsequent mechanical treatment (e.g. firm pipetting flushing the suspension 5–10 times through a narrow tipped pipette).
- Proteolytic enzymes can be used to release isolated cells from magnetic particles by selective cleavage of the protein epitope or Ab involved in the immunomagnetic binding.
- Application of an Ab that reacts with the Fab fragments of primary monoclonal Abs on magnetic beads and thus enables direct dissociation of the antigen–Ab binding, thereby producing cells without Abs remaining on the surface and with unchanged antigen expression (e.g. DETACHaBEAD from Invitrogen).
- Synthetic peptides that bind specifically to the antigen-binding site of primary Abs compete with the target cell–magnetic particles complexes and enable to obtain target cells with unchanged antigen expression (e.g. used by Baxter Healthcare).
- Carbohydrate units on the Fc part of the Abs allow reversible attachment of the Abs to the magnetic particles with immobilised $-B(OH)_3$ groups. After selective isolation of the target cells, sorbitol is added, which replaces the Ab on the magnetic bead.
- A complex primary Ab–DNA linker can be immobilised on magnetic particles, and after cell binding, the DNA linker can be splitted enzymatically using DNase.
- In specific cases, decrease of pH can cause immunomagnetic particles release.

Magnetic (nano)particles with immobilised annexin V have been employed for the simple and efficient separation of apoptotic cells from normal culture. This procedure is based on the fact that annexin V is a Ca^{2+}-dependent phospholipids-binding protein with high affinity for negatively charged phosphatidylserine (PS), which is redistributed from the inner to the outer plasma membrane leaflet in apoptotic or dead cells. Once on the cell surface, PS becomes available for binding to annexin V and any of its magnetic conjugates[93,94].

In addition to whole cells, cell organelles such as ectosomes[95] or peroxisomes[96] can also be efficiently isolated using immunomagnetic separation (IMS).

During cells transplantation, it is necessary to track and monitor the grafted cells in the transplant recipient. To screen cells both *in vitro* and *in vivo*, SPIO nanoparticles, such as MRI contrast agent Endorem or dextran-based magnetic nanoparticles MicroBeads (Miltenyi Biotec), have been used to label the stem cells; nanoparticles can often be taken up by cells during cultivation by endocytosis. The magnetically labelled cells enable either *in vitro* detection by staining for iron to produce ferric ferrocyanide (Prussian blue) or *in vivo* detection using MRI visualisation, due to the selective shortening of T_2-relaxation time, leading to a hypo-intense (dark) signal. MRI can be used to evaluate the cells engraftment, the time course of cell migration and their survival in the targeted tissue[97].

Magnetofection is a simple and highly efficient transfection method that uses magnetic fields to concentrate particles containing nucleic acid into target cells. Magnetofection is based on three steps: formulation of a magnetic vector, its addition to the medium covering cultured cells and application of a magnetic field in order to direct the vector towards the target cells. The simplest approach to form magnetic derivatives of nucleic acids employs magnetic nanocomposites covered with charged biocompatible polymers, which enable the formation of ionic complexes. Transfected cells can be separated from the non-transfected ones using an appropriate magnetic separation technique[98,99].

Also eukaryotic microbial cells have been efficiently magnetically modified, for example, using different types of water-based magnetic fluids. Living cells of baker's yeasts (*S. cerevisiae*) modified with tetramethylammonium hydroxide stabilised ferrofluid (see Figure 8.10) exhibited good magnetic response and enabled efficient decomposition of hydrogen peroxide and conversion of sucrose into glucose and fructose, due to the presence of active intracellular catalase and invertase.

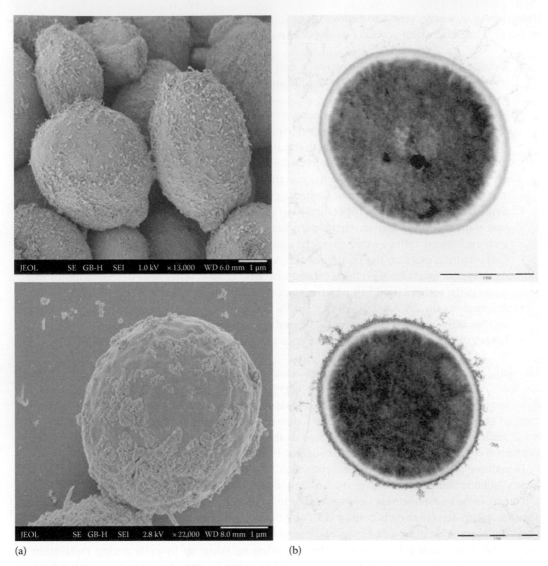

(a)   (b)

FIGURE 8.10   (a) SEM micrographs of ferrofluid modified *S. cerevisiae* cells showing attached magnetic nanoparticles and their aggregates on the cell surface (bars: 1 μm). (b) – TEM micrographs of *S. cerevisiae* cells (bars: 1 μm). Top – native cell, bottom – ferrofluid modified cell with attached magnetic iron oxide nanoparticles on the cell wall. (Reproduced from Safarikova, M., Maderova, Z., and Safarik, I., *Food Res. Int.*, 42, 521–524, 2009. With permission.)

Ferrofluid modification of yeast cells performed under the gentle conditions did not lead to the substantial changes in intracellular enzyme activities. This is an example of inexpensive, non-toxic, magnetically responsive whole cells biocatalyst, which could find interesting applications in biotechnology[46].

Dead baker's yeast cells (*S. cerevisiae*)[100], brewer's yeast cells (*S. cerevisiae* subsp. *uvarum*)[101], fodder yeast cells (*Kluyveromyces fragilis*)[102] and algae cells (*Chlorella vulgaris*)[103] were magnetically modified with perchloric acid stabilised magnetic fluid. Detailed analysis has confirmed the presence of isolated single-domain maghemite nanoparticles on the cell surface although a little amount of agglomerates was also present[104]. Magnetically responsive microbial cells could adsorb high amounts of organic xenobiotics (e.g. water-soluble dyes), as well as heavy metal ions[105] and radionuclides[106].

8.4.4.2 Magnetic Separation and Labelling of Prokaryotic Cells

One of the most important tasks in food, clinical, veterinary and environmental microbiology is the detection of important pathogenic bacteria in various matrices (e.g. food, clinical samples, soil, water, mud, etc.). Standard microbiology procedures for their detection usually require four stages and at least four different growth media; hence, the total time from sampling the analysed material to obtaining a result can be measured in days. One of the possibilities for shortening the isolation and detection period is to replace the selective enrichment stage (usually taking 24 h) with a non-growth-related procedure. This can be achieved by specific IMS of the target bacteria directly from the sample or the pre-enrichment medium (Figure 8.11). Isolated cells can then be identified by standard microbiology, molecular biology and microscopy techniques. IMS can be effectively combined with the PCR; the main purpose of IMS is to remove the PCR-inhibitory compounds from a sample without loss of sensitivity through dilution and the concentration of target cells. The oligonucleotide primers should be specific either for the target genus (e.g. detection of different strains of *Salmonella*) or for the individual strain of interest[38].

IMS is not only faster but also usually gives higher number of positive samples. Also, sublethally injured microbial cells can be isolated using IMS. Several types of commercially available immunomagnetic particles for pathogens detection are available, such as Dynabeads anti-*Salmonella*, Dynabeads anti-*E. coli* O157, Dynabeads EPEC/VTEC O26, Dynabeads EPEC/VTEC O103, Dynabeads EPEC/VTEC O111, Dynabeads EPEC/VTEC O145, Dynabeads anti-*Legionella* and Dynabeads anti-*Listeria* (all from Invitrogen).

8.4.5 BIOASSAYS BASED ON MAGNETIC NANOPARTICLES

In the area of bioassays, magnetically responsive (nano)composite particles have several advantages over the use of conventional solid-phase supports, such as a microplate. First, magnetic particles provide a relatively large surface area compared to a single microwell. This allows for the utilisation of a higher concentration of capture Abs enhancing the sensitivity of the assay. Second, Ab-coated magnetic particles are freely suspended in the antigen-containing solution and thus contact between Abs and antigens is facilitated by agitation of the mixture; this results in more rapid reaction rates and reduced assay time. Third, isolation of the magnetic particles containing antigen–Ab complexes can be easily and rapidly performed using a simple magnetic separator; this approach allows for the direct separation of the target analyte from the test sample, which is extremely beneficial when complex mixtures are analysed. By sequestering the analyte with the magnetic particles, contaminated materials can be removed by washing with minimal loss of

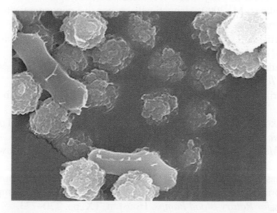

FIGURE 8.11 Electron microscopy of *Legionella pneumophila* bound to immunomagnetic beads (Dynabeads My One Streptavidin (Invitrogen) with bound biotinylated polyclonal anti-*Legionella* Ab). (Reproduced from Reidt, U., Geisberger, B., Heller, C., and Friedberger, A., *J. Lab. Autom.*, 16, 157–164, 2011. With permission.)

the target material. The overall effect of this procedure will reduce the non-specific background reactivity leading to enhanced sensitivity and specificity. Finally, immunomagnetic particles are capable of capturing small amounts of analytes from a large, dilute sample volume. Thus, when the magnetic particles are collected the total amount of analyte will be effectively concentrated[39]. Detailed information about bioassays based on the use of magnetic nanoparticles is given in another chapter of this book.

8.5   CONCLUSIONS

Different types of nanomaterials, among them magnetically responsive nanomaterials, have enormous potential to significantly influence different areas of biosciences, medicine, biotechnology, environmental technology, etc. where dozens of extremely important applications, both *in vitro* and *in vivo*, have already been described. Further progress, especially in *in vitro* area could be expected if cost-effective biocompatible magnetic particles would become available. Extremely inexpensive magnetic nanocomposites and microparticles for large-scale biotechnology and environmental applications (e.g. separation of biologically active compounds from agricultural wastes or removal of xenobiotics from wastewater) are necessary. Specific applications require specific magnetic particles covered by appropriate biocompatible molecules in a precisely predefined way (e.g. quantity, orientation, geometric arrangement). Safety and biocompatibility studies of magnetically responsive materials, in particular long-term toxicity studies, have to be carried out. The collaboration of scientists from different fields (e.g. materials and polymer science, chemistry, biochemistry and molecular biology, cell biology, biotechnology, magnetic characterisation, etc.) is necessary. It can be expected that the potential of magnetic nanomaterials will expand in the future.

ACKNOWLEDGEMENTS

This research was supported by the Ministry of Education of the Czech Republic (project OC 09052 – Action COST MP0701) and by the Ministry of Industry and Trade of the Czech Republic (Project No. 2A-1TP1/094).

REFERENCES

1. Gao, Y. 2005. Biofunctionalization of magnetic nanoparticles. In: *Nanotechnologies for the Life Sciences Vol 1* (Kumar, C.S.S.R., Ed.). Wiley-VCH, Weinheim, Germany, pp. 72–98.
2. Safarik, I. and Safarikova, M. 2009. Magnetically responsive nanocomposite materials for bioapplications. *Solid State Phenomena* 151, 88–94.
3. Berry, C.C. and Curtis, A.S.G. 2003. Functionalisation of magnetic nanoparticles for applications in biomedicine. *Journal of Physics D: Applied Physics* 36, R198–R206.
4. Arruebo, M., Fernandez-Pacheco, R., Ibarra, M.R. and Santamaria, J. 2007. Magnetic nanoparticles for drug delivery. *Nano Today* 2, 22–32.
5. Safarik, I. and Safarikova, M. 2009. Magnetic nano- and microparticles in biotechnology. *Chemical Papers* 63, 497–505.
6. Safarik, I. and Safarikova, M. 2004. Magnetic techniques for the isolation and purification of proteins and peptides. *Biomagnetic Research and Technology* 2, Article No. 7.
7. Svoboda, J. 2004. *Magnetic Methods for the Treatment of Minerals*. Kluwer Academic Publishers, New York.
8. Mitamura, Y., Takahashi, S., Kano, K., Okamoto, E., Murabayashi, S., Nishimura, I. and Higuchi, T.A. 2008. Sealing performance of a magnetic fluid seal for rotary blood pumps. *16th Congress of the International Society for Rotary Blood Pumps*. Houston, TX, pp. 770–773.
9. Lubbe, A.S., Alexiou, C. and Bergemann, C. 2001. Clinical applications of magnetic drug targeting. *Journal of Surgical Research* 95, 200–206.
10. Latorre, M. and Rinaldi, C. 2009. Applications of magnetic nanoparticles in medicine: Magnetic fluid hyperthermia. *Puerto Rico Health Sciences Journal* 28, 227–238.

11. Safarik, I. and Safarikova, M. 2002. Magnetic nanoparticles and biosciences. *Monatshefte für Chemie* 133, 737–759.
12. Kciuk, M. and Turczyn, R. 2006. Properties and application of magnetorheological fluids. *Journal of Achievements in Materials and Manufacturing Engineering* 18, 127–130.
13. Johnsen, S. and Lohmann, K.J. 2008. Magnetoreception in animals. *Physics Today* 61, 29–35.
14. Schüler, D. (Ed.). 2007. *Magnetoreception and Magnetosomes in Bacteria.* Springer, Berlin, Germany.
15. Lowenstam, H.A. 1962. Magnetite in denticle capping in recent chitons (Polyplacophora). *Bulletin of the Geological Society of America* 73, 435–438.
16. Waren, A., Bengtson, S., Goffredi, S.K. and Van Dover, C.L. 2003. A hot-vent gastropod with iron sulfide dermal sclerites. *Science* 302, 1007.
17. Bazylinski, D.A., Frankel, R.B. and Konhauser, K.O. 2007. Modes of biomineralization of magnetite by microbes. *Geomicrobiology Journal* 24, 465–475.
18. Arakaki, A., Nakazawa, H., Nemoto, M., Mori, T. and Matsunaga, T. 2008. Formation of magnetite by bacteria and its application. *Journal of the Royal Society Interface* 5, 977–999.
19. Medeiros, S.F., Santos, A.M., Fessi, H. and Elaissari, A. 2011. Stimuli-responsive magnetic particles for biomedical applications. *International Journal of Pharmaceutics* 403, 139–161.
20. Laurent, S., Forge, D., Port, M., Roch, A., Robic, C., Elst, L.V. and Muller, R.N. 2008. Magnetic iron oxide nanoparticles: Synthesis, stabilization, vectorization, physicochemical characterizations, and biological applications. *Chemical Reviews* 108, 2064–2110.
21. Lin, C.R., Chu, Y.M. and Wang, S.C. 2006. Magnetic properties of magnetite nanoparticles prepared by mechanochemical reaction. *Materials Letters* 60, 447–450.
22. Krishnan, K.M. 2010. Biomedical nanomagnetics: A spin through possibilities in imaging, diagnostics, and therapy. *IEEE Transactions on Magnetics* 46, 2523–2558.
23. Gervald, A.Y., Gritskova, I.A. and Prokopov, N.I. 2010. Synthesis of magnetic polymeric microspheres. *Russian Chemical Reviews* 79, 219–229.
24. Philippova, O., Barabanova, A., Molchanov, V. and Khokhlov, A. 2011. Magnetic polymer beads: Recent trends and developments in synthetic design and applications. *European Polymer Journal* 47, 542–559.
25. Zheng, B.Z., Zhang, M.H., Xiao, D., Jin, Y. and Choi, M.M.F. 2010. Fast microwave synthesis of Fe_3O_4 and Fe_3O_4/Ag magnetic nanoparticles using Fe^{2+} as precursor. *Inorganic Materials* 46, 1106–1111.
26. Berger, P., Adelman, N.B., Beckman, K.J., Campbell, D.J., Ellis, A.B. and Lisensky, G.C. 1999. Preparation and properties of an aqueous ferrofluid. *Journal of Chemical Education* 76, 943–948.
27. Massart, R. 1981. Preparation of aqueous magnetic liquids in alkaline and acidic media. *IEEE Transactions on Magnetics* 17, 1247–1248.
28. Kroll, E., Winnik, F.M. and Ziolo, R.F. 1996. In situ preparation of nanocrystalline γ-Fe_2O_3 in iron(II) cross-linked alginate gels. *Chemistry of Materials* 8, 1594–1596.
29. Salazar-Alvarez, G., Muhammed, M. and Zagorodni, A.A. 2006. Novel flow injection synthesis of iron oxide nanoparticles with narrow size distribution. *Chemical Engineering Science* 61, 4625–4633.
30. Kluchova, K., Zboril, R., Tucek, J., Pecova, M., Zajoncova, L., Safarik, I., Mashlan, M., Markova, I., Jancik, D., Sebela, M., Bartonkova, H., Bellesi, V., Novak, P. and Petridis, D. 2009. Superparamagnetic maghemite nanoparticles from solid-state synthesis – Their functionalization towards peroral MRI contrast agent and magnetic carrier for trypsin immobilization. *Biomaterials* 30, 2855–2863.
31. Lockman, P.R., Mumper, R.J., Khan, M.A. and Allen, D.D. 2002. Nanoparticle technology for drug delivery across the blood-brain barrier. *Drug Development and Industrial Pharmacy* 28, 1–13.
32. Molday, R.S. and Mackenzie, D. 1982. Immunospecific ferromagnetic iron-dextran reagents for the labeling and magnetic separation of cells. *Journal of Immunological Methods* 52, 353–367.
33. Korneva, G., Ye, H.H., Gogotsi, Y., Halverson, D., Friedman, G., Bradley, J.C. and Kornev, K.G. 2005. Carbon nanotubes loaded with magnetic particles. *Nano Letters* 5, 879–884.
34. Fonnum, G., Johansson, C., Molteberg, A., Morup, S. and Aksnes, E. 2005. Characterisation of Dynabeads® by magnetization measurements and Mossbauer spectroscopy. *Journal of Magnetism and Magnetic Materials* 293, 41–47.
35. Safarik, I., Safarikova, M., Weyda, F., Mosiniewicz-Szablewska, E. and Slawska-Waniewska, A. 2005. Ferrofluid-modified plant-based materials as adsorbents for batch separation of selected biologically active compounds and xenobiotics. *Journal of Magnetism and Magnetic Materials* 293, 371–376.
36. Guisan, J.M. (Ed.). 2006. *Immobilization of Enzymes and Cells.* Humana Press, Totowa, NJ.
37. Safarik, I., Safarikova, M. and Forsythe, S.J. 1995. The application of magnetic separations in applied microbiology. *Journal of Applied Bacteriology* 78, 575–585.
38. Safarik, I. and Safarikova, M. 1999. Use of magnetic techniques for the isolation of cells. *Journal of Chromatography B* 722, 33–53.

39. Yu, H., Raymonda, J.W., McMahon, T.M. and Campagnari, A.A. 2000. Detection of biological threat agents by immunomagnetic microsphere-based solid phase fluorogenic- and electro-chemiluminescence. *Biosensors and Bioelectronics* 14, 829–840.

40. Lee, J.H., Yigit, M.V., Mazumdar, D. and Lu, Y. 2010. Molecular diagnostic and drug delivery agents based on aptamer-nanomaterial conjugates. *Advanced Drug Delivery Reviews* 62, 592–605.

41. Chiu, T.C. and Huang, C.C. 2009. Aptamer-functionalized nano-biosensors. *Sensors* 9, 10356–10388.

42. Hermanson, G.T. 1996. *Bioconjugate Techniques*. Academic Press, New York.

43. Safarik, I. and Safarikova, M. 1997. Overview of magnetic separations used in biochemical and biotechnological applications. In: *Scientific and Clinical Applications of Magnetic Carriers* (Hafeli, U., Schutt, W., Teller, J. and Zborowski, M., Eds.). Plenum Press, New York, pp. 323–340.

44. Qi, D., Deng, Y., Liu, Y., Lin, H., Deng, C., Li, Y., Zhang, X., Yang, P. and Zhao, D. 2008. Development of core–shell magnetic mesoporous SiO_2 microspheres for the immobilization of trypsin for fast protein digestion. *Journal of Proteomics and Bioinformatics* 1, 346–358.

45. De Cuyper, M. and Joniau, M. 1990. Immobilization of membrane enzymes into magnetizable, phospholipid bilayer-coated, inorganic colloids. *Progress in Colloid and Polymer Science* 82, 353–359.

46. Safarikova, M., Maderova, Z. and Safarik, I. 2009. Ferrofluid modified *Saccharomyces cerevisiae* cells for biocatalysis. *Food Research International* 42, 521–524.

47. Liao, M.H. and Chen, D.H. 2001. Immobilization of yeast alcohol dehydrogenase on magnetic nanoparticles for improving its stability. *Biotechnology Letters* 23, 1723–1727.

48. Namdeo, M. and Bajpai, S.K. 2009. Immobilization of α-amylase onto cellulose-coated magnetite (CCM) nanoparticles and preliminary starch degradation study. *Journal of Molecular Catalysis B: Enzymatic* 59, 134–139.

49. Liao, H.D., Chen, D., Yuan, L., Zheng, M., Zhu, Y.H. and Liu, X.M. 2010. Immobilized cellulase by polyvinyl alcohol/Fe_2O_3 magnetic nanoparticle to degrade microcrystalline cellulose. *Carbohydrate Polymers* 82, 600–604.

50. Kuroiwa, T., Noguchi, Y., Nakajima, M., Sato, S., Mukataka, S. and Ichikawa, S. 2008. Production of chitosan oligosaccharides using chitosanase immobilized on amylose-coated magnetic nanoparticles. *Process Biochemistry* 43, 62–69.

51. Heegaard, N.H.H. 1999. Microscale characterization of the structure-activity relationship of a heparin-binding glycopeptide using affinity capillary electrophoresis and immobilized enzymes. *Journal of Chromatography A* 853, 189–195.

52. Pan, C.L., Hu, B., Li, W., Sun, Y., Ye, H. and Zeng, X.X. 2009. Novel and efficient method for immobilization and stabilization of β-D-galactosidase by covalent attachment onto magnetic Fe_3O_4-chitosan nanoparticles. *Journal of Molecular Catalysis B: Enzymatic* 61, 208–215.

53. Rossi, L.M., Quach, A.D. and Rosenzweig, Z. 2004. Glucose oxidase–magnetite nanoparticle bioconjugate for glucose sensing. *Analytical and Bioanalytical Chemistry* 380, 606–613.

54. Zhu, H., Pan, J., Hu, B., Yu, H.L. and Xu, J.H. 2009. Immobilization of glycolate oxidase from *Medicago falcata* on magnetic nanoparticles for application in biosynthesis of glyoxylic acid. *Journal of Molecular Catalysis B: Enzymatic* 61, 174–179.

55. Mak, K.H., Lu, C.Y., Kuan, I.C. and Lee, S.L. 2009. Immobilization of *Pseudomonas cepacia* lipase onto magnetic nanoparticles for biodiesel production. *ISESCO Science and Technology Vision* 5, 19–23.

56. Xie, W.L. and Ma, N. 2009. Immobilized lipase on Fe_3O_4 nanoparticles as biocatalyst for biodiesel production. *Energy and Fuels* 23, 1347–1353.

57. Saiyed, Z.M., Sharma, S., Godawat, R., Telang, S.D. and Ramchand, C.N. 2007. Activity and stability of alkaline phosphatase (ALP) immobilized onto magnetic nanoparticles (Fe_3O_4). *Journal of Biotechnology* 131, 240–244.

58. Xu, X.Q., Deng, C.H., Yang, P.Y. and Zhang, X.M. 2007. Immobilization of trypsin on superparamagnetic nanoparticles for rapid and effective proteolysis. *Journal of Proteome Research* 6, 3849–3855.

59. Berensmeier, S. 2006. Magnetic particles for the separation and purification of nucleic acids. *Applied Microbiology and Biotechnology* 73, 495–504.

60. Dias, A.M.G.C., Hussain, A., Marcos, A.S. and Roque, A.C.A. 2011. A biotechnological perspective on the application of iron oxide magnetic colloids modified with polysaccharides. *Biotechnology Advances* 29, 142–155.

61. Dunnill, P. and Lilly, M.D. 1974. Purification of enzymes using magnetic bio-affinity materials. *Biotechnology and Bioengineering* 16, 987–990.

62. Horak, D., Babic, M., Jendelova, P., Herynek, V., Trchova, M., Pientka, Z., Pollert, E., Hajek, M. and Sykova, E. 2007. D-Mannose-modified iron oxide nanoparticles for stem cell labeling. *Bioconjugate Chemistry* 18, 635–644.

63. Fu, B., Zhao, P., Yu, H.C., Huang, J.W., Liu, J. and Ji, L.N. 2009. Magnetic polymer nanospheres immobilizing metalloporphyrins. Catalysis and reuse to hydroxylate cyclohexane with molecular oxygen. *Catalysis Letters* 127, 411–418.

64. Ke, S.J. and Ji, J. 2007. Preparation and bacterial separation function of norvancomycin hydrochloride-immobilized magnetic nanoparticles. *Chemical Journal of Chinese Universities-Chinese* 28, 26–28.

65. Babic, M., Horak, D., Trchova, M., Jendelova, P., Glogarova, K., Lesny, P., Herynek, V., Hajek, M. and Sykova, E. 2008. Poly(L-lysine)-modified iron oxide nanoparticles for stem cell labeling. *Bioconjugate Chemistry* 19, 740–750.

66. Choi, J., Lee, J.I., Lee, Y.B., Hong, J.H., Kim, I.S., Park, Y.K. and Hur, N.H. 2006. Immobilization of biomolecules on biotinylated magnetic ferrite nanoparticles. *Chemical Physics Letters* 428, 125–129.

67. Muller-Schulte, D. and Brunner, H. 1995. Novel magnetic microspheres on the basis of poly(vinyl alcohol) as affinity medium for quantitative detection of glycated hemoglobin. *Journal of Chromatography A* 711, 53–60.

68. Inada, Y., Matsushima, A., Takahashi, K. and Saito, Y. 1990. Polyethylene glycol-modified lipase soluble and active in organic solvents. *Biocatalysis* 3, 317–328.

69. Inada, Y., Matsushima, A., Kodera, Y. and Nishimura, H. 1990. Polyethylene glycol(PEG)-protein conjugates: Application to biomedical and biotechnological processes. *Journal of Bioactive and Compatible Polymers* 5, 343–364.

70. Takahashi, F., Sakai, Y. and Mizutani, Y. 1997. Immobilized enzyme reaction controlled by magnetic heating: γ-Fe_2O_3-loaded thermosensitive polymer gels consisting of *N*-isopropylacrylamide and acrylamide. *Journal of Fermentation and Bioengineering* 83, 152–156.

71. Sakai, Y., Tamiya, Y. and Takahashi, F. 1994. Enhancement of ethanol formation by immobilized yeast containing iron powder or Ba-ferrite due to eddy current or hysteresis. *Journal of Fermentation and Bioengineering* 77, 169–172.

72. Safarik, I. and Safarikova, M. 2000. Biologically active compounds and xenobiotics: Magnetic affinity separations. In: *Encyclopedia of Separation Science* (Wilson, I.D., Adlard, R.R., Poole, C.F. and Cook, M.R., Eds.). Academic Press, London, U.K., pp. 2163–2170.

73. Yavuz, C.T., Prakash, A., Mayo, J.T. and Colvin, V.L. 2009. Magnetic separations: From steel plants to biotechnology. *Chemical Engineering Science* 64, 2510–2521.

74. Schuster, M., Wasserbauer, E., Ortner, C., Graumann, K., Jungbauer, A., Hammerschmid, F. and Werner, G. 2000. Short cut of protein purification by integration of cell-disrupture and affinity extraction. *Bioseparation* 9, 59–67.

75. Franzreb, M., Siemann-Herzberg, M., Hobley, T.J. and Thomas, O.R.T. 2006. Protein purification using magnetic adsorbent particles. *Applied Microbiology and Biotechnology* 70, 505–516.

76. Hanyu, N., Nishio, K., Hatakeyama, M., Yasuno, H., Tanaka, T., Tada, M., Nakagawa, T., Sandhu, A., Abe, M. and Handa, H. 2009. High-throughput bioscreening system utilizing high-performance affinity magnetic carriers exhibiting minimal non-specific protein binding. *Journal of Magnetism and Magnetic Materials* 321, 1625–1627.

77. Safarik, I. and Safarikova, M. 2010. Magnetic affinity separation of recombinant fusion proteins. *Hacettepe Journal of Biology and Chemistry* 38, 1–7.

78. Shi, R.B., Wang, Y.C., Hu, Y.L., Chen, L. and Wan, Q.H. 2009. Preparation of magnetite-loaded silica microspheres for solid-phase extraction of genomic DNA from soy-based foodstuffs. *Journal of Chromatography A* 1216, 6382–6386.

79. Park, J.S., Park, J.H., Na, S.Y., Choe, S.Y., Choi, S.N. and You, K.H. 2001. Rapid isolation of genomic DNA from normal and apoptotic cells using magnetic silica resins. *Journal of Microbiology and Biotechnology* 11, 890–894.

80. Fritz, J.S., Dumont, P.J. and Schmidt, L.W. 1995. Methods and materials for solid-phase extraction. *Journal of Chromatography A* 691, 133–140.

81. Berrueta, L.A., Gallo, B. and Vicente, F. 1995. A review of solid phase extraction: Basic principles and new developments. *Chromatographia* 40, 474–483.

82. Safarikova, M. and Safarik, I. 1999. Magnetic solid-phase extraction. *Journal of Magnetism and Magnetic Materials* 194, 108–112.

83. Chen, L., Wang, T. and Tong, J. 2011. Application of derivatized magnetic materials to the separation and the preconcentration of pollutants in water samples. *TrAC Trends in Analytical Chemistry*, 30, 1095–1108.

84. Safarikova, M., Kibrikova, I., Ptackova, L., Hubka, T., Komarek, K. and Safarik, I. 2005. Magnetic solid phase extraction of non-ionic surfactants from water. *Journal of Magnetism and Magnetic Materials* 293, 377–381.

85. Safarik, I. and Safarikova, M. 2002. Detection of low concentrations of malachite green and crystal violet in water. *Water Research* 36, 196–200.
86. Komarek, K., Safarikova, M., Hubka, T., Safarik, I., Kandelova, M. and Kujalova, H. 2009. Extraction of alkylphenols and nonylphenol mono- and diethoxylates from water using magnetically modified adsorbents. *Chromatographia* 69, 133–137.
87. Safarikova, M., Lunackova, P., Komarek, K., Hubka, T. and Safarik, I. 2007. Preconcentration of middle oxyethylated nonylphenols from water samples on magnetic solid phase. *Journal of Magnetism and Magnetic Materials* 311, 405–408.
88. Safarik, I. and Safarikova, M. 2007. Magnetically modified microbial cells: A new type of magnetic adsorbents. *China Particuology* 5, 19–25.
89. Apel, M., Heinlein, U.A.O., Miltenyi, S., Schmitz, J. and Campbell, J.D.M. 2007. Magnetic cell separation for research and clinical applications. In: *Magnetism in Medicine* (Andra, W. and Nowak, H., Eds.). Wiley-VCH, Weinheim, Germany, pp. 571–595.
90. Chen, W., Shen, H.B., Li, X.Y., Jia, N.Q. and Xu, J.M. 2006. Synthesis of immunomagnetic nanoparticles and their application in the separation and purification of CD34$^+$ hematopoietic stem cells. *Applied Surface Science* 253, 1762–1769.
91. Miltenyi, S., Muller, W., Weichel, W. and Radbruch, A. 1990. High gradient magnetic cell separation with MACS. *Cytometry* 11, 231–238.
92. Shao, J.-Y., Ting-Beall, H.P. and Hochmuth, R.M. 1998. Static and dynamic lengths of neutrophil microvilli. *Proceedings of the National Academy of Sciences of the United States of America* 95, 6797–6802.
93. Dirican, E.K., Ozgun, O.D., Akarsu, S., Akin, K.O., Ercan, O., Ugurlu, M., Camsari, C., Kanyilmaz, O., Kaya, A. and Unsal, A. 2008. Clinical outcome of magnetic activated cell sorting of non-apoptotic spermatozoa before density gradient centrifugation for assisted reproduction. *Journal of Assisted Reproduction and Genetics* 25, 375–381.
94. Makker, K., Agarwal, A. and Sharma, R.K. 2008. Magnetic activated cell sorting (MACS): Utility in assisted reproduction. *Indian Journal of Experimental Biology* 46, 491–497.
95. Hess, C., Sadallah, S., Hefti, A., Landmann, R. and Schifferli, J.-A. 1999. Ectosomes released by human neutrophils are specialized functional units. *The Journal of Immunology* 163, 4564–4573.
96. Luers, G.H., Hartig, R., Mohr, H., Hausmann, M., Fahimi, H.D., Cremer, C. and Volkl, A. 1998. Immunoisolation of highly purified peroxisomes using magnetic beads and continuous immunomagnetic sorting. *Electrophoresis* 19, 1205–1210.
97. Sykova, E. and Jendelova, P. 2005. Magnetic resonance tracking of implanted adult and embryonic stem cells in injured brain and spinal cord. *Annals of the New York Academy of Sciences* 1049, 146–160.
98. Sanchez-Antequera, Y., Mykhaylyk, O., van Til, N.P., Cengizeroglu, A., de Jong, J.H., Huston, M.W., Anton, M., Ian, C.D., Johnston, I.C.D., Pojda, Z., Wagemaker, G. and Plank, C. 2011. Magselectofection: An integrated method of nanomagnetic separation and genetic modification of target cells. *Blood* 117, e171–e181.
99. Schillinger, U., Brill, T., Rudolph, C., Huth, S., Gersting, S., Krotz, F., Hirschberger, J., Bergemann, C. and Plank, C. 2005. Advances in magnetofection – Magnetically guided nucleic acid delivery. *Journal of Magnetism and Magnetic Materials* 293, 501–508.
100. Safarik, I., Ptackova, L. and Safarikova, M. 2002. Adsorption of dyes on magnetically labeled baker's yeast cells. *European Cells and Materials* 3 (Suppl. 2), 52–55.
101. Safarikova, M., Ptackova, L., Kibrikova, I. and Safarik, I. 2005. Biosorption of water-soluble dyes on magnetically modified *Saccharomyces cerevisiae* subsp. *uvarum* cells. *Chemosphere* 59, 831–835.
102. Safarik, I., Rego, L.F.T., Borovska, M., Mosiniewicz-Szablewska, E., Weyda, F. and Safarikova, M. 2007. New magnetically responsive yeast-based biosorbent for the efficient removal of water-soluble dyes. *Enzyme and Microbial Technology* 40, 1551–1556.
103. Safarikova, M., Pona, B.M.R., Mosiniewicz-Szablewska, E., Weyda, F. and Safarik, I. 2008. Dye adsorption on magnetically modified *Chlorella vulgaris* cells. *Fresenius Environmental Bulletin* 17, 486–492.
104. Mosiniewicz-Szablewska, E., Safarikova, M. and Safarik, I. 2010. Magnetic studies of ferrofluid-modified microbial cells. *Journal of Nanoscience and Nanotechnology* 10, 2531–2536.
105. Yavuz, H., Denizli, A., Gungunes, H., Safarikova, M. and Safarik, I. 2006. Biosorption of mercury on magnetically modified yeast cells. *Separation and Purification Technology* 52, 253–260.
106. Ji, Y.Q., Hu, Y.T., Tian, Q., Shao, X.Z., Li, J.Y., Safarikova, M. and Safarik, I. 2010. Biosorption of strontium ions by magnetically modified yeast cells. *Separation Science and Technology* 45, 1499–1504.
107. Reidt, U., Geisberger, B., Heller, C. and Friedberger, A. 2011. Automated immunomagnetic processing and separation of *Legionella pneumophila* with manual detection by sandwich ELISA and PCR amplification of the *ompS* gene. *Journal of Laboratory Automation* 16, 157–164.

Ivo Safarik was born in 1954 and studied at the Institute of Chemical Technology in Prague, Czech Republic, from which he graduated in 1978. He obtained his PhD and associate professor degrees from the same Institute in 1984 and 1993, respectively. He was awarded Doctor of Science Degree in 2001 and became full professor of biochemistry in 2008. He is the deputy director of the Institute of Nanobiology and Structural Biology of GCRC the Academy of Sciences of the Czech Republic. He published over 140 scientific journal papers and book chapters. In recent years, his research interest was mainly focussed on different aspects of biomagnetic research and technology, nano-technology and nanobiotechnology. Detailed information can be found on the author's web page (http://www.nh.cas.cz/people/safarik/).

Mirka Safarikova graduated from the Institute of Chemical Technology, Prague, in 1980, where she also obtained a PhD degree in Biochemistry in 1999. In 2001, she was appointed senior scientist and since then has been the Head of the Department of Nanobiotechnology (formerly Department of Biomagnetic Techniques) in the Institute of Nanobiology and Structural Biology of GCRC (formerly Institute of Systems Biology and Ecology) of the Academy of Sciences of the Czech Republic. Her main field of interest is the preparation of biocompatible magnetic nano- and microparticles, magnetically responsive biocomposites, adsorbents and microbial cells and application of these materials for the isolation, determination, detection and immobilisation of biologically active compounds, xenobiotics and microbial cells.

9 Magnetic Nanoparticles in Immunoassays

Peter Hawkins and Richard Luxton

CONTENTS

9.1 IMMUNOASSAYS

9.1.1 INTRODUCTION

Since the early 1960s, antibodies have been used as an analytical reagent to capture a target molecule (antigen) from a sample forming an antibody–antigen complex also known as an immune complex that is then quantified either directly or through the detection of a label (or reporter molecule) conjugated to the antibody or antigen. The specific binding properties of antibody to antigen molecules have driven the development and implementation of many thousands of antibody-based analytical methods called immunoassays. There are three critical factors in the development of an immunoassay: the antibody, the label and the detection technology; and in this chapter, we will introduce the concept of immunoassays before discussing how nanometre- or micrometre-sized magnetic particles (or beads) and magnetic detection technologies can be used in an immunoassay.

9.1.2 ANTIBODIES AND ANTIGENS

Antibodies are generated by the immune system to provide protection against specific infections and toxins (antigens). In humans, there are five classes of antibody: immunoglobulin (Ig) G, IgA, IgM, IgE and IgD. The IgG class of antibody is most commonly used in an immunoassay, and Figure 9.1 shows a schematic diagram of an IgG molecule.

An antibody contains two identical light chains and two identical heavy chains held together by disulphide bonds to form a Y-shaped molecule. Each light chain and heavy chain is made from several domains with two domains making the light chains and four domains making each heavy chain. IgM and IgE have an additional domain in the heavy chain. It is the structure of the heavy chain that gives rise to the classification of the different types of antibodies. The Y-shaped part containing the two arms of the molecule is called the Fab region, and the terminal domains of the heavy and light chains forming the arms have variable domains at the tips with an antigen-binding site on each tip. The remaining protein domains on the heavy and light chains are constant for all antibody molecules of that class and form the 'tail' of the antibody called the Fc fragment. This describes the basic structure of IgG, IgE and IgD antibodies with two antigen-binding sites. The IgM antibody has five basic antibody units linked together, giving a total of 10 antigen-binding sites on the antibody, and the IgA antibody has two basic antibody units with four antigen-binding sites. Each antigen-binding site is formed from a small number of amino acids found on the variable region of the light chain and variable region of the heavy chain and binds specifically to a region on the target antigen that has a complementary shape and charge structure known as an epitope. One antigen-binding site will bind with one epitope. Large molecules may have many different epitopes to which different antibodies can bind and because an antibody has at least two antigen-binding sites results in the antibody being able to cross-link two antigen molecules, and at the optimum

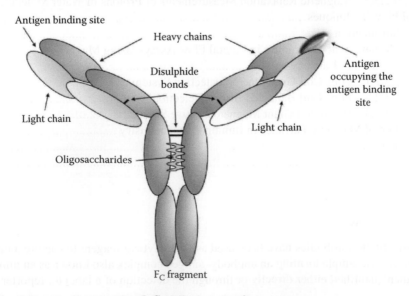

FIGURE 9.1 Schematic diagram of an IgG antibody molecule.

ratio of antibody to antigen concentration large networks of cross-linked molecules form insoluble precipitates, which is the basis of several label-free antibody-based tests and some techniques using magnetic particle labels.

Traditionally, antibodies are raised in animals through the inoculation of the target antigen mixed with an adjuvant to stimulate the host immune response. After a few weeks, a blood sample is taken from the animal, and the antibody purified from the serum. As many different immune cells are recruited into the immune response within the host animal, a wide range of antibody molecules directed against the target are produced with different chemical and physical characteristics resulting in a spectrum of antibodies with different binding energy against the target, and this type of antibody is known as polyclonal antibody.

In the 1970s, the introduction of monoclonal antibody techniques enabled the production of highly specific antibody from a single clone of antibody-producing cells. This allowed the production of identical antibody with high specificity and well-characterised binding properties. The use of protein engineering has enabled the development of techniques to produce antibodies and antibody fragments in the laboratory.

In most assays, the antigen is the analyte, and for larger molecules, such as proteins, the production of antibodies to the analyte molecule is generally not a problem, but small molecules, such as small hormones or drugs, do not generally provoke an immune response. In this case, the small molecule is conjugated to a larger protein such as bovine serum albumin that is immunogenic and causes an immune response producing antibodies, some of which would be directed against the target molecule on the protein. The antibody of interest is obtained after several purification stages to remove antibody directed against the carrier protein.

The binding of antibody (Ab) to antigen (Ag) is an equilibrium reaction governed by the law of mass action.

$$Ab + Ag \underset{k_d}{\overset{k_a}{\rightleftharpoons}} AbAg$$

Antibodies interact with a particular antigen through the formation of non-covalent bonds such as electrostatic bonds, hydrophobic bonds and dispersion forces known as Van der Waals and London forces. The rate of formation of immune complex is related to the association (k_a) and dissociation (k_d) constants, and the ratio k_a/k_d is the affinity of the antibody to the particular antigen. Antibody affinity is a measure of the strength of the interaction between antibody and antigen. High-affinity antibody binds to the antigen strongly with less dissociation of the immune complex, and low-affinity antibody binds less strongly with higher levels of dissociation of the immune complex.

9.1.3 IMMUNOASSAYS

Immunoassay is the term used to describe analytical techniques that use antibodies to confer specificity to the assay and usually refer to techniques employing a label. The label refers to any physical, chemical or biochemical entity that can be easily detected and quantified, which is incorporated into an antibody molecule (or antigen molecule for some assay systems). In many immunoassay systems, it is essential to separate label that has been incorporated into the antibody–antigen complex (bound label) from label that has not reacted (unbound or free label). Assays systems requiring a separation step are known as heterogeneous assays, whereas those that do not require the separation of free from bound label prior to measurement are known as homogeneous assays. The choice of antibody used in an immunoassay is important as it dictates the specificity and dynamic range of the assay, and the sensitivity of the assay is a function of antibody affinity and the ability to detect the label that depends on the type of label used in the system and on the instrumentation used to detect and quantify the signal coming from the label. Using high-affinity antibody and appropriate labels with high-sensitivity detection technologies, many of the newer immunoassay systems can obtain attomolar (1×10^{-18} M) sensitivies with some assays reporting zeptomolar (1×10^{-21} M) sensitivities (a concentration of 1 zM is equivalent to about 600 molecules L^{-1}).

The development of immunoassays has enabled a plethora of highly sensitive assays to be devised for medical applications. The use of high-affinity antibodies has allowed highly selective methods to be developed that can detect small numbers of molecules in complex mixtures such as serum or whole blood samples. The only criterion required in the development of a new assay is that an antibody can be made against the target antigen (analyte). Much research is undertaken to discover new biomarkers for a disease that will then be used to develop a new assay for the disease: for example, new biomarkers for the detection of bladder cancer[1] and ovarian cancer[2] have been reported.

Several hundreds of immunoassays have been developed to measure molecules in biological samples, including hormones, cytokines, enzymes, signalling molecules, cancer proteins and infectious agents. Many of the new immunoassay systems are designed to give rapid results at the point-of-care (POC). Immunoassay protocols have been incorporated into biosensor technology, which have also been designed to facilitate POC testing. In some recent examples, Scott[3] describes a POC test for tetanus, and Caygill et al.[4] review the use of biosensors to detect human pathogens. Also, many assays have been described to detect cancer, for example, the measurement of prostate-specific antigen (PSA) for prostate cancer[5]. In addition to the measurement of many soluble molecules found in blood, the use of labelled antibodies can be applied to the detection of cell surface proteins and intracellular proteins using tissue slices or cell preparations as described in Section 9.4.

9.1.4 IMMUNOASSAY FORMATS

When immunoassays were first introduced, the reactions took place in a test tube, and the antibody and target antigens were in solution. Following the reaction between the antibody and antigen, larger complexes were left in suspension in the test tube, but often the complex was soluble. In order to measure the label that was associated with the immune complex, it was necessary to perform labour-intensive separation stages to remove the unreacted components from the complex and associated label. This was often achieved using a second antibody to react with the first antibody and form an insoluble complex that was centrifuged to the bottom of the test tube, and the unreacted components were discarded. The insoluble complex was washed by re-suspending it in a wash solution and then centrifuged again, and the label associated with the immune complex quantified after the wash solution had been discarded. This process often took many hours, and there was a possibility of losing some of the immune complex during the washing steps. In order to overcome the problems associated with separating the bound and free label, solid-phase techniques were introduced. Almost all modern immunoassay techniques rely on solid-phase technology that has the antibody immobilised on a solid surface (solid phase), the most common type of solid phase being the inside of a plastic test tube or the well of a microtiter plate, and other types of materials used as a solid phase include nitrocellulose membranes, glass slides, polystyrene beads, latex particles and magnetic particles. The advantage of forming the immune complex on a surface is that un-reacted components of the assay can be easily washed away and new reagents added.

Solid-phase immunoassays for larger molecules use two antibodies where one antibody immobilised on the reaction surface is called the capture antibody and the other antibody in solution is conjugated with a label and is called the detector antibody. This type of immunoassay is called a sandwich (or non-competitive) assay (Figure 9.2), because the target antigen (analyte) is captured between the immobilised antibody and the detector antibody forming a sandwich on the surface.

The sample is mixed with labelled antibody and placed in contact with the capture antibody immobilised on a surface (Figure 9.2a). After a period of incubation, the capture antibody binds with antigen, capturing it on the surface, and the detector antibody also binds to the antigen forming the top layer of the 'sandwich' (Figure 9.2b). The un-reacted components are washed away, leaving the label on the detector antibody, associated with the captured antigen to be quantified (Figure 9.2c). The greater the concentration of target in the sample the greater the signal generated giving a positive slope dose–response curve.

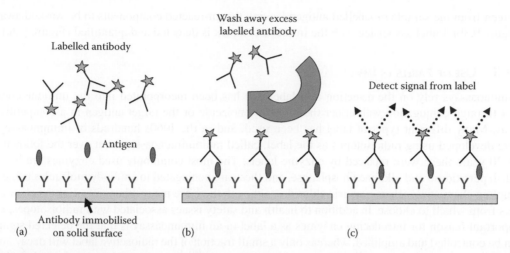

FIGURE 9.2 Diagram showing the steps in a sandwich or non-competitive immunoassay. (a) A solution containing analyte (antigen) and labelled capture antibody is introduced to a solid surface with immobilised capture antibody. (b) A sandwich assay forms when the immobilised antibody bonds to one epitope of the analyte and the labelled antibody to another and excess labelled antibody is removed. (c) The analyte is quantified by determining the amount of bound label.

Smaller molecules have only a single epitope and cannot cross-link antibody molecules, and hence the 'sandwich' structure described earlier cannot be formed. In this case, a competitive technique is used where a labelled antigen is used to compete for a limited number of antigen-binding sites. In this assay, a fixed concentration of labelled antigen is mixed with the sample containing an unknown concentration of antigen, and the mixture is applied to a surface on which an antibody has been immobilised. In this type of assay, the number of antigen-binding sites is optimised to allow competition between the sample antigen and labelled antigen. In the situation where there is little antigen in the sample, the majority of antigen captured by the immobilised antibody will be the labelled antigen. Conversely, if there is a high concentration of the target antigen in the sample, the immobilised antibody will bind predominantly target antigen and low levels of labelled antigen. Consequently, the typical dose–response for a competitive assay is an inverse sigmoidal curve. Figure 9.3 shows a diagram of the assay protocol.

Sample is mixed with a fixed amount of labelled antigen and added to antibody immobilised on a surface (Figure 9.3a), and after an incubation period, the immobilised antibody captures either

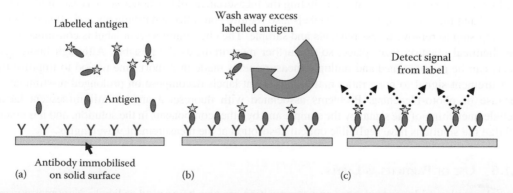

FIGURE 9.3 Diagram showing the steps in a competitive immunoassay. (a) A solution containing analyte (antigen) and labelled antigen is introduced to a solid surface with immobilised capture antibody. (b) Antibody binds to the antigen and the labelled antigen through the single epitope of the antigen and excess labelled antigen is removed. (c) The concentration of analyte is inversely related to the intensity of signal from the bound label.

antigen from the sample or labelled antigen leaving the un-reacted components to be washed away (Figure 9.3b). Label associated with the immune complex is detected and quantified (Figure 9.3c).

9.1.5   Use of Labels in Immunoassays

Immunoassays rely on the detection of a label that has been incorporated into the immune complex through conjugation with either the antibody molecule or the target antigen in a competitive assay. Many different types of label has been used, and in the 1960s hundreds of immunoassays were developed using radioisotopes as the label, called radioimmunoassays, but over the following 10–20 years these were replaced by enzyme labels. The most commonly used enzymes are horse radish peroxidase and alkaline phosphatase, and these are conjugated to antibody molecules using a coupling chemistry such as cyanuric chloride[6], although there are many other cross-linking chemistries from which to choose. In addition to health and safety issues associated with radioisotopes, an important reason for introducing enzymes as a label in an immunoassay is that the detected signal can be controlled and amplified, whereas only a small fraction of the radioactive label will decay and be counted in a typical measurement period leading to measurement errors. Enzymes on the other hand can be detected through a biochemical reaction changing a colourless substrate to a coloured product, and one enzyme molecule can generate many product molecules, bringing an amplification factor into the detection process. They are also very versatile as many different substrates can be used to create products of varying colours or other substrates that generate fluorescent and luminescent products. The main disadvantage of an enzyme label is that enzymes are protein molecules with some being of a similar size to an antibody molecule, which may result in steric hindrance affecting the immunoreactivity of the labelled antibody to the antigen, and the conjugation chemistry can also inactivate the protein (enzyme or antibody) involved in the cross-linking reaction in some cases[7]. The enzyme reaction needs to be performed under controlled conditions with fixed pH and temperature, and as an enzyme is a biological molecule, there may be issues of stability and environmental effects such as the presence of metal ions or other molecules that inhibit enzyme activity. Radioisotopes, however, do not suffer from problems with steric hindrance or environmental effects.

In recent years, fluorescent and chemiluminescent molecules have been introduced as alternative labels to overcome the limitations of enzyme labels because they are much smaller than enzymes and can be covalently coupled to the antibody or antigen molecule. The labels produce easily detectable, characteristic photons when the fluorescent label is exposed to short wavelength light and the chemiluminescent labels undergoes a chemical reaction usually with hydrogen peroxide. Several fluorescent labels have been described, for example, fluorescein, rhodamine derivatives and europium chelates, and examples of chemiluminescent labels include luminol and acridinium esters. These labels are used widely in medical diagnostics, including the measurement of hormones such as estadiol[8], cancer makers[9] and infectious agents[10]. Both labels also require controlled environments and often include a washing step to remove excess reagents and sample. The chemiluminescent label is consumed when the chemical reaction takes place, so no further measurements are possible. Although fluorescent labels can be re-stimulated and multiple measurements made in a short time period to improve the measurement signal to noise ratio, many fluorescent labels decompose on prolonged re-stimulation because of photo-bleaching. Problems associated with fluorescent and chemiluminescent labels include quenching of the signal by the sample and by other components in the solution, and it is essential that the solutions have negligible optical absorption at the measurement wavelengths.

9.1.6   Use of Particles as Labels

Particles with a wide range of sizes and composition are also being used as labels. Nanometre-sized particles such as quantum dots are smaller than an antibody molecule and can be conjugated to an antibody in a similar way[11,12]. In contrast, many immunoassay systems use larger particles with sizes from hundreds of nanometres to tens of micrometres with many antibody molecules immobilised on their surface, and detection of the immune reaction can be through antibody–antigen interactions, causing agglutination of the particles or through cross-linking of the particle to a sensor surface. Many different materials have been used as particles in immunoassays, and the most common are latex particles, colloidal metals such as gold, polystyrene beads and magnetic particles. Latex particles coated with suitable antibody are used in a simple and rapid technique for detecting and typing bacteria in which the antibody binds to different sites on the bacteria causing cross-linking between the particles and bacteria, the formation of agglutinates and a visible precipitate. Similarly, red blood cells act as particles in haemagglutination techniques used for blood group typing when soluble antibodies cross-link the red cells by binding to antigens on the surface of the red cell.

9.1.7 Coupling Proteins to Particle Surfaces

An advantage of using particles is that there are many techniques and materials available to enable the surface of the particles to be functionalised, and, as they have a large surface area, many antibody molecules can be immobilised on their surfaces, which will enhance the sensitivity of the assay. There are three commonly used chemistries to immobilise proteins to the surface of a particle, and these are to surfaces that have been modified to contain tosyl, carboxy or amino groups (Figure 9.4).

Whichever chemistry is chosen, it is important that the labelling methods do not inactivate the biological reactivity of the antibody. Strategies for antibody attachment to solid phases have been reviewed by Nisnevitch and Firer[13]. Tosyl activated particles allow coupling of proteins through the formation of a covalent bond between the tosyl group and a primary amino group or through sulfhydryl groups on the protein. This is a simple method which is well suited to immobilising fragile proteins as binding takes place at a neutral to high pH at room temperature or 37°C without

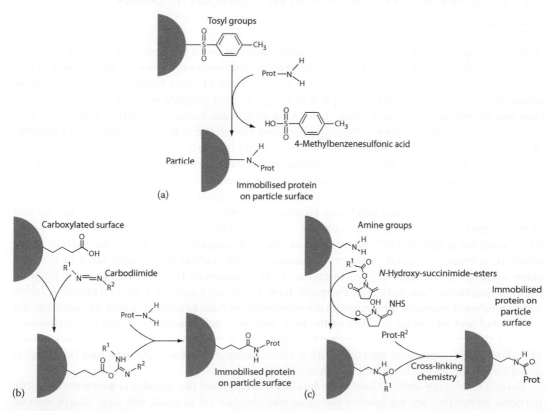

FIGURE 9.4 Diagram showing main methods for immobilising antibodies to the surface of a particle. (a) Tosyl activated surface, (b) carboxylated surface, (c) amino surface.

further chemistry. Figure 9.4a shows the reaction chemistry and the formation of the covalent link through an amine bond. In comparison, other methods of covalent binding require the addition of a cross-linking agent or initial surface activation chemistry. Carboxylated particles form a covalent bond with the antibody through a primary amino group as shown in Figure 9.4b. Although the use of carboxylated particles allows a rapid binding chemistry, they require activation using a carbodiimide derivative before the protein can be immobilised on the surface. Particles containing amino groups on their surface can be activated using a range of chemistries to introduce various cross-linking reagents, allowing covalent coupling through a variety of chemical groups such as aldehyde and ketone groups using sodium periodate oxidation to immobilise glycoproteins. Activation of the particle surface using a cross-linking molecule such as a (N-hydroxy-succinimidyl)-ester is commonly used to immobilise proteins, but the particular chemical group through which the bond is formed will depend on the ester used; cross-linking the amino group on the particle surface to sulfhydryl, carboxyl and hydroxyl groups on the protein. Figure 9.4c shows a generalised reaction scheme for particles with surface amino groups.

Many chemical techniques are available to cross-link antibodies directly to the surface of a particle, and other techniques use indirect biochemical immobilisation where a linker molecule such as protein A or protein G is covalently bonded to the surface and the antibody is captured through a high-affinity bond to its Fc fragment, allowing the antigen-binding sites to be in the correct orientation to facilitate capturing the analyte. Another example of a biological binding pair is biotin–streptavidin that has a binding energy close to that of a covalent bond. Streptavidin is immobilised onto the particle surface which then bonds with the biotinylated antibody and binds it to the surface.

9.2   IMMUNOASSAYS USING MAGNETIC PARTICLES AS LABELS

9.2.1   INTRODUCTION

Antibody-coated magnetic particles have many applications in biomedical investigations, such as the extraction of tumour cells from bone marrow, the isolation of viruses and organelles and in molecular genetics and immunoassays[14]. An advantage of using magnetic particles in these applications is that they can be manipulated using external magnets to assist in analyte extraction and sample purification because there is no significant magnetic material in most samples to interfere with the operations. With the particles held at the side of container by an external magnet, the liquid in the container can be removed and the particles washed leaving the purified target on the particle. In extraction processes, where the target molecule is required for further investigations, the analyte can be released from the particle surface by breaking the immunological bond using a low-pH environment or a chaotropic reagent such as ammonium thiocyanate. In immunoassays, traditional labels can be used to quantify the antigen captured on the surface of the magnetic particle such as an enzyme[15], a fluorescent or a chemiluminescent label as used in an assay for α-fetoprotein[16]. Many commercial multi-analyte analysers use the surface of the magnetic particle as a solid phase in immunoassays and a chemiluminescent label in the final stage to quantify the analyte (e.g. 'Immulite' from Diagnostics Products Corporation, Siemens Medical Solutions, New York and 'Fastpack' from Qualigen, Carlsbad, CA). In a recent development, combined magnetic and quantum dot nanoparticles have been used, where the quantum dot is the label and advantage is taken of the high luminescence quantum efficiency and photostability of the quantum dots[17], and in another development by Argento Diagnostics, Twickenham, Middlesex, U.K., magnetic particles with a silver nanoparticle label are used and the assay is quantified using an electrochemical technique[18,19].

Several measuring methods have been developed over the last two decades in which the magnetic particles themselves are the label in the assay and this has led to rapid, one step, assays with no washing step. The absence of magnetic material in most samples also means that there is nothing to interfere in the determination of bound magnetic labels in the assay so no, or very little, sample

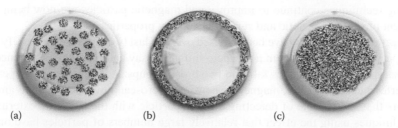

(a)   (b)   (c)

FIGURE 9.5 Diagram showing how magnetic nanoparticles are imbedded into the structures of the three main types of micrometre-sized paramagnetic particles used in immunoassays. In (a) the magnetic nanoparticles are uniformly distributed throughout the polymer matrix, in (b) they are concentrated near the surface and in (c) they are concentrated in the core of the polymer matrix.

preparation is necessary, and in sandwich assays, the magnetic particle label together with detector antibody and analyte can be pulled towards the capture antibody on the reaction surface using an external magnetic field to speed up the assay[20]. Another advantage is that, unlike chemiluminescent and radioactive labels, the magnetic particles labels are not consumed in the assay, and unlike fluorescent labels, they do not readily degrade so the samples may be stored and retested later if necessary. An important requirement of the magnetic particles is that they do not have significant permanent magnetic moments that would cause clusters or clumps of particles to form in the absence of an external field. Magnetic particles are available from several manufacturers and typically contain ferromagnetic nanoparticles (e.g. 20–30nm diameter magnetite, Fe_3O_4) embedded in a polymer matrix. Nanoparticles of magnetite with this diameter are single domain, lose their ferromagnetic properties and become superparamagnetic. Figure 9.5 shows the three main ways the magnetic nanoparticles are distributed in the micrometre-sized paramagnetic particles used in immunoassays.

The particle shown in Figure 9.5a has a uniform distribution of magnetic material throughout a polymer matrix forming a regular-sized sphere, similar to the fruit distributed in a fruitcake. The particle shown in Figure 9.5b has magnetic material distributed on the surface of a polymer sphere and then stabilised in a polymer coat, similar to the skin of an orange. The particle shown in Figure 9.5c has a large single mass of magnetic material embedded in polymer forming a sphere, similar to the stone in a plum. Many particles with diameters about less than 100nm have very irregular shapes.

Manufacturing particles with consistent magnetic, shape and surface-chemistry properties is difficult, particularly for particles with diameters of less than 1 µm, and this can seriously affect the reproducibility of measurements. Although the magnetic properties of bulk quantities of particular particles from a manufacturer might be consistent, the magnetic properties of individual particles may vary widely with some particles actually being non-magnetic[21]. Particles with the same nominal diameter from different manufacturers and particles of various sizes from the same manufacturer often produce widely different responses in a detector[22]. In addition, changes in the shape and surface chemistry of the particles can result in differing numbers of antibody molecules binding to the surface and reacting with the antigen. Variations in the properties of individual particles are less likely to lead to poor reproducibility in measurement techniques using relatively large numbers of magnetic particles because the average magnetic property and number of capture antibody molecules per particle will be similar to those of the particles in the bulk. Poorer reproducibility may arise in techniques that use a few particles, and for this reason, larger, micrometre-sized, particles are preferred in some immunoassays because of their more uniform magnetic characteristics and for their ability to immobilise a larger amount of capture antibody on the surface[23]. A problem experienced, however, in sandwich assays using larger particles as labels that have been coated with many capture antibody molecules is that each particle can capture several analyte molecules that then bind to a large number of immobilised capture antibody on the reaction surface resulting in a rapid depletion of available antibody on the surface and a non-linear response of the measurement system, so in these applications it is usually preferable to use smaller magnetic particles.

Manufacturing techniques continue to improve, and magnetic particles are now being made with more consistent physical, magnetic and surface chemistry properties.

Several ingenious techniques have been devised over the last two decades to quantify the amount of magnetic particles used as label in a magneto-immunoassay and from this, the concentration of analyte in a sample. The techniques fall into two broad groups: techniques using superconducting quantum interference, Hall effect, magnetoresistive or micro-cantilever-based force-amplified biological sensors that are capable of detecting a single particle with the aid of an external magnetic field; and techniques using the effect that relatively large numbers of particles have on the inductance of a coil or on the nuclear magnetic resonance (NMR) spectrum of hydrogen nuclei in water. An immunoassay format that has become very popular because of its ease of use is the lateral flow, and investigations are underway in using magnetic particles as labels in these assays along with suitable measurement techniques.

9.2.2 Techniques Capable of Detecting a Single Magnetic Particle Label

9.2.2.1 Superconducting Quantum Interference Device

The superconducting quantum interference device (SQUID) has the highest sensitivity and largest dynamic range of all magnetic field measuring devices and is capable of detecting a single particle if it is at the same temperature as the SQUID (77 K for a high-temperature SQUID). The concept of using a SQUID in magnetically labelled immunoassays was introduced by Kotitz et al.[24] A SQUID responds only to magnetic fields so at some point in the measurement an external magnetic field has to be applied when working with superparamagnetic nanoparticles, because they do not have a permanent magnetic moment. In an example of the application of a SQUID (Figure 9.6), a sandwich assay is formed between a capture antibody immobilised on a reaction surface, an analyte molecule and a second capture antibody immobilised on a superparamagnetic particle.

Excess particles remain unbound and suspended in the solution. Applying an external magnetic field causes the particles to align along the magnetic field. When the field is removed, the particles remain aligned for a brief period before randomising again. This short, but decaying, period of self-magnetisation is described as magnetic remanence and the field produced by the remanence can be

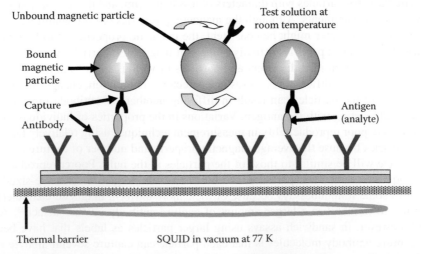

FIGURE 9.6 Schematic diagram illustrating how the magnetisation in unbound particles is rapidly randomised through Brownian motion, and the slower decay in remanence measured by the SQUID is from Néel relaxation in the bound particles.

detected by a SQUID. The decay in the remanence in superparamagnetic nanoparticles was studied by Néel, and the relaxation time for the decay is given by

$$\tau_N = \tau_0 \exp\left(\frac{KV}{kT}\right),$$

where

$\tau_0 \approx 1\,\text{ns}$ is the damping or excitation time

K is the anisotropy constant of the bulk material with volume V

k is the Boltzmann's constant

T is the absolute temperature

The particles are also subject to thermal motion, and the unbound, but not the bound particles, undergo Brownian rotations that cause a rapid randomisation of the magnetised particles. The decay constant for the Brownian rotation is given by

$$\tau_B = \frac{3V_H\eta}{kT},$$

where

V_H is the hydrodynamic particle volume

η is the viscosity of the medium

Typically, $\tau_N \approx 1\,\text{s}$ and $\tau_B \approx 1\,\text{ms}$ for superparamagnetic particles at room temperature, so the magnetic field measured by the SQUID is almost entirely from the decaying remanence in the bound particles involved in the assay, and the technique is capable of distinguishing between bound and unbound magnetic label, which means that it can be used in one-step immunoassays where very little or no sample preparation is required.

The sensitivity of a SQUID falls rapidly with distance, so the sample has to be placed close to it, which creates an enormous technical problem because of the large temperature difference that has to be maintained between the SQUID and the sample. Chemla et al.[25] reported a SQUID 'microscope' in which a high-temperature SQUID in a vacuum at a temperature of 77 K was separated by 40 μm from a sample at room temperature and had a detection limit of $5 \pm 2 \times 10^3$ particles for particles with core sizes of $35 \pm 5\,\text{nm}$. In the development of this technique, Tsukamoto et al.[26] detected 30 pg of Fe_3O_4 particles, and Enpuku et al.[27] reported a detection limit of 2 amol for an immunoassay for the antibody IgE using a 25 nm diameter Fe_3O_4 magnetic label. Despite these major developments, the SQUID still remains an expensive technique.

9.2.2.2 Hall Effect Sensors

In the classic Hall effect, an external magnetic field is applied at right angles to the direction of flow of current I in a conductor, causing the charge carriers carrying the current to be deflected to the side of the conductor producing an electric field and measurable Hall voltage, V_H at right angles to both I and H.

For a thin film probe, the Hall voltage is given by[28]

$$V_H = \frac{IHR_H}{t},$$

where

t is the effective thickness of the probe across which V_H is measured

R_H is the Hall coefficient for the conductor

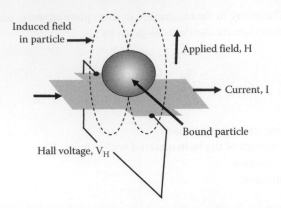

FIGURE 9.7 Schematic diagram showing how a single magnetic particle bound to the surface of a Hall sensor can be detected if the diameter of the particle is about the same as the dimensions of the sensor.

The equation shows that reducing t increases the sensitivity of the probe. Besse et al.[29] reported a Hall probe made using standard complementary metal oxide semiconductor (CMOS) technology with an effective area of $2.4 \times 2.4\,\mu m^2$ that was able to detect a single $2.8\,\mu m$ diameter magnetic particle because its cross-sectional area was comparable to that of the probe, so its presence on the probe significantly changed the magnetic field, H, reaching the probe (Figure 9.7).

The Hall voltage produced by the probe is very small, so synchronous electronic circuitry and pulsed magnetic fields are necessary to improve the signal-to-noise ratio. In the development of the technique, Aytur et al.[30] described a 1024-element array of micrometre-sized Hall sensors that were coated with capture antibody. An immunological assay for the antigen capture of purified mouse IgG and the detection of human anti-dengue virus IgG in clinical serum were presented. A controlled magnetic field was used to remove non-specifically bound magnetic particles from the sensor surface.

9.2.2.3 Magnetoresistive Sensors

Several systems for use in magnetically labelled immunoassays and DNA analysis and based on magnetoresistance (MR) sensors have been reported in recent years. MR is the property of a material to change its electrical resistance when an external magnetic field is applied to it. The sensitivity of an MR material is given by the ratio:

$$MR_R\,(\%) = \frac{|R_0 - R_S| \times 100}{R_S}$$

where
   R_0 is the resistance of the material before a field has been applied
   R_S is the resistance of the material in a saturation magnetic field

Although the MR effect was discovered over 150 years ago, it is only in recent years that major applications have been found for it, which include readout sensors in magnetically recorded tapes and discs, non-volatile computer memories and proximity sensors, and these applications have encouraged investigations into ways of improving the sensitivity of the devices. Thin film technology is used now to make MR sensors with large values of MR_R using several layers of different combinations of metals, alloys and magnetic materials. The MR effect depends on both the strength of the magnetic field and the direction of the field relative to the current. There are now four distinct types of MR: ordinary, anisotropic, giant and colossal, which use different materials and have different mechanisms.

Ordinary MR applies mainly to non-magnetic metals, and the effect at low magnetic fields is very small, and although it can be large for high fields, there are currently no applications of sensors based on ordinary MR in magneto-immunoassays. Colossal MR occurs in La-Ca-Mn-O films with $MR_R \approx 99.9\%$ near room temperature[31], but as an intense field of 10 kOe is required to achieve this effect, it too has no applications at present in sensors used in magneto-immunoassays.

In anisotropic MR materials, the resistance of the material increases when the magnetic field is parallel to the current flow and decreases when the field is perpendicular. A commonly used construction for anisotropic MR devices has several thin layers containing Ni, Fe and other elements, resulting in sensors with values of $MR_R \approx 2\%$ at a saturation field of only 5–10 Oe[32], which makes them well suited for use in magnetic-recording read heads.

A method of detecting magnetic particles using an anisotropic MR ring-shaped sensor with an outer diameter of 5 μm and an inner diameter of 3.2 μm was described by Miller et al.[33] in what appears to an adaptation of the well-known Corbino anisotropic MR disc (Figure 9.8).

Electrical connections are made to the inner and outer edges of the MR ring, and a current, I, passes through the ring. When a magnetic field is applied along the axis of the ring, the current takes a spiral path through the ring because of the interaction between this field and the field created by the current. Placing a magnetic particle at the centre of the ring changes the field experienced by the ring because of the induced field in the particle, and the path taken by the current becomes radial, causing the resistance of the ring to decrease. The change in resistance was measured using an ac bridge. Maximum change in resistance is observed when the diameter of the particle is about the same as the inner diameter of the ring.

Giant magnetoresistive (GMR) sensors have been constructed with MR_R values of 10%–30% at saturation fields as low as 1 Oe[34]. The GMR effect was first described independently by Bailbich et al.[35] and Binasch et al.[36], and it occurs between two or more thin magnetic layers separated by a thin non-magnetic layer. Because spin-dependent scattering of the conduction electrons occurs at the interfaces between the layers, the resistance is a maximum when the magnetic moments of the layers are anti-parallel and a minimum when the magnetic moments are parallel. Different constructions have been used to make GMR sensors. The first and simplest design of GMR sensor was multilayered and consisted of a non-magnetic conducting layer (e.g. Cu) sandwiched between two exchange-coupled antiferromagnetic layers. When the directions of the magnetisations of the two magnetic layers are anti-parallel, electrons with spin orientations in one particular direction are scattered by one of the magnetic layers and electrons with spins in the opposite direction are scattered by the other magnetic layer, so all electrons are scattered as they pass through the layers and the sensor has a high resistance. When the magnetic layers are parallel, only those electrons with spins in one of the orientations are now scattered so the total number of scattered electrons reduces and the electrical resistance falls. Multilayer GMR sensors have values of MR_R of about 4%–9%[34].

Spin valve GMR sensors are similar in construction to the multilayer sandwich sensors but include additional layers. Several different constructions of spin valve sensors for use with

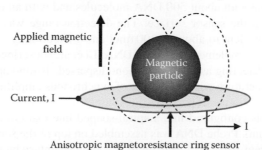

FIGURE 9.8 The current, I, flowing between the outer and inner edges of the MR ring takes a spiral path when the field is applied. The presence of a magnetic particle in the ring produces a radial flow of current and a fall in resistance.

magneto-immunoassays have been described, and some examples are presented in articles by Graham et al.[37–39] and Li et al.[40,41] The main difference between spin valve and multilayer GMR sensors is the addition of another magnetic layer on the top or the bottom of the pile, which couples with the adjacent magnetic layer and pins its magnetisation to a fixed direction but the magnetisation of the other layer is still free to rotate. These materials do not require the field from a current to achieve anti-parallel alignment or a strong antiferromagnetic exchange coupling to adjacent layers. Spin valve GMR sensors have values of MR_R of 10%–30% or higher and saturation fields of about 1 Oe[34].

A third type of GMR sensor is the magnetic tunnel junction or tunnelling MR, which was first reported by Miyazaki and Tezuka[42] and Moodera et al.[43] The main difference between this type of sensor and the spin-valve MR sensor is that the conducting Cu layer between the magnetic layers is replaced by a very thin insulating layer of an oxide of Al or Mg that acts as a tunnel barrier for the electrons and the sensitivity can be varied by changing the thickness of the insulating layer. The tunnel junction MR sensors are the most sensitive of the GMR sensors with typical MR_R of 20%–50%[32] but up to 220% in some cases[44]. The relatively high resistance of the sensors means that they are better suited to use in portable battery-operated instruments.

Several systems using GMR sensors have been developed for use in magneto-immunoassays. Baselt et al.[45] introduced the concept of adapting MR computer memory technology to fabricate millions of sensors on a chip to detect thousands of analytes using magnetic nanoparticles in immunoassays and DNA–DNA interactions, and this gave rise to the Bead ARay Counter (BARC), which was developed by the Naval Research Laboratory, Washington, and then evolved into BARC-III with 64 sensing zones[21]. Each multilayer GMR sensing zone is a trace 1.6 μm wide on a 4.0 μm pitch that follows a serpentine path within a 200 μm diameter circular sensing area, so that the total length of each MR trace is 8 mm. The magnetisation of the multilayer lies in the plane of the film. An alternating field with a frequency of 200 Hz and intensity of 80–400 Oe is applied normally to the sensor. A magnetic nanoparticle binding to a reaction surface on a sensing zone perturbs the magnetic field affecting the MR and changes its resistance, which is detected using a Wheatstone bridge and a synchronous amplifier. Although it is possible to make a small MR sensor capable of detecting a single magnetic nanoparticle, the approach used in the BARC is to have larger sensing areas so that more particles are detected because of the difficulty in targeting analyte molecules and magnetic particle labels to a very small sensor surface[23] and because a large number of particles captured on the surface will avoid problems caused by inconsistencies in the properties of individual particles as described earlier. A detailed description of the integration of the BARC into a complete instrument (called cBASS), including the fluidic system required, has been given in Reference [23]. A similar system is also being developed by Diagnostic Biosensors LLC, Minneapolis, MN[46].

To facilitate the rapid focusing and hybridisation of magnetically labelled target DNA with complementary probe DNA on a GMR chip bound on the surface of high-sensitivity spin-valve sensors $(2 \times 6 \mu m^2)$, current-carrying metallic lines have been added that tapered from 150 to 5 μm to produce local magnetic field gradients to guide the magnetic particles to the sensors[39]. Each 250 nm particle was functionalised with about 500 DNA molecules and with an estimated 70 DNA–DNA interactions per particle at the sensor surface. The detection range was about 140–14,000 DNA molecules per sensor equivalent to about 2–200 fmol cm^{-2}. No binding signals were observed for magnetically labelled non-complementary target DNA. Li et al.[41] described a submicron spin-valve GMR that is capable of detecting as few as 23 monodispersed, 16 nm-superparamagnetic magnetite particles at room temperature without the use of synchronous amplifiers, and the sensor signal increased linearly with the number of particles. Schotter et al.[47] described a GMR multilayer system consisting of 206 elements configured into a spiral-shaped line that covered the area of a typical DNA spot (diameter, 70 μm). Probe DNA was assembled on top of the sensor elements in different concentrations ranging from 16 pg μL^{-1} to 10 ng μL^{-1}. Complementary biotin-labelled analyte DNA was hybridised to the probe DNA at a concentration of 10 ng μL^{-1}. Several different commercially available magnetic particles were investigated as markers, and an assay was developed that was superior to a fluorescent marker technique at low levels of DNA.

A system for use in magnetic-particle labelled immunoassays and DNA analysis and based on spin-valve GMR technology is being developed by MagArray Inc., Sunnyvale, CA[48–50]. To overcome some of the problems experienced in earlier systems, this one uses very small magnetic particles (50 nm) with small magnetic signatures, which they have called NanoTags, in conjunction with MR sensors that have a very thin passivation layer of only 30 nm[51]. The small particles exhibit long-term suspension stability and excellent binding selectivity and do not clump because of their small magnetic signature. The very thin passivation layer means that particles bound to the surface of the MR are easily detected, whereas unbound particles are not, and this will lead to simple, one-step, homogeneous assays with no washing steps.

9.2.2.4 Micro-Cantilever-Based Force-Amplified Biological Sensor

A micro-cantilever-based force-amplified biological sensor uses a sandwich assay in which capture antibody to an analyte is bound to a piezoresistive micro-cantilever and is able to detect a single magnetic nanoparticle[52]. Particles coated with antibody are added to a test solution, and analyte molecules are captured by the antibody. The particles plus captured analyte are passed over the antibody on the cantilever, and the particles become attached to it through a sandwich assay formed between the analyte and the antibodies (Figure 9.9).

The tethered particles are attracted to a solenoid positioned below the cantilever by a diverging magnetic field created by a solenoid, which causes the cantilever to bend, and the assay is quantified by the change in resistance of the piezoresistive cantilever. Arrays of piezoresistive micro-cantilevers have been made using micro-engineering techniques and with each cantilever coated with different antibodies, so that several analytes can be quantified at the same time. Using this technique, Weizmann et al.[53] reported a detection limit of 7.1×10^{-20} M (71 zM) in a DNA analysis. There are several problems with this technique, including difficulties in coating the antibody on the cantilevers, directing the magnetic particle and analyte to the cantilevers, and the cantilevers are sensitive to vibrations and temperature changes.

9.2.3 Techniques Using Relatively Large Numbers of Magnetic Particles as Label

9.2.3.1 Use of a Maxwell–Wein Bridge to Detect Changes in Magnetic Permeability

One of the first measurement systems for use in immunoassays using a relatively large number of magnetic particles as label was reported by Kriz et al.[54], which was based on a Maxwell–Wein bridge (Figure 9.10).

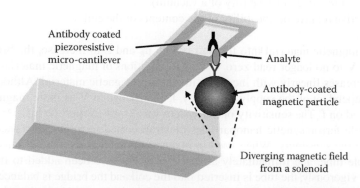

Antibody coated
piezoresistive
micro-cantilever

Analyte

Antibody-coated
magnetic particle

Diverging magnetic field
from a solenoid

FIGURE 9.9 Schematic diagram showing a micro-machined piezoresistive cantilever coated with a layer of capture antibody forming a sandwich assay with the analyte and an antibody-coated magnetic particle. The particle is attracted to an external solenoid, and a force is exerted on the cantilever producing a change in resistance.

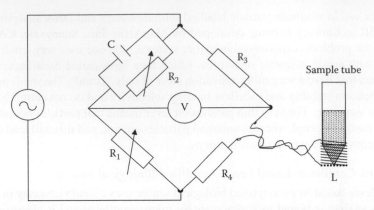

FIGURE 9.10   The basic Maxwell–Wein bridge circuit is balanced by adjusting R_1 and R_2. The immunoassay taking place in the sample tube results in magnetic label sinking to the bottom of the tube, causing the bridge to go off balance and producing a reading on V that is directly related to the quantity of bound label.

A sample coil with self-inductance, L, forms a part of the bridge circuit, which is balanced by adjusting resistors R_1 and R_2 until V reads zero, and under this condition, the impedances of the capacitive and inductive arms of the bridge must match so,

$$R_4 + j2\pi fL = R_1 R_3 \left\{ \frac{1}{R_2} + j2\pi fC \right\},$$

where f is the frequency of the sinusoidal supply. Separating the resistive (real) and reactive (imaginary) components, $R_4 R_2 = R_1 R_3$, which is the balance condition for a basic Wheatstone bridge, and $L = R_1 R_3 C$. This bridge circuit works well with coils having low values of Q but not at all with those having high values of Q. The inductance of the coil is given by

$$L = \left(\frac{\mu_R \mu_0 A}{l} \right) N^2,$$

where
   N is the number of turns in the coil of length l and cross sectional area A
   $\mu_0 = 4\pi \times 10^{-7}$ H m^{-1} (the permeability of a vacuum)
   μ_R is the effective relative permeability of the contents of the coil

Introducing magnetic material into the coil causes μ_R and L to increase, the bridge to become unbalanced and V to no longer read zero. For small additions of magnetic material in the coil, the reading of V increases linearly with increases in mass of magnetic material. Although the balance condition is independent of the supply frequency, f, the change in V as more magnetic material is added will depend on f. The sensitivity of the system was $21 \pm 4\,\mu$V μg^{-1} Fe mL^{-1}[54].

Specially made ferromagnetic nanoparticles (dextran-coated ferrofluid) are used in the instrument to improve the sensitivity. When used in magneto-immunoassays, the affinity reactions take place in a sample tube, and immediately after the reagents have been added to the tube and the contents shaken vigorously, the tube is inserted into the coil and the bridge is balanced. Power losses will occur in the coil caused by the high conductivity of the ionic solution in the tube and so reduce the Q of the coil and introduce an equivalent resistance that will contribute to R_4 in the bridge circuit, so for this reason, the bridge must be balanced for each test solution at the start of the assay. The bound particles sink to the bottom of the tube, which is positioned near the centre of the coil, causing L to increase and the bridge to go out of balance and producing a reading of V that is related to the number of magnetic particles involved in the assay. Using this technique, a sandwich assay for Concanavalin A has been demonstrated[55] with a range of 0–1.9 μM, a limit of detection at 250 nM and a relative standard deviation of 6.2% (n = 3). Comparative measurements using whole blood samples containing the inflammation marker C-reactive protein (CRP) between this technique and a standard hospital laboratory analyser produced a good linear correlation over the CRP concentration range 0–260 mg L^{-1} (y = 1.001x + 0.42, R^2 = 0.982, n = 50) with a limit of detection of 3 mg L^{-1} and a total imprecision (coefficient of variation) of 10.5%[56]. A more rapid assay for CRP in whole blood (5.5 min) was obtained by using micrometre-sized silica particles to conjugate with the magnetic nanoparticles to reduce the sedimentation time[57]. An instrument based on this technique is manufactured by LifeAssays, Lund, Sweden.

9.2.3.2 Resonant Coil Magnetometer

This device measures the change in resonant frequency of a coil with a self inductance L in parallel with a capacitor of capacitance C when a relatively large number of magnetic particles are placed inside the coil. The LC circuit (Figure 9.11a) has a resonant frequency of $f_0 = (2\pi(LC)^{1/2})^{-1}$.

The presence of the magnetic particles in the coil causes L to increase and f_0 to decrease, and the decrease is directly related to the number of magnetic particles in the coil[58]. When the LC circuit is incorporated into a Colpitts oscillator (Figure 9.11b), the output frequency of the oscillator is governed by the resonant frequency of the LC circuit, f_0, and can be conveniently read using a frequency meter. Putting magnetic particles in the coil causes the reactance of the coil to increase and the frequency of the oscillator to fall by an amount directly related to the number of magnetic particles. Using a helical coil, this arrangement is able to determine the quantity of magnetic particles in suspension of buffer solution over the range 0–9 mg of particles per gram of buffer solution to an accuracy of 10%. However, when used in an immunoassay using the magnetic particles as label, the Colpitts oscillator was found to be not stable enough to measure the small changes in frequency produced by the particles. An improvement in this circuit is to use a voltage-controlled oscillator to drive the LC circuit in conjunction with a phase-locked loop, making use of the fact that at resonance, the applied voltage and the current flowing into the circuit are in phase (Figures 9.11a and 9.12).

A phase-sensitive detector compares the phase of the voltages across the resistor, R, and produces an output related to the difference which, after passing through a low-pass filter, provides a dc error signal that is fed back to the oscillator and changes the output frequency until there is no phase difference in the voltages across R so the output frequency of the oscillator equals the resonant frequency of the LC circuit, f_0. This circuit works well with different designs of coils. In one design, a coil with an oval cross-section was used to determine the number of particles immobilised on polyethylene terephthalate strips, and this had a good linear response for 1×10^5 to at least 3.33×10^6 magnetic particles with diameters of 2.8 μm[59]. In a sandwich assay for the blood protein, human transferrin, a linear dose response was achieved with a detection limit of 260 fM[60].

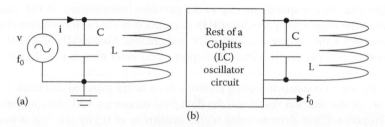

FIGURE 9.11 (a) When the source frequency matches the resonant frequency of the LC resonant circuit f_0, the current I is in phase with the applied voltage, V. Magnetic material placed in the coil causes L to increase and f_0 to decrease. (b) The output frequency of a Colpitts oscillator is controlled by an LC circuit.

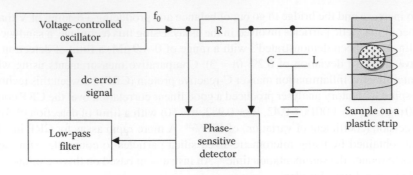

FIGURE 9.12 The phase-locked loop ensures that the frequency of the output of the oscillator always matches the resonant frequency of the LC circuit.

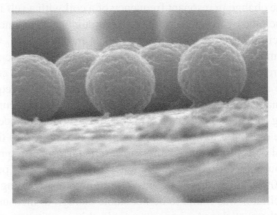

FIGURE 9.13 An EM image showing immobilised 2.8 μm diameter particles on a plastic strip in a sandwich assay.

Figure 9.13 shows an electron microscope image of the 2.8 μm diameter particles immobilised on the strip at the end of the sandwich assay.

The strands of antibody and antigen protein conjugate tethering the particles to the strip are clearly visible and show that many antibody–analyte–antibody complex links are used to immobilise each magnetic particle that can result in rapid saturation of the capture antibody on the reaction surface and a non-linear response.

A flat spiral measuring coil has been used so that the reaction surface at the bottom of a vessel can be placed on top of the coil and external magnets moved above and below the reaction vessel to pull unbound magnetic particle alternately towards and away from the surface in order to encourage mixing of the particles and to speed up the reaction at the surface[20]. Another advantage of using flat spiral coils in this arrangement is that the sensitivity of the measuring coil to the particles falls off rapidly with distance, so it responds mainly to the particles immobilised on the adjacent reaction surface and not to the un-reacted particles still in suspension in the buffer solution (Figure 9.14).

Sandwich assays for the cardiac markers, CRP and creatine kinase-MB (CKMB), were demonstrated with good responses over the clinically important concentration ranges. In a further development, flat spiral coils have been incorporated into the base of a reaction vessel[22,61], and this has led to a disposable biosensor constructed using printed circuit board material and thick film processing techniques to create the reaction vessel and the flat spiral measuring coil. An immunoassay for cardiac marker, Troponin I, was demonstrated with a sensitivity of 0.5 ng mL^{-1}. It is possible to have more than one sensing coil and associated reaction surfaces in a reaction vessel for simultaneous multi-analyte analysis in a sample, and Figure 9.15 shows an experimental vessel containing five separate coils.

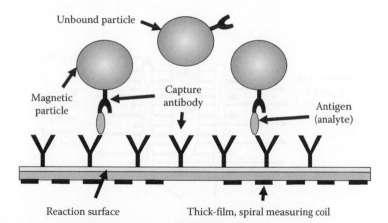

FIGURE 9.14 A schematic diagram showing the arrangement for a sandwich assay on a reaction surface on top of a measuring coil.

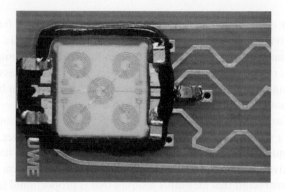

FIGURE 9.15 Photograph of an experimental reaction vessel containing five sensing coils for simultaneous multi-analyte analysis.

The whole of the electronics for the sensing circuitry can be obtained by adapting a readily available frequency modulated radio chip. An instrument based on this design is being developed by Randox Laboratories, County Antrim, U.K.

9.2.3.3 Methods Based on Magnetic Saturation Measurements

In a technique described by Meyer et al.[62] for determining a relatively large number of magnetic particles in an immunoassay, the sample is subjected to an intense alternating magnetic field so that the magnetisation of the particles become saturated, or nearly saturated, at the maxima of the field. Superparamagnetic particles based on magnetite become saturated at applied field strengths greater than about 2000 Oe[22]. The immunoassay is performed in a specially designed tube that is then placed at the centre of three sets of coils (Figure 9.16).

One set of coils provides an intense alternating magnetic field with a frequency, f_2, of 61 Hz, a second set produces a weaker field at a frequency, f_1, of 49.382 kHz and the third forms a differential pickup coil that is carefully balanced to about one part in 1000, so there is negligible direct induction at 49.382 kHz in the absence of a sample. The sample is positioned between one half of the pickup coil, and an out-of-balance signal is induced in the coil depends on the number of magnetic particles in the sample. When the intense field is at a maximum, the particles are saturated and the induced signal in the pickup coil falls to zero, or near zero, and when the field is zero, the induced signal is at a maximum. As the intense field has one positive and one negative maximum per cycle, the induced current in the pickup coil has a combined frequency of $f_1 + 2f_2$ with an intensity that

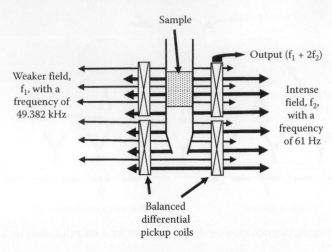

FIGURE 9.16   Simplified diagram of the equipment described by Meyer et al.[62]

depends on the number of particles. The driving currents for the coils at frequencies f_1 and f_2 are derived by frequency dividing the output of a common oscillator so they are in phase thus enabling a synchronous demodulating circuit to be used to extract the signal. The output of the device has a linear response over the range 0.12–1300 mg L^{-1} iron concentration. In an assay for CRP, the device had a linear detection range from 25 ng mL^{-1} to 2.5 μg mL^{-1}. An instrument based on this technique is manufactured by Senova GmbH, Weimar, Germany.

In an approach similar to the one described earlier, but using a simpler coil arrangement, Nikitin et al.[63] have developed a magnetic reader that can detect 100–1000 particles depending on the particular particles, that equates to 3 ng of Fe_3O_4 in a volume of 0.1 cm^3 and has a dynamic measurement range of almost five orders of magnitude[64]. In a further development of this approach[65], an instrument has been constructed that measures a signal that is proportional to the second derivative of the magnetisation curve of the magnetic particles. An ependorf tube containing the particles is positioned at the centre of two coils of wire powered by individual signal generators so the particles are exposed simultaneously to two alternating magnetic fields produced by the coils, one with a frequency of 0.025 Hz and the other with a frequency of 24.4 kHz. The amplitude of the high-frequency field is set at a fixed value between ±8.88 Oe, and the amplitude of the other field, H, is varied between ±452.4 Oe. A signal, proportional to the second derivative of the magnetisation, M, curve of the material, d^2M/dH^2, is then recorded when the amplitude of the high-frequency magnetic field is varied. It was shown that large magnetosomes biosynthesised by magnetotactic bacteria produce a different signal from small nanoparticles synthesised chemically. A change in the signal was also observed when interactions occurred between the magnetosomes and the nanoparticles. A detection scheme, capable of simultaneously detecting two biological entities, is described. An instrument based on this technique is being manufactured by Magnisense, San Francisco, CA.

9.2.3.4   Magnetic Relaxation Measurements of Protons in Water Molecules

NMR spectroscopy has been a scientific analytical tool for over 60 years and is used to determine the structure of organic molecules, and it is also the basis of magnetic resonance imaging (MRI). It is currently being adapted to work with magnetic particles to make a rapid homogeneous assay system on raw samples, including turbid samples and whole cell lysates without protein purification or other sample cleaning using spin–spin relaxation times (T_2) of protons in the nuclei of water molecules[66].

NMR is observed in any atom with an odd number of protons or neutrons in its nucleus, but the strongest effect is observed in the ^1H nucleus that has just one proton. The protons in the hydrogen nuclei spin with two possible quantised spin states of +1/2 and −1/2, and the spinning protons have magnetic moments because they are also charged. In the absence of a magnetic field, the two spin

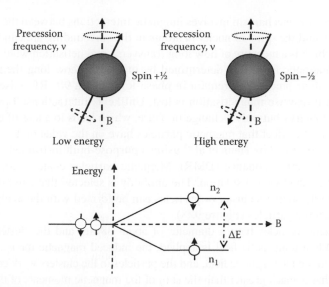

FIGURE 9.17 Protons have a precession frequency of ν in a magnetic field B and those with spin of +1/2 have a lower energy than those with spins of −1/2.

states have the same energy, but in a strong field B, a proton with a spin of +1/2 aligns in the direction of the field and has a Larmor precession about it, and a proton with a spin of −1/2 aligns in the opposite direction to the field (Figure 9.17).

The protons with spins of +1/2 have lower energies in the field than those with spins of −1/2. The Larmor frequency is given by $\nu = \gamma_p B/(2\pi)$, where $\gamma_p/(2\pi) = 42.57$ MHz T^{-1} (the gyromagnetic ratio for a proton in a H_2O molecule), and the difference in energy between the spin +1/2 and spin −1/2 nuclei is $\Delta E = h\nu$, where h = 6.626×10^{-34} J s (Planck's constant). If a proton with a spin of +1/2 is excited with radio frequency (RF) electromagnetic radiation having a frequency equal to the Larmor frequency, then it changes state and becomes aligned in a direction opposite to the field. Thus, when a sample of pure water is placed in a magnetic field of 1 T (10,000 Oe), an absorption is observed when an RF signal with a frequency of 42.57 MHz is applied as protons with a spin +1/2 move to the higher quantum energy state, 2.82×10^{-26} J. As ΔE is very small, many protons will have sufficient thermal energy to move between the energy states, and the relative populations of the higher (n_2) and lower (n_1) energy levels at room temperature are given by the Boltzmann law, $n_2/n_1 = \exp(-\Delta E/kT) \approx 0.99999$ where k = 1.38×10^{-23} J K^{-1} (Boltzmann's constant) and T is the absolute temperature, which means that at room temperature, the probability of observing an upward transition is only slightly greater than that for a downward transition—and the overall probability of observing a transition is quite small, so relatively large samples are usually required to obtain a measurable signal. RF radiation is emitted as the excited protons fall back to the lower energy level.

After excitation, there are two main relaxation processes that can return the populations of proton spins in the water molecules in the energy levels to their thermal equilibrium values. The first of these involves interactions between the proton spins and the lattice and has a (longitudinal) time constant of T_1, which is the time taken for the spin magnetisations to return to 63% of its equilibrium value and is fastest when the motion of the nucleus (rotations, translations or tumbling rates) matches that of the Larmor frequency. As a result, T_1 relaxation is dependent on B, which also determines the Larmor frequency, so longer T_1 times are associated with higher fields. It also determines the time required for a substance to become magnetised after first being placed in a magnetic field. Values of T_1 vary considerably depending on the nature of the material surrounding the protons in the water molecules, and determining these values is an important part of MRI (e.g. T_1 ranges from 1.5 to 2 s in fluids, from 400 ms to 1.2 s in water-based tissues and from 100 to 150 ms in fat-based tissues), and the values are used to produce T_1-weighted images.

The second relaxation mechanism involves magnetic interactions between the spins of protons in the water molecules and the spins of nuclei in atoms in the surrounding area and has a (transverse) time constant T_2, which is a measure of how long transverse magnetisation would last in a perfectly uniform external magnetic field. It is determined by measuring how long the resonating protons remain coherent or their precessions remain in phase following a 90° RF pulse when the protons lose coherence and transverse magnetisation is lost. Unlike T_1 interactions, T_2 interactions do not involve a transfer of energy but only a change in phase, which leads to a loss of coherence.

A technique using the effect that magnetic particles have on the value of T_2 for protons in water molecules has been described by Haun et al.[67] using a purpose-built instrument, and this they have called diagnostic magnetic resonance (DMR). Magnetic particles coated with antibody or other capture species are added to a test solution. The antibody is selected that can bind to two different epitopes on the analyte so that a cluster of particles can be formed with the analyte molecules providing links between the particles (Figure 9.18).

Individual magnetic particles have a diameter of about 38 nm, and the clusters have a diameter of about 300 nm. When a magnetic field is applied, the induced magnetic moments in the magnetic nanoparticles align along the magnetic field, and the particles in the clusters work collectively to have a magnetic moment that is much greater than the sum of the magnetic moments of the particles making up the cluster[66]. This is due in part because the bound particles in the cluster cannot move freely so that their induced magnetic dipoles remain aligned with the field, whereas unbound particles can still move freely and are subject to Brownian motion that tends to randomise their dipoles. The presence of the clusters in the test solution decreases the value of T_2 for the proton resonance coming from the water molecules because they disturb the homogeneity in the external magnetic field that destroys the coherent precession of nuclear spins of neighbouring water protons, and the decrease in T_2 is related to the number of particles in the clusters from which the concentration of analyte in the test solution can be deduced. Unbound particles make a negligible contribution to the decrease in T_2 so do not have to be removed from the test solution prior to measurements being made. The time constant T_1 is also measured as the decrease in T_1 is related to the total number of particles. A variation of this technique is to form the clusters first using the magnetic particles and the analyte, add a suitable agent such as an enzyme to cleave the analyte–particle bonds and then observe the increase in T_2 as the particles disperse[68].

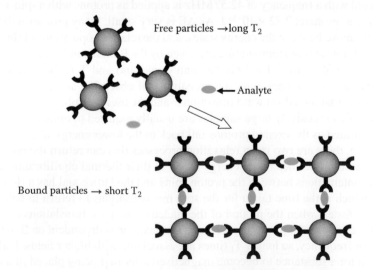

FIGURE 9.18 The antibody on the particles can bind with two epitopes on the analyte and form clusters. The clusters reduce T_2 for the water protons.

TABLE 9.1

Summary of Detection Methods for Magnetic Labels in Immunoassays Described in Section 9.2

Detection Method	Features
The superconducting quantum interference device (SQUID) (see Section 9.2.2.1)	Single particle detection is possible if particle at same temperature as SQUID
	Possible to have sample at room temperature and SQUID at 77 K but with reduced sensitivity (5×10^4 particles with diameter of 56 nm)
	Sensitivity of SQUID falls rapidly with distance so not necessary to remove unbound magnetic label from liquid sample at room temperature
	Shielding from electromagnetic interference is required
	Expensive and relatively large instrumentation
Hall effect sensors (see Section 9.2.2.2)	Single particle detection is possible if sensor is about the same size as the particle
	Arrays of sensors can be made using integrated circuit fabrication techniques
	Multiplexing electronics is required to address a large number of sensors
	Substantial dead space between sensors in the arrays
	Difficult to deposit capture layer on very small sensors and diffusion limits must be considered
	Difficult to remove unbound and non-specifically bound particles
Magnetoresistive sensors (see Section 9.2.2.3)	Possible to detect a single magnetic particle under favourable conditions
	Several different sensor arrays and geometries are possible using integrated circuit fabrication techniques
	Multiplexing electronics required to address a large number of sensors
	Works with solid- or liquid-phase samples
	Recent developments using sensors with very thin passivation layers in conjunction with 50 nm diameter particles should overcome problems of earlier systems and lead to simple, one-step homogeneous assays with no washing steps and a handheld device
Micro-cantilever-based force-amplified biological sensor (see Section 9.2.2.4)	Cantilevers can be fabricated and measurements made using established micro-electromechanical systems technology and measurement electronics
	Multi-analyte assays possible
	Suitable for use in bench top or portable instruments
	Difficult to coat the cantilevers with capture layers
	Cantilevers sensitive to temperature and vibrations
Use of a Maxwell–Wein bridge to detect changes in magnetic permeability (see Section 9.2.3.1)	Uses simple instrumentation that is well suited for handheld instrumentation
	Uses specially coated ferromagnetic particles that are easier to detect
	Liquid phase assay with no preparation of substrate surface
	Single analyte with a relatively large sample volume and response time
	Shown that the method can determine important cardiac markers over the clinically important range
Resonant coil magnetometer (see Section 9.2.3.2)	Simple instrumentation using readily available electronics and simple thick-film technology to make the coils
	Multi-coils can be used to make simultaneous measurements on solutions containing many analytes
	Uses commercially available magnetic particles
	External magnets can be used to reduce the response time and lead to rapid, one-step homogeneous assays with no washing steps
	Simple single-chip detection electronics
	Could be made into a handheld device

(continued)

TABLE 9.1 (continued)
Summary of Detection Methods for Magnetic Labels in Immunoassays
Described in Section 9.2

Detection Method	Features
Methods based on magnetic saturation measurements (see Section 9.2.3.3)	Good linear response with a wide operating range using liquid samples
	No sample preparation necessary, although system marketed by Senova uses specially constructed sample tubes containing a filter
	In present form, seems better suited for use in a bench analyser
	The approach developed by Magnisense uses the second derivative of the magnetisation and is capable of distinguishing between different magnetic particles used in one-step, multi-analyte analysis
Magnetic relaxation measurements of protons in water molecules (see Section 9.2.3.4)	Based on proton nuclear magnetic resonance in water molecules and the effect that clusters of magnetic particle labels and analyte have on the relaxation time T_2
	Able to distinguish between bound and unbound particles
	A handheld instrument has been developed that uses $1\,\mu L$ samples
	Can be used in DNA analysis as well as immunoassays
	Instrument requires a powerful permanent magnet with a uniform field across the sample
	Technique uses expensive equipment and signal processing

Details of a miniaturised, handheld instrument using this technique has been presented that uses very small samples with a volume of $1\,\mu L$ that fit miniature coils[68–71]. The NMR circuitry is integrated on a single complementary metal-oxide semiconductor (CMOS) chip. A permanent magnet with a field of less than $1\,T$ is used to provide the field, B, which is not completely uniform because of the small size of the magnet. Imperfections in the field produce faster decay times ($T_2{}^*$) than expected, and a spin-echo technique is used to correct this in order to determine a value for T_2. A further problem arising from a non-uniform field, and not commented on by the authors, is that both the free and clustered magnetic particles will move along any magnetic field gradients and be held on the side of the sample tube so they will not be uniformly distributed throughout the sample: it may be that the particles do not move far during the time of the measurements because of their small size. The technique has been applied to a wide range of analytes including DNA, RNA, proteins, enzymes, small molecules, viruses, bacteria and cancer cells, and a list is given by Haun et al.[67] Table 9.1 summarizes the detection methods described in Section 9.2 for magnetic labels used in immunoassays.

9.3 LATERAL FLOW TECHNIQUES

9.3.1 INTRODUCTION

The lateral flow immunoassay (or immunochromatography) test strip is one of the simplest immunoassay formats and is the basis of many over-the-counter, one-step tests requiring little or no sample preparation including the very successful pregnancy test originally developed by May et al.[72], which detects the pregnancy marker hormone, human chorionic gonadotrophin, in a urine sample. In addition to home diagnostic test kits, there are applications of these devices in such diverse areas as veterinary medicine, farming, food quality control, market gardening and environmental monitoring.

A typical lateral flow device (Figure 9.19) is made from a series of overlapping layers of fibrous material arranged into a single continuous strip.

A drop of the sample is placed on the sample pad that serves as a filter to remove unwanted solid matter such as blood cells in blood samples or particulate proteins in urine samples. The whole of the lateral flow strip is backed with plastic so that the liquid in the sample diffuses by capillary

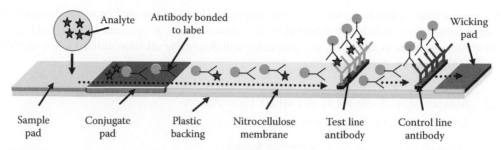

FIGURE 9.19 Schematic diagram of a lateral flow immunoassay.

action along the length of the strip and eventually reaches the wicking pad. The analyte is carried by the liquid into a conjugate pad made from fibre glass containing an excess of dried antibody bonded to a label and usually stabilised with sucrose. The liquid rehydrates the antibody and attached label, and the analyte bonds with the antibody. The bound analyte, antibody and label, together with surplus antibody and attached label, are carried by the liquid into a nitrocellulose membrane and diffuse along the membrane. The membrane contains two narrow lines of dried antibodies immobilised onto the membrane, which are rehydrated when the sample liquid reaches them. The first (test) line contains a second antibody to the analyte so that when the label, first antibody and analyte encounter it, a sandwich assay is formed and the label becomes immobilised on the test line. Surplus antibody and attached label are carried by the liquid to the second (control) line containing an anti-species antibody to the first (e.g. if one antibody is derived from sheep then the other might be from mouse), and the label becomes immobilised on the line. The first lateral flow devices used blue-coloured latex particles as label, so a simple visual outcome of the test was displayed, and a blue control line showed that the assay had finished and a blue test line showed that the test was positive. More recent devices use gold particles that produce lines that are clearer to see and may be used with a reader to avoid visual interpretation of the results[73], and fluorescent labels have also been used.

Although lateral flow devices are easy to use, there are myriad problems in developing an assay that gives consistent results. The choice of nitrocellulose membrane is important, and the average pore size of the membrane should be about 10 times the diameter of the particulate label. A membrane with a pore size that is too small will cause antibody and its label to become trapped, whereas a membrane with a pore size that is too large will allow the sample liquid to travel too quickly along the length of the strip and there will not be enough time for the antibody reactions at the capture lines. The width of the capture lines is also important as a line that is too narrow could mean that there are not enough sites for reactions to occur with the antibody in the capture lines, whereas lines that are too wide could result in too much undesirable non-specific binding. Lateral flow immunoassay devices are better suited to tests in which the concentration of the analyte in the sample is above or below a threshold value so the result of the test is only either positive or negative. It is much more difficult to construct lateral flow devices where the number of immobilised label in the test strip is directly related to the concentration of analyte in the sample. This is mainly because of the difficulty of depositing sufficient capture antibody in the lines so as to give a wide operating concentration range and for the flowing analyte–antibody–label complex to interact with the antibody immobilised on the surface of the membrane. In an attempt to give some indication of the concentration of the analyte, lateral flow devices have been made that have two or more test lines to indicate whether the concentration of analyte is within a particular range. For example, a lateral flow device is available for determining the concentration of PSA in blood that is used to assist in the management of prostate cancer and has three test lines to show if the concentration of PSA is less than 4 ng mL^{-1}, between 4 and 25 ng mL^{-1} or over 25 ng mL^{-1}, but this device is not considered sufficiently reliable for the result to be accepted without a further test on a hospital laboratory analyser. A problem with lateral flow devices using optical methods for determining the quantity of label in the capture lines is that these methods only respond to label on the surface of the nitrocellulose membrane and do not include any label below the surface. A way of overcoming this problem is to use magnetic particles as label because a detector will determine all immobilised label, including any immobilised below the capture line.

9.3.2 Measurement Techniques in Lateral Flow Assays Using Magnetic Particles as Label

9.3.2.1 Four-Coil Gradiometer Detection System

The original version of this measurement system used two circular, planar, spiral coils[74], and the later version, which is being developed by MagnaBioSciences LLC, San Diego, CA, uses four square coils, A, B, C and D (Figure 9.20), arranged as a gradiometer to determine the number of magnetic particle label in a lateral flow immunoassay.

The four coils are identical with the same self-inductance except that alternate coils are wound in opposite directions[75]. The coils are placed between the pole pieces of an electromagnet that produces a uniform alternating magnetic field. A mechanical system accurately positions and moves the lateral flow strip at a constant rate between the coils and the pole pieces. The coils are connected in such a way that there is no output from the gradiometer when there is no lateral flow strip in place because the currents induced into coils A and C are exactly balanced by the currents induced into coils B and D. In order to have consistent results, the strips are specially constructed to fit the mechanical system, the capture lines are accurately deposited with narrow widths and the intensity of the magnetic field is strong enough to saturate the magnetic particles. When the lateral flow strip moves beneath the coils and a capture line is under coil A, the presence of magnetic particles in the line produces a localised increase in magnetic flux, which in turn increases the induced

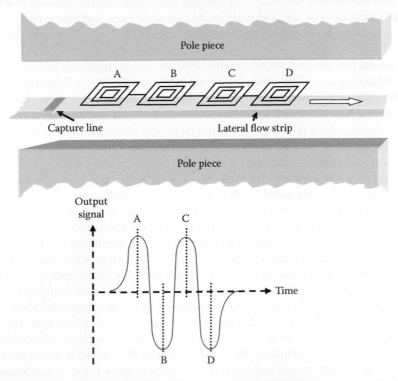

FIGURE 9.20 Simplified diagram of the instrument described by LaBorde and O'Farrell[75]. Coils A, B, C and D are wired in series, and alternate coils are wound in opposite directions so the output of the coils changes as a capture line moves beneath them.

current in this coil and disturbs the balance of the gradiometer producing an output. The output of the gradiometer moves in the opposite direction when the capture line is under coil B, back again when under C and finally in the opposite direction when under D. The output of the gradiometer is integrated and compared with previously stored outputs of the gradiometer to known numbers of particles on calibration strips. Correction is also made for the background signal caused by magnetic particles that inevitably become trapped in the membrane. An application of this measurement system for human papillomavirus 16 E6 has been described[76] in which the strip-to-strip reproducibility had a CV less than 10%, and the reader had a reproducibility with a CV of 1.5%.

9.3.2.2 Four-Coil Balanced Bridge Detection System

The measurement system being developed by Magnasense, Survontie, Finland, uses four planar spiral coils, A, B, C and D, arranged in an impedance bridge circuit (Figure 9.21).

The coils are identical with the same self-inductance, but coils A and C are wound in the opposite direction to coils B and D. The lateral flow test strip is placed under coils A and B with the capture line under coil A. The bridge circuit is energised by a high-frequency source, and the coils have been designed so that the field created by coil A is affected by the presence of the magnetic particles in the capture line[77]. With no lateral flow strip placed below the coils, the bridge is balanced because all the inductances are equal and V reads zero, and changes in ambient temperature will not affect the balance point as the inductances of the four coils will change by the same amount. When a strip is placed below the coils, the inductance of coil A increases because of the presence of the paramagnetic particles in the capture line, causing the bridge to become unbalanced and producing a reading on V that is related to the number of particles in the capture line. The inductance of coil B is affected by the presence of paramagnetic particles trapped in the membrane away from the capture line so its inductance will increase and, because of its position in the bridge, will reduce the reading of V, thus compensating for background readings. Advantages of this system are that it does not require an external magnet or a mechanism to move the lateral flow strip below the coils.

9.3.2.3 Integrated Giant Magnetoresistive Strip

Diagnostic Biosensors LLC of Minneapolis, MN, is developing a lateral flow system in which a GMR strip is integrated into the membrane. A GMR chip has been demonstrated[78] to detect the presence of magnetic labels in capture spots whose volume is approximately $150 \times 150 \times 150\,\mu m^3$. The range of linear detection is better than two orders of magnitude, and the total range is up to four orders of magnitude. The system was demonstrated with protein detection of rabbit IgG and interferon-γ achieving detection of $12\,pg\,mL^{-1}$ protein. Ultimately, the goal

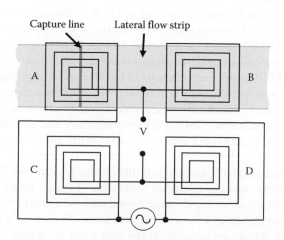

FIGURE 9.21 Simplified diagram of the balanced inductance bridge described by Mäkiranta and Lekkala[77].

is for the detector to be fully integrated into the lateral flow strip backing to form a single consumable item that is interrogated by a handheld device.

9.4   APPLICATION OF MAGNETIC PARTICLES IN IMMUNOHISTOCHEMISTRY

Immunoassays are used not simply to determine the concentration of an analyte in a solution but in other quite different biomedical investigations, and this section describes how a magneto-immunoassay can be used in imaging in histopathology. Histology is the study of the anatomy of cells using optical or electron microscopy, and the medical application of histology is histopathology which, in cancer investigations, is the search for abnormal cells in a thin slice of a biopsy taken from the patient, the presence of which is final confirmation of cancer. Many molecular pathways in the cell are altered in cancer, and some of the alterations can be targeted in cancer therapy. Immunohistochemistry can give an indication of the state of the disease and a likely prognosis as well as assessing which tumours will probably respond to therapy by detecting the presence or elevated levels of the molecular target, and this is achieved by using an immunoassay technique to look for certain antigens (e.g. proteins) in the cells using antibodies. A label is attached to the antibody so that the amount of molecular target in the cells can be quantified or scored. Several different labels are used in immunohistochemistry investigations, such as enzymes and fluorescent molecules.

An investigation into the use of a magnetic particle label in immunohistochemistry applied to a breast cancer was reported by Mitchels et al.[79] The human epidermal growth factor receptor-2 (HER2 or HER2/neu) is over-expressed in approximately 20%–30% of breast cancers. Over-expression of the HER2 protein has been shown to correlate with amplification of the HER2 gene and is associated with an aggressive disease course and a poor prognosis. HER2 over-expression is also a strong predictor of response to the humanised monoclonal antibody, trastuzumab, which has demonstrated a significant survival benefit in patients with HER2-positive breast cancer. Assessment of HER2 status has therefore become a standard practice in women with breast cancer in order to identify those patients most likely to benefit from treatment with trastuzumab. Most laboratories in North America and Europe use immunohistochemistry to determine the status of the HER2 protein expression on the cell surface. A common technique uses an anti-human HER2 antibody with an enzyme such as horse radish peroxidase as label, which when hydrogen peroxide is added causes a reaction with an organic molecule such as 3,3′-diaminobenzidine to produce a brown stain around the edges of the cells on a microscope slide with an intensity that is related to the amount of molecular target. The intensity and score of the stain is judged subjectively by a pathologist looking through a microscope at the slide. Attempts have been made to automate the process by measuring the optical density of the stain, but these have not proved reliable enough for even a preliminary screening of the slides. Samples with scores of 2+ or 3+ undergo another test using a fluorescent label [fluorescent *in situ* hybridisation (FISH)], which determines the actual degree of HER2 gene amplification. Both tests are recommended in the HER2 testing algorithm, as they are highly specific and reproducible methodologies when performed using standardised and validated protocols, and patients with tumours having scores of 3+ or FISH positive are considered HER2 positive and eligible for treatment with trastuzumab. However, both tests have certain limitations. While the staining technique is readily available in most laboratories, results are susceptible to variations in methodology and operator interpretation between laboratories. FISH is a highly sensitive and specific method for determination of HER2 gene amplification but requires special skills using expensive equipment that is not available in most pathology laboratories. In addition, FISH cannot readily distinguish between invasive carcinoma and carcinoma *in situ* and is limited by the fact that the fluorescent signal decays rapidly at room temperature, making it difficult to preserve slides for future reference.

Using magnetic particles as labels in these immunohistochemistry tests could have several advantages. The particles appear as a brown stain when immobilised on a tissue slice on a microscope slide without any further preparation so the slides can be scored in the same way as the current method by a pathologist using a microscope. The particles on the slide could be scanned

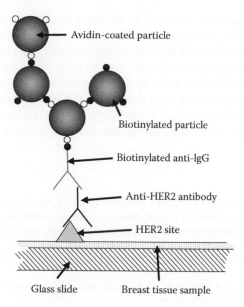

FIGURE 9.22 Schematic diagram showing an amplification method to enhance the signal in an immunohistochemistry investigation of HER2 breast cancer using magnetic particles as label.

using a suitable magnetic scanner, and this would provide the same information as FISH. If this is considered to be as reliable as FISH, then it could be used in an automated system to make preliminary scores of slides and samples with high scores investigated again using conventional methods. In addition, magnetic labels do not deteriorate quickly like fluorescent labels. Mitchels et al.[79] used an amplification technique to bind several magnetic particles to the HER2 sites on sections of breast tissue (Figure 9.22) so as to obtain a detectable signal.

Well-characterised samples of breast tissue with known scores of HER2 were obtained from commercial suppliers, and sections of the tissue, about 2 μm thick, were fixed to glass microscope slides. The cells in the slide were about 10–30 μm long, and the HER2 sites, 2.5 nm wide, were situated in the plasma membrane of the cell, which was 10 nm thick. Anti-human-HER2 antibody was added to the section that bound to the HER2 sites, followed by biotinylated anti-IgG antibody that bound to the first antibody. Magnetic particles coated with avidin were added that bound to the biotin on the antibody, followed by magnetic particles coated with biotin that bound to unused sites on the avidin-coated particles. Several magnetic particles could be attached to the HER2 site by adding alternately biotin- and avidin-coated magnetic particles until the required numbers were attached. The slides were then inspected using a microscope, and the HER2 scores were determined, and these were in agreement with the known scores for the breast tissues, indicating that the change in intensity of the staining produced by the particles was the same as that produced by the conventional method. A magnetic field was applied to the slide, and a magnetic force microscope was used to scan the slides in order to quantify the amount of immobilised magnetic particles. After applying a suitable statistical analysis of the results, it was demonstrated that the quantities of magnetic particles immobilised on the slides were in good agreement with the score for the breast tissue on the sample. This technique could be the basis for a fully automated system that could be used in the rapid testing of routine samples in histopathology.

9.5 CONCLUSIONS

Immunoassays have had a major impact on medical diagnostics and will continue to do so particularly with the demand for rapid clinical chemistry tests and the increase in near patient monitoring. Magnetic particles have played an important part in the development of immunoassays in sample extraction and purification, and now as a label in the direct quantification of analytes. Several techniques for quantifying the magnetic particle label in immunoassays have been developed, but it remains to be seen which of these will become widely accepted. All commercially available magnetic particles are based on superparamagnetic nanoparticles of magnetite, and the measurement techniques would be enhanced in some cases by the introduction of new particles based on materials with stronger magnetic signatures and more consistent properties. Immunoassays using magnetic particles as label are finding other medical applications, such as in histopathology, and these applications are bound to increase, so the future of magnetic particles in immunoassays looks very bright.

REFERENCES

1. Chade D C, Shariat S F, Godoy G, Meryn S and Dalbagni G, Critical review of biomarkers for the early detection and surveillance of bladder cancer, *Journal of Men's Health*, 6, 2009, 368–382.
2. Husseinzadeh N, Status of tumor markers in epithelial ovarian cancer has there been any progress? A review, *Gynecologic Oncology*, 120, 2011, 152–157.
3. Scott T, Point-of-care tetanus immunoassay, *International Journal of Orthopaedic and Trauma Nursing*, 14, 2010, 222–223.
4. Caygill R L, Blair G E and Millner P A, A review on viral biosensors to detect human pathogens, *Analytica Chimica Acta*, 681, 2010, 8–15.
5. Oh S W, Kim Y M, Kim H J, Kim S J, Cho J-S and Choi E Y, Point-of-care fluorescence immunoassay for prostate specific antigen, *Clinica Chimica Acta*, 406, 2009, 18–22.
6. Abuknesha R A, Luk C Y, Griffith H H M, Maragkou A and Iakovaki D, Efficient labelling of antibodies with horseradish peroxidase using cyanuric chloride, *Journal of Immunological Methods*, 306, 2005, 211–217.
7. Vira S, Mekhedov E, Humphrey G and Blank P S, Fluorescent-labeled antibodies: Balancing functionality and degree of labeling, *Analytical Biochemistry*, 402, 2010, 146–150.
8. Xin T-B, Chen H, Lin Z, Liang S-X and Lin J-M, A secondary antibody format chemiluminescence immunoassay for the determination of estradiol in human serum, *Talanta*, 82, 2010, 1472–1477.
9. Yang X, Guo Y and Wang A, Luminol/antibody labeled gold nanoparticles for chemiluminescence immunoassay of carcinoembryonic antigen, *Analytica Chimica Acta*, 666, 2010, 91–96.
10. Martins T B, Jaskowski T D, Tebo A and Hill H R, Development of a multiplexed fluorescent immunoassay for the quantitation of antibody responses to four *Neisseria meningitidis* serogroups, *Journal of Immunological Methods*, 342, 2009, 98–105.
11. Algar W R, Tavares A J and Krull U J, Beyond labels: A review of the application of quantum dots as integrated components of assays, bioprobes, and biosensors utilizing optical transduction, *Analytica Chimica Acta*, 673, 2010, 1–25.
12. Jie G, Li L, Chen C, Xuan J and Zhu J-J, Enhanced electrochemiluminescence of CdSe quantum dots composited with CNTs and PDDA for sensitive immunoassay, *Biosensors and Bioelectronics*, 24, 2009, 3352–3358.
13. Nisnevitch M and Firer M A, The solid phase in affinity chromatography: Strategies for antibody attachment, *Journal of Biochemical and Biophysical Methods*, 49, 2001, 467–480.
14. Hawkanes B I and Kvam C, Application of magnetic beads in bioassays, *Biotechnology*, 11, 1993, 60–64.
15. Liu Z, Zhang L, Yang H, Zhu Y, Jin W, Song Q and Yang X, Paramagnetic microbead-based enzyme-linked immunoassay for detection of *Schistosoma japonicum* antibody in human serum, *Analytical Biochemistry*, 404, 2010, 127–134.
16. Zhang Q, Wang X, Li Z and Lin J-M, Evaluation of α-fetoprotein (AFP) in human serum by chemiluminescence enzyme immunoassay with paramagnetic particles and coated tubes as solid phases, *Analytica Chimica Acta*, 631, 2009, 212–217.
17. Selvan S T, Silica-coated quantum dots and magnetic nanoparticles for bioimaging applications (minireview), *Biointerphases*, 5(3), 2010, FA110–FA115.
18. Porter R, Kabil A, Forstern C, Slevin C, Kouwenberg K, Szymanski M and Birch B, A novel, generic, electroanalytical immunoassay format utilising silver nano-particles as a bio-label, *Journal of Immunoassay and Immunochemistry*, 30, 2009, 428–440.
19. Wilson P K, Szymanski M and Porter R, Poster: Point-of-care instrument platform based on magnetic beads and silver nanoparticle metalloimmunoassay, *Oak Ridge Conference, American Association for Clinical Chemistry*, 22 and 23 April 2010, San José, CA.

20. Luxton R, Badesha J, Kiely J and Hawkins P, Use of external magnetic fields to reduce reaction times in an immunoassay using micrometer-sized paramagnetic particles as labels (magnetoimmunoassay), *Analytical Chemistry*, 76, 2004, 1715–1719.
21. Rife J C, Miller M M, Sheehan P E, Tamanaha C R, Tondra M and Whitman L J, Design and performance of GMR sensors for the detection of magnetic microbeads in biosensors, *Sensors and Actuators A*, 107, 2003, 209–218.
22. Eveness J, Kiely J, Hawkins P, Wraith P and Luxton R, Evaluation of paramagnetic particles for use in a resonant coil magnetometer based magneto-immunoassay, *Sensors and Actuators B*, 139, 2009, 538–542.
23. Tamanaha C R, Mulvaney J C, Rife J C and Whitman L J, Magnetic labelling, detection, and system integration, *Biosensors and Bioelectronics*, 24, 2008, 1–13.
24. Kotitz R, Matz H, Trahms L, Koch H, Weitschies W, Rheinlander T, Semmler W and Bunte T, SQUID based remanence measurements for immunoassays, *IEEE Transactions in Applied Superconductivity*, 7, 1997, 3678–3681.
25. Chemla Y R, Grossman H L, Poon Y, McDermott R, Stevens R, Alper M D and Clarke J, Ultrasensitive magnetic biosensor for homogeneous immunoassay, *Proceeding of the National Academy of Science of USA*, 97, 2000, 14268–14272.
26. Tsukamoto A, Saitoh K, Suzuki D, Sugita N, Seki Y, Kandori A, Tsukada K, Sugiura Y, Hamaoka S, Kuma H, Hamasaki N and Enpuku K, Development of multisample biological immunoassay system using HTSSQUID and magnetic nanoparticles, *IEEE Transactions in Applied Superconductivity*, 15, 2005, 656–659.
27. Enpuku K, Inoue K, Soejima K, Yoshinaga K, Kuma H and Hamasaki N, Magnetic immunoassays utilizing magnetic markers and a high-T_c SQUID, *IEEE Transactions in Applied Superconductivity*, 15, 2005, 660–663.
28. Marinace J C, High sensitivity hall effect probe, US Patent No. 3202913, 1965.
29. Besse P-A, Boero G, Demierre M, Pott V and Popovic R, Detection of a single magnetic microbead using a miniaturized silicon Hall sensor, *Applied Physics Letters*, 80, 2002, 4199–4201.
30. Aytur T, Foley J, Anwar M, Boser B, Harris E and Beatty P R, A novel magnetic bead bioassay platform using a microchip-based sensor for infectious disease diagnosis, *Journal of Immunological Methods*, 314, 2006, 21–29.
31. Jin S, Tiefel T H, McCormack M, Fastnacht R A, Ramesh R and Chen L H, Thousandfold change in resistivity in magnetoresistive La–Ca–Mn–O films, *Science*, 264, 1994, 413–415.
32. Slaughter J M, Chen E Y, Whig R, Engel B N, Janesky J and Tehrani S, Magnetic tunnel junction materials for electronic applications, *JOM-e*, 52(6), 2000. http://www.tms.org/pubs/journals/JOM/0006/Slaughter/Slaughter-0006.html
33. Miller M M, Prinz G A, Cheng S-F and Bounnak S, Detection of a micron-sized magnetic sphere using a ring-shaped anisotropic magnetoresistance-based sensor: A model for a magnetoresistance-based biosensor, *Applied Physics Letters*, 81, 2002, 2211–2213.
34. Schneider R W and Smith C H, Low magnetic field sensing with GMR sensors, part 1: The theory of solid-state magnetic sensing, *Sensors*, 1999. http://www.sensorsmag.com/sensors/electric-magnetic/low-magnetic-field-sensing-with-gmr-sensors-part-1-the-theor-928
35. Bailbich M N, Broto J M, Fert A, Van Dau F N, Petroff F, Etienne P, Creuzet G, Friederich A and Chazelas J, Giant magnetoresistance of (001)Fe/(001) Cr magnetic superlattices, *Physical Review Letters*, 61, 1988, 2472–2475.
36. Binasch G, Grünberg P, Saurenbach F and Zinn W, Enhanced magnetoresistance in layered magnetic structures with antiferromagnetic interlayer exchange, *Physical Review B*, 39, 1989, 4828–4830.
37. Graham D L, Ferreira H, Bernardo J, Freitas P P and Cobra J M S, Single magnetic microsphere placement and detection on-chip using current line designs with integrated spin valve sensors: Biotechnological applications, *Journal of Applied Physics*, 91, 2002, 7786–7788.
38. Graham D L, Ferreira H A and Freitas P P, Magnetoresistive-based biosensors and biochips, *Trends in Biotechnology*, 22, 2004, 455–462.
39. Graham D L, Ferreira H A, Feliciano N, Freitas P P, Clarke L A and Amaral M D, Magnetic field-assisted DNA hybridisation and simultaneous detection using micron-sized spin-valve sensors and magnetic nanoparticles, *Sensors and Actuators B: Chemical*, 107, 2005, 936–944.
40. Li G, Joshi V, White R L, Wang S X, Kemp J T, Webb C, Davis R W and Sun S, Detection of single micron-sized magnetic bead and magnetic nanoparticles using spin valve sensors for biological applications, *Journal Applied Physics*, 93, 2003, 7557–7559.
41. Li G, Sun S, Wilson R J, White R L, Pourmand N and Wang S X, Spin valve sensors for ultrasensitive detection of superparamagnetic nanoparticles for biological applications, *Sensors and Actuators A: Physical*, 126, 2006, 98–106.

42. Miyazaki T and Tezuka N, Giant magnetic tunneling effect in Fe/Al$_2$O$_3$/Fe junction, *Journal of Magnetism and Magnetic Materials*, 139, 1995, L231–L234.

43. Moodera J S, Kinder L R, Wong T M and Meservey R, Large magnetoresistance at room temperature in ferromagnetic thin film tunnel junctions, *Physical Review Letters*, 74, 1995, 3273–3276.

44. Parkin S S P, Kaiser C, Panchula A, Rice P M, Hughes B, Samant M and Yang S-H, Giant tunnelling magnetoresistance at room temperature with MgO (100) tunnel barriers, *Nature Materials*, 3, 2004, 862–867.

45. Baselt D R, Lee G U, Natesan M, Metzger S W, Sheehan P E and Colton R J, A biosensor based on magnetoresistance technology, *Biosensors and Bioelectronics*, 13, 1998, 731–739.

46. Tondra M, Microchip-based sensors for detection of magnetic microbead label, *2006 Oak Ridge Conference of the American Association of Clinical Chemistry*, 21 April 2006, San Jose, CA.

47. Schotter J, Kamp P B, Becker A, Pühler A, Reiss G and Brückl H, Comparison of a prototype magnetoresistive biosensor to standard fluorescent DNA detection, *Biosensors and Bioelectronics*, 19, 2004, 1149–1156.

48. Xu L, Yu H, Akhras M S, Han S-J, Osterfeld S J, White R L, Pourmand N and Wang S X, Giant magnetoresistive biochip for DNA detection and HPV genotyping, *Biosensors and Bioelectronics*, 24, 2008, 99–103.

49. Gaster R S, Hall D A, Nielsen C H, Osterfeld S J, Yu H, Mach K E, Wilson R J, Murmann B, Liao J C, Gambhir S S and Wang S X, Matrix-insensitive protein assays push the limits of biosensors in medicine, *Nature Medicine*, 15, 2009, 1327–1332.

50. Fishbein I and Levy R J, The matrix neutralized, *Nature*, 461, 2009, 890–891.

51. Osterfeld S J, Yu H, Gaster R S, Caramuta S, Xu L, Han S-J, Hall D A, Wilson R J, Sun S, White R L, Davis R W, Pourmand N and Wang S X, Multiplex protein assays based on real-time magnetic nanotag sensing, *Proceedings of the National Academy of Sciences of USA*, 105, 2008, 20637–20640.

52. Baselt D R, Lee G U, and Colton R J, Biosensor based on force microscope technology, *Journal of Vacuum Science and Technology B*, 14, 1996, 789–793.

53. Weizmann Y, Patolsky F, Lioubashevski O and Willner I, Magneto-mechanical detection of nucleic acids and telomerase activity in cancer cells, *Journal of the American Chemical Society*, 126, 2004, 1073–1080.

54. Kriz C B, Rådevik K and Kriz D, Magnetic permeability measurements in bioanalysis and biosensors, *Analytical Chemistry*, 68, 1996, 1966–1970.

55. Kriz K, Gehrke J and Kriz D, Advancements toward magneto immunoassays, *Biosensors and Bioelectronics*, 13, 1998, 817–882.

56. Kriz K, Ibraimi F, Lu M, Hansson LO and Kriz D, Detection of C-reactive protein utilizing magnetic permeability detection based immunoassays, *Analytical Chemistry*, 15, 2005, 5920–5924.

57. Ibraimi F, Kriz D, Lu M, Hansson L O and Kriz K, Rapid one-step whole blood C-reactive protein magnetic permeability immunoassay with monoclonal antibody conjugated nanoparticles as superparamagnetic labels and enhanced sedimentation, *Analytical and Bioanalytical Chemistry*, 384, 2006, 651–657.

58. Hawkins P, Luxton R, and Macfarlane J, Measuring system for the rapid determination of the concentration of coated micrometer-sized paramagnetic particles suspended in aqueous buffer solutions, *Review of Scientific Instruments*, 72, 2001, 237–242.

59. Richardson J, Hill A, Luxton R and Hawkins P, A novel measuring system for the determination of paramagnetic particle labels for use in magneto-immunoassays, *Biosensors and Bioelectronics*, 16, 2001, 1127–1132.

60. Richardson J, Hawkins P and Luxton R, The use of coated paramagnetic particles as a physical label in a magneto-immunoassay, *Biosensors and Bioelectronics*, 16, 2001, 989–993.

61. Kiely J, Hawkins P, Wraith P and Luxton R, Paramagnetic particle detection for use with an immunoassay based biosensor, *IET Science, Measurement & Technology*, 1, 2007, 270–275.

62. Meyer M H F, Hartmann M, Krause H-J, Blankenstein G, Mueller-Chorus, Oster B J, Miethe P and Keusgen M, CRP determination based on a novel magnetic biosensor, *Biosensors and Bioelectronics*, 22, 2007, 973–979.

63. Nikitin P I, Vetoshko P M and Ksenevich T I, New type of biosensor based on magnetic nanoparticle detection, *Journal of Magnetism and Magnetic Materials*, 311, 2007, 445–449.

64. Lenglet L, Nikitin P and Péquignot C, Magnetic immunoassays: A new paradigm in POCT, *IVD Technology*, July 2008.

65. Alphandéry E, Lijeour L, Lalatonne Y and Motte L, Different signatures between chemically and biologically synthesized nanoparticles in a magnetic sensor: A new technology for multiparametric detection, *Sensors and Actuators B*, 147, 2010, 786–790.

66. Perez J M, Josephson L, O'Loughlin T, Högemann D and Weissleder R, Magnetic relaxation switches capable of sensing molecular interactions, *Nature Biotechnology*, 20, 2002, 816–820.

67. Haun J B, Yoon T-J, Lee H and Weissleder R, Magnetic nanoparticle biosensors, *Wiley Interdisciplinary Reviews, Nanomedicine and Nanobiotechnology*, 2, 2010, 291–304.
68. Lee H, Sun E, Ham D and Weissleder R, Chip-NMR biosensor for detection and molecular analysis of cells, *Nature Medicine*, 14, 2008, 869–874.
69. Lee H, Yoon T-J and Weissleder R, Ultrasensitive detection of bacteria using core–shell nanoparticles and an NMR-filter system, *Angewandte Chemie International Edition*, 48, 2009, 5657–5660.
70. Lee H, Yoon T-J, Figueiredo J L, Swirski F K and Weissleder R, Rapid detection and profiling of cancer cells in fine-needle aspirates, *Proceedings of the National Academy of Sciences of USA*, 106, 2009, 12459–12464.
71. Liu Y, Sun N, Lee H, Weissleder R and Ham D, CMOS mini nuclear magnetic resonance system and its application for biomolecular sensing, *ISSCC Digest of Technical Papers*, 1, 2008, 140–141.
72. May K, Prior M E and Richards I, U.K. Patent No. GB 2204398, 1988.
73. Kim S and Park J-K, Development of a test strip reader for a lateral flow membrane based immunochromatographic assay, *Biotechnology and Bioprocess Engineering*, 9, 2004, 127–131.
74. Simmonds M B, Method and apparatus for making qualitative measurements of localized accumulations of target particles having magnetic particles bound thereto, US Patent No. 6046585, 2000.
75. LaBorde R T and O'Farrell B, Paramagnetic labeling offers an alternative method for analyte detection, *IVD Technology*, April 2002. http://www.ivdtechnology.com/article/paramagnetic-labeling-offers-alternative-method-analyte-detection
76. Peck R B, Schweizer Weigl B H, Somoza C, Silver J, Sellors J W and Lu P S, A magnetic immunochromatographic strip test for detection of human papillomavirus 16 E6, *Clinical Chemistry*, 52, 2006, 2170–2172.
77. Mäkiranta J and Lekkala J, Optimization of a novel magnetic nanoparticles sensor, *XVIII IMEKO World Conference, Metrology for a Sustainable Development*, Rio de Janeiro, Brazil, 2006.
78. Taton K, Johnson D, Guire P, Lange E and Tondra M, Lateral flow immunoassay using magnetoresistive sensors, *Journal of Magnetism and Magnetic Materials*, 321, 2009, 1679–1682.
79. Mitchels J, Hawkins P, Luxton R and Rhodes A, Quantification in histopathology—Can magnetic particles help? *Journal of Magnetism and Magnetic Materials*, 311, 2007, 264–268.

Peter Hawkins has over 40 years of research and teaching experience in the area of sensor science and technology. He has published numerous papers on a wide range of sensors, including electrochemical, optical and magnetic sensors, and is the named inventor on 15 patents for devices used in electro-luminescent display panels, a transcutaneous carbon dioxide blood gas analyser, a blood pressure measuring system and a magnetic particle measuring system for use in immunoassays. Now happily retired, he is engaged in a variety of activities including restoring a 1936 Austin 10 car, maintaining an 1837 church turret clock and visiting his grandchildren.

Professor Richard Luxton studied clinical chemistry in the National Health Service for 13 years in Bristol before moving to the Institute of Neurology in London to investigate antibody affinity in the cerebral spinal fluid of patients with multiple sclerosis. His current research interests in the University of the West of England are in developing new rapid detection technologies for point of care diagnostics, environmental analysis, food safety and homeland defence applications. He is currently the director of the Institute of Bio-Sensing Technology in the University of the West of England that seeks to develop new collaborations between industry and academia through inter-disciplinary research.

10 Magnetic Nanoparticles in Lab-on-a-Chip Devices

Nicole Pamme

CONTENTS

In this chapter, the potential of microfluidic lab-on-a-chip devices for applications involving magnetic nanoparticles will be reviewed. In Section 10.1, the concept of microfluidics is introduced together with some underlying physical phenomena. Section 10.2 is dedicated to on-chip synthesis of magnetic nanoparticles as well as on-chip synthesis of magnetite-containing microspheres. Lab-on-a-chip devices may provide an alternative or complementary approach to the synthesis methods outlined in Part II of this book. Section 10.3 of this chapter concerns on-chip applications with magnetic nanoparticles. These include physical applications where magnetic nanoparticles act as pumps, valves, mixers or switches as well as bioanalytical applications such as immunoassays or sorting of magnetically labelled cells.

10.1 LAB-ON-A-CHIP DEVICES/MICROFLUIDIC DEVICES

Microfluidic lab-on-a-chip devices feature small fluid-carrying channels, typically a few tens or hundreds of micrometres in width and depth through which small amounts of fluids (nanolitre or below) can be manipulated in a precise manner. Such microfluidic chips are often fabricated from glass or polymeric materials (Figure 10.1).

Lab-on-a-chip devices have found applications in a wide range of disciplines, including chemical synthesis, biomedical assays and analytical separations. By performing fluid handling on the microfluidic scale, a number of useful capabilities can be exploited: the required sample and reagent volumes are greatly reduced; the amount of waste generated is reduced; chemical reactions and analytical separations can be performed in shorter times and often with better efficiencies due to favourable scaling laws and finally the footprint of the device is greatly reduced, which may allow for point-of-care analysis.

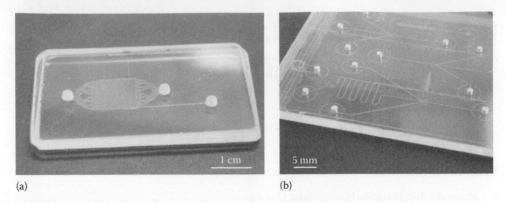

(a)   (b)

FIGURE 10.1 Microfluidic devices feature channels with tens or hundreds of micrometre dimensions. They can be fabricated from (a) glass or (b) polymeric materials such as polymethyl methacrylate (PMMA).

One important feature of microfluidic devices is the flow regime. Fluid flow in microchannels is laminar, and therefore, mixing between fluid streams occurs solely by diffusion. This means that mixing between streams is predictable and thus controllable. An example is shown in Figure 10.2. Flow streams of iron chloride and potassium thiocyanate are pumped into a microfluidic chamber. As the streams diffuse into each other, red iron thiocyanate is formed. The extent of diffusion depends on the residence time within the microfluidic chamber that can be visualised by the appearance of the red reaction product. According to experimental requirements, mixing can be tailored by changing the length of the chamber or by changing the applied flow rate. Therefore, local concentrations can be predicted and are controllable, that is, there is spatial and temporal control.

Many researchers have taken advantage of magnetic forces to manipulate materials inside microfluidic channels[1–3]. One motivation for this approach is that magnets can be placed externally to the microchannel and do not need to be in direct contact with the liquid. Furthermore, magnetic forces on particles inside the microchannel are generally independent of pH, ionic strength and temperature. Most commonly, so far, commercially available superparamagnetic microparticles doped with iron oxide have been used for on-chip applications. Often researchers have utilised small permanent NdFeB magnets

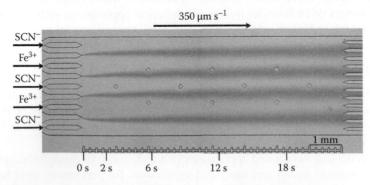

FIGURE 10.2 Flow regimes within microfluidic devices are laminar, and mixing occurs based on diffusion. The extent of diffusion can be controlled by adjusting flow rates and therefore residence times of molecules within the microchannel. In this figure, diffusion is visualised by a chemical reaction. Five inlets merge into a chamber that is 3 mm wide, 8 mm long and 20 μm deep. The flow velocity in the chamber is 350 μm s^{-1}. Two inlet streams contain iron (III) chloride solution, and the other three inlet streams contain potassium thiocyanate solution. When these streams diffuse into each other, a red product, iron thiocyanate, is formed. As the fluid flows along the chamber, more and more inter-diffusion occurs between the streams as seen by the widening red bands.

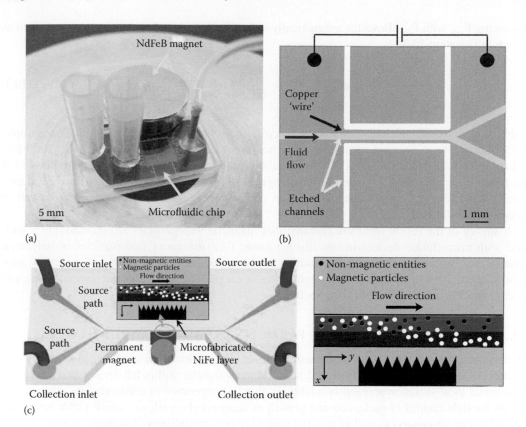

FIGURE 10.3 Magnetic forces can be used to manipulate material inside a microchannel by means of an external magnetic field. (a) Magnetic fields can be generated from external permanent magnets such as NdFeB magnets. (Courtesy of S. A. Peyman.) (b) Microfabrication techniques can be used to create microwires[4]. Here, channels were etched into a copper plate. Some channels carried fluids while others acted as a barrier for an applied electric field. In this way, a small copper wire was created along a 5 mm length of the microchannel. (c) Microfabricated permalloy features in close proximity to a microchannel can be magnetised by an external larger magnet and lead to a concentration of field lines near the channel. (Reprinted with kind permission from Springer Science + Business Media: *Microfluid. Nanofluid.*, Local control of magnetic objects in microfluidic channels, 8, 2010, 123–130, Derec, C., Wilhelm, C., Servais, J., and Bacri, J.C.; *Biomed. Microdev.*, Combined microfluidic-micromagnetic separation of living cells in continuous flow, 8, 2006, 299–308, Xia, N., Hunt, T.P., Mayers, B.T., Alsberg, E., Whitesides, G.M., Westervelt, R.M., and Ingber, D.E.)

(Figure 10.3a). Alternatively, microfabricated electromagnets (Figure 10.3b) or microfabricated permalloy features that concentrate field lines from an external magnet (Figure 10.3c) can be used.

The magnetic force $\mathbf{F}_{\mathrm{mag}}$ on a magnetic particle in a microchannel, as shown in Equation 10.1, is a function of the volume of magnetic material inside the particle (V_{m}), the difference in magnetic susceptibility between the particle and its surrounding medium ($\Delta\chi$) as well as the strength and gradient of the applied magnetic field ($\mathbf{B}$), with $\mu_0 = 4\pi \times 10^{-7}$ H m^{-1}:

$$\mathbf{F}_{\mathrm{mag}} = \frac{V_{\mathrm{m}} \cdot \Delta\chi}{\mu_0}(\mathbf{B} \cdot \nabla)\mathbf{B} \qquad (10.1)$$

In some applications, magnetic particles are moved through a liquid, which leads to the generation of a viscous drag force $\mathbf{F}_{\mathrm{vis}}$ (Equation 10.2), depending on the viscosity (η) of the medium and the radius (r), as well as the magnetically induced velocity ($\mathbf{u}_{\mathrm{mag}}$) of the moving particle:

$$\mathbf{F}_{\mathrm{vis}} = 6\pi\,\eta\,r\,\mathbf{u}_{\mathrm{mag}} \qquad (10.2)$$

Balancing $\mathbf{F}_{mag}$ with $\mathbf{F}_{vis}$ allows the magnetically induced velocity on the particle to be expressed as shown in Equation 10.3:

$$\mathbf{u}_{mag} = \frac{\mathbf{F}_{mag}}{6\pi\eta r} = \frac{V_m \Delta\chi}{6\mu_0 \pi\eta r}(\mathbf{B} \cdot \nabla)\mathbf{B} \tag{10.3}$$

It should be noted that the magnetic force scales with the volume of magnetic material. For example, a 10-fold reduction in size of an iron oxide particle from 100 to 10 nm would result in a 1000-fold reduction of the force acting on it. The manipulation of individual magnetic nanoparticles within microfluidic devices therefore remains challenging, and most strategies outlined in this chapter involve the manipulation of larger numbers of nanoparticles, for example, attached to a cell or incorporated within a droplet.

A survey of the recent literature revealed that magnetic nanoparticles are increasingly associated with microfluidic devices in two distinct areas: (i) *synthesis* of nanoparticles or polymeric spheres with embedded nanoparticles and (ii) *applications* that employ the functionality of magnetic nanoparticles to achieve pumping, movement of reagents or separation and sorting. These will be explained in detail in the following sections.

10.2   SYNTHESIS OF MAGNETIC PARTICLES IN LAB-ON-A-CHIP DEVICES

Microreactor systems allow for precise control of reaction conditions. The small length scales and low reaction volumes together with enhanced heat and mass transfer within lab-on-a-chip devices lead to minimal local variation of reaction conditions, that is, temperature or concentrations. This in turn allows for tight control of nucleation and growth in nanoparticle synthesis, which yields better size control (mono-dispersity) as well as material control (mono-crystallinity). Synthesis of nanoparticles in microreactors is usually carried out in continuous flow (Figure 10.4a). Reagent streams are merged,

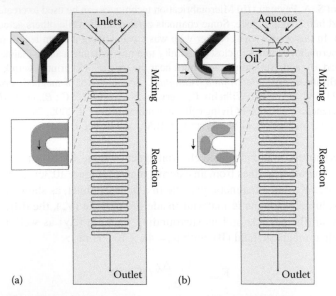

(a)   (b)

FIGURE 10.4   Microfluidic reactors for continuous flow synthesis of nanoparticles can operate based on two principles. (a) Single-phase system: Two reagent streams are merged and flowed through a serpentine channel. The reagents mix *via* diffusion and the product is collected at the outlet. (b) Two-phase system: Water-in-oil microdroplets can be generated on-chip. These act as small reaction vessels with picolitre volumes. Mixing inside these droplets is based on diffusion and re-circulating streamlines and is therefore faster than in the single-phase system.

allowed to flow side by side and to diffuse into each other. Sometimes an active mixer is employed to speed up mixing of the two streams. Control of flow rates allows for control of diffusion and reaction times, and therefore, any further reagents or quenchers can be introduced at the optimum point in time. An alternative approach is to generate droplets, small aqueous picolitre vesicles in a surrounding oil phase (Figure 10.4b). Fast mixing of precise volumes and concentrations can be achieved in these isolated reaction vessels, and therefore, an even higher control of synthesis conditions is obtained compared with single-phase systems.

10.2.1 SYNTHESIS OF MAGNETIC NANOPARTICLES

Magnetic nanoparticles of various materials have been synthesised within microfluidic devices. *Iron oxide particles* are commonly prepared *via* co-precipitation of Fe^{2+} and Fe^{3+} in a basic environment. Reaction conditions can be optimised by varying concentrations of the reagents as well as relative flow rates:

$$Fe^{2+}_{(aq)} + 2Fe^{3+}_{(aq)} + 8OH^-_{(aq)} \rightarrow Fe_3O_{4(s)} + 4H_2O_{(l)}$$

Abou Hassan et al.[6] devised a reactor with a 3D sheath-flow regime where an inner capillary of 150 μm diameter merged into an outer capillary of 1.7 mm diameter. An acidic mixture of iron(II) chloride and iron(III) chloride was pumped through the inner capillary, while the outer stream contained tetramethyl-ammonium hydroxide (NMe_4OH). Iron oxide nanoparticles were formed as the two streams diffused into each other. Since the precipitate formed inside the channel, away from the walls, clogging problems were minimised. Residence times within the reactor were less than 1 min. The particles were collected into a vessel with a cationic surfactant in cyclohexane to yield a stable colloidal suspension. Transmission electron microscopy (TEM) measurements showed fairly spherical particles with a size around 7 nm. Diffraction patterns suggested that maghemite, γ-Fe_2O_3, was present. A series of three such sheath-flow type microreactors was employed for the encapsulation of magnetic nanoparticles to form *fluorescent core/shell magnetic/silica nanoparticles* in continuous flow[7]. In the first microreactor, citrated γ-Fe_2O_3 particles were reacted with aminopropyl triethoxy silane (APTES). The obtained particles were then injected into the central stream of the second microreactor, and precursors for sol–gel formation, namely, tetraethoxy silane (TEOS) and fluorescently labelled APTES, were introduced *via* the outer stream. This mixture was pumped into the third microreactor together with ammonia. The resulting hydrolysis and condensation yielded magnetic nanoparticles encapsulated in a fluorescent silica shell. The reaction was quenched off-chip. Residence times in each microreactor were 2–3 min, and the three reaction steps were accomplished within 7 min as opposed to the 24 h required in conventional systems.

Iron oxide nanoparticles were also synthesised in a polydimethylsiloxane (PDMS) microchannel with integrated mixing elements[8]. Initially, acetic solutions of $FeCl_2$ and $FeCl_3$ were introduced and mixed for 1 min. Then an aqueous sodium hydroxide solution was slowly added over a period of 12 min. Finally, an organic acid was added to stabilise the suspension. Under optimised conditions, spherical Fe_3O_4 particles were formed with an average size of around 5 nm and standard deviations of about 20%.

Yujin Song et al. have published extensively on the synthesis of magnetic nanoparticles in microfluidic devices. They generally employed a microfluidic set-up with two inlets merging into a serpentine channel of about 20–50 cm in length. Solutions were kept under a nitrogen atmosphere and temperatures could be controlled as required. Reactions were either quenched off-chip or an additional quenching inlet was included on the chip itself. By varying reaction temperature, time for ripening and quenching conditions, it was possible to tailor crystal structure, magnetic properties and size of the nanoparticles. Initially, *worm-like iron nanoparticles* were synthesised at room temperature by pumping $FeCl_2$ in tetrahydrofuran (THF) solution through the first inlet and a solution with

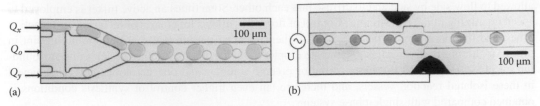

(a)   (b)

FIGURE 10.5 Fusion of droplet pairs with iron chloride and a base allows for a controlled initiation of the reaction within a very confined reaction vessel. (a) Two aqueous phases, Q_x and Q_y, were synchronously broken into droplets by a central oil stream, Q_o. (b) Pairs of droplets were fused *via* electrocoalescence by applying an electrical voltage U between two electrodes. (Copyright Wiley-VCH Verlag. Reproduced with permission.)

the reducing agent, LiBEt$_3$H, and a stabiliser, polyvinylpyrrolidone (PVP), in THF through the second inlet[9]. *Cobalt nanoparticles* were synthesised by pumping CoCl$_2$ in THF through the first inlet and a mixture of LiBEt$_3$H and stabiliser 3-(*N*,*N*-dimethyldodecylammonium) propane-sulphonate in THF through the second inlet[10]. Finally, *CoSm alloy nanoparticles* were synthesised by introducing a CoCl$_2$ and SmCl$_3$ mixture into the first inlet and the reducing agent LiBEt$_3$H together with PVP through the second inlet[11].

A *droplet-based microreactor* was investigated for the synthesis of *iron oxide nanoparticles* (Figure 10.5)[12]. Two types of aqueous droplets were produced, containing an Fe^{2+}/Fe^{3+} salt mixture and ammonium hydroxide solution, respectively. Droplet pairs were pumped past a set of electrodes. A voltage applied between the electrodes enabled controlled merging of two of the alternate droplets by electrocoalescence. Iron oxide precipitate was formed within 2 ms of droplet fusion, and monocrystalline particles with (220) spinel planes of 4 ± 1 nm diameter were obtained, compared with 9 ± 3 nm in a bulk experiment.

10.2.2 SYNTHESIS OF POLYMER PARTICLES WITH INCORPORATED MAGNETIC NANOPARTICLES

Droplets generated in microfluidic devices can also be *polymerised on-chip via* UV-radiation or *via* heat to form microparticles (Figure 10.6). Several research groups have incorporated magnetic nanoparticles into such droplets to achieve magnetic functionality.

In the simplest set-up, magnetic nanoparticles can be added to the dispersed phase and thus be incorporated into the droplets and finally the polymerised microparticles (Figure 10.6a). For example, polycaprolactone (PCL) microcapsules were formed on-chip with encapsulated quantum dots, iron oxide nanoparticles and the anti-cancer drug tamoxifen[13]. Such capsules could act as multifunctional drug delivery vesicles allowing for fluorescence imaging as well as magnetic targeting. Oil droplets of PCL, tamoxifen, quantum dots and 5 nm magnetic iron oxide nanoparticles in chloroform were formed in a continuous phase of aqueous poly(vinyl alcohol) (PVA). A cross-linking reaction between PCL and PVA occurred. Average droplet size could be controlled *via* the applied flow rates to between 270 and 440 μm with relative standard deviation (RSD) < 4%. The droplets were collected off-chip, and the chloroform was evaporated, leading to shrinkage and cross-linking of the PCL matrix. The obtained capsules were between 50 and 200 μm in diameter and showed superparamagnetic behaviour. In another example, polylactide (PLA) microspheres were synthesised in a coaxial microfluidic device[14]. An inner stream of PLA in dichloromethane including oleic acid–coated magnetic nanoparticles was pumped into a sheath of aqueous PVA solution. Droplets were formed and collected off-chip to complete cross-linking. The 300–500 μm diameter spheres showed superparamagnetic behaviour.

Droplets can also be formed from gases. Microbubbles loaded with magnetic nanoparticles are potential contrast enhancing agents. Park et al.[15] formed such gas droplets in a continuous aqueous phase containing a protein (lysozyme), a polysaccharide (alginate) and anionic magnetic

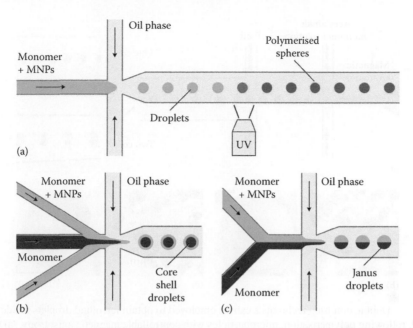

FIGURE 10.6 Magnetic nanoparticles (MNPs) can be incorporated into polymeric microspheres by on-chip continuous flow droplet generation and downstream polymerisation: (a) magnetic nanoparticles dispersed throughout the microparticle, (b) synthesis of core–shell microparticles with magnetic functionality and (c) synthesis of Janus particles with magnetic functionality.

nanoparticles. The gaseous stream contained the water-soluble gas CO_2 as well as a small amount of low water-soluble gases such as nitrogen, oxygen, helium and neon. After droplet formation, the carbon dioxide dissolved into the continuous phase and the bubbles shrank, while the medium around the bubbles became acidic. This in turn led to a process of co-deposition of the protein, polysaccharide and nanoparticles at the bubble/water interface due to electrostatic interactions. The shrunk and stabilised bubbles featured diameters down to 5 µm with a narrow size distribution of RSD < 6% and were shown to be stable for up to 3 months.

With a slightly more complex channel design, *core/shell particles* can be produced (Figure 10.6b). Gong et al.[16] demonstrated this approach for the synthesis of core/shell particles for drug delivery. At a first on-chip junction, an aqueous aspirin solution was enveloped between two streams of an aqueous solution of magnetite nanoparticles and chitosan. Downstream, at a second junction, hexadecane was introduced, and water-in-oil droplets were formed. The droplets featured an aspirin core and chitosan shell with embedded magnetite. Depending on flow rates, the droplet sizes could be adjusted to between 20 and 400 µm. In order to solidify the outer layer, *n*-butanol and glutaraldehyde were added at a junction further downstream. Collection off-chip was then followed by baking at 60°C for 2 h to fully cure the spheres. By applying AC magnetic fields, the spheres were found to extend and compress, and the embedded aspirin could be released.

Such a two-junction approach can also be used to form *double emulsions* (Figure 10.7)[17]. At the first junction, a ferrofluid suspension with styrene and divinylbenzene monomers as well as initiators was dispersed into a continuous aqueous phase of acrylamide monomer solution. The ferrofluid droplets in aqueous acrylamide solution were then guided towards a second junction, where a fluorocarbon oil formed an oil/water/oil double emulsion. The obtained droplets, depending on flow rates, could contain one or two of the smaller ferrofluid droplets. The droplets were collected off-chip, UV cured and thermally treated to fully polymerise them.

Laminar flow regimes within microfluidic devices can also be employed for the formation of *Janus particles*, that is, particles with two distinct halves (Figure 10.6c). Several groups have

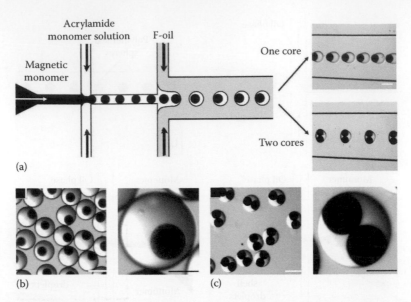

(a)

(b)   (c)

FIGURE 10.7 Double emulsion techniques can be employed to obtain ferrofluid droplets inside acrylamide droplets and, following polymerisation, microparticles with controllable magnetic anisotropy: (a) Chip design for the formation of double emulsion droplets. (b) Images of particles with one core and (c) images of particles with two cores. The white scale bar is 50 μm, and the black scale bar is 20 μm. (From Chen, C.H., Abate, A.R., Lee, D.Y., Terentjev, E.M., and Weitz, D.A.: Microfluidic assembly of magnetic hydrogel particles with uniformly anisotropic structure. *Adv. Mater.* 2009. 21. 3201. Copyright Wiley-VCH Verlag GmbH & Co. KGaA. Reproduced with permission.)

recently shown the incorporation of magnetic nanoparticles into such Janus particles. Magnetic functionality to allow for controlled movement is therefore localised in one-half of the particle, whereas any biofunctionality can be contained within the other half, separated from the magnetic particles to avoid chemical interference and to maintain optical performance for detection.

Yuet et al.[18] employed a commercial aqueous ferrofluid with 10 nm Fe_3O_4 particle for this purpose (Figure 10.8). A magnetic pre-polymer solution with ferrofluid, poly(ethylene glycol) diacrylate, 2-hydroxy-2-methylpropiophenone, glycerol and water was merged with a non-magnetic pre-polymer solution that contained a fluorescent dye such as rhodamine or biomolecules such as DNA. From these two co-flowing laminar streams, droplets were formed at a downstream junction with mineral oil. The droplets were polymerised with UV radiation before any mixing between the liquids in the two halves could occur. The resulting Janus particles were 48 μm in diameter. The authors showed how these Janus particles could be assembled into chain-like structures and hinted at their possible use for bottom-up assembly in tissue engineering applications.

Lewis et al.[19] synthesised hydrogel Janus particles with magnetic nanoparticles in one-half and a genetically modified tobacco mosaic virus doped with catalytic Pd particles in the other half. The aqueous virus and magnetic nanoparticle streams both contained poly(ethylene glycol) and diacrylate. They were flowed side by side and guided towards a junction with a mineral oil, where droplets with two distinct halves were formed. UV-photo-polymerisation led to Janus particles with diameters of about 40 μm.

Seiffert et al.[20] synthesised hydrogel particles from poly(N-isopropylacrylamide) (pNIPAAm) with a red fluorescent dye in one-half, a green fluorescent dye in the other half and chains of magnetic nanoparticles aligned perpendicular to the colour pattern. The resulting spheres were essentially mini-magnets, with red and green poles and permanent magnetic moment due to the particle chains. To achieve this, a chip design with two junctions was employed (Figure 10.9). At the first junction, three flow streams were merged. A central stream contained a suspension of 300 nm diameter ferromagnetic particles. One of the side channels contained a solution of

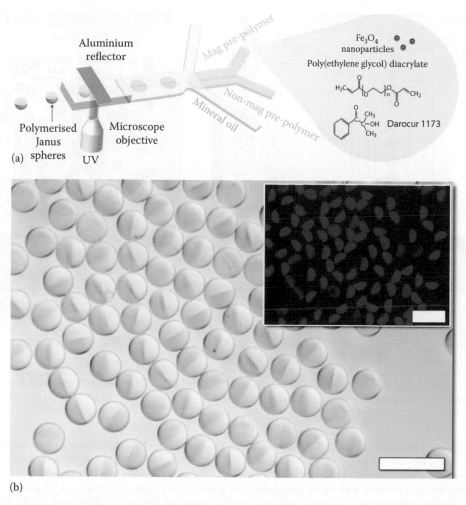

FIGURE 10.8 Magnetic Janus particle can be synthesised on-chip by exploiting laminar flow behaviour, controlled droplet formation and on-chip polymerisation: (a) schematic of chip design and set-up, (b) optical microscopy images of particles with magnetic and fluorescent halves. The scale bar is 100 μm wide in the larger image and 25 μm wide for the inserted fluorescent image. (Reprinted with permission from Yuet, K.P., Hwang, D.K., Haghgooie, R., and Doyle, P.S., Multifunctional superparamagnetic Janus particles, *Langmuir*, 26, 4281–4287, 2010. Copyright 2010 American Chemical Society.)

pNIPAAm functionalised with a red fluorescent dye, the other side channel contained a similar solution with green fluorescent dye. Downstream, two oil streams were introduced to enable droplet formation. The viscous droplets were collected in a wide channel and UV polymerised in an applied magnetic field.

Zhao et al.[21] incorporated cells into magnetic alginate Janus particles. Two laminar flow streams of aqueous sodium alginate solution, one containing 520 nm Ademtech particles and the other HeLa cells, were formed on-chip. Water droplets were generated from these two streams at a junction with an oil phase. Gelation of the alginate was induced further downstream by the addition of calcium chloride droplets. Depending on the applied flow rates, Janus particles of 45–60 μm were formed. Cell viability was comparable, albeit somewhat lower than in conventional alginate sheets.

The examples outlined earlier demonstrate how spatial and temporal control in microfluidic reactors can be exploited to control local reaction conditions and therefore enable the synthesis of materials with more confined properties than is typically achievable on the larger scale.

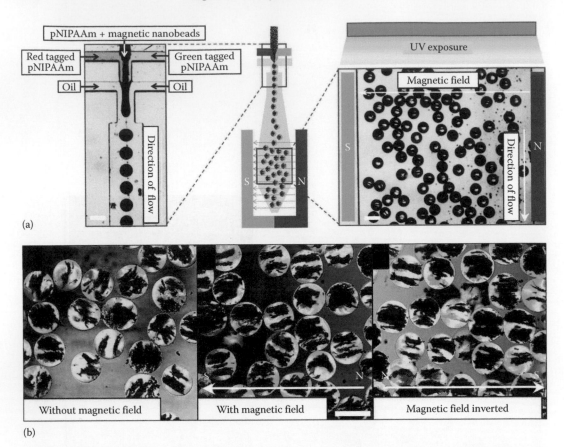

FIGURE 10.9 Janus particles with red and green halves and magnetic particle chains, oriented perpendicular to the boundary, were produced. (a) Three inlets featured aqueous solutions of pNIPAAm, the central stream contained magnetic nanoparticles, while the two outer streams contained red and green fluorescent labels. Droplets were obtained at a junction with an oil phase and were guided into a wider microchannel for polymerisation with UV under application of a magnetic field. The scale bar represents 150 μm. (b) Micrographs of the produced Janus particles under three experimental conditions: without application of magnetic field, the magnetic bead chains were randomly oriented; with magnetic field, the chains were oriented along the field lines; with inverted magnetic field, the particles switched their orientation by 180°. The scale bar equals 200 μm. (Reprinted with permission from Seiffert, S., Romanowsky, M.B., and Weitz, D.A., Janus microgels produced from functional precursor polymers, *Langmuir*, 26, 14842–14847, 2010. Copyright 2010 American Chemical Society.)

10.3 APPLICATIONS OF MAGNETIC NANOPARTICLES IN LAB-ON-A-CHIP DEVICES

Magnetic nanoparticles are not only synthesised in microfluidic lab-on-a-chip devices, they are also employed in a range of useful applications such as enhancing mixing, enabling pumping or to perform bioanalysis with magnetic particle–based assays or magnetically labelled cells. Forces on individual nanoparticles are usually too small for noticeable effects; therefore, in most applications, large numbers of nanoparticles are employed in combination with strong magnetic fields, typically provided by permanent NdFeB magnets.

10.3.1 PUMPING

Plugs of ferrofluid inside a microchannel can be employed as pistons to push liquids. The ferrofluid needs to be immiscible with the liquid to be pumped; therefore, hydrophobic ferrofluids are used to pump aqueous solutions. The ferrofluid plugs can be moved with an external magnet, and

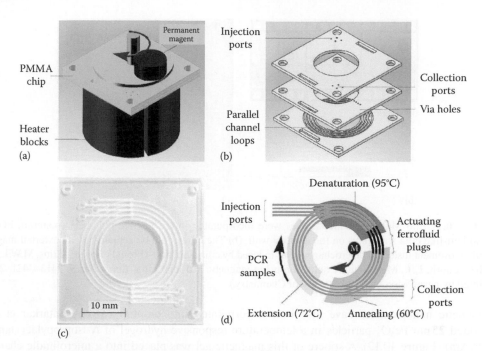

FIGURE 10.10 Plugs of ferrofluid can be employed for on-chip circular pumping with external magnets to perform the PCR in continuous flow. (a) Schematic of the set-up featuring three heater blocks, a PMMA chip and a rotating permanent magnet. (b) The PMMA chip was constructed from three layers. (c) Photograph of the fabricated microfluidic chip showing three concentric PCR loops. (d) Samples and ferrofluids were injected into the three different loops. The moving external magnet resulted in movement of the ferrofluid plugs and therefore pumping of the PCR mixtures in the three sample loops over the heating blocks. (Reprinted with permission from Sun, Y., Nguyen, N.T., and Kwok, Y.C., High-throughput polymerase chain reaction in parallel circular loops using magnetic actuation, *Anal. Chem.*, 80, 6127–6130, 2008. Copyright 2008 American Chemical Society.)

this in turn results in displacement of liquid[2,22]. Such a pumping scheme is particularly attractive for circular channels as has been shown for polymerase chain reactions (PCRs) in such channels (Figure 10.10)[23–25]. A heat-sensitive ferrofluid was employed by Love et al.[26] for magnetocaloric pumping. Metal-reducing bacteria were used to obtain iron oxide nanoparticles with incorporated manganese and zinc, which featured Curie temperatures below 80°C. A uniform magnetic field and a temperature gradient were applied over a plug of ferrofluid made from such particles. When the Curie temperature was reached, the fluid lost its magnetic attraction and was displaced by cooler fluid. With the resulting pressure gradient, a flow of $2\,mm\,s^{-1}$ was achieved.

Magnetic nanoparticles have also been incorporated into polymeric materials such as PDMS. Fahrni et al.[27] assembled a microchannel with magnetic artificial cilia on the channel wall (Figure 10.11). Iron nanoparticles with carbon shells of 70 nm diameter were mixed with the PDMS precursors. The microfabricated cilia were 300 μm long, 100 μm wide and 15 μm thick and could be actuated by rotating magnetic fields of a few tens of millitesla. The cilia movement led to net movement of the liquid inside the microchannel.

10.3.2 MIXERS, VALVES AND SWITCHES

Magnetic nanoparticles can also be used as passive entities to bring about *mixing* of fluid streams in microfluidic channels, which may be desired when diffusion-based mixing is too slow. Wen et al.[28] showed mixing between two laminar flow streams containing a ferrofluid and fluorescent rhodamine B dye, respectively. An AC electromagnetic field was applied along the microchannel, which induced movement of the ferrofluid particles and led to 95% mixing between the two streams in 2.0 s at a distance of 3.0 mm from the merging point.

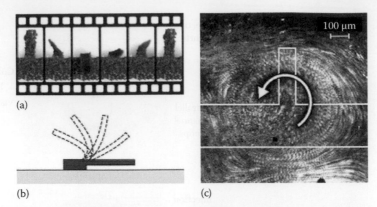

FIGURE 10.11 (a) Magnetic nanoparticles were incorporated into a flexible polymeric material, PDMS, to yield hair-like cilia structures on the channel wall. (b) The cilia could be actuated by an external magnet. (c) Fluid movement inside the microchannel was induced by cilia actuation. (From Fahrni, F., Prins, M.W.J., and Van Ijzendoorn, L.J., Micro-fluidic actuation using magnetic artificial cilia, *Lab Chip*, 9, 3413–3421, 2009. Reprinted by permission of Royal Society of Chemistry.)

Magnetic nanoparticles have also been incorporated into on-chip valves[2]. Satarkar et al.[29] embedded 25 nm Fe_3O_4 particles in a temperature responsive hydrogel of N-isopropylacrylamide (NIPAAm) (Figure 10.12). A sphere of this magnetic gel was placed into a microfluidic channel. In its swollen, cool state, it blocked fluid from flowing through the channel. When an alternating magnetic field (AMF) was applied, the magnetic particles become hotter and in turn heated the gel. At a temperature of about 30°C, the gel started to shrink, fluid could pass the gel ball, and therefore, the valve was in its open state. The process was reversible. A similar concept was also demonstrated by Ghosh et al.[30] They employed a 300 µm diameter cylindrical gel of N-isopropylacrylamide with embedded magnetic Fe_3O_4 particles of 25–35 nm diameter. The gel cylinder blocked a channel in its swollen state and shrank when heated to allow liquid to flow past. In both examples, the shrinking process happened within a few seconds, while the swelling took in the order of minutes.

Latham et al.[31] presented a method for *switching flow streams*. Two orthogonal microfluidic channels were fabricated in a PDMS chip device. Magnetic nanoparticles such as Fe_2O_3 and $MnFe_3O_4$, as well as non-magnetic gold nanoparticles, were pumped through one of the channels. In the absence of a magnetic field, the particles remained in their original channel when passing the orthogonal junction. When a magnet was placed at the junction, magnetic nanoparticles were dragged from one channel into the other, while the gold particles remained in their original stream. The transfer of particles was not complete. However, the set-up did allow for multiple injection of magnetic particle plugs into a desired channel by periodic application of the magnetic field.

10.3.3 Magnetic Droplets

As mentioned earlier in this chapter, uniformly sized droplets can be generated within microfluidic devices and utilised as nano- or picolitre reaction vessels. It is often desirable to merge, split or move droplets to certain locations. For such droplet manipulation, electric forces are used most commonly, as seen in Figure 10.5. There are some examples in the literature of droplet manipulation based on magnetic forces; here two examples are shown for droplets with incorporated magnetic nanoparticles.

Zhang et al.[32] synthesised iron oxide nanoparticles and dispersed them as an aqueous ferrofluid. Droplets were generated on-chip with an oil as the continuous phase (Figure 10.13a). The researchers guided the droplets into a chamber and applied a magnetic field perpendicular to the direction of flow, which resulted in the droplets being deflected from their laminar flow path. Depending on the amount of magnetic nanoparticles inside the droplet and depending on the magnetic field applied, the droplets could be guided to one of the three outlets. Up to 10 droplets could be manipulated per second.

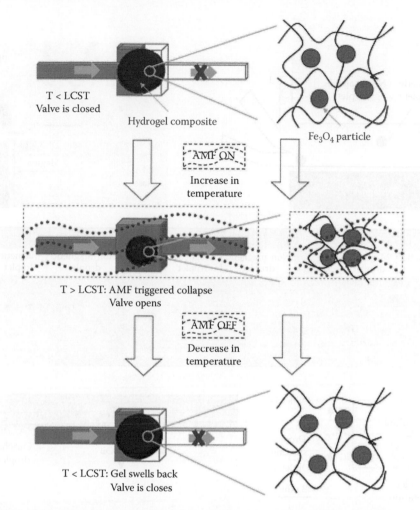

FIGURE 10.12 Magnetic nanoparticles were incorporated into a heat-sensitive hydrogel sphere that was placed in a microchannel to act as a valve. When an AMF was applied, the hydrogel was heated to a temperature (T) above its lower critical solution temperature (LCST) and collapsed, thereby allowing fluid flow. Upon cooling, the hydrogel swelled back to its original size and blocked the fluid flow. (From Satarkar, N.S., Zhang, W.L., Eitel, R.E., and Hilt, J.Z., Magnetic hydrogel nanocomposites as remote controlled microfluidic valves, *Lab on a Chip*, 9, 1773–1779, 2009. Reprinted by permission of Royal Society of Chemistry.)

Al-Hetlani et al.[33] employed a commercially available aqueous ferrofluid and generated water-in-oil droplets. This group also demonstrated the deflection of magnetic droplets, in this case to one of the four outlets, depending on nanoparticle loading and applied magnetic field. Furthermore, the splitting of droplets into magnetically enriched and depleted daughter droplets was demonstrated (Figure 10.13b). Both methods were continuous flow manipulations of droplets with external magnetic fields. The magnetic nanoparticles either could either act as passive entities simply fulfilling the purpose of manoeuvring droplets as desired or could be functionalised for specific molecular recognition of sample components.

10.3.4 PARTICLE BEDS

Magnetic particles can be packed inside a microfluidic channel to form particle beds that perform functions depending on the particles' surface functionalisation (Figure 10.14).

Chen et al.[34] employed a bed of magnetic nanoparticles for *solid-phase extraction of heavy metals*. Silica-coated magnetic nanoparticles were functionalised with γ-mercapto propyl trimethoxy silane

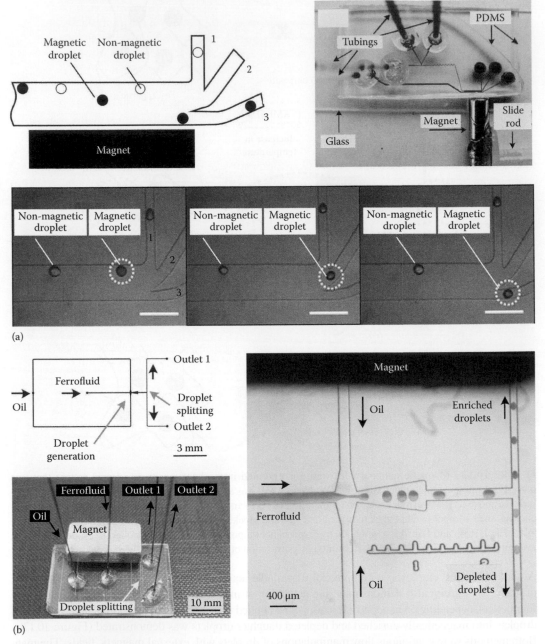

FIGURE 10.13 (a) Aqueous droplets containing varying amounts of ferrofluid were generated on-chip and sorted into one of the three outlets depending on magnetite content and magnetic field strength. (From Zhang, K., Liang, Q.L., Ma, S., Mu, X.A., Hu, P., Wang, Y.M., and Luo, G.A., On-chip manipulation of continuous picoliter-volume superparamagnetic droplets using a magnetic force, *Lab on a Chip*, 9, 2992–2999, 2009. Reprinted by permission of Royal Society of Chemistry.) (b) Magnetic droplets of ferrofluid were generated on-chip and split into magnetically enriched and depleted daughter droplets. (Reprinted with permission from Al-Hetlani, E., Hatt, O.J., Vojtíšek, M., Tarn, M.D., Iles, A., and Pamme, N., Sorting and manipulation of magnetic droplets in continuous flow, *AIP Conf. Proc.*, 1311, 167–175, 2010. Copyright 2011 American Institute of Physics.)

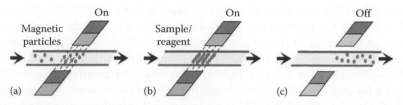

FIGURE 10.14 Trapping of magnetic particles to form a plug inside a microfluidic channel. (a) Magnetic particles are pumped into a microchannel with a magnetic field applied and become trapped. (b) Reagents or sample can be flushed over the particle bed as required. (c) The particles can be released by switching off the magnetic field.

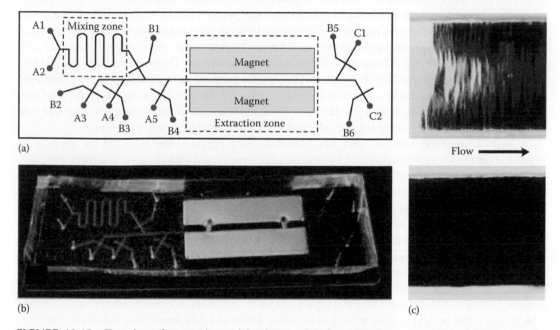

FIGURE 10.15 Trapping of magnetic particles into a plug for solid-phase extraction of heavy metals: (a) chip design, (b) photograph of the set-up, (c) enlarged photographs of the magnetic particle bed being formed inside the microchannel (top) and after completed bed formation (bottom). (Adapted from Chen, B.B., Heng, S.J., Peng, H.Y., Hu, B., Yu, X., Zhang, Z.L., Pang, D.W., Yue, X., and Zhu, Y., Magnetic solid phase microextraction on a microchip combined with electrothermal vaporization-inductively coupled plasma mass spectrometry for determination of Cd, Hg and Pb in cells, *J. Anal. Atom. Spectrom.*, 25, 1931–1938, 2010. Reprinted by permission of Royal Society of Chemistry.)

(γ-MPTS). These 50–70 nm particles were trapped inside in a microchannel by external NdFeB magnets to form a particle bed that acted as a solid-phase extraction matrix for metals due to the thiol functional groups on the particles' surface (Figure 10.15). Samples containing Cd, Pb and Hg were flushed through the particle bed, and the metals were trapped on the particles. After washing of the bed, the metals were eluted with acidic thiourea solution and analysed off-chip *via* inductively coupled plasma–mass spectrometry (ICP–MS). Enrichment factors of >40 were achieved, and limits of detection were 0.72 ng L^{-1} for Cd, 0.86 ng L^{-1} for Hg and 1.12 ng L^{-1} for Pb. The chip design also allowed for metal analysis of cells. HepG2 cells that had been incubated with Cd, Pb and Hg were introduced *via* one inlet and a lysing agent *via* another inlet. The cell stream and lysing stream were allowed to mix, and the cells were thus ruptured on-chip before being pumped through the particle bed for metal trapping. The detectable values were several orders of magnitude lower than IC50 values; therefore, this miniaturised and integrated method could have potential for the analysis of real samples.

Fe$_3$O$_4$ nanoparticles coated with silica and octadecyl silane (ODS) were used for *solid-phase extraction of organic molecules* followed by downstream electrophoretic separation[35]. Using electroosmotic flow for pumping, the ODS-coated particles were captured within a microfluidic chip *via* a small electromagnet. A mixture of the organic molecules, rhodamine 110 and sulforhodamine B, was flushed over the particle bed and adsorbed on the surface. In a subsequent elution step, the two components were removed from the particle bed into a separation channel and separated in less than 1 min.

Qu et al.[36] employed a bed of magnetic particles with molecular imprints on their surface as a stationary phase for on-chip *capillary electrochromatography of enantiomers*. Magnetic nanoparticles of 25 nm were functionalised with 3-methacryloyloxy propyl trimethoxy silane, and a polymer shell was produced around the particles *via* co-polymerisation of methacrylic acid and ethylene glycol dimethacrylate in the presence of the template molecule S-ofloxacin. The 200 nm diameter particles were pumped into a microchannel and held in place with an external magnet to form the stationary-phase bed. When a mixture of ofloxacin enantiomers was pumped through this bed, S-ofloxacin was retained while the R-ofloxacin passed un-retained and thus the two enantiomers could be separated.

Tryptic digestion is a further application for beds formed from magnetic nanoparticles[37]. Commercially available beads of 290 and 570 nm were functionalised with trypsin. The particles were trapped in a microchannel with external NdFeB magnets. The smaller particles were, however, found to form too dense a plug that caused high back pressure. Results for the 570 nm particles were shown for the digestion of five proteins ranging from 4.2 to 150 kDa. The authors stated that the microfluidic approach with the large surface to volume ratio led to a reduction in digestion time from hours to minutes.

A bed of magnetic *nanoparticles functionalised with enzymes* was reported for triglyceride detection on-chip[38]. Fe_3O_4 particles of 80 nm diameter were coated with chitosan and three enzymes: lipase, glycerokinase and glycerol-3-phosphate oxidase. The resulting particles were 120–140 nm in size. They were pumped through a microfluidic device *via* electroosmotic flow and trapped with an external NdFeB magnet. Triglyceride solution was then flushed through the particle bed. The triglyceride reacted with the three enzymes to form hydrogen peroxide, which could be detected electrochemically downstream. A linear range of 0–10 mM was reported, with a limit of detection of 0.6 μM. Analysis of serum samples was also shown.

Trapping of magnetic nanoparticles can be enhanced by either incorporating some form of barrier into the microchannel design[39] or by concentrating the magnetic field gradients[40,41]. Larcharme et al.[39] designed a zipper-like structure for the trapping of chains of 500 nm Ademtech beads (Figure 10.16A). The chains could be held in place as long as magnetic forces and drag forces were balanced. The slight movement of the particle chains as a result of drag forces was found to increase particle-reagent interactions when assays for murine monoclonal antibodies were performed on the surface of the beads. The set-up allowed for a limit of detection of 1 ng mL^{-1} from a sample volume of 31 nL. Teste et al.[40] employed magnetic iron beads of 6–8 μm diameter that were packed into a microfluidic chamber to focus the field lines of an external magnetic and thus to create high local field gradients that would enhance magnetic forces on 30 nm particles flushed through the bed (Figure 10.16B). They reported trapping/release of the nanoparticles within 20 s, with a pre-concentration factor of 4000. Chen et al.[41] employed a packing of iron particles of 25–75 μm for concentrating external magnetic field lines (Figure 10.16C). Their analyte, plasma samples with HIV Type 1 virions, was incubated on-chip with magnetic nanoparticles functionalised with anti-CD44. The magnetic particles with bound HIV were then trapped and therefore concentrated in the iron packing chamber. Finally, a lysing agent was pumped through the chamber and extraction efficiencies of up to 62% with up to 80-fold concentration were reported.

10.3.5   SEPARATION OF MAGNETIC NANOPARTICLES

While there are many examples of the separation of magnetic *micro*particles employing the micrometre spatial control within microfluidic devices[1,2], the differential separation of magnetic nanoparticles is more challenging.

Beveridge et al.[42] reported on a separation device similar to a chromatographic column for 8–17 nm diameter $CoFe_2O_4$ nanoparticles. They wrapped a 250 μm i.d. fused silica capillary around a spacer that was placed between the poles of two electromagnets. By carefully adjusting

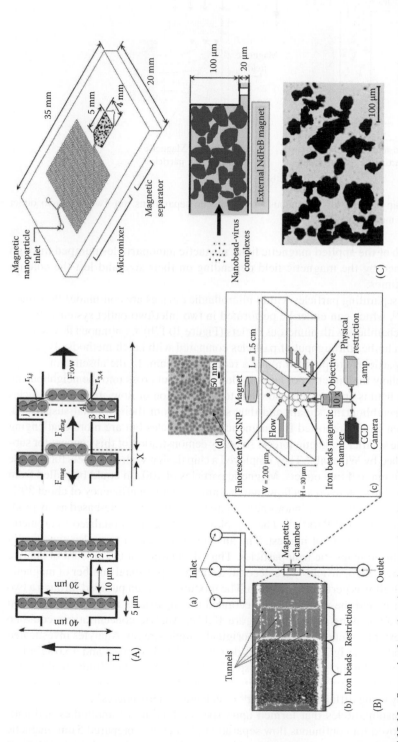

FIGURE 10.16 Strategies for enhancing the trapping of magnetic nanoparticles within microfluidic channels. (A) A zipper-like structure was employed to form barriers that helped retaining chains of 500 nm magnetic particles against the drag forces caused by fluid movement. (Reprinted with permission from Lacharme, F., Vandevyver, C., and Gijs, M.A.M., Full on-chip nanoliter immunoassay by geometrical magnetic trapping of nanoparticle chains, *Anal. Chem.*, 80, 2905–2910, 2008. Copyright 2008 American Chemical Society.) (B) A packing of 30 nm magnetic iron beads of 6–8 μm diameter was used to concentrate field lines from an external magnet and therefore to increase the trapping efficiency of 30 nm magnetic particles. (From Feste, B., Malloggi, F., Gassner, A.-L., Georgelin, T., Siaugue, J.-M., Varenne, A., Girault, H., and Descroix, S., Magnetic core shell nanoparticles trapping in a microdevice generating high magnetic gradient, *Lab on a Chip*, 11, 833–840, 2011. Reprinted by permission of Royal Society of Chemistry.) (C) Iron particles of 25–75 μm were filled into a microdevice to concentrate the field lines in order to improve the capture of magnetic nanoparticles for HIV analysis. (Reprinted with permission from Chen, G.D., Alberts, C.J., Rodriguez, W., and Toner, M., Concentration and purification of human immunodeficiency virus Type 1 virions by microfluidic separation of superparamagnetic nanoparticles, *Anal. Chem.*, 82, 723–728, 2011. Copyright 2011 American Chemical Society.)

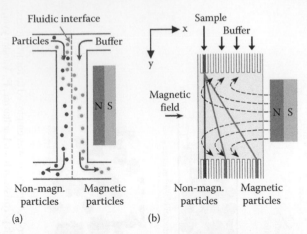

FIGURE 10.17 Typical chip designs for continuous flow magnetic separations. (a) Two inlet/two outlet design and (b) chamber with multiple outlets.

flow rates and the strength of the applied magnetic field, magnetic nanoparticles pumped through this capillary were retained by the magnetic field depending on their size and left the column after different residence times.

More efficient ways of separating particles within microfluidic devices are continuous flow magnetic separation methods[43], which can either be performed in two inlet/two outlet systems (Figure 10.17a) or in microfluidic chambers with numerous outlets (Figure 10.17b). Continuous flow separations generally allow for a higher throughput of particles compared with batch methods. Typically, suspensions of magnetic particles are pumped along a receiving stream. In the absence of a magnetic field, the particles follow the direction of laminar flow and do not cross into the neighbouring stream. When a magnetic field is applied perpendicular to the direction of flow, magnetic particles can be dragged into the neighbouring stream and thus separated from their original suspension. Such approaches have been widely reported for magnetic *micro*particles but are more challenging for nanoparticles due to the smaller magnetic forces. An early demonstration of this concept for sub-microparticles was published by Wu et al.[44] They employed a chip design with two inlets for particle suspension and buffer solution and two outlets. Magnetic particles of 100 nm could be pulled from the particle stream into the buffer stream at a flow rate of 1 mm s^{-1} with an efficiency of about 30%.

A considerable body of research for continuous flow separation of nanoparticle-based assay products has been published by Kim and Park[45–47]. They employed surface-functionalised micrometre polymer particles that bind an analyte of interest. The analyte of interest in turn also specifically binds to surface-functionalised magnetic nanoparticles. Thus, depending on the amount of analyte present in the assay mixture, microparticles were decorated with a proportional number of magnetic nanoparticles. The assay itself was performed off-chip. The mixture was then pumped through a two inlet–two outlet chip device, and the magnetically decorated microparticles could be pulled from their original stream into a receiving stream (see Figure 10.17a). Analyte concentrations could be determined based on the extent of deflection from the original sample stream. Analytes investigated included (i) rabbit IgG and mouse IgG with limits of detection of 244 pg mL^{-1} and 15.6 ng mL^{-1}, respectively[45], (ii) streptavidin-biotin[46], (iii) allergen-specific IgE in serum with limits of detection of a few hundred femtomolar[47] and (iv) prostate-specific antigen (PSA) at 45 fg mL^{-1} [48]. Such assays can also be multiplexed by using different fluorescent[45] or coloured microspheres[49].

Clusters of magnetic nanoparticles that formed upon specific binding of antibodies and antigens have also been employed for continuous flow separations. Lai et al.[50] prepared 5 nm magnetic particles that aggregated and disaggregated by the action of a pH-responsive polymer coating of poly(tert-butyl methacrylate-co-N-isopropylacrylamide). The particle surface was further functionalised with biotin. The particles were mixed off-chip with a sample of streptavidin, and nanoclusters

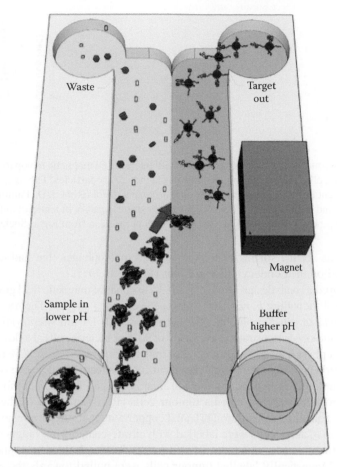

FIGURE 10.18 Transfer of a bioanalyte from one reagent stream into a neighbouring reagent stream was achieved by embedding the analytes in clusters of magnetic nanoparticles that disaggregated at high pH. (From Lai, J.J., Nelson, K.E., Nash, M.A., Hoffman, A.S., Yager, P., and Stayton, P.S., Dynamic bioprocessing and microfluidic transport control with smart magnetic nanoparticles in laminar-flow devices, *Lab on a Chip*, 9, 1997–2002, 2009. Reprinted by permission of Royal Society of Chemistry.)

with incorporated streptavidin were formed. The cluster mixture was then pumped into a two inlet–two outlet chip (Figure 10.18). The stream with magnetic nanoclusters featured a pH of 7.3, while the neighbouring stream featured a higher pH of 8.4. The magnetic aggregates experienced a sufficient magnetic force to be pulled into the neighbouring stream by means of an external magnet. Due to the higher pH, the clusters disaggregated and released the protein of interest into the stream collected at the second outlet with 81% efficiency.

10.3.6 Cells Labelled with Magnetic Nanoparticles

Biological cells labelled with magnetic nanoparticles can also be manipulated within microfluidic lab-on-a-chip devices. Only a few researchers have so far published in this field, and their work is summarised here. Pamme and Wilhelm[51] reported the sorting of human ovarian cancer (HeLa) cells as well as mouse macrophages that had been incubated with 8 nm magnetic maghemite nanoparticles, which led to the internalisation of millions of nanoparticles per cell depending on cell type and incubation time. The cells were then pumped into a microfluidic chamber similar to that shown in Figure 10.17b and were deflected from their direction of laminar flow by means of an external NdFeB magnet. Flow rates of 0.4–2 mm s^{-1} were employed, and it was shown that the extent of deflection depended on magnetite loading and cell size. This work was recently extended to a flow

FIGURE 10.19 Continuous flow separation of cells with internalised magnetic nanoparticles. Depending on endocytotic capacity, cells take up varying amounts of magnetic nanoparticles. The magnetic cells were then sorted into five different fractions of increasing magnetite content. (From Robert, D., Pamme, N., Conjeaud, H., Gazeau, F., Iles, A., and Wilhelm, C. Cell sorting by endocytotic capacity in a microfluidic magnetophoresis device. *Lab on a Chip*, 11, 1902–1910, 2011. Reprinted with permission from Royal Society of Chemistry.)

cell with five outlets for the sorting of monocytes and macrophages that had taken up magnetic nanoparticles based on their endocytotic capacities (Figure 10.19)[52].

Xia et al.[5] employed a comb design (see Figure 10.3c) to enhance magnetic field gradients in the vicinity of a microchannel for pulling *E. coli* cells from one stream into another. The bacteria cells had been reacted with biotinylated anti-*E. coli* antibody and mixed with 130nm streptavidin-coated magnetic nanoparticles. The authors demonstrated the transfer of 90% of bacteria cells at a flow rate of 8.4mm s^{-1}. When mixed with a physiologically relevant concentration of red blood cells, 80% of the magnetically functionalised bacteria cells were transferred into the received stream at a flow rate of 6.9mm s^{-1}.

Derec et al.[4] employed embedded microelectromagnets (see Figure 10.3b) to sort magnetically labelled human prostatic adenocarcinoma tumour cells (PC 3) into one of the two outlets. The magnetic field was generated by a microfabricated copper wire that ran parallel to the microchannel for a length of 5mm. Cancer cells were labelled with citrate-coated 8nm magnetic particles. A 1:1 mixture of magnetic and non-magnetic cells was pumped through the device and at a flow rate of about 500μm s^{-1}. Magnetically labelled tumour cells were pulled towards the electromagnet and exited *via* the outlet near the magnet with 93% efficiency. The different approaches for cell separation illustrate clearly that higher magnetic field gradients and larger magnetic nanoparticles allow for faster flow rates and thus larger volume throughput.

Kim et al.[53] demonstrated the trapping of magnetically labelled cells in a microfluidic channel. Magnetic nanoparticles of about 50nm were isolated from *Magnetospirillum sp. AMB-1* and internalised into endothelial progenitor cells (EPCs). These cells were pumped into a microchannel and were trapped with an external NdFeB magnet. A flow rate of 6.7mm s^{-1} allowed for a trapping efficiency of about 80%.

A quite different method for cell manipulation was demonstrated by Kose et al.[54] They employed a ferrofluid with citrate-coated cobalt-ferrite nanoparticles of 11nm. Patterned electrodes at the bottom of a microfluidic channel were used to generate an alternating travelling wave magnetic field. Diamagnetic repulsion forces could thus act on non-magnetic living cells inside the fluidic channel. The cells were pushed upwards and towards the gap between the electrodes and were found to roll along the channel ceiling. The resulting continuous translation along the wall was dependent on cell shape, as shown for the separation of live red blood cells from sickle cells. Seventy-five percent of cell viability was reported.

10.4 SUMMARY AND CONCLUSION

In this chapter, applications of magnetic nanoparticles within microfluidic lab-on-a-chip devices have been described. The development of clinical applications involving magnetic nanoparticles may well benefit from microfluidic technology. Microfluidic chips offer unique control over fluid flow and particle movement. This has been exploited for the synthesis of magnetic nanoparticles with mono-crystalline structure and narrow size distribution. Such homogeneous particle populations should be favourable for clinical applications such as contrast enhancement in magnetic resonance imaging (MRI), hyperthermia or drug delivery for cancer therapy. Microfluidic channel networks have also been utilised for the formation of monomer-containing droplets and their subsequent polymerisation to yield microspheres with magnetic functionality. These could find clinical applications *in vitro* for analyte extraction or *in vivo* as drug delivery vesicles. Other researchers have demonstrated the readout of particle-based assays with low limits of detection as well as the manipulation and sorting of magnetically labelled cells. Cell populations with very similar magnetic properties can thus be obtained, which is of interest for cell imaging as well as cell therapy. Much progress has been made with systems for the on-chip detection of magnetic nanoparticles. Magnetoresistive sensors[55-57], as well as Doppler tomography[58], have all been reported for the detection of magnetic sub-micron or nanoparticles within microfluidic channels. With more development in this field anticipated, we are likely to see many more integrated microfluidic platforms performing several procedural steps in an automated fashion. Ultimately, it would be desirable to have robust microfluidic platforms on-site that can be operated by a clinician for targeted clinical applications.

REFERENCES

1. Gijs, M. A. M. 2004. Magnetic bead handling on-chip: new opportunities for analytical applications. *Microfluidics and Nanofluidics*, 1, 22–40.
2. Pamme, N. 2006. Magnetism and microfluidics. *Lab on a Chip*, 6, 24–38.
3. Ganguly, R. and Puri, I. K. 2010. Microfluidic transport in magnetic MEMS and bioMEMS. *Wiley Interdisciplinary Reviews – Nanomedicine and Nanobiotechnology*, 2, 382–399.
4. Derec, C., Wilhelm, C., Servais, J. and Bacri, J. C. 2010. Local control of magnetic objects in microfluidic channels. *Microfluidics and Nanofluidics*, 8, 123–130.
5. Xia, N., Hunt, T. P., Mayers, B. T., Alsberg, E., Whitesides, G. M., Westervelt, R. M. and Ingber, D. E. 2006. Combined microfluidic-micromagnetic separation of living cells in continuous flow. *Biomedical Microdevices*, 8, 299–308.
6. Abou Hassan, A., Sandre, O., Cabuil, V. and Tabeling, P. 2008. Synthesis of iron oxide nanoparticles in a microfluidic device: Preliminary results in a coaxial flow millichannel. *Chemical Communications*, 1783–1785.
7. Abou-Hassan, A., Bazzi, R. and Cabuil, V. 2009. Multistep continuous-flow microsynthesis of magnetic and fluorescent gamma-Fe_2O_3@SiO_2 core/shell nanoparticles. *Angewandte Chemie International Edition*, 48, 7180–7183.
8. Lee, W. B., Weng, C. H., Cheng, F. Y., Yeh, C. S., Lei, H. Y. and Lee, G. B. 2009. Biomedical microdevices synthesis of iron oxide nanoparticles using a microfluidic system. *Biomedical Microdevices*, 11, 161–171.
9. Song, Y. J., Jin, P. Y. and Zhang, T. 2010. Microfluidic synthesis of Fe nanoparticles. *Materials Letters*, 64, 1789–1792.
10. Song, Y. J., Henry, L. L. and Yang, W. T. 2009. Stable amorphous cobalt nanoparticles formed by an in situ rapidly cooling microfluidic process. *Langmuir*, 25, 10209–10217.
11. Song, Y. J. and Henry, L. 2009. Nearly monodispersion CoSm alloy nanoparticles formed by an in-situ rapid cooling and passivating microfluidic process. *Nanoscale Research Letters*, 4, 1130–1134.
12. Frenz, L., El Harrak, A., Pauly, M., Begin-Colin, S., Griffiths, A. D. and Baret, J. C. 2008. Droplet-based microreactors for the synthesis of magnetic iron oxide nanoparticles. *Angewandte Chemie International Edition*, 47, 6817–6820.
13. Yang, C. H., Huang, K. S., Lin, Y. S., Lu, K., Tzeng, C. C., Wang, E. C., Lin, C. H., Hsu, W. Y. and Chang, J. Y. 2009. Microfluidic assisted synthesis of multi-functional polycaprolactone microcapsules: Incorporation of CdTe quantum dots, Fe_3O_4 superparamagnetic nanoparticles and tamoxifen anticancer drugs. *Lab on a Chip*, 9, 961–965.
14. Zhu, L. P., Li, Y. G., Zhang, Q. H., Wang, H. Z. and Zhu, M. F. 2010. Fabrication of monodisperse, large-sized, functional biopolymeric microspheres using a low-cost and facile microfluidic device. *Biomedical Microdevices*, 12, 169–177.
15. Park, J. I., Jagadeesan, D., Williams, R., Oakden, W., Chung, S. Y., Stanisz, G. J. and Kumacheva, E. 2010. Microbubbles loaded with nanoparticles: A route to multiple imaging modalities. *ACS Nano*, 4, 6579–6586.

16. Gong, X. Q., Peng, S. L., Wen, W. J., Sheng, P. and Li, W. H. 2009. Design and fabrication of magnetically functionalized core/shell microspheres for smart drug delivery. *Advanced Functional Materials*, 19, 292–297.

17. Chen, C. H., Abate, A. R., Lee, D. Y., Terentjev, E. M. and Weitz, D. A. 2009. Microfluidic assembly of magnetic hydrogel particles with uniformly anisotropic structure. *Advanced Materials*, 21, 3201–3204.

18. Yuet, K. P., Hwang, D. K., Haghgooie, R. and Doyle, P. S. 2010. Multifunctional superparamagnetic Janus particles. *Langmuir*, 26, 4281–4287.

19. Lewis, C. L., Lin, Y., Yang, C. X., Manocchi, A. K., Yuet, K. P., Doyle, P. S. and Yi, H. 2010. Microfluidic fabrication of hydrogel microparticles containing functionalized viral nanotemplates. *Langmuir*, 26, 13436–13441.

20. Seiffert, S., Romanowsky, M. B. and Weitz, D. A. 2010. Janus microgels produced from functional precursor polymers. *Langmuir*, 26, 14842–14847.

21. Zhao, L. B., Pan, L., Zhang, K., Guo, S. S., Liu, W., Wang, Y., Chen, Y., Zhao, X. Z. and Chan, H. L. W. 2009. Generation of Janus alginate hydrogel particles with magnetic anisotropy for cell encapsulation. *Lab on a Chip*, 9, 2981–2986.

22. Hatch, A., Kamholz, A. E., Holman, G., Yager, P. and Bohringer, K. F. 2001. A ferrofluidic magnetic micropump. *Journal of Microelectromechanical Systems*, 10, 215–221.

23. Sun, Y., Kwok, Y. C. and Nguyen, N. T. 2007. A circular ferrofluid driven microchip for rapid polymerase chain reaction. *Lab on a Chip*, 7, 1012–1017.

24. Sun, Y., Nguyen, N. T. and Kwok, Y. C. 2008. High-throughput polymerase chain reaction in parallel circular loops using magnetic actuation. *Analytical Chemistry*, 80, 6127–6130.

25. Sun, Y., Kwok, Y. C., Lee, P. F. P. and Nguyen, N. T. 2009. Rapid amplification of genetically modified organisms using a circular ferrofluid-driven PCR microchip. *Analytical and Bioanalytical Chemistry*, 394, 1505–1508.

26. Love, L. J., Jansen, J. F., Mcknight, T. E., Roh, Y. and Phelps, T. J. 2004. A magnetocaloric pump for microfluidic applications. *IEEE Transactions on NanoBioscience*, 3, 101–110.

27. Fahrni, F., Prins, M. W. J. and Van Ijzendoorn, L. J. 2009. Micro-fluidic actuation using magnetic artificial cilia. *Lab on a Chip*, 9, 3413–3421.

28. Wen, C. Y., Yeh, C. P., Tsai, C. H. and Fu, L. M. 2009. Rapid magnetic microfluidic mixer utilizing AC electromagnetic field. *Electrophoresis*, 30, 4179–4186.

29. Satarkar, N. S., Zhang, W. L., Eitel, R. E. and Hilt, J. Z. 2009. Magnetic hydrogel nanocomposites as remote controlled microfluidic valves. *Lab on a Chip*, 9, 1773–1779.

30. Ghosh, S., Yang, C., Cai, T., Hu, Z. B. and Neogi, A. 2009. Oscillating magnetic field-actuated microvalves for micro- and nanofluidics. *Journal of Physics D: Applied Physics*, 42(13), 135501.

31. Latham, A. H., Tarpara, A. N. and Williams, M. E. 2007. Magnetic field switching of nanoparticles between orthogonal microfluidic channels. *Analytical Chemistry*, 79, 5746–5752.

32. Zhang, K., Liang, Q. L., Ma, S., Mu, X. A., Hu, P., Wang, Y. M. and Luo, G. A. 2009. On-chip manipulation of continuous picoliter-volume superparamagnetic droplets using a magnetic force. *Lab on a Chip*, 9, 2992–2999.

33. Al-Hetlani, E., Hatt, O. J., Vojtíšek, M., Tarn, M. D., Iles, A. and Pamme, N. 2010. Sorting and manipulation of magnetic droplets in continuous flow. *AIP Conference Proceedings*, 1311, 167–175.

34. Chen, B. B., Heng, S. J., Peng, H. Y., Hu, B., Yu, X., Zhang, Z. L., Pang, D. W., Yue, X. and Zhu, Y. 2010. Magnetic solid phase microextraction on a microchip combined with electrothermal vaporization-inductively coupled plasma mass spectrometry for determination of Cd, Hg and Pb in cells. *Journal of Analytical Atomic Spectrometry*, 25, 1931–1938.

35. Tennico, Y. H. and Remcho, V. T. 2010. In-line extraction employing functionalized magnetic particles for capillary and microchip electrophoresis. *Electrophoresis*, 31, 2548–2557.

36. Qu, P., Lei, J. P., Zhang, L., Ouyang, R. Z. and Ju, H. X. 2010. Molecularly imprinted magnetic nanoparticles as tunable stationary phase located in microfluidic channel for enantioseparation. *Journal of Chromatography A*, 1217, 6115–6121.

37. Bilkova, Z., Slovakova, M., Minc, N., Futterer, C., Cecal, R., Horak, D., Benes, M., Le Potier, I., Krenkova, J., Przybylski, M. and Viovy, J. L. 2006. Functionalized magnetic micro- and nanoparticles: Optimization and application to mu-chip tryptic digestion. *Electrophoresis*, 27, 1811–1824.

38. Chen, S.-P., Yu, X.-D., Xu, J.-J. and Chen, H.-Y. 2010. Lab-on-a-chip for analysis of triglycerides based on a replaceable enzyme carrier using magnetic beads. *Analyst*, 135, 2979–2986.

39. Lacharme, F., Vandevyver, C. and Gijs, M. A. M. 2008. Full on-chip nanoliter immunoassay by geometrical magnetic trapping of nanoparticle chains. *Analytical Chemistry*, 80, 2905–2910.

40. Teste, B., Malloggi, F., Gassner, A.-L., Georgelin, T., Siaugue, J.-M., Varenne, A., Girault, H. and Descroix, S. 2011. Magnetic core shell nanoparticles trapping in a microdevice generating high magnetic gradient. *Lab on a Chip*, 11, 833–840.

41. Chen, G. D., Alberts, C. J., Rodriguez, W. and Toner, M. 2010. Concentration and purification of human immunodeficiency virus type 1 virions by microfluidic separation of superparamagnetic nanoparticles. *Analytical Chemistry*, 82, 723–728.

42. Beveridge, J. S., Stephens, J. R., Latham, A. H. and Williams, M. E. 2009. Differential magnetic catch and release: Analysis and separation of magnetic nanoparticles. *Analytical Chemistry*, 81, 9618–9624.

43. Pamme, N. 2007. Continuous flow separations in microfluidic devices. *Lab on a Chip*, 7, 1644–1659.

44. Wu, L. Q., Zhang, Y., Palaniapan, M. and Roy, P. 2009. Magnetic nanoparticle migration in microfluidic two-phase flow. *Journal of Applied Physics*, 105(12), 123909.

45. Kim, K. S. and Park, J. K. 2005. Magnetic force-based multiplexed immunoassay using superparamagnetic nanoparticles in microfluidic channel. *Lab on a Chip*, 5, 657–664.

46. Kim, K. S. and Park, J. K. 2006. Superparamagnetic nanoparticle-based nanobiomolecular detection in a microfluidic channel. *Current Applied Physics*, 6, 976–981.

47. Hahn, Y. K., Jin, Z., Kang, J. H., Oh, E., Han, M. K., Kim, H. S., Jang, J. T., Lee, J. H., Cheon, J., Kim, S. H., Park, H. S. and Park, J. K. 2007. Magnetophoretic immunoassay of allergen-specific IgE in an enhanced magnetic field gradient. *Analytical Chemistry*, 79, 2214–2220.

48. Jin, Z., Hahn, Y. K., Oh, E., Kim, Y. P., Park, J. K., Moon, S. H., Jang, J. T., Cheon, J. and Kim, H. S. 2009. Magnetic nanoctusters for ultrasensitive magnetophoretic assays. *Small*, 5, 2243–2246.

49. Hahn, Y. K., Chang, J. B., Jin, Z., Kim, H. S. and Park, J. Y. 2009. Magnetophoretic position detection for multiplexed immunoassay using colored microspheres in a microchannel. *Biosensors and Bioelectronics*, 24, 1870–1876.

50. Lai, J. J., Nelson, K. E., Nash, M. A., Hoffman, A. S., Yager, P. and Stayton, P. S. 2009. Dynamic bioprocessing and microfluidic transport control with smart magnetic nanoparticles in laminar-flow devices. *Lab on a Chip*, 9, 1997–2002.

51. Pamme, N. and Wilhelm, C. 2006. Continuous sorting of magnetic cells via on-chip free-flow magnetophoresis. *Lab on a Chip*, 6, 974–980.

52. Robert, D., Pamme, N., Conjeaud, H., Gazeau, F., Iles, A. and Wilhelm, C. 2011. Cell sorting by endocytotic capacity in a microfluidic magnetophoresis device. *Lab on a Chip*, 11, 1902–1910.

53. Kim, J. A., Lee, H. J., Kang, H. J. and Park, T. H. 2009. The targeting of endothelial progenitor cells to a specific location within a microfluidic channel using magnetic nanoparticles. *Biomedical Microdevices*, 11, 287–296.

54. Kose, A. R., Fischer, B., Mao, L. and Koser, H. 2009. Label-free cellular manipulation and sorting via biocompatible ferrofluids. *Proceedings of the National Academy of Sciences of the United States of America*, 106, 21478–21483.

55. Martins, V. C., Germano, J., Cardoso, F. A., Loureiro, J., Cardoso, S., Sousa, L., Piedade, M., Fonseca, L. P. and Freitas, P. P. 2010. Challenges and trends in the development of a magnetoresistive biochip portable platform. *Journal of Magnetism and Magnetic Materials*, 322, 1655–1663.

56. Germano, J., Martins, V. C., Cardoso, F. A., Almeida, T. M., Sousa, L., Freitas, P. P. and Piedade, M. S. 2009. A portable and autonomous magnetic detection platform for biosensing. *Sensors*, 9, 4119–4137.

57. Schotter, J., Shoshi, A. and Brueckl, H. 2009. Development of a magnetic lab-on-a-chip for point-of-care sepsis diagnosis. *Journal of Magnetism and Magnetic Materials*, 321, 1671–1675.

58. Kim, J., Oh, J., Milner, T. E. and Nelson, J. S. 2007. Imaging nanoparticle flow using magneto-motive optical Doppler tomography. *Nanotechnology*, 18, 035504.

Nicole Pamme obtained her first degree, a Diploma in Chemistry, at the University of Marburg, Germany. For her PhD, she went to Imperial College London, where she worked on single particle analysis in microfluidic chips under the supervision of Prof. Andreas Manz. This was followed by a stay in Tsukuba, Japan, as an independent research fellow in the International Centre for Young Scientists (ICYS) at the National Institute for Materials Science (NIMS). In December 2005, she took up a lectureship position at the University of Hull, Yorkshire, United Kingdom. She has conducted extensive research on magnetic applications in microfluidic devices.

11 Separation and Characterisation of Magnetic Particulate Materials

P. Stephen Williams

CONTENTS

In this chapter, we focus on the analytical separation and characterisation technique of magnetic field-flow fractionation (MgFFF) and the preparative separation technique of continuous magnetic split-flow thin channel (SPLITT) fractionation. The general concepts for both field-flow fractionation (FFF) and SPLITT fractionation originated with J. Calvin Giddings of the University of Utah in 1966 and 1985, respectively. Both techniques achieve separation within thin, parallel-walled channels carrying a flow of fluid with a field applied across the channel thickness, perpendicular to the direction of flow. The mechanism of separation differs, however. Separation takes place in the direction of flow in the case of FFF, while separation occurs across the channel thickness in the case of SPLITT fractionation.

The channels have uniform, microfluidic thickness and have cross sections of high aspect ratio. Fluid flow therefore has a laminar parabolic velocity profile across the microfluidic thickness that is uniform across the channel breadth. This flow profile is essential for particle separation, and the breadth of the channel also allows for relatively large sample size or throughput as compared to most microfluidic systems. Giddings was a pioneer in the use of microfluidics for particle separation.

11.1 FIELD-FLOW FRACTIONATION

The concept for the technique of FFF was presented in 1966[1,2]. It is a method for separation and characterisation of macromolecules, polymers, colloids and particles up to tens of microns in diameter. It is a separation technique similar to chromatography in operation in which a small sample is introduced into the flow of a carrier solution entering a separation channel and is separated into its different components as they are carried along the length of the channel at differing velocities. As in chromatography, a detector is placed at the channel outlet to register the elution of the sample components from the channel. Unlike chromatography, the separation takes place within the mobile liquid phase alone, rather than with partition between mobile and stationary phases. FFF exploits the differing, non-uniform distributions of sample materials across a gradient in fluid velocity rather than differing distributions between the mobile and stationary phases.

In its original conception, illustrated in Figure 11.1, separation takes place within a fluid driven along a thin, planar, parallel-walled channel with a field of some type applied across the thin channel dimension, perpendicular to the direction of flow. In the thin channel, the fluid takes on a parabolic velocity profile with maximum velocity at the centre and zero at the walls. For channels having cross sections of high aspect ratio, we can ignore the small perturbations to the parabolic profile at the channel edges. It is the differing interaction of the sample components with the applied field and/or their differing diffusion coefficients that leads to their different distributions across the fluid velocity profile that in turn leads to their separation.

A variety of fields have been used to bring about separation, including the earth's gravitational field, sedimentation (or centrifugation), a thermal gradient and an electrical potential gradient. A secondary flow of fluid across the channel thickness entering and exiting *via* semipermeable

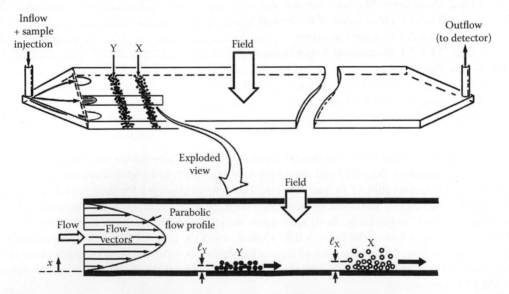

FIGURE 11.1 Schematic of a FFF channel. The expanded edge-on view illustrates the mechanism of differential elution velocity. The more diffuse distribution for the particles X can sample faster flow vectors as well as the slow vectors close to the wall, and the zone has a higher mean elution velocity than zone Y. (From Giddings, J.C., 1993. Field-flow fractionation: Analysis of macromolecular, colloidal, and particulate materials, *Science*, 260, 1456–1465. Reprinted with permission of AAAS.)

walls, resulting in a cross-channel drag force on sample materials, has also been used. These different applied fields and cross flow correspond to different forms of FFF: gravitational FFF (GrFFF), sedimentation FFF (SdFFF), thermal FFF (ThFFF), electrical FFF (ElFFF) and flow FFF (FlFFF, or sometimes F4). Other forms of FFF using different fields and field gradients have also been proposed and investigated[3–5].

Different applied fields interact with different sample properties, and it is often the sample property of interest that determines the particular FFF technique used. The majority of FFF instruments still utilise channels of planar geometry, with one major exception being hollow fibre FFF. This is a form of FlFFF carried out using a semipermeable ultrafiltration fibre rather than a flat channel. Sample materials are carried to the inner wall of the fibre by an outward flow of a fraction of the carrier fluid through the wall[6–9]. In SdFFF, the channel is not planar but is curved around a centrifuge basket, but the field still acts across the thickness of the channel that is essentially parallel-walled[10]. The optimum channel geometry depends on the nature of the field, as we shall see in the following.

The mechanism of particle separation is illustrated in Figure 11.1, which shows the separation of two monodisperse components, X and Y, which interact differently with the applied field and/or have different diffusion coefficients. The particles are driven across the channel thickness by the field towards the so-called accumulation wall. The accumulation of particles next to the wall is opposed by diffusion, and dynamic, steady-state distributions are quickly obtained in the thin channel. Channels are typically between 100 and 300 μm in thickness, and the distributions of particles, or particle zones, adjacent to the wall occupy a small fraction of the channel thickness. The zone thicknesses are exaggerated in Figure 11.1 to more clearly illustrate the separation mechanism. Due to the small distances involved, there is a rapid exchange of particle positions within each zone thickness, and all particles spend time in different fluid streamlines in proportion to the local particle concentration. The particles of a monodisperse sample consequently move along the channel with very nearly the same average velocities. With the imposition of the fluid velocity gradient on the thin sample zone, there is inevitably a small departure from the equilibrium distribution, and a small variation in particle mean velocities. This leads to a broadening of the particle zone as it migrates along the channel. This is known as the non-equilibrium contribution to band spreading[11,12]. In a well-constructed and operated FFF system, this is the major contribution to band spreading.

A polydisperse sample is made up of particles having a wide range of mean velocities that elute from the channel over a wide range of time. The response of the detector at the channel outlet to the eluting particles as a function of time is known as the elution profile. The flow of fluid in the thin channel has a parabolic velocity profile due to the drag of the channel walls, and particles having different steady state distributions are therefore separated by virtue of the differences in the velocities of the fluid streamlines in which they are entrained. Particles that interact strongly with the field and/or have low diffusion coefficients form thin layers adjacent to the accumulation wall while those that interact less strongly and/or have higher diffusion coefficients form thicker layers. Particles occupying thicker layers sample faster streamlines further from the wall as well as those close to the wall, and their averaged velocities are therefore higher. In Figure 11.1, component X, forming the thicker layer, migrates more quickly than component Y. The result is that particles elute in the order of increasing elution time with increasing strength of interaction with the field and/or with decreasing diffusion coefficient. This is equally true for polydisperse samples where the breadth of the elution profile is a reflection of the polydispersity. In fact, for polydisperse samples, the breadth of the elution curve tends to be dominated by the polydispersity and the additional contribution to breadth caused by non-equilibrium band spreading and other effects is negligible.

For polydisperse samples, it is generally necessary to initially apply a high field strength and gradually reduce the field strength during the elution. This is because the particles having weak interaction with the field (often the smallest ones) may be sufficiently retained at the initial high field strength, and those having the strongest interaction are eluted as the field decays. If the high initial field was maintained, the strongly retained particles would take far too long to elute. Programmed field decay, or field programming, and its optimisation have been the subject of many studies over the years[13–22]. The field may be reduced in an infinite number of ways. It is, however, essential for the analysis of polydisperse samples that the field should approach zero asymptotically.

The discussion thus far has concerned the mechanism for the so-called normal mode of elution. Particles larger than around 1 μm in diameter are eluted in the so-called steric mode[23]. In this mode, the back-diffusion from the accumulation wall is negligible for these larger particles and it is principally the size of the particles and how far they project into the flow that determines their migration velocity along the channel. The larger particles project further into the fluid stream and are driven more quickly along the channel. The order of elution for the steric mode is reversed in comparison to the normal mode with the larger particles eluting before smaller particles. Because of this reversal in order, experimental conditions should ideally be arranged so that a sample will elute in either the normal mode or the steric mode, but not both, to avoid co-elution of different sized particles[24,25]. It has been found that in the steric mode, the particles are not driven along the channel in contact with the wall but are raised to positions above the wall where the force due to interaction with the applied field is exactly opposed by hydrodynamic lift forces[26–31].

As mentioned earlier, most implementations of FFF have employed plane, parallel-plate channel geometry. The first FFF channels were constructed using two flat-faced blocks, clamped or bolted around a thin spacer (usually Teflon, Mylar or polyimide) out of which the channel outline had been cut. Ducts through the blocks conveyed the carrier solution to and from the ends of the channel. Planar channels are suited to those fields that can be applied directly across the thickness of the channel, and such channels are still in use for gravitational, thermal, electrical and flow FFF. The quadrupole MgFFF, to be described later, uses a quadrupole magnetic field in which the magnitude of the field increases linearly from the axis of the quadrupole aperture in all directions—the magnitude of the field is therefore axisymmetrical. Such a field symmetry calls for a different channel geometry as explained in Section 11.2.2.

11.1.1 GENERAL RETENTION THEORY

One of the great advantages of FFF is that, for a given set of experimental conditions, the retention time of a sample may be predicted from its properties. Conversely, sample particle properties may be determined from particle retention times. For the ideal model of normal mode FFF retention, it is assumed that there are no particle–particle or particle–wall interactions, and the particles are assumed to have negligible size[4]. In this case, the steady state concentration distribution next to the accumulation wall is obtained by setting the net flux J away from the accumulation wall to zero:

$$J = uc - D\frac{dc}{dx} = 0 \tag{11.1}$$

where
    u is the velocity in the direction away from the wall due to interaction with the applied field (u is therefore negative)
    c is the local concentration
    D is the particle diffusion coefficient
    dc/dx is the local gradient in concentration with distance from the wall x (dc/dx is negative)

Solving Equation 11.1, we obtain the concentration profile:

$$c = c_0 \exp\left(-\frac{|u|x}{D}\right) = c_0 \exp\left(-\frac{|F|x}{kT}\right) \tag{11.2}$$

where
  c_0 is the concentration at the accumulation wall
  F is the force on a single particle in the direction away from the wall due to its interaction with
    the applied field (F is negative)
  k is the Boltzmann constant
  T the absolute temperature

Note that u and F are always negative for this coordinate set-up, and the absolute values are included in Equation 11.2 together with the negative signs, as is the convention in the FFF literature. The second form on the right-hand side of Equation 11.2 is obtained by replacing D with the Nernst–Einstein expression kT/f, where f is the particle friction coefficient, and replacing the field-induced velocity towards the wall $|u|$ by $|F|/f$. Equation 11.2 may be written in the following form:

$$c = c_0 \exp\left(-\frac{x}{\ell}\right) = c_0 \exp\left(-\frac{x}{\lambda w}\right) \tag{11.3}$$

in which
  $\ell = kT/|F|$ and is approximately equal to the mean particle layer thickness
  λ is the so-called FFF retention parameter, which is equal to ℓ/w, where w is the channel thickness

In a thin, parallel-plate channel, the fluid velocity profile is essentially parabolic, ignoring the small perturbations at the channel edges, and is described by

$$v = 6\langle v \rangle \frac{x}{w}\left(1 - \frac{x}{w}\right) \tag{11.4}$$

where
  $\langle v \rangle$ is the mean fluid velocity
  the angle-brackets indicate the averaged value of some enclosed quantity q across the channel thickness, $\langle q \rangle = (1/w)\int_0^w q\, dx$

The retention ratio R is defined as the ratio of particle zone velocity to mean fluid velocity, and is given by

$$R = \frac{\langle cv \rangle}{\langle c \rangle \langle v \rangle} \tag{11.5}$$

in which the numerator integrates the concentration weighted velocity that is then normalised by $\langle c \rangle$ in the denominator. Solving the integrals in Equation 11.5 after substituting in Equations 11.3 and 11.4 results in the following equation:

$$R = 6\lambda\left(\coth\left(\frac{1}{2\lambda}\right) - 2\lambda\right) \approx 6\lambda(1 - 2\lambda) \tag{11.6}$$

where the approximate form on the right is accurate to within 0.37% for λ up to 0.15, or to within 0.20% for R up to 0.6. In practice, retention ratio is determined *via* the ratio of void time to retention time.

11.2 MAGNETIC FIELD-FLOW FRACTIONATION

11.2.1 INITIAL APPROACHES

The early approaches to the development of MgFFF were not very successful for various reasons. Some made use of tubular channels with transverse magnetic field gradients[32–34]. This is far from ideal for FFF as discussed by Giddings[35].

The use of a parallel-plate channel with transverse magnetic field was explored by Schunk et al.[36] The channel was simply placed on top of the core of a fairly large electromagnet. The separation of singlet 0.8 μm rod-shaped iron oxide particles used in the recording industry from doublets of the same particles was demonstrated. The channel was exposed to a maximum field of 0.0275 T and field gradient of 2.1 T/m, and it may be estimated that the magnetic force on the particles was around an order of magnitude greater than that due to gravity.

Others have considered parallel-plate channels with ferromagnetic wires embedded in one or both channel walls, generally aligned along the channel length in the direction of fluid flow[37–46]. The wires were to be magnetised by the application of an external magnetic field. The problem with this approach is that the wires would generate very short-range high-gradient magnetic fields (HGMFs), extending for distances on the order of the wire's diameter. Such high-gradient fields are more suited to particle capture, as in high-gradient magnetic separation (HGMS) systems[47–49] than to FFF. However, even if the particles are not captured, they would be subjected to a magnetic force that varied strongly across the channel breadth. This would lead to a large contribution to band spreading and poor separation.

11.2.2 QUADRUPOLE MgFFF

The quadrupole magnetic field has the property that the magnitude of the field is axisymmetrical and increases linearly with distance from the axis. Such symmetry is most usefully exploited for FFF by a cylindrical or annular channel mounted symmetrically around the quadrupole axis. Quadrupole magnets having apertures of capillary dimensions are not feasible. A practical approach uses a quadrupole magnet having an aperture of 1–2 cm diameter, with a channel occupying an annular region axisymmetrical with the aperture. For the initial design of quadrupole MgFFF[50], a simple annular channel was constructed. Fluid occupied the full annular circumference and flowed in the longitudinal direction. However, such a channel requires a uniform distribution of fluid around the circumference at the inlet and uniform collection at the outlet. This has to be accomplished using multiple ports or slots, but the uniform distribution of flow is unlikely to be realised in practice. Any interference in flow at one or more inlet or outlet ports would disrupt the flow pattern and degrade the separation. Also, the cylindrical rod that forms the inner wall of the annulus would have to be very precisely centred in the tube forming the outer wall. The channels are typically only 250 μm in thickness and any slight misalignment or curvature in the rod or tube would have a major impact on the flow pattern.

In a better design, the channel takes a helical path through the annular region. This is achieved by taking a polymer rod that fits tightly into a non-magnetic stainless steel tube and machining the helical channel into its surface, as shown in Figure 11.2. The fluid is introduced at a single inlet and collected at a single outlet, so there is no possibility of uneven flow distribution between multiple ports.

FIGURE 11.2 The helical channel is machined into the surface of a Delrin cylinder that fits tightly into a stainless steel tube. The inlet and outlet ports are seen close to the ends of the channel. (Reprinted from *J. Magn. Magn. Mater.*, 293, Carpino, F., Moore, L.R., Zborowski, M., Chalmers, J.J., and Williams, P.S., Analysis of magnetic nanoparticles using quadrupole magnetic field-flow fractionation, 546–552, Copyright 2005, with permission from Elsevier.)

After assembly, the unmachined surface of the rod remains in tight contact with the tube and this maintains a uniform channel thickness so that across the channel breadth the fluid flow velocity is uniform. Finally, any small variation in field around the circumference of the quadrupole aperture would degrade the separation in the case of a full annular channel because some fractions of the sample would be exposed to different fields than others. With the helical channel, the whole sample is carried around the circumference several times, through the variations in field, and the separation would not be degraded.

The need for the gradual reduction of applied field during the analysis of polydisperse particulate materials has already been mentioned in Section 11.1, and it is therefore necessary to use a quadrupole electromagnet with computer control of the current supply.

11.2.2.1 Description of the System

The quadrupole electromagnet was designed and assembled in the laboratory[51,52]. Four large copper wire coils (American Wire Gauge 18 coated copper wire, each nominally of 1900 m length) were wound around low-carbon steel plates of high magnetic permeability, 15.24 cm tall and 2.54 cm thick. One end of each plate was tapered and machined to the hyperbolic profile necessary to generate an axisymmetrical quadrupole field[53]. The four plates were bolted into a square magnetic flux return yoke, of the same steel plate, 15.24 cm tall and 1.91 cm thick. The yoke completes the magnetic circuit and maximises the field in the quadrupole aperture. The hyperbolic profiles came together as the four pole pieces of the quadrupole magnet with an aperture of 16 mm diameter. A schematic of the cross section of the electromagnet plates and coils is shown in Figure 11.3.

The coils were wired in parallel and current supplied by a regulated power supply capable of a nominal maximum current of 5 A at 60 V. After assembly, the numbers of turns on the coils were fine-tuned to equate the field at the four pole pieces. The power supply was controlled by a computer and the current could be programmed to generate any desired field as a function of time. A photograph of the system is shown in Figure 11.4.

It was mentioned in Section 11.1 that it is necessary for the field to asymptotically approach zero for the analysis of polydisperse magnetic nanoparticle samples. When all four coils are connected in parallel and current is reduced to zero, the steel plate forming the solenoid cores and flux return yoke has a remnant field that prevents approach to zero field in the aperture. A method has been

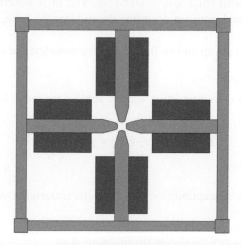

FIGURE 11.3 Schematic of the quadrupole electromagnet cross section. The low-carbon steel pole pieces and yoke are shown in pale grey, and the copper wire coils in darker grey. The stainless steel tube that encloses the helical channel fits snugly between the four pole tips at the centre of the system. (Reprinted from *J. Magn. Magn. Mater.*, 293, Carpino, F., Moore, L.R., Zborowski, M., Chalmers, J.J., and Williams, P.S., Analysis of magnetic nanoparticles using quadrupole magnetic field-flow fractionation, 546–552, Copyright 2005, with permission from Elsevier.)

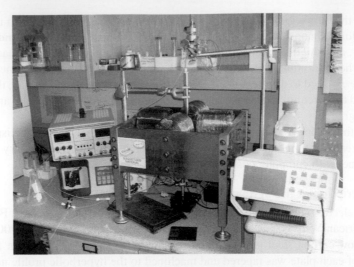

FIGURE 11.4 The quadrupole MgFFF system, showing the quadrupole electromagnet with the channel and sample injector held above by the clamps. The Gauss/Tesla meter is to the right. The constant current supply is on the left on top of the UV detector, and the programmed current supply is to the right of the detector.

implemented in which two of the coils are supplied with a low constant current sufficient to counter-act this remnant field. The current to only the other two coils is controlled by the computer[54]. The field is recorded during sample analysis using a Gauss meter placed adjacent to the channel, and a correction applied to obtain the field B_o at the accumulation wall. The output of the Gauss meter is recorded to a data file along with the output from the UV detector using an analogue to digital converter.

The channels were machined to a depth of 250 μm into the surface of a precision-turned Delrin (DuPont) rod of 1.488 cm diameter and 15.5 cm length. The length of a typical helical channel was 23.5 cm with a breadth of 1.6 cm, and it made four full revolutions around the rod. The inlet and outlet ports were carried to the ends of the rod and these were threaded to accept tube fittings. The rods were inserted into internally polished, non-magnetic stainless steel tubes (316 grade), outer diameter of 1.588 cm and wall thickness of 0.051 cm. The fit is watertight and accomplished by cooling the two parts in liquid nitrogen, inserting the rod into the tube, and then allowing them to warm to room temperature.

A liquid chromatography pump and an injection valve complete the system.

11.2.2.2 Retention Theory

The force F_m on a particle due to interaction with an applied magnetic field is given by

$$F_m = \chi V_m \nabla \left(\frac{1}{2} HB \right) = \chi V_m \nabla \left(\frac{B^2}{2\mu_0} \right) \tag{11.7}$$

in which
   χ is the volume magnetic susceptibility of the magnetic material of volume V_m in the particle
   ∇ is the gradient operator
   H is the magnetic field strength
   B is the magnetic flux density, or simply the magnetic field

The second form on the right-hand side of Equation 11.7 is obtained *via* the relationship between H and B given by $H = B/\mu_0$, where μ_0 is the magnetic permeability of free space, equal to $4\pi \times 10^{-7}$ Tm/A. Note that the particles may be composed of multiple materials. In this case it is assumed that there is one magnetic component that interacts with the magnetic field, while the other components are assumed to have negligible interaction. The magnetisation M of a magnetisable material is related to the applied magnetic field strength by the equation $M = \chi H$, and the equation for the force on a particle may be written as

$$F_m = V_m M \nabla B \tag{11.8}$$

where

V_m is the volume of magnetic material in the particle having an induced magnetisation M

∇B is the gradient in magnetic field B

Paramagnetic and diamagnetic materials interact relatively weakly with magnetic fields, and for these materials, M is directly proportional to H, so that χ is constant. Paramagnetic materials typically have volume susceptibilities between 10^{-5} and 10^{-2}, while diamagnetic materials have susceptibilities of the order of -10^{-5}. For example, water is diamagnetic and has a susceptibility of -9.04×10^{-6} at 20°C. When a paramagnetic or diamagnetic particle is suspended in a medium, the net magnetic force on the particle is given by

$$F_m = \Delta\chi V_m \nabla \left(\frac{1}{2} HB \right) = \Delta\chi V_m \nabla \left(\frac{B^2}{2\mu_0} \right) \tag{11.9}$$

where $\Delta\chi$ is the difference in susceptibility between the magnetic component of the particle and the medium, and is constant.

Ferromagnetic materials have much higher, positive susceptibilities and the susceptibility of the suspending medium can often be ignored. The magnetisation approaches a saturation level and χ is therefore not constant but is a function of H. If the ferromagnetic material is sufficiently finely divided, it becomes energetically favourable for the magnetic spins to be aligned with one another. Such particles, generally less than 100 nm in diameter, effectively become single magnetic domains and therefore magnetically saturated along a so-called easy axis. At lower magnetic fields, the directions of magnetisation might not be aligned with the field because Brownian rotation will tend to orient the particles in random directions, but as the field is increased, the particles are brought into alignment with the field. Because such particles have permanent magnetisation, there is also a possibility for them to interact with one another to form chains or clusters. For particles that are even smaller (of the order of 10 nm in diameter), the orientation of the magnetisation may spontaneously flip directions along the easy axis. This will occur if the energy barrier between magnetisation directions is much lower than the thermal energy. In this case, the collection of particles behaves like a paramagnetic substance except that the particles have a much larger magnetic moment than the molecules of a typical paramagnetic substance. For this reason, these materials have become known as superparamagnetic materials[55]. Superparamagnetic particles exhibit zero coercivity (magnetisation falls to zero when the applied field is removed) and they magnetically saturate at relatively low magnetic fields. These properties make them particularly useful for biotechnology such as for magnetic sorting of biological cells and for magnetic drug targeting[56,57]. Such particles may contain a volume of magnetite that is equivalent to a larger, multi-domain nanoparticle, but this may be finely divided into superparamagnetic nanoparticles so that the overall behaviour is superparamagnetic.

The nanoparticles analysed by quadrupole MgFFF have all been superparamagnetic. Paramagnetic and diamagnetic nanoparticles would not interact with the field of the current electromagnet to a sufficient degree to be retained and cannot be analysed. We therefore use the general magnetic force equation corresponding to Equation 11.8. The retention parameter λ is then given by

$$\lambda = \frac{kT}{F_m w} = \frac{kTr_o}{V_m M B_o w} \tag{11.10}$$

Magnetisation M varies with B, which in turn varies with the radial distance from the quadrupole axis. The channel is very thin compared to the aperture, however, and this small variation across the channel thickness may be ignored. We can therefore assume that M corresponds to the magnetisation at field B_o, and λ is then simply a function of B_o and of M at B_o.

The fluid velocity profile in the helical channel departs slightly from parabolic, but again, because the channel is thin, we can assume the departure from parabolic is negligible. The derivations of R for longitudinal and angular flow in an annulus have been presented[58,59], and as expected, the retention ratio in each case approaches the parallel-plate solution as the ratio of inner to outer wall radii approaches unity. Note that for the later consideration of annular SPLITT fractionation in Sections 11.4.2.2.1 and 11.4.2.2.2, we cannot make this assumption regarding velocity profile because the channels tend to be considerably thicker.

The retention ratio for quadrupole MgFFF may be assumed to be given by Equation 11.6. From the definition of the retention ratio, we have

$$v_p = R \langle v \rangle \tag{11.11}$$

where v_p is the zone velocity for a narrow fraction of particles of a polydisperse sample. Generally, the magnetic field is gradually reduced during an analysis, and so R changes with time for this narrow fraction and for every other fraction of the sample. The retention time t_r for the narrow fraction corresponds to the time taken to migrate along the full length L of the channel. Integrating the velocity over time up to t_r must result in the channel length L:

$$L = \int_0^{t_r} v_p dt = \int_0^{t_r} R \langle v \rangle dt \tag{11.12}$$

If $\langle v \rangle$ is held constant, this reduces to

$$\int_0^{t_r} R dt - t^0 = 0 \tag{11.13}$$

where t^0 is the void time, or the time for non-retained material to be carried through the channel, so that $t^0 = L/\langle v \rangle$. Therefore, if T, r_o, w, t^0, and the relationship between M and B_o are known, and if B_o as a function of time is known (it is recorded during an analysis), then the retention time may be calculated for any V_m by numerically solving Equation 11.13.

11.2.2.3 Data Reduction

A general algorithm for data reduction in FFF has been presented in the literature[60]. In most forms of FFF, the particle properties do not vary with the strength of the applied field. For example, particle density does not vary with centrifugal field strength. For MgFFF, however, the algorithm requires the inclusion of the dependence of induced magnetisation on the applied field[61]. During the sample analysis, the values of detector response $h(i)$ (measured as voltage) and applied field $B_o(i)$ (measured in T) are recorded to a data file at discrete times $t(i)$. The field is actually monitored at a point adjacent to the channel exterior and an adjustment made for the displacement from the accumulation wall inside the tube. The retention time for a particle of known V_m is given by the solution of the integral equation

$$\int_0^{t_r} R(B_o(t), M(B_o(t)), V_m) dt - t^0 = 0 \tag{11.14}$$

where the dependence of M on B_o and the dependence of R on B_o, M and V_m are shown explicitly. The magnetisation M may have some crystal size dependence as well as B_o dependence[62], but this cannot be taken into account because multiple crystals may be incorporated into particles. If a bulk magnetisation curve is available for the material used to prepare the nanoparticles then it may be used. Alternatively, a magnetisation curve may be taken from the literature. For example, a function relating M to B_o has been fitted to a typical magnetisation curve for bare magnetite nanoparticles[63,64]:

$$M = \frac{9.152 \times 10^6 B_o}{1 + 27.30 B_o - 0.9229 B_o^2}$$

(11.15)

The algorithm requires that Equation 11.14 be numerically solved for t_r for a series of discrete values of V_m, taking the actual recorded values of $B_o(i)$ and the measured or assumed dependence of M on B_o. The computer program ensures that there is a uniform distribution of these discrete V_m across the range of eluted material. Interpolation is then used to associate a value of $V_m(i)$ with every discrete retention time $t(i)$ in the data file for which a solution is found. These $t(i)$ are equivalent to retention times $t_r(i)$ when a solution for $V_m(i)$ exists. The $V_m(i)$ may be transformed to equivalent spherical magnetic core diameters $d_m(i)$ if desired $(d_m(i) = (6V_m(i)/\pi)^{1/3})$. The final step is to transform the elution curve of $h(i)$ versus $t(i)$ into a mass distribution in d_m.

Suppose $m(d_m(i))$ represents the mass of particles having equivalent spherical core diameters between $d_m(i)$ and $d_m(i) + \delta d_m$, it follows that

$$m(d_m(i)) = c(i)Q\frac{\delta t_r(i)}{\delta d_m}$$

(11.16)

where

$c(i)$ is the mass concentration of particles eluting at time $t_r(i)$
$\delta t_r(i)$ is the difference in retention time for particles of diameters $d_m(i)$ and $d_m(i) + \delta d_m$
Q is the volumetric channel flow rate

Generally, a UV detector is used and the exact relationship between the detector response $h(i)$ and mass concentration $c(i)$ is not known. We have to assume that there is a simple linear relationship at this stage. Once the $d_m(i)$ have been associated with $t_r(i)$, it is possible to find the derivative $dt_r/dd_m(i)$ at every discrete point corresponding to $t_r(i)$, and carry out the transformation of Equation 11.16. This yields the particle core diameter distribution of $m(d_m(i))$ as a function of $d_m(i)$. The result may be normalised as desired.

11.2.2.4 Non-Idealities

It was mentioned earlier that the ideal theory for FFF requires the absence of both particle–particle and particle–wall interactions. It has been explained that MgFFF differs from other forms of FFF in that a magnetic moment is induced in the particles by the applied field, and this magnetic moment interacts with the field gradient to induce particle migration. The particles' magnetic moments have the potential to interact with each other as well as with the field gradient. If the interaction is strong enough, particles may behave as clusters that will perturb their ideal elution; they will behave as larger particles. The types of particles that have been analysed are not bare magnetic materials. They are composite materials that are coated with some polymer or surfactant to stabilise them in aqueous suspension. They may also carry antibodies to cell surface antigens for cell labelling. Magnetic nanoparticle drug carriers will carry a drug component and possibly a non-polar material to carry the drug. The presence of all of these components tends to reduce the strength of the interactions between the particles' magnetic dipoles. The potential for deviations of particle retention times from the ideal model remains, however. The phenomenon has been discussed in the literature[61,65]. It was also suggested that particles may interact with one another and be captured on the channel wall at the high magnetic field that is initially applied, and be subsequently released and dissociated as the field decays.

11.2.2.5 Biomedical Applications

Quadrupole MgFFF has been applied to the separation and characterisation of magnetic nanoparticles used for the magnetic labelling of biological cells[51,52,61,63]. It has been shown that these nanoparticle samples are invariably polydisperse in their magnetic properties. The technique of MgFFF was shown to be extremely reproducible[51], and samples eluted under different flow rate and field decay conditions yielded consistent results following data reduction[61]. The analysis of magnetic nanoparticle labels by MgFFF will be useful for quality control purposes.

MgFFF has also been used to characterise magnetic nanoparticle drug carriers to be used in magnetically targeted chemotherapy[65]. Chemotherapeutic drugs are generally highly toxic, and the targeting of the drug to a localised area results in therapeutic concentrations at the targeted site while reducing systemic toxicity. For a polydisperse magnetic drug carrier reagent, the least magnetic particles will contribute most to systemic toxicity because they are least likely to be captured at the target site. The magnetically weaker carrier particles in the formulation contribute most to the side effects and limit the drug dose that can be safely administered. This potentially reduces the effectiveness of the treatment and may lead to relapse. The characterisation of magnetic drug carriers by MgFFF provides a direct measure of their response to a magnetic field gradient while in a sheared fluid flow, and it does not matter if this is influenced by particle magnetic dipole–dipole interaction. Characterisation of magnetic drug carriers by MgFFF would be invaluable for quality control and would contribute significantly to the safety and effectiveness of these formulations.

11.3   SPLIT-FLOW THIN CHANNEL FRACTIONATION

The technique of split-flow thin channel (SPLITT) fractionation was also invented by Calvin Giddings[66,67]. Like FFF, it employs thin, parallel-walled channels with a field applied across the channel thickness. However, in SPLITT fractionation, the separation takes place across the channel thickness rather than along the channel length. While FFF is able to separate a small polydisperse sample into a continuous distribution in elution time, SPLITT fractionation is able to continuously separate much larger samples into two, or potentially more, fractions. When operated in a continuous manner, it is known as continuous SPLITT fractionation, or CSF. The most common configuration implements the so-called transport mode[66,67], although other modes of operation involving diffusion[68,69] and hydrodynamic lift forces[70] have been used. A modified form of transport mode in which there is no inlet splitter and the feed stream occupies the full channel thickness at the inlet has also been used. This is known as the full feed depletion mode in which some fraction of the sample is driven across the channel to exit at outlet b'[71,72]. This can be a useful method to remove unwanted materials without dilution of the feed stream. The principles of the transport mode are illustrated in Figure 11.5.

In the transport mode, the sample is continuously introduced as a thin laminar stream adjacent to one of the channel walls. This is accomplished by merging the sample flow with a higher flow of pure carrier solution introduced on opposite sides of a stream divider placed between the channel walls at the channel entrance. The channel therefore has two inlets, a' and b', one on each side of this stream divider. By convention, the sample enters at a'. It is important that the stream divider be parallel to the channel walls in order to obtain a uniform initial thickness for the sample lamina next to the wall. Optimum separation depends on a uniform range of starting positions for particle migration across the channel thickness. The virtual surface in the main body of the channel dividing fluid elements that enter on the sample side from those that enter on the pure carrier side is known as the inlet splitting surface (ISS). The ISS is represented by the upper dashed line in Figure 11.5. The position of the surface between the channel walls is determined by the ratio of the two inlet flow rates.

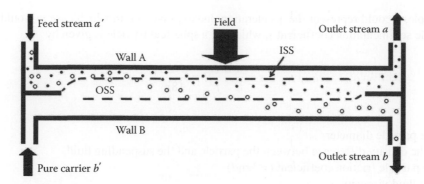

FIGURE 11.5 Schematic of the cross section of a split-flow thin (SPLITT) channel operated in transport mode. The channel thickness is exaggerated to more clearly illustrate the mechanism of particle separation.

Under the influence of the field, the sample components are driven across the channel thickness while they are carried along the length of the channel by the flow. An outlet stream splitter, having the same construction as the inlet stream divider, divides the flow of carrier solution at the outlet end of the channel and conveys it to two outlets, a and b, with outlet a being on the same side of the channel as inlet a'. The outlet stream splitter must also be parallel to the channel walls so that the virtual surface within the channel flow, dividing the fluid that further downstream passes to each side of the splitter, is also parallel to the walls. This virtual surface is known as the outlet splitting surface (OSS). The inlet flow rate ratio $Q(a'):Q(b')$ (typically 0.1) is generally arranged to be lower than the ratio of outlet flow rates $Q(a):Q(b)$ (typically > 0.2) in order to obtain a finite distance from the ISS to the OSS.

The fluidic region between the ISS and the OSS is known as the transport zone or transport lamina, and it is the differing rates of transport of particles across this transport lamina that are exploited to achieve separation. In Figure 11.5, the OSS corresponds to the lower dashed line. The sample components that migrate more quickly across the channel thickness (the open circles in Figure 11.5), and pass beyond the OSS, are conveyed to outlet b. Those that migrate more slowly (the filled circles) are retained in the carrier solution that exits the channel at outlet a. The channels are typically between 100 and 500 μm in thickness, and transverse transport distances are necessarily smaller than the channel thickness. Separation therefore tends to be achieved relatively quickly within the channel. The throughput depends on the transverse migration velocities that in turn determine the flow rates that allow sufficient residence time in the channel for migration.

11.3.1 GENERAL OPERATION

The general principles of transport mode operation and its optimisation have been described in the literature[73]. As mentioned before, the inlet flow rate ratio $Q(a'):Q(b')$ is generally arranged to be lower than the outlet flow rate ratio $Q(a):Q(b)$, so that particles must traverse a fluid lamina known as the transport lamina in order to exit at outlet b. The flow rate $Q(t)$ within this transport lamina is given by

$$Q(t) = Q(a) - Q(a') = Q(b') - Q(b) \tag{11.17}$$

In fact, particles starting their migration from the wall A must traverse the flow stream $Q(a')$ as well as $Q(t)$ to exit at outlet b. Consider a particle having a transverse velocity u given by

$$u = Sm \tag{11.18}$$

where
 S is a measure of the applied field
 m is the particle mobility

For example, S could represent the acceleration due to gravity G, in which case m would represent the particle sedimentation coefficient s, which for a spherical particle is given by

$$s = \frac{\pi}{6}\frac{d^3 \Delta\rho}{f} = \frac{d^2 \Delta\rho}{18\eta} \qquad (11.19)$$

where
  d is the particle diameter
  $\Delta\rho$ is the density difference between the particle and the suspending fluid
  f is the particle friction coefficient ($= 3\pi\eta d$)
  η is the fluid viscosity

The time δt taken for the particle to traverse a distance δx at velocity u is given by

$$\delta t = \frac{\delta x}{u} \qquad (11.20)$$

and during this time it is transported along the length of the channel, a distance δz given by

$$\delta z = v\delta t = v\frac{\delta x}{u} \qquad (11.21)$$

in which v is the local fluid velocity. The volumetric flow δQ that passes along this fluid lamina of thickness δx corresponds to

$$\delta Q = bv\delta x \qquad (11.22)$$

where b is the breadth of the channel. Combining Equations 11.21 and 11.22, we see that

$$\delta z = \frac{\delta Q}{bu} \qquad (11.23)$$

There is seen to be a simple linear relationship between the incremental distance gained by the particle along the channel length and the volumetric flow rate traversed. All particles must travel the full length of the channel L from splitter edge to splitter edge, and integration over length elements yields the simple relationship:

$$\Delta Q = bLu = bLSm \qquad (11.24)$$

where ΔQ is the volumetric flow rate corresponding to the thickness of the flow lamina traversed by the particle as it is carried along the length of the channel. This relationship makes the set-up and operation of parallel-plate SPLITT fractionation particularly easy. For example, for a monodisperse set of particles:

$$\text{if } \Delta Q \leq Q(t) \quad \text{then } F_b = 0 \qquad (11.25)$$

$$\text{if } Q(t) \leq \Delta Q \leq Q(a') + Q(t) \quad \text{then } F_b = \left(\frac{\Delta Q - Q(t)}{Q(a')}\right) \qquad (11.26)$$

$$\text{if } Q(a') + Q(t) \leq \Delta Q \quad \text{then } F_b = 1 \qquad (11.27)$$

where F_b is the fraction of the particles that will be collected at outlet b. A critical mobility m_0 is associated with the condition $\Delta Q = Q(t)$, so that the particles having mobilities $m \leq m_0$ are all carried to outlet a, and $F_b = 0$. A second critical mobility m_1 is associated with the condition $\Delta Q = Q(a') + Q(t)$, so that the particles having mobilities $m \geq m_1$ are all carried to outlet b, and $F_b = 1$. It follows that

$$m_0 = \frac{Q(t)}{bLS} = \frac{Q(a) - Q(a')}{bLS} \tag{11.28}$$

$$m_1 = \frac{Q(a') + Q(t)}{bLS} = \frac{Q(a)}{bLS} \tag{11.29}$$

$$\Delta m = m_1 - m_0 = \frac{Q(a')}{bLS} \tag{11.30}$$

and

$$\text{if } m_0 < m < m_1 \quad \text{then } F_b = \frac{m - m_0}{\Delta m} \tag{11.31}$$

The resolution of the separation may be expressed in terms of $m_1/\Delta m$[73], and it is apparent that

$$\frac{m_1}{\Delta m} = \frac{Q(a)}{Q(a')} \tag{11.32}$$

Then, if a fractionation is required for some given m_1 with a required $m_1/\Delta m$, the suitable flow rates $Q(a')$ and $Q(a)$ may be calculated from Equations 11.29 and 11.30, respectively. The throughput TP is of course directly related to the concentration of the feed stream and the feed flow rate $Q(a')$. It is then seen from Equation 11.30 that

$$TP = CQ(a') = CbLS \frac{m_1}{m_1/\Delta m} \tag{11.33}$$

where C is the feed concentration. For a given field strength S, cut-off mobility m_1 and resolution $m_1/\Delta m$, the throughput may be increased by increasing the feed stream concentration C, channel length L and/or the channel breadth b. The increase of C is limited by particle–particle interactions, which can interfere with particle migration across streamlines. High particle concentrations can also lead to instabilities in the flow. This has been shown in gravitational SPLITT fractionation, where unstable density layering can cause instabilities and degradation of separation efficiency[74–76]. The same mechanism can disrupt separation for any applied field. For example, a magnetic separation could be disrupted when the magnetic particles are at too high a concentration and they become influenced as a group rather than as individual particles. Another concentration-dependent phenomenon that may reduce separation efficiency is that of viscous resuspension that is related to shear-induced diffusion[77–81].

The increase in $Q(a')$ in proportion to L at fixed b would be limited by the onset of turbulent flow because the separation depends on laminar flow. In fact, it has been shown that the degradation of separation efficiency occurs well before the expected onset of turbulence[82]. Using computational fluid dynamics simulations, it was shown that vortices may form in the region of merging inlet flows even at relatively low Reynolds numbers of the order of 10, and this tendency for vortex formation is exacerbated at high inlet flow rate ratios. (The Reynolds number Re is a dimensionless number that gives a ratio of inertial to viscous forces, and for fluid flow in a parallel-plate channel is equal to $2\rho\langle v \rangle w/\eta$, where ρ is the fluid density and η is the fluid viscosity. Fluid flow is generally considered laminar for Re of up to around 1000.) It was suggested that the presence of a vortex could cause some mixing of the inlet flows and the experiments seemed to support the assertion.

Note that the conditions selected for some fixed m_1 and $m_1/\Delta m$ do not specify the $Q(b)$ and $Q(b')$. These may be defined by considering the mobility of particles that approach the wall B before they exit the channel. The approach of particles to wall B may not mean that they will be retained in the channel; they could still be swept out of outlet b. Some samples, such as biological cells, may be disturbed by wall contact, however, and this may put a lower limit on the total channel flow rate Q. Again, considering a monodisperse collection of particles,

$$\text{if } \Delta Q \leq Q(b') \quad \text{then } F_w = 0 \tag{11.34}$$

$$\text{if } Q \leq \Delta Q \leq Q(b') \quad \text{then } F_w = \left(\frac{\Delta Q - Q(b')}{Q(a')} \right) \tag{11.35}$$

$$\text{if } Q \leq \Delta Q \quad \text{then } F_w = 1 \tag{11.36}$$

where Q is the total flow rate:

$$Q = Q(a') + Q(b') = Q(a) + Q(b) \tag{11.37}$$

and F_w is the fraction of the particles that impact the wall B before either exiting at outlet b or remaining in the channel. A third critical mobility m_2 may be associated with the condition $\Delta Q = Q(b')$, so that no particles having mobilities $m \leq m_2$ are carried to wall B and $F_w = 0$. A fourth critical mobility m_3 may be associated with the condition $\Delta Q = Q$, so that particles having mobilities $m \geq m_3$ will all be carried to the wall B. It follows that

$$m_2 = \frac{Q(b')}{bLS} = \frac{Q - Q(a')}{bLS} \tag{11.38}$$

$$m_3 = \frac{Q}{bLS} \tag{11.39}$$

It is also true that

$$\Delta m = m_1 - m_0 = m_2 - m_3 \tag{11.40}$$

and

$$\text{if } m_2 < m < m_3 \quad \text{then } F_b = \frac{m_3 - m}{\Delta m} \tag{11.41}$$

If the requirement is that no particles of some mobility m_2, where $m_2 \geq m_1$, are to impact wall B then the total minimum flow rate Q can be calculated from Equation 11.38, as $Q(a')$ is already known. Once Q is known, then $Q(b')$ and $Q(b)$ can be obtained.

11.4 MAGNETIC SPLITT FRACTIONATION

11.4.1 APPROACHES EMPLOYING PARALLEL-PLATE CHANNELS

There have been attempts to implement SPLITT fractionation using parallel-plate channels with a transverse magnetic field applied across the thickness[83–87]. Fuh and co-workers generated a magnetic field along a long, narrow interpolar gap, using neodymium–iron–boron (NdFeB) permanent magnets with soft iron pole pieces. Various magnet assemblies were constructed with lengths of 7, 10 and 14 cm and with gap widths of 2, 5 and 10 mm. The field generated increases as the gap is reduced. Fields of 1.2 and 0.72 T were measured for 5 and 10 mm interpolar gaps, respectively. Channels were made to suit gap widths and lengths, and were all 250 μm in thickness. With this design, there will be a variation of field and field gradient across the channel breadth, particularly when the channel is placed relatively close to the interpolar gap. While this variation would cause problems with band spreading if used for FFF, SPLITT fractionation is a comparatively low-resolution technique and typically used only for binary separations. The system was successful in separating a number of different samples containing magnetic and non-magnetic species. Some materials, including silica beads, yeasts and red blood cells (RBCs), were made magnetic by labelling with various paramagnetic metal ions[83,84,86].

An alternative approach employing a parallel-plate channel has also been presented. This was termed HGMF-SPLITT fractionation[88]. This used a deposit of powdered steel wool laid over the surface of one of the channel walls, coated with an aerosolised Teflon mould release substance. The whole channel was placed in the bore of a superconducting magnet in a region of homogeneous magnetic field of 7 T. Channels varied from 0.005 to 0.007 in. (127–178 μm) in thickness. This approach has similar drawbacks to the attempts at implementing magnetic FFF using magnetised wires embedded in the channel wall. It is more suited to magnetic nanoparticle capture than operation in continuous mode. This was indeed found to be the case in the experiments reported.

In a similar vein, a theoretical model of a magnetic SPLITT device using magnetisable wires embedded in one of the channel walls has been presented[89]. The wires were of 300×300 μm cross section placed across the channel breadth at intervals of 300 μm. It was shown that deoxygenated RBCs could theoretically be separated from white blood cells (WBCs) in a channel of 120 μm thickness. WBCs as well as oxygenated and deoxygenated RBCs are diamagnetic (negative susceptibility), but deoxygenated RBCs are less diamagnetic and, importantly, less diamagnetic than aqueous medium. WBCs and oxygenated RBCs therefore exhibit negative $\Delta\chi$ in aqueous suspension, and deoxygenated RBCs positive $\Delta\chi$[90]. Deoxygenated RBCs and WBCs will therefore be driven in opposite directions by a magnetic field gradient.

Finally, a new approach to parallel-plate SPLITT fractionation was recently presented[91]. This uses a linear Halbach array of transversely magnetised cylindrical NdFeB magnets to provide the field. We shall return to this in Section 11.5.

11.4.2 QUADRUPOLE MAGNETIC SEPARATOR

11.4.2.1 Description of the System

A typical device makes use of a quadrupole magnet constructed using permanent NdFeB magnet blocks and soft iron pole pieces. The pole pieces should ideally be machined to the hyperbolic profile necessary to generate an axisymmetrical quadrupole field[53]. An example of a quadrupole magnet that has been constructed has a length of 7.6 cm and an aperture radius of 4.82 mm and generates a field of 1.42 T at the poles[92–96]. Inserted into the aperture of the quadrupole magnet is an annular channel made up of an inner polyetherimide (Ultem®) rod of 2.38 mm radius, and a non-magnetic stainless steel tube having an internal radius of 4.53 mm and wall thickness of 0.25 mm. The annular channel thickness was therefore 2.15 mm. The twin inlet and outlet tubes for the fluid were connected *via* inlet and outlet manifolds incorporating the flow splitters. The stainless steel splitters were cylindrical, and concentric to the outer tube and inner rod. The concentricity is quite critical for good separation performance. This has been demonstrated both by computational fluid dynamics modelling and by experiment[97,98]. The channel was exposed to a field gradient of 295 T/m in the aperture and S_{mo} of 156×10^6 TA/m^2 for $r_o = 4.53$ m.

A larger system has also been constructed[99]. This magnet was 15.2 cm long, and had an aperture radius of 8.00 mm, and also generated a field of 1.42 T at its poles. The field gradient was therefore 178 T/m in the aperture and S_{mo} 93.1×10^6 TA/m^2. It will be shown later in Sections 11.4.2.2.1 and 11.4.2.2.2 that the potential throughput is a proportional to L and to either B_o^2 or B_o. The lower field gradient and S_{mo} for the larger system do not result in a lower throughput. In fact the longer length gives a potentially higher throughput.

A commercial system called the Quadrasep QMS™ is now available from IKOTECH LLC (New Albany, IN). A larger system called the Quadrasep/LP QMS™ is also available for separation of larger cells, clusters and particles.

11.4.2.2   General Operation

The description of the separation process is more complicated for the annular magnetic SPLITT than for the parallel-plate system considered in Section 11.3.1. This is because the force on a magnetised particle is generally a function of the magnetic field and therefore of position in the annulus, and because of the annular channel geometry. It is necessary to consider the trajectories taken by particles of critical mobilities m_0, m_1, m_2 and m_3 as they are carried through the system. It has been explained that particles may be paramagnetic or they may be superparamagnetic. Paramagnetic behaviour has been considered in the literature[100,101], but for the case of biological cells labelled with superparamagnetic nanoparticles, the magnetically saturated model would be more applicable. We discuss the two limiting behaviours in the following.

11.4.2.2.1   Paramagnetic Model

The net force on a paramagnetic particle suspended in some medium and subjected to a magnetic field was given by Equation 11.9. The magnetic field-induced migration velocity u_m may then be written as

$$u_m = \frac{F_m}{f} = \frac{\Delta\chi V_m}{f} \nabla\left(\frac{B^2}{2\mu_0}\right) = m_m S_m \tag{11.42}$$

in which
the magnetophoretic mobility $m_m = \Delta\chi V_m / f$
S_m is the magnetic energy density gradient, a measure of the magnetomotive field strength, equal to $\nabla(B^2/2\mu_0)$

In the quadrupole field, B increases linearly with distance from the axis, and S_m is given by

$$S_m = \frac{B_o^2}{\mu_0 r_o^2} r = \frac{B_o^2}{\mu_0 r_o}\rho = S_{mo}\rho \tag{11.43}$$

where
B_o is the magnetic field at the radius r_o of the outer annulus wall
r is the radial distance from the axis
ρ is equal to r/r_o
S_{mo} is the value of S_m at r_o

It follows that the field-induced migration velocity in the aperture of the quadrupole is given by

$$u_m(\rho) = m_m S_{mo}\rho \tag{11.44}$$

The time t_w for a particle to migrate across the full channel thickness is given by

$$t_w = r_o \int_{\rho_i}^{1} \frac{d\rho}{u_m(\rho)} = \frac{r_o}{m_m S_{mo}} \ln\left(\frac{1}{\rho_i}\right) = \frac{\mu_0 r_o^2}{m_m B_o^2} \ln\left(\frac{1}{\rho_i}\right) \tag{11.45}$$

where $\rho_i = r_i/r_o$ is the ratio of inner to outer wall radii, and t_w therefore increases rapidly as $\rho_i \rightarrow 0$. This is the reason for the use of an annular channel that excludes particles from the region around the quadrupole axis where the field falls to zero.

The fluid velocity profile in an annulus is given by

$$v(\rho) = \frac{2\langle v \rangle}{A_1}(1 - \rho^2 + A_2 \ln\rho) \tag{11.46}$$

where $\langle v \rangle$ is the mean fluid velocity, and A_1 and A_2 are functions of ρ_i:

$$A_1 = 1 + \rho_i^2 - A_2 \tag{11.47}$$

$$A_2 = \frac{\left(1 - \rho_i^2\right)}{\ln\left(\frac{1}{\rho_i}\right)} \tag{11.48}$$

It is now possible to derive an expression for the particle trajectory *via* the integral equation:

$$\int_0^z dz = r_o \int_{\rho_1}^{\rho} \frac{v(\rho)}{u_m(\rho)}d\rho \tag{11.49}$$

where
ρ_1 corresponds to the initial radial position of the particle at $z = 0$
ρ corresponds to the radial position at arbitrary distance z from the inlet splitter

Carrying out the integration, we obtain the following equation:

$$z = \frac{r_o\langle v \rangle}{2A_1 m_m S_{mo}}[4\ln\rho - 2\rho^2 + 2A_2(\ln\rho)^2]_{\rho_1}^{\rho} \tag{11.50}$$

The total volumetric flow rate in the annulus Q is related to $\langle v \rangle$ by

$$Q = \pi r_o^2 \langle v \rangle \left(1 - \rho_i^2\right) \tag{11.51}$$

and on substituting for $\langle v \rangle$ in Equation 11.50 and rearranging, we can obtain the following equation:

$$2\pi r_o m_m S_{mo} A_1 \left(1 - \rho_i^2\right) z = QI_1[\rho_1, \rho] \tag{11.52}$$

in which

$$I_1[\rho_1, \rho] = [4\ln\rho - 2\rho^2 + 2A_2(\ln\rho)^2]_{\rho_1}^{\rho} \tag{11.53}$$

If the particle does not come into contact the outer wall at $\rho = 1$ before reaching the outlet splitter at $z = L$ then the final radial position of the particle ρ_2 may be determined by numerical solution of the equation

$$2\pi r_o m_m S_{mo} A_1 (1-\rho_i^2)L = QI_1[\rho_1, \rho_2] \tag{11.54}$$

The critical mobilities are then defined just as for the parallel-plate SPLITT fractionator, but they are not so simply related to volumetric flow rates. The critical mobilities are defined in terms of Equation 11.54 with the respective limits of transport:

$$m_0 = \frac{Q}{2\pi r_o L S_{mo}} \frac{I_1[\rho_{ISS}, \rho_{OSS}]}{A_1(1-\rho_i^2)} \tag{11.55}$$

$$m_1 = \frac{Q}{2\pi r_o L S_{mo}} \frac{I_1[\rho_i, \rho_{OSS}]}{A_1(1-\rho_i^2)} \tag{11.56}$$

$$m_2 = \frac{Q}{2\pi r_o L S_{mo}} \frac{I_1[\rho_{ISS}, 1]}{A_1(1-\rho_i^2)} \tag{11.57}$$

$$m_3 = \frac{Q}{2\pi r_o L S_{mo}} \frac{I_1[\rho_i, 1]}{A_1(1-\rho_i^2)} = \frac{Q}{\pi r_o L S_{mo} A_2} \tag{11.58}$$

$$\Delta m = m_1 - m_0 = \frac{Q}{2\pi r_o L S_{mo}} \frac{I_1[\rho_i, \rho_{ISS}]}{A_1(1-\rho_i^2)} \tag{11.59}$$

and the resolution of the separation is given by

$$\frac{m_1}{\Delta m} = \frac{I_1[\rho_i, \rho_{OSS}]}{I_1[\rho_i, \rho_{ISS}]} \tag{11.60}$$

The selection of the flow rate conditions needed to achieve some desired fractionation is essentially the same as for the parallel-plate system except that some numerical solutions are required for the equations[101]. It is also necessary to relate the positions of the ISS and OSS to the relative volumetric flow rates at the inlets and outlets. The splitting surfaces are ideally cylindrical and concentric with the annular channel walls. Integrating fluid flow from the inner radius to the ISS must correspond to $Q(a')$. Therefore, we have

$$Q(a') = r_o^2 \int_{\rho_i}^{\rho_{ISS}} 2\pi\rho v(\rho)d\rho \tag{11.61}$$

Substituting for $v(\rho)$ using Equation 11.46 and integrating, we obtain the result

$$Q(a') = \frac{\pi r_o^2 \langle v \rangle}{A_1} I_2[\rho_i, \rho_{ISS}] \tag{11.62}$$

in which

$$I_2[\rho_i, \rho_{ISS}] = [2\rho^2 - \rho^4 + 2A_2\rho^2 \ln \rho - A_2\rho^2]_{\rho_i}^{\rho_{ISS}} \tag{11.63}$$

The total flow rate Q corresponds to the integral of flow velocity over the whole annular cross section, and this of course reduces to Equation 11.51. It follows that

$$\frac{Q(a')}{Q} = \frac{I_2[\rho_i, \rho_{ISS}]}{I_2[\rho_i, 1]} = \frac{I_2[\rho_i, \rho_{ISS}]}{A_1(1 - \rho_i^2)} \tag{11.64}$$

We can similarly derive an expression for the position of the OSS:

$$\frac{Q(a)}{Q} = \frac{I_2[\rho_i, \rho_{OSS}]}{I_2[\rho_i, 1]} = \frac{I_2[\rho_i, \rho_{OSS}]}{A_1(1 - \rho_i^2)} \tag{11.65}$$

The throughput is again directly related to the concentration C of the feed stream and the feed flow rate $Q(a')$, and it can be shown[101] that

$$TP = 2\pi CL \frac{B_o^2}{\mu_0} \frac{I_2[\rho_i, \rho_{ISS}]}{I_1[\rho_i, \rho_{ISS}]} \frac{m_1}{\frac{m_1}{\Delta m}} \tag{11.66}$$

Again, we see that the throughput may be increased by increasing both the length of the channel and the flow rates, but this approach is limited by the requirement for laminar flow and a smooth merging of the inlet flow streams without vortex formation. Throughput may also be limited by potential shear stress on biological cells. Interestingly, the throughput is seen to be dependent on B_o^2 and independent of r_o, provided there is laminar flow and vortices are avoided. There may be a temptation to reduce channel dimensions simply because one can generate a given B_o at a smaller r_o using smaller permanent magnet blocks. However, Equations 11.55 through 11.58 show that if ρ_i is held constant, the volumetric flow rates scale with the product of L and B_o^2. Turbulent flow, vortex formation or shear stress, and their consequent disadvantages, will be observed at a lower volumetric flow rate for a smaller radius channel. In addition, for smaller radius quadrupole apertures and channels, tolerances on the concentricity of the inner rod, the splitters and the outer tube become more critical.

11.4.2.2.2 Magnetically Saturated Model

The force on a magnetically saturated particle is given by Equation 11.8 in which magnetisation is constant at the saturation level M_s:

$$F_m = V_m M_s \nabla B \tag{11.67}$$

and in the quadrupole aperture, the gradient in magnetic field is constant and equal to B_o/r_o. In this case, the field-induced migration velocity is constant and given by

$$u_m = \frac{V_m M_s}{f} \frac{B_o}{r_o} = m'_m S'_m \tag{11.68}$$

in which m'_m is a different measure of magnetophoretic mobility that we shall term gradient mobility ($m'_m = V_m M_s / f$), and S'_m is simply the gradient in magnetic field ($S'_m = B_o / r_o$). In the quadrupole field, a magnetically saturated particle therefore migrates radially at constant velocity and the time t_w for migration across the channel thickness is given by

$$t_w = \frac{r_o}{m'_m S'_m}(1 - \rho_i) = \frac{r_o^2}{m'_m B_o}(1 - \rho_i) \tag{11.69}$$

This may be compared with Equation 11.45 for the paramagnetic particle. Note that Equation 11.69 must break down as $\rho_i \to 0$ because the particle cannot be magnetically saturated where the field falls to zero.

The trajectory is determined as before using Equation 11.49, where u_m is now constant and given by Equation 11.68:

$$2\pi r_o m'_m S'_m A_1 \left(1 - \rho_i^2\right) z = Q I'_1[\rho_1, \rho] \tag{11.70}$$

in which

$$I'_1[\rho_1, \rho] = \left[4\rho - \frac{4}{3}\rho^3 - 4A_2\rho + 4A_2\rho \ln\rho \right]_{\rho_1}^{\rho} \tag{11.71}$$

and ρ_1 is the initial radial position of the particle at $z = 0$.

Again, if the particle does not come into contact with the outer wall at $\rho = 1$ before reaching the outlet splitter at $z = L$ then the final radial position of the particle ρ_2 may be determined by numerical solution of the equation

$$2\pi r_o m'_m S'_m A_1 \left(1 - \rho_i^2\right) L = Q I'_1[\rho_1, \rho_2] \tag{11.72}$$

The critical gradient mobilities are now defined by the equations

$$m'_0 = \frac{Q}{2\pi r_o L S'_m} \frac{I'_1[\rho_{ISS}, \rho_{OSS}]}{A_1 \left(1 - \rho_i^2\right)} \tag{11.73}$$

$$m'_1 = \frac{Q}{2\pi r_o L S'_m} \frac{I'_1[\rho_i, \rho_{OSS}]}{A_1 \left(1 - \rho_i^2\right)} \tag{11.74}$$

$$m'_2 = \frac{Q}{2\pi r_o L S'_m} \frac{I'_1[\rho_{ISS}, 1]}{A_1 \left(1 - \rho_i^2\right)} \tag{11.75}$$

$$m'_3 = \frac{Q}{2\pi r_o L S'_m} \frac{I'_1[\rho_i, 1]}{A_1 \left(1 - \rho_i^2\right)} \tag{11.76}$$

$$\Delta m' = m_1' - m_0' = \frac{Q}{2\pi r_o L S_m'} \frac{I_1'[\rho_i, \rho_{ISS}]}{A_1 \left(1 - \rho_i^2\right)} \tag{11.77}$$

$$\frac{m_1'}{\Delta m'} = \frac{I_1'[\rho_i, \rho_{OSS}]}{I_1'[\rho_i, \rho_{ISS}]} \tag{11.78}$$

The flow velocity profile remains the same and the Equations 11.61 through 11.65 remain valid. The selection of suitable flow rate conditions follows the same approach as for the paramagnetic model. In this case, provided there is laminar flow and vortices are avoided, the throughput scales with the product of channel length L and B_o, and is again independent of r_o:

$$TP = 2\pi C L B_o \frac{I_2[\rho_i, \rho_{ISS}]}{I_1'[\rho_i, \rho_{ISS}]} \frac{m_1'}{m_1'/\Delta m'} \tag{11.79}$$

11.4.2.3 Biomedical Applications

The QMS has been used to demonstrate the separation of various magnetically labelled, cultured cells from non-labelled cells[92,95,96,99,100,102–109]. The objective was often to demonstrate the higher potential throughput of the QMS in comparison to fluorescence-activated cell sorting (FACS), and the gentleness of the separation by QMS so that cell viability was maintained. Recently the QMS has been applied to the detection of rare head and neck tumour cells circulating in the peripheral blood stream[110,111]. A negative selection technique was used where the normal leukocytes are magnetically labelled and are removed[112]. The larger commercial system from IKOTECH LLC has been used to isolate pancreatic islets[113,114].

For most of the studies, it is necessary to determine the response of magnetically labelled biological cells to a magnetic field. This allows the correct set-up of flow rate conditions and also allows comparison of predicted fractionations with the experimental results. The measurement of magnetophoretic mobilities is carried out using an instrument known as a cell tracking velocimeter (CTV)[94,115–122]. It follows the migration of up to several hundred cells at a time in a well-characterised magnetic field gradient using a microscope and video capture. The software distinguishes the individual tracks and produces a histogram of the velocities and the statistical quantities describing the distributions. The detailed description of this particular instrument is beyond the scope of this chapter. Recently, a commercial CTV, the High-Definition Magnetic Cell-Tracking Velocimeter (HD-MCTV™), has been made available from IKOTECH LLC.

11.5   FUTURE DIRECTIONS

11.5.1   USE OF HIGHER ORDER AXISYMMETRICAL FIELDS

The quadrupole magnetic field has been used for both MgFFF and magnetic SPLITT fractionation, but there is no reason why a hexapole, octopole or even higher order field may not be used. In the case of MgFFF where an electromagnet must be used to provide the axisymmetrical field, it may be advantageous to use only four coils and pole pieces. However, for magnetic SPLITT fractionation where the separation is achieved with a balance of field-induced migration against time for migration (*via* flow rate adjustment) and permanent magnets are generally used, there may be advantages in the use of higher order fields.

Circular arrangements of magnet blocks, known as Halbach ring magnets, may be constructed to generate dipole, quadrupole or higher order fields inside the aperture. The order of the field generated depends on the number of complete rotations made by the magnetisation vector of the magnet blocks around the circumference. The idea originated with an observation made by Mallinson[123] in 1973, but they have become known as Halbach arrays due to the considerable contributions to the field made by Klaus Halbach[124–127]. Coey proposed the use of cylindrical magnets that are magnetised across their diameter to generate different types of field[128,129]. The advantage associated with the cylindrical magnets is the freedom to physically rotate them relative to one another to alter the type of field generated. Hence, they have become to be known as magnetic mangles.

Take, for example, a circular array of N cylindrical magnets arranged around an origin. The centres of the magnets will be at angular intervals of $2\pi/N$ around the origin. Suppose the centres fall at angles $\theta_i = 2\pi(i - 1)/N$, for $i = 1$ to N. The direction of magnetisation of the successive magnets must progress at regular intervals around the array and make 2, 3, 4 or more rotations (a single rotation of 2π is not useful). If two full rotations are made then the internal field is a uniform dipole field; three full rotations generate a quadrupole field, four full rotations generate a hexapole field, etc. It is obvious that there is a minimum number of magnets required to generate each field pattern. The direction of magnetisation of the magnets should correspond to $\varphi_i = \varphi_0 + 2\pi(i - 1)(1 + (n/2))/N$, where $i = 1$ to N, and $n = 2$ for a uniform dipole field, $n = 4$ for a quadrupole, $n = 6$ for a hexapole, etc. The angle φ_0 is arbitrary; it is only necessary that $\Delta\theta = 2\pi/N$ and $\Delta\varphi = 2\pi(1 + (n/2))/N$.

The time for paramagnetic and magnetically saturated particles to migrate across the thickness of an annular channel in a quadrupole field has been given in Equations 11.45 and 11.69, respectively. For an annular channel in an ideal hexapole magnetic field, where field increases with the square of distance from the axis ($B = B_o\rho^2$). It follows that $S_m = S_{mo}\rho^3$, where $S_{mo} = 2B_o^2/\mu_o r_o$ and $S'_m = 2(B_o/r_o)\rho$. A paramagnetic particle then takes a time given by

$$t_w = \frac{r_o}{2m_m S_{mo}}\left(\frac{1}{\rho_i^2} - 1\right) = \frac{\mu_o r_o^2}{4m_m B_o^2}\left(\frac{1}{\rho_i^2} - 1\right) \tag{11.80}$$

to migrate across the channel thickness, and a magnetically saturated particle takes a time given by

$$t_w = \frac{r_o}{2m'_m S'_m}\ln\left(\frac{1}{\rho_i}\right) = \frac{r_o^2}{2m'_m B_o}\ln\left(\frac{1}{\rho_i}\right) \tag{11.81}$$

For an annular channel in an ideal octopole magnetic field, where field increases with the cube of distance from the axis ($B = B_o\rho^3$). It follows that $S_m = S_{mo}\rho^5$, where $S_{mo} = 3B_o^2/\mu_o r_o$ and $S'_m = 3(B_o/r_o)\rho^2$. The migration time for a paramagnetic particle is given by

$$t_w = \frac{r_o}{4m_m S_{mo}}\left(\frac{1}{\rho_i^4} - 1\right) = \frac{\mu_o r_o^2}{12m_m B_o^2}\left(\frac{1}{\rho_i^4} - 1\right) \tag{11.82}$$

and a magnetically saturated particle's migration time is

$$t_w = \frac{r_o}{3m'_m S'_m}\left(\frac{1}{\rho_i} - 1\right) = \frac{r_o^2}{3m'_m B_o}\left(\frac{1}{\rho_i} - 1\right) \tag{11.83}$$

The magnetic field B_o at the radius r_o may vary to some extent for the different configurations of magnets, but not a great deal. If it is assumed that B_o is not greatly different for the quadrupole, hexapole and octopole fields, then it follows that as $\rho_i \rightarrow 1$, the time for migration across the annulus for a hexapole field approaches half that of a quadrupole field, and the time for an octopole field approaches one-third that of a quadrupole field. This is true for both the paramagnetic and the magnetically saturated models. The terminal migration velocity as $\rho \rightarrow 1$ also increases by a factor of 2 for the hexapole field and a factor of 3 for the octopole field, when compared to the quadrupole. This suggests that it

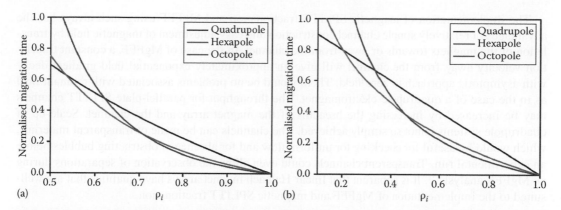

FIGURE 11.6 Normalised time for migration from ρ_i to $\rho = 1$ for (a) the paramagnetic particle model and (b) the magnetically saturated particle model in the quadrupole, hexapole and octopole magnetic fields.

could be advantageous to use higher order fields for fractionation in batch mode or in transport mode of magnetic SPLITT fractionation. Figure 11.6 shows plots of normalised migration times as functions of ρ_i for quadrupole, hexapole and octopole magnetic fields, for (a) the paramagnetic particle model and (b) the magnetically saturated particle model. For the paramagnetic particle model, the times are normalised by $\mu_0 r_o^2/m_m B_o^2$, and for the magnetically saturated model by $r_o^2/m_m' B_o$.

It can be seen that in the case of the paramagnetic model, the migration time in the hexapole field is shorter than in the quadrupole field when $\rho_i > 0.54$, shorter in the octopole field than in the quadrupole when $\rho_i > 0.63$ and shorter in the octopole than in the hexapole when $\rho_i > 0.71$.

In the case of the magnetically saturated model, the migration time is shortest in the octopole field for all $\rho_i > 0.5$, followed by the hexapole and finally the quadrupole. For the saturated model, the hexapole results in a faster migration time than the quadrupole for $\rho_i > 0.21$, the octopole gives a faster migration than the quadrupole for $\rho_i > 0.34$ and the octopole bests the hexapole for $\rho_i > 0.47$. It must be emphasised that we have not taken into account the channel wall thickness or small distortions in the field close to the surface of the magnets. In the case of a finite channel wall thickness, it must be realised that the field drops off more rapidly from a point close to the surface of the magnets for a hexapole field than it does for a quadrupole, and the drop off is even faster for the octopole. Therefore, for a given channel wall thickness, a hexapole field generated by a fixed number of magnets might generate a lower B_o than the same number of magnets generating a quadrupole field, and B_o may be even lower for an octopole field. This could reduce to some extent the advantage of higher migration velocity associated with the higher order fields. The results indicate that there could still be a significant improvement to throughput with the use of higher order axisymmetrical magnetic fields, and this area warrants further investigation.

11.5.2 Linear Halbach Magnet Arrays

It was mentioned in Section 11.4.1 that a linear Halbach array of transversely magnetised NdFeB magnets had been used to generate a cross-channel magnetic field gradient for SPLITT fractionation using a parallel-plate channel. The effectiveness of the separation was successfully demonstrated using a mixture of magnetic (6.2 μm micromod, Partikeltechnologie GmbH, Rostock-Warnemuende, Germany) and non-magnetic (15.8 μm polystyrene, Duke Scientific, Palo Alto, CA) microparticles[91].

These were preliminary, proof-of-principle experiments, and the linear Halbach magnet array was not optimised for the purpose. However, the approach holds great promise for both magnetic SPLITT fractionation and MgFFF. An ideal linear Halbach array generates a magnetic field and a field gradient that decay exponentially with distance from the magnets. The rate of their decay is a function of the rotation rate of the magnetisation vector[130]. These relationships, together with magnetic field modelling, can be used to optimise the design of the magnet array.

The implementation of magnetic SPLITT fractionation and MgFFF using such arrays has the advantages of relatively simple channel construction and easy adjustment of magnetic field by translation of the magnets towards or away from the channel. In the case of MgFFF, a constant translation velocity away from the channel will give an approximately exponential field gradient decay with asymptotic approach to zero field. There would be no problems associated with remnant field as in the case of a quadrupole electromagnet. The throughput for parallel-plate SPLITT channels may be increased by increasing the breadths of the magnet array and the channel. Scale-up of quadrupole systems is not so simply achieved. The channels can be made of transparent materials, which would be useful for checking for uniform flow and for absence of obstructing bubbles before an experimental run. Transparent channels could even allow the observation of separations during an MgFFF analysis[131]. It is apparent that linear Halbach magnet arrays have qualities that are well-suited to the implementation of MgFFF and magnetic SPLITT fractionation.

REFERENCES

1. Giddings, J. C. 1966. A new separation concept based on a coupling of concentration and flow nonuniformities. *Separation Science*, 1, 123–125.
2. Thompson, G. H., Myers, M. N. and Giddings, J. C. 1967. An observation of a field-flow fractionation effect with polystyrene samples. *Separation Science*, 2, 797–900.
3. Giddings, J. C. 1993. Field-flow fractionation: Analysis of macromolecular, colloidal, and particulate materials. *Science*, 260, 1456–1465.
4. Martin, M. and Williams, P. S. 1992. Theoretical basis of field-flow fractionation. In: Dondi, F. and Guiochon, G. (eds.) *Theoretical Advancement in Chromatography and Related Separation Techniques*. Dordrecht, the Netherlands: Kluwer Academic Publisher, pp. 513–580.
5. Schimpf, M. E., Caldwell, K. and Giddings, J. C. (eds.) 2000. *Field-Flow Fractionation Handbook*. New York: John Wiley & Sons, Inc.
6. Jönsson, J. Å. and Carlshaf, A. 1989. Flow field flow fractionation in hollow cylindrical fibers. *Analytical Chemistry*, 61, 11–18.
7. Wijnhoven, J. E. G. J., Koorn, J. P., Poppe, H. and Kok, W. T. 1995. Hollow-fibre flow field-flow fractionation of polystyrene sulphonates. *Journal of Chromatography A*, 699, 119–129.
8. Kang, D. and Moon, M. H. 2005. Hollow fiber flow field-flow fractionation of proteins using a microbore channel. *Analytical Chemistry*, 77, 4207–4212.
9. Reschiglian, P., Zattoni, A., Roda, B., Cinque, L., Parisi, D., Roda, A., Piaz, F. D., Moon, M. H. and Min, B. R. 2005. On-line hollow-fiber flow field-flow fractionation-electrospray ionization/time-of-flight mass spectrometry of intact proteins. *Analytical Chemistry*, 77, 47–56.
10. Giddings, J. C., Yang, F. J. F. and Myers, M. N. 1974. Sedimentation field-flow fractionation. *Analytical Chemistry*, 46, 1917–1924.
11. Giddings, J. C. 1968. Nonequilibrium theory of field-flow fractionation. *Journal of Chemical Physics*, 49, 81–85.
12. Giddings, J. C., Yoon, Y. H., Caldwell, K. D., Myers, M. N. and Hovingh, M. E. 1975. Nonequilibrium plate height for field-flow fractionation in ideal parallel plate columns. *Separation Science*, 10, 447–460.
13. Yau, W. W. and Kirkland, J. J. 1981. Retention characteristics of time-delayed exponential field-programmed sedimentation field-flow fractionation. *Separation Science and Technology*, 16, 577–605.
14. Kirkland, J. J. and Yau, W. W. 1985. Thermal field-flow fractionation of polymers with exponential temperature programming. *Macromolecules*, 18, 2305–2311.
15. Giddings, J. C., Williams, P. S. and Beckett, R. 1987. Fractionating power in programmed field-flow fractionation: Exponential sedimentation field decay. *Analytical Chemistry*, 59, 28–37.
16. Williams, P. S. and Giddings, J. C. 1987. Power programmed field-flow fractionation: A new program form for improved uniformity of fractionating power. *Analytical Chemistry*, 59, 2038–2044.
17. Williams, P. S., Giddings, J. C. and Beckett, R. 1987. Fractionating power in sedimentation field-flow fractionation with linear and parabolic field decay programming. *Journal of Liquid Chromatography*, 10, 1961–1998.
18. Williams, P. S., Kellner, L., Beckett, R. and Giddings, J. C. 1988. Comparison of experimental and theoretical fractionating power for exponential field decay sedimentation field-flow fractionation. *The Analyst*, 113, 1253–1259.

19. Williams, P. S. and Giddings, J. C. 1991. Comparison of power and exponential field programming in field-flow fractionation. *Journal of Chromatography*, 550, 787–797.

20. Williams, P. S. and Giddings, J. C. 1994. Theory of field-programmed field-flow fractionation with corrections for steric effects. *Analytical Chemistry*, 66, 4215–4228.

21. Williams, P. S. 2000. Programmed field-flow fractionation – Retention. In: Schimpf, M. E., Caldwell, K. and Giddings, J. C. (eds.) *Field-Flow Fractionation Handbook*. New York: Wiley-Interscience, pp. 145–165.

22. Williams, P. S. 2000. Programmed field-flow fractionation – Fractionating power and optimization. In: Schimpf, M. E., Caldwell, K. and Giddings, J. C. (eds.) *Field-Flow Fractionation Handbook*. New York: Wiley-Interscience, pp. 167–182.

23. Giddings, J. C. and Myers, M. N. 1978. Steric field-flow fractionation: A new method for separating 1 to 100 μm particles. *Separation Science and Technology*, 13, 637–645.

24. Myers, M. N. and Giddings, J. C. 1982. Properties of the transition from normal to steric field-flow fractionation. *Analytical Chemistry*, 54, 2284–2289.

25. Lee, S. and Giddings, J. C. 1988. Experimental observation of steric transition phenomena in sedimentation field-flow fractionation. *Analytical Chemistry*, 60, 2328–2333.

26. Caldwell, K. D., Nguyen, T. T., Myers, M. N. and Giddings, J. C. 1979. Observations on anomalous retention in steric field-flow fractionation. *Separation Science and Technology*, 14, 935–946.

27. Williams, P. S., Koch, T. and Giddings, J. C. 1992. Characterization of near-wall hydrodynamic lift forces using sedimentation field-flow fractionation. *Chemical Engineering Communications*, 111, 121–147.

28. Williams, P. S., Moon, M. H. and Giddings, J. C. 1992. Fast separation and characterization of micron size particles by sedimentation/steric field-flow fractionation: Role of lift forces. In: Stanley-Wood, N. G. and Lines, R. W. (eds.) *Particle Size Analysis. Proceedings of 25th Anniversary Conference of the Particle Characterization Group*. Cambridge, U.K.: Royal Society of Chemistry, pp. 280–289.

29. Williams, P. S., Lee, S. and Giddings, J. C. 1994. Characterization of hydrodynamic lift forces by field-flow fractionation. Inertial and near-wall lift forces. *Chemical Engineering Communications*, 130, 143–166.

30. Williams, P. S., Moon, M. H. and Giddings, J. C. 1996. Influence of accumulation wall and carrier solution composition on lift force in sedimentation/steric field-flow fractionation. *Colloids and Surfaces A: Physicochemical and Engineering Aspects*, 113, 215–228.

31. Williams, P. S., Moon, M. H., Xu, Y. and Giddings, J. C. 1996. Effect of viscosity on retention time and hydrodynamic lift forces in sedimentation/steric field-flow fractionation. *Chemical Engineering Science*, 51, 4477–4488.

32. Vickrey, T. M. and Garcia-Ramirez, J. A. 1980. Magnetic field-flow fractionation: Theoretical basis. *Separation Science and Technology*, 15, 1297–1304.

33. Mori, S. 1986. Magnetic field-flow fractionation using capillary tubing. *Chromatographia*, 21, 642–644.

34. Latham, A. H., Freitas, R. S., Schiffer, P. and Williams, M. E. 2005. Capillary magnetic field flow fractionation and analysis of magnetic nanoparticles. *Analytical Chemistry*, 77, 5055–5062.

35. Giddings, J. C. 2000. The field-flow fractionation family: Underlying principles. In: Schimpf, M. E., Caldwell, K. and Giddings, J. C. (eds.) *Field-Flow Fractionation Handbook*. New York: John Wiley & Sons, pp. 3–30.

36. Schunk, T. C., Gorse, J. and Burke, M. F. 1984. Parameters affecting magnetic field-flow fractionation of metal oxide particles. *Separation Science and Technology*, 19, 653–666.

37. Semenov, S. N. 1986. Flow fractionation in a strong transverse magnetic field. *Russian Journal of Physical Chemistry*, 60, 729–731.

38. Ohara, T., Mori, S., Oda, Y., Yamamoto, K., Wada, Y. and Tsukamoto, O. 1994. FFF using high gradient and high intensity magnetic field: Process analysis. *Fourth International Symposium on Field-Flow Fractionation (FFF94)*. Lund, Sweden.

39. Ohara, T., Mori, S., Oda, Y., Wada, Y. and Tsukamoto, O. 1995. Feasibility of using magnetic chromatography for ultra-fine particle separation. *Proceedings of the IEE Japan—Power and Energy '95*, 161–166.

40. Tsukamoto, O., Ohizumi, T., Ohara, T., Mori, S. and Wada, Y. 1995. Feasibility study on separation of several tens nanometer scale particles by magnetic field-flow-fractionation technique using superconducting magnet. *IEEE Transactions on Applied Superconductivity*, 5, 311–314.

41. Ohara, T., Mori, S., Oda, Y., Wada, Y. and Tsukamoto, O. 1996. Feasibility of magnetic chromatography for ultra-fine particle separation. *Transactions of the IEE Japan*, 116-B, 979–986.

42. Wang, X., Ohara, T., Whitby, E. R., Karki, K. C. and Winstead, C. H. 1997. Computer simulation of magnetic chromatography system for ultra-fine particle separation. *Transactions of the IEE Japan*, 117-B, 1466–1474.

43. Ohara, T. 1997. Feasibility of using magnetic chromatography for ultra-fine particle separation. In: Schneider-Muntau, H. J. (ed.) *High Magnetic Fields: Applications, Generations, Materials*. Hackensack, NJ: World Scientific, pp. 43–55.

44. Ohara, T., Wang, X., Wada, H. and Whitby, E. R. 2000. Magnetic chromatography: Numerical analysis in the case of particle size distribution. *Transactions of the IEE Japan*, 120-A, 62–67.

45. Karki, K. C., Whitby, E. R., Patankar, S. V., Winstead, C., Ohara, T. and Wang, X. 2001. A numerical model for magnetic chromatography. *Applied Mathematical Modelling*, 25, 355–373.

46. Mitsuhashi, K., Yoshizaki, R., Ohara, T., Matsumoto, F., Nagai, H. and Wada, H. 2002. Retention of ions in a magnetic chromatograph using high-intensity and high-gradient magnetic fields. *Separation Science and Technology*, 37, 3635–3645.

47. Oberteuffer, J. A. 1973. High gradient magnetic separation. *IEEE Transactions on Magnetics*, 9, 303–306.

48. Gerber, R., Takayasu, M. and Friedlaender, F. J. 1983. Generalization of HGMS theory: The capture of ultra-fine particles. *IEEE Transactions on Magnetics*, 19, 2115–2117.

49. Kramer, A. J., Janssen, J. J. M. and Perenboom, J. A. A. J. 1990. Single-wire HGMS of colloidal particles: The evolution of concentration profiles. *IEEE Transactions on Magnetics*, 26, 1858–1860.

50. Williams, P. S., Moore, L. R., Chalmers, J. J. and Zborowski, M. 2002. The potential of quadrupole magnetic field-flow fractionation for determining particle magnetization distributions. *European Cells and Materials*, 3, 203–205.

51. Carpino, F., Moore, L. R., Chalmers, J. J., Zborowski, M. and Williams, P. S. 2005. Quadrupole magnetic field-flow fractionation for the analysis of magnetic nanoparticles. *Journal of Physics: Conference Series*, 17, 174–180.

52. Carpino, F., Moore, L. R., Zborowski, M., Chalmers, J. J. and Williams, P. S. 2005. Analysis of magnetic nanoparticles using quadrupole magnetic field-flow fractionation. *Journal of Magnetism and Magnetic Materials*, 293, 546–552.

53. Zborowski, M. 1997. Physics of the magnetic cell sorting. In: Häfeli, U., Schütt, W., Teller, J. and Zborowski, M. (eds.) *Scientific and Clinical Applications of Magnetic Carriers*. New York: Plenum Press, pp. 205–231.

54. Williams, P. S., Carpino, F., Moore, L. R. and Zborowski, M. 2010. Magnetic field programming in quadrupole magnetic field-flow fractionation. *Physics Procedia*, 9, 91–95.

55. Bean, C. P. and Livingston, J. D. 1959. Superparamagnetism. *Journal of Applied Physics*, 30, S120–S129.

56. Pankhurst, Q. A., Connolly, J., Jones, S. K. and Dobson, J. 2003. Applications of magnetic nanoparticles in biomedicine. *Journal of Physics D: Applied Physics*, 36, R167–R181.

57. Trahms, L. 2009. Biomedical applications of magnetic nanoparticles. In: Odenbach, S. (ed.) *Colloidal Magnetic Fluids: Basics, Development and Application of Ferrofluids*. Berlin, Germany: Springer-Verlag, pp. 327–358.

58. Williams, P. S., Carpino, F. and Zborowski, M. 2009. Theory for nanoparticle retention time in the helical channel of quadrupole magnetic field-flow fractionation. *Journal of Magnetism and Magnetic Materials*, 321, 1446–1451.

59. Williams, P. S., Carpino, F. and Zborowski, M. 2010. Erratum to: 'Theory for nanoparticle retention time in the helical channel of quadrupole magnetic field-flow fractionation' [*Journal of Magnetism and Magnetic Materials* 321 (2009) 1446–1451]. *Journal of Magnetism and Magnetic Materials*, 322, 3605.

60. Williams, P. S., Giddings, M. C. and Giddings, J. C. 2001. A data analysis algorithm for programmed field-flow fractionation. *Analytical Chemistry*, 73, 4202–4211.

61. Williams, P. S., Carpino, F. and Zborowski, M. 2010. Characterization of magnetic nanoparticles using programmed quadrupole magnetic field-flow fractionation. *Philosophical Transactions of the Royal Society of London, Series A*, 368, 4419–4437.

62. Mukadam, M. D., Yusuf, S. M., Sharma, P. and Kulshreshtha, S. K. 2004. Particle size-dependent magnetic properties of γ-Fe_2O_3 nanoparticles. *Journal of Magnetism and Magnetic Materials*, 272–276, 1401–1403.

63. Carpino, F., Zborowski, M. and Williams, P. S. 2007. Quadrupole magnetic field-flow fractionation: A novel technique for the characterization of magnetic nanoparticles. *Journal of Magnetism and Magnetic Materials*, 311, 383–387.

64. Yamaura, M., Camilo, R. L., Sampaio, L. C., Macêdo, M. A., Nakamura, M. and Toma, H. E. 2004. Preparation and characterization of (3-aminopropyl) triethoxysilane-coated magnetite nanoparticles. *Journal of Magnetism and Magnetic Materials*, 279, 210–217.

65. Williams, P. S., Carpino, F. and Zborowski, M. 2009. Magnetic nanoparticle drug carriers and their study by quadrupole magnetic field-flow fractionation. *Molecular Pharmaceutics*, 6, 1290–1306.

66. Springston, S. R., Myers, M. N. and Giddings, J. C. 1987. Continuous particle fractionation based on gravitational sedimentation in split-flow thin cells. *Analytical Chemistry*, 59, 344–350.

67. Gao, Y., Myers, M. N., Barman, B. N. and Giddings, J. C. 1991. Continuous fractionation of glass microspheres by gravitational sedimentation in split-flow thin (SPLITT) cells. *Particulate Science and Technology*, 9, 105–118.

68. Williams, P. S., Levin, S., Lenczycki, T. and Giddings, J. C. 1992. Continuous SPLITT fractionation based on a diffusion mechanism. *Industrial and Engineering Chemistry Research*, 31, 2172–2181.

69. Levin, S. and Giddings, J. C. 1991. Continuous separation of particles from macromolecules in split-flow thin (SPLITT) cells. *Journal of Chemical Technology and Biotechnology*, 50, 43–56.

70. Zhang, J., Williams, P. S., Myers, M. N. and Giddings, J. C. 1994. Separation of cells and cell-sized particles by continuous SPLITT fractionation using hydrodynamic lift forces. *Separation Science and Technology*, 29, 2493–2522.

71. Contado, C., Dondi, F., Beckett, R. and Giddings, J. C. 1997. Separation of particulate environmental samples by SPLITT fractionation using different operating modes. *Analytica Chimica Acta*, 345, 99–110.

72. Lee, S., Cho, S. K., Yoon, J. W., Choi, S.-H., Chun, J.-H., Eum, C. H. and Kwen, H. 2010. Removal of aggregates from micron-sized polymethyl methacrylate (PMMA) latex beads using full feed depletion mode of gravitational SPLITT fractionation (FFD-GSF). *Journal of Liquid Chromatography and Related Technologies*, 33, 27–36.

73. Giddings, J. C. 1992. Optimization of transport-driven continuous SPLITT fractionation. *Separation Science and Technology*, 27, 1489–1504.

74. Gupta, S., Ligrani, P. M. and Giddings, J. C. 1997. Investigations of performance characteristics including limitations due to flow instabilities in continuous SPLITT fractionation. *Separation Science and Technology*, 32, 1629–1655.

75. Ligrani, P. M., Gupta, S. and Giddings, J. C. 1998. Onset and effects of instabilities from unstable stratification of density on mass transfer in channel shear layers at low Reynolds numbers. *International Journal of Heat and Mass Transfer*, 41, 1667–1679.

76. Gupta, S., Ligrani, P. M. and Giddings, J. C. 1999. Characteristics of flow instabilities from unstable stratification of density in channel shear layers at low Reynolds numbers. *International Journal of Heat and Mass Transfer*, 42, 1023–1036.

77. Leighton, D. and Acrivos, A. 1987. Measurement of shear-induced self-diffusion in concentrated suspensions of spheres. *Journal of Fluid Mechanics*, 177, 109–131.

78. Leighton, D. and Acrivos, A. 1986. Viscous resuspension. *Chemical Engineering Science*, 41, 1377–1384.

79. Contado, C. and Hoyos, M. 2007. SPLITT cell analytical separation of silica particles. Non-specific crossover effects: Does the shear-induced diffusion play a role? *Chromatographia*, 65, 453–462.

80. Callens, N., Hoyos, M., Kurowski, P. and Iorio, C. S. 2008. Particle sorting in a mini step-split-flow thin channel: Influence of hydrodynamic shear on transversal migration. *Analytical Chemistry*, 80, 4866–4875.

81. Williams, P. S., Hoyos, M., Kurowski, P., Salhi, D., Moore, L. R. and Zborowski, M. 2008. Characterization of nonspecific crossover in split-flow thin channel fractionation. *Analytical Chemistry*, 80, 7105–7115.

82. Fuh, C. B., Trujillo, E. M. and Giddings, J. C. 1995. Hydrodynamic characterization of SPLITT fractionation cells. *Separation Science and Technology*, 30, 3861–3876.

83. Fuh, C. B. and Chen, S. Y. 1998. Magnetic split-flow thin fractionation: New technique for separation of magnetically susceptible particles. *Journal of Chromatography A*, 813, 313–324.

84. Fuh, C. B. and Chen, S. Y. 1999. Magnetic split-flow thin fractionation of magnetically susceptible particles. *Journal of Chromatography A*, 857, 193–204.

85. Fuh, C. B., Lai, J. Z. and Chang, C. M. 2001. Particle magnetic susceptibility determination using analytical split-flow thin fractionation. *Journal of Chromatography A*, 923, 263–270.

86. Fuh, C. B., Tsai, H. Y. and Lai, J. Z. 2003. Development of magnetic split-flow thin fractionation for continuous particle separation. *Analytica Chimica Acta*, 497, 115–122.

87. Tsai, H., Fang, Y. S. and Fuh, C. B. 2006. Analytical and preparative applications of magnetic split-flow thin fractionation on several ion-labeled red blood cells. *Biomagnetic Research and Technology*, 4, 6.

88. Wingo, R. M., Prenger, F. C., Johnson, M. D., Waynert, J. A., Worl, L. A. and Ying, T. -Y. 2004. High-gradient magnetic field split-flow thin channel (HGMF-SPLITT) fractionation of nanoscale paramagnetic particles. *Separation Science and Technology*, 39, 2769–2783.

89. Furlani, E. P. 2007. Magnetophoretic separation of blood cells at the microscale. *Journal of Physics D: Applied Physics*, 40, 1313–1319.

90. Zborowski, M., Ostera, G. R., Moore, L. R., Milliron, S., Chalmers, J. J. and Schechter, A. N. 2003. Red blood cell magnetophoresis. *Biophysical Journal*, 84, 2638–2645.

91. Hoyos, M., Moore, L., Williams, P. S. and Zborowski, M. 2011. The use of a linear Halbach array combined with a step-SPLITT channel for continuous sorting of magnetic species. *Journal of Magnetism and Magnetic Materials*, 323, 1384–1388.

92. Hoyos, M., McCloskey, K. E., Moore, L. R., Nakamura, M., Bolwell, B. J., Chalmers, J. J. and Zborowski, M. 2002. Pulse-injection studies of blood progenitor cells in a quadrupole magnet flow sorter. *Separation Science and Technology*, 37, 745–767.

93. Moore, L. R., Williams, P. S., Chalmers, J. J. and Zborowski, M. 2002. Magnetic flow sorting using susceptibility-modified carrier fluids. *European Cells and Materials*, 3, 56–59.

94. Moore, L. R., Milliron, S., Williams, P. S., Chalmers, J. J., Margel, S. and Zborowski, M. 2004. Control of magnetophoretic mobility by susceptibility-modified solutions as evaluated by cell tracking velocimetry and continuous magnetic sorting. *Analytical Chemistry*, 76, 3899–3907.

95. Nakamura, M., Decker, K., Chosy, J., Comella, K., Melnik, K., Moore, L., Lasky, L. C., Zborowski, M. and Chalmers, J. J. 2001. Separation of a breast cancer cell line from human blood using a quadrupole magnetic flow sorter. *Biotechnology Progress*, 17, 1145–1155.

96. Zborowski, M., Moore, L. R., Williams, P. S. and Chalmers, J. J. 2002. Separations based on magnetophoretic mobility. *Separation Science and Technology*, 37, 3611–3633.

97. Williams, P. S., Moore, L. R., Chalmers, J. J. and Zborowski, M. 2003. Splitter imperfections in annular split-flow thin separation channels: Effect on nonspecific crossover. *Analytical Chemistry*, 75, 1365–1373.

98. Williams, P. S., Decker, K., Nakamura, M., Chalmers, J. J., Moore, L. R. and Zborowski, M. 2003. Splitter imperfections in annular split-flow thin separation channels: Experimental study of nonspecific crossover. *Analytical Chemistry*, 75, 6687–6695.

99. Moore, L. R., Rodriguez, A. R., Williams, P. S., McCloskey, K., Bolwell, B. J., Nakamura, M., Chalmers, J. J. and Zborowski, M. 2001. Progenitor cell isolation with a high-capacity quadrupole magnetic flow sorter. *Journal of Magnetism and Magnetic Materials*, 225, 277–284.

100. Zborowski, M., Williams, P. S., Sun, L., Moore, L. R. and Chalmers, J. J. 1997. Cylindrical SPLITT and quadrupole magnetic field in application to continuous-flow magnetic cell sorting. *Journal of Liquid Chromatography and Related Technologies*, 20, 2887–2905.

101. Williams, P. S., Zborowski, M. and Chalmers, J. J. 1999. Flow rate optimization for the quadrupole magnetic cell sorter. *Analytical Chemistry*, 71, 3799–3807.

102. Chalmers, J. J., Zborowski, M., Sun, L. and Moore, L. 1998. Flow through, immunomagnetic cell separation. *Biotechnology Progress*, 14, 141–148.

103. Sun, L., Zborowski, M., Moore, L. R. and Chalmers, J. J. 1998. Continuous, flow-through immunomagnetic cell sorting in a quadrupole field. *Cytometry*, 33, 469–475.

104. Zborowski, M., Sun, L., Moore, L. R. and Chalmers, J. J. 1999. Rapid cell isolation by magnetic flow sorting for applications in tissue engineering. *ASAIO Journal*, 45, 127–130.

105. Zborowski, M., Sun, L., Moore, L. R., Williams, P. S. and Chalmers, J. J. 1999. Continuous cell separation using novel magnetic quadrupole flow sorter. *Journal of Magnetism and Magnetic Materials*, 194, 224–230.

106. Lara, O., Tong, X., Zborowski, M. and Chalmers, J. J. 2004. Enrichment of rare cancer cells through depletion of normal cells using density and flow-through, immunomagnetic cell separation. *Experimental Hematology*, 32, 891–904.

107. Lara, O., Tong, X., Zborowski, M., Farag, S. S. and Chalmers, J. J. 2006. Comparison of two immunomagnetic separation technologies to deplete T cells from human blood samples. *Biotechnology and Bioengineering*, 94, 66–80.

108. Jing, Y., Moore, L. R., Williams, P. S., Chalmers, J. J., Farag, S. S., Bolwell, B. and Zborowski, M. 2007. Blood progenitor cell separation from clinical leukapheresis product by magnetic nanoparticle binding and magnetophoresis. *Biotechnology and Bioengineering*, 96, 1139–1154.

109. Tong, X., Xiong, Y., Zborowski, M., Farag, S. S. and Chalmers, J. J. 2007. A novel high throughput immunomagnetic cell sorting system for potential clinical scale depletion of T cells for allogeneic stem cell transplantation. *Experimental Hematology*, 35, 1613–1622.

110. Tong, X., Yang, L., Lang, J. C., Zborowski, M. and Chalmers, J. J. 2007. Application of immunomagnetic cell enrichment in combination with RT-PCR for the detection of rare circulating head and neck tumor cells in human peripheral blood. *Cytometry Part B: Clinical Cytometry*, 72B, 310–323.

111. Yang, L., Lang, J. C., Balasubramanian, P., Jatana, K. R., Schuller, D., Agrawal, A., Zborowski, M. and Chalmers, J. J. 2009. Optimization of an enrichment process for circulating tumor cells from the blood of head and neck cancer patients through depletion of normal cells. *Biotechnology and Bioengineering*, 102, 521–534.

112. Jing, Y., Moore, L. R., Schneider, T., Williams, P. S., Chalmers, J. J., Farag, S. S., Bolwell, B. and Zborowski, M. 2007. Negative selection of hematopoietic progenitor cells by continuous magnetophoresis. *Experimental Hematology*, 35, 662–672.

113. Kennedy, D. J., Todd, P., Logan, S., Becker, M., Papas, K. K. and Moore, L. R. 2007. Engineering quadrupole magnetic flow sorting for the isolation of pancreatic islets. *Journal of Magnetism and Magnetic Materials*, 311, 388–395.

114. Rizzari, M. D., Suszynski, T. M., Kidder, L. S., Stein, S. A., O'Brien, T. D., Sajja, V. S. K., Scott, W. E., Kirchner, V. A., Weegman, B. P., Avgoustiniatos, E. S., Todd, P. W., Kennedy, D. J., Hammer, B. E., Sutherland, D. E. R., Hering, B. J. and Papas, K. K. 2010. Surgical protocol involving the infusion of paramagnetic microparticles for preferential incorporation within porcine islets. *Transplantation Proceedings*, 42, 4209–4212.

115. Chalmers, J. J., Haam, S., Zhao, Y., McCloskey, K., Moore, L., Zborowski, M. and Williams, P. S. 1999. Quantification of cellular properties from external fields and resulting induced velocity: Cellular hydrodynamic diameter. *Biotechnology and Bioengineering*, 64, 509–518.

116. Chalmers, J. J., Haam, S., Zhao, Y., McCloskey, K., Moore, L., Zborowski, M. and Williams, P. S. 1999. Quantification of cellular properties from external fields and resulting induced velocity: Magnetic susceptibility. *Biotechnology and Bioengineering*, 64, 519–526.

117. Chalmers, J. J., Zhao, Y., Nakamura, M., Melnik, K., Lasky, L., Moore, L. and Zborowski, M. 1999. An instrument to determine the magnetophoretic mobility of labeled, biological cells and paramagnetic particles. *Journal of Magnetism and Magnetic Materials*, 194, 231–241.

118. Zhang, H., Nakamura, M., Comella, K., Moore, L., Zborowski, M. and Chalmers, J. 2002. Characterization/quantification of the factors involved in the imparting a magnetophoretic mobility on cells and particles. *European Cells and Materials*, 3, 34–36.

119. Moore, L. R., Fujioka, H., Williams, P. S., Chalmers, J. J., Grimberg, B., Zimmerman, P. A. and Zborowski, M. 2006. Hemoglobin degradation in malaria-infected erythrocytes determined from live cell magnetophoresis. *FASEB Journal*, 20, 747–749.

120. Zhang, H., Williams, P. S., Zborowski, M. and Chalmers, J. J. 2006. Binding affinities/avidities of antibody-antigen interactions: Quantification and scale-up implications. *Biotechnology and Bioengineering*, 95, 812–829.

121. Jin, X., Zhao, Y., Richardson, A., Moore, L., Williams, P. S., Zborowski, M. and Chalmers, J. J. 2008. Differences in magnetically induced motion of diamagnetic, paramagnetic, and superparamagnetic microparticles detected by cell tracking velocimetry. *The Analyst*, 133, 1767–1775.

122. Jing, Y., Mal, N., Williams, P. S., Mayorga, M., Penn, M. S., Chalmers, J. J. and Zborowski, M. 2008. Quantitative intracellular magnetic nanoparticle uptake measured by live cell magnetophoresis. *FASEB Journal*, 22, 4239–4247.

123. Mallinson, J. C. 1973. One-sided fluxes – A magnetic curiosity? *IEEE Transactions on Magnetics*, 9, 678–682.

124. Halbach, K. 1979. Strong rare earth cobalt quadrupoles. *IEEE Transactions on Nuclear Science*, NS-26, 3882–3884.

125. Halbach, K. 1980. Design of permanent multipole magnets with oriented rare earth cobalt material. *Nuclear Instruments and Methods*, 169, 1–10.

126. Halbach, K. 1981. Physical and optical properties of rare earth cobalt magnets. *Nuclear Instruments and Methods*, 187, 109–117.

127. Halbach, K. 1985. Application of permanent magnets in accelerators and electron storage rings. *Journal of Applied Physics*, 57, 3605–3608.

128. Cugat, O., Hansson, P. and Coey, J. M. D. 1994. Permanent magnet variable flux sources. *IEEE Transactions on Magnetics*, 30, 4602–4604.

129. Coey, J. M. D. 2002. Permanent magnet applications. *Journal of Magnetism and Magnetic Materials*, 248, 441–456.

130. Hayden, M. E. and Häfeli, U. O. 2006. 'Magnetic bandages' for targeted delivery of therapeutic agents. *Journal of Physics: Condensed Matter*, 18, S2877–S2891.

131. Melucci, D., Guardigli, M., Roda, B., Zattoni, A., Reschiglian, P. and Roda, A. 2003. A new method for immunoassays using field-flow fractionation with on-line, continuous chemiluminescence detection. *Talanta*, 60, 303–312.

P. Stephen Williams obtained his PhD in 1980 under the direction of Prof. J. Howard Purnell at Swansea University, Wales, United Kingdom, who was one of the pioneers in gas chromatography. He later joined Prof. J. Calvin Giddings at the University of Utah, USA where he spent 13 years carrying out research on various theoretical aspects of FFF and SPLITT fractionation. For the past 10 years he has worked in the Department of Biomedical Engineering at the Lerner Research Institute, Cleveland Clinic, Clevland, Ohio, on magnetic separations. He has developed quadrupole magnetic FFF for characterising magnetic nanoparticles used for cell labelling and drug targeting.

12 Nanomagnetic Gene Transfection

Angeliki Fouriki and Jon Dobson

CONTENTS

12.1 BRIEF OVERVIEW OF GENE THERAPY

Genes are heritable units controlling identifiable traits in every organism. Living beings rely on genes coding for all proteins, following the central dogma of molecular biology to ensure specific transcription from DNA to RNA and translation from RNA to protein[1,2]. In September 1966, Joshua Lederberg talked about the necessity of evolutionary theory to model a self-modifying system and the potential of genetic therapy to repair any 'genetic–metabolic' disease[3]. Nowadays, gene therapy is considered as a method to insert genetic material into an individual's cells and tissues in order to replace a defective gene, or by introducing a new function into cells to attempt to treat more complex inherited or acquired genetic disorders such as cancer, immunodeficiency syndrome, cystic fibrosis (CF) and haemophilia[4–9]. Gene therapy as a potential therapeutic tool requires effective gene delivery *via* 'vectors' such as viruses that are used as vessels to carry foreign genetic material into the target cells[10,11].

There are two main types of gene therapy:

> *Somatic gene therapy (SGT)* is the only technique that is currently considered appropriate for use in humans. It is the insertion of a single gene into somatic cells of an individual with a life-threatening genetic disease that is intended only to eliminate the clinical consequences of the disease. The genetic intervention affects only a subset of patient's cells and passes no genetic information to the next generation. Therefore, it does not directly affect any descendents[12].

Human germ line gene therapy (HGLGT) is fundamentally different from SGT with respect to long-term consequences. HGLGT involves the insertion of a healthy gene into the fertilised egg of an animal that has a specific genetic defect. Therefore, the transgene is acquired by all cells of the body as the embryo develops and has the potential to be passed on to future generations[13]. One of the primary constraints limiting the use of HGLGT in humans is the fact that it is not possible to diagnose genetic diseases within the fertilised egg nor to manipulate the transgene inside the egg[14].

In SGT, genes can be delivered by two different approaches. The *ex vivo* approach is the most common and involves the removal of physiologically accessible target cells from the patient. These cells are manipulated in the laboratory in order to incorporate the vector containing the transgene or recombinant vector and then are introduced back into the patient[15–17]. This approach is more efficient than other methods due to the higher vector ratio to target cell present in cell culture, and also because it allows the reintroduction only of those cells back into the patient that have incorporated the transgene[18]. When the recombinant vector is injected directly into the target tissue, the method is called *in situ*. The disadvantages of this technique include invasive procedures for the extraction and reintroduction of cells from and back into the patient and patient specificity in terms of immunogenicity[17].

The second approach, known as *in vivo*, allows direct administration of gene transfer vectors to patients and therefore permits the movement of the vector throughout the body, into cells that are still part of a living organism, eliminating immunogenicity issues[17,19]. This method is less invasive compared to *ex vivo* and in addition is advantageous for working with less accessible cells or cell types that do not easily survive for long outside the body[20]. This technique's limitations include (i) the non-specific gene targeting, since the vector moves around the body until it reaches the target tissue, (ii) the possibility of the vector infecting the germ line instead of the target tissue and (iii) the potential for insertional mutagenesis, placing the gene in the wrong place within the patient's chromosome to trigger cancer[18].

Gene therapy by its definition, and due to the ability of acting on a single-cell basis, has a wide therapeutic potential that could provide more effective treatment in some existing therapeutic options or alleviate and even cure a currently untreatable disease. A greater understanding of the technical barriers and true benefits or harms of gene therapy is essential to provide optimised clinical treatments. In fact, there is a great need for improvement of the current gene delivery systems, and one of the simplest but most daunting barriers that scientists have to overcome is the insufficient vector accumulation at the target sites that lead to insufficient therapeutic gene delivery into the cells[21–23]. Multiple methods for gene transfer have been considered, trying to resolve this limitation, and the major currently used systems for *in vitro* and *in vivo* studies will be discussed in the following sections.

12.2   OVERVIEW OF GENE DELIVERY APPROACHES

An ideal gene therapy vector has to be specifically targeted, biodegradable or easy to remove, stable in the bloodstream and non-toxic. Additionally, it has to be able to protect DNA against degradation during transport, to initiate no inflammatory or immunogenic response and be stable for storage[11,24–26].

Many different gene delivery methods have been developed. These protocols are initially being tested *in vitro* for their gene transfer efficiency and have proved to be an invaluable tool for understanding genetic pathways, diseases and drug screening[27–29]. In addition, reporter DNA constructs, such as firefly luciferase and green fluorescent protein (GFP), are often incorporated into the vector sequence to facilitate sensitive monitoring of gene transfer and expression into the target cells[30,31]. Further *in vivo* experiments using many of these gene transfer methods have been done in animal models[32–34] to extend our understanding and ensure safety prior to human testing[6,35]. The currently available methods for gene transfer are divided into two main classes: viral and non-viral gene delivery.

12.2.1 Viral Delivery

Viruses (Latin for toxin or poison) are microscopic infectious agents unable to grow or reproduce outside a host cell. They can be considered as packages used to transfer viral genes into host cells[23]. Once these small parasites infect a cell by receptor-mediated endocytosis, they employ its cellular machinery to maintain their survival in terms of metabolism and nutrient supply and to target the nucleus for the delivery of their genes and regulatory proteins for initiation of transcription and replication[17,25]. Each viral particle, also known as virion, consists of nucleic acid genome (DNA or RNA) that may be single- or double-stranded[2].

Increasing knowledge about the viral replication mechanism has enabled researchers to modify viruses for distinct scientific uses such as vaccine development[36,37]. For gene transfer purposes, all viral DNA has been disabled to minimise the replication of the virus into the target cell and the initiation of virally caused diseases. Scientists introduce modified DNA carrying a therapeutic gene inside the viral particles, in place of the viral genome. Then, in a process known as transduction, these viral vectors are directed into the target cells and express their functional genetic information[25] (Figure 12.1).

Viral vectors appear to be important for SGT due to their increased overall transduction rate compared to non-viral vectors[19,38,39]. Further molecular engineering of these systems, including protein engineering approaches to generate viral vectors with novel gene delivery capabilities, have been employed trying to meet complicated human therapeutic needs[40].

The most commonly used viral vectors for gene delivery are as follows:

1. *Retroviruses*, including recently developed lentiviruses, which have high transduction efficiency in many cell types (including non-dividing) and stably integrate the transformed single-stranded RNA to double-stranded DNA into the host genome[18,41–43]. They have been used in many *in vitro* and *in vivo* studies[18,19,25,41,44–47] and more recently in clinical trials for

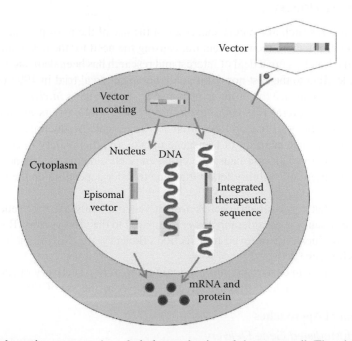

FIGURE 12.1 Schematic representation of viral transduction of the target cell. The viral vector carrying a therapeutic gene binds specifically to the surface of the host cell *via* a receptor and enters the target cell. The viral genome is inserted into the nucleus and following specific modifications can remain as an episome or be integrated into the host genome. Gene expression follows. (Adapted from Kay, M. et al., *Nat. Med.*, 7, 33, 2001.)

the successful infusion of CD34$^+$ bone marrow cells transduced with a retroviral vector containing the adenosine deaminase (ADA) gene, into 10 children with severe combined immunodeficiency (SCID) due to lack of the ADA gene[6].

2. *Adenoviruses* are similar to retroviruses in terms of the wide variety of cell types that are able to transduce as well as their efficiency, and they are considered more advantageous due to their capacity to accommodate relatively large pieces for foreign DNA (7–8 kb)[17]. Adenoviruses have been successfully used in many cancer therapy studies and clinical trials[17,48–50], neurodegenerative and cardiovascular diseases and pulmonary hypertension[17,34].

3. *Adeno-associated viruses* (AAV) are used quite successfully for viral gene delivery[51]. They transduce a wide variety of cell types; so far they have caused no known human disease and have a good safety record[28,51,52]. They selectively integrate into the host genome and are generally stable[15,51,53]. The successes of AAVs *in vitro* and *in vivo* among others include: the microsphere-mediated delivery of recombinant AAV2 for improved vector transduction efficiency[38,54], the self-complementary (sf) AAV for intra-articular gene delivery[52] and the AVV6 use for gene transfer in skeletal muscle[51].

12.2.1.1   Limitations of Viral Vectors

Despite the fact that viral vectors have provided invaluable results for gene delivery, some general disadvantages associated with their use include (i) safety issues (excluding the AAVs)[55–57], (ii) the host inflammatory or immunogenic response that viruses initiate from the infusion of foreign materials[17,19,58–60], (iii) the target tissue specificity that can cause tissue damage in undesired tissues and organs (when delivery and transduction is non-specific)[55], (iv) recombination incidents that can lead to the production of a replicating virus causing a direct threat[23,57], (v) the limited capacity to accommodate large transgenes and manufacture sufficient stable vector stocks[17,51,60–63] and (vi) the inefficiency in transferring the engineered gene into suspension cell lines[39].

12.2.2   NON-VIRAL DELIVERY

Despite the extensive research and recent successes in the use of the most popular viral vectors for gene delivery, some limitations and risks remain, causing the need for the development of non-viral alternative methods. Lately, a great deal of interest and research has been dedicated to non-viral vector gene delivery, leading to the first non-viral gene therapy clinical trial in 1992 using direct gene transfer to successfully enhance the immune response against tumours *in vivo*[24].

In general, there are several reasons to explain why non-viral carriers are advantageous and promising: (i) the simplicity of use and potential for large-scale production, (ii) the low cost of manufacturing, (iii) the large capacity of transgene uptake, (iv) their safety in comparison to viral vectors since they contain no disease-causative viral genes, eliminating any chance of viral replication[26,64,65] and (v) the high cell viability following gene transfer depending on the specific non-viral technique used.

The main reason that non-viral approaches are not yet commonly used for gene delivery is that their transfection efficiency is considerably lower compared to the viral vectors. Recent advances in the field showing significantly improved transfection efficiency are shifting the focus towards the potential of non-viral gene delivery.

Non-viral techniques can be divided into two main categories: (i) delivery of encapsulated genes mediated by chemical approaches[66] and (ii) delivery of DNA by the use of physical methods[67].

12.2.2.1   Chemical Approaches

12.2.2.1.1   Lipid-Mediated Gene Delivery

Liposome-mediated gene transfer, also known as lipofection, was first developed in 1987. It occurs *via* endocytosis[68], and its principle relies on the ability of liposomes to encapsulate DNA, fuse with the cell membrane and thereby deliver it into the target cell[69]. Among other mentioned benefits that many

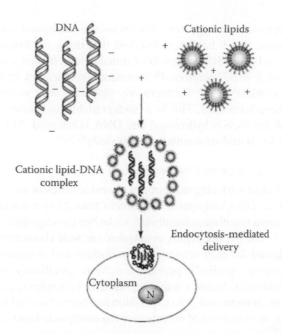

FIGURE 12.2 Schematic representation of cationic lipid/DNA complexes. Cationic lipids mix and associate with DNA to form spherical liposomes. Once the complexes have entered the cell by endocytosis, the DNA is released and transported into the nucleus of the target cell. (Adapted from Schmidt-Wolf, G.D. et al., *Trends Mol. Med.*, 9, 67, 2003.)

non-viral vectors have in common, this method protects the DNA against enzymatic degradation[70] as it is released from the intracellular vesicles before being broken down by lysosomal compartments[71]. Lipid-mediated gene delivery (Figure 12.2) has been successful with *in vitro* and *in vivo* gene therapy experimental models[26,68] and has been tested in clinical trials for the treatment of cancer and CF[26,70].

Cationic-lipid transfection has many advantages but there are drawbacks that require improvement such as (i) further increase of transfection efficiency (a common drawback among non-viral systems), (ii) their transient expression that requires repeated doses, (iii) the extended required time for treatment and incubation of cells (4–6 h), (iv) the toxic side effects reported with high doses of lipocomplex administration, (v) the administration route-dependent transfection efficiency *in vivo,* requiring individualised optimisation of the vector for distinct target cells and (vi) the potentially high clearance rate from bloodstream when lipocomplexes interact with serum proteins resulting in their volume increase[65,70,72,73].

Trying to resolve some of these concerns, multiple studies have been done to improve plasmid encapsulation, several cationic lipids have been used and many cationic headgroup manipulations of monovalent lipids have considerably increased gene transfer efficiency[65,66,68,74]. Lipofection remains one of the most popular non-viral transfection delivery methods but modification of cationic lipid molecules has been saturated and its research is focused on specific *in vitro* applications.

12.2.2.1.2 Polymer-Mediated Gene Delivery

Another actively studied class of synthetic vectors is the polymer-based gene delivery system. Generally, cationic polymers are more efficient in condensing DNA compared to cationic lipids[64,66]. The cationic polymer poly-L-lysine (PLL) is one of the first polymers used for gene transfer *in vivo* that form electrostatic complexes with negatively charged DNA, known as polyplexes, which then are internalised by the cells[26,71].

Many types of polymers have been developed such as polyamidoamine, cationic dextran, chitosan[75], cationic proteins and peptides[71]. Another well known and quite promising example of cationic polymer is polyethylenimine (PEI) that has been used *in vivo* for efficient gene transfer *via* lung instillation or kidney perfusion, without using the endosome-disruption component[64,68,71]. PEI has a non-biodegradable nature, and it has been observed that using its different isoforms (linear or branched) complexed with DNA, different levels of transfection efficiency and cell viability can be obtained both *in vitro* and *in vivo*. Therefore, PEI transfection efficiency and toxicity are molecular weight and structure dependent[64,71,74,76]. Furthermore, polymers such as poly[α-(4-aminobutyl)-L-glycolic acid] (PAGA) have been used. This is a biodegradable and water-soluble polymer, which initially condenses DNA but then is hydrolysed and DNA is released. This method demonstrated satisfactory transfection levels and substantially low toxicity[68,74,77].

12.2.2.1.3   *Peptide-Mediated Gene Delivery*

Peptides are covalently linked with oligonucleotides to enable efficient and cell-specific gene delivery *in vitro*. These peptide–DNA conjugates use protein transduction domains or cell-penetrating peptides that can easily cross the plasma membrane to deliver the oligonucleotide sequence or gene of interest into the cell[68]. The peptide–DNA conjugates are well characterised molecules[26]. Since thiols have been investigated for their sensitivity in reduction and oxidation to control lipid/DNA complex formation they appear similar in peptides too. More specifically, when peptides condense DNA, thiol groups are oxidised to create a stable peptide/DNA complex capable of high transfection levels *in vitro*. In addition, an increased gene expression has been observed without any alteration in DNA uptake by the cells, an indication of reduction of the disulphide bond, essential in intracellular release of DNA[74].

12.2.2.2   **Physical Methods**

12.2.2.2.1   *Direct Injection of Naked DNA*

This method is the simplest non-viral gene delivery system currently in use for gene therapy purposes. It involves the injection of naked DNA directly into target tissues (e.g. skeletal muscle, thyroid, liver, heart muscle, skin and tumour) or into the systemic circulation *via* a needle or jet[26,71,74]. The benefit of jet injection relies on the high-pressure stream of liquid that penetrates the skin and targets the tissue of interest. As shown recently, this minimises patients' discomfort and increases gene expression[67,78]. The advantages of naked DNA injection comprise the absence of a biological vector and immunogenicity. Finally, the limitations include the low transfection efficiency, the possibility of some tissue damage or increase of the pressure on the surface of the target cell following injection[64,67,69] and during systemic circulation, the possibility of naked DNA not escaping enzymatic degradation due to the action of nucleases and uptake of phagocytes[26].

12.2.2.2.2   *Gene Gun*

During gene gun transfection, the targeted tissue is bombarded with heavy metal particles (e.g. gold particles) with plasmid DNA deposited on their surface[79]. The particle acceleration, generally mediated by a gas discharge in a gun, is required for the penetration of the particles through the cell membrane into the cytoplasm, offering the advantage of bypassing the endosomal compartment[26,67,74]. Using the gene gun approach, DNA can be directly delivered into the nucleus and skin, mucosa, liver or muscle and also long-lasting transgene expression has been achieved[67–74]. The main limitation of this technique involves the low penetration of the metal particles and subsequently the low efficiency in reaching the entire tissue, leading to the requirement for surgery to access non-superficial tissue[67]. For this reason, the application of the method is restricted to mainly skin.

12.2.2.2.3   *Electroporation*

Electroporation (or electropermeabilisation) is a versatile, well-studied method by which DNA and other molecules can be introduced into cells by a pulsed electric field[26]. The short and intense electric shock of electrocompressive forces makes the cell membrane temporarily permeable allowing the DNA to enter the cell[67,74]. Electroporation has achieved high transfection efficiency in many tissues and prevents enzymatic degradation of the DNA. The method is limited by the use of electrodes with restricted range that cause difficulties in transfection of cells in tissues for therapeutic purposes. In addition, the use of high-intensity electric fields can cause permanent changes in the cell membrane structure, resulting in apoptosis or necrosis[67,71].

12.2.2.2.4   Ultrasound

Ultrasound, also known as sonoporation, enables gene transfection by the generation of irradiating ultrasonic waves that temporarily increase cell membrane permeability and allow the internalisation of large macromolecules such as plasmid DNA[74]. Ultrasound is a flexible and approved method for clinical research due to its safety, non-invasive nature and low levels of energy delivered. The system has demonstrated a 10- to 20-fold increase in transgene expression over that of naked DNA transfection[71] and has been used for gene delivery in foetal mouse studies[26], muscle[74] and for the delivery of naked DNA into the carotid artery, liver, kidney and solid tumours[67]. However, it has been reported that it can induce the breakdown of the cytoskeleton causing significant alterations in cellular mechanisms such as DNA trafficking[67,76].

12.2.2.2.5   Hydrodynamic Gene Delivery

This method allows naked DNA to be introduced into the cells in highly perfused internal organs such as the liver, and it has been applied only *in vivo* in animal models with encouraging results[67]. With this technique, molecules such as small interfering RNA (siRNA), dye molecules and proteins have been delivered into the target cell by direct transfer into the cytoplasm, overcoming endocytosis[71]. Hydrodynamic gene delivery yields high transfection efficiency that depends on the target organ and the speed and volume of injection[71]. The limitation of the method is the high volume of DNA solution required for injection and the frequent lethal effects of rapid injection[71]. Recent work has developed an injection device to ensure safe and controlled hydrodynamic injection that might provide a method for the future in clinical use.

12.2.2.3   Magnetic Drug Delivery as the Basis for MNP-Based Gene Transfection and Targeting

It has been shown that some non-viral transfection methods, due to their chemical or physical characteristics as well as their acquired technical advances, can offer impressive results and clinical potential. The current limitations of these systems, such as toxicity and low transfection efficiency combined with the challenge of specific targeting to sites of interest in the body, maintain the need to develop suitable carriers for gene delivery for *in vitro* investigations and *in vivo* therapeutic applications.

Magnetic micro- and nanoparticle-based drug delivery is a chemical and physical combination that offers potential for *in vitro* and *in vivo* gene delivery. Coated and functionalised magnetic nanoparticles can be used as gene carriers and targeted to specific sites *in vitro* and *in vivo* by the application of external, focused magnetic fields[80], forming the basis of magnetic nanoparticle-based gene transfection and targeting. In addition, since systemic drug delivery requires larger doses of drug circulating and associated deleterious side effects in healthy cells, the potential of specific magnetically targeted drug and gene delivery attracts growing interest[80–82].

The first clinical application of magnetic particles was demonstrated in 1957 by Gilchrist et al. when maghemite (γFe_2O_3) particles were injected and selectively heated using an AC magnetic field to treat lymph nodes and metastases[83]. The principle of magnetically guiding particles into specific target areas was introduced in 1963, when intravenously injected iron particles were accumulated into the veins of dog legs by the application of an externally applied magnet[84]. A few years later, in 1979, the first well-defined magnetic microspheres were constructed for the purpose of magnetic targeting and the theory was developed[85].

Magnetic particles are capable of coupling and responding to an applied magnetic field. In magnetically targeted therapy, a cytotoxic drug or a gene is bound to biocompatible magnetic micro- or nanoparticles, which then are injected into the patient's bloodstream or, in some cases, directly into the target tissue. By the use of externally applied magnetic fields, the drug/particle

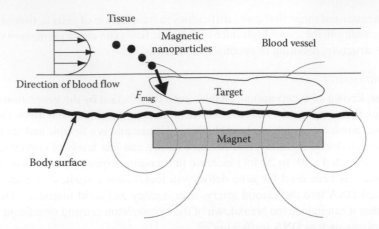

FIGURE 12.3 Schematic side view section of magnetic nanoparticle-based gene targeting *in vivo*. Grey rings represent the lines of magnetic flux due to externally applied magnetic field. F_{mag} represents the magnetic force vector applied on the particles as they flow through the bloodstream. (From Dobson, J., *Gene Ther.*, 13, 283–287, 2006.)

complexes can be pulled from circulation and targeted to a region within the body or held in place at the target[81] (Figure 12.3).

Subsequently, the therapeutic agent is slowly released from the magnetic carrier—by heat, enzymatic activity or degradation of the linker compounds–and a local effect of the agent released is initiated[29,87]. Studies have shown that magnetically targeted drug delivery can successfully replace the large amounts of non-specific, systemic delivery with lower amounts of the therapeutic agent concentrated at the target tissue of interest. In these cases, the level of drug accumulation at the disease sites is significantly increased, while systemic distribution is reduced[87].

Although multiple small animal studies have successfully used magnetically targeted drug delivery, the clinical potential of the method has not yet been realised. Technical barriers preventing this include physical constraints, such as the rapid decrease of magnetic field strength when the target tissue is deep into the body and limitations of bypassing intervening vasculature[82,86,88].

In order to minimise these limitations and enhance attachment of the drugs, many studies have worked on the development of novel magnetic nanoparticles that vary size, multifunctional coatings and use high-moment materials and thermoresponsive hydrogels and particles[23,29,80,89,90].

In summary, magnetic drug targeting must be safe and effective and be capable of delivering the maximum amount of therapeutic agent with the least possible amount of magnetic particles[91]. Additionally, the most desirable magnetic carriers possess characteristics such as (i) small size to allow insertion into the target area, (ii) a wide range of particle sizes to allow their selection depending on the specific application, (iii) a large biocompatible surface area and high capacity for transgene or therapeutic agent uptake and (iv) the ability to biodegrade, eliminating toxic by-products[92].

12.3 NANOMAGNETIC GENE TRANSFECTION

12.3.1 OVERVIEW AND PRINCIPLES

Nanomagnetic gene transfection follows on from the principles of magnetically targeted gene delivery where magnetic nanoparticles are coupled with DNA, and in response to an externally applied magnetic field are specifically directed to the site of interest or onto the cell culture. The principle of nanomagnetic gene transfection was introduced in 2000 by Mah et al.[38,54] who first demonstrated the use of magnetic microparticles for gene transfection purposes by linking those to viral vectors delivering GFP reporter gene to targeted areas of C12S cells *in vitro* and in mouse models *in vivo*. The method has generated great interest and based on this idea many subsequent studies have applied the principle to non-viral systems that will be discussed in the following section.

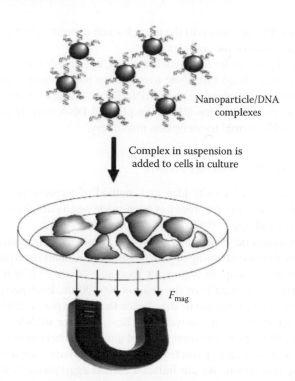

Nanoparticle/DNA complexes

Complex in suspension is added to cells in culture

F_{mag}

FIGURE 12.4 Schematic representation of gene delivery by nanomagnetic gene transfection *in vitro*. The vector is attached to magnetic nanoparticles to form complexes, which are added to the cell culture. The high-gradient rare-earth magnet is placed below the culture dish and pulls the particles towards the magnetic field source. F_{mag} is the force vector exerted on the particles by the magnetic field. (From Dobson, J., *Gene Ther.*, 13, 283–287, 2006.)

During *in vitro* nanomagnetic gene transfection, magnetic nanoparticle/DNA complexes are introduced into the cell culture, where the field gradient produced by the applied rare-earth magnets positioned underneath the cell culture, result in an increase of complex sedimentation and transfection rate[86] (Figure 12.4).

Ultimately, in future *in vivo* applications, following the magnetically targeted gene delivery model illustrated in the previous section (Figure 12.3), the particles carrying the therapeutic gene could be intravenously injected and circulated by the bloodstream until captured by externally applied magnets. These magnets could be focused close to the target tissue, to promote transfection and targeting of therapeutic genes to a specific site or organ of the body[88].

Nanomagnetic gene transfection relies on the selection of the appropriate type, size and surface coating of magnetic nanoparticles necessary for binding with the plasmid DNA or siRNA (Figure 12.5). The particle/DNA complex is pulled into contact with the target cells, which are 'sandwiched' between the particles and high-gradient magnetic fields positioned below the cell culture plates

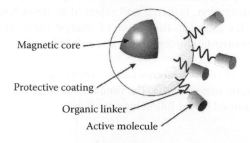

Magnetic core

Protective coating

Organic linker

Active molecule

FIGURE 12.5 Schematic representation of a typical magnetic nanoparticle for biotechnology. (From McBain, S.C., Yiu, H.H.P., Dobson, J., *Intl. J Nanomed.*, 3, 169–180, 2008.)

(for *in vitro* transfection). The magnets pull the particles into contact with the cells where endocytosis results in the internalisation of the complex.

By selecting a charged polymer for the particle coating, it is possible to exploit the changes in pH, which occur within the internalised endosome to activate a proton sponge mechanism, which ruptures the internalised endosome, releasing the particle/DNA complex into the cell's cytoplasm. The change in pH within the cytoplasm then allows the plasmid DNA to be released from the particle and transcribed by the cell's normal transcription machinery.

12.3.2 Studies Using Static Magnet Arrays

Over the past decade, and based on the idea and method of nanomagnetic gene transfection as presented by Mah et al., significant progress has been made on the development of non-viral nanomagnetic transfection technologies[22,93-95].

In these studies, nanomagnetic gene transfection, also known as 'magnetofection' when used with static magnet arrays, involved the coupling of genetic material to small superparamagnetic iron oxide (magnetite core) nanoparticles with an organic or inorganic coating[96] (Figure 12.4). The hydrodynamic radius of the particles is generally on the order of 50–100 nm. Each particle consists of clusters of magnetite crystallites less than <10 nm in size (hence their superparamagnetic nature) coated with a charged polymer, generally PEI or its derivatives. These DNA (or siRNA)/nanoparticle complexes are targeted to specific cell types by the application of a magnetic field as described earlier.

In magnetofection, iron oxide nanoparticles coated with PEI have been used and associated with vectors by electrostatic interaction and salt-induced colloid aggregation[22]. To avoid attachment of these nanoparticles with other external elements, they can be spread into a polymer matrix such as silica or they can be enclosed within a polymer or metallic coat[93]. Their therapeutic genes can be released by enzymatic cleavage of cross-linking molecules, pH charge interactions and polymer matrix degradation[86].

Some recently developed magnetic nanoparticles can be embedded within a matrix such as a hydrogel, and release of the therapeutic gene is mediated by heating of the hydrogel carrier[82]. Other particles such as magnetoliposomes are similar to liposomes with a magnetic ferrite core structure enclosed within a spherical lipid membrane[97]. Also, mesoporous silica nanoparticles, which may consist of up to 80% iron oxide, have been used along with several molecules to promote uptake by the target cells and have been associated with *N*-[1-(2,3-dioleoyloxy)propyl]-*N,N,N*-trimethylammonium chloride (DOTAP chloride) to both prevent particle aggregation and promote uptake[86,98].

The size of the nanoparticle/DNA complex is important in transfection efficiency. A comparison between small (30–60 nm) and large (300–600 nm) nanoparticles has shown that both types were transfected, but the large nanoparticles were able to achieve higher transfection efficiencies[94].

Once the magnetic nanoparticle formation and consistency has been optimised, it has to be coupled to the corresponding gene of interest. This may be linked to naked DNA or enclosed into a vector. Often naked DNA is attached to magnetic nanoparticles by electrostatic interactions or salt-induced colloid aggregation, targeted and released as described in the previous section. Also, magnetic nanoparticles linked to their gene of interest *via* a viral or non-viral vector can be coupled through both electrostatic interactions and salt-induced colloid aggregation as well as cross-linking[93].

Finally, the applied magnetic field is another factor to consider when optimising transfection efficiency. For superparamagnetic nanoparticles, the fundamental mechanism by which nanomagnetic transfection works is determined by the following equation:

$$F_{mag} = (\chi_2 - \chi_1)V \frac{1}{\mu_0} B(\nabla B)$$

where

F_{mag} is the force on the magnetic particle

χ_2 is the volume of magnetic susceptibility of the magnetic particle

χ_1 is the volume of magnetic susceptibility of the surrounding medium

μ_0 is the magnetic permeability of free space

V is the particle volume

B is the magnetic flux density in Tesla (T)

∇B is the field gradient and can be reduced to $\partial B/\partial x$, $\partial B/\partial y$, $\partial B/\partial z$, of which $\partial B/\partial z$ plays the most critical role due to the geometry of the system[86]

Generally, neodymium iron boron (NdFeB) magnets are used and placed directly underneath the cell culture. The main advantage of this system is the ability to increase the speed of particle sedimentation onto the cell surface and therefore significantly reduce the required time for cell transfection[93].

Though magnetofection is a rapid and efficient non-viral technique for transfection, particularly *in vitro*, improvements to transfection efficiency are still required if it is to compete with viral systems.

12.3.3 Advances Using Oscillating Magnet Arrays

Since 2005, a lot of research has been focused on the optimisation of the technique by the use of oscillating magnet arrays to enhance overall transfection efficiency of magnetofection and decrease transfection duration maintaining the high cell viability that nanomagnetic gene transfection offers.

In this system, magnetic nanoparticle/DNA complexes are added into the cell culture, and the presence of the oscillating magnet arrays, positioned beneath the cell culture plate, introduce a lateral motion to the particle/DNA complexes (Figure 12.6).

This lateral motion results in more efficient, mechanical stimulation of endocytosis and an increase in transfection efficiency in most cell types when compared to static magnetofection[100–102]. The additional energy transferred from the oscillating field to the cells promotes particle uptake and increases transfection efficiency and protein expression levels by up to 10-fold in comparison to the use of both static field nanomagnetic gene transfection, as well as the most widely used cationic lipid–based technique in various cell types and primary cells[88,102] (Figure 12.7).

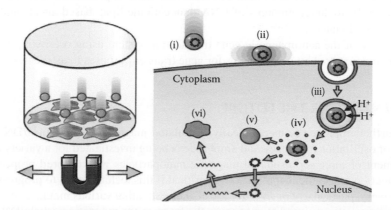

FIGURE 12.6 Principle of oscillating nanomagnetic transfection: Plasmid DNA or siRNA is attached to magnetic nanoparticles and incubated with cells in culture (left). An oscillating magnet array below the surface of the cell culture plate pulls the particles into contact with the cell membrane (i) and drags the particles from side-to-side across the cells (ii), mechanically stimulating endocytosis (iii). Once the particle/DNA complex is endocytosed, proton sponge effects rupture the endosome (iv) releasing the DNA, (v) which then transcribes the target protein (vi). (From Fouriki, A., Farrow, N., Clements, M., Dobson, J., *Nanorev.*, 1, 5167, 2010.)

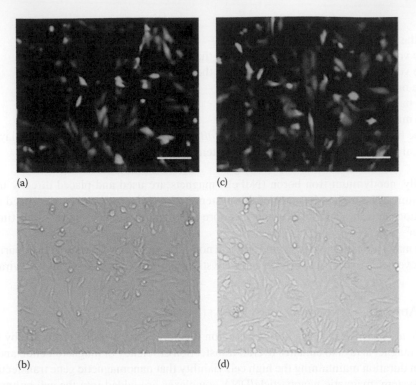

FIGURE 12.7   Enhanced GFP expression in human MG-63 cells; (a) transfected with 100 nm nTMag magnetic nanoparticles (MNPs) coated with pEGFPN1 plasmid DNA in the presence of static magnetic field, for 30 min (c) and in the presence of an oscillating field (nanoTherics magnefect-nano array at f = 2 Hz and amplitude = 200 μm). (b) The corresponding light image from Figure 12.7a, and (d) from Figure 12.7c. Scale bar = 200 μm. MG-63: human osteosarcoma fibroblasts; GFP; nTMag MNPs: nanoTherics nTMag magnetic nanoparticles; F: oscillation frequency.

Further, experimental work has repeatedly demonstrated that the oscillating magnet array clearly outperforms the most efficient cationic lipid–based transfection agents currently available. Furthermore, these results have been obtained from transfections performed at shorter transfection times (30 min), utilising lower amounts of DNA than cationic lipid–based agents and showing no adverse effect in cell viability[60,86,102–104].

The superiority of the nanomagnetic gene transfection system using oscillating magnet arrays has also been demonstrated by other studies for gene transfer purposes to astrocytes[60].

12.4   POTENTIAL FOR THE FUTURE

In order to further improve the mechanically stimulated magnetic nanoparticle/DNA uptake, an optimisation of oscillation frequency and amplitude is being investigated for a variety of cell types. The development of more sophisticated magnetic nanoparticles with tailored properties for both *in vitro* transfection and clinical applications is critical if this technology is to progress and replace viral gene delivery, particularly *in vivo*. We are also investigating various magnetic field geometries and magnet configurations aimed at enhancing the force on the magnetic particle/DNA complex.

Nanomagnetic gene transfection has great potential for use as an effective, non-viral transfection agent for *in vitro* and *in vivo* applications, offering an efficient gene delivery system that does not affect cell viability, morphology and proliferation. Oscillating nanomagnetic transfection has demonstrated further improvements in this technology and promises future advances, perhaps leading to clinical application of nanomagnetic transfection for diseases such as CF, where up to now it has not been possible to treat due to the failure of current techniques to penetrate the thick mucous layer that lines the lungs of CF patients[86,102].

REFERENCES

1. Crick, F. 1970. Central dogma of molecular biology. *Nature*, 227, 561–563.
2. Alberts, B., Johnson, A., Lewis, J. et al. 2002. *Molecular Biology of the Cell*. Garland Science, Taylor & Francis Group, New York and Milton Park, Abingdon, UK.
3. Lederberg, J. 1966. Experimental genetics and human evolution. *The American Naturalist*, 100, 519–531.
4. Herrmann, F. 1995. Cancer gene therapy: Principles, problems, and perspectives. *Journal of Molecular Medicine*, 73, 157–163.
5. Vile, R.G., Russell, S.J., Lemoine, N.R. 2000. Cancer gene therapy: Hard lessons and new courses. *Gene Therapy*, 7, 2–8.
6. Aiuti, A., Cattaneo, F., Galimberti, S. et al. 2009. Gene therapy for immunodeficiency due to adenosine deaminase deficiency. *The New England Journal of Medicine*, 360:5, 447–458.
7. Ferrari, S., Geddes, D.M., Alton, E.W. 2002. Barriers to and new approaches for gene therapy and gene delivery in cystic fibrosis. *Advances in Drug Delivery Reviews*, 54, 1373–1393.
8. Griesenbach, U., Geddes, D.M., Alton, E.W.F.W. 2006. Gene therapy progress and prospects: Cystic fibrosis. *Gene Therapy*, 13, 1061–1067.
9. Walsh, C.E. 2003. Gene therapy progress and prospects: Gene therapy for the hemophilias. *Gene Therapy*, 10, 999–1003.
10. Giannoukakis, N., Thomson, A.W. 1999. Gene therapy in transplantation. *Gene Therapy*, 6, 1499–1511.
11. Grigsby, C.L., Leong, K.W. 2010. Balancing protection and release of DNA: Tools to address a bottleneck of non-viral gene therapy. *Journal of Royal Society Interface*, 7, S67–S82.
12. Mountain, A. 2000. Gene therapy: The first decade. *Trends in Biotechnology*, 18, 119–128.
13. Nielsen, T.O. 1997. Human germline gene therapy. *McGill Journal of Medicine*, 3, 126–132.
14. Fletcher, J.C., Anderson, W.F. 1992. Germ-line gene therapy: A new stage of debate. *Law, Medicine and Health Care*, 20, 26–30.
15. Mulligan, R.C. 1993. The basic science of gene therapy. *Science*, 260, 926–932.
16. Friedmann, T. June 1997. Overcoming the obstacles to gene therapy. *Scientific American*, 276, 96–101.
17. Breyer, B., Jiang, W., Cheng, H. et al. 2001. Adenoviral vector-mediated gene transfer for human gene therapy. *Current Gene Therapy*, 1, 149–162.
18. Lee, J.H., Klein, H.G. 1995. Cellular gene therapy. *Hematology/Oncology Clinics of North America*, 9, 91–113.
19. Lu, Y. 2001. Viral based gene therapy for prostate cancer. *Current Gene Therapy*, 1, 183–200.
20. Culver, K.W. 1996. Gene therapy. *Journal of Insurance Medicine*, 2, 121–124.
21. Luo, D., Saltzman, W.M. 2000. Enhancement of transfection by physical concentration of DNA at the cell surface. *Nature Biotechnology*, 18, 893–895.
22. Scherer, F., Anton, M., Schillinger, U. et al. 2002. Magnetofection: Enhancing and targeting gene delivery by magnetic force in vitro and vivo. *Gene Therapy*, 9, 102–109.
23. McBain, S.C., Yiu, H.H.P., Dobson, J. 2008. Magnetic nanoparticles for gene and drug delivery. *International Journal of Nanomedicine*, 3:2, 169–180.
24. Nabel, E.G. 1992. Direct gene transfer into the arterial wall. *Journal of Vascular Surgery*, 15, 931–932.
25. Kay, M., Glorioso, J.C., Naldini, L. 2001. Viral vectors for gene therapy: The art of turning infectious agents into vehicles of therapeutics. *Nature Medicine*, 7, 33–40.
26. Schmidt-Wolf, G.D., Schmidt-Wolf, I.G.H. 2003. Non-viral and hybrid vectors in human gene therapy: An update. *Trends in Molecular Medicine*, 9, 67–72.
27. Dykxhoorn, D.M., Novina, C.D., Sharp, P.A. 2003. Killing the messenger: Short RNAs that silence gene expression. *Nature Reviews Molecular Cell Biology*, 4, 457–467.
28. Ostedgaard, L.S., Rokhlina, T., Karp, P.H. et al. 2005. A shortened adeno-associated virus expression cassette for CFTR gene transfer to cystic fibrosis airway epithelia. *Proceedings of the National Academy of Sciences of the United States of America*, 102:8, 2952–2957.
29. Nishijima, S., Mishima, F., Terada, T., Takeda, S. 2007. A study on magnetically targeted drug delivery system using a superconducting magnet. *Physica C: Superconductivity and its Applications*, 463–465, 1311–1314.

30. Baldwin, T.O. 1996. Firefly luciferase: The structure is known, but the mystery remains. *Structure*, 15, 223–228.

31. Welsh, S., Kay, S.A. 1997. Reporter gene expression for monitoring gene transfer. *Current Opinion in Biotechnology*, 8, 617–622.

32. Mikata, K., Uemura, H., Ohuchi, H. et al. 2002. Inhibition of growth of human prostate cancer xenograft by transfection of *p53* gene: Gene transfer by electroporation 1. *Molecular Cancer Therapeutics*, 1, 247–252.

33. Shi, E., Jiang, X., Kazui, T. et al. 2006. Nonviral gene transfer of hepatocyte growth factor attenuates neurologic injury after spinal cord ischemia in rabbits. *The Journal of Thoracic and Cardiovascular Surgery*, 132:4, 941–947.

34. Horimoto, S., Horimoto, H., Mieno, S. et al. 2006. Implantation of mesenchymal stem cells overexpressing endothelial nitric oxide synthase improves right ventricular impairments caused by pulmonary hypertension. *Circulation*, 4, I181–I185.

35. Kalka, C., Baumgartner, I. 2008. Gene and stem cell therapy in peripheral arterial occlusive disease. *Vascular Medicine*, 13, 157–172.

36. Randrianarison-Jewtoukoff, V., Perricaudet, M. 1995. Recombinant adenoviruses as vaccines. *Biologicals*, 23, 145–157.

37. Goedegebuure, P.S., Eberlein, T.J. 1997. Vaccine trials for the clinician: Prospects for viral and non-viral vectors. *The Oncologist*, 2, 300–310.

38. Mah, C., Fraites, T.J., Zolotukhin, I. et al. 2002. Improved method of recombinant AAV2 delivery for systemic targeted gene delivery. *Molecular Therapy*, 6:1, 106–112.

39. Bhattarai, S.R., Kim, S.Y., Jang, K.Y. et al. 2008. Laboratory formulated magnetic nanoparticles for enhancement of viral gene expression in suspension cell line. *Journal of Virological Methods*, 147, 213–218.

40. Schaffer, D.V., Koerber, J.T., Lim, K. 2008. Molecular engineering of viral gene delivery vehicles. *Annual Review of Biomedical Engineering*, 10, 169–194.

41. Naldini, L., Blomer, U., Gallay, P. et al. 1996. In vivo gene delivery and stable transduction of nondividing cells by a lentiviral vector. *Science*, 272, 263–267.

42. Barton, G., Medzhitov, R. 2002. Retroviral delivery of small interfering RNA into primary cells. *Proceedings of the National Academy of Sciences of the United States of America*, 99:23, 14943–14945.

43. Anson, D.S. 2004. The use of retroviral vectors for gene therapy—What are the risks? A review of retroviral pathogenesis and its relevance to retroviral vector-mediated gene delivery. *Genetic Vaccines and Therapy*, 2, 9.

44. Miller, A.D., Jolly, D.J., Friedmann, T., Verma, I.M. 1983. A transmissible retrovirus expressing human hypoxanthine phosphoribosyltransferase (HPRT): Gene transfer into cells obtained from humans deficient in HPRT. *Proceedings of the National Academy of Science of the United States of America*, 80, 4709–4713.

45. Joyner, A., Keller, G., Phillips, R.A., Bernstein, A. 1983. Retrovirus transfer of a bacterial gene into mouse haematopoietic progenitor cells. *Nature*, 305, 556–558.

46. Willis, R.C., Jolly, D.J., Miller, A.D. et al. 1984. Partial phenotypic correction of human Lesch–Nyhan (hypoxanthine–guanine phosphoribosyltransferase-deficient) lymphoblast's with a transmissible retroviral vector. *Journal of Biological Chemistry*, 259, 8742–8749.

47. Swift, S., Lorens, J., Achacoso, P., Nolan, G.P. 2001. Rapid production of retroviruses for efficient gene delivery to mammalian cells using 293T cell-based systems. *Current Protocols in Immunology*, 10.17.14–10.17.29.

48. Molnar-Kimber, K.L., Sterman, D.H., Chang, M., 1998. Impact of preexisting and induced humoral and cellular immune responses in adenovirus-based gene therapy phase I clinical trial for localized mesothelioma. *Human Gene Therapy*, 9, 2121–2133.

49. Jie, J., Wang, J., Qu, J., Hung, T. 2007. Suppression of human colon tumor growth by adenoviral vector-mediated NK4 expression in an athymic mouse model. *World Journal of Gastroenterology*, 13:13, 1938–1946.

50. Yanagie, H., Tanabe, T., Sumimoto, H. et al. 2009. Tumor growth suppression by adenovirus-mediated introduction of a cell growth suppressing gene tob in a pancreatic cancer model. *Biomedicine & Pharmacotherapy*, 63, 275–286.

51. Qiao, C., Zhang, W., Yuan, Z. et al. 2010. Adeno-associated virus serotype 6 caspid tyrosine-to-phenylalanine mutations improve gene transfer to skeletal muscle. *Human Gene Therapy*, 21:10, 1343–1348.

52. Kay, J.D., Gouze, E., Oligino, T.J. et al. 2009. Intra-articular gene delivery and expression of interleukin-1Ra mediated by self-complementary adeno-associated virus. *The Journal of Gene Medicine*, 11:7, 605–614.

53. Carter, P.J., Samulski, R.J. 2001. Adeno-associated viral vectors as gene delivery vehicles. *International Journal of Molecular Medicine*, 6:1, 17–44.

54. Mah, C., Zolotukhin, I., Fraites, T.J., Dobson, J., Batich, C., Bryne, B.J. 2000. Microsphere-mediated delivery of recombinant AVV vectors *in vitro* and *in vivo*. *Molecular Therapy*, 1, S239.

55. Verma, I.M. 2000. A tumultuous year for gene therapy. *Molecular Therapy*, 2, 415–416.

56. Hacein-Bey, S. 2003. A serious adverse event after successful gene therapy for X-linked severe combined immunodeficiency. *New England Journal of Medicine*, 348, 255–256.

57. Klink, D., Schindelhauer, D., Laner, A. et al. 2004. Gene delivery systems—Gene therapy vectors for cystic fibrosis. *Journal of Cystic Fibrosis*, 3, 203–212.

58. Spack, E.G., Sorgi, F.L. 2001. Developing non-viral DNA delivery systems for cancer and infectious disease. *Drug Delivery Today*, 6:4, 186–197.

59. Davies, J.C. 2006. Gene and cell therapy for cystic fibrosis. *Paediatric Respiratory Reviews*, 7S, S163–S165.

60. Pickard, M., Chari, D. 2010. Enhancement of magnetic nanoparticle-mediated gene transfer to astrocytes by 'magnetofection': Effects of static and oscillating fields. *Nanomedicine*, 5:2, 217–232.

61. Smith, A.E. 1995. Viral vectors in gene therapy. *Annual Review of Microbiology*, 49, 807–838.

62. Flotte, T.R., Carter, B.J. 1995. Adeno-associated viruses as vectors for gene therapy. *Gene Therapy*, 2, 357–362.

63. Schiedner, G., Hertel, S., Kochanek, S. 2000. Efficient transformation of primary human amniocytes by E1 functions of ad5: Generation of new cells for adenoviral vector production. *Human Gene Therapy*, 11, 2105–2116.

64. Li, S., Ma, Z. 2001. Non viral gene therapy. *Current Gene Therapy*, 1, 201–226.

65. Martin, B., Sainlos, M., Aissaoui, A. et al. 2005. The design of cationic lipids for gene delivery. *Current Pharmaceutical Design*, 11, 375–394.

66. Midoux, P., Pichon, C., Yaouanc, J., Jaffres, P. 2009. Chemical vectors for gene delivery: A current review on polymers, peptides and lipids containing histidine or imidazole as nucleic acids carriers. *British Journal of Pharmacology*, 157, 166–178.

67. Villemejane, J., Mir, L.M. 2009. Physical methods of nucleic acid transfer: General concepts and applications. *British Journal of Pharmacology*, 157, 207–219.

68. Mintzer, M.A., Simanek, E.E. 2009. Nonviral vectors for gene delivery. *Chemical Reviews*, 109, 259–302.

69. Felgner, P.L., Gadek, T.R., Holm, M. et al. 1987. Lipofection: A highly efficient, lipid-mediated DNA-transfection procedure. *Proceedings of the National Academy of Sciences of the United States of America*, 84, 7413–7417.

70. Li, S., Huang, L. 2000. Nonviral gene therapy: Promises and challenges. *Gene Therapy*, 7, 31–34.

71. Gao, X., Kim, K.S., Liu, D. 2007. Nonviral gene delivery: What we know and what is next. *The AAPS Journal*, 9, E92–E104.

72. Faneca, H., Simoes, M.C., Lima, P. 2002. Evaluation of lipid-based reagents to mediate intracellular gene delivery. *Biochimica et Biophysica Acta*, 1567, 23–33.

73. Templeton, N.S. 2004. Liposomal delivery of nucleic acids *in vivo*. *DNA and Cell Biology*, 21:12, 857–867.

74. Niidome, T., Huang, L. 2002. Gene therapy progress and prospects: Nonviral vectors. *Gene Therapy*, 9, 1647–1652.

75. Koping-Hoggard, M., Tubulekas, I., Guan, H. et al. 2001. Chitosan as a nonviral gene delivery system. Structure–property relationships and characteristics compared with polyethylenimine *in vitro* and after lung administration *in vivo*. *Gene Therapy*, 8, 1108–1121.

76. Skorpikova, J., Dolnikova, M., Hrazdira, I., Janisch, R. 2001. Changes in microtubules and microfilaments due to a combined effect of ultrasound and cytostatics in HeLa cells. *Journal of Cellular and Molecular Biology*, 47:4, 143–147.

77. Wightman, L., Kircheis, R., Carotta, S. et al. 2001. Different behavioral of branched and linear polyethylenimine for gene delivery *in vitro* and *in vivo*. *Journal of Gene Medicine*, 3, 362–372.

78. Andre, F.M., Cournil-Henrionnet, C., Vernery, D., Opolon, P., Mir, L.M. 2006. Variability of naked DNA expression after direct local injection: The influence of the injection speed. *Gene Therapy*, 13, 1619–1627.

79. Yaron, Y., Kramer, R.L., Johnson, M.P., Evans, M.I. 1997. Gene therapy: Is the future here yet? *Obstetrics and Gynecology Clinics of North America*, 24, 179–199.

80. Corchero, J.L., Villaverde, A. 2009. Biomedical applications of distally controlled magnetic nanoparticles. *Trends in Biotechnology*, 27:8, 468–476.
81. Pankhurst, Q.A., Connolly, J., Jones, S.K., Dobson, J. 2003. Applications of magnetic nanoparticles in biomedicine. *Journal of Physics D: Applied Physics*, 36, R167–R181.
82. Pankhurst, Q.A., Thanh, N.K.T., Jones, S.K., Dobson, J. 2009. Progress in applications of magnetic nanoparticles in biomedicine. *Journal of Physics D: Applied Physics*, 42, 224001.
83. Gilchrist, R.D., Medal, R., Shorey, W.D. et al. 1957. Selective inductive heating of lymph nodes. *Annals of Surgery*, 4, 596–606.
84. Meyers, P.H., Cronic, F., Nice, C.M. 1963. Experimental approach in the use and magnetic control of metallic iron particles in the lymphatic and vascular system of dogs as a contrast and isotopic agent. *American Journal of Roentgenology, Radium Therapy and Nuclear Medicine*, 90, 1068–1077.
85. Widder, K.J., Senyei, A.E., Ranney, D.F. 1979. Magnetically responsive microspheres and other carriers for the biophysical targeting of antitumor agents. *Advances in Pharmacology and Chemotherapy*, 16, 213–271.
86. Dobson, J. 2006. Gene therapy progress and prospects: Magnetic nanoparticle-based gene delivery. *Gene Therapy*, 13, 283–287.
87. Hafeli, U.O. 2004. Magnetically modulated therapeutic systems. *International Journal of Pharmaceutics*, 277, 19–24.
88. Dobson, J. 2006. Magnetic micro and nano-particle based targeting for drug and gene delivery. *Nanomedicine*, 1, 31–37.
89. Sanvicens, N., Marco, M.P. 2008. Multifunctional nanoparticles—Properties and prospects for their use in human medicine. *Trends in Biotechnology*, 26:8, 425–433.
90. Xie, J., Lee, S., Chen, X. 2010. Nanoparticle-based theranostic agents. *Advanced Drug Delivery Reviews*, 62, 1064–1079.
91. Lubbe, A.S., Bergemann, C., Brock, J., McClure, D.G. 1999. Physiological aspects in magnetic drug-targeting. *Journal of Magnetism and Magnetic Materials*, 194, 149–155.
92. Silva, G.A., Ducheyne, P., Reis, R.L. 2007. Materials in particulate form for tissue engineering. 1. Basic concepts. *Journal of Tissue Engineering and Regenerative Medicine*, 1, 4–24.
93. Plank, C., Schillinger, U., Scherer, F. et al. 2003. The magnetofection method: Using magnetic force to enhance gene delivery. *Biological Chemistry*, 384, 737–747.
94. Schillinger, U., Brill, T., Rudolph, C. et al. 2005. Advances in magnetofection—Magnetically guided nucleic acid delivery. *Journal of Magnetism and Magnetic Materials*, 293, 501–508.
95. Mykhaylyk, O., Zelphati, O., Rosenecker, J., Plank, C. 2008. siRNA delivery by magnetofection. *Current Opinion in Molecular Therapeutics*, 10, 493–505.
96. Neuberger, T., Schopf, B., Hofmann, H., Hofmann, M., Rechenberg, B. 2005. Superparamagnetic nanoparticles for biomedical applications: Possibilities and limitations of a new drug delivery system. *Journal of Magnetism and Magnetic Materials*, 293, 483–496.
97. Gonzales, M., Kirshnan, K.M. 2005. Synthesis of magnetoliposomes with monodisperse iron oxide nanocrystal cores for hyperthermia. *Journal of Magnetism and Magnetic Materials*, 293, 265–270.
98. Yiu, H.P.H., Dobson, J., El Haj, A.J. 2005. Internalization of functionalized magnetic nanoparticles into human cells with the use of DOTAP as a transfecting agent. *Vienna Magnetics Group Report*, 176–177. ISBN: 3-902105-00-1.
99. Fouriki, A., Farrow, N., Clements, M., Dobson, J. 2010. Evaluation of the magnetic field requirements for nanomagnetic gene transfection. *Nanoreviews*, 1, 5167.
100. Dobson, J., Batich, C. 2005. Gene delivery. Patent Pending No. WO2006111770.
101. Kamau, S.W., Hassa, P.O., Steitz, B. et al. 2006. Enhancement of the efficiency of non-viral gene delivery by application of pulsed magnetic field. *Nucleic Acids Research*, 34:5, e40.
102. McBain, S.C., Griesenbach, U., Xenariou, S., Keramane, A., Batich, C.D., Alton, E.W.F.W., Dobson, J. 2008. Magnetic nanoparticles as gene delivery agents: Enhanced transfection in the presence of oscillating magnet arrays. *Nanotechnology*, 19. doi: 10.1088/0957–4484/19/40/405102.
103. Kim, J.S., Yoon, T., Yu, K.N. et al. 2006. Toxicity and tissue distribution of magnetic nanoparticles in mice. *Toxicological Sciences*, 89:1, 338–347.
104. Dobson, J., McBain, S.C., Farrow, N., Batich, C.D. 2008. Oscillating magnet arrays for enhanced magnetic nanoparticle-based gene transfection. *European Cells and Materials*, 16:3, 48.

Angeliki Fouriki received a BSc in molecular biology from the University of Wales, Bangor in 2007. She obtained an MSc in cell and tissue engineering from the University of Keele in 2008, and in the same year continuing at the University of Keele, she began her PhD studies under the direction of Professor J. Dobson. Her research focuses on biomedical engineering and nanomagnetic gene transfection for gene delivery.

Jon Dobson is professor of biomedical engineering and biomaterials at the University of Florida and professor of biomedical engineering at Keele University in the United Kingdom. He has a BSc and an MSc from the University of Florida and a PhD from the Swiss Federal Institute of Technology (ETH-Zurich, 1991). Professor Dobson received a Royal Society of London Wolfson Research Merit Award in 2004, the UK Medical Research Council's César Milstein Award in 2008 and was co-recipient of The Wellcome Trust's Sir Henry Wellcome Showcase Award 2002. He was elected as Fellow of the Royal Society of Medicine and the Institute for Nanotechnology in 2007 and is on the editorial board of seven journals, including *IEEE Transactions in NanoBioscience*. He has more than 170 publications, including more than 20 invited book chapters and articles for journals such as *Nature Nanotechnology* and *Journal of Physics D* (the second most highly cited paper in this journal over the past decade), and has been an invited speaker at more than 45 institutes and scientific meetings, including the recent *Gordon Conference on Magnetic Nanostructures* (2010). He is also the founder of two spin-off companies, nanoTherics and MICA BioSystems.

Part V

In Vivo *Application of MNPs*

Part V

In Vivo Application of MNPs

13 Imaging and Manipulating Magnetically Labelled Cells

Florence Gazeau and Claire Wilhelm

CONTENTS

Labelling living cells with magnetic nanoparticles (MNPs) creates opportunities for numerous imaging and therapeutic applications such as cell manipulation, cell patterning for tissue engineering, magnetically assisted cell delivery or magnetic resonance imaging (MRI)-assisted cell tracking. The unique advantage of magnetic-based methods is to activate or monitor cell behaviour by a remote stimulus, the magnetic field. Cell labelling methods using superparamagnetic nanoparticles (NPs) have been widely developed, showing no adverse effect on cell proliferation and functionalities while conferring magnetic properties to various cell types. This chapter describes a controlled method for MNP cellular labelling and gives an overview of the most important applications of 'magnetic cells', focusing particularly on the medical promises.

13.1 CELL LABELLING WITH MNP: INTRACELLULAR TRAFFICKING AND LONG-TERM OUTCOME

In the last decade, a variety of strategies have been developed to label living cells with MNPs. The main requirement is to supply cells with sufficient magnetisation to be detectable by MRI and/or manipulated by external magnetic field, while preserving cell viability and functionalities. Dextran-coated NPs (also clinically approved as MRI contrast agent[1]) were first evaluated for cell labelling *in vitro*[2-4]. However, cellular uptake of these NPs was poorly efficient, motivating the research for other strategies, relying on modification of the NP surface to trigger receptor-mediated cell uptake. For instance, monoclonal antibodies[5,6] or human transferrins[7] were used to target cell membrane–specific receptors, with the possible drawback of an insufficient numbers of receptor and of species specificity. HIV-Tat peptide conjugated to the NPs was also shown to improve cell uptake[8,9]. However, the most popular method was to associate a transfection agent, like, for example, a highly cationic peptide, polyamine[10,11] or dendrimer[12], with polymer-coated NPs to improve NPs' internalisation. This non-specific strategy, inspired from gene transfection techniques, has shown good uptake efficiency in a variety of cells for incubation time of a minimum of 6 h[13]. Other techniques based on instant electroporation have also been tested to incorporate nanomagnets in cells with improved efficiency[14,15]. All these advanced techniques lead to intracellular uptake in the range of several picogram of iron, corresponding to millions of 8 nm iron oxide NPs per cell. In an attempt to unravel the mechanism of interactions between MNPs and cells, we and others have varied the

NPs surface properties and quantified their subsequent uptake in different cell types[16]. A key issue is the stability of the colloid in the medium used for incubation with cells. The cells react in a different manner depending on whether the NPs remain dispersed in suspension or become aggregated and flocculated. In particular, cell toxicity can arise from NPs aggregates, whereas the same NPs have no deleterious effect when correctly dispersed. Also, a quantitative control of cell labelling is only possible for disperse NPs. In this review, we focus on MNPs conserving their colloidal stability during incubation with cells like the commonly used dextran-coated MNPs, albumin-coated MNPs or anionic citrate-coated MNPs (AMNPs), providing the culture medium is supplemented with free citrate. In these conditions, the presence of a polymer, surface charge and degree of hydrophilicity were shown to strongly affect the affinity of NPs for the cell plasma membrane[17]. In particular, dextran- or albumin-coated MNPs show a low affinity for the cell plasma membrane, whereas AMNPs spontaneously adsorb on it[17]. In general, electrostatic interactions favour the membrane adsorption of NPs, while polymers tend to restrict interactions with cells due to steric effects. As a consequence, the cell uptake mechanism also differs: non-interacting NPs will be ingested together with the extracellular fluid (pinocytosis) with a low efficiency, while adsorbed MNPs enter the cell through the active endocytosis pathway[16]. Cell uptake can be a hundred times larger for AMNP, despite very short incubation time (typically less than 1 h) and no use of transfection agent.

The internalisation of AMNPs has been fully characterised[18] both qualitatively and quantitatively. The uptake mechanism consists in a two-step process. The first step is the non-specific adsorption of AMNPs on the plasma membrane, following a generic Langmuir kinetics. This step can be investigated separately if cells are maintained at 4°C, thus inhibiting the internalisation process. We have verified that the affinity of AMNPs for cell membrane is almost constant whatever the cell type ($1.6-4 \times 10^7$ M^{-1}) and that the binding capacity (the maximum number of NPs that can be adsorbed on the cell membrane) only depends on the cell size (see Ref. [16] for details of the model). Typically, the binding capacity per unit surface is about 0.03 pg/µm^2 or 2.4×10^4 NPs/µm^2. Therefore, the larger the cell, the higher the number of NPs that can be adsorbed and further internalised.

The second step is the internalisation of the plasma membrane: the plasma membrane invaginates and encloses the NPs into vesicles, which fuse with pre-existing intracellular compartments, successively in early endosomes, late endosomes and finally lysosomes (Figure 13.1A). This process can be described as a first-order kinetic law, limited by a maximal fraction of surface that can be internalised (internalisation capacity) and a characteristic time for internalisation of the order of 1 h. It must be noted that the adsorption and internalisation occur concomitantly at 37°C, but the binding sites (cationic sites) for NP adsorption are continuously recycled. Remarkably, the internalisation capacity, as well as the internalisation time, are quite homogenous among the different cell types, with the notable exception of macrophages, which have an outstanding internalisation capacity in agreement with their specialised function for phagocytosis. This general model succeeds in describing the AMNP uptake by endothelial cells, tumour cells, lymphocytes, smooth muscle cells (SMCs), etc and provides a tight control on cell labelling. A maximum mass of iron can be predicted for any cell type according to its size. For example, small-sized T lymphocyte will uptake a maximum of 2–3 pg of iron, whereas large hepatocytes can concentrate more than 50 pg. The model also helps to optimise the labelling procedure in terms of incubation time and extracellular iron concentration, allowing very short incubation time to be used (as low as 10 min) in contrast with protocols involving transfection agents.

Whatever the labelling procedure, the internalisation process tends to confine MNPs at high concentration in endosomes or lysosomes[19] (Figure 13.1A). The endosomal sequestration of MNPs has several important consequences. Firstly, it confers strong magnetic properties to the endosomes, which are then able to elongate and attract each other to form chaplets when the cell is submitted to a magnetic field[20] (Figure 13.1B). Therefore, magnetic endosomes can be manipulated inside the living cell to probe the microrheology of the cell interior[21,22] or to confine them in a specified area of the cell[23]. Secondly, endosomes and lysosomes protect the cell from a potential toxicity of MNPs, hampering the release of any free iron species in the cytoplasm. The lysosomes are used by cells to handle MNPs and to degrade them at long term[24,25]. Recent data suggest that the *in vivo*

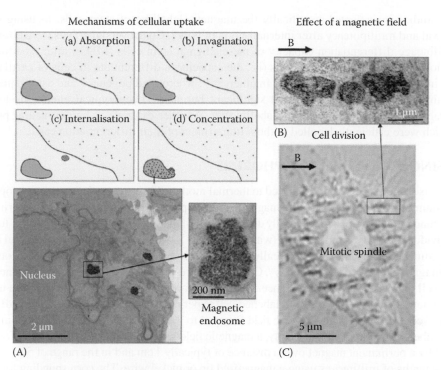

Mechanisms of cellular uptake

Effect of a magnetic field

(a) Absorption (b) Invagination

(c) Internalisation (d) Concentration

(B) Cell division

Nucleus

Mitotic spindle

200 nm

Magnetic
endosome

2 µm

5 µm

(A) (C)

FIGURE 13.1 Citrate-coated AMNPs are uptaken by cell following a universal mechanism of endocytosis. (A) They are first adsorbed (A.a) as small clusters on the cell plasma membrane, which invaginates (A.b) to form intracellular vesicles containing the NPs (A.c, early endosomes). These vesicles are subsequently transported through the cytosol and fuse with pre-existing late endosomes and lysosomes, concentrating the NPs in them (A.d). (From *Biomaterials*, 29, Wilhelm, C. and Gazeau, F., Universal cell labelling with anionic magnetic nanoparticles, 3161–3174, Copyright 2008, with permission from Elsevier.) The electron microscopy picture shows a lymphocyte having internalised the NPs into intracellular endosomes. Note the high concentration of NPs within the endosomes[29]. (B) Upon application of a magnetic field (100 mT) on the living cell, the intracellular endosomes attract each other to form small chaplets and deform in the direction of the field. From the analysis of this field-induced deformation, one can deduce the mechanical properties of the endosomal membrane. (From Wilhelm, C. et al., *Eur. Biophys. J.*, 32, 655, 2003.) (C) Chains of endosomes can be observed by optical microscopy. They rotate following the direction of the field. From the monitoring of their motion, one can deduce a cartography of the viscoelasticity of the cell interior. (Adapted from Wilhelm, C., *Phys. Rev. Lett.*, 101, 028101, 2008; Robert, D. et al., *PLoS One*, 5, e10046, 2010.) Here the cell is observed during its division, as assessed by the presence of a mitotic spindle. The magnetic endosomes will be shared between the two daughter cells.

biotransformation of MNPs occurs intracellularly within the lysosomes of macrophages and that the iron, coming from the degradation of NPs, is locally stored within the ferritin protein[26,27]. Beyond the particular case of macrophages, which are professional scavengers, we tested AMNP labelling on a wide variety of cell types (e.g. immune cells[28–31], endothelial cells[32,33], cancer cells[18,34], primary culture or established cell lines, progenitors and stem cells[35,36]) and never observed any detrimental effect on cell proliferation and cell functions at short and long terms, *in vitro* or *in vivo* (see Ref. [16] for a review). When a cell undergoes division (Figure 13.1C), it shares the magnetic endosomes between the two daughter cells. The iron load is thus reduced by a factor of 2 at each division. In normal conditions, there is no exocytosis of AMNPs. However, under stress conditions, some magnetically labelled cells can release microvesicles in the extracellular medium, containing NPs as cargo[37,38]. These cell-released vesicles can transport MNPs in the extracellular space, including body fluids and transfer NPs to other naïve cells[38] especially macrophages. This process, if confirmed *in vivo*, could participate to an horizontal intercellular transfer of NPs, challenging to some extent the initial specificity of cell labelling[39,40].

Some studies investigated specifically the magnetic labelling of stem cells, focusing on their self-renewal and multipotency after internalisation of NPs[41]. Controversial effects were observed on the multilineage differentiation capacity of mesenchymal stem cells (MSCs) after labelling using transfection agents. The chondrogenesis (i.e. the capacity to differentiate in cells of cartilage) was partially inhibited in one study[42] but not in others[43-46], whereas adipogenesis and osteogenesis were not impaired. Gene expression showed significant but subtle phenotypical alterations following superparamagnetic iron oxide (SPIO) labelling[44,47]. Another example is human neural precursor cells, which were efficiently labelled without impairment of their differentiation capacity[48,49].

13.2   SINGLE CELL MAGNETOPHORESIS

Nanosized isolated particles are submitted to thermal motion. Their magnetic moment is proportional to their volume. As a consequence, the magnetic force applied to one single NP using experimentally available magnetic field gradient generally does not exceed thermal forces. It is thus unlikely to manipulate individual NPs. By contrast, a cell (with diameter of 10–30 µm) can concentrate several millions of NPs confined in lysosomes (with typically 10^3–5×10^4 NPs/lysosome)[20]. Therefore, it becomes easy to apply magnetic forces to labelled cells[16]. Cells become responsive to inhomogeneous magnetic field, allowing cell manipulation from a distance. In a non-uniform magnetic field B, defined by a unidirectional magnetic field gradient gradB, a labelled cell experiences a magnetic force M(B)gradB, where M(B) is the cell magnetic moment in the field B (equal to the magnetic moment of one NP multiplied by the number of NPs per cell). Typically, a magnetic field gradient in the range of 10–50 T/m can be generated by a permanent magnet over a distance of typically 1 cm and in the range of 500–1500 T/m over a few tenths of millimetres using a magnetised tip or nickel wire. The corresponding force experienced by a cell (with a typical iron load of 10 pg) may vary from 1 pN to a few nN[50].

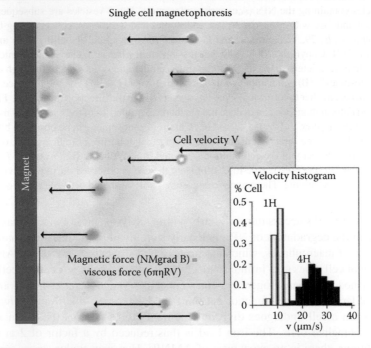

FIGURE 13.2   Magnetically labelled cells are put into motion in the magnetic field gradient (gradB), created by a magnet. In a cellular suspension, the magnetic force (MgradB) experienced by each cell is balanced by the viscous force ($6\pi\eta RV$), where R is the cell radius, V the cell velocity and the η viscosity of carrier medium. From the measurements of R and V by video microscopy, one can deduce the magnetic moment of each cell, proportional the number N of uptaken NPs[51]. The figure displays an example of velocity histograms for macrophages incubated for 1 h and 4 h with AMNP at an iron concentration of [Fe] = 1 mM.

The first direct application measured the cell's magnetic mobility to quantify NP uptake[51].

For cells in suspension, the magnetic force is simply balanced by the viscous force $6\pi\eta RV$, where η is the viscosity of the medium, R the cell radius and V the cell velocity. In a set-up with calibrated B and gradB (18 T/m), it becomes easy to deduce N from the determination of V and R by video microscopy for each cell (Figure 13.2). This experiment, called single cell magnetophoresis, allows determining the whole distribution of NPs uptake in a cell population. More generally, it provides quantitative data on the cell magnetic motilities that can be achieved in any type of magnetic device designed for cell sorting or for controlling cell migration. Typically, magnetic forces in the range of 1–10 pN will be sufficient to move circulating cells with magnetophoretic mobility of 10–100 μm/s. Magnetic control of cell transport offers opportunities for selective cell sorting on a microfluidic platform as a function of iron load[52,53] or of endocytotic capacity[54] (see also Chapter 10).

13.3 CELLULAR MAGNETOTAXIS *IN VITRO* AND ITS APPLICATION IN TISSUE ENGINEERING

Magnetic forces manipulate cells in different ways, whether the cells adhere on a substrate or not. If the magnetic force is lower than adhesion constraint (which is usually the case on a 2D substrate), the cell cannot move as a whole and the magnetic force acts on intracellular endosomes carrying the NPs. Such intracellular constraints have been used to deform the cell in a controlled direction and observe the subsequent effects on cell migration[50,55]. Figure 13.3 exemplifies the case of endothelial cells, which spontaneously associate and elongate themselves to form a capillary-like network[32]. In the vicinity of a magnetic tip, some pseudopods containing magnetic endosomes are progressively dragged by magnetic force along the magnetic field gradient (in the range of 500–1500 T/m), pulling the cell towards the magnet. From the force needed to disrupt a cell line, one can estimate the force linking adjacent cells in the lattice. This type of experiment is useful to unravel the role of mechanical forces during cell migration or self-organisation on a substrate[36].

In contrast to adherent cells, suspended cells move as a whole when submitted to remote magnetic forces. Very promising applications are developing in the field of tissue engineering based on the magnetic attraction concept. As seen in Figure 13.4, the design of magnetic attractor with prescribed geometry and magnetic field gradient enables to construct 3D aggregates of cells[57] (Figure 13.4A) or to pattern the cells on 2D substrate[58] (Figure 13.4B). The volume and cellular compacity of the 3D cell assembly as well as the spatial resolution of 2D patterning are directly controlled by the magnetic force, depending both on the magnetic attractor and on the cell iron load[57]. Using a similar

Magnetic deformation of endothelial cell capillaries

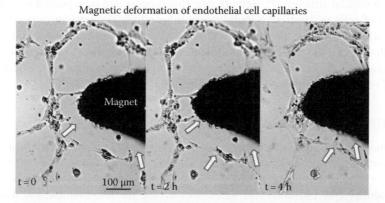

FIGURE 13.3 EPCs self-organise to form capillary-like structures, which are precursors of blood capillaries[32]. In the vicinity of a magnetic tip, the magnetically labelled EPCs (2 h incubation with [Fe] = 5 mM) are attracted *via* magnetic forces on their intracellular endosomes. Real-time monitoring of the cells shows the extension of cell pseudopods in the direction of the tip, driven by the magnetic endosomes. The force linking the cells in a junction forming the network can be evaluated from the magnetic force needed to disrupt the junction.

Magnetically assisted tissue engineering

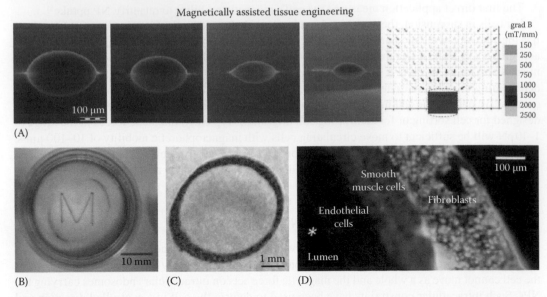

FIGURE 13.4 Remotely applied magnetic forces on labelled cells can improve tissue engineering techniques. (A) 3D cell aggregates with controlled shape, geometry and cell compacity can be constructed using specially designed magnetic attractors. Here, magnetically labelled EPCs (cell number from 1.4×10^4 to 1.4×10^5) were attracted by the magnetic device on the right[57]. (B) A variety of cell patterns can be obtained by using magnetic forces. Magnetically labelled cells are seeded into a culture dish placed on steel plates with prescribed geometry and submitted to a magnet. The width of the cell line can be varied, and cord-like cell structure can be harvested from cell culture non-adherent surface and further manipulated. (From Ino, K. et al., *Biotechnol. Bioeng.*, 97, 1309, 2007. With permission.) (C,D) Vascular tubular constructs are fabricated by means of magnetic cell sheet technology. (From Ito, A. et al., *Tissue Eng.*, 11, 1553, 2005. With permission.) 2D cell sheet is first obtained from ultra-low attachment surface submitted to a magnet. To design a tubular geometry, the detached cell sheets are then rolled over a cylindrical magnet enclosed into a silicone tube (C). To mimic the cellular organisation of a native vessel, endothelial cell sheet, SMC sheet and fibroblast sheet are successively enrolled. This magnetically assisted tissue engineering methodology does not require the use of any scaffold. (From Ito, A. et al., *Tissue Eng.*, 11, 1553, 2005. With permission.)

approach, sheets of magnetic cells grown on removable substrates have been enrolled on cylindrical magnet to form tubular structures[59]. Remarkably, this process can be repeated several times using different types of cells to construct a multilayered tube, which elegantly mimics the native hierarchical organisation of a native vessel (Figure 13.4D). Other strategies combine the advantage of a 3D scaffold to enhance cell seeding and engraftment, with the possibility to control cell organisation by magnetic means[60]. Magnetic manipulation of cells opens fascinating prospects for the *in vitro* elaboration of tissue substitute comprising multicellular organised assemblies.

13.4   IMAGING CELL MIGRATION *IN VIVO*

The most developed application of magnetic cell labelling is MRI cell tracking[41]. Iron oxide NPs are known as potent MR contrast agents creating a positive or negative contrast depending on their concentration and on the MR sequences used[61,62]. They are characterised by their relaxivities r_1 and r_2, that is, their ability (per millimolar of agent) to increase the longitudinal and transverse relaxation rate of proton magnetisation. However, when MNPs are internalised into cellular endosomes or lysosomes, their contrast properties are drastically changed[63]: their longitudinal relaxivity is strongly diminished due to the poor accessibility of water protons to the highly concentrated NPs in the endosomes or lysosomes. Magnetic interactions between MNPs into endosomes also likely play a role in the change of relaxivities. The r_2/r_1 ratio is subsequently increased by cell internalisation.

An important consequence is that the cell, when magnetised in the MRI uniform magnetic field, creates a very local but strong magnetic inhomogeneity. Typically, the field increment is about 10^{-4} T at the surface of a cell (containing 5 pg of iron) and falls to 10^{-7} T at 50 μm from the cell. This cell-induced magnetic artefact is seen by protons as static and can be easily detected using T_2^*-sensitive gradient echo sequences[64]. Consequently, as soon as the spatial resolution is sufficiently high (typically <100 μm), it is possible to detect single magnetic cells as a focal signal void. Single cell detection has been proven using high-field MRI[64] or clinical 1.5 T field equipped with a low noise superconducting detection coil[65]. As an illustration, Figure 13.5A shows 9.4 T MR images of

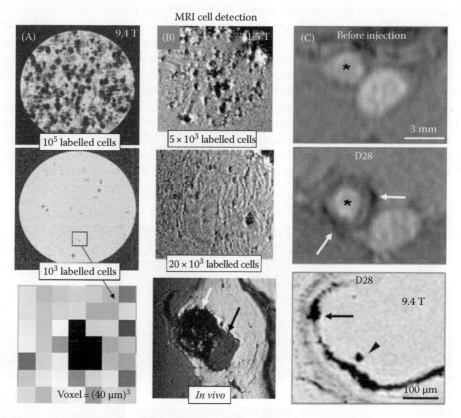

FIGURE 13.5 High-resolution MRI can detect single magnetically labelled cells. In (A), magnetically labelled HeLa cells were dispersed in agarose gel at different cell densities and observed in a 9.4 T MRI device using gradient echo sequence. When magnetised by the static magnetic field, each cell creates a signal void whose spatial extension depends on the echo time TE (TE = 10 ms). If the spatial resolution is higher than 120 μm in the three directions, a single cell can be detected. (From Smirnov, P., Gazeau, F., Beloeil, J. C., Doan, B.T., Wilhelm, C., and Gillet, B.: Single-cell detection by gradient echo 9.4 T MRI: A parametric study. *Contrast Media Mol Imaging.* 2006. 1. 165–174. Copyright Wiley-VCH Verlag GmbH & Co. KGaA. With permission.) In (B), MSCs were first labelled with AMNP and then seeded into lamellar polysaccharide-based scaffolds designed for tissue engineering. (From Poirier-Quinot, M. et al., *Tissue Eng. Part C Methods*, 16, 185, 2009.) High-resolution MRI was performed in a 1.5 T MRI set-up using a superconducting coil. The density and spatial distribution of cells into the scaffold can be directly assessed by MRI. The same scaffolds seeded with magnetic (white arrow) or non-magnetic MSC (black arrows) are imaged *in vivo* after subcutaneous implantation in mouse (bottom image). (C) Magnetically labelled SMCs were directly implanted in an aortic aneurysm in a rat. SMCs were used here as therapeutic cells to participate to the *in situ* regeneration of the aortic wall. 1.5 T *in vivo* MRI permitted to visualise the short-term delivery and long-term implantation of SMCs in the aorta and in the same time the evolution of the aorta diameter. High-resolution 9.4 T *ex vivo* MRI allowed assessing the distribution of SMCs in the regenerated aortic wall, with resolution closed to histology. (From Deux, J.F. et al., *Radiology*, 246, 185, 2008.)

agarose gels containing different densities of magnetically labelled cells[64]. At low density, one can observe signal voids, each of them being associated with a single cell. Of note, the MR apparent cell size is much larger than the real size of the cell and depends on the MR sequence parameters, such as the echo time. At high density, isolated cells cannot be resolved, but a global signal loss will be observed independently of the resolution.

MRI cell detection has been recently shown to be useful for assessing the distribution of stem cells seeded in a scaffold to construct tissue substitutes[36] (Figure 13.5B). The short-term engraftment of the cells in the scaffold as well as the *in vivo* implantation of the tissue construct could be monitored by MRI[36]. MRI cell tracking has also become the method of choice to evaluate cell therapies, which involve direct (local or intravenous) administration of labelled cells[41]. As an example, AMNP-labelled SMCs were injected in an aortic aneurysm in a rat model, and the success of endovascular cell therapy could be documented by the MRI follow-up (Figure 13.5C): Low-resolution 1.5 T MRI allowed delineating the immediate cell delivery as well as the cell engraftment and the evolution of aorta diameter over a month, while high-resolution *ex vivo* MRI confirmed histology to demonstrate the involvement of the injected cells in the aortic wall reconstruction[66].

In a model of cancer cell therapy, we also used MRI to monitor the migration of lymphocytes injected intravenously into tumour-bearing mice[29] (Figure 13.6). Lymphocytes were targeted to tumour cells through immune recognition. Surprisingly, the MRI follow-up showed a complex migration pathway of lymphocytes. They first homed to the spleen to multiply and become activated, but only then infiltrated the tumour and made it regress. This study was important from a

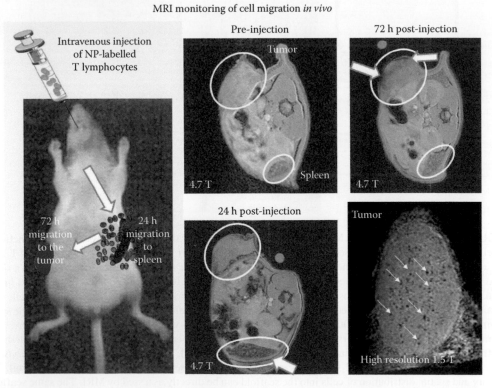

FIGURE 13.6 MRI follow-up of cell migration *in vivo*. To monitor cancer cell therapy using immune cells, cytotoxic T lymphocytes were first labelled with AMNP and injected intravenously to mice bearing a tumour. The lymphocytes were specifically targeted to the tumour cells thanks to ovalbumin receptors. From *in vivo* 7 T MRI follow-up, we could deduce that the lymphocyte first migrated to the spleen, where they divided (as assessed by a darkening of the spleen and enhancement of spleen volume) and only in the second time infiltrated the tumour. T lymphocyte tumour infiltration conducted to the tumour regression[29]. Using high-resolution 1.5 T MRI, single T lymphocytes could be detected *in vivo* in the tumour as punctual signal voids[65].

methodological point of view, because it was the first time that single cells could be detected by MRI directly *in vivo* in a tumour[65]. This *'tour de force'* was even complicated by the fact that the lymphocytes were poorly labelled after several divisions in the spleen (cell iron mass 0.2 pg) and that we used a clinical 1.5 T MRI device. Together with many other studies by different groups, it shows the great potential of MRI for cellular tracking, which was also illustrated in a clinical assay monitoring the role of dendritic cells in melanoma[67,68]. Beyond the magnetic labelling of cells, AMNP labelling of tissue explants was explored and proved useful to monitor muscle cell therapy using muscle explants as reservoir of myogenic stem cells[69].

13.5 MAGNETIC CELL TARGETING FOR CELL THERAPIES

Stem cell transplantation is considered a promising strategy for regeneration of injured tissue. It was proposed for the treatment of acute or chronic cardiac ischaemia, but showed moderate therapeutic benefit in preclinical trials. Stem cell transplantation fails to improve heart function, because the majority of injected cells escape the injured site due to the combination of tissue blood flow washing the cells out and cardiac contractions squeezing the cells out. Consequently, the cells retained in the injured area represent less than 10% of the originally injected cells, and the subsequent cell engraftment is poor.

It has been recently proposed that magnetic targeting of therapeutic cells could enhance cell retention in the injured sites and improve cell implantation. The pioneering studies in this field have concerned liver cell transplantation. It has been shown that active magnetic guiding of cells could influence cell trafficking to the liver parenchyma and enhance the liver concentration of transplanted cells[70,71]. However, perivascular aggregates of cells, probably created by magnetic interactions, also limited the integration of cells into the liver parenchyma. This raised the issue of the time and duration of magnet application with respect to the cell injection in order to optimise cell retention while minimising the formation of large aggregates. Interestingly, the importance of cell aggregation on magnetic cell targeting was recently highlighted both theoretically and experimentally in a model system of arterial bifurcation under the MRI field gradient[72].

Up to now, more potent developments have been achieved in the field of heart and vascular diseases[56]. The key concern in magnetically assisted cell delivery is whether magnetic forces exerted on cells can overcome the bloodstream in arteries, arterioles or capillaries. Microfluidic channels were designed to model physiological blood flow *in vitro* and simulate the attraction of magnetic cells[73]. Synthetic vascular grafts were successfully covered with endothelial cells by means of magnetic sheet annealed to the surface of the graft[74,75]. Similar strategies were directly tested *in vivo* to localise endothelial cells on the surface of intra-arterial stents in the presence of blood flow. High magnetic gradients were created in the steel stent by coating it with a layer of nickel[75] or by applying a strong uniform magnetic field during cell delivery[76]. Using a magnetic device that could be positioned outside the body, the engraftment of endothelial progenitor cells (EPCs) at a site of artery injury was increased by a factor of 5 compared with cell delivery without a magnet[77]. In an interesting study, Hofmann et al.[78] (Figure 13.7A) combined magnetically assisted transfection of endothelial cells with magnetic positioning of these transduced cells in an injured mouse carotid artery. By this way, gene therapy could be associated with cell-based therapy to improve vascular function. All these studies highlight the potential of magnetic cell targeting for vascular repair providing the basis for future clinical applications.

Magnetically assisted cell delivery was recently assessed for cardiac cell transplantation (Figure 13.7B). AMNP-labelled EPCs were injected in the infracted myocardium upon application of a magnet, which resulted in an increased concentration of cells. The short-term effect on cell retention was monitored by *in vivo* MRI and quantified by RT-PCR[33].

Remarkably, the work by Cheng et al.[79] was the first to evaluate the long-term engraftment and functional benefits of magnetically assisted cell retention. Under the action of magnetic forces, the injected stem cells clustered around the heart ischaemic zone and the short-term cell retention

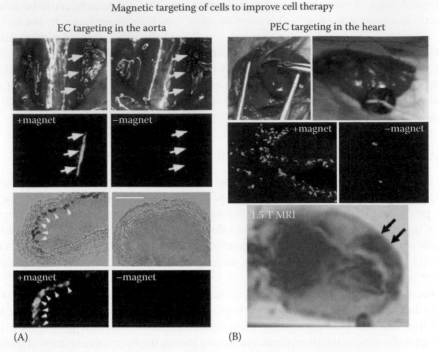

FIGURE 13.7 Magnetic cells can also be manipulated *in vivo* to enhance the efficiency of cell therapy or of gene-based therapy. The principle is that magnetic forces can increase cell retention by countering the efflux of transplanted cells due to blood flow or cardiac contractions. (A) Endothelial cells were genetically modified using a combination of MNPs and lentiviral vectors (LVs). Cell transduction with lentiviral vectors was significantly enhanced in the presence of a magnetic field gradient both *in vitro* and *in vivo*. Furthermore, the local implantation of LV-GFP-transduced endothelial cells to the intima of injured carotid arteries (top images) was improved in the presence of magnets. Images were taken 10 min after application of the cells and restoration of the blood flow. Bottom images show histological analysis of the carotid arteries. Accumulations of NP-loaded cells and of GFP fluorescence are seen at the vessel wall exposed to the magnet, but not in the control vessel. (From Hofmann, A. et al., *Proc. Natl. Acad. Sci. USA*, 106, 44, 2009. With permission.) (B) EPCs have been proposed as therapeutic cells to regenerate cardiac function. In a recent work, we tested the strategy of magnetically assisted cell delivery to enhance myocardial engraftment of human cord blood-derived EPCs after intra-coronary injection in a rat model of occlusion–reperfusion. The application of a magnet during the injection of labelled EPCs resulted in an increased concentration of cells in the myocardium as assessed by immunohistology (medium images, injected cells appear in green) or by 1.5 T MRI (bottom image, the area containing the magnetically labelled EPCs (black arrows) shows attenuated signal)[33].

tripled. Long-term engraftment of cells was also enhanced, and adverse remodelling of the heart was attenuated. As a result, ventricular function was improved with magnetic targeting. According to the researchers of Cedars-Sinai Heart Institute, the method is simple, readily scalable and could easily be coupled with current stem cell treatments to enhance their effectiveness. Although magnetically assisted cell or gene therapy is still in its infancy, it clearly offers a great potential for translation to clinical applications. As it is readily combined with a non-invasive MRI evaluation of the therapy[80,81], the use of magnetically labelled and magnetically manipulated cells in cell-based therapy will undoubtedly increase in future medicine.

13.6   CONCLUSION

The control of the interactions between live cells and nanomaterials is the corner stone of every advance in nanomedicine. Due to intensive researches in the last years, MNPs – mainly iron oxide – are now remarkably well commanded in their interactions with cells. Magnetic labelling of living cells has given rise to a variety of biomedical applications ranging from diagnostic imaging to regenerative medicine. The essential reasons boosting the use of 'magnetic cells' are the biocompatibility of iron oxide NPs and the unique opportunity to both image and manipulate cells using remote magnetic stimuli. While MRI cell tracking has already become a method of choice for monitoring migration and fate of administrated cells, magnetic cell manipulation is forecast to be more developed to improve tissue engineering technique and to assist cell-based therapies. Owing to the clinical approval of iron oxide NPs as MRI contrast agent, there is no major obstacle in the way of translation to human clinics of the magnetic methods summarised in this chapter. Different issues will require careful consideration for future clinical assessment of magnetically assisted cell therapies, especially the spatial scale change compared to small animal studies, the large tissue depths and the potentially higher physiologically relevant flow rates. More extensive studies will also be needed to optimise magnetic devices for cell attraction *in vivo* and for 3D control of cell organisation within *in vitro*-designed tissue construct.

REFERENCES

1. Weissleder, R., Elizondo, G., Wittenberg, J., Lee, A. S., Josephson, L. and Brady, T. J. 1990. Ultrasmall superparamagnetic iron oxide: An intravenous contrast agent for assessing lymph nodes with MR imaging. *Radiology*, 175, 494–498.
2. Yeh, T. C., Zhang, W., Ildstad, S. T. and Ho, C. 1993. Intracellular labeling of T-cells with superparamagnetic contrast agents. *Magn Reson Med*, 30, 617–625.
3. Dodd, S. J., Williams, M., Suhan, J. P., Williams, D. S., Koretsky, A. P. and Ho, C. 1999. Detection of single mammalian cells by high-resolution magnetic resonance imaging. *Biophys J*, 76, 103–109.
4. Weissleder, R., Cheng, H. C., Bogdanova, A. and Bogdanov, A., Jr. 1997. Magnetically labeled cells can be detected by MR imaging. *J Magn Reson Imaging*, 7, 258–263.
5. Ahrens, E. T., Feili-Hariri, M., Xu, H., Genove, G. and Morel, P. A. 2003. Receptor-mediated endocytosis of iron-oxide particles provides efficient labeling of dendritic cells for in vivo MR imaging. *Magn Reson Med*, 49, 1006–1013.
6. Daldrup-Link, H. E., Rudelius, M., Oostendorp, R. A., Settles, M., Piontek, G., Metz, S., Rosenbrock, H., Keller, U., Heinzmann, U., Rummeny, E. J., Schlegel, J. and Link, T. M. 2003. Targeting of hematopoietic progenitor cells with MR contrast agents. *Radiology*, 228, 760–767.
7. Bulte, J. W., Zhang, S., Van Gelderen, P., Herynek, V., Jordan, E. K., Duncan, I. D. and Frank, J. A. 1999. Neurotransplantation of magnetically labeled oligodendrocyte progenitors: Magnetic resonance tracking of cell migration and myelination. *Proc Natl Acad Sci USA*, 96, 15256–15261.
8. Lewin, M., Carlesso, N., Tung, C. H., Tang, X. W., Cory, D., Scadden, D. T. and Weissleder, R. 2000. Tat peptide-derivatized magnetic nanoparticles allow in vivo tracking and recovery of progenitor cells. *Nat Biotechnol*, 18, 410–414.
9. Josephson, L., Tung, C. H., Moore, A. and Weissleder, R. 1999. High-efficiency intracellular magnetic labeling with novel superparamagnetic-Tat peptide conjugates. *Bioconjug Chem*, 10, 186–191.
10. Kalish, H., Arbab, A. S., Miller, B. R., Lewis, B. K., Zywicke, H. A., Bulte, J. W., Bryant, L. H., Jr. and Frank, J. A. 2003. Combination of transfection agents and magnetic resonance contrast agents for cellular imaging: Relationship between relaxivities, electrostatic forces, and chemical composition. *Magn Reson Med*, 50, 275–282.
11. Arbab, A. S., Yocum, G. T., Kalish, H., Jordan, E. K., Anderson, S. A., Khakoo, A. Y., Read, E. J. and Frank, J. A. 2004. Efficient magnetic cell labeling with protamine sulfate complexed to ferumoxides for cellular MRI. *Blood*, 104, 1217–1223.
12. Bulte, J. W., Douglas, T., Witwer, B., Zhang, S. C., Strable, E., Lewis, B. K., Zywicke, H., Miller, B., Van Gelderen, P., Moskowitz, B. M., Duncan, I. D. and Frank, J. A. 2001. Magnetodendrimers allow endosomal magnetic labeling and in vivo tracking of stem cells. *Nat Biotechnol*, 19, 1141–1147.
13. Montet-Abou, K., Montet, X., Weissleder, R. and Josephson, L. 2007. Cell internalization of magnetic nanoparticles using transfection agents. *Mol Imaging*, 6, 1–9.
14. Walczak, P., Kedziorek, D. A., Gilad, A. A., Lin, S. and Bulte, J. W. 2005. Instant MR labeling of stem cells using magnetoelectroporation. *Magn Reson Med*, 54, 769–774.
15. Xie, D., Qiu, B., Walczak, P., Li, X., Ruiz-Cabello, J., Minoshima, S., Bulte, J. W. M. and Yang, X. 2010. Optimization of magnetosonoporation for stem cell labeling. *NMR Biomed*, 23, 480–484.

16. Wilhelm, C. and Gazeau, F. 2008. Universal cell labelling with anionic magnetic nanoparticles. *Biomaterials*, 29, 3161–3174.

17. Wilhelm, C., Billotey, C., Roger, J., Pons, J. N., Bacri, J. C. and Gazeau, F. 2003. Intracellular uptake of anionic superparamagnetic nanoparticles as a function of their surface coating. *Biomaterials*, 24, 1001–1011.

18. Wilhelm, C., Gazeau, F., Roger, J., Pons, J. N. and Bacri, J. C. 2002. Interaction of anionic superparamagnetic nanoparticles with cells: Kinetic analyses of membrane adsorption and subsequent internalization. *Langmuir*, 18, 8148–8155.

19. Riviere, C., Wilhelm, C., Cousin, F., Dupuis, V., Gazeau, F. and Perzynski, R. 2007. Internal structure of magnetic endosomes. *Eur Phys J E Soft Matter*, 22, 1–10.

20. Wilhelm, C., Cebers, A., Bacri, J. C. and Gazeau, F. 2003. Deformation of intracellular endosomes under a magnetic field. *Eur Biophys J*, 32, 655–660.

21. Wilhelm, C. 2008. Out-of-equilibrium microrheology inside living cells. *Phys Rev Lett*, 101, 028101.

22. Wilhelm, C., Gazeau, F. and Bacri, J. C. 2003. Rotational magnetic endosome microrheology: Viscoelastic architecture inside living cells. *Phys Rev E Stat Nonlin Soft Matter Phys*, 67, 06190801-12.

23. Gao, J., Zhang, W., Huang, P., Zhang, B., Zhang, X. and Xu, B. 2008. Intracellular spatial control of fluorescent magnetic nanoparticles. *J Am Chem Soc*, 130, 3710–3711.

24. Levy, M., Lagarde, F., Maraloiu, V. A., Blanchin, M. G., Gendron, F., Wilhelm, C. and Gazeau, F. 2010. Degradability of superparamagnetic nanoparticles in a model of intracellular environment: Follow-up of magnetic, structural and chemical properties. *Nanotechnology*, 21, 395103.

25. Arbab, A. S., Wilson, L. B., Ashari, P., Jordan, E. K., Lewis, B. K. and Frank, J. A. 2005. A model of lysosomal metabolism of dextran coated superparamagnetic iron oxide (SPIO) nanoparticles: Implications for cellular magnetic resonance imaging. *NMR Biomed*, 18, 383–389.

26. Pawelczyk, E., Arbab, A. S., Pandit, S., Hu, E. and Frank, J. A. 2006. Expression of transferrin receptor and ferritin following ferumoxides–protamine sulfate labeling of cells: Implications for cellular magnetic resonance imaging. *NMR Biomed*, 19, 581–592.

27. Levy, M., Luciani, N., Alloyeau, D., Elgrabli, D., Deveaux, V., Pechoux, C., Chat, S., Wang, G., Vats, N., Gendron, F., Factor, C., Lotersztajn, S., Luciani, A., Wilhelm, C. and Gazeau, F. 2011. Long term in vivo biotransformation of iron-oxide nanoparticles. *Biomaterials*, 32, 3988–3999.

28. Smirnov, P., Gazeau, F., Lewin, M., Bacri, J. C., Siauve, N., Vayssettes, C., Cuenod, C. A. and Clement, O. 2004. In vivo cellular imaging of magnetically labeled hybridomas in the spleen with a 1.5-T clinical MRI system. *Magn Reson Med*, 52, 73–79.

29. Smirnov, P., Lavergne, E., Gazeau, F., Lewin, M., Boissonnas, A., Doan, B. T., Gillet, B., Combadiere, C., Combadiere, B. and Clement, O. 2006. In vivo cellular imaging of lymphocyte trafficking by MRI: A tumor model approach to cell-based anticancer therapy. *Magn Reson Med*, 56, 498–508.

30. Billotey, C., Aspord, C., Beuf, O., Piaggio, E., Gazeau, F., Janier, M. F. and Thivolet, C. 2005. T-cell homing to the pancreas in autoimmune mouse models of diabetes: In vivo MR imaging. *Radiology*, 236, 579–587.

31. Luciani, N., Gazeau, F. and Wilhelm, C. 2009. Reactivity of the monocyte/macrophage system to superparamagnetic anionic nanoparticles. *J Mater Chem*, 19, 6373–6380.

32. Wilhelm, C., Bal, L., Smirnov, P., Galy-Fauroux, I., Clement, O., Gazeau, F. and Emmerich, J. 2007. Magnetic control of vascular network formation with magnetically labeled endothelial progenitor cells. *Biomaterials*, 28, 3797–3806.

33. Chaudeurge, A. C. W., Chen-Tournoux, A., Farahmand, P. V. B., Autret, G., Larghéro, J., Desnos, T., Hagège, A., Gazeau, F., Clement and Menasché, P. in press. Can magnetic targeting of magnetically-labeled endothelial progenitor circulating cells optimize intramyocardial cell engraftment? *Cell Transplant*.

34. Wilhelm, C., Fortin, J. P. and Gazeau, F. 2007. Tumour cell toxicity of intracellular hyperthermia mediated by magnetic nanoparticles. *J Nanosci Nanotechnol*, 7, 2933–2937.

35. Naveau, A., Smirnov, P., Menager, C., Gazeau, F., Clement, O., Lafont, A. and Gogly, B. 2006. Phenotypic study of human gingival fibroblasts labeled with superparamagnetic anionic nanoparticles. *J Periodontol*, 77, 238–247.

36. Poirier-Quinot, M., Frasca, G., Wilhelm, C., Luciani, N., Ginefri, J. C., Darrasse, L., Letourneur, D., Le Visage, C. and Gazeau, F. 2009. High resolution 1.5T magnetic resonance imaging for tissue engineering constructs: A non invasive tool to assess 3D scaffold architecture and cell seeding. *Tissue Eng Part C Methods*, 16, 185–200.

37. Wilhelm, C., Lavialle, F., Pechoux, C., Tatischeff, I. and Gazeau, F. 2008. Intracellular trafficking of magnetic nanoparticles to design multifunctional biovesicles. *Small*, 4, 577–582.

38. Luciani, N., Wilhelm, C. and Gazeau, F. 2010. The role of cell-released microvesicles in the intercellular transfer of magnetic nanoparticles in the monocyte/macrophage system. *Biomaterials*, 31, 7061–7069.

39. Pawelczyk, E., Arbab, A. S., Chaudhry, A., Balakumaran, A., Robey, P. G. and Frank, J. A. 2008. In vitro model of bromodeoxyuridine or iron oxide nanoparticle uptake by activated macrophages from labeled stem cells: Implications for cellular therapy. *Stem Cells*, 26, 1366–1375.

40. Pawelczyk, E., Jordan, E. K., Balakumaran, A., Chaudhry, A., Gormley, N., Smith, M., Lewis, B. K., Childs, R., Robey, P. G. and Frank, J. A. 2009. In vivo transfer of intracellular labels from locally implanted bone marrow stromal cells to resident tissue macrophages. *PLoS ONE*, 4, e6712. doi:10.1371/journal.pone.0006712.

41. Arbab, A. S. and Frank, J. A. 2008. Cellular MRI and its role in stem cell therapy. *Regen Med*, 3, 199–215.

42. Kostura, L., Kraitchman, D. L., Mackay, A. M., Pittenger, M. F. and Bulte, J. W. 2004. Feridex labeling of mesenchymal stem cells inhibits chondrogenesis but not adipogenesis or osteogenesis. *NMR Biomed*, 17, 513–517.

43. Farrell, E., Wielopolski, P., Pavljasevic, P., Van Tiel, S., Jahr, H., Verhaar, J., Weinans, H., Krestin, G., O'Brien, F. J., Van Osch, G. and Bernsen, M. 2008. Effects of iron oxide incorporation for long term cell tracking on MSC differentiation in vitro and in vivo. *Biochem Biophys Res Commun*, 369, 1076–1081.

44. Farrell, E., Wielopolski, P., Pavljasevic, P., Kops, N., Weinans, H., Bernsen, M. R. and van Osch, G. J. V. M. 2009. Cell labelling with superparamagnetic iron oxide has no effect on chondrocyte behaviour. *Osteoarthritis Cartilage*, 17, 961–967.

45. Arbab, A. S., Yocum, G. T., Rad, A. M., Khakoo, A. Y., Fellowes, V., Read, E. J. and Frank, J. A. 2005. Labeling of cells with ferumoxides–protamine sulfate complexes does not inhibit function or differentiation capacity of hematopoietic or mesenchymal stem cells. *NMR Biomed*, 18, 553–559.

46. Henning, T. D., Sutton, E. J., Kim, A., Golovko, D., Horvai, A., Ackerman, L., Sennino, B., McDonald, D., Lotz, J. and Daldrup-Link, H. E. 2009. The influence of ferucarbotran on the chondrogenesis of human mesenchymal stem cells. *Contrast Media Mol Imaging*, 4, 165–173.

47. Kedziorek, D. A., Muja, N., Walczak, P., Ruiz-Cabello, J., Gilad, A. A., Jie, C. C. and Bulte, J. W. M. 2010. Gene expression profiling reveals early cellular responses to intracellular magnetic labeling with superparamagnetic iron oxide nanoparticles. *Magn Reson Med*, 63, 1031–1043.

48. Neri, M., Maderna, C., Cavazzin, C., Deidda-Vigoriti, V., Politi, L. S., Scotti, G., Marzola, P., Sbarbati, A., Vescovi, A. L. and Gritti, A. 2008. Efficient in vitro labeling of human neural precursor cells with superparamagnetic iron oxide particles: Relevance for in vivo cell tracking. *Stem Cells*, 26, 505–516.

49. Cohen, M. E., Muja, N., Fainstein, N., Bulte, J. W. M. and Ben-Hur, T. 2010. Conserved fate and function of ferumoxides-labeled neural precursor cells in vitro and in vivo. *J Neurosci Res*, 88, 936–944.

50. Wilhelm, C., Riviere, C. and Biais, N. 2007. Magnetic control of *Dictyostelium* aggregation. *Phys Rev E Stat Nonlin Soft Matter Phys*, 75, 041906.

51. Wilhelm, C., Gazeau, F. and Bacri, J. C. 2002. Magnetophoresis and ferromagnetic resonance of magnetically labeled cells. *Eur Biophys J*, 31, 118–125.

52. Pamme, N. and Wilhelm, C. 2006. Continuous sorting of magnetic cells via on-chip free-flow magnetophoresis. *Lab Chip*, 6, 974–980.

53. Adams, J. D., Kim, U. and Soh, H. T. 2008. Multitarget magnetic activated cell sorter. *Proc Natl Acad Sci USA*, 105, 18165–18170.

54. Robert, D., Pamme, N., Conjeaud, H., Gazeau, F., Iles, A. and Wilhelm, C. 2011. Cell sorting by endocytotic capacity in a microfluidic magnetophoresis device. *Lab Chip*, 11, 1902–1910.

55. Riviere, C., Marion, S., Guillen, N., Bacri, J. C., Gazeau, F. and Wilhelm, C. 2007. Signaling through the phosphatidylinositol 3-kinase regulates mechanotaxis induced by local low magnetic forces in *Entamoeba histolytica*. *J Biomech*, 40, 64–77.

56. Gazeau, F. and Wilhelm, C. 2010. Magnetic labeling, imaging and manipulation of endothelial progenitor cells using iron oxide nanoparticles. *Future Med Chem*, 2, 397–408.

57. Frasca, G., Gazeau, F. and Wilhelm, C. 2009. Formation of a three-dimensional multicellular assembly using magnetic patterning. *Langmuir*, 25, 2348–2354.

58. Ino, K., Ito, A. and Honda, H. 2007. Cell patterning using magnetite nanoparticles and magnetic force. *Biotechnol Bioeng*, 97, 1309–1317.

59. Ito, A., Ino, K., Hayashida, M., Kobayashi, T., Matsunuma, H., Kagami, H., Ueda, M. and Honda, H. 2005. Novel methodology for fabrication of tissue-engineered tubular constructs using magnetite nanoparticles and magnetic force. *Tissue Eng*, 11, 1553–1561.

60. Robert, D., Fayol, D., Le Visage, C., Frasca, G., Brule, S., Menager, C., Gazeau, F., Letourneur, D. and Wilhelm, C. 2010. Magnetic micro-manipulations to probe the local physical properties of porous scaffolds and to confine stem cells. *Biomaterials*, 31, 1586–1595.

61. Corot, C., Robert, P., Idee, J. M. and Port, M. 2006. Recent advances in iron oxide nanocrystal technology for medical imaging. *Adv Drug Deliv Rev*, 58, 1471–1504.

62. Laurent, S., Forge, D., Port, M., Roch, A., Robic, C., Vander Elst, L. and Muller, R. N. 2008. Magnetic iron oxide nanoparticles: Synthesis, stabilization, vectorization, physicochemical characterizations, and biological applications. *Chem Rev*, 108, 2064–2110.

63. Billotey, C., Wilhelm, C., Devaud, M., Bacri, J. C., Bittoun, J. and Gazeau, F. 2003. Cell internalization of anionic maghemite nanoparticles: Quantitative effect on magnetic resonance imaging. *Magn Reson Med*, 49, 646–654.

64. Smirnov, P., Gazeau, F., Beloeil, J. C., Doan, B. T., Wilhelm, C. and Gillet, B. 2006. Single-cell detection by gradient echo 9.4T MRI: A parametric study. *Contrast Media Mol Imaging*, 1, 165–174.

65. Smirnov, P., Poirier-Quinot, M., Wilhelm, C., Lavergne, E., Ginefri, J. C., Combadiere, B., Clement, O., Darrasse, L. and Gazeau, F. 2008. In vivo single cell detection of tumor-infiltrating lymphocytes with a clinical 1.5 Tesla MRI system. *Magn Reson Med*, 60, 1292–1297.

66. Deux, J. F., Dai, J., Riviere, C., Gazeau, F., Meric, P., Gillet, B., Roger, J., Pons, J. N., Letourneur, D., Boudghene, F. P. and Allaire, E. 2008. Aortic aneurysms in a rat model: In vivo MR imaging of endovascular cell therapy. *Radiology*, 246, 185–192.

67. De Vries, I. J., Lesterhuis, W. J., Barentsz, J. O., Verdijk, P., Van Krieken, J. H., Boerman, O. C., Oyen, W. J., Bonenkamp, J. J., Boezeman, J. B., Adema, G. J., Bulte, J. W., Scheenen, T. W., Punt, C. J., Heerschap, A. and Figdor, C. G. 2005. Magnetic resonance tracking of dendritic cells in melanoma patients for monitoring of cellular therapy. *Nat Biotechnol*, 23, 1407–1413.

68. Bulte, J. W. 2009. In vivo MRI cell tracking: Clinical studies. *AJR Am J Roentgenol*, 193, 314–325.

69. Riviere, C., Lecoeur, C., Wilhelm, C., Pechoux, C., Combrisson, H., Yiou, R. and Gazeau, F. 2009. The MRI assessment of intraurethrally—Delivered muscle precursor cells using anionic magnetic nanoparticles. *Biomaterials*, 30, 6920–6928.

70. Arbab, A. S., Jordan, E. K., Wilson, L. B., Yocum, G. T., Lewis, B. K. and Frank, J. A. 2004. In vivo trafficking and targeted delivery of magnetically labeled stem cells. *Hum Gene Ther*, 15, 351–360.

71. Luciani, A., Wilhelm, C., Bruneval, P., Cunin, P., Autret, G., Rahmouni, A., Clement, O. and Gazeau, F. 2009. Magnetic targeting of iron-oxide-labeled fluorescent hepatoma cells to the liver. *Eur Radiol*, 19, 1087–1096.

72. Riegler, J., Wells, J. A., Kyrtatos, P. G., Price, A. N., Pankhurst, Q. A. and Lythgoe, M. F. 2010. Targeted magnetic delivery and tracking of cells using a magnetic resonance imaging system. *Biomaterials*, 31, 5366–5371.

73. Kim, J. A., Lee, H. J., Kang, H. J. and Park, T. H. 2009. The targeting of endothelial progenitor cells to a specific location within a microfluidic channel using magnetic nanoparticles. *Biomed Microdevices*, 11, 287–296.

74. Pislaru, S. V., Harbuzariu, A., Agarwal, G., Witt, T., Gulati, R., Sandhu, N. P., Mueske, C., Kalra, M., Simari, R. D. and Sandhu, G. S. 2006. Magnetic forces enable rapid endothelialization of synthetic vascular grafts. *Circulation*, 114, I314–I318.

75. Pislaru, S. V., Harbuzariu, A., Gulati, R., Witt, T., Sandhu, N. P., Simari, R. D. and Sandhu, G. S. 2006. Magnetically targeted endothelial cell localization in stented vessels. *J Am Coll Cardiol*, 48, 1839–1845.

76. Polyak, B., Fishbein, I., Chorny, M., Alferiev, I., Williams, D., Yellen, B., Friedman, G. and Levy, R. J. 2008. High field gradient targeting of magnetic nanoparticle-loaded endothelial cells to the surfaces of steel stents. *Proc Natl Acad Sci USA*, 105, 698–703.

77. Kyrtatos, P. G., Lehtolainen, P., Junemann-Ramirez, M., Garcia-Prieto, A., Price, A. N., Martin, J. F., Gadian, D. G., Pankhurst, Q. A. and Lythgoe, M. F. 2009. Magnetic tagging increases delivery of circulating progenitors in vascular injury. *JACC Cardiovasc Interv*, 2, 794–802.

78. Hofmann, A., Wenzel, D., Becher, U. M., Freitag, D. F., Klein, A. M., Eberbeck, D., Schulte, M., Zimmermann, K., Bergemann, C., Gleich, B., Roell, W., Weyh, T., Trahms, L., Nickenig, G., Fleischmann, B. K. and Pfeifer, A. 2009. Combined targeting of lentiviral vectors and positioning of transduced cells by magnetic nanoparticles. *Proc Natl Acad Sci USA*, 106, 44–49.

79. Cheng, K., Li, T. S., Malliaras, K., Davis, D. R., Zhang, Y. and Marban, E. 2010. Magnetic targeting enhances engraftment and functional benefit of iron-labeled cardiosphere-derived cells in myocardial infarction. *Circ Res*, 106, 1570–1581.

80. Kedziorek, D. A. and Kraitchman, D. L. 2010. Superparamagnetic iron oxide labeling of stem cells for MRI tracking and delivery in cardiovascular disease. *Methods Mol Biol*, 660, 171–183.

81. Kraitchman, D. L., Kedziorek, D. A. and Bulte, J. W. M. 2011. MR imaging of transplanted stem cells in myocardial infarction. *Methods Mol Biol*, 680, 45–52.

82. Robert, D., Nguyen, T. H., Gallet, F. and Wilhelm, C. 2010. In vivo determination of fluctuating forces during endosome trafficking using a combination of active and passive microrheology. *PLoS One*, 5, e10046.

Dr. Florence Gazeau (born in 1970) completed her PhD degree in 1997 at the University Paris 7 – Denis Diderot, focusing on the magnetic and hydrodynamic properties of ferrofluids in the group of Professor Jean-Claude Bacri. She joined the centre national de la Recherche scientifique (CNRS) as a staff scientist in 1998 to develop biomedical applications of MNPs. Her main research interest includes the physics of nanomagnetism applied to nanomedicine: interactions of NPs with cells, MRI of cell migration, NPs-mediated hyperthermia and magnetic targeting. She is a CNRS senior scientist since 2009 and animates, together with Dr. Claire Wilhelm, the group NanoBioMagnetism in the laboratory MSC of the University Paris Diderot.

Dr. Claire Wilhelm (born in 1975), CNRS staff scientist since 2003, is a young scientist in physics known to develop new applications of MNPs for medical diagnosis (MRI) with the use of MNPs as contrast agent, *in vitro* diagnosis test based on magneto-optical detection, magnetic cell sorting on lab-on-chip, intracellular magnetic tweezers and therapy (monitoring of cell migration by MRI, cell therapy and tissue engineering assisted by magnetic force, magnetic guiding of drug vectors, magnetically induced hyperthermia). She has authored more than 50 publications in international journals and about 20 proceedings.

Claire Wilhelm will receive the CNRS Bronze medal in 2011, which recognises a researcher's first work, making that person a specialist with talent in a particular field.

14 Non-Invasive Magnetically Targeted tPA Delivery for Arterial Thrombolysis*

Haitao Chen, Michael D. Kaminski, Xianqiao Liu,
Patricia Caviness Stepp, Yumei Xie, and Axel J. Rosengart[†]

CONTENTS

14.1 INTRODUCTION

Many research groups are interested in the development of drug-loaded magnetic spheres in the nano- and micro-size range for non-invasive targeted drug delivery in humans. We combined experience in nanotechnology-based drug delivery and expertise in engineering to improve the treatment of a common disease affecting many individuals at any stage and with devastating consequences—acute strokes. Brain strokes or ischaemic brain injuries are caused by acute (or acute-on-chronic) obstruction of a brain artery and stand out as the third leading cause of death and disability in the industrialised world. In this chapter, we introduce the reader to a conceptual approach of applying modern nanotechnology to a clinically common disease and discuss the initial building blocks towards a novel drug delivery method for reopening acutely occluded brain arteries.

* Argonne National Laboratory, operated by UChicago Argonne, LLC, for the U.S. Department of Energy under Contract No. DE-AC02-06CH11357.
† Kaminski and Rosengart's photographs and biographies are featured in Chapter 7.

Currently, medication-induced reopening of an occluded brain artery, called thrombolysis, is achieved in about one out of three stroke victims for patients arriving at a specialised centre in less than 5 h after symptom onset. However, this rather narrow treatment time window is only met by <10% of all strokes as many victims are either not immediately aware of the urgency of their functional deficits (i.e. arm weakness or numbness, unsteadiness, etc.) or are waking up in the morning only to realise that they had suffered a stroke during sleep with the exact timing of stroke onset remaining unknown ('strokes generally do not hurt'; in contrast, heart attacks, also caused by an arterial occlusion but within the heart, are commonly excruciatingly painful, and patients immediately seek medical advice).

Progress has been made to increase the treatment time window and, hence, accessibility to thrombolysis by providing local (catheter-based, intra-arterial) thrombolysis. This costly approach is only available for a minority of patients but uses ~1/10 of the systemic dose of the commonly employed clot lysis drug called tissue plasminogen activator (tPA). However, this catheter-based procedure has the added advantage of mechanical clot disruption by the catheter guide wire. In addition, clinical studies point to a small time benefit in terms of extending the treatment window by ~1–2 h for certain subtypes of strokes without an increased bleeding risk. Therefore, the majority of stroke patients cannot receive thrombolytic therapy either because it is too risky (given systemically past 5 h) or targeted, catheter-based therapy is not available due to its high procedural complexities.

We reasoned that a novel drug delivery system with precise, intravenous delivery of tPA directly into the blood clot, shielded or partly shielded against enzymatic degradation, and with on-demand drug release once at the target clot location (Figure 14.1) is without the need for a risky, invasive, catheter-based technology. Such development could, therefore, be of help to a large number of stroke victims. This aim could be achieved by employing systemic, intravenous injection of magnetically responsive and, hence, externally (from outside the body) guidable, tPA-loaded spheres responsive to an external trigger to unload and release tPA at the target clot location. With such an approach, lysis could be obtained at low concentrations (targeted delivery into the clot) and without the need for invasiveness. Of note, tPA is an enzyme that cleaves plasminogen, which is regularly present within the bloodstream, into plasmin, the major enzyme responsible for clot breakdown. Hence, tPA is currently the preferred drug employed for vascular clot lysis (venous and arterial).

The concept of magnetically guided spheres represents a unique drug delivery strategy, as traditional pharmacotherapy is almost entirely dependent on physiological properties such as blood flow, tissue barriers (e.g. blood–brain barrier) and cellular receptors for passive delivery. Passive delivery, however, is exceedingly difficult into a region without flow of blood, such as the occluded brain artery. The benefit of non-invasive targeting thrombolytics to vascular occlusions leading has been previously suggested

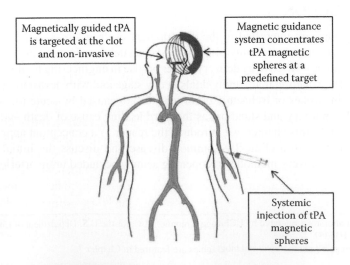

FIGURE 14.1   Proposed magnetically guided thrombolysis.

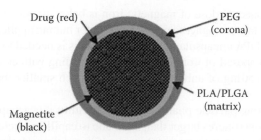

FIGURE 14.2   A diagram of a tPA encapsulated magnetic sphere (not to scale).

employing non-magnetic liposomes and microparticles[1-3]. The uniqueness of the discussed drug delivery strategy stems from the specific physical and chemical characteristics of our 'key ingredient', the individual, biodegradable and non-toxic magnetic sphere (Figure 14.2). The sphere matrix is designed with the stability of a polymerised shell to control biochemical degradation and permit manipulations (i.e. guidance) within fast-flowing blood. Its surface layer consists of poly(ethylene glycol) (PEG) to control pharmacokinetic properties (reducing immune uptake from the bloodstream). The sphere core contains the drug load of choice, (here tPA), and magnetic nanoparticles, the latter providing the sphere with superparamagnetic properties. As the human body is transparent to magnetic fields (as the MRI demonstrates), externally applied magnetic field can be used to trap injected magnetic spheres within the human body. In future applications, the magnetic fields needed to exactly localise freely blood-circulating tPA-loaded nanospheres at the blood clot side could ultimately be set up by a 3D array of electromagnetic coils, the position and field strength of which are determined by computer-based imaging data of the anatomic area in question. Indeed, the non-trivial engineering problem of targeting magnetic particles *in vivo* has been developed slowly over the years, and progress has been reviewed recently[4]. The reader is also referred to Chapter 15, which discusses the use of blood cells as magnetically targeted delivery system. In addition, Chapter 16 reviews the design and magnetic control of magnetic systems to guide therapeutic nanoparticles, while Chapter 19 introduces the reader to the use of magnetic microbubbles.

14.2   NOVEL NON-INVASIVE YET TARGETED DELIVERY SYSTEM: CONCEPTUAL APPROACH

14.2.1   BIOCOMPATIBILITY AND BIODEGRADABILITY OF THE CARRIER SYSTEM

For future human use, the sphere should be non-toxic and biodegradable. Poly(lactic acid) (PLA) and poly(lactic-*co*-glycolic acid) (PLGA) have been extensively studied as biomaterials since the 1960s[5-9], being used in suture material and also as polymers for PLA/PLGA-based microsphere and nanospheres[10-19]. PLA and PLGA degrade *in vivo*[20] by hydrolysis (esterase activity) into lactic acid and glycolic acid, which are then incorporated into the tricarboxylic acid cycle and excreted. In addition, PLA and PLGA are considered to be immunologically inert[8].

PEG is a biocompatible, water-soluble polymer that resists recognition by the immune system[21], possibly *via* static hindrance between PEG and blood proteins, thus delaying the removal of the spheres from the bloodstream *via* the reticuloendothelial system (liver, spleen, etc.)[22]. Because of these remarkable *in vivo* properties, PEGs have been used extensively in controlled, targeted and sustained drug delivery systems[23], and their use approved by the Food and Drug Administration (FDA). For instance, activated PEGs have been linked to proteins and enzymes[24,25] or lipids and liposomes[26] and are used in drug spheres[15,27].

The presence of magnetic nanocrystals incorporated within the polymer matrix introduces another source of potential toxicity. The most commonly used magnetic materials are magnetite (Fe_3O_4) and maghemite (Fe_2O_3)[28-33], although iron cobalt oxides[34], passivated iron metals[35,36] and cobalt metals[37,38] may become more readily available. Magnetite nanocrystals are, in part, metabolised, thereby increasing hepatic and splenic ferritin stores and, in part, incorporated into red blood cells[39,40].

Thus, provided that the injected dose of magnetic iron is below the estimated toxic dose threshold (750 mg iron for standard man), magnetite seems suitable for human applications. If we estimate that a single dose of 100 mg of tPA encapsulated magnetic spheres is needed to treat a typical stroke subject, a fraction will be composed of iron. Assuming 50% loading with elemental iron, this injection will expose the patient to 50 mg of unbound iron, a dose much smaller than a single injection dose leading to toxic effects.

Initial work investigating particle pharmacokinetics within our group and by others identified serious adverse effects from spheres larger than 5 µm due to capillary occlusion[41,42]. In contrast, very small spheres, that is, below approximately 90 nm, are filtered through the Bowman membrane and quickly excreted by the kidney[35,42]. Consequently, designer tPA encapsulated spheres should ideally range in diameter between 90 nm and 3 µm to avoid vascular occlusions as well as rapid excretion.

14.2.2 'MEDICATION' OF THE DESIGNED MAGNETIC SPHERES

Drug encapsulation is the 'enclosure' of the drug such as tPA into the PLA/PLGA magnetic matrix during particle synthesis. Once introduced into a fluid medium, the medication remains 'shielded' until released from the matrix by spontaneous[43–46] or active release by some trigger mechanisms[47]. Encapsulated drugs are characteristically released in several defined stages. First, a burst effect releases some percentage of drug immediately upon exposure to fluids[45,46] from drug adsorbed superficially. A second release phase of untriggered drug-loaded spheres lasts days to weeks and is attributed to both the degradation of the matrix with slow release of the encapsulated material and diffusion of aqueous fluid within the porous network of the spheres. In contrast, triggered drug burst release can be caused by pH-responsive polymers, thermo-responsive polymers or by adsorbing external energy resulting in thermal, mechanical or chemical degradation of the polymer and is initiated when the spheres are exposed to conditions that induce rapid degradation or conformation changes[48–50].

14.2.3 INTRAVASCULAR CIRCULATION OF MAGNETIC NANOSPHERES

Prolonged *in vivo* vascular circulation times of similar designed spheres have previously been described with the conjugation of sorption of amphipathic polyethylene glycol (so-called PEGylation). Starting with Klibanov et al.[51], who recognised similar studies in the extended vascular circulation of PEGylated proteins, PEGylation has been adapted to artificial carrier systems: first with liposomes[52], then with polymeric nanospheres. For example, Gref et al. provide a good discussion regarding surface PEGylation, that is, changing rate of antibody formation, opsonisation by phagocytes and protein adsorption[14], therefore influencing pharmacological behaviour and improving vascular circulation of spheres within the bloodstream[14,15]. The half-life of PEGylated spheres depends not only on the total molecular weight, but also increases with increasing PEG chain length as longer chain PEG demonstrated a vascular circulation of up to 20 h[53,54]. An increase in blood circulation times was found with an increase in the molecular weight of covalently linked PEG. For instance, 5 h after systemic injection, only one-third of 20 kDa PEG-conjugated PLGA nanospheres (140 nm) had been captured by the liver as compared to uncoated spheres[14,54].

14.2.4 MAGNETIC PROPERTIES OF MAGNETIC NANOSPHERES

The magnetic moment of the polymeric-based nanospheres is derived from the magnetic nanocrystals encapsulated in the polymer matrix during synthesis. For the need of manipulating and externally controlling or influencing sphere movements within the body, the spheres are required to have a high magnetic moment. Options for magnetically responsive inclusions are nickel, cobalt or iron magnetic nanocrystals. A marked limitation in the use of some of these materials is their tendency towards oxidation at room temperature and, hence, their decreased biocompatibility.

To reduce oxidation, passive layers may be coated onto the magnetic materials[55]. In addition to magnetic material loading of the individual spheres, the retention of carriers at a particular vascular target site depends on multiple additional parameters such as magnetic field gradient and strength, agglomeration and build-up over time, distance between the target site, the magnetic field source and local blood flow velocities, among others.

We investigated theoretically the retention of magnetic drug carriers at the human carotid artery bifurcation (Figure 14.3)[56]. Three magnetic drug targeting systems were evaluated in terms of their collection efficiency and found that using a magnetic field plus a magnetisable conductor (i.e. wire) positioned close to the bifurcation provides a strong, magnetic field gradient suitable to magnetically retain spheres against high-blood flow velocities[56]. Other investigations have demonstrated that retention of magnetic carriers can range from 10% in the rat head[57] to 70% liver targeting[58]. We demonstrated in a primate model that after both proximal intra-arterial and intravenous injections, the intravascular sequestration of magnetic polystyrene particles (mean diameter 4.5 μm, magnetite loading 50%) against arterial blood flow was achieved selectively at the site next to an externally placed permanent magnet (Figure 14.4)[59]. More elegantly, Shapiro[60] has shown that dynamic magnetic arrays can create local maxima in the magnetic field that can more generally target internal targets in the body. This is quite important, because static magnetic fields create force vectors that point towards the highest value of the magnetic field, which occurs at the surface of the magnet. Since the magnets are generally located outside the body, superficial targeting of magnetic spheres is viable but

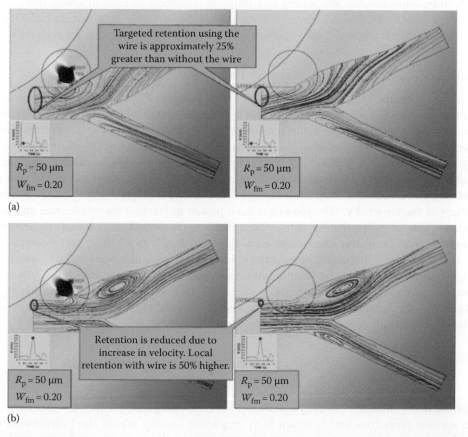

FIGURE 14.3 Simulation of magnetic drug targeting at human carotid artery bifurcation using a magnetisable wire adjacent to the bifurcation [wire diameter equal to one-half the diameter of the common carotid artery; particles are assumed to be aggregates with mean radius (R_p) equal to 50 μm and composed of 20% by mass (W_{fm}) in magnetite]. (a) identifies trapping of particles during diastole, whereas (b) identifies aggregation during systole.

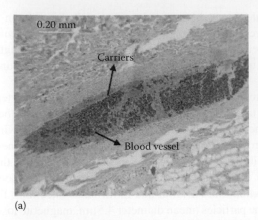
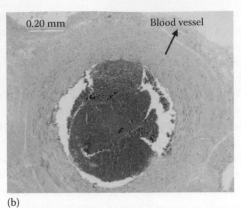

FIGURE 14.4 (a) Histology verified particle concentration at the arterial region under the surface magnetic after injection. (b) The particles were not found at the control side. (Reprinted with permission from Chen, H., Kaminski, M.D., Pytel, P., Macdonald, L., and Rosengart, A.J., Capture of magnetic carriers within large arteries using external magnetic fields, *J. Drug Target*, 16, 262–268, 2008. Copyright 2009 Informa Healthcare.)

targeting to internal organs is very difficult or impossible. But, by placing magnetisable wires local to the target or by producing local maxima with dynamic magnetic arrays, targeting to internal organs appears to be quite feasible. We believe that future development of focal magnetic fields to target distant sites within the body and current development of magnetisable drug carriers make it reasonable to propose non-invasive delivery of tPA spheres to a target vascular occlusion within the body.

14.2.5 Drug Release from Magnetic Spheres

Drug release from the magnetic spheres can be a sole function of natural degradation of the individual spheres within a specific fluid medium[61] that can be controlled by synthesis parameters such as variations in polymer matrix, molecular weight, etc. In contrast to passive degradation and drug release, we pursue for our technology an external trigger for burst release of drug to achieve high (lytic) concentrations of tPA directly at the clot location. To achieve this, we are utilising externally applied ultrasound with a focused beam directed to the area of vascular occlusion, that is, the carotid artery in the neck or at the skull base, in order to trigger magnetically localised tPA spheres selectively at the target clot. Ultrasound has been proven to increase the release rate and polymer degradation rate of polymer drug delivery systems[62] as ultrasound energy leads to heat-induced rupture of the particle matrix, liberating the encapsulated drug[62]. In addition, ultrasound also accelerates tPA diffusion into the clot as the kinetic energy increases clot porosity and leads to build-up of cavitations within the clot resulting in deeper penetration and increased tPA lysis rate[63–65]. Furthermore, as ultrasound simultaneously measures the existence of blood flow, this approach could also provide a real-time monitoring technique for vascular reopening allowing concurrent clot visualisation, lysis enhancement and detection of clot lysis[66].

14.3   SYNTHESIS AND PHYSICOCHEMICAL CHARACTERISATION OF NON-MEDICATED MAGNETIC SPHERES

One of the main challenges is to synthesise spheres with magnetic compounds that have hydrophilic surface properties and to attain high magnetite loading within the hydrophobic biodegradable polymers. One method of encapsulating hydrophilic magnetite is to prepare a double emulsion where an aqueous suspension of the magnetite is emulsified into the organic solvent containing the polymer. This is then introduced again into a water solution to harden the spheres. However, a considerable portion of the magnetite inevitably partitions from the primary emulsion into the second water solution resulting in poor encapsulation of the magnetic material. Resulting from a double emulsion, the spheres are composed of both hydrophilic and lipophilic compartments. If the magnetic material is contained within the hydrophilic compartment, then there is little space left to accommodate a high loading of hydrophilic drug like tPA. Compared with this water-in-oil-in-water double emulsion solvent evaporation method, a single emulsion solvent evaporation protocol was reported to obtain biodegradable and biocompatible magnetic microspheres that are 1–2 μm in mean diameter and contain a high concentration of magnetic material by encapsulating a hydrophobic magnetite material[67].

14.3.1 Synthesis of Hydrophobic Oleic Acid–Coated Magnetite Gel

Hydrophobic oleic acid–coated magnetite gel was prepared by alkaline precipitation of iron (II, III) salts in the presence of oleic acid[68]. A simplified method was developed to efficiently prepare the highly concentrated hydrophobic magnetite gel mainly by adjusting the amount of ammonium hydroxide and oleic acid and the time of oleic acid addition. The resultant magnetite nanoparticles in the magnetite gel were in a range of 5–15 nm in diameter (Figure 14.5) and exhibited superparamagnetic characteristics at room temperature with a saturation magnetisation of about 46 emu/g[69], about one-half the saturation magnetisation of pure magnetite (90 emu/g). The key to the successful preparation of such hydrophobic magnetite gel is to add an appropriate amount of ammonium hydroxide and oleic acid, so that the final solution remains neutral and the magnetite gel precipitates spontaneously. Herein, the oleic acid, as a reactant, is added immediately after the formation of magnetite crystal, simultaneously with the crystal growth. We demonstrated that the oleic acid efficiently coated the iron oxide crystals at the growth stage, creating highly concentrated hydrophobic magnetite gel.

14.3.2 Synthesis of Copolymers and Magnetic Spheres

Poly(lactic acid-ethylene glycol) (PLA–PEG) diblock copolymers were synthesised according to ring-opening polymerisation[67]. Purified L-lactide was combined with PEG and stannous octoate, as solutions in dry toluene. The polymerisation was then carried out under moisture-free high-purity argon atmosphere at 110°C for 2 h.

For magnetic sphere synthesis, we introduced hydrophobic magnetite gel into a solution of PLGA and methoxy PEG (5,000 Da)–PLA (40,000 Da) in dichloromethane. In addition, a single-control batch (pure PLGA/PLA–PEG spheres) was prepared by the same procedure except the magnetite

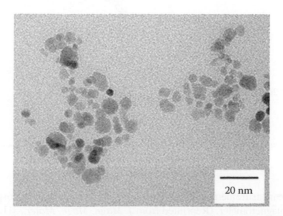

20 nm

FIGURE 14.5 TEM image of synthesised magnetite gel. (Reprinted from *J. Magn. Magn. Mater.*, 306, Liu, X., Kaminski, M.D., Guan, Y., Chen, H., Liu, H., and Rosengart, A.J., Preparation and characterization of hydrophobic superparamagnetic magnetite gel, 248–253, Copyright 2006, with permission from Elsevier B.V.)

was omitted. The synthesised magnetic microspheres were spherical in geometry (Figure 14.6a). The surface was generally smooth, although some roughness could be identified in certain areas of some spheres. Magnetite (dark domains shown in Figure 14.6b) appeared to be encapsulated heterogeneously within the microspheres. The experimental saturation magnetisation of oleic acid magnetite nanocrystals was 60–65 emu/g, which is less than that of the uncoated bulk magnetite (90 emu/g). This reduction is likely due to the existence of disordered spins at the surface of nanosized magnetic particles[69]. The magnetisations of microspheres at 10 kOe (1 T) were 17 and 27 emu/g for various synthesis samples with zero residual magnetisation at zero applied fields (Figure 14.7). The oleic acid

(a)   (b)

FIGURE 14.6 SEM (a: 28% magnetite loading) and TEM (b: 45% magnetite loading) images of synthesised magnetic PLGA/PLA–PEG composite microspheres. (Reprinted from *J. Magn. Magn. Mater.*, 311, Liu, X., Kaminski, M.D., Riffle, J.S., Chen, H., Finck, M.R., Torno, M., Taylor, L., and Rosengart, A.J., Preparation and characterization of PEGylated biodegradable magnetic nanocomposite carriers by single emulsion-solvent evaporation, 84–87, Copyright 2006, with permission from Elsevier B.V.)

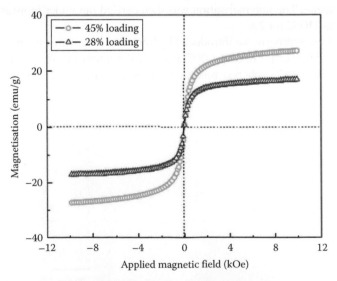

FIGURE 14.7 Magnetic measurement of synthesised magnetic composite microspheres measured at 300 K using a vibrating sample magnetometer. (Reprinted from *J. Magn. Magn. Mater.*, 311, Liu, X., Kaminski, M.D., Riffle, J.S., Chen, H., Finck, M.R., Torno, M., Taylor, L., and Rosengart, A.J., Preparation and characterization of PEGylated biodegradable magnetic nanocomposite carriers by single emulsion-solvent evaporation, 84–87, Copyright 2006, with permission from Elsevier B.V.)

magnetite nanocrystals had a magnetisation of 60 emu/g at 10 kOe[68]. Therefore, the calculated magnetite loading was about 28% and 45% by mass, close to our aim of 50% magnetite loading based on theoretical models in support of the proposed system[56,70,71].

14.4 DRUG LOADING OF MAGNETIC SPHERES

14.4.1 PREPARATION AND CHARACTERISATION OF MEDICATED MAGNETIC SPHERES

Using the FDA-approved polymer PLA–PEG, we report here on the successful co-encapsulation of tPA with an osmolyte (1 M trehalose or 0.5 M trimethylamine N-oxide-TMAO) and oleic acid magnetite[72]. The spheres of our initial synthesis series had an unusual appearance (Figure 14.8a). Regardless of the presence or type of osmolyte, which was added to the concentrated tPA solution in order to increase the stability of the enzyme, the carrier surface was heavily dimpled with numerous craters. This condition is likely the characteristic of an osmotic imbalance between the inner and outer microsphere volumes. The mean carrier size was about 3 μm. The larger carriers were odd shaped (Figure 14.8b) with an average diameter of 53.5 μm. However, the zeta potential of the carriers was 0 to −5.0 mV from pH 5.5–9.5, indicating good PEG coverage.

Further, there were observed differences in magnetisation between the magnetic carriers containing trehalose and those without trehalose (the carriers with TMAO were not tested) as shown in Figure 14.9. The magnetisation at 10 kOe was about 9.9 emu/g for the magnetic carriers containing trehalose and about 7.4 emu/g for those without trehalose.

The tPA encapsulation efficiency was 59%–93% with loadings from 3.3 to 9.4 wt% with significant improvement in tPA encapsulation efficiency with TMAO osmolyte compared with trehalose.

As mentioned, the magnetic spheres prepared from the procedure described previously did not have well-controlled surface morphology, though the tPA encapsulation efficiency was acceptable. Hence, we developed a modified encapsulation procedure using earlier described oleic acid–based magnetite and avoiding vortexing during the synthesis procedure. Due to the hydrophilicity of the surface and the relative small nanosize of individual magnetite crystals, our next-generation magnetic spheres had smoother surfaces and more controlled size (Figure 14.10).

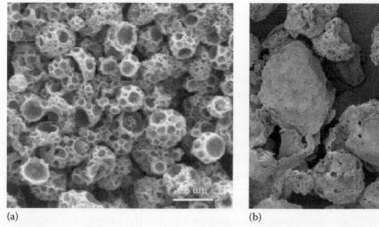

(a)   (b)

FIGURE 14.8   (a) Typical view of the smaller carriers containing tPA, scale bar: 2.5 μm and (b) large magnetic microcarriers containing tPA, scale bar: 50 μm. (Reprinted from *Eur. J. Pharm. Sci.*, 35, Kaminski, M.D., Xie, Y., Mertz, C.J. et al., Encapsulation and release of plasminogen activator from biodegradable magnetic microcarriers, 96–103, Copyright 2008, with permission from Elsevier B.V.)

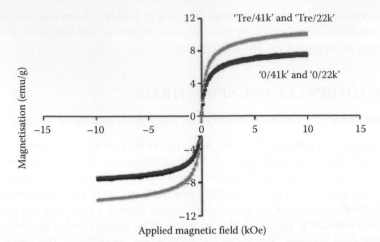

FIGURE 14.9 Magnetisation of various magnetic carrier formulations. The samples included the spheres with trehalose (Tre/41 K and Tre/22 K) and those without trehalose (0/41 K and 0/22 K). (Reprinted from *Eur. J. Pharm. Sci.*, 35, Kaminski, M.D., Xie, Y., Mertz, C.J. et al., Encapsulation and release of plasminogen activator from biodegradable magnetic microcarriers, 96–103, Copyright 2008, with permission from Elsevier B.V.)

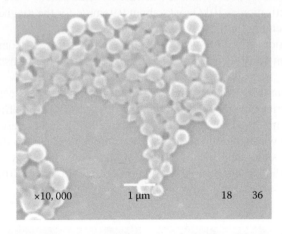

FIGURE 14.10 SEM of tPA-loaded magnetic spheres from the improved preparation procedure.

14.4.2 tPA Release from Magnetic Spheres

Enhanced tPA delivery at the desired target location is achieved by ultrasound insonation of the drug carriers, which concomitantly enhances the penetration of the tPA and plasma into the thrombus. Both must enter the clot as the plasma provides the plasminogen needed for tPA action and clot lysis[73]. We used theoretical modelling to demonstrate that thrombolytic tPA concentrations can be achieved in humans using the developed nanospheres even when adjusted for currently existing bioengineering constraints and applying strict clinical boundary conditions[74].

The tPA-loaded magnetic microspheres used for *in vitro* drug release experiments performed at body temperature and while employing ultrasound trigger stimulus (i.e. 20 kHz sonication probe with a 4 mm tip size at 50 W for pulses of 10 s each). An ultrasound pulse was given on the carrier suspension at the 25, 40 or 55 min time point, depending on whether the sample was randomised to receive one, two or three pulses. After triggering, the tPA magnetic carriers were separated from the solution, and the supernatant was sampled to determine the tPA concentration (quantified by human tPA total antigen assay). We compared the short-term and long-term release of tPA under natural release and after ultrasound insonation.

Under natural conditions, tPA eluted to a concentration of 58–74 µg/mL in 20–360 min from the microcarriers without the stabiliser trehalose, which corresponded to 13%–16% of its tPA content. Most of this tPA was released within 20 min ($p > 0.15$). In contrast, the concentration of tPA released from the microcarriers with trehalose was significantly lower (by about a factor of 2).

After ultrasound insonation, the release of active enzyme from the sample was depressed compared to controls (no insonation; Figure 14.11). We assume that the reduction in tPA release was a result of misfolding or partial denaturation of the enzyme after exposure to ultrasound energy. As an enzyme, tPA is sensitive not only to changes in temperature, pressure, acidity, light, etc.[75], but also to sonication[76]. In order to confirm this, we quantified free tPA activity by the human tPA activity assay. As reasoned before, we found that longer ultrasound exposure times significantly decreased the tPA activity (Figure 14.12). The primary cause seems to be the temperature rise within the small-volume suspension of tPA-carriers during the sonication experiments, and we reason that

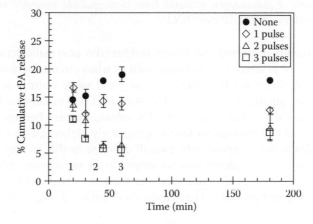

FIGURE 14.11 Effect of insonation (1–3 pulses) on active tPA release. Samples received no ultrasound ('none'), 1 ultrasound pulse at 25 min, 2 ultrasound pulses at 25 and 40 min, or 3 ultrasound pulses at 25, 40 and 55 min. (Reprinted from *Eur. J. Pharm. Sci.*, 35, Kaminski, M.D., Xie, Y., Mertz, C.J. et al., Encapsulation and release of plasminogen activator from biodegradable magnetic microcarriers, 96–103, Copyright 2008, with permission from Elsevier B.V.)

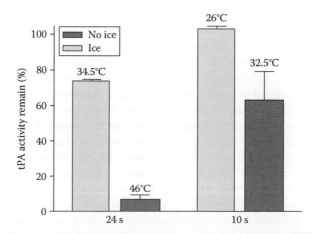

FIGURE 14.12 tPA activity following 10 and 24 s insonation for free tPA in solution[65]. Values are the average of single batches sampled in duplicate. (Reprinted from *Eur. J. Pharm. Sci.*, 35, Kaminski, M.D., Xie, Y., Mertz, C.J. et al., Encapsulation and release of plasminogen activator from biodegradable magnetic microcarriers, 96–103, Copyright 2008, with permission from Elsevier B.V.)

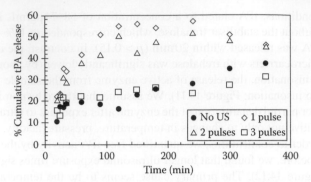

FIGURE 14.13 Effect of insonation on active tPA from larger drug carriers. Values are the average of single batches sampled in duplicate. (Reprinted from *Eur. J. Pharm. Sci.*, 35, Kaminski, M.D., Xie, Y., Mertz, C.J. et al., Encapsulation and release of plasminogen activator from biodegradable magnetic microcarriers, 96–103, Copyright 2008, with permission from Elsevier B.V.)

temperature is more readily dispersed and, hence, residual tPA enzymatic activity more preserved when spheres are exposed to larger fluid volumes such as when circulating within the vasculature. Studies utilising varied ultrasound frequency, intensity and mode settings to optimise triggered release and to reduce enzymatic injury are needed. As a next step to improve tPA release activity, we increased the carrier size. Interestingly, the tPA release kinetics became distinctly different (Figure 14.13) as one and two insonations led to significantly higher release of tPA than that with no insonation. In addition, we observed only a small decrease in tPA activity, suggesting that the tPA entrapped in the larger microspheres was not subject to the same denaturing effects as what we observed in Figure 14.11[72]. We speculate that this improvement is the direct result of a different, less harmful temperature distribution within larger individual carriers during insonation.

14.5 *IN VITRO* AND *IN VIVO* THROMBOLYSIS EXPERIMENTS

Successful intravascular blood clot lysis induced by freely circulating tPA depends on sufficient local tPA concentrations at and within the clot site for activation of plasminogen to plasmin, ultimately achieving clot lysis. An *in vitro* clot lysis study was executed to further investigate this approach[77].

The clot lysis experiments were performed using a static model as well as a dynamic flow model at room temperature. In the static lysis model, a single blood clot was placed at the bottom of a tube, which allowed the clot to occupy the complete inner circumference of the tube, and a small permanent magnet was placed underneath. The various treatments were placed on the top of the clot to remain in contact primarily with the clot surface. The dynamic flow model[77] was built to mimic the physiological conditions related to a single occluded artery in humans. Individual blood clots were placed at a reduction section within a Tygon® lab tubing where the inner diameter decreased from 3.8 to 3.1 mm. A flow rate of 2.1 mL/s was used to simulate the blood volume rate found in arteries. Different treatments such as tPA, magnetic spheres, magnetic field and ultrasound were introduced into the flow system as proposed at time zero. Red blood clots and ^{125}I-radiolabelled red blood clots were produced using a previously described procedure[77]. The reperfusion time for dissolving the clots in the different treatment groups, as well as the radioactivity loss of the clot before and after the treatments, was measured to determine the lysis efficiency.

The combination of tPA with magnetic field and microspheres gave a considerable increase in tPA clot penetration, predominantly along the passage trajectories of the magnetic spheres. Lysis front and clot fragmentation increased with increasing exposure time as seen on the sections prepared immediately after as well as 1 and 2 min post-treatment with tPA and magnetically driven magnetic spheres. Visual inspection of the sections showed the tPA concentration remarkably increased near the deposited magnetic spheres, along the clot length and following the magnetic field gradients. The greatest lysis efficiency improvements in the dynamic lysis studies were noted for the combined treatment with tPA, magnetic spheres, magnetic field and ultrasound (87%). The time period for first reperfusion decreased from 87 s for the use of magnetic spheres with magnetic field, 21 s when tPA was added and 11 s for the combined use of tPA, magnetic spheres, magnetic field and ultrasound. Therefore, the addition of ultrasound, magnetic field and tPA resulted in both greatest lysis efficiency and shortest reperfusion time among the treatment groups, which corresponded to an approximate twofold increase in lysis efficiency and sevenfold decrease in reperfusion time compared to the control group with magnetic spheres alone.

14.6   *IN VIVO* TARGETED THROMBOLYTIC THERAPY

So far, there are few published studies that have successfully demonstrated *in vivo* thrombolysis[78–80]. Ma et al. developed an *in vivo* model using male Sprague Dawley (SD) rats[78–80]. Here, the right iliac artery of the rats was cannulated with a catheter consisting of a needle and a three-way connector. The catheter tip reached the bifurcation of aorta and iliac arteries. A piece of whole blood clot prepared from the blood of another SD rat 24 h before the experiment was injected into the three-way connector and flushed into the left iliac artery. The clot was reproducibly lodged in the left iliac artery before branching into the pubic epigastric and femoral arteries. The exact site of clot lodging was determined by the colour of the artery. Bolus injection of recombinant tPA (rtPA) to the right iliac artery was preformed with a syringe inserted into the three-way connector connected with tubing for continuous infusion of saline. Two types of rtPA magnetic nanoparticles were tested in this rat embolic model. In the first study, rtPA was immobilised to polyacrylic acid-coated magnetite nanoparticles (PAA-MNP) with a hydrodynamic diameter of 246 ± 11 nm. 1-ethyl-3-(3-dimethyl-aminopropyl) carbodiimide hydrochloride (EDC) and *N*-hydroxysulphosuccinimide (NHS) were used as the cross-linking agents to promote formation of amide bonds between activated carboxyl group of PAA-MNP and amino groups of rtPA. A whole blood clot was introduced to almost completely block the left iliac blood flow in the SD rats. PAA-MNP–(rtPA) was administered into the right iliac artery *via* a catheter. A magnet was attached to the clotting site in the left iliac artery. As early as 25 min after injection, aortic blood flow, iliac blood flow and hind limb perfusion reversed to significantly higher levels than those in the free rtPA (0.2 mg/kg) group (p 0.05)[78,79]. The results demonstrated thrombolysis in response to PAA-MNP–(rtPA) under direct magnetic guidance in the embolic rat model.

In another study, chitosan-coated magnetite nanoparticles were used as the carriers for rtPA[80]. Administration of a dosage of chitosan-MNP–(rtPA) with activity equivalent to 0.2 mg/kg of free rtPA plus magnetic guidance significantly increased tissue perfusion from 29% ± 4% to 77% ± 15% of the basal levels before clot introduction[80]. In contrast, vehicle administration did not improve the hind limb perfusion in the control group ($n = 8$). As restriction in blood supply caused by introduced blood clot will lead to damages of blood vessels, subsequent reperfusion of blood after rtPA induced thrombolysis may therefore provide a full recovery of the haemodynamics. Nonetheless, the recovery of 70%–80% blood flow with magnetically guided chitosan-MNP–(rtPA) at a dosage of 0.2 mg/kg rtPA represents 20% the dose (1 mg/kg) required for thrombolysis by free rtPA[78].

14.7   CONCLUSION AND OUTLOOK

Novel methods are in early development for non-invasive yet targeted clot lysis aiming at improving the treatment for acute vascular occlusion such as seen in stroke patients. The technology is based on magnetically guidable, plasminogen activator encapsulated magnetic spheres that are systemically injected, circulated throughout the bloodstream and magnetically trapped at the vascular occlusion site. Triggered release of the thrombolytic can be achieved over natural release utilising focused ultrasound. However, the technology is still in its infancy as improvements are needed in each of the technology areas including improved loading of thrombolytic, better control of the size distribution and magnetisation of magnetic spheres, improved methods of releasing drug on-demand either using external or internal triggers and novel approaches to magnetic targeting either through magnetic implants or dynamic magnetic arrays. Yet, we envision that improvements in such technologies will reduce the inherent risk of current thrombolysis techniques and improve the thrombolysis rate, both of which would ultimately lead to reduced morbidity and mortality from acute vascular syndromes.

REFERENCES

1. Culp W. C., Porter T. R., Lowery J., Xie F., Roberson P. K., Marky L. 2004. Intracranial clot lysis with intravenous microbubbles and transcranial ultrasound in swine. *Stroke*, 35, 2407–2411.
2. Molina C. A., Ribo M., Rubiera M., Montaner J., Santamarina E., Delgado-Mederos R., Arenillas J. F., Huertas R., Purroy F., Delgado P., Alvarez-Sabín J. 2006. Microbubble administration accelerates clot lysis during continuous 2-MHz ultrasound monitoring in stroke patients treated with intravenous tissue plasminogen activator. *Stroke*, 37, 425–429.
3. Orekhova N. M., Akchurin R. S., Belyaev A. A., Smirnov M. D., Ragimov S. E., Orekhov A. N. 1990. Local prevention of thrombosis in animal arteries by means of magnetic targeting of aspirin-loaded red cells. *Thromb Res*, 57, 611–616.
4. Devineni D., Klein-Szanto A., Gallo J. M. 1995. Tissue distribution of methotrexate following administration as a solution and as a magnetic microsphere conjugate in rats bearing brain tumors. *J Neurooncol*, 24, 143–152.
5. Athanasiou K. A., Niederauer G. G., Agrawal C. M. 1996. Sterilization, toxicity, biocompatibility and clinical applications of polylactic acid/polyglycolic acid copolymers. *Biomaterials*, 17, 93–102.
6. Kobayashi H., Shiraki K., Ikada Y. 1992. Toxicity test of biodegradable polymers by implantation in rabbit cornea. *J Biomed Mater Res*, 26, 1463–1476.
7. Robert P., Mauduit J., Frank R. M., Vert M. 1993. Biocompatibility and resorbability of a polylactic acid membrane for periodontal guided tissue regeneration. *Biomaterials*, 14, 353–358.
8. Santavirta S., Konttinen Y. T., Saito T., Grönblad M., Partio E., Kemppinen P., Rokkanen P. 1990. Immune response to polyglycolic acid implants. *J Bone Joint Surg Br*, 72, 597–600.
9. Matsusue Y., Yamamuro T., Oka M., Shikinami Y., Hyon S. H., Ikada Y. 1992. In vitro and in vivo studies on bioabsorbable ultra-high-strength poly(L-lactide) rods. *J Biomed Mater Res*, 26, 1553–1567.
10. Cui C., Schwendeman S. P. 2001. Surface entrapment of polysine in biodegradable poly(DL-lactide-*co*-glycolide) microparticles. *Macromolecules*, 34, 8426–8433.
11. Mosqueira V. C., Legrand P., Pinto-Alphandary H., Puisieux F., Barratt G. 2000. Poly(D,L-lactide) nanocapsules prepared by a solvent displacement process: influence of the composition on physicochemical and structural properties. *J Pharm Sci*, 89, 614–626.
12. Mosqueira V. C., Legrand P., Morgat J. L., Vert M., Mysiakine E., Gref R., Devissaguet J. P., Barratt G. 2001. Biodistribution of long-circulating PEG-grafted nanocapsules in mice: effects of PEG chain length and density. *Pharm Res*, 18, 1411–1419.
13. Dunn S. E., Coombes A. G. A., Garnett M. C., Davis S. S., Davies M. C., Illum L. 1997. In vitro cell interactions and in vivo biodistribution of poly(lactide-*co*-glycolide) nanospheres surface modified by poloxamer and poloxamine copolymers. *J Control Release*, 44, 65–76.
14. Gref R., Minamitake Y., Peracchia M. T., Trubetskoy V., Torchilin V., Langer R. 1994. Biodegradable long-circulating polymeric nanospheres. *Science*, 263, 1600–1603.
15. Gref R., Domb A., Quellec P., Blunk T., Müller R. H., Verbavatz J. M., Langer R. 1995. The controlled intravenous delivery of drugs using PEG-coated sterically stabilized nanospheres. *Adv Drug Deliv Rev*, 16, 215–233.
16. Morita T., Sakamura Y., Horikiri Y., Suzuki T., Yoshino H. 2000. Protein encapsulation into biodegradable microspheres by a novel S/O/W emulsion method using poly(ethylene glycol) as a protein micronization adjuvant. *J Control Release*, 69, 435–444.
17. O'Hagan D. T., Rahman D., McGee J. P., Jeffery H., Davies M. C., Williams P., Davis S. S., Challacombe S. J. 1991. Biodegradable microparticles as controlled release antigen delivery systems. *Immunology*, 73, 239–242.
18. Li X., Zhang Y., Yan R., Jia W., Yuan M., Deng X., Huang Z. 2000. Influence of process parameters on the protein stability encapsulated in poly-DL-lactide-poly(ethylene glycol) microspheres. *J Control Release*, 68, 41–52.

19. Liu Y., Deng X. 2002. Influences of preparation conditions on particle size and DNA-loading efficiency for poly(DL-lactic acid-polyethylene glycol) microspheres entrapping free DNA. *J Control Release*, 83, 147–155.
20. Arshady R., Monshipouri M. 1999. Targeted delivery of microparticulate carriers. In: R. Arshady (Ed.), *Microspheres Microcapsules and Liposomes*. Citrus Books, London.
21. Harris J. M. 1992. *Poly(Ethylene Glycol) Chemistry: Biotechnical and Biomedical Applications*. New York: Plenum Press.
22. Yoshioka H. 1991. Surface modification of haemoglobin-containing liposomes with polyethylene glycol prevents liposome aggregation in blood plasma. *Biomaterials*, 12, 861–864.
23. Bhadra D., Bhadra S., Jain P., Jain N. K. 2002. Pegnology: a review of PEG-ylated systems. *Pharmazie*, 57, 5–29.
24. Zalipsky S. 1995. Chemistry of polyethylene-glycol conjugates with biologically-active molecules. *Adv Drug Deliv Rev*, 16, 157–182.
25. Caliceti P., Schiavon O., Veronese F. M., Chaiken I. M. 1990. Effects of monomethoxypoly(ethylene glycol) modification of ribonuclease on antibody recognition, substrate accessibility and conformational stability. *J Mol Recognit*, 3, 89–93.
26. Woodle M. C., Lasic D. D. 1992. Sterically stabilized liposomes. *Biochim Biophys Acta*, 1113, 171–199.
27. Matsumoto J., Nakada Y., Sakurai K., Nakamura T., Takahashi Y. 1999. Preparation of nanoparticles consisted of poly(L-lactide)–poly(ethylene glycol)–poly(L-lactide) and their evaluation in vitro. *Int J Pharm*, 185, 93–101.
28. Huang Z. B., Tang F. Q., Zhang L. 2005. Morphology control and texture of Fe_3O_4 nanoparticle-coated polystyrene microspheres by ethylene glycol in forced hydrolysis reaction. *Thin Solid Films*, 471, 105–112.
29. Bruce I. J., Taylor J., Todd M., Davies M. J., Borioni E., Sangregorio C., Sen T. 2004. Synthesis, characterisation and application of silica-magnetite nanocomposites. *J Magn Magn Mater*, 284, 145–160.
30. Yamaura M., Camilo R. L., Felinto M. C. F. C. 2002. Synthesis and performance of organic-coated magnetite particles. *J Alloy Compd*, 344, 52–156.
31. Sudakar C., Kutty T. R. N. 2004. Structural and magnetic characteristics of cobalt ferrite-coated nanofibrous γ-Fe_2O_3. *J Magn Magn Mater*, 279, 363–374.
32. Itoh H., Sugimoto T. 2003. Systematic control of size, shape, structure, and magnetic properties of uniform magnetite and maghemite particles. *J Colloid Interface Sci*, 265, 283–295.
33. Gilbert I., Millan A., Palacio F., Falqui A., Snoeck E., Serin V. 2003. Magnetic properties of maghemite nanoparticles in a polyvinylpyridine matrix. *Polyhedron*, 22, 2457–2461.
34. Naik R., Kroll E., Rodak D., Tsoi G. M., Mccullen E., Wenger L. E., Suryanarayanan R., Naik V. M., Vaishnava P. P., Tao Q., Boolchand P. 2004. Magnetic properties of iron-oxide and (iron, cobalt)-oxide nanoparticles synthesized in polystyrene resin matrix. *J Magn Magn Mater*, 272–276 (Suppl 1), E1239–E1241.
35. Kelberg E. A., Grigoriev S. Y., Okorokov A. I., Eckerlebe H., Grigorieva N. A., Eliseev A. A., Lukashin A. V., Vertegel A. A., Napolskii K. S. 2004. Magnetic properties of iron nanoparticles in mesoporous silica. *Physica B: Condens Matter*, 15, E305–E308.
36. Psarras G. C., Manolakaki E., Tsangaris G. M. 2003. Dielectric dispersion and AC conductivity in iron particles-loaded polymer composites. *Compos Pt A—Appl Sci Manuf*, 34, 1187–1198.
37. Capek I. 2004. Preparation of metal nanoparticles in water-in-oil (w/o) microemulsions. *Adv Colloid Interface Sci*, 110, 49–74.
38. Park I. W., Yoon M., Kim Y. M., Kim Y., Kim J. H., Kim S., Volkov V. 2004. Synthesis of cobalt nanoparticles in polymeric membrane and their magnetic anisotropy. *J Magn Magn Mater*, 272–276, 1413–1414.
39. Okon E., Pouliquen D., Okon P., Kovaleva Z. V., Stepanova T. P., Lavit S. G., Kudryavtsev B. N., Jallet P. 1994. Biodegradation of magnetite dextran nanoparticles in the rat. A histologic and biophysical study. *Lab Invest*, 71, 895–903.
40. Pouliquen D., Le Jeune J. J., Perdrisot R., Ermias A., Jallet P. 1991. Iron oxide nanoparticles for use as an MRI contrast agent: pharmacokinetics and metabolism. *Magn Reson Imaging*, 9, 275–283.
41. Slack J. D., Kanke M., Simmons G. H., DeLuca P. P. 1981. Acute hemodynamic effects and blood pool kinetics of polystyrene microspheres following intravenous administration. *J Pharm Sci*, 70, 660–664.
42. Yoshioka T., Hashida M., Muranishi S., Sezaki H. 1981. Specific delivery of Mitomycin-C to the liver, spleen and lung: nanospherical and microspherical carriers of gelatin. *Int J Pharm*, 8, 131–141.

43. Zhang Y., Zhuo R. X. 2005. Synthesis and drug release behavior of poly(trimethylene carbonate)–poly(ethylene glycol)–poly(trimethylene carbonate) nanoparticles. *Biomaterials*, 26, 2089–2094.

44. Bilati U., Allemann E., Doelker E. 2005. Strategic approaches for overcoming peptide and protein instability within biodegradable nano- and microparticles. *Eur J Pharm Biopharm*, 59, 375–388.

45. Teixeira M., Alonso M. J., Pinto M. M., Barbosa C. M. 2005. Development and characterization of PLGA nanospheres and nanocapsules containing xanthone and 3-methoxyxanthone. *Eur J Pharm Biopharm*, 59, 491–500.

46. Mingxing L., Jing D., Yajiang Y., Xiangliang Y., Huibi X. 2005. Characterization and release of triptolide-loaded poly(D,L-lactic acid) nanoparticles. *Eur Polym J*, 41, 375–382.

47. Frinking P. J., Bouakaz A., de Jong N., Ten Cate F. J., Keating S. 1998. Effect of ultrasound on the release of micro-encapsulated drugs. *Ultrasonics*, 36, 709–712.

48. Marin A., Muniruzzaman M., Rapoport N. 2001. Acoustic activation of drug delivery from polymeric micelles: effect of pulsed ultrasound. *J Control Release*, 71, 239–249.

49. Husseini G. A., Myrup G. D., Pitt W. G., Christensen D. A., Rapoport N. Y. 2000. Factors affecting acoustically triggered release of drugs from polymeric micelles. *J Control Release*, 69, 43–52.

50. Lavon I., Kost J. 1998. Mass transport enhancement by ultrasound in non-degradable polymeric controlled release systems. *J Control Release*, 54, 1–7.

51. Klibanov A. L., Maruyama K., Torchilin V. P., Huang L. 1990. Amphipathic polyethyleneglycols effectively prolong the circulation time of liposomes. *FEBS Lett*, 268, 235–237.

52. Torchilin V. P., Klibanov A. L., Huang L., O'Donnell S., Nossiff N. D., Khaw B. A. 1992. Targeted accumulation of polyethylene glycol-coated immunoliposomes in infarcted rabbit myocardium. *FASEB J*, 6, 2716–2719.

53. Unezaki S., Maruyama K., Ishida O., Suginaka A., Hosoda J., Iwatsuru M. 1995. Enhanced tumor targeting and improved antitumor activity of doxorubicin by long-circulating liposomes containing amphipathic poly(ethylene glycol). *Int J Pharm*, 126, 41–48.

54. Li Y., Pei Y., Zhang X., Gu Z., Zhou Z., Yuan W., Zhou J., Zhu J., Gao X. 2001. PEGylated PLGA nanoparticles as protein carriers: synthesis, preparation and biodistribution in rats. *J Control Release*, 71, 203–211.

55. Qiang Y., Anthony J., Marino M. G., Pendyala S. 2004. Synthesis of core-shell nanoclusters with high magnetic moment for biomedical applications. *IEEE Trans Magn*, 40, 3538–3540.

56. Aviles M. O., Ebner A. D., Chen H., Kaminski M. D., Rosengart A. J., Ritter J. A. 2005. Theoretical analysis of transdermal ferromagnetic implants for retention of magnetic drug carrier particles. *J Magn Magn Mater*, 293, 605–615.

57. Ovadia H., Paterson P. Y., Hale J. R. 1983. Magnetic microspheres as drug carriers: factors influencing localization at different anatomical sites in rats. *Isr J Med Sci*, 19, 631–637.

58. Goodwin S., Peterson C., Hoh C., Bittner C. 1999. Targeting and retention of magnetic targeted carriers (MTCs) enhancing intra-arterial chemotherapy. *J Magn Magn Mater*, 194, 132–139.

59. Chen H., Kaminski M. D., Pytel P., Macdonald L., Rosengart A. J. 2008. Capture of magnetic carriers within large arteries using external magnetic fields. *J Drug Target*, 16, 262–268.

60. Shapiro B. 2009. Towards dynamic control of magnetic fields to focus magnetic carriers to targets deep inside the body. *J Magn Magn Mater*, 321, 1594.

61. Lin S. Y., Chen K. S., Teng H. H., Li M. J. 2000. In vitro degradation and dissolution behaviours of microspheres prepared by three low molecular weight polyesters. *J Microencapsul*, 17, 577–586.

62. Kost J., Leong K., Langer R. 1989. Ultrasound-enhanced polymer degradation and release of incorporated substances. *Proc Natl Acad Sci USA*, 86, 7663–7666.

63. Francis C. W., Blinc A., Lee S., Cox C. 1995. Ultrasound accelerates transport of recombinant tissue plasminogen activator into clots. *Ultrasound Med Biol*, 21, 419–424.

64. Akiyama M., Ishibashi T., Yamada T., Furuhata H. 1998. Low-frequency ultrasound penetrates the cranium and enhances thrombolysis in vitro. *Neurosurgery*, 43, 828–832; discussion 832–833.

65. Pfaffenberger S., Devcic-Kuhar B., El-Rabadi K., Gröschl M., Speidl W. S., Weiss T. W., Huber K., Benes E., Maurer G., Wojta J., Gottsauner-Wolf M. 2003. 2 MHz ultrasound enhances t-PA-mediated thrombolysis: comparison of continuous versus pulsed ultrasound and standing versus travelling acoustic waves. *Thromb Haemost*, 89, 583–589.

66. Devcic-Kuhar B., Pfaffenberger S., Gherardini L., Mayer C. H., Groschl M., Kaun C. H., Benes E., Tschachler E., Huber K., Maurer G., Wojta J., Gottsauner-Wolf M. 2003. Ultrasound changes distribution and localization of fibrinolytic enzymes within blood clots during thrombolytic treatment. In: *Third International Symposium on Therapeutic Ultrasound*, Inserm Unite' 556, Lyon, France, pp. 101–106.

67. Liu X., Kaminski M. D., Riffle J. S., Chen H., Finck M. R., Torno M., Taylor L., Rosengart A. J. 2006. Preparation and characterization of PEGylated biodegradable magnetic nanocomposite carriers by single emulsion–solvent evaporation. *J Magn Magn Mater*, 311, 84–87.
68. Liu X., Kaminski M. D., Guan Y., Chen H., Liu H., Rosengart A. J. 2006. Preparation and characterization of hydrophobic superparamagnetic magnetite gel. *J Magn Magn Mater*, 306, 248–253.
69. Shimizu K., Ito A., Lee J. K., Yoshida T., Miwa K., Ishiguro H., Numaguchi Y., Murohara T., Kodama I., Honda H. 2007. Construction of multi-layered cardiomyocyte sheets using magnetite nanoparticles and magnetic force. *Biotechnol Bioeng*, 96, 803–809.
70. Chen H., Ebner A. D., Rosengart A. J., Kaminski M. D., Ritter J. A. 2004. Analysis of magnetic drug carrier particle capture by a magnetizable intravascular stent: 1. Parametric study with single wire correlation. *J Magn Magn Mater*, 284, 181–194.
71. Chen H., Ebner A. D., Rosengart A. J., Kaminski M. D., Ritter J. A. 2005. Analysis of magnetic drug carrier particle capture by a magnetizable intravascular stent: 2. Parametric study with multiple-wire two-dimensional model. *J Magn Magn Mater*, 293, 616–632.
72. Kaminski M. D., Xie Y., Mertz C. J., Finck M. R., Chen H., Rosengart A. J. 2008. Encapsulation and release of plasminogen activator from biodegradable magnetic microcarriers. *Eur J Pharm Sci*, 35, 96–103.
73. Rosengart A. J., Chen H., Xie Y., Kaminski M. D. 2005. Magnetically guided plasminogen activator loaded designer spheres for acute stroke lysis. *Med Hypotheses Res*, 2, 413–424.
74. Xie Y., Kaminski M. D., Guy S. G., Rosengart A. J. 2005. Plasminogen activator loaded magnetic carriers for stroke therapy: a mass balance feasibility evaluation. *J Biomed Nanotechnol*, 1, 410–415(6).
75. Wang W. 1999. Instability, stabilization, and formulation of liquid protein pharmaceuticals. *Int J Pharm*, 185, 129–188.
76. Özbeka B., Ülgenb K. Ö. 2000. The stability of enzymes after sonication. *Process Biochem*, 35, 1037–1043.
77. Torno M. D., Kaminski M. D., Xie Y., Meyers R. E., Mertz C. J., Liu X., O'Brien W. D. Jr., Rosengart A. J. 2008. Improvement of in vitro thrombolysis employing magnetically-guided microspheres. *Thromb Res*, 121, 799–811.
78. Ma Y. H., Wu S. Y., Wu T., Chang Y. J., Hua M. Y., Chen J. P. 2009. Magnetically targeted thrombolysis with recombinant tissue plasminogen activator bound to polyacrylic acid-coated nanoparticles. *Biomaterials*, 30, 3343–3351.
79. Ma Y. H., Hsu Y., Chang Y. J., Hua M. Y., Chen J. P., Wu T. 2007. Intra-arterial application of magnetic nanoparticles for targeted thrombolytic therapy: a rat embolic model. *J Magn Magn Mater*, 311, 342–346.
80. Chen J. P., Yang P. C., Ma Y. H., Wu T. 2011. Characterization of chitosan magnetic nanoparticles for in situ delivery of tissue plasminogen activator. *Carbohydr Polymer*, 84, 364–372.

15 Blood Cells as Carriers for Magnetically Targeted Delivery of Drugs

Nadine Sternberg, Kristin Andreas,
Hans Bäumler, and Radostina Georgieva

CONTENTS

In recent years, cell-based new therapeutic strategies have been intensively explored for tissue regeneration and targeted drug delivery. In this context, cells loaded with magnetic nanoparticles (MNPs) gain increasing interest as MNP migration can easily be followed *in vivo* by magnetic resonance imaging (MRI). Moreover, an external magnetic field enables the direction and concentration of cells loaded with MNPs at the desired sites for imaging, tissue repair, drug delivery or cancer therapy by hyperthermia. Several cell types have been used to test this strategy. This chapter focuses on loading blood cells with MNPs additionally providing an overview on general strategies for entrapment of drugs and nanoparticles in cells, examples of promising results and applications.

15.1   BLOOD CELLS

Blood is a natural transport and delivery system in the body. It is also the most important instrument for defence against infections, for clearance of toxic compounds and for tissue repair in case of injury. All these functions are managed by the blood cells (Figure 15.1), supported by unique compounds of the blood plasma[1].

The most abundant blood cells are the erythrocytes, also called red blood cells (RBCs). Due to the fact that these cells are accessible in large quantities, they have been already described in the seventeenth century[2]. RBCs and their membranes have been intensively studied and preferably used as a model system to investigate the transport processes across biological membranes for more than 80 years[3–5]. As a consequence, the structure, physiology and membrane properties of RBCs are well known. They are anucleated, biconcave, disk-like cells containing predominantly haemoglobin that accounts approximately to 90% of their dry mass. Their main function is the transport of oxygen and partly of carbon dioxide between the lungs and the other tissues. The lack of nucleus allows more space for haemoglobin, and the biconcave shape raises the surface/volume ratio, ensuring large binding capacity and optimal conditions for efficient diffusion of oxygen. RBCs live about 120 days, completing one whole cycle through the circulatory system of the body every 20 s. RBCs are also studied as potential drug carriers, since last 40 years[6–8].

The leucocytes, also known as white blood cells (WBC), are infection-fighting cells and thus essential components of the immune defence. They are classified in two main groups: granular and agranular. The agranulocytes are lymphocytes or monocytes. Lymphocytes are mainly responsible for production of antibodies and cytokines, whereas monocytes have multiple responsibilities. Their main functions are phagocytosis, antigen presentation and cytokine production. In response to inflammation signals, monocytes can move to sites of infection in the tissues, where they divide or differentiate into macrophages and dendritic cells to elicit an immune response. Granulocytes are classified and named according to their staining properties: neutrophils, basophils and eosinophils. Neutrophils make up 60%–70% of total leucocytes count. They defend against bacterial or fungal infection and inflammatory processes by very active phagocytosis. Eosinophils primarily fight parasitic infections. Furthermore, they are the predominant inflammatory cells in allergic reactions. Basophils are involved in allergic response by releasing histamine[1].

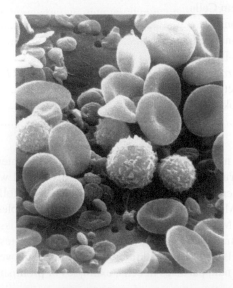

FIGURE 15.1   Scanning electron microscopy (SEM) image of peripheral blood. Erythrocytes, leucocytes and small disc-shaped platelets are displayed. (From Bruce Wetzel and Hary Schaefer (photographers), National Cancer Institute, Bethesda, MD, http://visualsonline.cancer.gov/details.cfm?imageid = 2129, public domain.)

Thrombocytes, more commonly named platelets, are the smallest blood cells with a diameter of 2–3 µm. Strictly speaking, platelets are not real cells. They are rather cell fragments derived from precursor megakaryocytes in the bone marrow (BM). The fundamental role of platelets is the initiation of blood clot formation in haemostasis. In addition, they are a natural source of growth factors[1]. Upon activation at sites of injury, platelets release a multiplicity of growth factors that play a significant role in tissue repair and regeneration[9,10].

The replenishment of blood cells occurs predominantly in the BM due to differentiation of hae-matopoietic stem and progenitor cells (HSCs) into any blood cell lineages. Almost 50 years ago, HSCs were first recognised in the BM[11]; mesenchymal stem and progenitor cells (MSCs) were described shortly after[12–14]. HSCs have been shown to leave the BM and to circulate in a very small number in the bloodstream[15–18]. Both HSCs and differentiated blood cells circulate within the BM and the periphery and play an important role in the homeostasis of the haematopoietic system. Actually, HSCs reside in the BM stroma and are only released to the circulation upon need in the peripheral tissues or in response to an intravenous administration of a stimulus. Under ischaemic conditions, for example, injured tissues secrete stroma cell-derived factor-1alpha (SDF-1α/CXCL12) to reverse the chemokine gradient between BM and the periphery to therefore release the stem cells from the BM to the circulation[19]. Subsequently, the circulating stem cells migrate to the injured tis-sue to give rise to differentiated cells and to secrete trophic factors for tissue regeneration. In this regard, multipotent adult stem cells, such as HSCs and MSCs, have the primary function to main-tain the steady-state functioning of a tissue—the tissue homeostasis—as reservoir of reparative cells that mobilise from the BM to the periphery to regenerate damaged tissue. The release of HSCs into circulation can be further stimulated by potent mobilising agents, such as chemotherapy, granu-locyte colony-stimulating factor (G-CSF) and haematopoietic growth factors[20]. This prompted the use of mobilised HSCs as an ideal cell source for both allogeneic and autologous transplantations. Nowadays, haematopoietic stem cell transplantation has emerged to a successful clinical procedure performed for patients with cancer and other disorders of the blood system.

BM is also the main source for MSC isolation. Although representing only 0.001%–0.01% of the total amount of nucleated cells in the marrow, these cells can easily be isolated and expanded with high efficiency[21]. Caplan and Owen were the first to introduce the concept of a mesenchymal stem cell to generate tissues such as bone, cartilage, fat, tendon and muscle[21–25]. Furthermore, MSCs help to provide the supportive connective tissue microenvironment for HSC proliferation and differen-tiation in the BM. When MSCs were delivered intravenously, these cells were capable of migrating along the chemokine gradient to the BM or to the site of injury such as bone fractures, infarcted myocardial and ischaemic nerve tissues, to promote tissue regeneration[26–28]. The ease of isolation from the BM, the low immune response of stem cells, the high proliferative capacity, the potential to migrate towards the site of injury after intravenous injection and the potential to differentiate into various tissues (such as bone[29,30], cartilage[31–33] and tendon[25,34,35]) make MSCs an attractive therapeu-tic tool capable of playing a role in a wide range of clinical applications[36–38].

Regarding these attributes and functions of the different cells in the blood, one can realise their immense potential as carriers. Since these carriers are actual endogenous cells, they will produce little or no antigenic response. In particular, loading nanoparticles like MNPs into blood cells could circumvent several side effects and disadvantages like rapid clearance from the circulation by the liver, spleen or kidneys as well as toxicity due to production of reactive oxygen species[39,40].

15.2   STRATEGIES FOR LOADING CELLS

From a biophysical point of view, cells are open thermodynamic systems able to exchange energy and materials with the surrounding. The cell membrane is organised in a way that allows highly selective material exchange through different mechanisms summarised as membrane transport. Small molecules and ions can permeate through the lipid bilayer or specialised membrane channels and pumps; large molecules and particles can be internalised by more complex processes mediated by membrane vesicles like endo-, phago- or pinocytosis. The pathways relevant for internalisation of nanoparticles have been shown to not only depend on the cell type (professional phagocytes vs. other cell types), but also on the physicochemical properties of the nanoparticles (size, geometry, surface charge and chemical composition, aggregation etc.) and on the parameters of the actual microenvironment[41,42].

In addition to these natural cellular mechanisms, the barrier function of the membrane can be overcome by inducing short living pores in the lipid bilayer. For example, opening of such pores occurs in hypotonic solutions due to swelling of the cells and mechanical stretching of the membrane. Application of strong electric fields for a short time also induces reversible breakdown and permeabilisation of the cell membrane[43]. Other strategies apply chemicals and molecules of biological origin, which are able to incorporate in the lipid bilayer or to form complexes with membrane proteins leading to formation of channels or pores[44].

These mechanisms altogether represent the fundamental for loading cells with particles and in particular with MNPs.

15.2.1 Vesicle-Mediated Loading

15.2.1.1 Endocytosis

Endocytosis is a basic process used by cells to internalise a large variety of materials, including molecules and particles of different sizes (Figure 15.2)[41].

Large particles such as bacteria can be taken up by phagocytosis—an endocytic mechanism mediated by cup-like membrane extensions, pseudopodia. Phagocytosis of particles is initiated by specific membrane receptors (e.g. scavenger receptors), leading to the formation of an early phagosome[45]. Pinocytosis by contrast is a type of endocytosis in which small particles and external fluids are unspecifically internalised by invagination of vesicles from the cell membrane. In brief, the cell membrane forms pockets, which pinch off into the cell to form small vesicles filled with extracellular fluid including particles. These vesicles fuse and form endosomes. Both processes—phago- and pinocytosis—appear to be triggered by and are dependent on actin-mediated remodelling of the plasma membrane at a large scale.

Clathrin-mediated endocytosis is performed in specific membrane regions, clathrin-coated pits. Formation of a clathrin-coated vesicle is followed by its uncoating and clathrin monomer recycling.

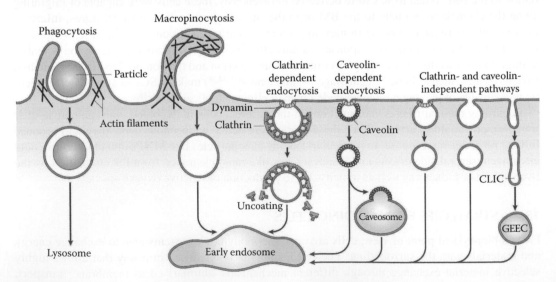

FIGURE 15.2 Pathways of entry into cells. (From Macmillan Publishers Ltd. *Nat. Rev. Mol. Cell Biol.*, Mayor, S. and Pagano, R. E., Pathways of clathrin-independent endocytosis, 8, 603–612, Copyright 2011.)

Subsequently, particles are processed by early endosomes, multivesicular bodies and late endosomes. This pathway has been confirmed mainly for particles with diameters below 200 nm. Particles with sizes up to 500 nm are predominantly taken up by another known mechanism of endocytosis *via* so-called lipid rafts, leading to the formation of caveosomes[46]. Endocytosis processes without involvement of clathrin or caveolin are referred to as non-clathrin non-caveolar-mediated endocytosis[41,42,47,48].

15.2.1.2 Drug-Induced Endocytosis

Endocytosis as a tool for loading cells with drugs or particles can be provoked in non-phagocytotic cells by exposure to some membrane-binding compounds[49,50] as a complex phenomenon occurring in several steps[51]. In RBCs, an amphipathic cationic drug induces the conversion of the discocytotic cells into stomatocytes. Thereafter, invagination of the membrane is followed by separation of an endocytotic vacuole from the bulk membrane[52]. Three distinct classes of drugs, primaquine, vinblastine and chlorpromazine, have been shown to be capable of inducing endocytosis in RBC[53,54].

15.2.1.3 Vesicle Fusion

Fusion of vesicles with cell membranes is another key process that facilitates the transport of materials between and within cells. Liposomes as carrier for the delivery of drugs have been studied extensively. A variety of fusion proteins, peptides and drugs have the potentials to promote controlled fusion of carrier liposomes with the desired cell type[55]. RBCs normally do not undergo endovesiculation. However, the fusion of small unilamellar liposomes with isolated erythrocyte membranes (ghosts) has been shown[56]. Moreover, by this method inositol monophosphate was entrapped into intact human RBCs to improve the oxygen carrying capacity of the cells[57]. In another approach, RBCs were also loaded with liposomes composed of vectamidine (a double-chain cationic amphiphile)[58].

15.2.2 CHEMICAL MEMBRANE PERMEABILISATION

The loading of cells with normally not permeating molecules can be mediated by using chemical compounds that are capable of causing specific or unspecific permeability increase[44]. Some antibiotics belong to this class of membrane active substances of which the antifungal and antibiotic polyene amphotericin B (AmB) is the most investigated. It binds to cholesterol in the eukaryotic plasma membrane to enhance the uptake of ions and non-permeating molecules from the surrounding medium. Applying low concentrations of AmB entrapment of daunomycin or halothane in RBCs has been shown[59-62].

Another approach, the so-called osmotic pulse method, is based on colloid osmotic haemolysis of RBCs. Highly membrane permeable compounds like dimethyl sulphoxide (DMSO), polyethylene glycol (PEG)[63], urea[64] or ammonium chloride[65] are added to the surrounding medium and diffuse into the cell due to a concentration gradient. The large transient gradient across the cell membrane is followed by influx of water to maintain the osmotic equilibrium leading to swelling and haemolysis. In this manner, DMSO was used to facilitate the incorporation of inositol hexaphosphate[66,67] and insulin[68] into RBCs. However, this procedure is associated by major changes in the membrane structure and significantly decreases the survival of the loaded cells *in vivo*.

15.2.3 REVERSIBLE MEMBRANE BREAKDOWN

15.2.3.1 Electroporation

Electroporation, also called electropermeabilisation, is a phenomenon that is observed upon application of high voltage to cells. The increase in transmembrane potential leads to the formation of nanoscale pores in the cell membrane (Figure 15.3) allowing molecules to be transported into and out of the cells[69]. Pore formation can be reversible or irreversible, depending on electric

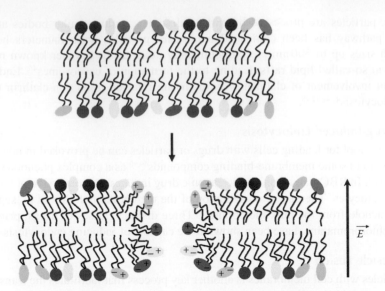

FIGURE 15.3 Scheme of electroporation: Top—membrane lipid domain in the absence of electric field; Bottom—the same lipid domain with a pore, stabilised by lipid head group dipole interactions with an external electric field.

pulse parameters and cell characteristics. By varying the pulse duration, the intensity of the electric field and the ionic strength of the surrounding solution characteristic pore diameters can be created[70]. The mechanisms behind electroporation have yet to be elucidated. There is a commonly accepted concept about pore creation: the pores form on a timescale of nanoseconds overcoming an energy barrier of approximately 45–50 kT. Electroporation techniques are generally divided into two groups: bulk or batch electroporation and single-cell electroporation[71].

Electroporation was first applied to load substances into cells by Zimmermann and co-workers in the early 1970s. Using an electric shock, they irreversibly impaired the RBC membrane[72]. A few years later, Kinosita and Tsong[73] exposed RBC suspensions in isotonic media to electric pulses, resulting in a dielectric breakdown of the cell membrane followed by transient electrolysis. By using this method, sucrose and methotrexate that were present in the surrounding medium were already successfully entrapped into RBCs[74–76]. Even encapsulation of latex beads with a diameter of 0.2 μm was successful[43]. Jain et al.[77] described an encapsulation efficiency of approximately 35%. Carrier RBCs generated by electroencapsulation have been shown to survive in the circulation *in vivo* as long as normal cells[43,74].

15.2.3.2 Hypotonic Swelling and Lysis

Perturbation of the osmotic equilibrium between cell and surrounding medium results in a change of cell volume caused by the high water permeability of the plasma membrane. Maintenance of a constant cell volume in the face of osmotic stress is an evolutionarily ancient homeostatic process[78]. To avoid the potential deleterious consequences of cell shrinkage or swelling, most eukaryotic cells have developed compensatory mechanisms that involve the activation of ion channels and/or transporters in the plasma membrane[79].

Mammalian erythrocytes are highly specialised cells with specialised physiology lacking mechanisms of volume control. In these cells, the physical permeability for cations is very low[80,81]. As a consequence, RBCs swell until a critical volume is reached, which corresponds to the state of maximal membrane stretching. The process of hypotonic swelling is accompanied by a shape change from biconcave to spherical with a volume gain of approximately 25%–50%. Further input of water is giving rise to transient holes (pores) in the lipid bilayer. The size of these pores ranges from 10 up to 50 nm with a certain lifetime depending on the temperature, thus allowing permeation of large molecules and nanoparticles by simple diffusion driven by the concentration gradient[82]. These processes are usually reversible, and the pores can be resealed completely by raising the salt concentration to its physiological value and warming the RBC sample to 37°C. Thereafter, the cells reassume their normal biconcave shape, while the substance remains encapsulated[83–86]. These erythrocytes are usually called carrier RBCs[83–85,87,88]. Several procedures have been reported for encapsulation of different drugs, other bioactive agents and nanoparticles in RBCs based on hypotonic swelling and haemolysis, for example, hypotonic dilution, hypotonic pre-swelling and hypotonic dialysis.

The hypotonic dilution procedure represents the simplest and fastest encapsulation method. The compound to be loaded is dissolved in water or a buffer with low ionic strength and added to the RBCs in a volume ratio of 1: 2–20. The final osmotic pressure of the RBC suspension is set below the critical value of approximately 150 mOsm/kg allowing pore generation. After incubation, the RBCs are resealed by increasing the osmolarity and the temperature. The supernatant is discarded and the loaded RBCs are washed in a physiological solution. Successful encapsulation of enzymes like β-galactosidase, β-glucosidase[6], asparaginase[89,90] or arginase[91] by means of hypotonic dilution was reported. The main drawbacks of this procedure are the relatively low entrapment efficiency[6,92–95] and the excessive loss of haemoglobin and other constituents of the RBC, which causes immense membrane damage and thus reduces the *in vivo* circulation time of the carrier RBCs[43,96,97]. However, we could recently show that a careful handling of the cells during the whole procedure, a dilution factor of 50% and the usage of a nutritious reversal buffer has the potential to reduce the membrane damage[98].

A special loading device, the red cell loader, based on the hypotonic dilution principle, was developed by Magnani et al.[99] The red cell loader allows working under standardised blood banking conditions. Two haemolysis steps are followed by the concentration of the RBC suspension in a haemofilter reducing haemoglobin and energy depletion of the loaded RBCs.

The pre-swell procedure[100] is based on RBC swelling in a slightly hypotonic solution, followed by centrifugation and haemolysis of the pellet by addition of small amounts of an aqueous solution of the drug to be encapsulated. The membrane pores are closed by restoration of isotonicity adding a hypertonic buffer and resealing at 37°C. Asparaginase[101], methotrexate and insulin[68,93] have been encapsulated successfully by using the pre-swell procedure. Here, the carrier RBCs have been reported to exhibit a strong similarity to native cells with an analogue lifetime *in vivo*[83,87,93,102]. However, it should be considered that RBC damage may be stimulated by the additional centrifugation steps.

The hypotonic dialysis method is the most promising and gentle procedure for RBC loading. Here, isotonic RBC suspensions are placed in a conventional dialysis tube that is loaded in a hypotonic buffer reservoir. The compound that has to be encapsulated is added into the dialysis tube prior or during the hypotonic swelling[83,103,104]. Compounds with a molecular weight below the cut-off of the dialysis membrane can be added directly to the dialysis buffer[84,94,105]. Resealing is performed by either addition of concentrated potassium chloride solution to achieve isotonicity or by replacing the hypotonic buffer against iso-osmotic solution (reverse dialysis). Standard haemodialysis equipment allows continuous flow dialysis with high loading capacity and a cell recovery of 70%–80%[106]. The main disadvantages of this procedure are the need of special equipment and the relatively long preparation time[43,83,92,94].

15.2.4 SURFACE CONJUGATION

Conjugation of compounds to the surface of cells represents an alternative to the entrapment strategies described earlier. Initially, therapeutic substances have been attached to the RBC surface by non-specific cross-linkers like tannic acid[107], chromium chloride[108,109] or glutaraldehyde[110] in analogy to the surface modification of colloidal particles. Later, the controlled biotinylation of lysine residues using N-Hydroxysuccinimide (NHS) esters of biotin became most popular[99,111–115]. The biotin–avidin bridge was extensively used for modular anchoring of diverse biotinylated amino acids[116–118] sulfhydryl groups[119], sugars[120] and lipids[121,122] to defined functional groups on the RBC membrane. RBCs with up to 10^5 molecules of protein conjugated *via* avidin have been shown to circulate *in vivo* without enhanced clearance, haemolysis or organ uptake[123,124]. The activity of uricase bound to the RBC surface was found to be even higher than that of the same enzyme entrapped inside the cells[125].

Recently, gentle conjugation methods of therapeutic substances to cell membranes have been introduced[126]. The novel strategy is based on engineered fused molecular constructs consisting of a cell-specific antibody and a drug or pro-drug.

For example, a single-chain antibody (scFv Ter-119) binding to mouse glycophorin A was fused with a variant human single-chain low-molecular-weight urokinase construct that can be activated selectively by thrombin. The fused protein interacts specifically with mouse RBCs *in vivo* without altering their biocompatibility. In the presence of thrombin, it converted into an active two-chain molecule and caused thrombin-induced fibrinolysis[127].

Furthermore, artificial immune complexes (heteropolymers) were designed consisting of a monoclonal antibody against the RBC complement receptor type I (CRI) conjugated to another antibody against a known pathogen or toxin in blood. CRI binds C3b component of the activated complement system or immune complexes containing this protein. Macrophages of the reticuloendothelial system (RES) recognise these complexes from RBCs in circulating bloodstream without cell damage[128]. These heteropolymers then were injected *in vivo* into blood, causing immune clearance of the pathogens from the bloodstream by macrophages without RBC damage[129]. Binding to CRI does not target RBCs for phagocytosis, only the extracellular CRI domain with the pathological cargo is cleaved off and eliminated[129–132]. Heteropolymers of C3b linked to anti-toxin or anti-pathogen antibodies were created and used for the same purpose. This method was used for elimination of bacterial[133] or viral[134–137] pathogens. Elimination of cytokines as anti-inflammatory intervention[138] was possible applying the same strategy. Generally, there is a potentially wide biomedical application of this concept[130].

The conjugation of particles to the surface of RBCs was also shown recently. Particles with a diameter above 450 nm in diameter were adsorbed onto RBCs *via* adsorption mimicking *Bartonella* bacteria. These microorganisms adhere to the RBC surface and thus remain in circulation. Modified RBCs were not eliminated by the immune system[139]. In a different approach, nanoparticles have been conjugated to the surface of RBCs *via* peptide ligands designed with high affinity to specific regions of the RBC membrane[140].

15.3 LOADING BLOOD CELLS WITH MNPs

Cell-based therapies and targeted drug delivery are nowadays seen as the most promising strategies to improve treatment of cancer, degenerative diseases and infections. Due to their natural origin, cells have many advantages as drug carriers combining low immunogenicity, large loading capacity and simultaneous incorporation of different compounds for therapy, diagnostic and targeting. Additionally, their specific functions can be exploited to support treatment and tissue repair. In this context, tracking injected carrier cells *in vivo* to monitor distribution and engraftment into host tissues is essential for improvement of therapy and in clinical trials. Stimulated by the development of novel engineered MNPs and improvement of MRI sensitivity and resolution offering new opportunities for tracking labelled cells *in vivo*, the interest of loading different cell types with MNPs steadily increases. Superparamagnetic iron oxide particles appear currently to be the preferred material since they are biocompatible and provide a high signal contrast change on T_2- and T_2^*-weighted images. Today, iron oxide–based MNPs are synthesised with different size, shape, surface charge, coatings and functionalities. They are usually encapsulated by organic polymers like dextrans, proteins, lipids, styrenes and other polymers, which increase their stability and biocompatibility and allow the chemical modification of their surface for further functionalisation[141].

As mentioned earlier, blood cells have a great potential in these novel diagnostic and therapeutic strategies due to their unique functions and properties. Depending on the specific cell physiology of the different blood cells, appropriate particles and loading approach have to be considered.

15.3.1 RED BLOOD CELLS

RBCs are among the widely studied particulate delivery systems for biopharmaceuticals as they possess several properties that make them unique and useful carriers[88,142,143]. They are available in huge amount, their preparation and storage is uncomplicated, they are non-immunogenic, biodegradable with a large capacity to carry drugs and have long circulation half-time. Numerous *in vivo* studies already confirmed efficacy and safety of pharmaceutical agents loaded into RBCs[144].

Almost every imaginable loading strategy has been tested with erythrocytes, including hypotonic haemolysis procedures (Figure 15.4), which have been considered to be the most applicable methods for the encapsulation of substances into erythrocytes and in particular for loading them with MNPs.

Magnetic RBCs, resulting from the co-encapsulation of drugs with ferrofluids, have been introduced almost 30 years ago[43]. Since then several research groups have reported successful encapsulation of MNPs in animal and human RBCs with or without co-encapsulated drugs by application of different hypotonic haemolysis protocols[43,145–153]. Generally, the loading efficiency and the physiological status of the carrier RBCs appear to be strongly dependent on the loading conditions and on the surface coating of the MNPs.

Earlier studies applied mainly electroporation[43] and pre-swell procedures[72], dealing with uncoated or silicon oil–coated ferrofluids. These carrier RBCs have been reported to have properties similar to that of native RBCs.

More recently, citrate-coated MNPs with a mean diameter of 10–15 nm have been loaded into RBCs with high efficiency applying a hypotonic dilutional procedure[150]. The obtained carriers responded strongly to an externally applied magnetic field as shown in Figure 15.5C. Transmission electron microscopy (TEM) imaging of loaded and control RBCs confirmed that MNPs are located mainly inside the cells forming aggregates of different size being also strongly incorporated in the membrane (Figure 15.5A and B). Citrate-coated MNPs have the tendency to destabilise at physiological ionic strengths[155]. During loading, the low salt concentration of 20 mM benefits the fine dispersion and stability of the MNPs. Most probably, the aggregation occurs during resealing caused by the reconstitution of the physiological ionic strength. Furthermore, the carrier RBCs prepared following this protocol are haemoglobin depleted to a relatively high degree.

In contrast, RBCs loaded with commercially available dextran- and carboxydextran-coated MNPs by hypotonic dialysis appear with a natural biconcave shape and uniform non-aggregated distribution of the MNPs inside the cell and absence of significant clustering or accumulation at the erythrocyte membrane as demonstrated by TEM imaging (Figure 15.5, lower panel)[151,153]. Very recently, Antonelli et al.[154] evaluated 19 commercially available and newly developed MNPs with different hydrodynamic diameters and surface coatings concerning efficient loading in RBCs versus integrity of the cellular carriers. They concluded that careful engineering of size, shape and surface modification is required to produce MNPs suitable for loading into RBCs.

A very promising approach for surface modification and functionalisation of nanoparticles has been recently developed applying atom transfer radical polymerisation (ATRP). This provides immense flexibility to incorporate different functional segments into one polymer chain and to add functional groups on one end of the polymer anchored on a surface. This procedure was applied for coating MNPs with thermosensitive polymer brushing random copolymers of 2-(2-methoxyethoxy)

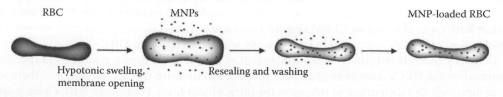

RBC   MNPs   MNP-loaded RBC

Hypotonic swelling,   Resealing and washing
membrane opening

FIGURE 15.4 Scheme of loading erythrocytes with MNP by hypotonic haemolysis.

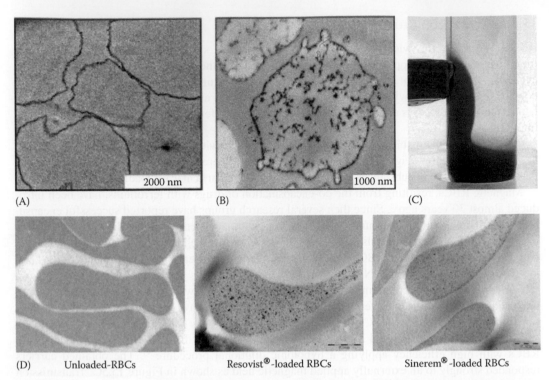

(A)   (B)   (C)

(D)   Unloaded-RBCs   Resovist®-loaded RBCs   Sinerem®-loaded RBCs

FIGURE 15.5 MNP-loaded RBCs. Upper panel: TEM images of (A) control RBCs; (B) RBCs loaded with citrate-coated MNPs by hypotonic dilution procedure; (C) photograph of RBCs loaded with citrate-coated superparamagnetic iron oxide nonoparticles (SPIONs) manipulated by a permanent magnet. (From Brahler, M., Georgieva, R., Buske, N., Muller, A., Muller, S., Pinkernelle, J., Teichgraber, U., Voigt, A., and Baumler, H., Magnetite-loaded carrier erythrocytes as contrast agents for magnetic resonance imaging, *Nano Lett.*, 6, 2505–2509, 2006. Copyright 2011 American Chemical Society.) (D) Lower panel: TEM images of unloaded and contrast agent loaded RBCs. (From Markov, D. E., Boeve, H., Gleich, B., Borgert, J., Antonelli, A., Sfara, C., and Magnani, M., Human erythrocytes as nanoparticle carriers for magnetic particle imaging, *Phys. Med. Biol.*, 55, 6461–6473, 2010, Copyright 2011 IOP Science.)

ethyl methacrylate (MEO$_2$MA) and oligo(ethylene glycol) methacrylate (OEGMA)[152]. This new class of thermosensitive polymers is of compelling interest for biomedical applications due to excellent biological compatibility arising from the oligo(ethylene glycol) side groups. Below the so-called lower critical solution temperature (LCST), these polymers are strongly hydrophilic and stabilise the MNPs in aqueous solutions. Above this temperature, the polymers perform conformational changes becoming hydrophobic and cause agglomeration of the MNPs. The copolymers are synthesised with different molar ratios of both components to allow controlling the temperature of particle agglomeration. In addition, they can be fluorescently labelled during synthesis. In this manner, agglomeration and disagglomeration of the MNPs can be tuned in a controlled and reversible manner by changing the temperature and visualised by means of confocal laser scanning microscopy (CLSM). It has been shown that these thermosensitive MNPs can be loaded simultaneously with fluorescein isothiocyanate (FITC)-albumin into RBCs *via* optimised hypotonic dilutional procedure (Figure 15.6A). The temperature-dependent reversible agglomeration of the MNPs inside the carrier RBCs was observed by CLSM (Figure 15.6B and C). This effect is potentially important for development of stimuli responsive *in vivo* imaging techniques applying MRI.

In a very recently published work, attachment of a viral spike fusion glycoprotein (haemagglutinin) on the RBCs' membrane has been proposed to improve specific targeting of therapeutic compounds and their efficient release at the intracellular level. Previously, RBCs were loaded with MNPs and a drug. These engineered RBCs have been shown to fuse with the cytoplasmic

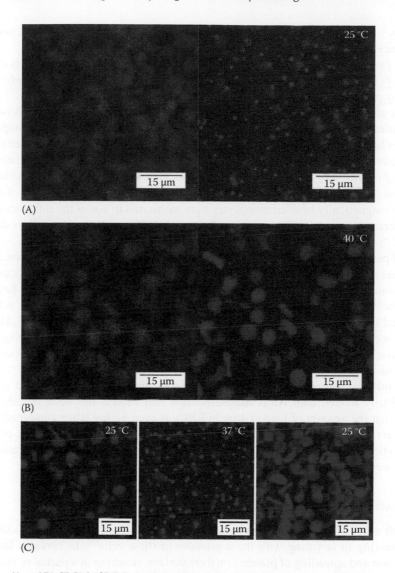

FIGURE 15.6 (A and B) CLSM of RBCs loaded with both Fe_3O_4@MEO_2MA_{90}-co-$OEGMA_{10}$-R nanoparticles (NPs) (LCST = 36°C in phosphate buffered saline (PBS)) and FITC-bovine serum albumin (FITC-BSA) labelling the inner volume of RBCs at 25°C (A) and 40°C (B). Note that FITC–BSA was used as the control to label RBCs. The left panel was recorded using the FITC channel and the right using the rhodamine one. (C) Confocal fluorescence microscopy images of RBCs loaded with both Fe_3O_4@MEO_2MA_{90}-co-$OEGMA_{10}$-R NPs and FITC-BSA when the environmental temperature first increased from 25°C to 37°C and then decreased back to 25°C. (From Chanana, M., Jahn, S., Georgieva, R., Lutz, J.F., Baumler, H., and Wang, D.Y., Fabrication of colloidal stable, thermosensitive, and biocompatible magnetite nanoparticles and study of their reversible agglomeration in aqueous milieu, *Chem. Mater.*, 21, 1906–1914, 2009. Copyright 2011 American Chemical Society.)

membrane of target HeLa human breast tumour cells in a very short time and to release MNPs and therapeutic compound inside these cells[156].

15.3.2 WHITE BLOOD CELLS (LEUCOCYTES)

Loading procedures of MNPs to different leucocyte populations are generally divided into unspecific phagocytosis and receptor-mediated endocytosis depending on the capability of the particular cell type to perform phagocytosis. A comprehensive review on uptake mechanisms of nanoparticles is provided by Unfried et al.[42] Existing MRI studies with MNP-loaded phagocytic and non-phagocytic leucocytes are extensively reviewed by Bulte and Kraitchman[141].

Although the uptake mechanism in macrophages, monocytes and granulocytes is clearly identified as phagocytosis that is independent on opsonisation of the MNPs, different studies also identified uptake by pinocytosis in case of non-opsonised particles[157].

Uptake of anionic maghemite nanoparticles by human leucocytes has been shown to depend on the differentiation state of the cells. The internalisation capacity of macrophages was more than 10 times higher than the one of monocytes[158]. An iron uptake in non-stimulated monocytes of up to 6 pg Fe/cell was measured without impairment of their capacity to differentiate into macrophages, giving opportunity for tracking labelled monocytes, recruitment at inflammation sites, up to differentiation into macrophages. Moreover, the morphology and gene expression of macrophages derived from magnetically labelled monocytes were monitored to assess the influence of the MNPs on the differentiation process. Thus, uptake efficacy of MNPs in phagocytic cells is defined by the properties of the particle and physiological characteristics of the cells.

Uptake of particles in non-phagocytic lymphocytes in contrast is dependent on specific receptor-mediated mechanisms of endocytosis. The Bulte group was the first to achieve specific magnetic labelling of human lymphocytes with anti-lymphocyte-directed monoclonal antibody-linked, dextran-coated MNPs[159]. Optimal labelling of activated rat myelin reactive T-cells was achieved using MNPs in combination with poly-L-lysine as an unspecific transfection agent (TA)[160]. However, size, charge and concentration of MNPs have been shown to strongly influence the loading efficiency of non-phagocytic T lymphocytes. A preferable size of approximately 100 nm was found for uptake without the use of cell-penetrating peptides and TAs[161].

Very recently, an interesting approach for *in vitro* loading of granulocytes and monocytes *via* RBCs, previously loaded with AmB nanosuspension and MNPs, has been introduced[162]. It has been shown that the amount of loaded drug in RBCs is even higher when combined loading with MNPs was carried out (unpublished own results). After a resting period of 24 h, the cells recovered their phagocytic activity and were able to release AmB to the culture medium successfully fighting *Candida albicans*.

15.3.3 PLATELETS (THROMBOCYTES)

Particles with a diameter up to 300 nm have been shown to enter the open canalicular system of platelets appearing in or fusing with the α granules of these cells[163]. In contrast, larger particles caused adhesion and spreading of platelets on their surface, resulting in a partial or complete covering of the particles by one or several cells; hence, the cell reaction is identical to their response to flat surfaces. Still, little is known about the mechanisms behind the interaction between small NPs and platelets.

Uptake of fluorescently labelled polymeric nanoparticles with a diameter from 50 to 150 nm by platelets has been studied *in vitro* in samples of whole blood by flow cytometry and fluorescence microscopy[164]. Smaller particles were internalised by platelets more effectively than larger particles. In addition, PEG surface modification prevented the uptake of the small particles. It has been suggested that strong interaction with the polymethacrylate corona is a prerequisite for effective internalisation of the polymeric nanoparticles by thrombocytes. Furthermore, a two-step mechanism of interaction between nanoparticles and platelets was proposed—a passive and reversible interaction between the particle surface and the thrombocyte surface and, subsequently, internalisation through an active process resulting in an intracellular granular localisation.

In our group, similar polymeric nanoparticles have been recently shown to increase only slightly the percentage of activated platelets in platelet-rich plasma. The aggregation kinetics of platelets loaded with nanoparticles was studied upon addition of collagen, adenosine diphosphate (ADP) and epinephrine as aggregation promoters. In general, the loading with nanoparticles did not influence platelet activation and aggregation. Nevertheless, larger particles without PEG coating were observed to cause a delay in the collagen-provoked aggregation (unpublished own results).

However, the potential of platelets as carriers for MNPs and their possible applications in imaging, image-guided therapies and drug targeting are not investigated until now.

15.3.4 CIRCULATING STEM CELLS

The intravenous injection of stem and progenitor cells is very promising for widely ranged therapies of diseased tissues, that is, bone fractures, myocardial infarction and ischaemic nerve tissue[26–28]. Crucial to the efficiency and safety of these cell transplantations is the ability of the injected stem and progenitor cells to migrate from the circulation to the injured tissue and to survive, differentiate and/or secrete factors that promote tissue regeneration. For this purpose, cells can be labelled with MNPs prior to infusion or transplantation. MRI can then be used to track repeatedly injected cells *in vivo* in a sensitive and non-invasive manner. MNPs with an iron oxide core unit, were preferred to gadolinium chelates for stem cell labelling. These MNPs provide a high signal contrast change on T_2- and T_2*-weighted images, are non-toxic and biodegradable as the iron is recycled by cells using natural biochemical pathways of iron metabolism and can be coated with functional groups and ligands for a more efficient and/or specific internalisation.

Among the different MNPs that are already commercialised for clinical use, surface coating with dextran (i.e. Endorem® from Guerbet and Feridex® from Berlex) and carboxydextran (Resovist® from Bayer HealthCare Pharmaceuticals) is most accomplished[165,166]. These MNPs with a hydrodynamic diameter between 50 and 200 nm were developed to achieve liver and spleen imaging, as they are easily phagocytosed by the Kupffer cells of the RES upon intravenous administration. Any type of cell that is capable of endocytosis can unspecifically internalise these MNPs. However, due to the lack of substantial phagocytic capacity, there is no efficient cellular uptake by stem and progenitor cells. The efficiency of MNP internalisation by stem and progenitor cells can be improved by TAs such as polycationic amines, dendrimers and lipid-based agents. These large, highly charged macromolecules are able to form complexes with MNPs adsorbing to the cell membrane *via* electrostatic interactions inducing membrane bending, which triggers endocytosis[167]. The complexation of commercial dextran-coated MNPs with TA has been reported to be an efficient and effective technique to magnetically label MSCs and HSCs[168,169]. However, unbound TAS are toxic to cells at high concentrations[170]. The ratio of TA and MNPs needs to be carefully titrated to provide a stable and non-toxic complex that can be efficiently internalised by the cells. Besides the use of TA, the magnetic labelling efficiency can furthermore be improved by linking the MNP surface to a translocation agent that facilitates unspecific MNP uptake. The probably most prominent membrane translocating signal peptide is HIV-1 Tat, which when coupled to dextran-coated MNPs, increases significantly the magnetic labelling efficiency of stem and progenitor cells[171]. However, stem cell loading with dextran-coated MNPs still remains a complex problem.

Almost all stem cells are non-phagocytotic and internalise MNPs by pinocytosis. In this regard, stem cells were magnetically labelled by simply adding MNPs to the culture medium. During cell culture, MNPs are taken up by the cells and accumulated in the endosomes. This type of particle uptake is quite inefficient for the internalisation of MNPs to yield sufficient MRI contrast of the labelled cells.

Another type of not yet commercialised MNPs has emerged in the last few years: dextran-free MNPs that are coated with anionic monomers or anionic chelating acids and thus, are relatively small, negatively charged and electrostatically stabilised in colloidal suspensions. Interestingly, the negative surface charge strikingly induces the internalisation into cells without the need of TA or MNP linking to additional peptides.

Figure 15.7A shows that the intracellular iron content is highly increased when MSCs were loaded with citrate-coated anionic MNPs compared to labelling with the same Fe concentration of commercial dextran-coated MNPs (Endorem) and carboxydextran-coated (Resovist) MNPs of similar size. High magnetic loading of MSCs can be visualised under the light microscope

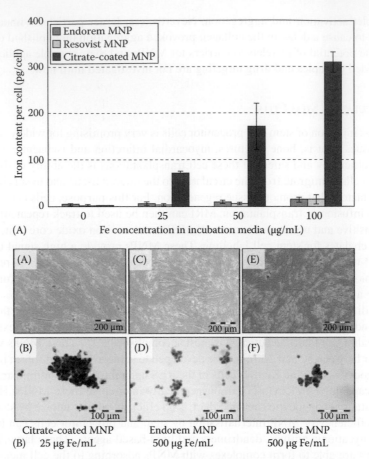

FIGURE 15.7 Uptake of MNPs in MSCs. (A) The intracellular iron content is highly increased when MSCs were loaded with citrate-coated anionic MNPs compared to labelling with dextran-coated (Endorem) and carboxydextran-coated (Resovist) commercial MNPs of similar size and concentration in the incubation medium, (B) Uptake of MNPs into MSC can be visualised in cell culture and after staining with Prussian Blue. Loading of MSC with Endorem and Resovist requires almost 20-fold MNP concentration to yield similar intracellular MNP accumulation to anionic MNPs.

giving a brownish colour in transmission mode and a blue staining with Prussian Blue reagent (Figure 15.7B). To achieve equivalent loading of MSCs, the commercial MNPs Endorem and Resovist must be employed with about 20-fold higher concentration in the incubation medium than the anionic MNPs.

Due to the highly negative surface charge, the anionic MNPs interact strongly with the cell plasma membrane and are efficiently captured through a non-specific adsorptive endocytosis pathway into endosomal compartments and are not internalised through the inefficient non-adsorptive pinocytosis[172,173]. Wilhelm and co-workers hypothesise that anionic MNPs repulsively interact with the large negatively charged domains of the cell membrane, non-specifically adsorbing to the membrane inducing a local neutralisation of the membrane with subsequent bending that seems to effectively promote particle endocytosis (see Chapter 13). Furthermore, the group of Safi revealed that MNPs stabilised with anionic chelating agents are effectively internalised due to the destabilisation of the MNP suspension in culture medium[155]. They propose that the initially fine dispersed MNPs destabilise in physiological solutions, form large aggregates, settle down to the bottom of the cell culture flask, accumulate and concentrate at the cell membrane that effectively promotes their internalisation. Altogether, these results suggest that anionically charged MNPs represent an effective alternative to commercial dextran- and carboxydextran-coated nanoparticles to label stem and progenitor cells magnetically.

For specific cell targeting, the MNP surface can be covalently linked with distinct high-affinity ligands that enter specific cells through receptor-mediated endocytosis[174]. Such receptor-mediated strategy involves, for example, the linkage of MNPs with antibodies that specifically bind to a distinct cell type and facilitate their internalisation. Recently, the specific labelling of CD133+ human stem and progenitor cells with antibody-linked MNPs has been reported[175]. However, the number of cell membrane receptors limits the internalisation efficiency.

It is known that high concentrations of free iron ions can cause cytotoxic effects[176]. Free iron has the capacity to exchange electrons interconverting between the ferric (Fe^{3+}) and the ferrous (Fe^{2+}) status and, thus, can produce free radicals inducing oxidative stress. Normally, only trace amounts of intracellular free iron can be detected due to protection mechanisms from iron-mediated oxidative damage. Therefore, high loading of stem cells with iron oxide MNPs may disturb the fine regulation of cellular iron homeostasis. MNPs may be degraded within the lysosomes into free iron ions that can potentially pass the nucleus and mitochondrial membrane and produce free radicals leading to DNA, protein and membrane damage. Thus, there is urgent need to identify any potential cellular perturbations to ensure that MSC still have the potential to differentiate after magnetic loading.

Until now, the potential cytotoxicity of MNPs on stem and progenitor cells remains an issue of debate. Several studies have been published so far demonstrating that MNPs with varying physicochemical characteristics possess no cytotoxicity at low iron concentrations but have cytotoxic effects at higher concentrations[177,178]. Thus, detrimental effects of stem cell loading with MNPs seem to be dose dependent. Labelling of human MSCs with MNPs does not affect the exposition of the typical cell surface antigens (CD44+, CD73+, CD105+, CD166+, CD14-, CD34- and CD45-), and magnetically labelled MSCs perform differentiation into the adipogenic and osteogenic lineage as efficiently as unlabelled cells (Figure 15.8A).

In contrast, high-intracellular MNP concentrations have been reported to interfere with the actin cytoskeleton and, thus, to hamper the spindle apparatus[179,180]. This causes detrimental effects on processes that involve the remodelling of the cytoskeleton, such as cell proliferation, chondrogenic differentiation and cell migration[178,180–182]. Hence, highly MNP-loaded MSCs show a decreased potential to differentiate into the chondrogenic lineage (Figure 15.8A) and to migrate along chemotactic gradients (Figure 15.8B). Contradictory, other studies have reported that magnetic labelling does not alter the differentiation and migration potential of MSCs[183,184]. Since the impact of MSC loading on MSC function is dose dependent, the respective studies may have employed cells that have not been sufficiently labelled with MNPs to observe the inhibitory effect on migration and chondrogenic differentiation. Furthermore, high levels of reactive oxygen species as evidence for oxidative stress are measured in MNP-loaded cells, although no acute cytotoxic effect is apparent. However, this may have long-term effects that have to be considered[180].

Although sufficient intracellular iron content is required for tracking MNP-labelled MSCs by MRI, very high MNP loading should be avoided to maintain full viability and stem cell function. The careful labelling of MSCs with low MNP concentrations can yield both sufficient MRI contrast for *in vivo* tracking and maintenance of cell viability and function.

15.4 BIOMEDICAL APPLICATIONS

15.4.1 Magnetic Resonance Imaging

15.4.1.1 MNP-Loaded RBCs as a Contrast Agent

A significant problem concerning biomedical applications of iron oxide–based contrast agents in MRI is associated with their relatively fast recognition and clearance from the blood circulation by the RES. Despite positive developments mainly related to particle surface modifications and coatings with biocompatible polymers, the circulation half-time of MNPs still remains in the order of

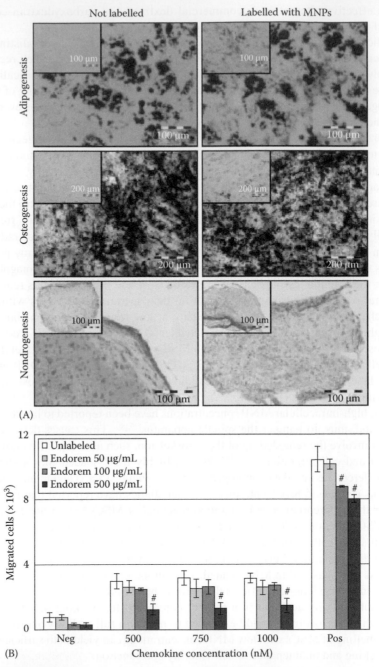

FIGURE 15.8 Differentiation and migration of MNP-loaded MSCs. (A) Compared with unlabelled MSCs (left) adipogenic and osteogenic differentiation is not affected by MNP labelling. Lipid accumulation is observed as a result of the adipogenic induction as shown by Oil Red O staining, and calcium nodules are formed after osteogenic induction as visualised by von-Kossa stain. Human MSCs failed to generate the typical chondrogenic extracellular matrix in a micromass culture shown by Alcian Blue staining of proteoglycans. Controls (labelled and unlabelled cells cultured in non-differentiation media) showed—as expected—no positive staining (small images). (B) Results (mean ± SD) of chemokine-induced migration of unlabelled MSCs and magnetically labelled MSCs. Increasing loading with dextran-coated MNPs (Endorem) resulted in a decreased potential of labelled MSCs to migrate towards 500, 750 and 1000 nM chemokine concentrations and towards the positive control (culture medium; #, $p < 0.05$).

some hours. RBCs loaded with MNPs are one of the most promising candidates to succeed as a long circulating iron oxide–based agent. The natural ability of RBCs to stay in circulation for up to 120 days is unique and exceeds by orders of magnitude than that of any other known carrier systems. They can be obtained easily in a large quantity and can be loaded with MNPs under sterile blood bank conditions (e.g. red cell loader, see 15.2.3.2). Moreover, the use of autologous RBCs will possibly provide the best biocompatibility at all.

The ability of RBCs loaded with citrate-coated MNPs to reduce relaxation times T_1 and T_2 was measured by means of relaxiometry (Figure 15.9D, upper panel)[150]. A suspension of MNP-loaded RBCs with a haematocrit of 25% induced a strong reduction of both T_1 and T_2 compared to the values measured in a suspension of native RBCs with the same haematocrit. In phantoms, MNP-loaded RBCs were added to whole blood at different concentrations and examined by MRI. The obtained magnetic resonance (MR) images are shown in Figure 15.9A and B. Even 2.5% of loaded cells added to native cells significantly enhance the contrast (Figure 15.9A). A concentration of 4% in a phantom with a diameter of 1 mm that simulates the conditions in small blood vessels produced clearly enhanced contrast (Figure 15.9B). Finally, it was possible to visualise individual loaded RBCs using a clinical 3T scanner (Figure 15.9C). The *in vivo* detection of single cells is a quality criterion for the loading procedure. This high *in vitro* resolution indicates that a relatively small number of loaded RBCs will be sufficient to contrast specific sites of interest *in vivo*.

Similar studies by other groups provided useful data that revealed a strong correlation between loading efficiency and changes in relaxation times and relaxation rates by RBCs loaded with different MNPs[151,153,154].

The behaviour of RBCs loaded with thermosensitive MNPs as a contrast-enhancing agent has to be mentioned. These MNPs have been shown to form aggregates above their LCST maintaining this ability when encapsulated in RBCs as well (Figure 15.6). Phantom tubes for MRI have been prepared by fixation of RBCs loaded with thermosensitive MNPs (LCST = 20°C) at temperatures below (4°C) and above (37°C) their LCST. The signal intensity in T_2 imaging (Figure 15.9A, lower panel) was considerably decreased by 44% for the sample maintained at 37°C (d) as compared to that obtained for the sample prepared at 4°C most likely due to the reduced inter-particle distances in agglomerated state. In addition, upon increase of surrounding temperature, agglomeration aids a stronger attraction of MNP-loaded RBCs to a magnet (Figure 15.9B, lower panel). Taken together, these observations demonstrate that magnetic retention and MRI contrast of RBCs loaded with thermosensitive MNPs can be controlled by changing the temperature[152].

Furthermore, *in vivo* MRI of thermosensitive particle was performed in Wistar Rats[185]. For imaging, the contrast enhancement in the liver, a T_2*-weighted 2D-gradient echo sequence fast low angle shot-(FLASH) in coronal view with respiration triggering was used. In this study, bare 5-methacryloylamino fluorescein (MAF)-labelled Fe_3O_4@MEO_2MA_{85}-*co*-$OEGMA_{15}$ (Fe_3O_4@MEO_2MA_{85}-*co*-$OEGMA_{15}$-MAF) and Fe_3O_4@MEO_2MA_{90}-*co*-$OEGMA_{10}$ MNPs and human RBCs loaded with Fe_3O_4@MEO_2MA_{90}-*co*-$OEGMA_{10}$ MNPs were investigated. Commercially available Resovist MNPs served as a control. The total amount of injected iron was the same in all four animals. After sample administration, the MRI scans of the liver were performed every 10 min for 1 h.

Figure 15.10A through D shows T_2*-weighted MRI images in coronal view of the liver before injection ($t = 0$) and after injection ($t = 10$ and 60 min) of the investigated MNPs. The mean intensity in the liver of all animals decreases with time (Figure 15.10, right). In the case of Resovist, the liver becomes immediately black after a few minutes (Figure 15.10D), suggesting clearly that the Resovist particles are immediately taken up by the RES. The normalised mean intensity here decreases immediately to the minimum value of 10% of the initial intensity at $t = 0$.

For both bare thermosensitive MNP samples the mean intensity in the liver decreases to roughly 60% of the initial intensity 60 min after injection. However, for the animal injected with human RBCs loaded with Fe_3O_4@MEO_2MA_{90}-*co*-$OEGMA_{10}$ MNPs, the mean liver intensity decreases only by 14% of the respective initial liver intensity. Consequently, these results confirm a significantly prolonged circulation time of thermosensitive MNPs loaded into RBCs compared to bare particles.

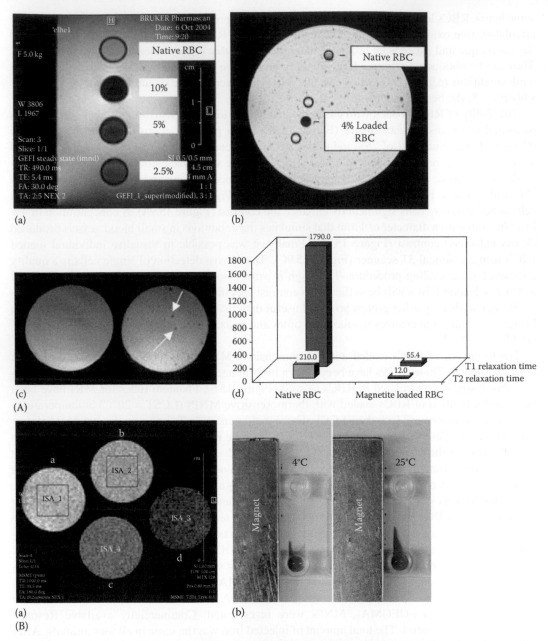

FIGURE 15.9 MRI of MNP-loaded RBCs. (A) (a) RBCs loaded with citrate-coated MNPs. Top down: whole blood and loaded RBCs in whole blood, 10, 5 and 2.5%; (b) MRI of native RBC and 4% RBCs loaded with citrate-coated MNPs; (c) single-cell MRI at 3.0 T of native RBCs (left) and RBCs loaded with citrate-coated MNPs (see arrows, right) (1000 cells/mL). (d) Relaxation times (T_1 and T_2) of native and RBCs loaded with citrate-coated MNPs. (From Brahler, M., Georgieva, R., Buske, N., Muller, A., Muller, S., Pinkernelle, J., Teichgraber, U., Voigt, A., and Baumler, H., Magnetite-loaded carrier erythrocytes as contrast agents for magnetic resonance imaging, *Nano Lett.*, 6, 2505–2509, 2006. Copyright 2011 American Chemical Society.) (B) (a) Phantom tubes containing 1×10^5 native RBCs (a and b) and RBCs loaded with Fe_3O_4@MEO_2MA thermosensitive MNPs (LCST 20°C) (C and D). RBCs in tubes a and c were fixed at 4°C and RBCs in tubes b and d at 37°C. (b) Photographs of using a magnet to manipulate RBCs loaded with the thermosensitive MNPs at 4°C (left) and 25°C (right). (From Chanana, M., Jahn, S., Georgieva, R., Lutz, J.F., Baumler, H. and Wang, D.Y., Fabrication of colloidal stable, thermosensitive, and biocompatible magnetite nanoparticles and study of their reversible agglomeration in aqueous milieu, *Chem. Mater.*, 21, 1906–1914, 2009. Copyright 2011 American Chemical Society.)

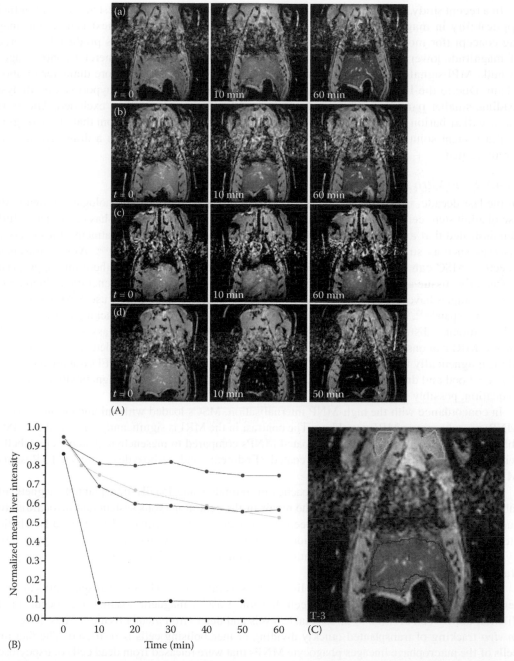

FIGURE 15.10 *In vivo* distribution of thermosensitive MNPs[185]. Left: T_2*-weighted 2D gradient-echo sequence (FLASH) in coronal view of four different Wistar Rats. The animals were first scanned at $t = 0$ for blank value. Then, the iron oxide NPs were injected into the tail vein of the rat and MRI scans of the liver were performed every 10 min for 1 h. (A) Fe_3O_4@MEO_2MA_{85}-*co*-$OEGMA_{15}$-MAF NPs (200 μL, $c = 3$ mg Fe/mL). (B) Fe_3O_4@ MEO_2MA_{90}-*co*-$OEGMA_{10}$ NPs (200 μL, $c = 3$ mg Fe/mL). (C) 2.4 mL 40 v%-suspension of human RBCs loaded with Fe_3O_4@MEO_2MA_{90}-*co*-$OEGMA_{10}$ NPs (NP, $c = 3$ mg Fe/mL). (D) Resovist (10 μL, $c = 28$ mg/mL). Right: Normalised mean liver intensity measured by MRI as a function of time. Fe_3O_4@MEO_2MA_{85}-*co*-$OEGMA_{15}$-MAF (red), Fe_3O_4@MEO_2MA_{90}-*co*-$OEGMA_{10}$ NPs (cyan) and FITC-labelled RBCs loaded with Fe_3O_4@ MEO_2MA_{90}-*co*-$OEGMA_{10}$ MNPs (blue) were injected into rats. As a control, Resovist (Bayer-Schering Pharma AG) was applied (black). Normalised mean intensity liver = signal-to-noise ratio (SNR) of the liver (red ROI, inset) divided by the SNR of the limb muscles (green region of interests (ROIs), inset).

In a recent study, RBCs loaded with Resovist and Sinerem® have been tested concerning their applicability in magnetic particle imaging (MPI)[186], which represents a next generation imaging concept (for more information, see Chapter 20)[153]. MNP-loaded RBCs produced one order of magnitude lower signal compared to the signal produced by the commercial contrast agent in bulk. MPI signals are mainly generated by particles with a magnetic core diameter of about 30 nm. Due to the limited size of the membrane pores generated during hypo-osmotic dialysis loading, smaller particles will penetrate faster and larger particles will be excluded. Therefore, the size distribution of the MNPs incorporated in RBCs probably differs from that of the original contrast agent solution. However, optimising the loading procedure, such a drawback may be circumvented.

15.4.1.2   *In Vitro* and *In Vivo* Tracking of MNP-Loaded MSCs

In the last decade, increasing advances were made in the field of cell technologies, allowing the use of adult stem cells for the regeneration of diseased tissues. In animals, it has been successfully demonstrated that *ex vivo*-expanded MSCs are a promising tool for the treatment of a variety of diseases such as stroke, spinal cord injury and myocardial infarction[26,187,188]. After intravenous injection, MSC can migrate to the injured tissue, engraft and differentiate there into appropriate regenerative tissue and/or secrete trophic molecules that promote tissue regeneration. Stem cell – based therapies have already entered the clinics or are underway, like for the use of MSCs for bone and joint repair[30,33]. To address clinical standards of cell-based therapies, tracking of transplanted cells to monitor distribution and engraftment into host tissues is essential to assess safety and efficiency. MRI can ensure very sensitive, repeatedly and non-invasive *in vivo* detection and tracking of the magnetically labelled cells. For this purpose, several commercial and FDA-approved MNPs (FDA = Food and drug administration, USA) can be used to label MSCs magnetically prior transplantation, possibly with the help of FDA-approved TAs.

In concordance with the high-MNP internalisation, MSCs loaded with anionic citrate – coated MNPs reveal a strong MRI signal loss. The contrast in the MRI is significantly enhanced for MSCs that are labelled with anionic citrate–coated MNPs compared to mesenchymal stem cells labelled with same concentrations of dextran-coated (Endorem) and carboxydextran-coated (Resovist) MNPs (Figure 15.11).

Most prominently, therapeutic approaches in animals using locally transplanted magnetically labelled stem and progenitor cells for the repair of the central nervous system and myocardial tissue have been described[189–191]. It has been shown that locally transplanted magnetically labelled stem cells can be tracked at the injection site in the brain, the spinal cord and the myocardium by MRI[192,193]. Other very promising studies applied magnetically labelled HSCs systemically and tracked cell homing to the BM[194].

However, tracking labelled MSCs is limited by several factors. The internalised MNPs are segregated into the daughter cells during cell division. Loss of magnetic loading does occur due to dilution during cell division to undetectable levels and due to iron metabolisation. Thus, long-term *in vivo* tracking of transplanted quickly dividing or metabolising cells is restricted. The fact that cells of the macrophage lineages phagocyte MNPs that were released from dead cells or exocytosed from cells after transplantation presents a further drawback. This causes false-positive signalling if labelled MSCs were injected locally into the target tissue. MNPs that were released into the blood after intravenous infusion of loaded cells are likely to be taken up by the RES, resulting in false-positive contrasts in the liver and spleen. Last but not least, *in vivo* detection of labelled cells does not discriminate live from dead cells. Here, a double labelling of transplanted cells — for example, a co-transfection with the *LacZ* gene encoding the β-galactosidase enzyme — to validate vitality and stem cell existence seems to be adequate. However, tracking of transplanted stem and progenitor cells not only in the animal but also during clinical studies is a very attractive tool to guide future stem cell–based therapeutic approaches.

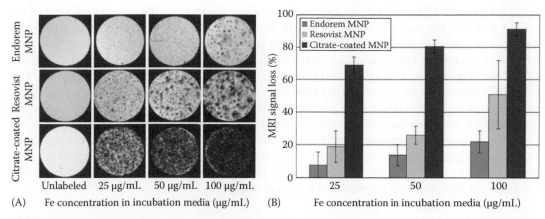

(A)   Fe concentration in incubation media (µg/mL)   (B)   Fe concentration in incubation media (µg/mL)

FIGURE 15.11 (A) MRI phantoms of MNP-loaded MSCs. MSCs loaded with anionic citrate–coated MNPs reveal a strong MRI signal loss. (B) The contrast in the MRI was significantly enhanced compared to cells labelled with same concentrations of dextran-coated (Endorem) and carboxydextran-coated MNPs (Resovist).

15.4.2 MAGNETIC TARGETING

The ability of MNP-loaded RBC to move towards a gradient of an external magnetic field has been demonstrated[150,152]. Hence, these cells are supposed to be highly appropriate for magnetic targeting. Applied intravenously, they circulate through the body and can be concentrated at a target by a strong inhomogeneous external magnetic field. Successful targeting of MNP-loaded RBC ghosts and magnetically guided delivery of drugs using carrier RBCs has been reported previously[145–149]. However, in these earlier studies, the carriers were only localised by chemical and histological analysis of isolated tissues after scarifying the experimental animals.

Magnetic targeting of RBCs loaded with citrate-coated MNPs was observed *in vitro* in an artificial circulation system simulating shear stress conditions comparable with these in venous blood (unpublished own results). The MNP-loaded RBCs were suspended in blood plasma at a concentration of 2% v/v. The flow chamber was fixed directly on an electromagnet connected to a power supply. The magnetic field intensity was adjusted by changing the current strength and photographs were taken every 10 min. Visible attachment of MNP-loaded RBCs occurred already after 15 min (Figure 15.12A).

Furthermore, the targeting potential was tested under similar conditions with RBCs loaded with thermosensitive MNPs[152] at temperatures below and above the critical precipitation point

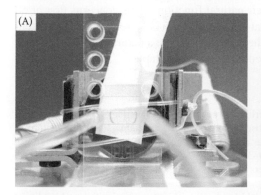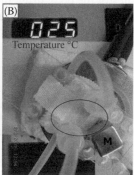

FIGURE 15.12 Photographs of the artificial circulation system for *in vitro* magnetic targeting of (A) RBCs loaded with citrate-coated MNPs by an electromagnet; (B) RBCs loaded with thermosensitive MNPs Fe₃O₄@MEO₂MA₉₀-*co*-OEGMA₁₀ (LCST in PBS = 36°C) by a permanent magnet (M) on a heating plate for temperature control at 25°C below LCST of the MNPs; (C) same suspension as in (B) at 40°C above LCST.

(LCST; Figure 15.12B and C). Here, the temperature was essential for the efficient retention of the carrier RBCs to the magnet. Below LCST, the MNPs are dispersed uniformly inside the RBCs and their magnetic moment is small. Under shear stress in the flow chamber, the attractive forces towards the magnetic field gradient are not sufficient for cell retention (Figure 15.12B). Above the LCST, the magnetic moment of MNP agglomerates is large and magnetic forces acting on the carrier RBCs become larger than the driving force of the flow resulting in retention and accumulation of the cells at the magnet (Figure 15.12C). This effect is reversible — the carrier RBCs are dispersed again as soon as the temperature is downregulated below the LCST.

However, magnetic targeting of MNP-loaded blood cells *in vivo* still remains a challenge and is limited to peripheral regions of the body. For magnetic targeting in deeper tissues, further development of instrumentation for focusing of magnetic fields with high accuracy is required to provide the technical equipment for magnetically guided drug delivery.

15.5   CHALLENGES AND OUTLOOK

Searching the appropriate MNPs for each specific cell type and particular medical application, scientists with different background discover more and more exciting opportunities. The fast development in synthesis of nanoparticles of different composition, size, shape and surface properties has the potential to match the high requirements for modern medical applications. Additionally, improvement of technical equipment and instrumentation allows measurements and visualisation of processes with enhanced accuracy and resolution both *in vitro* and *in vivo*.

Although extensive work has already been done on loading MNPs into blood cells and their potential for biomedical applications, there are still many open questions and challenges. In particular, the potential application of different blood cells in combination with novel multifunctional MNPs remains a great field of further research. Open questions concerning the influence of each newly synthesised type of particles on specific cellular functions and the mechanisms behind have to be addressed prior medical applications. Since intravenous application of MNPs is needed for MRI, interactions between MNPs and blood cells have to be explored intensively.

Various possible medical applications of MNP-loaded blood cells are still not explored but seem to be very promising. MNP-loaded RBCs as long circulating contrast agent for MRI and MPI have yet to be validated for clinical applications. The surface of MNP-loaded RBCs can be modified easily by attachment of peptides or antibodies to promote uptake by cells of targeted tissues. In hyperthermia, treatment of cancer, MNP-loaded RBCs are promising to be of great advantage especially in combination with thermosensitive MNPs. In addition to magnetic targeting, different blood cells can be targeted to specific sites of the body using their natural functions. This concerns first of all circulating stem cells, leucocytes and platelets. A great unexplored area is represented by the potential application of platelets loaded with MNPs since these blood cells are a natural source of growth factors and could find a broad application in cell-based regenerative therapies.

REFERENCES

1. Mueller-Eckhardt, C. 2004. *Transfusionsmedizin*, Berlin, Germany, Springer Verlag.
2. Van Leeuwenhoek, A. 1675. Other microscopical observations made by the same, about the texture of the blood, the sap of some plants, the figures of sugar and salt, and the probable cause of the difference of their tastes. *Philos Trans Royal Soc Lond*, 10, 380–385.
3. Gorter, E. and Grendel, F. 1925. On bimolecular layers of lipoids on the chromocytes of the blood. *J Exp Med*, 41, 439–443.
4. Bessis, M. and Delpech, G. 1981. Discovery of the red blood cell with notes on priorities and credits of discoveries, past, present and future. *Blood Cells*, 7, 447–480.
5. Mohandas, N. and Gallagher, P. G. 2008. Red cell membrane: Past, present, and future. *Blood*, 112, 3939–3948.
6. Ihler, G. M., Glew, R. H. and Schnure, F. W. 1973. Enzyme loading of erythrocytes. *Proc Natl Acad Sci USA*, 70, 2663–2666.

7. Zimmermann, U., Pilwat, G. and Riemann, F. 1974. Dielectric breakdown of cell membranes. *Biophys J*, 14, 881–899.

8. Gothoskar, A. V. 2004. Resealed erythrocytes: A review. *Pharm Technol*, 140–158.

9. Knighton, D. R., Ciresi, K. F., Fiegel, V. D., Austin, L. L. and Butler, E. L. 1986. Classification and treatment of chronic nonhealing wounds. Successful treatment with autologous platelet-derived wound healing factors (PDWHF). *Ann Surg*, 204, 322–330.

10. Celotti, F., Colciago, A., Negri-Cesi, P., Pravettoni, A., Zaninetti, R. and Sacchi, M. C. 2006. Effect of platelet-rich plasma on migration and proliferation of SaOS-2 osteoblasts: Role of platelet-derived growth factor and transforming growth factor-beta. *Wound Repair Regen*, 14, 195–202.

11. Becker, A. J., Mc, C. E. and Till, J. E. 1963. Cytological demonstration of the clonal nature of spleen colonies derived from transplanted mouse marrow cells. *Nature*, 197, 452–454.

12. Friedenstein, A. J., Piatetzky, S., II and Petrakova, K. V. 1966. Osteogenesis in transplants of bone marrow cells. *J Embryol Exp Morphol*, 16, 381–390.

13. Friedenstein, A. J., Chailakhjan, R. K. and Lalykina, K. S. 1970. The development of fibroblast colonies in monolayer cultures of guinea-pig bone marrow and spleen cells. *Cell Tissue Kinet*, 3, 393–403.

14. Friedenstein, A. J. 1976. Precursor cells of mechanocytes. *Int Rev Cytol*, 47, 327–359.

15. Siena, S., Bregni, M., Brando, B., Ravagnani, F., Bonadonna, G. and Gianni, A. M. 1989. Circulation of CD34+ hematopoietic stem cells in the peripheral blood of high-dose cyclophosphamide-treated patients: Enhancement by intravenous recombinant human granulocyte–macrophage colony-stimulating factor. *Blood*, 74, 1905–1914.

16. Shields, L. E. and Andrews, R. G. 1998. Gestational age changes in circulating CD34+ hematopoietic stem/progenitor cells in fetal cord blood. *Am J Obstet Gynecol*, 178, 931–937.

17. Erices, A., Conget, P. and Minguell, J. J. 2000. Mesenchymal progenitor cells in human umbilical cord blood. *Br J Haematol*, 109, 235–242.

18. He, Q., Wan, C. and Li, G. 2007. Concise review: Multipotent mesenchymal stromal cells in blood. *Stem Cells*, 25, 69–77.

19. Ceradini, D. J., Kulkarni, A. R., Callaghan, M. J., Tepper, O. M., Bastidas, N., Kleinman, M. E., Capla, J. M., Galiano, R. D., Levine, J. P. and Gurtner, G. C. 2004. Progenitor cell trafficking is regulated by hypoxic gradients through HIF-1 induction of SDF-1. *Nat Med*, 10, 858–864.

20. Mayani, H., Alvarado-Moreno, J. A. and Flores-Guzman, P. 2003. Biology of human hematopoietic stem and progenitor cells present in circulation. *Arch Med Res*, 34, 476–488.

21. Pittenger, M. F., Mackay, A. M., Beck, S. C., Jaiswal, R. K., Douglas, R., Mosca, J. D., Moorman, M. A., Simonetti, D. W., Craig, S. and Marshak, D. R. 1999. Multilineage potential of adult human mesenchymal stem cells. *Science*, 284, 143–147.

22. Beresford, J. N., Bennett, J. H., Devlin, C., Leboy, P. S. and Owen, M. E. 1992. Evidence for an inverse relationship between the differentiation of adipocytic and osteogenic cells in rat marrow stromal cell cultures. *J Cell Sci*, 102 (Pt 2), 341–351.

23. Caplan, A. I. 1991. Mesenchymal stem cells. *J Orthop Res*, 9, 641–650.

24. Owen, M. and Friedenstein, A. J. 1988. Stromal stem cells: Marrow-derived osteogenic precursors. *Ciba Found Symp*, 136, 42–60.

25. Pittenger, M., Vanguri, P., Simonetti, D. and Young, R. 2002. Adult mesenchymal stem cells: Potential for muscle and tendon regeneration and use in gene therapy. *J Musculoskelet Neuronal Interact*, 2, 309–320.

26. Shake, J. G., Gruber, P. J., Baumgartner, W. A., Senechal, G., Meyers, J., Redmond, J. M., Pittenger, M. F. and Martin, B. J. 2002. Mesenchymal stem cell implantation in a swine myocardial infarct model: Engraftment and functional effects. *Ann Thorac Surg*, 73, 1919–1925; discussion 1926.

27. Otsuru, S., Tamai, K., Yamazaki, T., Yoshikawa, H. and Kaneda, Y. 2008. Circulating bone marrow-derived osteoblast progenitor cells are recruited to the bone-forming site by the CXCR4/stromal cell-derived factor-1 pathway. *Stem Cells*, 26, 223–234.

28. Wang, Y., Deng, Y. and Zhou, G. Q. 2008. SDF-1alpha/CXCR4-mediated migration of systemically transplanted bone marrow stromal cells towards ischemic brain lesion in a rat model. *Brain Res*, 1195, 104–112.

29. Bruder, S. P., Jaiswal, N., Ricalton, N. S., Mosca, J. D., Kraus, K. H. and Kadiyala, S. 1998. Mesenchymal stem cells in osteobiology and applied bone regeneration. *Clin Orthop Relat Res*, S247–S256.

30. Horwitz, E. M., Gordon, P. L., Koo, W. K., Marx, J. C., Neel, M. D., Mcnall, R. Y., Muul, L. and Hofmann, T. 2002. Isolated allogeneic bone marrow-derived mesenchymal cells engraft and stimulate growth in children with osteogenesis imperfecta: Implications for cell therapy of bone. *Proc Natl Acad Sci USA*, 99, 8932–8937.

31. Wakitani, S., Imoto, K., Yamamoto, T., Saito, M., Murata, N. and Yoneda, M. 2002. Human autologous culture expanded bone marrow mesenchymal cell transplantation for repair of cartilage defects in osteoarthritic knees. *Osteoarthr Cartil*, 10, 199–206.

32. Ponticiello, M. S., Schinagl, R. M., Kadiyala, S. and Barry, F. P. 2000. Gelatin-based resorbable sponge as a carrier matrix for human mesenchymal stem cells in cartilage regeneration therapy. *J Biomed Mater Res*, 52, 246–255.

33. Murphy, J. M., Fink, D. J., Hunziker, E. B. and Barry, F. P. 2003. Stem cell therapy in a caprine model of osteoarthritis. *Arthritis Rheum*, 48, 3464–2474.

34. Young, R. G., Butler, D. L., Weber, W., Caplan, A. I., Gordon, S. L. and Fink, D. J. 1998. Use of mesenchymal stem cells in a collagen matrix for Achilles tendon repair. *J Orthop Res*, 16, 406–413.

35. Hoffmann, A., Pelled, G., Turgeman, G., Eberle, P., Zilberman, Y., Shinar, H., Keinan-Adamsky, K., Winkel, A., Shahab, S., Navon, G., Gross, G. and Gazit, D. 2006. Neotendon formation induced by manipulation of the *Smad8* signalling pathway in mesenchymal stem cells. *J Clin Investig*, 116, 940–952.

36. Caplan, A. I. and Bruder, S. P. 2001. Mesenchymal stem cells: Building blocks for molecular medicine in the 21st century. *Trends Mol Med*, 7, 259–264.

37. Barry, F. P. and Murphy, J. M. 2004. Mesenchymal stem cells: Clinical applications and biological characterization. *Int J Biochem Cell Biol*, 36, 568–584.

38. Minguell, J. J., Erices, A. and Conget, P. 2001. Mesenchymal stem cells. *Exp Biol Med (Maywood)*, 226, 507–520.

39. Berry, C. C. and Curtis, A. S. G. 2003. Functionalisation of magnetic nanoparticles for applications in biomedicine. *J Phys D – Appl Phys*, 36, R198–R206.

40. Apopa, P. L., Qian, Y., Shao, R., Guo, N. L., Schwegler-Berry, D., Pacurari, M., Porter, D., Shi, X., Vallyathan, V., Castranova, V. and Flynn, D. C. 2009. Iron oxide nanoparticles induce human microvascular endothelial cell permeability through reactive oxygen species production and microtubule remodeling. *Part Fibre Toxicol*, 6, 1.

41. Mayor, S. and Pagano, R. E. 2007. Pathways of clathrin-independent endocytosis. *Nat Rev Mol Cell Biol*, 8, 603–612.

42. Unfried, K., Albrecht, C., Klotz, L. O., Von Mikecz, A., Grether-Beck, S. and Schins, R. P. F. 2007. Cellular responses to nanoparticles: Target structures and mechanisms. *Nanotoxicology*, 1, 52–71.

43. Zimmermann, U. 1983. *Cellular Drug-Carrier Systems and their Possible Targeting*, New York, John Wiley & Sons.

44. Deuticke, B., Kim, M. and Zollner, C. 1973. Influence of amphotericin-B on permeability of mammalian erythrocytes to nonelectrolytes, anions and cations. *Biochim Biophys Acta*, 318, 345–359.

45. Aderem, A. and Underhill, D. M. 1999. Mechanisms of phagocytosis in macrophages. *Annu Rev Immunol*, 17, 593–623.

46. Rejman, J., Oberle, V., Zuhorn, I. S. and Hoekstra, D. 2004. Size-dependent internalization of particles via the pathways of clathrin- and caveolae-mediated endocytosis. *Biochem J*, 377, 159–169.

47. Mukherjee, S., Ghosh, R. N. and Maxfield, F. R. 1997. Endocytosis. *Physiol Rev*, 77, 759–803.

48. Gruenberg, J. 2001. The endocytic pathway: A mosaic of domains. *Nat Rev Mol Cell Biol*, 2, 721–730.

49. Schrier, S. L., Bensch, K. G., Johnson, M. and Junga, I. 1975. Energized endocytosis in human erythrocyte-ghosts. *J Clin Investig*, 56, 8–22.

50. Deloach, J. R. 1983. Encapsulation of exogenous agents in erythrocytes and the circulating survival of carrier erythrocytes. *J Appl Biochem*, 5, 149–157.

51. Schrier, S. L., Hardy, B. and Bensch, K. G. 1979. Endocytosis in erythrocytes and their ghosts. *Prog Clin Biol Res*, 30, 437–449.

52. Ginn, F. L., Hochstein, P. and Trump, B. F. 1969. Membrane alterations in hemolysis: Internalization of plasmalemma induced by primaquine. *Science*, 164, 843–845.

53. Ben-Bassat, I., Bensch, K. G. and Schrier, S. L. 1972. Drug-induced erythrocyte membrane internalization. *J Clin Investig*, 51, 1833–1844.

54. Schrier, S. L., Junga, I., Krueger, J. and Johnson, M. 1978. Requirements of drug-induced endocytosis by intact human erythrocytes. *Blood Cells*, 4, 339–359.

55. Marsden, H. R., Tomatsu, I. and Kros, A. 2011. Model systems for membrane fusion. *Chem Soc Rev*, 40, 1572–1585.

56. Papahadjopoulos, D., Mayhew, E., Poste, G., Smith, S. and Vail, W. J. 1974. Incorporation of lipid vesicles by mammalian cells provides a potential method for modifying cell behaviour. *Nature*, 252, 163–166.

57. Nicolau, C. and Gersonde, K. 1979. Incorporation of inositol hexaphosphate into intact red blood cells. I. Fusion of effector-containing lipid vesicles with erythrocytes. *Naturwissenschaften*, 66, 563–566.

58. Hagerstrand, H., Danieluk, M., Bobrowska-Hagerstrand, M., Pector, V., Ruysschaert, J., Kralj-Iglic, V. and Iglic, A. 1999. Liposomes composed of a double-chain cationic amphiphile (vectamidine) induce their own encapsulation into human erythrocytes. *Biochim Biophys Acta*, 1421, 125–130.

59. Yu, B. G., Okano, T., Kataoka, K. and Kwon, G. 1998. Polymeric micelles for drug delivery: Solubilization and haemolytic activity of amphotericin B. *J Control Release*, 53, 131–136.

60. Kitao, T., Hattori, K. and Takeshita, M. 1978. Agglutination of leukemic cells and daunomycin entrapped erythrocytes with lectin *in vitro* and *in vivo*. *Experientia*, 34, 94–95.

61. Kitao, T. and Hattori, K. 1980. Erythrocyte entrapment of daunomycin by amphotericin B without hemolysis. *Cancer Res*, 40, 1351–1353.

62. Lin, W., Mota de Freitas, D., Zhang, Q. and Olsen, K. W. 1999. Nuclear magnetic resonance and oxygen affinity study of cesium binding in human erythrocytes. *Arch Biochem Biophys*, 369, 78–88.

63. Billah, M. M., Finean, J. B., Coleman, R. and Michell, R. H. 1977. Permeability characteristics of erythrocyte ghosts prepared under isoionic conditions by a glycol-induced osmotic lysis. *Biochim Biophys Acta*, 465, 515–526.

64. Davson, H. and Danielli, J. F. 1970. *The Permeability of Natural Membranes*, Hafner Pub. Co. Darien, Connecticut, pp. 365.

65. Chernyshev, A. V., Tarasov, P. A., Semianov, K. A., Nekrasov, V. M., Hoekstra, A. G. and Maltsev, V. P. 2008. Erythrocyte lysis in isotonic solution of ammonium chloride: Theoretical modeling and experimental verification. *J Theor Biol*, 251, 93–107.

66. Franco, R. S., Weiner, M., Wagner, K. and Martelo, O. J. 1983. Incorporation of inositol hexaphosphate into red blood cells mediated by dimethyl sulfoxide. *Life Sci*, 32, 2763–2768.

67. Franco, R. S., Wagner, K., Weiner, M. and Martelo, O. J. 1984. Preparation of low-affinity red cells with dimethylsulfoxide-mediated inositol hexaphosphate incorporation: Hemoglobin and ATP recovery using a continuous-flow method. *Am J Hematol*, 17, 393–400.

68. Bird, J., Best, R. and Lewis, D. A. 1983. The encapsulation of insulin in erythrocytes. *J Pharm Pharmacol*, 35, 246–247.

69. Neumann, E., Schaefer-Ridder, M., Wang, Y. and Hofschneider, P. H. 1982. Gene transfer into mouse lyoma cells by electroporation in high electric fields. *EMBO J*, 1, 841–845.

70. Kinosita, K., Jr. and Tsong, T. Y. 1977. Formation and resealing of pores of controlled sizes in human erythrocyte membrane. *Nature*, 268, 438–441.

71. Wang, M., Orwar, O., Olofsson, J. and Weber, S. G. 2010. Single-cell electroporation. *Anal Bioanal Chem*, 397, 3235–3248.

72. Zimmermann, U. 1973. *Jahresbericht der Kernforschungsanlage Jülich GmbH*, Nuclear Research Center, Jülich, Annual report, pp. 55–58.

73. Kinosita, K., Jr. and Tsong, T. T. 1977. Hemolysis of human erythrocytes by transient electric field. *Proc Natl Acad Sci USA*, 74, 1923–1927.

74. Kinosita, K., Jr. and Tsong, T. Y. 1978. Survival of sucrose-loaded erythrocytes in the circulation. *Nature*, 272, 258–260.

75. Zimmermann, U., Riemann, F. and Pilwat, G. 1976. Enzyme loading of electrically homogeneous human red blood-cell ghosts prepared by dielectric-breakdown. *Biochim Biophys Acta*, 436, 460–474.

76. Tsong, T. Y. and Kinosita, K., Jr. 1985. Use of voltage pulses for the pore opening and drug loading, and the subsequent resealing of red blood cells. *Bibl Haematol*, 51, 108–114.

77. Jain, S., Jain, S. K. and Dixit, V. K. 1995. Erythrocytes based delivery of isoniazid: Preparation and *in vitro* characterization. *Indian Drugs*, 32, 471–476.

78. Strange, K., Emma, F. and Jackson, P. S. 1996. Cellular and molecular physiology of volume-sensitive anion channels. *Am J Physiol*, 270, C711–C730.

79. Lang, F., Busch, G. L., Ritter, M., Volkl, H., Waldegger, S., Gulbins, E. and Haussinger, D. 1998. Functional significance of cell volume regulatory mechanisms. *Physiol Rev*, 78, 247–306.

80. Raker, J. W., Taylor, I. M., Weller, J. M. and Hastings, A. B. 1950. Rate of potassium exchange of the human erythrocyte. *J Gen Physiol*, 33, 691–702.

81. Sheppard, C. W. and Martin, W. R. 1950. Cation exchange between cells and plasma of mammalian blood; methods and application to potassium exchange in human blood. *J Gen Physiol*, 33, 703–722.

82. Seeman, P. 1967. Transient holes in the erythrocyte membrane during hypotonic hemolysis and stable holes in the membrane after lysis by saponin and lysolecithin. *J Cell Biol*, 32, 55–70.

83. Ihler, G. M. and Tsang, H. C. 1987. Hypotonic hemolysis methods for entrapment of agents in resealed erythrocytes. *Methods Enzymol*, 149, 221–229.

84. Deloach, J. R., Harris, R. L. and Ihler, G. M. 1980. An erythrocyte encapsulator dialyzer used in preparing large quantities of erythrocyte ghosts and encapsulation of a pesticide in erythrocyte ghosts. *Anal Biochem*, 102, 220–227.

85. Ihler, G. M. 1983. Erythrocyte carriers. *Pharmacol Ther*, 20, 151–69.

86. Pierige, F., Serafini, S., Rossi, L. and Magnani, A. 2008. Cell-based drug delivery. *Adv Drug Deliv Rev*, 60, 286–295.

87. Jain, S. and Jain, N. K. 1997. Engineered erythrocytes as a drug delivery system. *Indian J Pharm Sci*, 59, 275–281.

88. Hamidi, M., Zarrin, A., Foroozesh, M. and Mohammadi-Samani, S. 2007. Applications of carrier erythrocytes in delivery of biopharmaceuticals. *J Control Release*, 118, 145–160.

89. Updike, S. J. and Wakamiya, R. T. 1983. Infusion of red blood cell-loaded asparaginase in monkey. Immunologic, metabolic, and toxicologic consequences. *J Lab Clin Med*, 101, 679–691.

90. Updike, S. J., Wakamiya, R. T. and Lightfoot, E. N., Jr. 1976. Asparaginase entrapped in red blood cells: Action and survival. *Science*, 193, 681–683.

91. Adriaenssens, K., Karcher, D., Lowenthal, A. and Terheggen, H. G. 1976. Use of enzyme-loaded erythrocytes in in-vitro correction of arginase-deficient erythrocytes in familial hyperargininemia. *Clin Chem*, 22, 323–326.

92. Jaitely, V., Kanaujia, P., Venkatesan, N., Jain, S. and Vyas, S. P. 1996. Resealed erythrocytes: Drug carrier potentials and biomedical applications. *Indian Drugs*, 33, 589–594.

93. Pitt, E., Johnson, C. M., Lewis, D. A., Jenner, D. A. and Offord, R. E. 1983. Encapsulation of drugs in intact erythrocytes: An intravenous delivery system. *Biochem Pharmacol*, 32, 3359–3368.

94. Deloach, J. and Ihler, G. 1977. A dialysis procedure for loading erythrocytes with enzymes and lipids. *Biochim Biophys Acta*, 496, 136–145.

95. Talwar, N. and Jain, N. K. 1992. Erythrocytes as carriers of metronidazole: In-vitro characterization. *Drug Dev Ind Pharm*, 18, 1799–1892.

96. Lewis, D. A. 1984. Red blood cells for drug delivery. *Pharm J*, 233, 384–385.

97. Baker, R. F. 1967. Entry of ferritin into human red cells during hypotonic haemolysis. *Nature*, 215, 424–425.

98. Sternberg, N., Georgieva, R., Duft, K. and Bäumler, H. 2011. Surface modified loaded human red blood cells for targeting and delivery of drugs. *J Microencapsul*, Submitted.

99. Magnani, M., Rossi, L., D'ascenzo, M., Panzani, I., Bigi, L. and Zanella, A. 1998. Erythrocyte engineering for drug delivery and targeting. *Biotechnol Appl Biochem*, 28 (Pt 1), 1–6.

100. Rechsteiner, M. C. 1975. Uptake of proteins by red blood cells. *Exp Cell Res*, 93, 487–492.

101. Alpar, H. O. and Lewis, D. A. 1985. Therapeutic efficacy of asparaginase encapsulated in intact erythrocytes. *Biochem Pharmacol*, 34, 257–261.

102. Field, W. N., Gamble, M. D. and Lewis, D. A. 1989. A comparison of the treatment of thyroidectomized rats with free-thyroxine and thyroxine encapsulated in erythrocytes. *Int J Pharm*, 51, 175–178.

103. Eichler, H. G., Gasic, S., Bauer, K., Korn, A. and Bacher, S. 1986. In vivo clearance of antibody-sensitized human drug carrier erythrocytes. *Clin Pharmacol Ther*, 40, 300–303.

104. Kravtzoff, R., Ropars, C., Laguerre, M., Muh, J. P. and Chassaigne, M. 1990. Erythrocytes as carriers for L-asparaginase. Methodological and mouse in-vivo studies. *J Pharm Pharmacol*, 42, 473–476.

105. Jrade, M., Villereal, M. C., Boynard, M., Dufeaut, J. and Ropars, C. 1987. Rheological properties of desferal loaded red blood-cells. *Clin Hemorheol*, 7, 872–872.

106. Zanella, A. 1987. Desferrioxamine loading of red cells for transfusion. *Adv Biosci*, (series) 67, 17–27.

107. Muzykantov, V. R., Smirnov, M. D., Zaltzman, A. B. and Samokhin, G. P. 1993. Tannin-mediated attachment of avidin provides complement-resistant immunoerythrocytes that can by lysed in the presence of activator of complement. *Anal Biochem*, 208, 338–342.

108. Muzykantov, V. R., Sakharov, D. V., Domogatsky, S. P., Goncharov, N. V. and Danilov, S. M. 1987. Directed targeting of immunoerythrocytes provides local protection of endothelial-cells from damage by hydrogen-peroxide. *Am J Pathol*, 128, 276–285.

109. Chiarantini, L., Droleskey, R., Magnani, M. and Deloach, J. R. 1992. In vitro targeting of erythrocytes to cytotoxic T-cells by coupling of Thy-1.2 monoclonal-antibody. *Biotechnol Appl Biochem*, 15, 171–184.

110. Mahan, D. E. and Copeland, R. L. 1978. Method using toluene-2,4-diisocyanate and glutaraldehyde to stabilize and conjugate antigens to erythrocytes for use in passive hemagglutination tests. *J Immunol Methods*, 19, 217–225.

111. Smirnov, M. D., Samokhin, G. P., Muzykantov, V. R., Idelson, G. L., Domogatsky, S. P. and Smirnov, V. N. 1983. Type-I and type-III collagens as a possible target for drug delivery to the injured sites of vascular bed. *Biochem Biophys Res Commun*, 116, 99–105.

112. Muzykantov, V. R., Sakharov, D. V., Smirnov, M. D., Domogatsky, S. P. and Samokhin, G. P. 1985. Targeting of enzyme immobilized on erythrocyte-membrane to collagen-coated surface. *FEBS Lett*, 182, 62–66.

113. Magnani, M., Chiarantini, L. and Mancini, U. 1994. Preparation and characterization of biotinylated red blood cells. *Biotechnol Appl Biochem*, 20, 335–345.

114. Muzykantov, V. R. and Murciano, J. C. 1996. Attachment of antibody to biotinylated red blood cells: Immuno-red blood cells display high affinity to immobilized antigen and normal biodistribution in rats. *Biotechnol Appl Biochem*, 24, 41–45.

115. Cowley, H., Wojda, U., Cipolone, K. M., Procter, J. L., Stroncek, D. F. and Miller, J. L. 1999. Biotinylation modifies red cell antigens. *Transfusion*, 39, 163–168.

116. Orr, G. A. 1981. The use of the 2-iminobiotin–avidin interaction for the selective retrieval of labeled plasma-membrane components. *J Biol Chem*, 256, 761–766.

117. Godfrey, W., Doe, B., Wallace, E. F., Bredt, B. and Wofsy, L. 1981. Affinity targeting of membrane-vesicles to cell-surfaces. *Exp Cell Res*, 135, 137–145.

118. Roffman, E., Mermosky, L., Benhur, H., Bayer, E. A. and Wilchek, M. 1986. Selective labeling of functional-groups on membrane-proteins or glycoproteins using reactive biotin derivatives and I-125 streptavidin. *Biochem Biophys Res Commun*, 136, 80–85.

119. Bayer, E. A., Safars, M. and Wilchek, M. 1987. Selective labeling of sulfhydryls and disulfides on blot transfers using avidin–biotin technology – Studies on purified proteins and erythrocyte-membranes. *Anal Biochem*, 161, 262–271.

120. Wilchek, M., Benhur, H. and Bayer, E. A. 1986. Para-diazobenzoyl biocytin – a new biotinylating reagent for the labeling of tyrosines and histidines in proteins. *Biochem Biophys Res Commun*, 138, 872–879.

121. Samokhin, G. P., Smirnov, M. D., Muzykantov, V. R., Domogatsky, S. P. and Smirnov, V. N. 1983. Red blood cell targeting to collagen-coated surfaces. *FEBS Lett*, 154, 257–261.

122. Muzykantov, V. R., Smirnov, M. D. and Klibanov, A. L. 1993. Avidin attachment to red blood cells via a phospholipid derivative of biotin provides complement-resistant immunoerythrocytes. *J Immunol Methods*, 158, 183–190.

123. Muzykantov, V. R., Murciano, J. C., Taylor, R. P., Atochina, E. N. and Herraez, A. 1996. Regulation of the complement-mediated elimination of red blood cells modified with biotin and streptavidin. *Anal Biochem*, 241, 109–119.

124. Chiarantini, L., Matteucci, D., Pistello, M., Mancini, U., Mazzetti, P., Massi, C., Giannecchini, S., Lonetti, I., Magnani, M. and Bendinelli, M. 1998. AIDS vaccination studies using an ex vivo feline immunodeficiency virus model: Homologous erythrocytes as a delivery system for preferential immunization with putative protective antigens. *Clin Diagn Lab Immunol*, 5, 235–241.

125. Magnani, M., Mancini, U., Bianchi, M. and Fazi, A. 1992. Comparison of uricase-bound and uricase-loaded erythrocytes as bioreactors for uric acid degradation. *Adv Exp Med Biol*, 326, 189–194.

126. Muzykantov, V. R. 2010. Drug delivery by red blood cells: Vascular carriers designed by mother nature. *Expert Opin Drug Deliv*, 7, 403–427.

127. Zaitsev, S., Spitzer, D., Murciano, J. C., Ding, B. S., Tliba, S., Kowalska, M. A., Marcos-Contreras, O. A., Kuo, A., Stepanova, V., Atkinson, J. P., Poncz, M., Cines, D. B. and Muzykantov, V. R. 2010. Sustained thromboprophylaxis mediated by an RBC-targeted pro-urokinase zymogen activated at the site of clot formation. *Blood*, 115, 5241–5248.

128. Hess, C. and Schifferli, J. A. 2003. Immune adherence revisited: Novel players in an old game. *News Physiol Sci*, 18, 104–108.

129. Taylor, R. P., Sutherland, W. M., Reist, C. J., Webb, D. J., Wright, E. L. and Labuguen, R. H. 1991. Use of heteropolymeric monoclonal-antibodies to attach antigens to the C3b receptor of human erythrocytes – a potential therapeutic treatment. *Proc Natl Acad Sci USA*, 88, 3305–3309.

130. Lindorfer, M. A., Hahn, C. S., Foley, P. L. and Taylor, R. P. 2001. Heteropolymer-mediated clearance of immune complexes via erythrocyte CR1: Mechanisms and applications. *Immunol Rev*, 183, 10–24.

131. Reist, C. J., Liang, H. Y., Denny, D., Martin, E. N., Scheld, W. M. and Taylor, R. P. 1994. Cross-linked bispecific monoclonal-antibody heteropolymers facilitate the clearance of human-IgM from the circulation of squirrel-monkeys. *Eur J Immunol*, 24, 2018–2025.

132. Ferguson, P. J., Martin, E. N., Greene, K. L., Kuhn, S., Cafiso, D. S., Addona, G. and Taylor, R. P. 1995. Antigen-based heteropolymers facilitate, via primate erythrocyte complement receptor-type-1, rapid erythrocyte binding of an autoantibody and its clearance from the circulation in rhesus-monkeys. *J Immunol*, 155, 339–347.

133. Lindorfer, M. A., Nardin, A., Foley, P. L., Solga, M. D., Bankovich, A. J., Martin, E. N., Henderson, A. L., Price, C. W., Gyimesi, E., Wozencraft, C. P., Goldberg, J. B., Sutherland, W. M. and Taylor, R. P. 2001. Targeting of *Pseudomonas aeruginosa* in the bloodstream with bispecific monoclonal antibodies. *J Immunol*, 167, 2240–2249.

134. Taylor, R. P., Sutherland, W. M., Martin, E. N., Ferguson, P. J., Reinagel, M. L., Gilbert, E., Lopez, K., Incardona, N. L. and Ochs, H. D. 1997. Bispecific monoclonal antibody complexes bound to primate erythrocyte complement receptor 1 facilitate virus clearance in a monkey model. *J Immunol*, 158, 842–850.

135. Taylor, R. P., Martin, E. N., Reinagel, M. L., Nardin, A., Craig, M., Choice, Q., Schlimgen, R., Greenbaum, S., Incardona, N. L. and Ochs, H. D. 1997. Bispecific monoclonal antibody complexes facilitate erythrocyte binding and liver clearance of a prototype particulate pathogen in a monkey model. *J Immunol*, 159, 4035–4044.

136. Hahn, C. S., French, O. G., Foley, P., Martin, E. N. and Taylor, R. P. 2001. Bispecific monoclonal antibodies mediate binding of dengue virus to erythrocytes in a monkey model of passive viremia. *J Immunol*, 166, 1057–1065.

137. Asher, D. R., Cerny, A. M. and Finberg, R. W. 2005. The erythrocyte viral trap: Transgenic expression of viral receptor on erythrocytes attenuates coxsackievirus B infection. *Proc Natl Acad Sci USA*, 102, 12897–12902.

138. Buster, B. L., Mattes, K. A. and Scheld, W. M. 1997. Monoclonal antibody-mediated, complement-independent binding of human tumor necrosis factor-alpha to primate erythrocytes via complement receptor 1. *J Infect Dis*, 176, 1041–1046.

139. Chambers, E. and Mitragoni, S. 2004. Prolonged circulation of large polymeric nanoparticles by noncovalent adsorption on erythrocytes. *J Control Release*, 100, 111–119.

140. Hall, S. S., Mitragoni, S. and Daugherty, P. S. 2007. Identification of peptide ligands facilitating nanoparticle attachment to erythrocytes. *Biotechnol Prog*, 23, 749–754.

141. Bulte, J. W. M. and Kraitchman, D. L. 2004. Iron oxide MR contrast agents for molecular and cellular imaging. *NMR Biomed*, 17, 484–499.

142. Magnani, M. 2003. *Erythrocyte Engineering for Drug Delivery and Targeting*, New York, Landes Bioscience/Eurekah.com and Kluwer Academic/Plenum Publishers.

143. Rossi, L., Serafini, S., Pierige, F., Antonelli, A., Cerasi, A., Fraternale, A., Chiarantini, L. and Magnani, M. 2005. Erythrocyte-based drug delivery. *Expert Opin Drug Deliv*, 2, 311–322.

144. Provotorov, V. M. and Ivanova, G. A. 2009. The role of erythrocytes in the system of controlled transport of pharmaceutical agents. *Klin Med (Mosk)*, 87, 4–8.

145. Sprandel, U., Lanz, D. J. and Von Horsten, W. 1987. Magnetically responsive erythrocyte ghosts. *Methods Enzymol*, 149, 301–312.

146. Orekhova, N. M., Akchurin, R. S., Belyaev, A. A., Smirnov, M. D., Ragimov, S. E. and Orekhov, A. N. 1990. Local prevention of thrombosis in animal arteries by means of magnetic targeting of aspirin-loaded red cells. *Thromb Res*, 57, 611–616.

147. Vyas, S. P. and Jain, S. K. 1994. Preparation and *in vitro* characterization of a magnetically responsive ibuprofen-loaded erythrocytes carrier. *J Microencapsul*, 11, 19–29.

148. Jain, S. K. and Vyas, S. P. 1994. Magnetically responsive diclofenac sodium-loaded erythrocytes: Preparation and *in vitro* characterization. *J Microencapsul*, 11, 141–151.

149. Jain, S., Jain, S. K. and Dixit, V. K. 1997. Magnetically guided rat erythrocytes bearing isoniazid: Preparation, characterization, and evaluation. *Drug Dev Ind Pharm*, 23, 999–1006.

150. Brahler, M., Georgieva, R., Buske, N., Muller, A., Muller, S., Pinkernelle, J., Teichgraber, U., Voigt, A. and Baumler, H. 2006. Magnetite-loaded carrier erythrocytes as contrast agents for magnetic resonance imaging. *Nano Lett*, 6, 2505–2509.

151. Antonelli, A., Sfara, C., Mosca, L., Manuali, E. and Magnani, M. 2008. New biomimetic constructs for improved in vivo circulation of superparamagnetic nanoparticles. *J Nanosci Nanotechnol*, 8, 2270–2278.

152. Chanana, M., Jahn, S., Georgieva, R., Lutz, J. F., Baumler, H. and Wang, D. Y. 2009. Fabrication of colloidal stable, thermosensitive, and biocompatible magnetite nanoparticles and study of their reversible agglomeration in aqueous milieu. *Chem Mater*, 21, 1906–1914.

153. Markov, D. E., Boeve, H., Gleich, B., Borgert, J., Antonelli, A., Sfara, C. and Magnani, M. 2010. Human erythrocytes as nanoparticle carriers for magnetic particle imaging. *Phys Med Biol*, 55, 6461–6473.

154. Antonelli, A., Sfara, C., Manuali, E., Bruce, I. J. and Magnani, M. 2011. Encapsulation of superparamagnetic nanoparticles into red blood cells as new carriers of MRI contrast agents. *Nanomedicine*, 6, 211–223.

155. Safi, M., Sarrouj, H., Sandre, O., Mignet, N. and Berret, J. F. 2010. Interactions between sub-10-nm iron and cerium oxide nanoparticles and 3T3 fibroblasts: The role of the coating and aggregation state. *Nanotechnology*, 21, 145103.

156. Cinti, C., Taranta, M., Naldi, I. and Grimaldi, S. 2011. Newly engineered magnetic erythrocytes for sustained and targeted delivery of anti-cancer therapeutic compounds. *PLoS One*, 6, e17132.

157. Moore, A., Weissleder, R. and Bogdanov, A. Jr. 1997. Uptake of dextran-coated monocrystalline iron oxides in tumor cells and macrophages. *J Magn Reson Imaging*, 7, 1140–1145.

158. Luciani, N., Gazeau, F. and Wilhelm, C. 2009. Reactivity of the monocyte/macrophage system to superparamagnetic anionic nanoparticles. *J Mater Chem*, 19, 6373–6380.

159. Bulte, J. W., Hoekstra, Y., Kamman, R. L., Magin, R. L., Webb, A. G., Briggs, R. W., Go, K. G., Hulstaert, C. E., Miltenyi, S., The, T. H. et al. 1992. Specific MR imaging of human lymphocytes by monoclonal antibody-guided dextran–magnetite particles. *Magn Reson Med*, 25, 148–157.

160. Baeten, K., Adriaensens, P., Hendriks, J., Theunissen, E., Gelan, J., Hellings, N. and Stinissen, P. 2010. Tracking of myelin-reactive T cells in experimental autoimmune encephalomyelitis (EAE) animals using small particles of iron oxide and MRI. *NMR Biomed*, 23, 601–609.

161. Thorek, D. L. J. and Tsourkas, A. 2008. Size, charge and concentration dependent uptake of iron oxide particles by non-phagocytic cells. *Biomaterials*, 29, 3583–3590.

162. Staedtke, V., Brahler, M., Muller, A., Georgieva, R., Bauer, S., Sternberg, N., Voigt, A., Lemke, A., Keck, C., Moschwitzer, J. and Baumler, H. 2010. *In vitro* inhibition of fungal activity by macrophage-mediated seques-tration and release of encapsulated amphotericin B nanosuspension in red blood cells. *Small*, 6, 96–103.

163. White, J. G. 2005. Platelets are covercytes, not phagocytes: Uptake of bacteria involves channels of the open canalicular system. *Platelets*, 16, 121–131.

164. Cartier, R., Kaufner, L., Paulke, B. R., Wustneck, R., Pietschmann, S., Michel, R., Bruhn, H. and Pison, U. 2007. Latex nanoparticles for multimodal imaging and detection in vivo. *Nanotechnology*, 18, 195102.

165. Wang, Y. X., Hussain, S. M. and Krestin, G. P. 2001. Superparamagnetic iron oxide contrast agents: Physicochemical characteristics and applications in MR imaging. *Eur Radiol*, 11, 2319–2331.

166. Weinmann, H. J., Ebert, W., Misselwitz, B. and Schmitt-Willich, H. 2003. Tissue-specific MR contrast agents. *Eur J Radiol*, 46, 33–44.

167. Bulte, J. W., Douglas, T., Witwer, B., Zhang, S. C., Strable, E., Lewis, B. K., Zywicke, H., Miller, B., Van Gelderen, P., Moskowitz, B. M., Duncan, I. D. and Frank, J. A. 2001. Magnetodendrimers allow endosomal magnetic labeling and in vivo tracking of stem cells. *Nat Biotechnol*, 19, 1141–1147.

168. Arbab, A. S., Yocum, G. T., Kalish, H., Jordan, E. K., Anderson, S. A., Khakoo, A. Y., Read, E. J. and Frank, J. A. 2004. Efficient magnetic cell labeling with protamine sulfate complexed to ferumoxides for cellular MRI. *Blood*, 104, 1217–1223.

169. Frank, J. A., Miller, B. R., Arbab, A. S., Zywicke, H. A., Jordan, E. K., Lewis, B. K., Bryant, L. H., Jr. and Bulte, J. W. 2003. Clinically applicable labeling of mammalian and stem cells by combining super-paramagnetic iron oxides and transfection agents. *Radiology*, 228, 480–487.

170. Arbab, A. S., Yocum, G. T., Wilson, L. B., Parwana, A., Jordan, E. K., Kalish, H. and Frank, J. A. 2004. Comparison of transfection agents in forming complexes with ferumoxides, cell labeling efficiency, and cellular viability. *Mol Imaging*, 3, 24–32.

171. Lewin, M., Carlesso, N., Tung, C. H., Tang, X. W., Cory, D., Scadden, D. T. and Weissleder, R. 2000. Tat peptide-derivatized magnetic nanoparticles allow in vivo tracking and recovery of progenitor cells. *Nat Biotechnol*, 18, 410–414.

172. Wilhelm, C., Billotey, C., Roger, J., Pons, J. N., Bacri, J. C. and Gazeau, F. 2003. Intracellular uptake of anionic superparamagnetic nanoparticles as a function of their surface coating. *Biomaterials*, 24, 1001–1011.

173. Wilhelm, C. and Gazeau, F. 2008. Universal cell labelling with anionic magnetic nanoparticles. *Biomaterials*, 29, 3161–3174.

174. Zhang, Y., Kohler, N. and Zhang, M. 2002. Surface modification of superparamagnetic magnetite nanoparticles and their intracellular uptake. *Biomaterials*, 23, 1553–1561.

175. Gamarra, L. F., Pavon, L. F., Marti, L. C., Pontuschka, W. M., Mamani, J. B., Carneiro, S. M., Camargo-Mathias, M. I., Moreira-Filho, C. A. and Amaro, E., Jr. 2008. *In vitro* study of CD133 human stem cells labeled with superparamagnetic iron oxide nanoparticles. *Nanomedicine*, 4, 330–339.

176. Emerit, J., Beaumont, C. and Trivin, F. 2001. Iron metabolism, free radicals, and oxidative injury. *Biomed Pharma*, 55, 333–9.

177. Ankamwar, B., Lai, T. C., Huang, J. H., Liu, R. S., Hsiao, M., Chen, C. H. and Hwu, Y. K. 2010. Biocompatibility of Fe(3)O(4) nanoparticles evaluated by *in vitro* cytotoxicity assays using normal, glia and breast cancer cells. *Nanotechnology*, 21, 75102.

178. Bulte, J. W., Kraitchman, D. L., Mackay, A. M. and Pittenger, M. F. 2004. Chondrogenic differentiation of mesenchymal stem cells is inhibited after magnetic labeling with ferumoxides. *Blood*, 104, 3410–3412; author reply 3412–3413.

179. Soenen, S. J., Nuytten, N., De Meyer, S. F., De Smedt, S. C. and De Cuyper, M. 2010. High intracellular iron oxide nanoparticle concentrations affect cellular cytoskeleton and focal adhesion kinase-mediated signaling. *Small*, 6, 832–842.

180. Soenen, S. J., Himmelreich, U., Nuytten, N. and De Cuyper, M. 2011. Cytotoxic effects of iron oxide nanoparticles and implications for safety in cell labelling. *Biomaterials*, 32, 195–205.

181. Nohroudi, K., Arnhold, S., Berhorn, T., Addicks, K., Hoehn, M. and Himmelreich, U. 2010. In vivo MRI stem cell tracking requires balancing of detection limit and cell viability. *Cell Transpl*, 19, 431–441.

182. Kostura, L., Kraitchman, D. L., Mackay, A. M., Pittenger, M. F. and Bulte, J. W. 2004. Feridex labeling of mesenchymal stem cells inhibits chondrogenesis but not adipogenesis or osteogenesis. *NMR Biomed*, 17, 513–517.

183. Balakumaran, A., Pawelczyk, E., Ren, J., Sworder, B., Chaudhry, A., Sabatino, M., Stroncek, D., Frank, J. A. and Robey, P. G. 2010. Superparamagnetic iron oxide nanoparticles labeling of bone marrow stromal (mesenchymal) cells does not affect their 'stemness'. *PLoS One*, 5, e11462.

184. Arbab, A. S., Yocum, G. T., Rad, A. M., Khakoo, A. Y., Fellowes, V., Read, E. J. and Frank, J. A. 2005. Labeling of cells with ferumoxides–protamine sulfate complexes does not inhibit function or differentiation capacity of hematopoietic or mesenchymal stem cells. *NMR Biomed*, 18, 553–559.

185. Chanana, M. 2010. *Synthesis of Stimuli-Responsive and Switchable Inorganic Nanoparticles for Biomedical Applications*. Dissertation. Potsdam University, Potsdam, Germany.

186. Gleich, B. and Weizenecker, R. 2005. Tomographic imaging using the nonlinear response of magnetic particles. *Nature*, 435, 1214–1217.

187. Zhao, L. R., Duan, W. M., Reyes, M., Keene, C. D., Verfaillie, C. M. and Low, W. C. 2002. Human bone marrow stem cells exhibit neural phenotypes and ameliorate neurological deficits after grafting into the ischemic brain of rats. *Exp Neurol*, 174, 11–20.

188. Hofstetter, C. P., Schwarz, E. J., Hess, D., Widenfalk, J., El Manira, A., Prockop, D. J. and Olson, L. 2002. Marrow stromal cells form guiding strands in the injured spinal cord and promote recovery. *Proc Natl Acad Sci USA*, 99, 2199–2204.

189. Kraitchman, D. L., Gilson, W. D. and Lorenz, C. H. 2008. Stem cell therapy: MRI guidance and monitoring. *J Magn Reson Imaging*, 27, 299–310.

190. Barbash, I. M., Chouraqui, P., Baron, J., Feinberg, M. S., Etzion, S., Tessone, A., Miller, L., Guetta, E., Zipori, D., Kedes, L. H., Kloner, R. A. and Leor, J. 2003. Systemic delivery of bone marrow-derived mesenchymal stem cells to the infarcted myocardium: Feasibility, cell migration, and body distribution. *Circulation*, 108, 863–868.

191. Jendelova, P., Herynek, V., Urdzikova, L., Glogarova, K., Kroupova, J., Andersson, B., Bryja, V., Burian, M., Hajek, M. and Sykova, E. 2004. Magnetic resonance tracking of transplanted bone marrow and embryonic stem cells labeled by iron oxide nanoparticles in rat brain and spinal cord. *J Neurosci Res*, 76, 232–243.

192. Bulte, J. W., Duncan, I. D. and Frank, J. A. 2002. In vivo magnetic resonance tracking of magnetically labeled cells after transplantation. *J Cereb Blood Flow Metab*, 22, 899–907.

193. Kraitchman, D. L., Heldman, A. W., Atalar, E., Amado, L. C., Martin, B. J., Pittenger, M. F., Hare, J. M. and Bulte, J. W. 2003. In vivo magnetic resonance imaging of mesenchymal stem cells in myocardial infarction. *Circulation*, 107, 2290–2293.

194. Niemeyer, M., Oostendorp, R. A., Kremer, M., Hippauf, S., Jacobs, V. R., Baurecht, H., Ludwig, G., Piontek, G., Bekker-Ruz, V., Timmer, S., Rummeny, E. J., Kiechle, M. and Beer, A. J. 2010. Non-invasive tracking of human haemopoietic CD34(+) stem cells in vivo in immunodeficient mice by using magnetic resonance imaging. *Eur Radiol*, 20, 2184–2193.

Nadine Sternberg studied medical biotechnology at Technical University of Berlin (TUB), Germany. After graduation, she joined Prof. Hans Bäumler's group at the Institute of Transfusion Medicine, Charité University Hospital as a PhD student and is now completing her thesis in cooperation with Prof. Roland Lauster of TUB. She is working on the development of natural and artificial drug carrier systems for targeted drug delivery.

Kristin Andreas studied biotechnology at the TUB. She graduated at the Tissue Engineering Group of Michael Sittinger and started a PhD thesis at the 'mesenchymal stem cell group' of Jochen Ringe in 2007. She investigated labelling of mesenchymal stem cells with superparamagnetic iron oxide nanoparticles to track cell migration *in vivo via* MRI. Current research is focused on *in situ* regeneration of degenerated joint structures using controlled delivery devices, differentiation factors and chemoattractants.

Hans Bäumler obtained his diploma in physics at the University of Leipzig, PhD as well as habilitation in biophysics at the Humboldt University of Berlin. He worked as a visiting researcher and guest professor at the medical academy in Petersburg (Russia) and at the University Paris 7 (France).

He is the head of the research department of the Institute of Transfusion Medicine and PI of the Berlin-Brandenburg Center for Regenerative Therapies of the Charité Universitätsmedizin Berlin.

Radostina Georgieva obtained her PhD at Humboldt University in Berlin. She joined Prof. Bäumler's team at the Institute of Transfusion Medicine, Charité and later Prof. Helmuth Möhwald's Department at the Max-Planck Institute of Colloids and Interface, Golm-Potsdam. Currently, she works in Prof. Bäumler's research group. Her research interests are artificial and natural carriers for targeted drug delivery and blood cells. She is teaching biophysics at the Medical Faculty, Trakia University, Stara Zagora, Bulgaria.

16 Putting Therapeutic Nanoparticles Where They Need to Go by Magnet Systems Design and Control

Arash Komaee, Roger Lee, Aleksander Nacev,
Roland Probst, Azeem Sarwar, Didier A. Depireux,
Kenneth J. Dormer, Isaac Rutel, and Benjamin Shapiro

CONTENTS

Earlier chapters in this book describe the synthesis and characterisation of magnetic nanoparticles, their biofunctionalisation and the medical applications they enable. This chapter focuses on magnet systems design and control: how should magnets be designed, where should they be placed and when should they be turned on and off so as to begin to understand how to better direct magnetic nanoparticles to *in vivo* targets? Shaping of magnetic fields and forces for directing therapy to disease locations is valuable for a wide range of clinical applications, and in this chapter, we will focus on just two: (1) magnetically pushing (injecting) nanoparticles to reach inner ear diseases and (2) development of methods for targeting deeper tissue tumours. Magnet design, optimisation and control will be informed by mathematical modelling of magnetic fields, forces and the resulting *in vivo* particle transport.

16.1 PHYSICAL FUNDAMENTALS OF MAGNETIC FIELDS AND FORCES

The first task is to quantify how magnetic fields vary around magnets, and then to describe and understand the resulting forces on magnetic nanoparticles from those fields. This aspect is standard and is well understood in the magnetic drug targeting literature[1-8]. Electromagnetic fields are classically described by Maxwell's equations[9]. Here, we will consider both permanent magnets and electromagnets whose current is changed slowly compared with radio frequencies, so that the magnetostatic portion of Maxwell's equations is appropriate:

$$\nabla \times \vec{H} = \vec{j}, \tag{16.1}$$

$$\nabla \times \vec{B} = 0, \tag{16.2}$$

$$\vec{B} = \mu_0(\vec{H} + \vec{M}) = \mu_0(\vec{H} + \chi\vec{H}), \tag{16.3}$$

where
  $\vec{B}$ is the magnetic field (in SI units of Tesla)
  $\vec{H}$ is the magnetic intensity (A/m)
  $\vec{j}$ is the current density (A/m^2)
  $\vec{M}$ is the material magnetisation (A/m), which can depend non-linearly on the applied magnetic field (material saturation and hysteresis)
  χ is the magnetic susceptibility (unitless)
  μ_0 is the permeability of a vacuum ($4\pi \times 10^{-7}$ N/A^2)

These equations hold true both in a vacuum and in materials. For a permanent magnetic material, the magnetisation $\vec{M} \neq 0$; for an electromagnet, the current density is non-zero $\vec{j} \neq 0$. Applied magnetic fields pass virtually unchanged through the human body because the magnetic susceptibility of blood and tissue is close to zero ($\chi \approx 10^{-6}$–10^{-4}). In contrast, ferromagnetic particles can have magnetic susceptibilities 5–7 orders of magnitude higher ($\chi \approx 20$) resulting in strong interactions with magnetic fields. This means magnetic particles can experience magnetic forces, while biological tissues remain unaffected.

The magnetic force on a single spherical nanoparticle depends on the magnetic field and field gradient (how the field varies in space) at that particle's location[10–13]:

$$\vec{F}_{mag} = \frac{4\pi a^3}{3} \frac{\mu_0\chi}{(1+\chi/3)} \left[\frac{\partial \vec{H}}{\partial \vec{x}}\right]^T \vec{H} = \frac{2\pi a^3}{3} \frac{\mu_0\chi}{(1+\chi/3)} \nabla(\|\vec{H}\|^2), \tag{16.4}$$

where
  a is the radius of the particle
  ∇ is the gradient operator
  $\vec{x} = (x, y, z)$ will denote the position of the particle

Equation 16.4 states that the magnetic force on a particle always points from low to high magnetic field intensity, up the gradient of $\|\vec{H}\|^2$. The middle expression is common in the magnetic drug delivery literature[7,14–18]: it illustrates that a spatially varying magnetic field with a non-zero Jacobian matrix ($\partial \vec{H}/\partial \vec{x} \neq 0$) is required to create a magnetic force, and it is equivalent to the last expression by the chain rule. Equation 16.4 also shows that the magnetic force scales with particle volume: a particle 10 times larger will experience a magnetic force a thousand times greater. Particle-to-particle interactions do often occur[19–23], but our interest here is in the direction of forces created by magnetic fields on either single particles or groups of particles moving together (e.g. a bolus of ferrofluid). For these cases, Equation 16.4 accurately describes the direction of magnetic forces. It further enables predictions of when magnetic targeting will or will not be successful, and these predictions match prior published *in vitro* and *in vivo* studies (see Refs. [24,25] and Section 16.2). Finally, control design based on Equation 16.4 has also allowed precise manipulation of ferrofluids in experiments[26].

16.2 MODELLING THE TRANSPORT OF NANOPARTICLES IN BLOOD AND TISSUE

Mathematical modelling of nanoparticle transport *in vivo* is needed to rationally choose magnet configurations and for choosing the dynamic actuation of electromagnets during precision ferro-fluid control. It is recognised that physical parameters, such as particle size and composition, as well as magnet placement, size, shape, strength and timing of actuation, are crucial for the success of magnetic targeting. However, the resulting behaviour of nanoparticles in blood and tissue has proved to be too complex to be effectively predicted by simple back-of-the-envelope estimates. At the same time, the number of design parameters is too large to be searched by animal experiments[24]. Instead, experimentally validated modelling can be used to fill the gap and inform magnet design and control.

A detailed survey of prior results for modelling magnetic particles *in vivo* is provided in Nacev et al.[25] Briefly, the modelling of single magnetic particles within a fluid flow and their resulting trajectories was introduced by Forbes et al.[11] and Voltairas et al.[27] The single particle case has been extended to evaluate the capture efficiency for various blood flow fields and for targeting techniques that utilise implanted magnets[2,7,14,16,17,28,29]. Further modelling advances have added magnetofection and other surface targeting techniques[6,8], have incorporated biologically relevant experimental data for ferrofluid targeting[1,30], have included biological effects into fluid viscosity and particle diffusion in blood[15,18] and have assessed the capturing of magnetically loaded cells or biomagnetic fluids[4,31]. More recent modelling research has added further biological phenomena, such as realistic blood flows, vessel edge effects and tissue properties and has considered a distributed ferrofluid instead of single particles[3,5].

Here, as in Refs. [24,25], we first focus our attention on ferrofluid behaviour in and around a single blood vessel. Distribution of magnetic particles in and around the vessel is considered after systemic injection of ferrofluid into the body (as done in the Lubbe clinical trials[32]). The selected blood vessel can be of any type, from a major artery or a vein to a minor capillary, fenestrated or not, at any depth and any applied magnetic force can be considered. Ferrofluid transport in and around this vessel can be described by the non-dimensionalised partial differential equations (PDEs):

$$\frac{\partial C_B}{\partial t} = -\nabla \left[\underbrace{-\frac{1}{Pe}\nabla C_B}_{\text{Diffusion}} + \left(\underbrace{\vec{V}_B}_{\text{Convection}} + \underbrace{(0,-\Psi)}_{\text{Magnetic forces}} \right) C_B \right], \tag{16.5}$$

$$\frac{\partial C_M}{\partial t} = -\nabla \mathcal{D} \left[-\frac{1}{Pe}\nabla C_M + (0,-\Psi)C_M \right],$$

$$\frac{\partial C_T}{\partial t} = -\nabla \mathcal{D}_T \left[-\frac{1}{Pe}\nabla C_T + (0,-\Psi)C_T \right]. \tag{16.6}$$

These equations quantify particle transport due to convection by blood flow, particle diffusion in tissue and blood (including particle scattering by collisions with blood cells[15]) and particle motion due to magnetic forces (see Ref. [25] for a detailed derivation of these equations). Here, these equations are written in two spatial dimensions: along the vessel and in a plane that cuts through it. The vessel is idealised to be straight and the magnetic force is pointed down, as shown in Figure 16.1 (curved vessels and other extensions are detailed in Ref. [25]—they do not substantially change the results). C_B, C_M and C_T are the concentration of the particles in space and time, in the blood vessel and the surrounding endothelial membrane and tissue, respectively; $\vec{V}_B$ is the blood velocity; t is time and

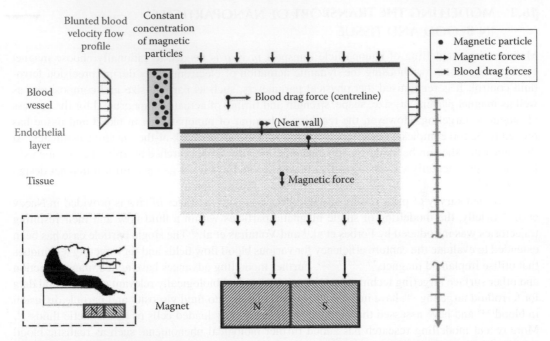

FIGURE 16.1 The considered blood vessel and surrounding tissue geometry. The blood vessel is idealised as a straight channel. Blood and, after systemic particle injection, an assumed constant concentration of magnetic nanoparticles enter from the left. The magnetic particles (black circles) within the blood vessel experience diffusion, migration under blood flow and magnetic forces. Magnetic particles in the surrounding endothelium and tissue experience diffusion and magnetic drift but no blood flow forces. The magnet can be at a long distance from the blood vessel (deep targeting) and here this is denoted by the break in the length bar on the right of the figure. (Figure taken from *J. Magn. Magn. Mater.*, 323, Nacev, A., Beni, C., Bruno, O., and Shapiro, B., The behaviors of ferro-magnetic nano-particles in and around blood vessels under applied magnetic fields, 651–668, Copyright 2011, with permission from Elsevier.)

$\nabla = (\partial/\partial x, \partial/\partial y)$ is the spatial gradient operator. Finally, Pe, Ψ, $\mathcal{D}$ and $\mathcal{D}_T$ are the non-dimensional mass Péclet number, magnetic-Richardson number and the endothelium and tissue Renkin reduced diffusion coefficients. These four non-dimensional numbers, discussed next, quantify the ratios between competing effects, for example, between blood viscosity and magnetic forces, and they uniquely specify *in vivo* particle behaviour.

The aforementioned PDE formulation is standard, see, for example, Refs. [15,33], and simply tracks the in and out flux of particles at each spatial location (net accumulation means an increased particle concentration in that infinitesimal volume). All the complexity and variation of human tissue must be represented in the choice of parameters. For example, particle extravasation (or lack thereof) is modelled by adjusting the diffusion coefficient in the endothelium $\mathcal{D}$ to a non-zero (or zero) value. Additional tissue-specific properties, for example, a decreased resistance to nanoparticle motion due to a compromised extracellular matrix with larger interstitial spaces in a tumour region, can be accounted for by increasing the magnetic drift coefficient (the magnetic-Richardson number) Ψ in that region.

The magnetic-Richardson number Ψ quantifies the ratio between the forces on a nanoparticle created by the magnet and those produced by the blood flowing in the vessel. It is defined as follows:

$$\Psi = \frac{\text{Magnetic force at vessel centre line}}{\text{Blood drag force at centre line}} = \frac{\|\vec{F}_M\|}{\|\vec{F}_S\|} = \frac{\|\vec{V}_R\|}{V_{\text{Bmax}}}. \tag{16.7}$$

The ratio of the two forces is equal to the ratio of the particle velocities they produce: the convection speed of the particle as it is carried by the blood flow V_{Bmax} versus the additional relative velocity $\vec{V}_R$ created by the applied magnetic force. As noted in Equation 16.4, the magnetic force depends both on the magnetic field $\vec{H}$ (units A/m) created by the magnet and its spatial gradient $[\mathrm{d}\vec{H}/\mathrm{d}\vec{x}]$ (A/m^2), and both of these quantities, as well as the particle's physical properties, must be known at the vessel location to determine Ψ.

The second key number that determines the success of magnetic drug delivery is the mass Péclet number. It quantifies the competition between particle movement (convection) by the blood flow versus particle diffusion in the blood. It is defined as:

$$Pe = \frac{\text{Blood vessel width} \times \text{Maximum blood velocity}}{\text{Total diffusion coefficient of particles}} = \frac{d_B V_{\text{Bmax}}}{D_{\text{Tot}}}. \qquad (16.8)$$

Here D_{Tot} (units m^2/s) takes into account the scattering of nanoparticles by collisions with blood cells[25], an effect that can be modelled as additional diffusion[15].

Finally, the Renkin reduced diffusion coefficients, for the endothelium and the surrounding tissue, are a description of the ease with which particles can move through these two tissue types compared to their diffusion through blood. These numbers are defined as follows:

$$\mathcal{D} = \frac{\text{Diffusion coefficient in membrane}}{\text{Total diffusion coefficient in blood}} = \frac{D_M}{D_B + D_S} = \frac{D_M}{D_{\text{Tot}}}, \qquad (16.9)$$

$$\mathcal{D}_T = \frac{\text{Diffusion coefficient in tissue}}{\text{Total diffusion coefficient in blood}} = \frac{D_T}{D_{\text{Tot}}}, \qquad (16.10)$$

where

D_B is the particle diffusion in blood due to thermal fluctuations

D_S is the additional diffusion caused by collisions with blood cells

D_M is the diffusion in the endothelial membrane (it is set to zero if there is no extravasation)

D_{Tot} is the particle diffusion in the surrounding tissue (all in units m^2/s)

Under physiological conditions, for vessel diameters and blood velocities ranging from a vessel width of $d_B = 6\,\mu\text{m}$ and blood velocity $V_{\text{Bmax}} = 0.1\,\text{mm/s}$ (for rat capillaries) to $d_B = 3\,\text{cm}$ and $V_{\text{Bmax}} = 40\,\text{cm/s}$ (a human aorta), for achievable particle magnetic parameters, for particle diameters ranging from 1 nm to 5 µm, and for magnet field strengths going up to 4 T [magnetic resonance imaging (MRI) strengths], the four non-dimensional parameters can range between $4 \times 10^{-18} \leq \Psi \leq 6 \times 10^3$, $1 \leq Pe \leq 1 \times 10^{12}$ and $0 \leq \min(\mathcal{D}, \mathcal{D}_T) \leq 1$. It suffices to consider the minimum of the two Renkin coefficients as only the smaller one determines particle behaviour[24,25].

The aforementioned model states that magnetic particles travel through blood and into surrounding tissue under three competing effects: under blood convection, under particle diffusion (which here includes both extravasation and particle scattering by collisions with blood cells) and due to the pull of the applied magnetic field. The Richardson, Péclet and Renkin numbers quantify the competition between these three effects.

Our simulations, across the range of achievable and physiologically relevant conditions, revealed three types of ferrofluid behaviours in and around a single vessel (Figure 16.2). In the velocity-dominated case, the created magnetic forces are weak compared to the blood flow forces, they cannot capture the particles, and so the nanoparticles are swept through the blood vessel and out the back of the region of interest. In the magnetic dominated case, the magnetic forces far exceed the ability of the vessel membrane and tissue to resist particle motion, to the point where the particles

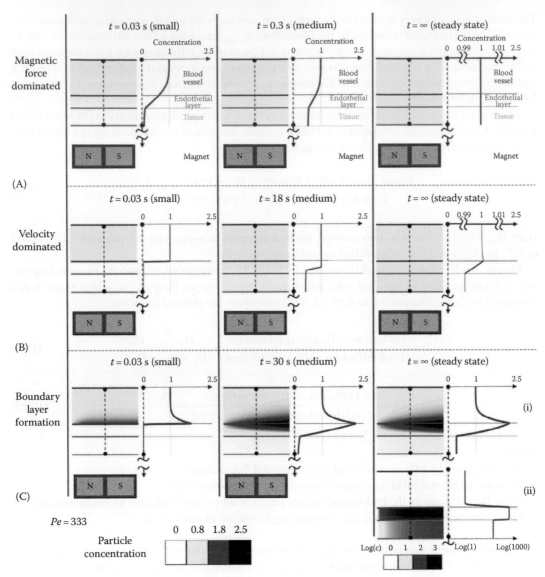

FIGURE 16.2 The three behaviours: (A) magnetic force dominated case ($\Psi = 10^{-3}$, $\mathcal{D} = 1$), (B) velocity-dominated case ($\Psi = 10^{-5}$, $\mathcal{D} = 10^{-3}$) and (C) boundary layer formation ($\Psi = 10^{-2}$, $\mathcal{D} = 10^{-3}$). (A) The magnetic force dominated case shows a cross-sectional concentration of the magnetic nanoparticles for three times, at $Pe = 333$. Particles are pulled towards the magnet and out through the bottom of the tissue resulting in a constant concentration equal to the blood inlet concentration. (B) The velocity-dominated case shows a cross-sectional concentration of the magnetic nanoparticles for three times, again at $Pe = 333$. Particles are washed out the back before they generate a significant boundary layer along the vessel wall. At long times, diffusion equilibrates the concentration between tissue and blood. (C) Boundary layer formation shows a cross-sectional magnetic nanoparticle concentration for three times, also at $Pe = 333$. (i) The steady-state profile for $\Psi = 10^{-2}$. Here, the particle concentration is shown on the same linear scale as in other time snap shots. (ii) The steady-state profile for a higher magnetic-Richardson number, for $\Psi = 10^{-1}$. Here, both the particle concentration and the cross-sectional plot are shown on a log scale. In both boundary layer cases ($\Psi = 10^{-2}$ and 10^{-1}), the particles build up along the vessel membrane on both the vessel side and within the membrane. The boundary layer forms very rapidly. In (ii), the membrane particle concentration is sufficiently high to cause a concentration in the tissue greater than the vessel inlet concentration. (Figure taken from *J. Magn. Magn. Mater.*, 323, Nacev, A., Beni, C., Bruno, O., and Shapiro, B., The behaviors of ferro-magnetic nano-particles in and around blood vessels under applied magnetic fields, 651–668, Copyright 2011, with permission from Elsevier.)

are pulled by the magnet out of the vessel and eventually also out of the region of tissue being considered. This case either requires exceedingly strong magnetic forces or a blood vessel membrane that does not substantially inhibit particle movement (such as a sufficiently 'leaky' tumour vessel). In the boundary layer case, nanoparticles accumulate in a layer at the vessel wall and, if extravasation is possible, are then in the correct location to enter the surrounding tissue. It is this last case that is most interesting and predicted to be effective for drug delivery as the applied magnetic field serves to concentrate the therapeutic magnetic particles[24,25].

As detailed in Refs. [24,25], to use this modelling, a magnetic targeting researcher should compute and estimate the non-dimensional magnetic-Richardson, Péclet and Renkin coefficients for his or her experiment or magnet design. Some terms in these numbers will depend on engineering properties such as particle size and magnet strength, others will depend on desired targeting depth (which blood vessel? how deep? with which blood flow velocity?) and yet others will require an educated best guess (diffusion coefficients for nanoparticles in tissue remain largely uncertain)[33]. Based on this magnetic-Richardson, Péclet and minimum Renkin number [Ψ, Pe, $\min(\mathcal{D}, \mathcal{D}_T)$] triplet, the resulting *in vivo* particle behaviour can be looked up using the domains marked in Figure 16.3. The boundaries of these domains are quantified in detail in Ref. [25], while the supplementary material in Ref. [24] provides a step-by-step workflow for determining the non-dimensional numbers and looking up the predicted particle behaviour. Using this workflow, a researcher can predict the success of magnetic targeting and so will be able to more rationally choose designs for improved drug targeting systems.

The results shown in Figure 16.3 have been compared to previously published magnetic drug delivery studies, for both *in vitro* and *in vivo* experiments. Our modelling was able to correctly predict both the occurrence and the amount of magnetic capture in *in vitro* experiments[30,34], and for *in vivo* studies we accurately predicted observed particle accumulation in rats[35–37] as well as the depth of focusing (~5 cm) in prior human clinical trials[32]. Variations and extensions of this PDE model are now being used to inform designs of dynamic magnet control for focusing of particles to deep tissue targets and for sweeping therapy into widespread hypoxic metastatic tumours[38].

The aforementioned modelling also indicates that there is a potential for external magnets to achieve deep tissue targeting. As noted in Ref. [25], a large (25 cm in diameter, 20 cm in length) and strong (2 Tesla) electromagnet acting on 100 nm diameter particles is predicted to create boundary layer behaviour in major vessels and minor capillaries to depths of ≤20 and ≤30 cm, respectively, although the nanoparticle boundary layer will be thinner for vessels deeper inside the body. With advances in particles properties and electromagnet designs, and with appropriate real-time imaging and control, magnetic particles in these thin boundary layers could be steered to deep tissue targets along the inside of blood vessel walls.

16.3 MAGNETIC PUSHING FOR TREATMENT OF INNER EAR DISEASES

Now that the physics of magnetic fields and particles and aspects of their transport *in vivo* have been summarised, we turn to magnet system designs to deliver particles to desired targets. We start with a simple push design for treatment of inner ear diseases.

Any single magnet, whether it is a permanent magnet or an electromagnet, will always attract paramagnetic, ferromagnetic, ferrimagnetic or superparamagnetic particles towards itself (Figure 16.4). This is apparent from Equation 16.4: the particles experience a force up the magnetic field intensity gradient ($\vec{F}_{mag} = k\nabla(\|\vec{H}\|^2)$) with $k = 4\pi a^3 \mu_0 \chi/(3+\chi) > 0$) and this field intensity is greatest nearest the magnet. A single magnet cannot push ferromagnetic particles away (diamagnetic particles with $\chi < 0$ can be pushed by a single magnet, but the magnetic forces they experience are too weak and most diamagnetic materials are not biocompatible making them impractical for magnetic drug targeting applications).

Yet there are a whole range of clinical needs, where it would be beneficial to magnetically push in particles: to drive them deeper into tumours, to non-invasively paint them into skin or open

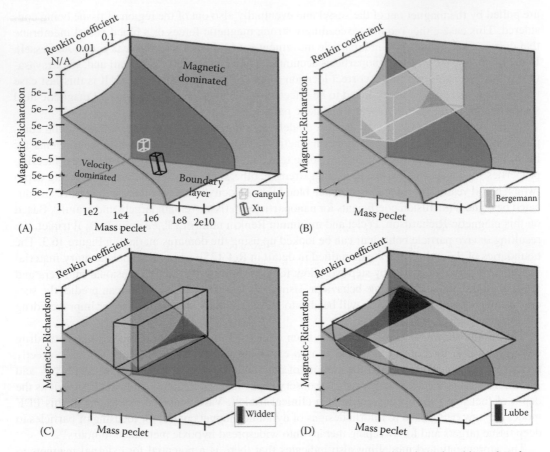

FIGURE 16.3 The three behaviour domains, velocity dominated, magnetic dominated and boundary layer behaviour, compared to prior magnetic targeting *in vitro* and *in vivo* experiments. The firmly shaded regions in green and purple denote the predicted magnetic and velocity behaviour domains. The boxed or curved wire-framed regions show the values spanned by each experiment. The dark shading (dark yellow in B, purple in C and dark red in D) shows the region of the corresponding experiment that exists in the velocity-dominated region. The light shading (light yellow in B, light purple in C and pink in D) shows the region of the experiments where the concentration created in the tissue is predicted to be greater than the vessel inlet concentration. The *in vitro* experiments (A)[30,34] exist entirely in the boundary layer regime. Widder (B)[36,37] and Bergemann (C)[39] have small portions in the velocity-dominated region, only at small magnetic-Richardson numbers and high Renkin coefficients. Lubbe (D)[32] extends into the velocity behaviour domain when mass Péclet numbers and magnetic-Richardson numbers are small, and this extent increases as the Renkin coefficient increases. (Figure taken from *J. Magn. Magn. Mater.*, 323, Nacev, A., Beni, C., Bruno, O., and Shapiro, B., The behaviors of ferro-magnetic nano-particles in and around blood vessels under applied magnetic fields, 651–668, Copyright 2011, with permission from Elsevier.)

wound bacterial infections and, as considered next, to deliver them to inner ear diseases. Acute noise-induced injury, uncontrolled labyrinthitis (dizziness) and profound tinnitus (loud ringing in the ears) affect millions of people[41] but injectable or ingestible drugs are ineffective or inadequate because the inner ear is behind the blood–brain barrier. Microcirculation vessels that supply the inner ear have walls that are impermeable even to small drug molecules[42]. Magnetic forces have been used to pull therapeutic nanoparticles into the inner ears of guinea pigs[43] and rats through the round window membrane (RWM). However, to pull nanoparticles into the inner ear of humans, a pulling magnet would have to be placed on the opposite side of the head at a long working distance (~12–15 cm). That magnet would need to be extremely strong (magnetic fields and forces drop off quickly with distance)[14,26,44,45] hence large, expensive and unwieldy. Instead, the ability to *push*

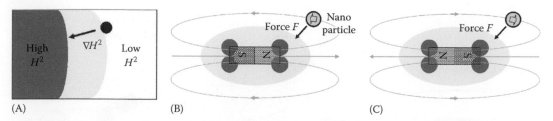

(A)   (B)   (C)

FIGURE 16.4 Magnetic particles are attracted to regions of highest magnetic field intensity. (A) A ferromagnetic particle (black dot) experiences a magnetic force (arrow) from low to high (white to pink) magnetic field intensity squared. Hence, a single permanent magnet (B) attracts particles to its corners where the magnetic field intensity is highest and (C) this force remains unchanged even if the polarity of the magnet is reversed (grey curves show the magnetic field lines; the open arrow shows the corresponding magnetisation direction inside the particle).

particles magnetically from the same side as the diseased ear, over a much shorter 3–5 cm working distance, would facilitate effective treatment with a small, safe and handheld magnetic device (Figure 16.5)[46,47].

Magnetic pushing can be created by a magnetic field cancellation node. From Equation 16.4, to create an outward push force, the magnetic field intensity (squared) $\left\lVert \vec{H} \right\rVert^2$ must be made to increase

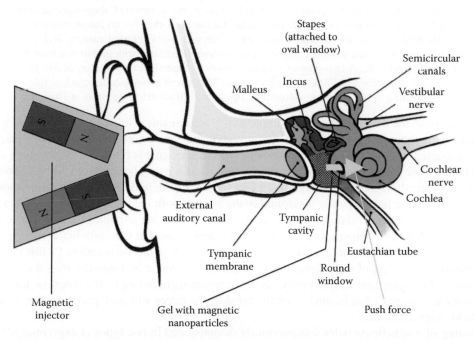

FIGURE 16.5 Microcirculation vessels that supply blood to the cochlea have tight capillary junctions that prevent therapies from reaching inner ear diseases. An ability to push therapeutic nanoparticles through the RWM would bypass this blood–brain barrier and transport therapy directly into the inner ear. The envisioned treatment is shown, from left to right: the handheld magnetic push system; human ear anatomy; a gel filled with magnetic nanoparticles that has been injected into the middle ear (light blue with black dots, in the tympanic cavity); the RWM (black oval) and a magnetic push force (yellow arrow) to deliver therapeutic particles through the RWM into the inner ear (the cochlea). (Reprinted with permission from Shapiro, B., Dormer, K., and Rutel, I.B., A two-magnet system to push therapeutic nanoparticles, in *8th International Conference on the Scientific and Clinical Applications of Magnetic Carriers,* American Institute of Physics, Conference Proceeding Series, Rostock, Germany. Copyright 2010 American Institute of Physics.)

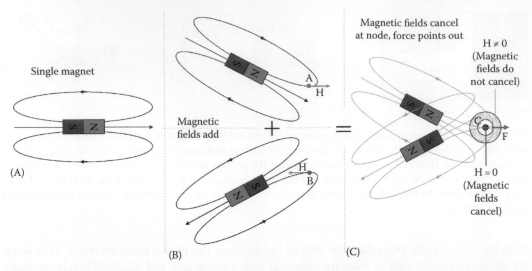

FIGURE 16.6 Unlike a single magnet, two permanent magnets correctly arranged can push nanoparticles away. (A) Schematic field lines around a single magnet magnetised along its length. (B) Two magnets. The top magnet is tilted down till the magnetic field at point A is purely to the right. The other magnet is flipped and tilted up, so that its magnetic field at B is purely to the left. (C) When these two magnets are brought together so that points A and B overlay, the magnetic field cancels at point C (black dot) but remains non-zero in an annulus around C. The resulting forces on particles radiate out from C—beyond C they create an outward push force. In this last panel, the grey curves are just guides for the eye—they are no longer magnetic field lines since magnetic field lines do not superpose and cannot intersect. For interested readers, Ref. [46] includes simulations of actual field lines around a magnetic push system. (Reprinted with permission from Shapiro, B., Dormer, K., and Rutel, I.B., A two-magnet system to push therapeutic nanoparticles, in *8th International Conference on the Scientific and Clinical Applications of Magnetic Carriers,* American Institute of Physics, Conference Proceeding Series, Rostock, Germany. Copyright 2010 American Institute of Physics.)

going away from the magnetic device. A simple way to achieve this is to create a displaced magnetic field minimum, a node. Figure 16.6 shows how such a node can be created with just two magnets. A single magnet will have the field lines shown. When the magnet is tilted clockwise, along a chosen field line there will be a location where the magnetic field is purely horizontal to the right (point A). A second identical magnet, flipped and tilted counter-clockwise, will have a like location (point B) where the magnetic field is purely horizontal to the left and with the same magnitude. If these two magnets are positioned as shown, so that points A and B coincide at point C, the magnetic fields add together (Maxwell's equations are linear) and exactly cancel at C to provide a zero magnetic field ($\vec{H} = 0$). Since the magnetic fields do not cancel at other points surrounding C, this point is a local minimum (a zero) of the magnetic field intensity $\|\vec{H}\|$. Again by Equation 16.4, the forces go from low to high magnetic field intensity, so in the region surrounding C, the magnetic forces will point outwards from C. Just beyond C, on the far side, the forces will push particles away from the handheld magnetic device.

Pushing of magnetic particles was previously demonstrated in two types of experiments[46]. Two magnets on wooden supports created an outward plume in an initially uniform solution of nanoparticles dispersed in water (Figure 16.7A and B). In this first experiment, it was conceivable that magnet pulling might have created fluid eddies that, through hydrodynamic effects, would have caused an outward fluid plume. To guard against this possibility, a second set of experiments was conducted, where a single droplet of ferrofluid was pushed away (Figure 16.7C).

Recently, we demonstrated magnetic pushing in live rat experiments. Red fluorescent 300 nm diameter Chemicell particles were first injected by a 1 cc 28.5 gage insulin syringe (Beckton–Dickson) into the middle ears of 15 rats (Long-Evans strain, Charles River) through their tympanic membranes. This filled the middle ear space with magnetic nanoparticles. The goal of the

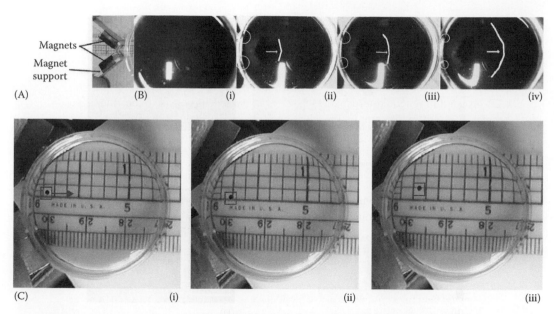

FIGURE 16.7 Experimental demonstrations of magnetic pushing. (A) The first experiment: two magnets held on wooden strips. (B, i) The initial uniform distribution of nanoparticles and the created outward plume at (ii) 5 s, (iii) 10 s and (iv) several minutes after introduction of the push device on the left (the plume leading edge is marked by the white curve). (C) A micro litre drop of ferrofluid being pushed away from the magnetic device. This series took place over 90 s; the magnetic device was again held on the left. Here, the ruler red grid lines are spaced 3.2 mm apart and the droplet was pushed 4 mm from its starting point. (Reprinted with permission from Shapiro, B., Dormer, K., and Rutel, I.B., A two-magnet system to push therapeutic nanoparticles, in *8th International Conference on the Scientific and Clinical Applications of Magnetic Carriers,* American Institute of Physics, Conference Proceeding Series, Rostock, Germany. Copyright 2010 American Institute of Physics.)

magnetic injection was to safely move the particles into the rat inner ears, as shown schematically in Figure 16.5 for a human ear. The magnetic injection system consisted of four 1.3 T (5.08 cm × 2.54 cm × 2.54 cm, Applied Magnets) magnets encased in a polymer holder fabricated by a 3D printer (Personal Portable 3D Printer, PP3DP). Two magnets were inserted side-by-side on each side of the holder in a 'V' geometry (Figure 16.8A). For these first push experiments in rats, it was easier to use four smaller magnets aligned side-by-side in pairs instead of two twice-as-large magnets, which would have been more difficult to handle. Likewise, this V geometry, where the push node has not yet been extended out to the 3–5 cm distance needed for human ear applications, was also a convenient starting point as it allowed an easier creation of strong push forces. In four rats, this magnetic push system was held next to the rat ears for 1 h, with the push node aligned with the long axis of the basal turn of the rats' cochlea. Four rats were used as negative controls and had no magnetic push applied. Both groups showed no decrease in their hearing abilities, as measured by auditory brainstem response (ABR) immediately after and 6 weeks after intra-tympanic injection.

The rats were sacrificed immediately at the end of the exposure to the magnetic field by CO_2 inhalation. The temporal bone was exposed with a ventral approach and the bulla was removed, quickly exposing the middle ear space. At this point, the oval window and the RWMs were exposed, as was a good part of the cochlea. The visible bone was gently cleaned mechanically with KimWipes and the space near the membranes was repeatedly rinsed with deionized water until no trace of the ferrofluid was visible. Then, a small break was made near the apex of the cochlea and a small cochleostomy was made near the RWM. The cochlear fluid was extracted by applying a capillary tube (400 μm inner diameter) to the cochleostomy. Further, a lateral cut was then made in the lateral wall of the cochlea, near the base, at the first half turn of the cochlea and near the apex. A surgical

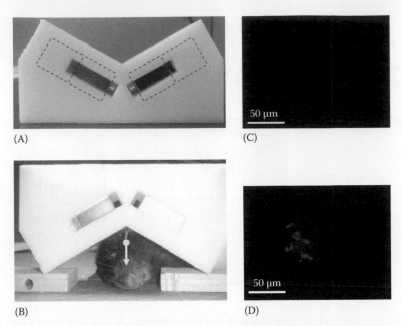

FIGURE 16.8   Magnetic pushing of 300 nm diameter Chemicell nano-screen MAG/R-D particles into the inner ears of rats. (A) The magnetic push system for these rat experiments. Two pairs of magnets are inside the white polymer holder, behind the two slots—their position is shown by the dashed blue outlines. (B) Experimental setup. The device is placed upside down above the rat's head in order to align the push node (yellow dot and down force arrow) on axis with the rat's RWM (here marked by the small, just visible, white rolled tissue paper above the yellow dot). Red fluorescence was measured from cochlea tissue scrapes, with blue and green channels subtracted to remove tissue autofluorescence. (C) No fluorescence from particles is visible in a cochlea tissue scrape for a rat where push was not applied versus (D) many red particles were seen in the cochlea of a rat where magnetic push was used. (We thank G. D. Marquart at the University of Maryland for helping us fabricate the polymer holder.)

spoon was inserted and tissue samples were extracted and placed in 0.9% saline. The cochlear fluids and the tissue samples were imaged by an Olympus IX51 Nikon microscope at 10× resolution with a red fluorescence channel to identify the red fluorescent particles, while green and blue channels were used to monitor and subtract any fluorescence not generated by the magnetic nanoparticles (tissue autofluorescence).

Figure 16.8C shows the result for a rat with no magnetic push—no nanoparticles are visible under fluorescence in the tissue scrape; versus Figure 16.8D where a magnetic push was applied and a high population of nanoparticles was observed in the rat cochlea tissue scrape. This was a first demonstration of successful magnetic push in animal experiments.

This system is now being optimised to provide sufficient force over the 3–5 cm distance necessary for treating humans, and this involves not only improving the design of the push system but also choosing the optimal particle size. Particles that are too small will experience small magnetic forces (magnetic force scales with volume)[5], while particles that are too large will experience too large tissue resistance forces (beyond a certain size, particles cannot move effectively through tissue)[33,48,49]. It is expected that optimally choosing particle sizes will reduce the magnetic gradients needed to drive particle through the RWM and will allow push at further distances with smaller magnets.

16.4   OPTIMISING PERMANENT MAGNET ARRAYS TO CREATE DEEPER FORCES

The reach of magnetic drug delivery—the distance from where the magnets are to where they can still effectively capture particles—depends on the magnetic force, which falls off quickly with distance from magnets[2,50]. Insufficient reach of magnetic force has limited the applicability of magnetic drug delivery. In cancer, it has limited treatment to shallow tumours[39,51]. Increasing the reach to greater depths would allow treatment of a greater number of disease profiles and patients. Improved magnet design can shape magnetic fields and forces to extend the reach of magnetic targeting[8,52,53]. Here, we consider composite permanent magnets built up of an array of sub-magnets, each magnetised in a specific direction (these are commonly referred to as Halbach arrays)[54], and we optimally choose the magnetisation directions to extend magnetic forces deeper into the body. In the following, we show how this can be achieved based on an optimisation over simulations (we have not yet demonstrated this optimal Halbach capability in experiments).

In previous research, permanent magnets and electromagnets have been designed to optimise magnetic fields for various applications, including motion actuators[55,56], MRI machines[57], magnetic levitation systems[58] and magnetic drug targeting[52,59–61]. Prior Halbach designs have been chosen based on both analytic models[62] and finite element simulations[63], and their optimisation has ranged from a comparison of different cases[8,64] to optimisation by sequential linear programming[65].

Figure 16.9 shows the currently considered Halbach optimisation problem in two spatial dimensions (the setup and method of solution is the same for three dimensions). The magnet is made up of a small number of sub-magnets, here nine, each of which is to be magnetised in an optimal direction to maximise the pull (or push) force on nanoparticles. If the strength of the magnets is unrestricted, then the magnetic force can be increased simply by making all the magnets stronger. However, there are safety limitations on the strength of the magnetic field that can be applied across the human body (the United States Food and Drug Administration currently considers 8 T fields safe for adults and up to 4 T appropriate for children)[66], as well as practical engineering constraints on the strength of permanent magnets (inexpensive magnets with strengths of around 1 T are commercially available in various shapes and sizes). Such strong magnets can be practically assembled into Halbach arrays, see, for example, Refs. [67,68]. Thus, a sensible optimisation problem is to maximise the magnetic force at a deep target location subject to a constraint on the maximum allowable magnetic field strength for each sub-magnet.

We phrase the optimisation of Figure 16.9 as a quadratic optimisation problem which, in contrast to using heuristic methods (such as neural networks or genetic algorithms), allows us to use an optimisation technique that has known rigorous bounds on the deviation away from true optimality. Meaning, when a solution is found, it is known that this solution is within a certain distance (the bound) from the true global optimum. In particular, we use semidefinite programming methods[70] and can guarantee that we find globally optimal solutions because our bounds converge to zero— thus, we can rigorously prove that we have found the very best Halbach array designs possible for the force maximisation problem shown in Figure 16.9[69].

The quadratic problem arises naturally as follows. The magnetic field around a uniformly magnetised rectangular magnet is known analytically[44], and we can use it to express the magnetic field

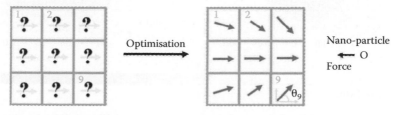

Unknown optimal magnetisation   Optimal magnetisation (sample)

FIGURE 16.9 Schematic of the Halbach array optimisation problem. The goal is to pick the magnetisation of each sub-magnet so as to maximise the pull (or push) force at a deep position. This translates to finding the optimal magnetisation angle θ for each of the sub-magnets to maximise the pull or push force at a deep location. (Figure taken from Sarwar, A., Nemirovski, A., and Shapiro, B., Optimal Halbach permanent magnet designs for maximally pulling and pushing nanoparticles, *J. Magn. Magn. Mater.*, in-press. With permission. Copyright 2010 IEEE.)

around any Halbach array. The magnetic field from two or more magnets can be added together to establish the net magnetic field[54]. Let $\vec{A}(x, y)$ represent the analytical expression for the magnetic field around a rectangular magnet that is uniformly magnetised along the positive horizontal axis, and let $\vec{B}(x, y)$ represent the analytical expression for the same magnet uniformly magnetised along the positive vertical axis. Then, the magnetic field when this magnet is uniformly magnetised at an angle θ is given by $\vec{A}(x, y)\cos\theta + \vec{B}(x, y)\sin\theta$. The magnetic field $\vec{H}$ around the entire Halbach array is generated by adding together similar expressions for all the sub-magnets in the Halbach array. For the ith magnet of the Halbach array with magnetisation direction θ_i, (which is yet to be determined), let $\alpha_i = \cos\theta_i$ and $\beta_i = \sin\theta_i$. Then, the magnetic field created at the location (x_0, y_0) by the Halbach array of N magnets is

$$\vec{H}(x_0, y_0) = \sum_{i=1}^{N} \alpha_i \vec{A}_i(x_0, y_0) + \beta_i \vec{B}_i(x_0, y_0), \tag{16.11}$$

which states that the magnetic field at (x_0, y_0) is the sum of the magnetic fields from all the sub-magnets of the array. The coefficients α_i and β_i are the unknown design variables.

In order to limit the strength of any given magnet in the Halbach array, here we choose 1 T as the limit, the constraint $\alpha_i^2 + \beta_i^2 \leq 1$ is imposed for all i. According to Equation 16.4, to maximise the horizontal force, we must maximise the horizontal component of the gradient of the magnetic field strength squared $\nabla \|\vec{H}\|^2$ at the desired location (x_0, y_0); either in the negative direction for pull or in the positive direction for maximal strength particle pushing. We can write the horizontal component of $\nabla \|\vec{H}(x_0, y_0)\|^2$ in matrix form as $\vec{q}^T P \vec{q}$ for an appropriate matrix P and a vector $\vec{q}$ that consists of the design variables α_i's and β_i's. Similarly, the constraint for limiting the strength of all the

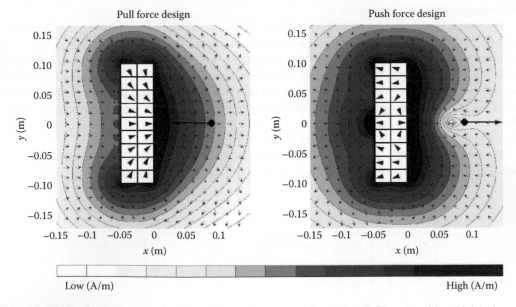

FIGURE 16.10 Optimal Halbach designs for maximum particle pulling (left) and pushing (right). Arrow heads inside each element show the optimal magnetisation directions of the 18 sub-magnets, thin arrows show the resulting external magnetic field and the contours show the strength of that field at each location. Both magnet arrays have a height of 20 cm, a width of 5 cm and a thickness (into the page) of 10 cm; and they maximise the force on nanoparticles at a distance of 9 cm away from the face (at the two black points). (Figure taken from Sarwar, A., Nemirovski, A., and Shapiro, B., Optimal Halbach permanent magnet designs for maximally pulling and pushing nanoparticles, *J. Magn. Magn. Mater.*, in-press. With permission. Copyright 2010 IEEE.)

sub-magnets below 1 T can be written in matrix form as $\vec{q}^T G_i \vec{q} \leq 1$ for appropriate G_i matrices. We can now rephrase the optimisation problem as: maximise (for push) or minimise (for pull) the quadratic cost function $\vec{q}^T P \vec{q}$ subject to the N quadratic constraints of $\vec{q}^T G_i \vec{q} \leq 1$, one for each Halbach array sub-magnet ($i = 1, 2, \ldots N$). This is a quadratic optimisation problem. We employ semidefinite quadratic programming[70] to solve this problem, finding the best possible values of α_i's and β_i's, so that the magnetic force (push or pull) is maximised at (x_0, y_0) (see Ref. [69] for mathematical details).

Sample globally optimal designs of an 18 element Halbach arrays are shown in Figure 16.10 to maximise pull (left panel) and push (right panel) forces on nanoparticles located 9 cm away from the right face of the array. The optimal pull Halbach array provides a pull force that is 80% greater than a magnet of same dimensions with uniform magnetisation across its width. Since it is not possible to compare pushing against a single magnet, as a single magnet will always pull particles in, we compare the optimal 9×2 Halbach pushing design against a minimal Halbach array consisting of just two rectangular magnets of identical size, one on top of the other. The optimal Halbach array provides a push force that is more than three times that provided by this reference two-element Halbach array.

16.5 ELECTROMAGNET CONTROL TO DIRECT NANOPARTICLES TO DEEP INTERNAL TARGETS

Focusing ferrofluids to deep tissue targets *in vivo* is a challenge. As part of an effort to work towards this goal, we have developed control algorithms and tested them in simulations[13,71,72], as well as verified them in initial *in vitro* experiments[26,73]. These control designs are now being extended to account for tissue properties and vasculature network geometries, are being tested in excised tissue samples and will be evaluated in animal experiments in the future.

Past research in control of particles and magnetisable objects has included magnetically assisted surgical procedures[74–79], MRI control of ferromagnetic cores[80–82] and implantable robots[83–85], ferrofluid droplet levitation[86], magnetic tweezers[87–92] and nanoparticle magnetic drug delivery in animal and human studies[32,39,93–95]. Means to manipulate a rigid permanent magnet through the brain with a view to guiding the delivery of hyperthermia to brain tumours are presented in Creighton and Ritter[45,96]. In these two references, a point-wise optimisation is stated for the magnetic force on the implant and example numerical solutions are shown, which display jumps and singularities similar to the ones we had to overcome in the following. Creighton and Ritter's focus changed to magnetically assisted cardiovascular surgical procedures and led to the formation of Stereotaxis (www.stereotaxis.com/). This company now uses magnetic control to guide catheters, endoscopes and other tools with magnetic tips for precision treatment of cardiac arrhythmias and other cardiovascular interventions. Control of magnetisable devices and ferromagnetic cores using an MRI machine as the actuator are presented by Martel and co-workers[80–82] who also discusses manipulation of implantable magnetic robots[83–85] and magnetic guidance of swimming magnetotactic bacteria[97,98]. In addition, recent work has demonstrated targeting and control of magnetised cells using a high-field MRI for potential treatments in cell transplantation surgeries[99,100]. In the current book, in Chapter 14, Rosengart and co-workers discuss recent efforts in targeting magnetic nanoparticles to occluded arteries in the brain. Overall, it is clear that improved techniques to control magnets to better direct magnetic implants, particles and focus ferrofluids to diseases locations will benefit a wide range of clinical applications.

To demonstrate the development of dynamic electromagnet control algorithms and to showcase challenges for deep manipulation of ferrofluids even in an *in vitro* setting, we start with optimal manipulation of a single droplet of ferrofluid[26]. Four electromagnets are arranged around a petri dish, which contains a single drop of ferrofluid. Control is achieved by feedback: an overhead camera and imaging software sense the real-time position of the droplet, the electromagnets are then optimally actuated at each moment so as to move the droplet from where it is towards where it

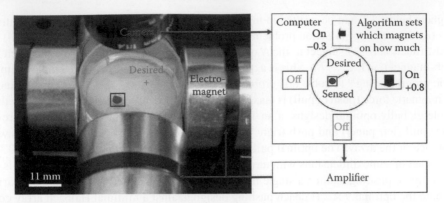

FIGURE 16.11 Four magnets around a petri dish containing a ferrofluid droplet create a test bed to develop and validate magnet control algorithms. The camera, a computer, an amplifier and the four electromagnets are connected in a feedback loop. The camera observes the current location of the droplet; the computer, using an optimal non-linear control algorithm[73], computes the electromagnet actuations required to move the droplet from where it is to where it should be; and the amplifier applies the needed voltages to do so. (Figure taken from *J. Magn. Magn. Mater.*, 323, Probst, R., Lin, J., Komaee, A., Nacev, A., Cummins, Z., and Shapiro, B., Planar steering of a single ferrofluid drop by optimal minimum power dynamic feedback control of four electromagnets at a distance, 885–896, Copyright 2011, with permission from Elsevier.)

should be (Figure 16.11). [In an *in vivo* setting, it is envisioned that deep real-time ferrofluid sensing could be achieved by advances in ultra-fast MRI[101]; or magnetic particles could be made radioactive, like imaging agents, and then sensed by fast gamma cameras[102,103] or by modified positron emission tomography (PET)[104].]

In order to choose control of the electromagnets to manipulate the droplet, we need to know how the motion of the droplet depends on the electromagnet actuation. Let $\vec{H}_1(x,y)$, $\vec{H}_2(x,y)$, $\vec{H}_3(x,y)$ and $\vec{H}_4(x,y)$ be the magnetic fields in the horizontal plane across the petri dish when each magnet is turned on with a 1 Ampere current. Each field is a 90 degree turn of the previous one, and the first field can be computed readily, for example, by COMSOL or by a semi-analytic method as in Ref. [26]. Denote the electrical currents in each of the four magnets by u_1, u_2, u_3 and u_4. Then, by the linearity of the magnetostatic Equations 16.1 through 16.3, and when operating the magnets below their saturation current, the total magnetic field inside the petri dish can be written as follows:

$$\vec{H}(x,y,t) = u_1(t)\vec{H}_1(x,y) + u_2(t)\vec{H}_2(x,y) + u_3(t)\vec{H}_3(x,y) + u_4(t)\vec{H}_4(x,y). \qquad (16.12)$$

During feedback control, we do not have direct access to the currents u_1, u_2, u_3 and u_4 because we cannot instantaneously charge a magnet to any desired strength. Instead, we control the vector of voltages $\vec{V}(t) = [V_1(t) \ V_2(t) \ V_3(t) \ V_4(t)]^T$. To first order, the current in each magnet is related to these voltages by simple time delay dynamics[105]. Our control corrects for this time delay by a temporal filter[73]. Substituting Equation 16.12 into the force Equation 16.4 gives the mapping from electromagnet currents to droplet velocity:

$$\frac{d}{dt}[x(t)\ y(t)] = k'\nabla\left\|\vec{H}(x,y,t)\right\|^2 = k'\nabla\left\|\sum_{i=1}^{4} u_i(t)\vec{H}_i(x,y)\right\|^2$$

$$= k'\left[\vec{u}^T(t)P_x(x,y)\vec{u}(t) \quad \vec{u}^T(t)P_y(x,y)\vec{u}(t)\right], \qquad (16.13)$$

where

$[x(t)\ y(t)]$ is the current location of the droplet in the petri dish

k' is a modified magnetic drift coefficient that reflects the balance of magnetic and viscous forces on a whole drop (as opposed to a single nanoparticle, see Ref. [26] for details)

The last equality is a compact matrix representation with superscript T denoting vector transpose and the matrices P_x and P_y defined as follows:

$$P_x(x,y) = \left[\frac{\partial \left(\vec{H}_i(x,y) \cdot \vec{H}_j(x,y) \right)}{\partial x} \right]_{4\times 4},$$

$$P_y(x,y) = \left[\frac{\partial \left(\vec{H}_i(x,y) \cdot \vec{H}_j(x,y) \right)}{\partial y} \right]_{4\times 4}. \tag{16.14}$$

Together with the time delay dynamics, this is a quadratic map from electromagnet currents to droplet motion, and this quadratic map depends non-linearly on space (magnetic forces fall off quickly with the droplet distance from magnets). The goal now is to optimally invert these dynamics: to choose actuations to manipulate the droplet with minimum electromagnet power.

The control operates by always directing the droplet from where it is sensed to be towards where it should go (Figure 16.11)—this enables both holding the ferrofluid at a target location (the control continually puts it back) or steering it along desired trajectories (the control is always moving it towards its next location). The task of the control algorithm is to decide how to optimally actuate the four magnets to create a droplet velocity $\vec{v} = k' \nabla \left\| \vec{H}(x,y,t) \right\|^2$ in Equation 16.13 that always points from the current droplet location to its next target location. This constraint on the droplet velocity set two degrees of freedom for the four control currents *via* two quadratic equations:

$$k'\ \vec{u}^T [P_x(x,y)]\ \vec{u} = v_x.$$
$$k'\ \vec{u}^T [P_y(x,y)]\ \vec{u} = v_y. \tag{16.15}$$

Hence, there are still two control degrees of freedom remaining. To achieve deep control, it is advantageous to choose the control that uses the minimum amount of electromagnet power to achieve a given velocity, and this is equivalent to choosing a control that minimises the quadratic cost function:

$$J = \| \vec{u} \|^2 = \vec{u}^T \vec{u}. \tag{16.16}$$

Such a control choice would use minimal magnet power to achieve maximal motion in the desired direction.

The solution of this constrained optimisation problem is detailed in Refs. [26,73]—it is a challenging problem with an interesting and intuitive solution. As the droplet traverses the petri dish, there are qualitative shifts in the kind of actuation strategy that is optimal, as shown in Figure 16.12 for moving a droplet from left to right. The optimal actuation switches from the top and bottom magnets to mainly the right magnet as the droplet is moved from left to right. This is because when the droplet is far away from the right magnet, it can be actuated to the right with lower current by using the closer top and bottom magnets; as the droplet approaches the right magnet, the optimal strategy switches to using only the right magnet. One difficulty comes from ensuring that these

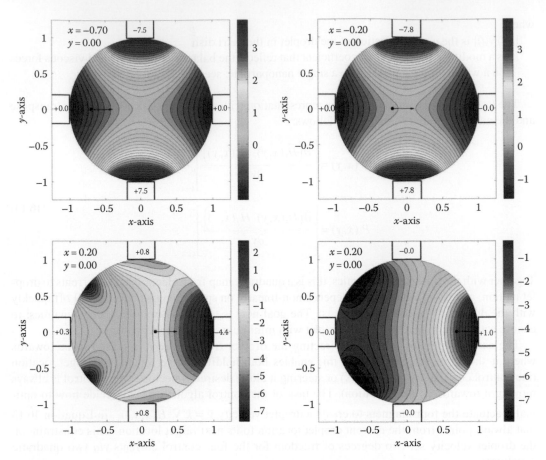

FIGURE 16.12 Magnetic energy and magnet actuation for control of a ferrofluid droplet from left to right. Each panel is for a different (x, y) droplet location (the black dot). The black arrow is the direction of the desired (and thus applied) magnetic force, the colour is the magnetic potential, which is equal to minus the magnetic field intensity squared (shown on a logarithmic scale), and the text inside each magnet states the current through that magnet (positive for clockwise current, negative for counter-clockwise). Notice how the optimal strategy switches from mainly the top and bottom magnets in the first two panels to only the right magnet in the last panel. (Figure taken from *J. Magn. Magn. Mater.,* 323, Probst, R., Lin, J., Komaee, A., Nacev, A., Cummins, Z., and Shapiro, B., Planar steering of a single ferrofluid drop by optimal minimum power dynamic feedback control of four electromagnets at a distance, 885–896, Copyright 2011, with permission from Elsevier.)

switches are done smoothly in time as the droplet traverses its domain, and this is achieved by a non-linear temporal filter that ensures the needed direction of the magnetic force is preserved, while the switch is smoothed in time[73].

A sample experimental result is shown in Figure 16.13. Here, a single droplet of ferrofluid was steered along a UMD path (for University of Maryland). The average error between the desired and the actual position of the ferrofluid droplet is defined as:

$$e_{\text{path}} = \frac{1}{T} \int_0^T \left\| \vec{x}_{\text{desired}}(t) - \vec{x}_{\text{measured}}(t) \right\| dt, \tag{16.17}$$

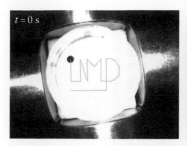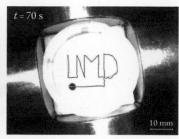

FIGURE 16.13 Minimum power manipulation of a 20 μL (1.7 mm radius) ferrofluid droplet along a UMD path with a 1.6 mm/s average velocity. (Figure taken from *J. Magn. Magn. Mater.,* 323, Probst, R., Lin, J., Komaee, A., Nacev, A., Cummins, Z., and Shapiro, B., Planar steering of a single ferrofluid drop by optimal minimum power dynamic feedback control of four electromagnets at a distance, 885–896, Copyright 2011, with permission from Elsevier.)

where T is the amount of time it took to traverse the entire path. In the experiment shown, this average error was 0.6 mm. Multiple additional experimental cases are presented in Ref. [26].

The experiment of Figure 16.13 shows precision control between magnets of a single object, a droplet of ferrofluid held together by surface tension. But in a patient, blood flow quickly distributes an injected ferrofluid throughout the body and there is a need to refocus that fluid to internal targets, for example, to a deep tissue tumour. Earnshaw's classic theorem[106], first published in 1842, applies to particles in a magnetic field and, *via* Maxwell's equations, states that no static magnetic field can create an internal stable energy trap for the particles[40]. One can imagine using barriers in the human body, such as blood vessels or aneurism walls, as local correctly oriented containment vessels to hold particles against opposed magnetic forces and so bypass the theorem. However, this is only practical for local targeting (a single vessel, a single aneurism). For deep targeting with external magnets, there is no guarantee that there will be barriers in the right places or that they will be correctly oriented.

Instead, we have been working on methods to move ferrofluid from concentrated locations at the edge of a domain to a central target inside the domain with minimal spreading. On average, this creates a ferrofluid hot spot at a deep location. This effort is still in its infancy: we have neither accounted for traversing the fluid through a dense vasculature network nor have we demonstrated our algorithms experimentally using the test bed shown in Figure 16.11 (the single droplet can be replaced by a distributed ferrofluid). So far we have only demonstrated theoretical results and have validated them in simulations. Our first result, reported in Ref. [13], shows a control strategy chosen by intuition that creates two ferrofluid hot spots at the edge of a circular domain and then moves them to the centre with minimal spreading. The sequence is repeated and creates hot spots on average at the centre and edges—in a patient this would correspond to high chemotherapy concentrations at the tumour and some skin locations but with a decreased systemic concentration. The difficult part is moving the ferrofluid a significant distance while limiting fluid spreading, and recently, we showed how this can be done in an optimal way[72]. We again used constrained quadratic programming[107], now to optimise the actuation of eight electromagnets, to move a droplet from edge to centre with minimal spreading (Figure 16.14, details in Komaee[72]).

Thus, we have shown preliminary theoretical results for optimal control of electromagnets to manipulate and focus ferrofluids to deep targets between magnets. Now these control schemes must be extended (e.g. to optimally control ferrofluids through blood vessel networks) and must be experimentally demonstrated, in gels, phantoms, excised tissue and in animals, to show the feasibility of focusing magnetic therapy to deep tissue targets with external electromagnets.

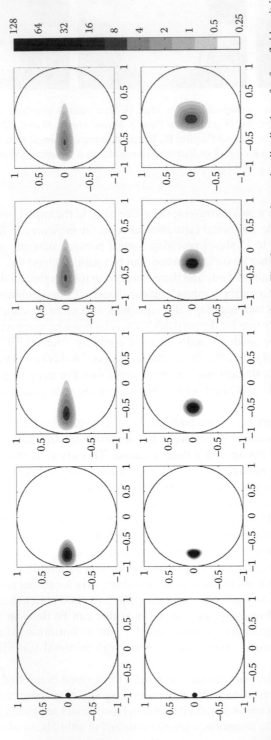

FIGURE 16.14 Moving a ferrofluid from edge to centre with minimal spreading (in simulations). The first row shows the distribution of a ferrofluid as it is pulled right by a single magnet — since the magnet pulls harder on the fluid closer to it, it acts to stretch the droplet out. The second row shows manipulation of the droplet to the centre by eight magnets optimally coordinated together to achieve the movement to centre with minimal spreading. (The bar on the right shows a non-dimensional logarithm concentration scale). (Figure taken from Komaee, A. and Shapiro, B., Magnetic steering of a distributed ferrofluid spot towards a deep target with minimal spreading, *IEEE Conference on Decision and Control*, Orlando, FL. With permission. Copyright 2011 IEEE.)

16.6 CONCLUSION

The magnetic drug community has expended tremendous (successful) effort in developing magnetic nanoparticles and other magnetic carriers, making them therapeutic and showing their biocompatibility and efficacy in cell cultures, animal models and even human clinical trials. In order to better direct such carriers to their intended *in vivo* targets, there is a need, and significant room, to improve the design and control of the magnetic systems that manipulate the carriers. The results described in this chapter show that simple arrangements of even two magnets can magnetically inject particles (Section 16.3); that magnetic forces, and hence the depth of focusing, can be extended deeper into the body by optimal design of magnet Halbach arrays (Section 16.4); and that magnets working in concert can steer a single droplet of ferrofluid or direct a ferrofluid to a deep target with minimal spreading (Section 16.5). The design of these magnet systems relied on first-principles modeling; numerical simulations of magnetic fields, gradients, forces and the resulting particle transport *in vivo*; a non-dimensional analysis and validation against experiments (Sections 16.1 and 16.2). This type of quantitative and validated modelling and analysis is necessary for rational and optimal design of next-generation magnetic systems to better manipulate nanoparticles in patients.

However, the reported results are only a start. There are multiple challenges to move these results from simulations, *in vitro* experiments and initial *in vivo* demonstrations (for the inner ear application) to clinical use. First, there is a lack of sufficient quantitative data on how nanoparticles behave *in vivo*. It is not enough to know biodistribution of particles to mouse organs. To design magnetic systems to successfully target particles to different parts of the body, it is necessary to know how quickly nanoparticles move through tissue, in blood, through vessel walls, the RWM, fat, muscle and through different types of tumours. This type of data has not been measured systematically; and without it, it is hard to do things like choose the best particle sizes or design controllers to manipulate nanoparticles through tissue effectively. Our group is currently building experimental systems that will measure *in vivo* particle properties, such as diffusion and magnetic velocities, in excised tissue, but that needs to happen on a broader scale—in other groups and for a variety of particles, tissue types and clinical needs.

Second, and related, there is a need to start designing magnetic carriers *for* specific clinical applications. What is the best carrier to safely and most easily penetrate deep into specific tumour types? Are flexible carriers that can deform past obstructions better than rigid particles? Beyond choosing the best size, are there optimal shapes, and coatings, to allow improved extravasation? It is likely that the carrier that is the best for reaching deep into tumours is not going to be the same as the one that is most suited to moving into diabetic wounds, through the RWM or into the skin.

This leads into the third and, possibly, most important point—the magnetic drug delivery community is rich in multidisciplinary research: we routinely bring together clinicians with experts in biology, chemistry and magnetism. But it is still not enough—we also need experts in microvasculature working with, and understanding, individuals who can mathematically model tissue resistance. We need optimisation experts who can take those equations and, together with material scientists and chemists, help design next-generation magnetic carriers that will move further, easier and more safely under magnetic fields. As such fields better interlink, there will be a stronger basis on which to design magnet delivery systems to more precisely direct magnetic carriers to target locations for ear and skin diseases, for strokes, cancer and diabetes.

ACKNOWLEDGEMENTS

A large number of collaborators have made this research possible. Thanks go to Andreas Lübbe, Catherine Beni, Oscar Bruno, Christian Bergemann, Michael Emmert-Buck, Jaime Rodriguez-Canales, Raymond Sedwick, Michael Tangrea and Irving Weinberg. Funding support was provided by the National Science Foundation (NSF) and the National Institute of Health (NIH) and is gratefully acknowledged.

REFERENCES

1. Ganguly, R., Zellmer, B. and Puri, I. K. 2005. Field-induced self-assembled ferrofluid aggregation in pulsatile flow. *Physics of Fluids*, 17, 097104 (1–8).
2. Rotariu, O. and Strachan, N. J. C. 2005. Modeling magnetic carrier particle targeting in the tumor microvasculature for cancer treatment. *Journal of Magnetism and Magnetic Materials – Proceedings of the Fifth International Conference on Scientific and Clinical Applications of Magnetic Carriers*, 293, 639–646.
3. David, A. E., Cole, A. J., Chertok, B., Park, Y. S. and Yang, V. C. 2011. A combined theoretical and in vitro modeling approach for predicting the magnetic capture and retention of magnetic nanoparticles in vivo. *Journal of Controlled Release*, 152(1), 67–75.
4. Huang, Z. Y., Pei, N., Wang, Y. Y., Xie, X. X., Sun, A. J., Shen, L., Zhang, S. N., Liu, X. B., Zou, Y. Z., Qian, J. Y. and Ge, J. B. 2010. Deep magnetic capture of magnetically loaded cells for spatially targeted therapeutics. *Biomaterials*, 31, 2130–2140.
5. Cherry, E. M., Maxim, P. G. and Eaton, J. K. 2010. Particle size, magnetic field, and blood velocity effects on particle retention in magnetic drug targeting. *Medical Physics*, 37, 175–182.
6. Furlani, E. P. and Ng, K. C. 2008. Nanoscale magnetic biotransport with application to magnetofection. *Physical Review E*, 77(6), D61914–22.
7. Cregg, P. J., Murphy, K. and Mardinoglu, A. 2008. Calculation of nanoparticle capture efficiency in magnetic drug targeting. *Journal of Magnetism and Magnetic Materials*, 320, 3272–3275.
8. Häfeli, U. O., Gilmour, K., Zhou, A., Lee, S. and Hayden, M. E. 2007. Modeling of magnetic bandages for drug targeting: Button vs. Halbach arrays. *Journal of Magnetism and Magnetic Materials*, 311, 323–329.
9. Feynman, R. P., Leighton, R. B. and Sands, M. 1964. *The Feynman Lectures on Physics*, Addison-Wesley Publishing Company, Boston, MA.
10. Fleisch, D. A. 2008. *A Student's Guide to Maxwell's Equations*, Cambridge, U.K., Cambridge University Press.
11. Forbes, Z. G., Yellen, B. B., Barbee, K. A. and Friedman, G. 2003. An approach to targeted drug delivery based on uniform magnetic fields. *IEEE Transactions on Magnetics*, 39, 3372–3377.
12. Forbes, Z. G., Yellen, B. B., Halverson, D. S., Fridman, G., Barbee, K. A. and Friedman, G. 2008. Validation of high gradient magnetic field based drug delivery to magnetizable implants under flow. *IEEE Transactions on Biomedical Engineering*, 55, 643–649.
13. Shapiro, B. 2009. Towards dynamic control of magnetic fields to focus magnetic carriers to targets deep inside the body. *Journal of Magnetism and Magnetic Materials*, 321, 1594–1599.
14. Furlani, E. P. and Ng, K. C. 2006. Analytical model of magnetic nanoparticle transport and capture in the microvasculature. *Physical Review E*, 73, 61919.
15. Grief, A. D. and Richardson, G. 2005. Mathematical modeling of magnetically targeted drug delivery. *Journal of Magnetism and Magnetic Materials – Proceedings of the Fifth International Conference on Scientific and Clinical Applications of Magnetic Carriers*, 293, 455–463.
16. Avilés, M. O., Ebner, A. D., Chen, H. T., Rosengart, A. J., Kaminski, M. D. and Ritter, J. A. 2005. Theoretical analysis of a transdermal ferromagnetic implant for retention of magnetic drug carrier particles. *Journal of Magnetism and Magnetic Materials*, 293, 605–615.
17. Iacob, G., Rotariu, O. and Chiriac, H. 2004. A possibility for local targeting of magnetic carriers. *Journal of Optoelectronics and Advanced Materials*, 6, 713–717.
18. Shaw, S., Murthy, P. V. S. N. and Pradhan, S. C. 2010. Effect of non-Newtonian characteristics of blood on magnetic targeting in the impermeable micro-vessel. *Journal of Magnetism and Magnetic Materials*, 322(8), 1037–1043.
19. Rosensweig, R. E. 1985. *Ferrohydrodynamics*, Mineola, NY, Dover Publications, Inc.
20. Zubarev, A. Y., Odenbach, S. and Fleischer, J. 2002. Rheological properties of dense ferrofluids. Effect of chain-like aggregates. *Journal of Magnetism and Magnetic Materials*, 252, 241–243.
21. Wu, M., Xiong, Y., Jia, Y., Niu, H., Qi, H., Ye, J. and Chen, Q. 2005. Magnetic field-assisted hydrothermal growth of chain-like nanostructure of magnetite. *Chemical Physics Letters*, 401, 374–379.
22. Mendelev, V. S. and Ivanov, A. O. 2004. Ferrofluid aggregation in chains under the influence of a magnetic field. *Physical Review E*, 70, 51502.
23. Wang, Z. and Holm, C. 2003. Structure and magnetic properties of polydisperse ferrofluids: A molecular dynamics study. *Physical Review E*, 68, 41401.
24. Nacev, A., Beni, C., Bruno, O. and Shapiro, B. 2010. Magnetic nanoparticle transport within flowing blood and into surrounding tissue. *Nanomedicine*, 5, 1459–1466.
25. Nacev, A., Beni, C., Bruno, O. and Shapiro, B. 2011. The behaviors of ferro-magnetic nano-particles in and around blood vessels under applied magnetic fields. *Journal of Magnetism and Magnetic Materials*, 323, 651–668.

26. Probst, R., Lin, J., Komaee, A., Nacev, A., Cummins, Z. and Shapiro, B. 2011. Planar steering of a single ferrofluid drop by optimal minimum power dynamic feedback control of four electromagnets at a distance. *Journal of Magnetism and Magnetic Materials*, 323, 885–896.

27. Voltairas, P. A., Fotiadis, D. I. and Michalis, L. K. 2002. Hydrodynamics of magnetic drug targeting. *Journal of Biomechanics*, 35, 813–821.

28. Yellen, B. B., Forbes, Z. G., Halverson, D. S., Fridman, G., Barbee, K. A., Chorny, M., Levy, R. and Friedman, G. 2005. Targeted drug delivery to magnetic implants for therapeutic applications. *Journal of Magnetism and Magnetic Materials*, 293, 647–654.

29. Avilés, M. O., Ebner, A. D. and Ritter, J. A. 2008. Implant assisted-magnetic drug targeting: Comparison of in vitro experiments with theory. *Journal of Magnetism and Magnetic Materials*, 320(21), 2704–2713.

30. Ganguly, R., Gaind, A. P., Sen, S. and Puri, I. K. 2005. Analyzing ferrofluid transport for magnetic drug targeting. *Journal of Magnetism and Magnetic Materials*, 289, 331–334.

31. Khashan, S. A. and Haik, Y. 2006. Numerical simulation of biomagnetic fluid downstream an eccentric stenotic orifice. *Physics of Fluids*, 18, 113601.

32. Lubbe, A. S., Bergemann, C., Riess, H., Schriever, F., Reichardt, P., Possinger, K., Matthias, M., Dorken, B., Herrmann, F., Gurtler, R., Hohenberger, P., Haas, N., Sohr, R., Sander, B., Lemke, A. J., Ohlendorf, D., Huhnt, W. and Huhn, D. 1996. Clinical experiences with magnetic drug targeting: A phase I study with 4'-epidoxorubicin in 14 patients with advanced solid tumors. *Cancer Research*, 56, 4686–4693.

33. Fournier, R. L. 2007. *Basic Transport Phenomena in Biomedical Engineering*, New York, Taylor & Francis.

34. Xu, H., Song, T., Bao, X. and Hu, L. 2005. Site-directed research of magnetic nanoparticles in magnetic drug targeting. *Journal of Magnetism and Magnetic Materials*, 293, 514–519.

35. Lubbe, A. S. 1997. Preclinical experiences with magnetic drug targeting: Tolerance and efficacy and clinical experiences with magnetic drug targeting: A phase I study with 4'-epidoxorubicin in 14 patients with advanced solid tumors—Reply. *Cancer Research*, 57, 3064–3065.

36. Widder, K. J., Morris, R. M., Poore, G., Howard, D. P. and Senyei, A. E. 1981. Tumor remission in Yoshida sarcoma-bearing rats by selective targeting of magnetic albumin microspheres containing doxorubicin. *Proceedings of the National Academy of Sciences of the United States of America*, 78, 579–581.

37. Widder, K. J., Senyei, A. E. and Ranney, D. F. 1979. Magnetically responsive microspheres and other carriers for the biophysical targeting of antitumor agents. *Advances in Pharmacology and Chemotherapy*, 16, 213.

38. Nacev, A., Kim, S. H., Rodriguez-Canales, J., Tangrea, M. A., Shapiro, B. and Emmet-Buck, M. R. in-press. A dynamic magnetic shift method to increase nanoparticle concentration in cancer metastases: A feasibility study using simulations on autopsy specimens. *International Journal of Nanomedicine*.

39. Lubbe, A. S., Bergemann, C., Huhnt, W., Fricke, T., Riess, H., Brock, J. W. and Huhn, D. 1996. Preclinical experiences with magnetic drug targeting: Tolerance and efficacy. *Cancer Research*, 56, 4694–4701.

40. Nacev, A., Komaee, A., Sarwar, A., Probst, R., Kim, S., Emmert-Buck, M., and Shapiro, B. to appear 2012. Towards control of magnetic fluids in patients: Directing therapeutic nanoparticles to disease locations *Control System Magazine*.

41. Swan, E. E., Mescher, M. J., Sewell, W. F., Tao, S. L. and Borenstein, J. T. 2008. Inner ear drug delivery for auditory applications. *Advanced Drug Delivery Reviews*, 60, 1583–1599.

42. Ganong, W. F. 2001. *Review of Medical Physiology*, New York, Lange/McGraw-Hill.

43. Kopke, R. D., Wassel, R. A., Mondalek, F., Grady, B., Chen, K., Liu, J., Gibson, D. and Dormer, K. J. 2008. Magnetic nanoparticles: Inner ear targeted molecule delivery and middle ear implant. *Audiology and Neurotology*, 11, 123–133.

44. Engel-Herbert, R. and Hesjedal, T. 2005. Calculation of the magnetic stray field of a uniaxial magnetic domain. *Journal of Applied Physics*, 97, 74504–74505.

45. Creighton, F. M. 1991. Control of magnetomotive actuators for an implanted object in brain and phantom materials. PhD, University of Virginia, Charlottesville, VA.

46. Shapiro, B., Dormer, K. and Rutel, I. B. 2010. A two-magnet system to push therapeutic nanoparticles. In: *8th International Conference on the Scientific and Clinical Applications of Magnetic Carriers*. American Institute of Physics, Conference Proceeding Series, Rostock, Germany.

47. Shapiro, B., Rutel, I. and Dormer, K. 2009. Therapeutic nano-particles magnetic injector system (MIS). *Frontiers in the Characterization and Control of Magnetic Carriers*. Clemson University, Clemson, SC.

48. Saltzman, W. M. 2001. *Drug Delivery: Engineering Principles for Drug Therapy*, New York, Oxford University Press.

49. Renkin, E. M. 1954. Filtration, diffusion, and molecular sieving through porous cellulose membranes. *Journal of General Physiology*, 38, 225–243.

50. Takeda, S.-I., Mishima, F., Fujimoto, S., Izumi, Y. and Nishijima, S. 2007. Development of magnetically targeted drug delivery system using superconducting magnet. *Journal of Magnetism and Magnetic Materials – Proceedings of the Sixth International Conference on the Scientific and Clinical Applications of Magnetic Carriers – SCAMC-06*, 311, 367–371.

51. Goodwin, S., Peterson, C., Hoh, C. and Bittner, C. 1999. Targeting and retention of magnetic targeted carriers (MTCs) enhancing intra-arterial chemotherapy. *Journal of Magnetism and Magnetic Materials*, 194, 132–139.

52. Fukui, S., Abe, R., Ogawa, J., Oka, T., Yamaguchi, M., Sato, T. and Imaizumi, H. 2007. Study on optimization design of superconducting magnet for magnetic force assisted drug delivery system. *Physica C: Superconductivity*, 463–465, 1315–1318.

53. Zablotskii, V., Pastor, J. M., Larumbe, S., Perez-Landazabal, J. I., Recarte, V. and Gomez-Polo, C. 2010. High-field gradient permanent micromagnets for targeted drug delivery with magnetic nanoparticles. *AIP Conference Proceedings*, 1311, 152–157.

54. Halbach, K. 1980. Design of permanent multipole magnets with oriented rare earth cobalt material. *Nuclear Instruments and Methods*, 169, 1–10.

55. Gutfrind, C., Jannot, X., Vannier, J. C., Vidal, P. and Sadarnac, D. 2010. Analytical and FEM magnetic optimization of a limited motion actuator for automotive application. In *XIX International Conference on Electrical Machines (ICEM)*, Rome, Italy, September 6–8.

56. Lee, M. G. and Gweon, D. G. 2004. Optimal design of a double-sided linear motor with a multi-segmented trapezoidal magnet array for a high precision positioning system. *Journal of Magnetism and Magnetic Materials*, 281(2–3), 336–346.

57. Wilson, J. L., Jenkinson, M., De Araujo, I., Kringelbach, M. L., Rolls, E. T. and Jezzard, P. 2002. Fast, fully automated global and local magnetic field optimization for fMRI of the human brain. *Neuroimage*, 17, 967–976.

58. Ham, C., Ko, W. and Han, Q. 2006. Analysis and optimization of a Maglev system based on the Halbach magnet arrays. *Journal of Applied Physics*, 99, 08P510.

59. Dames, P., Gleich, B., Flemmer, A., Hajek, K., Seidl, N., Wiekhorst, F., Eberbeck, D., Bittmann, I., Bergemann, C., Weyh, T., Trahms, L., Rosenecker, J. and Rudolph, C. 2007. Targeted delivery of magnetic aerosol droplets to the lung. *Nature Nanotechnology*, 2, 495–499.

60. Alexiou, C., Diehl, D., Henninger, P., Iro, H., Rockelein, R., Schmidt, W. and Weber, H. 2006. A high field gradient magnet for magnetic drug targeting. *IEEE Transactions on Applied Superconductivity*, 16, 1527–1530.

61. Slabu, I., Röth, A., Schmitz-Rode, T. and Baumann, M. 2009. Optimization of magnetic drug targeting by mathematical modeling and simulation of magnetic fields. In: *4th European Conference of the International Federation for Medical and Biological Engineering*, 22, 2309–2312.

62. Dwari, S. and Parsa, L. 2011. Design of Halbach array based permanent magnet motors with high acceleration. *IEEE Transactions on Industrial Electronics*, 58(9), 3768–3775.

63. Raich, H. and Blümler, P. 2004. Design and construction of a dipolar Halbach array with a homogeneous field from identical bar magnets: NMR Mandhalas. *Concepts in Magnetic Resonance Part B: Magnetic Resonance Engineering*, 23B(1), 16–25.

64. Hayden, M. E. and Hafeli, U. O. 2006. 'Magnetic bandages' for targeted delivery of therapeutic agents. *Journal of Physics: Condensed Matter*, 18, S2877–S2891.

65. Choi, J.-S. and Yoo, J. 2008. Design of a Halbach magnet array based on optimization techniques. *IEEE Transactions on Magnetics*, 44, 2361–2366.

66. Chakeres, D. W. and De Vocht, F. 2005. Static magnetic field effects on human subjects related to magnetic resonance imaging systems. *Progress in Biophysics and Molecular Biology*, 87, 255–265.

67. Zhang, X., Mahesh, V., Ng, D., Hubbard, R., Ailiani, A., O'hare, B., Benesi, A. and Webb, A. 2005. *Design, Construction and NMR Testing of a 1 Tesla Halbach. Permanent Magnet for Magnetic Resonance*, Boston, MA.

68. Chun, L. and Devine, M. 2009. Strong permanent magnet dipole with reduced demagnetizing effect. *IEEE Transactions on Magnetics*, 45, 4380–4383.

69. Sarwar, A., Nemirovski, A. and Shapiro, B. Optimal Halbach permanent magnet designs for maximally pulling and pushing nanoparticles. *Journal of Magnetism and Magnetic Materials*, in press.

70. Luo, Z.-Q., Ma, W.-K., So, A. M.-C., Ye, Y. and Zhang, S. 2010. Semidefinite relaxation of quadratic optimization problems. *IEEE Signal Processing Magazine*, 27, 20–34.

71. Shapiro, B., Probst, R., Potts, H. E., Diver, D. A. and Lubbe, A. 12–14 December 2007. Control to concentrate drug-coated magnetic particles to deep-tissue tumors for targeted cancer chemotherapy. In: *46th IEEE Conference on Decision and Control*, New Orleans, LA, 3901–3906.

72. Komaee, A. and Shapiro, B. 2011. Magnetic steering of a distributed ferrofluid towards a deep target with minimal spreading. In *50th IEEE Conference on Decision and Control and European Control Conference (CDC-ECC)*, Orlando, FL.

73. Komaee, A. and Shapiro, B. 2011. Steering a ferromagnetic particle by optimal magnetic feedback control. *IEEE Transactions on Control Systems Technology*, 99, 1–14.

74. Howard Iii, M. A., Ritter, R. C. and Grady, M. S. 26 September 1989. *Video tumor fighting system*. U.S. patent application.

75. Eyssa, Y. M. 2005. Apparatus and methods for controlling movement of an object through a medium using a magnetic field, patent application number 20020186526.

76. Ritter, R. C. 2001. Open field system for magnetic surgery. U.S. patent application, 20010038683.

77. Werp, P. R. 2002. Methods and apparatus for magnetically controlling motion direction of a mechanically pushed catheter, patent application number 20030125752.

78. Bova, F. J. and Friedman, W. A. 2003. Computer controlled guidance of a biopsy needle. U.S. patent application 09/975,200.

79. Ritter, R. C., Werp, P. R. and Lawson, M. A. 2000. Method and apparatus for rapidly changing a magnetic field produced by electromagnets. U.S. patent application 08/921,298.

80. Martel, S., Mathieu, J.-B., Felfoul, O., Chanu, A., Aboussouan, E., Yahia, L., Beaudoin, G., Soulez, G. and Mankiewicz, M. 2007. Automatic navigation of an untethered device in the artery of a living animal using a conventional clinical magnetic resonance imaging system. *Applied Physics Letters*, 90, 114105.

81. Mathieu, J.-B., Beaudoin, G. and Martel, S. 2006. Method of propulsion of a ferromagnetic core in the cardiovascular system through magnetic gradients generated by an MRI system. *IEEE Transactions on Biomedical Engineering*, 53, 292–299.

82. Tamaz, S., Gourdeau, R., Chanu, A., Mathieu, J.-B. and Martel, S. 2008. Real-time MRI-based control of a ferromagnetic core for endovascular navigation. *IEEE Transactions on Biomedical Engineering*, 55, 1854–1863.

83. Yesin, K. B., Vollmers, K. and Nelson, B. J. 2006. Modeling and control of untethered biomicrorobots in a fluidic environment using electromagnetic fields. *The International Journal of Robotics Research*, 25, 527–536.

84. Yesin, K. B., Vollmers, K. and Nelson, B. J. 2004. Analysis and design of wireless magnetically guided microrobots in body fluids. In: *Robotics and Automation, 2004. Proceedings. ICRA '04. 2004 IEEE International Conference on*. Vol. 2, 1333–1338.

85. Mathieu, J. B. and Martel, S. 2007. In vivo validation of a propulsion method for untethered medical microrobots using a clinical magnetic resonance imaging system. In *International Conference on Intelligent Robots and Systems (IROS)*, San Diego, CA.

86. Potts, H. E., Barrett, R. K. and Diver, D. A. 2001. Dynamics of freely-suspended drops. *Journal of Physics D: Applied Physics*, 34, 2629–2636.

87. Alenghat, F. J., Fabry, B., Tsai, K. Y., Goldmann, W. H. and Ingber, D. E. Analysis of cell mechanics in single vinculin-deficient cells using a magnetic tweezer. *Biochemical and Biophysical Research Communications*, 277, 93–99.

88. Hosu, B. G., Jakab, K., Banki, P., Toth, F. I. and Forgacs, G. Magnetic tweezers for intracellular applications. *Review of Scientific Instruments*, 74, 4158–4163.

89. Vries, A. H. B. D., Krenn, B. E., Driel, R. V. and Kanger, J. S. Micro magnetic tweezers for nanomanipulation inside live cells. *Biophysical Journal*, 88, 2137–2144.

90. Kanger, J. S., Subramaniam, V. and Van Driel, R. Intracellular manipulation of chromatin using magnetic nanoparticles. *Chromosome Research: An International Journal on the Molecular, Supramolecular and Evolutionary Aspects of Chromosome Biology*, 16, 511–522.

91. Amblard, F., Yurke, B., Pargellis, A. and Leibler, S. 1996. A magnetic manipulator for studying local rheology and micromechanical properties of biological systems. *Review of Scientific Instruments*, 67, 818–827.

92. Neuman, K. C. and Nagy, A. 2008. Single-molecule force spectroscopy: Optical tweezers, magnetic tweezers and atomic force microscopy. *Nature Methods*, 5, 491–505.

93. Wilson, M. W., Kerlan, R. K., Jr., Fidelman, N. A., Venook, A. P., Laberge, J. M., Koda, J. and Gordon, R. L. 2004. Hepatocellular carcinoma: Regional therapy with a magnetic targeted carrier bound to doxorubicin in a dual MR imaging/conventional angiography suite—Initial experience with four patients. *Radiology*, 230, 287–293.

94. Lemke, A. J., Von Pilsach, M. I. S., Lubbe, A., Bergemann, C., Riess, H. and Felix, R. 2004. MRI after magnetic drug targeting in patients with advanced solid malignant tumors. *European Radiology*, 14, 1949–1955.

95. Polyak, B. and Friedman, G. 2009. Magnetic targeting for site-specific drug delivery: Applications and clinical potential. *Expert Opinion on Drug Delivery*, 6, 53–70.

96. Meeker, D. C., Maslen, E. H., Ritter, R. C. and Creighton, F. M. 1996. Optimal realization of arbitrary forces in a magnetic stereotaxis system. *IEEE Transactions on Magnetics*, 32, 320–328.

97. Martel, S., Mohammadi, M., Felfoul, O., Zhao, L. and Pouponneau, P. 2009. Flagellated magnetotactic bacteria as controlled MRI-trackable propulsion and steering systems for medical nanorobots operating in the human microvasculature. *The International Journal of Robotics Research*, 28, 571–582.

98. Martel, S., Felfoul, O. and Mohammadi, M. 2008. Flagellated bacterial nanorobots for medical interventions in the human body. In: *Biomedical Robotics and Biomechatronics, 2008. BioRob 2008. 2nd IEEE RAS and EMBS International Conference*, 264–269.

99. Riegler, J., Allain, B., Cook, R. J., Lythgoe, M. F. and Pankhurst, Q. A. 2011. Magnetically assisted delivery of cells using a magnetic resonance imaging system. *Journal of Physics D: Applied Physics*, 44, 055001.

100. Riegler, J., Wells, J. A., Kyrtatos, P. G., Price, A. N., Pankhurst, Q. A. and Lythgoe, M. F. 2010. Targeted magnetic delivery and tracking of cells using a magnetic resonance imaging system. *Biomaterials*, 31, 5366–5371.

101. Allen, E. D. and Burdette, J. H. 2001. Questions and answers in MRI.

102. Parker, S. I., Kenney, C. J., Gnani, D., Thompson, A. C., Mandelli, E., Meddeler, G., Hasi, J., Morse, J. and Westbrook, E. M. 2006. 3DX: An x-ray pixel array detector with active edges. *IEEE Transactions on Nuclear Science*, 53, 1676–1688.

103. Kenney, C. J., Segal, J. D., Westbrook, E. M., Paker, S., Hasi, J., Davia, C., Watts, S. and Morse, J. 2006. Active-edge planar radiation sensors. *Nuclear Instruments and Methods in Physics Research A*, 565, 272–277.

104. Devaraj, N. K., Keliher, E. J., Thurber, G. M., Nahrendorf, M. and Weissleder, R. 2009. 18F labeled nanoparticles for in vivo PET-CT imaging. *Bioconjugate Chemistry*, 20, 397–401.

105. Desoer, C. A. and Kuh, E. S. 1969. *Basic Circuit Theory: Chapters 1 through 10*, U.S., McGraw-Hill Inc.

106. Earnshaw, S. 1842. On the nature of the molecular forces which regulate the constitution of the luminiferous ether. *Transactions of the Cambridge Philosophical Society*, 7, 97–112.

107. Nesterov, Y., Wolkowicz, H. and Ye, Y. 2000. Nonconvex quadratic optimization. In: Wolkowicz, H., Saigal, R. and Vandenberghe, L. (eds.) *Handbook of Semidefinite Programming, Theory, Algorithms, and Applications*. Boston, MA, Kluwer.

Arash Komaee received his PhD in electrical engineering from University of Maryland, College Park in 2008. His research interests are in the areas of optimal control, stochastic decision making and control, estimation theory and its applications and non-linear filtering. In the past, he has done research on applications of these fields in free-space optics and optimal management of inventory systems. His current research focuses on targeted magnetic drug delivery, a method of concentrating drug-coated ferromagnetic nanoparticles in a targeted region of a patient's body such as the area around a solid tumour.

Roger Lee received his BS in neurophysiology from the University of Maryland at College Park. He is currently a research scientist involved in developing a better way to deliver drugs to the inner ear to treat for tinnitus.

Aleksander Nacev received a BS degree in aerospace engineering at the A. James Clark School of Engineering, University of Maryland, College Park in 2008, where he is also currently obtaining a PhD in the Fischell Department of Bioengineering. His research interests include magnetically targeted distributed ferrofluid control for therapeutic applications, specifically developing the methodology by utilising simulated, table top, cell culture and live animal experiments.

Roland Probst received his diploma in computer engineering from the University of Applied Sciences (Mannheim, Germany) and a PhD in aerospace engineering from the University of Maryland (College Park, Maryland). He is currently a research associate with the Fischell Department of Bioengineering at the University of Maryland. His main interest is to invent and develop smart yet simple methods to control micro- and nanoscale systems by electrokinetic and magnetic actuation. He is currently involved in the control of magnetic nanoparticles in patients and the development of novel electrokinetic tweezers for 3D nanoassembly applications.

Azeem Sarwar received his PhD in mechanical engineering from the University of Illinois at Urbana-Champaign (Urbana-Champaign, Illinois) in 2009 with a focus on robustness, stability, identification and adaptation of distributed systems. He also received a master's degree in mechanical engineering and another master's degree in mathematics from the same institution. He is currently working as a research associate with the Fischell Department of Bioengineering at the University of Maryland, College Park, Maryland. His current research is focused towards maximising the reach of magnetic drug delivery by optimally designing permanent magnets arrays.

Isaac B. Rutel received a BS degree in engineering physics from Worcester Polytechnic Institute (WPI) and a PhD from the Experimental Condensed Matter Physics group at Florida State University and the National High Magnetic field Laboratory (NHMFL). He is currently an assistant professor in the Department of Radiological Sciences at OUHSC College of Medicine in Oklahoma City, Oklahoma, and his research interests have since focused on practical innovations in clinical diagnostic physics phantoms, radiation shielding measurement technology, graduate education methodology and magnetically assisted delivery of therapeutic nanoparticles.

Didier A. Depireux received a BS in physics from the Université de Liège and a PhD in theoretical physics (with Jim Gates as advisor) from the University of Maryland at College Park, Maryland. After completing postdoctoral fellowships in Canada, he changed fields to computational neuroscience, and in particular studied auditory cortex with Shihab Shamma. He then moved to the University of Maryland School of Medicine, Department of Anatomy. In 2009, he (re)joined the University of Maryland at College Park, at the Institute for Systems Research with an adjunct appointment with the Department of Bio-Engineering. His current research is in modelling the perception of natural sounds, particularly in noisy environments and in the aetiology and possible prevention of tinnitus induced by noise trauma.

Kenneth J. Dormer received his BS in marine biology from Cornell University then MS and PhD in physiology/biology from UCLA. He is presently a professor of physiology at the University of Oklahoma Health Sciences Center and his research interests are in magnetic targeting of therapeutics. He recently co-founded NanoMed Targeting Systems, Inc., to commercialise such drug deliveries including one into the heart to prevent or suppress atrial fibrillation.

Benjamin Shapiro received a BS in aerospace engineering from the Georgia Institute of Technology and a PhD from the Control and Dynamical Systems option at the California Institute of Technology. He joined the University of Maryland in 2000. He is currently an associate professor in the Fischell Department of Bioengineering and the Institute for Systems Research at the University of Maryland and his lab is focused on precisely manipulating bioparticles (cells, quantum dots, etc.) in microfluidic systems and on control of magnetic nanoparticles in patients. In 2009, he spent his sabbatical at NIH and NIST: half-time within the Advanced Technology Center at the National Cancer Institute, the other half at the Center for Nanoscale Science and Technology at the National Institute of Standards and Technology. He is a Fulbright scholar (Germany 2009).

17 Nanocrystalline Oxides in Magnetic Fluid Hyperthermia

Emil Pollert and Karel Závěta

CONTENTS

17.1 HYPERTHERMIA, THERAPEUTIC TOOL IN CARCINOLOGY

The thermal treatment at elevated temperatures, called hyperthermia, of carcinomas is an alternative or supplement therapeutic modality to potentiate the effects of chemotherapy and radiotherapy. Essentially, it is based on a difference between physiology, in particular the physical and chemical properties of tumourous and healthy tissues. While the vasculature of the tumourous tissues is chaotic and consequently contains hypoxic areas with low pH, they do not exist in the healthy tissues where the architecture of vasculature is undisturbed. Therefore, the tissues invaded by cancerous cells are predisposed more easily to deteriorate at ~41°C–46°C in comparison to more resistant normally oxygenated healthy tissues.

The classical hyperthermia exploits various sources of the thermal energy, for example, hot water and microwave radiation. It is applied directly on the whole body of the patient or alternatively on a selected part where the tumour is localised. Some results accomplished by these techniques are given, for example, in Refs. [1–3]. Nevertheless, their use is limited, in particular by the difficulty to achieve a localised heating of a selected area and by poor diffusion of heat through the tissue.

Therefore, the development of the magnetic fluid hyperthermia (MFH) was found to be more promising, as it can eliminate these substantial shortcomings. The method is based on the use of magnetic mediators, particles or seeds, placed inside or in a close vicinity of the tumour, and subjected to an interaction with external alternating magnetic field. The magnetic energy is dissipated and finally converted to heat by various loss processes like hysteresis, relaxation or friction.

17.2 FUNDAMENTALS OF MAGNETIC FLUID HYPERTHERMIA

17.2.1 Magnetism in Nanoparticles

A specific feature distinguishing the magnetic nanoparticles (MNPs) from their bulk analogues is the dependence of their properties on the size.

The problem can be illustrated on an example of an MNP of a size corresponding to a single magnetic domain. One can easily understand that the resulting state depends on a competition between two energy contributions, that is, of the terms $K_u V$ and $k_B T$; here K_u is the anisotropy constant of uniaxial magnetocrystalline anisotropy, V the MNP volume, k_B the Boltzmann constant and T the absolute temperature. According to Néel, the magnetic moment of the particle relaxes to its equilibrium position with a relaxation time τ_N given by the relation (Néel[4] and Brown[5]):

$$\tau_N = \tau_0 \exp\left(\frac{K_u V}{k_B T}\right), \tag{17.1}$$

where τ_0 is a constant in the range of 10^{-9}–10^{-12} s.

If the energy barrier $\Delta E = K_u V \gg k_B T$, the relaxation time τ_N becomes much longer than the time of observation of the moment even for DC magnetic measurements, where the relevant time is of the order of 1–10^2 s. In the given time, the magnetic moment of the individual particle is stable and aligned parallel to the easy axis of magnetisation, and the MNP exhibits ferromagnetic or ferrimagnetic ordering. The magnetisation curve of such a particle dramatically depends on the orientation of the external field with respect to the easy axis. If these two directions are parallel, the curve is rectangular and the moment flips from one orientation of the saturated magnetisation M_s to the opposite one at a coercive field H_C given by

$$H_C = \frac{2 K_u}{M_s}. \tag{17.2}$$

For the other geometrically limiting case, when the applied field is perpendicular to the easy axis, the magnetisation curve is linear and displays no hysteresis.

A system of such randomly oriented and magnetically non-interacting MNPs behaves in a way described by the Stoner–Wohlfarth model[6] exhibiting hysteretic behaviour with remanent magnetisation and coercivity derived from the parameters of the individual particle.

If the volume V of the MNP is sufficiently small, the thermal energy $k_B T$ prevails over the energy barrier $K_u V$, and τ_N decreases below the observation time. This leads to spontaneous fluctuations of the direction of the magnetic moment between the two orientations along the (uniaxial) easy axis, and the time average of the magnetic moment vanishes. In an applied magnetic field, the behaviour of a system of such particles resembles a paramagnet and the state of the system is called superparamagnetic. We must have in mind, however, that this state is displayed below the Curie temperature T_C of the material of the particles, and the usual paramagnetic behaviour starts at the transition temperature of the material. Let us remark at this point that the appearance of superparamagnetism is closely connected with the relevant time τ_m ('time window') of our measurement, observation or application of external magnetic fields.

The two limiting possibilities $\tau_N \gg \tau_m$ and $\tau_N \ll \tau_m$ thus lead either to magnetically ordered or stable state in the former case or superparamagnetism in the latter one. They are separated by the region where $\tau_N \sim \tau_m$. This condition may be fulfilled according to (17.1) by changing either the particle volume or temperature, assuming that the anisotropy constant does not essentially change in the considered temperature range.

For the given particle, volume V and the relevant time τ_m, we can then define the so-called blocking temperature T_B:

$$T_B = \frac{K_u V}{C\, k_B},$$ (17.3)

where the constant $C = \ln(\tau_m/\tau_0)$ is equal to 27.6, 16.1 and 6.9 for the relevant times of observation 100 s (DC measurement), 10^{-3} s (frequency of 1 kHz) and 10^{-7} s (e.g. Mössbauer spectroscopy) assuming $\tau_0 = 10^{-10}$ s. Relation (17.3) clearly illustrates the sensitivity of the blocking temperature to the time window. Let us note in particular that the blocking temperature derived from the DC magnetic measurements is about four times lower than that determined from the behaviour of the Mössbauer spectra if the anisotropy remains essentially constant.

In all this reasoning, we assumed that the magnetocrystalline anisotropy of the individual particle is of the uniaxial type, and the size of the particle is so small that it forms a single domain and its mechanical position remains fixed even in an applied magnetic field.

For a multi-domain particle, the magnetisation curve is always rounded even for specific orientations of the applied field, and the magnetisation process is realised by both the motion of the domain walls and rotation of magnetisation in the domains. The contribution of these two processes depends on the intrinsic parameters of the material and the intensity and frequency of the applied field. Though the domain wall motion is also partly irreversible, the corresponding coercive fields originating in this irreversibility are usually much smaller than for single-domain behaviour, and the hysteresis losses are thus also lower.

One phenomenon influencing also the behaviour of MNP is a difference between the magnetic properties of the inner part and the outer surface layer where a spin disorder arises due to a reduced coordination and broken exchange bonds. It becomes especially marked when the number of atoms in the surface layer approaches their number in the inner 'core' part of the particle[7-9]. As the properties of the particles are usually averaged over its whole volume, the effect results in the dependence of, for example, magnetisation and Curie temperature on the particle size. Then, employing a simple spheric model, the total saturation magnetisation M_s (measured) is given by the following relation:

$$M_s = \frac{(M_{s1} V_1 + M_{s2} V_2)}{(V_1 + V_2)} = M_{s1} - \frac{6\,t(M_{s1} - M_{s2})}{d},$$ (17.4)

where
$M_{s1} V_1$ are saturated magnetisation and volume of the 'core'
$M_{s2} V_2$ are saturated magnetisation and volume of the surface layer
t is the thickness of the outer layer
d is the diameter of the spherical particle

Supposing M_{s1} to be equal to the bulk value $M_{s(bulk)}$ and $M_{s2} = 0$, the desired data of the layer thickness can be evaluated[10,11].

17.2.2 Requirements of Magnetic Properties for Magnetic Fluid Hyperthermia

Magnetic heating realised by the conversion of magnetic energy to heat is a complex problem where the conversion is mediated by loss processes arising due to the changes of the MNP magnetisation under an alternating magnetic field.

The resulting effect depends on a number of various factors. First of all, there are parameters of the applied field, characterised by its maximum amplitude H_{max} and frequency ν, which, together with properties of the liquid medium, that is, viscosity η, specific heat capacity c_p and thermal conductivity λ can be classified as 'external parameters'. The relevant properties of the MNP can be called 'internal parameters', and in spite of the complexity of their interrelationships, two types of these properties can be distinguished. Namely, there are intrinsic especially magnetic properties of the cores as magnetisation M, coercivity H_C, magnetocrystalline anisotropy K, remanent magnetisation and Curie temperature, and the extrinsic properties related to the size and the shape of the cores, hydrophobicity/hydrophilicity of the shell and hydrodynamic volume of the MNP.

Prior to the subsequent discussion, let us enumerate the required restrictions of the applied alternating magnetic field parameters for applications on human objects, which should be respected in the search of suitable materials.

At first, it is the criterion for the whole-body exposure to magnetic field expressed as the product of the strength and frequency $H\nu$, which should be less than 4.85×10^8 A m^{-1} s^{-1} [12]. Nevertheless, for smaller body regions, this limitation is not so strict and, for example, parameters of $H_{max} \sim 10$ kA m^{-1} and $\nu \sim 400$ kHz were suggested for the treatment of breast cancer[13]. With respect to the medical demands, an important aspect is the applied frequency, which should be higher than 50 kHz to avoid stimulation of peripheral and skeletal muscles and possible cardiac stimulation and arrhythmia[14] and lower than 10 MHz preferably below 2 MHz, in order to achieve a reasonable penetration depth of the rf field into the tissue. Finally, because of technical reasons concerning the construction of the facility producing the magnetic field, there is a limit on the maximum amplitude of the applied field of about 16 kA m^{-1}.

Generally, the power dissipation originating from the cyclic increase of internal energy produced by the magnetic work on the system is given by the following relation:

$$P = \mu_0 \nu \oint_H M\, dH, \qquad (17.5)$$

where
  μ_0 is the permeability of free space
  ν is the frequency of the AC field
  H is the magnetic field
  M is the magnetisation of the measured sample

Subsequent rearrangement and integration give the following expression for the power dissipation:

$$P = \mu_0 \pi \chi''(\nu)\nu H^2, \qquad (17.6)$$

where
  H is the amplitude of the applied field
  χ'' is the out-of-phase (imaginary) component of the susceptibility[15]

In principle, three different mechanisms can be distinguished, namely hysteresis power losses, originating in the irreversibility of the magnetisation process, Néel relaxation, conditioned by the rotation of the magnetic moments of the particles and friction losses due to the Brownian rotation of the magnetic particles as a whole.

These mechanisms can act either separately or to a certain extent simultaneously in dependence on the contributions of the individual factors outlined earlier.

17.2.2.1 Hysteresis Losses

They are connected with the existence of irreversible magnetisation loops, and the power losses correspond to the relation (17.6) where χ'', the imaginary component of the susceptibility, is for polydisperse suspensions in a good approximation frequency independent in the range of 100 kHz–1 MHz currently used in MFH systems[16,17].

Two different magnetic states in ferro (ferri)-magnetically ordered systems should be distinguished, controlled by the size of particles. The single-domain particles of a smaller size possess high barrier for magnetisation reversal along the easy axis, which results in large coercivities, and the hysteresis loops are thus wide. Therefore, systems of such particles provide high hysteresis losses but because of the restriction of the field amplitude on the limit of ~16 kA m⁻¹, the full hysteresis loops cannot be utilised and the capability for heating is substantially reduced.

The properties markedly change with the transition to larger multi-domain particles where the displacement of the Bloch walls is energetically less demanding and typically takes place at lower magnetic fields. The hysteresis loops become considerably narrower possessing high initial permeabilities and low coercivities. Then, even a relatively low amplitude of the applied field, below the limit of ~16 kA m⁻¹, may reach the magnitude of the coercive field, and hysteresis losses promising for the heating effect can be achieved.

17.2.2.2 Néel Relaxation Losses

A destabilisation of the single-domain magnetic ordering due to a decrease of the particles size leads to a gradual transition to superparamagnetic behaviour where magnetic energy is converted to the thermal one by the rotation of the particle moment between two metastable antiparallel orientations.

In contrast to the previous case, the imaginary part of the susceptibility depends on the frequency as

$$\chi''(\nu) = \frac{\chi_0 \nu \tau_N}{[1 + (\nu \tau_N)^2]},$$
(17.7)

where

$$\chi_0 = \frac{\mu_0 M_s^2 V}{kT}$$

M_s is the saturation magnetisation.

Introduction of (17.7) into (17.6) gives for the power dissipation:

$$P = \frac{\mu_0 \pi \chi_0 \nu^2 H^2 \tau_N}{[1 + (\nu \tau_N)^2]}$$
(17.8)

and employing the relation for the Néel relaxation time $\tau_N = \tau_0 \exp(K_u V/k_B T)$, the relation between the materials properties and frequency of the applied alternating magnetic field can be analysed.

At low frequencies, that is, for $\nu \tau_N \ll 1$ the superparamagnetic relaxation is favoured and the magnetic losses, in other words the corresponding output of heating power, increase with the square of the frequency. Simultaneous influence of the anisotropy energy KV on the Néel relaxation time leads to a strong increase of the output heating power with a sharp maximum in a dependence on the particle size. Evidently, the effect is conditioned by a narrow size distribution of an ensemble of the particles and disappears continuously by gradual increase of the particles size leading to the blocked state. Therefore, a superposition of Néel relaxation and hysteresis losses has to be considered[18,19].

17.2.2.3 Brownian Friction Losses

They originate from the rotation of magnetic particles induced by the external alternating magnetic field on the particle magnetisation, and an expression analogous to (17.8) is obtained:

$$P = \frac{\mu_0 \pi \chi_0 v^2 H^2 \tau_B}{[1 + (v\tau_B)^2]}. \tag{17.9}$$

Brownian relaxation time τ_B is given as

$$\tau_B = \frac{3\eta V_H}{kT}, \tag{17.10}$$

where
 η is the viscosity of the surrounding liquid medium
 V_H is the hydrodynamic volume of the particle sensitive to the quality of the particles coating

As a rule, both Néel and Brownian relaxation mechanisms act simultaneously, and the effective relaxation time τ is given by

$$\frac{1}{\tau} = \frac{1}{\tau_N} + \frac{1}{\tau_B}. \tag{17.11}$$

Nevertheless, from a comparison of markedly different dependences of the relaxation times on the size, one can expect a prevailing contribution of Néel losses at higher frequencies and smaller sizes, characteristic for MFH application. On the other hand, the Brownian losses become essential for lower frequencies and bigger sizes of nanoparticles. They can be enhanced by a strong coupling of the magnetic moment direction with the particles themselves due to a strong magnetocrystalline anisotropy and a low viscosity of the suspension medium allowing easy reorientation of the particles.

Further, an attention should be paid to the temperature dependence of the existence and stability of the magnetic ordering in the specific range relevant to medical application at body temperature and slightly above it, that is, ~310–360 K. First, it is the ferro (ferri)–superparamagnetic transition characterised by the blocking temperature T_B, which decides about the mechanism of the heating. Then, according to the relation (17.3) for $T < T_B$, heating is effectuated by hysteresis losses while for $T \geq T_B$ by Néel relaxation losses. Second, it is the temperature of the disappearance of magnetic ordering and transition to the paramagnetic state, characterised by the Curie temperature T_C. Then, regardless of the mechanism, the magnetic heating in the vicinity of T_C gradually decreases and it is switched off when the transition temperature is achieved. Therefore, if the Curie temperature is suitably adjusted, this behaviour can be used as a mechanism of self-control and a local overheating of the healthy tissue can be prevented.

17.3 MAGNETIC CORES

A successful evolution of the cores towards the required behaviour, that is, to an optimum interaction with the applied alternating magnetic field, is qualified by detailed knowledge of their magnetic properties. From a general point of view, they depend on the composition, structural and microstructural properties; and this approach will be demonstrated on the appropriately selected types of magnetic oxides exhibiting typical distinct features, see also Ref. [20].

The two main aspects for the selection of the following types of materials are, on the one hand, the optimisation of the conversion of the energy from the AC magnetic field and, on the other hand, the effort to reach the required control of the temperature by suitable parameters of the used magnetic materials, in particular the temperature of their transition to paramagnetic behaviour.

17.3.1 FERRIMAGNETIC SPINELS

They possess the general formula $(Me)^A[Me'_2]^BO_4$ where Me denotes cations placed in tetrahedral sites (A), and Me' cations placed in octahedral sites (B). For a better understanding, the spinel structure is depicted in Figure 17.1.

The resulting magnetic ordering is controlled by the interactions between neighbouring magnetic cations *via* the intermediary anion, that is,

Me^A-O-Me'^B between cations in tetrahedral and octahedral sites with a bonding angle of $\sim$125°

Me^B-O-Me'^B between cations in octahedral sites with a bonding angle of $\sim$90°

Me^A-O-Me'^A between cations in tetrahedral sites with a bonding angle of $\sim$80°

Among them, the A–B interactions leading to collinear antiparallel ordering of the magnetic moments of the two sublattices are usually the strongest. The resultant total magnetic moment is thus given by the difference between these two magnetic moments; for details see, for example, Refs. [21,22].

17.3.1.1 Magnetite

It is a simple ferrimagnetic iron oxide spinel that due to a strong preference of Fe^{3+} for tetrahedral sites exhibits the inverse structure $(Fe^{3+})^A[Fe^{2+}Fe^{3+}]^BO_4$. Its bulk form possesses the saturated magnetisation $M_s = 91$ A m^2 kg^{-1} at 300 K and Curie temperature $T_C = 858$ K, for example, in Ref. [21]. Further with respect to its magnetocrystalline anisotropy, it can be characterised as a magnetically soft material with a low magnetic coercivity H_C.

The properties, however, can be modified by variation of the size of particles as it can be seen from the evolution of the hysteresis loops given in Figure 17.2.

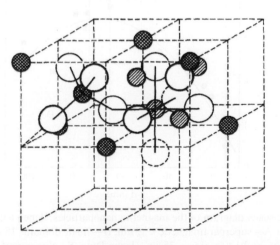

FIGURE 17.1 Elementary cell of the cubic spinel structure: ●—Me cations in tetrahedral sites (A), ◉—Me' cations in octahedral sites (B), ○—oxygen anions.

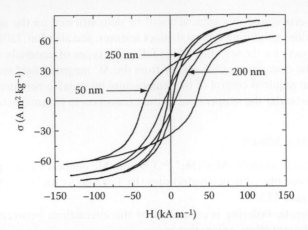

FIGURE 17.2   Saturation hysteresis loops at 300 K of samples with the crystallite size (XRD) 250 nm, 200 and 50 nm, respectively. (From Hergt, R.A., d'Ambly, W., Hilger, C.G., Kaiser, I., Richter, W.A., Schmidt, U. et al., Physical limits of hyperthermia using magnetite fine particles, *IEEE Trans. Magn.*, 34, 3745–3754. Copyright 1998 IEEE.)

Thus, the particles of the mean size $d_{XRD} \sim 50$ nm exhibit a large coercivity of 34 kA m^{-1} ascribed to their single-domain character, but the morphology seen in the TEM figure indicates maghemite phase[18]. The magnetisation reversal in the multi-domain particles of $d_{XRD} = 200$ and 250 nm is energetically less demanding and is manifested in high initial permeability and low coercivities of 7.5 and 3.3 kA m^{-1}, respectively.

A detailed study of this behaviour resulted in the magnetic states diagram, where the areas of the particular magnetic ordering are delimited with respect to the dependence of the coercivity on the particles sizes, measured in a static field at 300 K (Figure 17.3).

Simultaneously as an influence of the different magnetic properties of the surface layer, a decrease of the saturation magnetisation appears (Figure 17.4).

Analysing these results, it is possible to propose a suitable size of the magnetite particles for the MFH in the range of 25–40 nm.

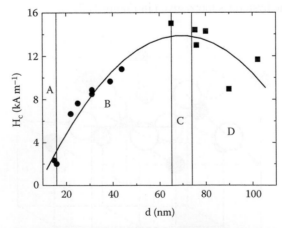

FIGURE 17.3   Magnetic states diagram of the magnetite nanoparticles, dependence of the coercivity H_c on the crystallite size (XRD): A—superparamagnetic < 15 nm, B—single domains 15–65 nm, C—pseudo-single domains 65–75 nm and D—multi-domains > 75 nm. (From Dutz, S., *Nanopartikel in der Medizin*, Verlag Dr. Kováč, Hamburg, Germany, 2008. Copyright Verlag Dr. Kováč e.K.)

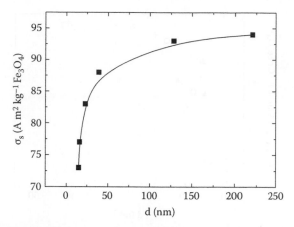

FIGURE 17.4 Dependence of the saturated magnetisation of Fe_3O_4 on the crystallite size (XRD), $T = 300\,K$. (Data from Műrbe, J. et al., *Mat. Chem. Phys.*, 110, 426, 2008.)

The reasons are evident as follows:

A decrease of the saturated magnetisation between 40 and 25 nm is relatively weak, about 6% only.

The magnetic coercivity values ranging approximately from 6 to 11 kA m^{-1} match well with the required limit of the maximum field amplitude ~16 kA m^{-1}.

With respect to the limit of the transition to the superparamagnetic behaviour, that is, 15 nm (static field, 300 K), the mechanism of hysteresis losses should prevail for heating with particles of the size above 20 nm, and the requirement on the uniform distribution of the particle size crucial in the case of the heating by Néel losses can be neglected. Let us note that to fulfil this last condition may be technologically difficult. The size of the particles in the range of 20–40 nm is sufficiently small and remains in a desirable limit for an application.

17.3.1.2 Cobalt Ferrites

It possesses mixed spinel structure of the formula $(Co_x^{2+}Fe_{1-x}^{3+})^A[Co_{1-x}^{2+}Fe_{1+x}^{3+}]^BO_4$. The site occupancy varies from $x = 0.20$ to $x = 0.07$ for rapidly quenched and slowly cooled samples from 1200°C, respectively[25]. One can thus suppose that $CoFe_2O_4$ nanoparticles prepared at substantially lower temperatures may have even more complete inverse structure. The magnetic properties of the bulk $CoFe_2O_4$ are characterised by the Curie temperature of $T_C \sim 496°C$ (769 K), and saturated magnetisation of 94 A m^2 kg^{-1} at 5 K and 81 A m^2 kg^{-1} at 300 K. Its high magnetocrystalline anisotropy, $K_1 = 270 \times 10^3$ J m^{-3} at 293 K decreasing to $K_1 = 90 \times 10^3$ J m^{-3} at 363 K, influences the resulting behaviour of the particles[26–28].

This effect is manifested in comparison with magnetite by an important shift of the superparamagnetic–ferrimagnetic transition down to the particle size of ~6 nm (static field, 300 K), relatively well established[29,30]. Simultaneously, the crossover from the single-domain to multi-domain behaviour is according to the magnitude of magnetocrystalline anisotropy and the dependence of the coercivity on the particle sizes expected to lie at ~70 nm[31]. Some more recent studies of cobalt ferrite nanoparticles with varying sizes due to annealing at various temperatures found a maximum in coercivity at ~30–40 nm[32,33]. The explanation of this dependence by the change to multi-domain character is considered improbable even by the authors themselves, and they admit that the particles are single domain in the whole range of sizes studied.

In spite of a dispersion of the magnetisation values in the vicinity of the ferrimagnetic–superparamagnetic transition, there is a slight decrease of the magnetisation of only 8% in the range from 40 nm down to 15 nm, see Figure 17.5. The undesirable differences in the data may be partially ascribed to the various modes of the synthesis influencing the distribution of cobalt ions between tetrahedral and octahedral sites.

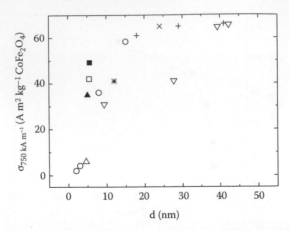

FIGURE 17.5 Dependence of the magnetisation $\sigma_{750\,kA\,m^{-1}}$ of $CoFe_2O_4$ on the crystallite size d(XRD), T = 300 K, + [34], × [32], ■ [35], □ [36], ○ [37], ▲ [38], ✳ [39], ▽ [33]. (Data from Veverka, M. et al., *Nanotechnology*, 18, 345704 (7pp), 2007; Maaz, K. et al., *J. Magn. Magn. Mater.*, 308, 289, 2007; Kim, D.H. et al., *J. Magn. Magn. Mater.*, 320, 2008; Ammar, S. et al., *J. Mater. Chem.*, 11, 186, 2001; Kim, Y.I. et al., *Physica. B*, 337, 42, 2003; Moumen, N. et al., *J. Magn. Magn. Mater.*, 149, 67, 1995; Vaidyanathan, G. and Sendhilnathan, S. *Physica B*, 403, 2157, 2008; El-Okr, M.M. et al., *J. Magn. Magn. Mater.*, 323, 920, 2011.)

TABLE 17.1

Mean Sizes and Fundamental DC Magnetic Data Measured at 300 K, Magnetisation $\sigma_{750\,kA\,m^{-1}}$ at 750 kA m⁻¹, Remanent Magnetisation σ_r and Coercivity H_C, of the Cobalt Ferrite Nanoparticles

d_{XRD} (nm)	σ_{750} (A m² kg⁻¹)	σ_r (A m² kg⁻¹)	H_C (kA m⁻¹)
18	61	14	18
29	65	25	43
41	66	26	50

Nanoparticles with relatively reasonable properties of the parameters required for MFH are shown in Table 17.1 and denoted in Figure 17.5 as '+'. Besides the smooth evolution of the magnetisation, there is a distinct decrease of the coercivity with decreasing size of the particles and nearly optimum fit of H_C = 18 kA m⁻¹ with the maximum field amplitude of ~16 kA m⁻¹ seems to be promising for an eventual clinical application. Nevertheless, let us mention that the DC characteristics are only a rough guide for attaining the optimum hysteresis losses, and the actual situation under an applied AC field can be essentially modified[34].

17.3.1.3 Cobalt–Zinc Ferrites

Properties of the cobalt ferrite can be further modified by a suitable compositional variation, for example, by a partial replacement of the cobalt by non-magnetic zinc cations. The mechanism of the substitution is described by the formula $(Zn_x^{2+}Fe_{(1-x)}^{3+})^A[Co_{(1-x)}^{2+}Fe_{(1+x)}^{3+}]^BO_4$, taking into account the strong preference of zinc ions for tetrahedral sites. It is valid for bulk materials, but for nanoparticles simultaneous presence of zinc ions in octahedral sites was reported on some parent spinels by several authors[40,41]. The total moment increases with the Zn doping practically linearly in the region $0 \leq x \leq 0.5$, as expected for decreasing population of magnetic ions in the A sublattice[42].

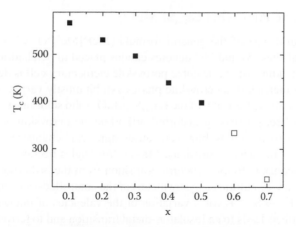

FIGURE 17.6 Dependence of the Curie temperature T_c of $Co_{1-x}Zn_xFe_2O_4$ nanoparticles on the composition, ■—mean size $d_{XRD} \sim 10\,nm$ [44], □—mean size $d_{XRD} \sim 14\,nm$ [45]. (Data from Arulmurugan, R.G. et al., *J. Magn. Magn. Mater.*, 303, 131, 2006; Veverka, M. et al., *J. Mag. Mag. Mat.*, 322, 2386, 2010.)

Nonetheless, the continuing decrease of the concentration of magnetic ions in the tetrahedral positions (x > 0.5) causes a weakening of the A–B interactions and finally destabilisation of the ferrimagnetic ordering[43], as also documented by a gradual decrease of the Curie temperature, see Figure 17.6.

According to this dependence, it seems to be possible to modify suitably the Curie temperature by a compositional variation in order to achieve the self-controlled heating mechanism in the range of ~315–335 K (42°C–60°C). This possibility is further elucidated by a comparison of the appropriate magnetic parameters displayed in Figure 17.7a and b.

One can thus conclude that the composition of x ~ 0.6 of the Curie temperature ~317–334 K and still relatively high specific magnetisation $\sigma_{750\,kA\,m^{-1}}$ of 30–35 A m^2 kg^{-1} $Co_{1-x}Zn_xFe_2O_4$ at 300 K exhibit adequate properties for the use in MFH. Let us note that particles exhibit superparamagnetic behaviour in the static regime at room temperature as it is evidenced by the blocking temperature lying around 200 K. However, under magnetic heating realised usually at frequencies above 100 kHz, the particles remain blocked up to the temperature of the application and exhibit hysteretic behaviour corresponding to the ferrimagnetic state.

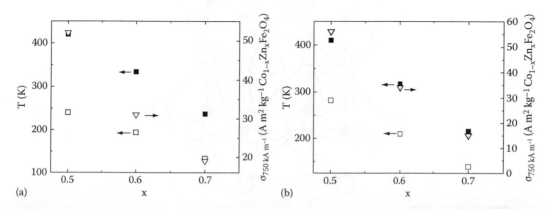

FIGURE 17.7 Evolution of the blocking temperature, Curie temperature and magnetisation σ_{750} of $Co_{1-x}Zn_x$-Fe_2O_4 with composition, mean crystallite size d(XRD), a ~ 14 nm, b ~ 23 nm [45], □—blocking temperature T_B, ■—Curie temperature T_c, ▽—magnetisation σ_{750} at T = 300 K. (Data from Veverka, M. et al., *J. Mag. Mag. Mat.*, 322, 2386, 2010.)

17.3.2 FERROMAGNETIC PEROVSKITES $La_{1-x}Sr_xMnO_3$

Perovskite is a structural type of the general formula $(Me)^A[Me']^BO_3$ where Me denotes cations placed in dodecahedral sites (A) and Me' denotes cations placed in octahedral sites [B].

For a better understanding, the basic cubic perovskite elementary cell is depicted in Figure 17.8. Nevertheless, let us mention that perovskite phases exhibit mostly various distorted structures.

Thus, the parent compound $LaMnO_3$ of the $La_{1-x}Sr_xMnO_3$ solid solutions is a single-valent (Mn^{3+}) antiferromagnetic insulator possessing distorted orthorhombic perovskite structure. The replacement of lanthanum by strontium ions, however, causes significant changes of its properties. It leads to a gradual decrease of the steric distortions, Mn–O–Mn angles become closer to the ideal 180° arrangement and consequently to the structural transition from the orthorhombic (*Pbnm*) to rhombohedral ($R\bar{3}c$) symmetry. Simultaneously, by means of the controlled-valence mechanism formally described as $La^{3+} + M^{3+} \leftrightarrow Sr^{2+} + Mn^{4+}$, variation of the valencies of manganese ions is induced. Action of both these effects leads to an insulator–metal transition and to ferromagnetic ordering due to double-exchange interactions above x ~ 0.1[46].

Thus, an appreciable dependence of the magnetisation and Curie temperature on the composition in the series $La_{1-x}Sr_xMnO_3$ exists. Herewith, as a consequence of an increasing influence of the outer magnetically 'dead layer', a gradual deterioration of the magnetic ordering with decreasing size of the crystallites appears[11,47]. Both these effects are shown in the established magnetic diagrams, Figure 17.9a and b.

Therefore, with respect to the required condition of the self-controlled heating mechanism, selection of the particles of the composition in the range of x = 0.25–0.3 and size of 20–30 nm seems to be reasonable. Let us note that the particles currently prepared exhibit at room temperature in the static regime very small hysteresis of about 1 kA m^{-1}. It can be thus assumed that they occur predominantly in the superparamagnetic state as it is evidenced by the blocking temperature lying around 260 K[48]. Nevertheless, under magnetic heating effectuated usually at frequencies above 100 kHz, the spins remain blocked up to the temperature of the application, and hysteretic behaviour corresponding to the ferromagnetic state is stabilised.

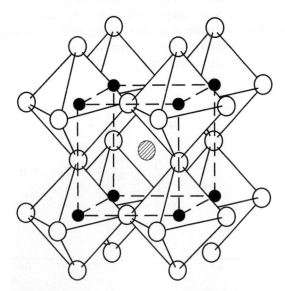

FIGURE 17.8 Elementary cell of the cubic perovskite structure, ●—cations in octahedral sites, ◉—cations in dodecahedral sites, ○—oxygen anions.

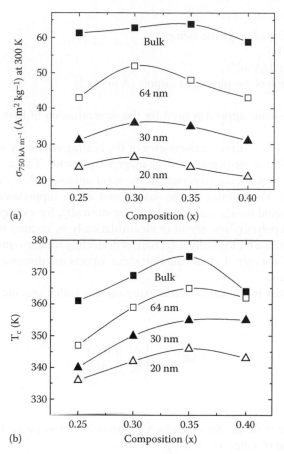

FIGURE 17.9 (a) Magnetisation σ_{750} at 300 K. (b) Curie temperature of the bulk and nanoparticles in the system $La_{1-x}Sr_xMnO_3$.

Alternatively, $La_{1-x}Na_xMnO_3$ and $La_{1-x}Ag_xMnO_3$ manganese perovskites, where lanthanum ions are replaced by monovalent sodium or silver, were investigated and ferromagnetic ordering with the Curie temperature around 45°C, that is, 318 K for x ~ 0.15–0.2, is reported. The results, however, seem to be affected by the difficulties of the synthesis of single-phase samples[49–51].

17.4 MAGNETIC HEATING

17.4.1 METHODOLOGY

Two different approaches can be used in order to quantify the heating efficiency. The first one is based on the determination of the AC hysteresis loops of the magnetic particles at constant temperature measured on dry compacted powder samples.

The produced heating power P (W g_y^{-1}) corresponds to the area of the hysteresis loop multiplied by the frequency:

$$P = 1.26 \times 10^{-3} \, y^{-1} \, v \oint_H M \, dH, \qquad (17.12)$$

where

    y is the weight fraction of magnetic elements
    ν is the frequency (Hz)
    H is the magnetic field (A m^{-1})
    M is the magnetisation of the measured sample (A m^2 kg^{-1})

The scheme of a home-made apparatus used for the determination of hysteresis loops is given in Figure 17.10.

The second approach is a 'direct' calorimetry of the heating efficiency realised by the application of an alternating field on suspensions of the magnetic particles. The measurements are carried out on model systems of particles dispersed in the solid agarose gel or on their water or saline suspensions of pH ~ 7. The stability of the suspensions and a suppression of a tendency to the agglomeration in the liquid media are achieved either sterically, for example, by coating the particles with dextran and polyethylene glycol or electrostatically by coating with silica. The created shell simultaneously forms a biologically inert barrier protecting the surrounding environment from the chemical effects of the core. Let us note that these aspects are discussed in more details elsewhere in several chapters.

The evaluated quantity in the calorimetric experiment is called specific absorption rate (SAR) defined as

$$\mathrm{SAR}^{\mathrm{T}}_{\mathrm{meas}} = \left(\frac{c^{\mathrm{T}}_{\mathrm{p}}}{\mathrm{x}} \right) \left(\frac{\mathrm{dT}}{\mathrm{dt}} \right)^{\mathrm{T}}, \tag{17.13}$$

where

    c_{p} is the specific heat of the medium in which the particles are suspended (usually roughly equal to the specific heat of water, i.e. 4.18 J g^{-1})
    $(\mathrm{dT}/\mathrm{dt})^{\mathrm{T}}$ is the slope of the temperature vs. time curve
    x is the weight fraction of magnetic elements in the medium

There are in principle two kinds of the facilities suitable for the study of the magnetic heating in the laboratory scale carried out on the nanoparticle suspensions themselves, experiments *in vitro* and eventually *in vivo* on small animals.

Devices where an alternating magnetic field of given amplitude is generated in an air gap (usually ~10 mm) of a ferrite core and an oscillation circuit including the capacitors allow continuous variation of the frequency, see Figure 17.11[53,54].

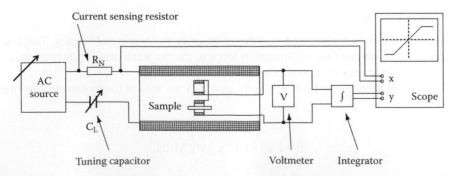

FIGURE 17.10 Scheme of the apparatus for AC magnetic hysteresis measurements. For more details about the measuring device (see Ref. [52]). (Data from Platil, A.J. et al., *Sens. Lett.*, 5, 311, 2007.)

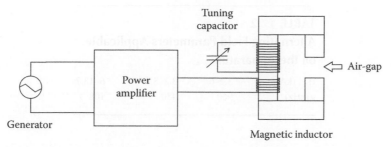

FIGURE 17.11 Scheme of the AC magnetic field application with air gap in magnetic circuit.

In a simpler type of device, the AC magnetic field is generated by an air exciting coil with continuous variation of the amplitude of the field linearly proportional to the current while the frequency is given by the generator. The circuit of the coil and auxiliary capacitors should be tuned to the frequency used by the proper selection of the capacitors.

The latter type of the devices is favoured and became more widely employed in the last years. Nevertheless, the use of these apparatuses still suffers in most cases by a limitation of the working space due to an effort to minimise the effects of magnetic field inhomogeneities[55–57]. The encountered problem inspired a design and construction of the home-made apparatus of a substantially bigger working space. Its scheme is given in Figure 17.12.

The exciting coil is made up of water-cooled copper tube of the outer diameter 8 mm, coil diameter 120 mm, length 223 mm, 18 turns. Homogeneity of the magnetic field in the transverse direction (inner diameter of the working volume 80 mm) and in the longitudinal direction of 80 mm centred in the position of the sensing coil is better than 5%. Applicable field parameters see Table 17.2.

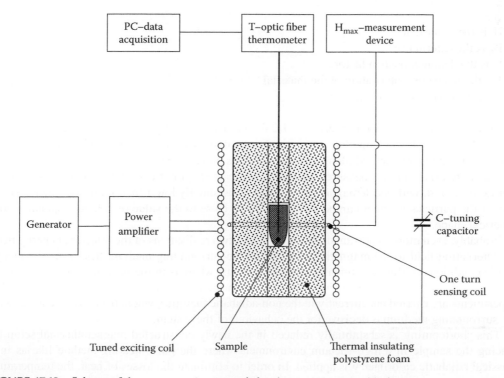

FIGURE 17.12 Scheme of the apparatus for magnetic heating experiments with air core.

TABLE 17.2
Alternating Field Parameters Applicable on the Apparatus

H_{max} (kA m^{-1})	3.2–6	3.2–8.8	6–13.7
ν (kHz)	960	480	108

The measured suspensions in plastic tubes (length 95 mm and diameter 14 mm) thermally insulated by polystyrene foam are placed into the exciting coil in the position of the sensing coil. The extent of the homogeneous magnetic field allows experiments on small animals (thermal insulating foam should be removed).

The described approaches, however, exploit different modalities so that the power losses P evaluated from the measurements of the hysteresis loops agree more likely only qualitatively with the SAR data obtained by calorimetric measurements.

First, while the measured hysteresis losses originate only from the magnetic properties of the ferrimagnetic and ferromagnetic nanoparticles possessing either single-domain or multi-domain character, the calorimetric data include the appropriate contributions of hysteresis losses, Néel relaxation and Brownian friction losses.

The other reason is an influence of the demagnetisation factor, which strongly depends on the shape of the measured sample according to the following relation:

$$H_i = H_0 - \frac{D \times J}{\mu_0}, \qquad (17.14)$$

where
   H_i is the inner field
   H_0 is the external field
   D is the demagnetisation factor
   J is the magnetic polarisation of the material
   μ_0 is the permeability of vacuum

Let us recall that D = 1/3 for a sphere, D = 1/2 for a thin and/or long cylinder magnetised normal to its length and D = 1 for a thin plate magnetised normal to its surface; when the thin (long) cylinder or the thin plate is magnetised along its length or in its plane, D = 0.

The hysteresis loops are measured on slightly compacted cylindrical samples of the length/diameter ratio ~3.5–4, with the demagnetisation factor substantially lower than for dilute water suspension of the particles where it can be considered to be close to the value of 1/3, corresponding to a spherical shape.

Probably, even more important difference in these two realisations of the transfer of energy from the alternating field to the magnetic particles and the surrounding medium lies in the thermodynamic conditions of the experiment and measurements; whereas in the measurement of hysteresis losses, we directly determine the hysteresis curves in the essentially isothermic regime; the SAR experiments are carried out currently in the non-adiabatic facilities, where the thermal contact with the surrounding medium is decisive for the behaviour of the system.

This shortcoming is substantially reduced in the newly constructed magnetothermal setup by placing the sample in a high-vacuum environment where the pulse heating method like as in a classical adiabatic calorimetry is applied. In order to eliminate the losses of heat, the temperature gradient between the sample and its environment is minimised. It is achieved by a specific adiabatic shield heated continuously to the temperature equal to that of the sample[58,59].

A simpler possibility to deal with the problem of the adiabatic vs. non-adiabatic conditions of the equipment is based on the independent determination of the thermal losses, that is, heat flow out of a preheated liquid sample during its spontaneous cooling[60] given as

$$\left(\frac{dQ}{dt}\right)^T = m \times c_p \left(\frac{dT}{dt}\right)^T,$$

(17.15)

where
   m is the mass of the liquid sample
   c_p is its specific heat capacity

And the corrected value of the heating efficiency is

$$SAR_{corr}^T = \frac{1}{x}\left[-\left(\frac{dQ}{dt}\right)^T\right] + SAR_{meas}^T.$$

(17.16)

Therefore, the isothermal arrangement of the measurements, where the inductive coil is kept at the body temperature, seems to reflect better the real situation[56].

The actual condition in the application of MFH, however, may be even more complicated when the particles are placed in a moving medium, for example, blood flow.

Finally, let us note that with respect to the dependence of the SAR values on the frequency and amplitude of the applied AC magnetic field, a direct comparison between experiments carried out under various magnetic conditions is not easy.

In the case of the existence of irreversible magnetisation loops, the power losses that correspond generally to the relation (17.6) can be rewritten in a more simple form:

$$P = \mu_0 \pi \chi'' v H^2,$$

(17.17)

where χ'', the imaginary component of the susceptibility, is in the range of frequencies of 100 kHz–1 MHz currently used in MFH systems for polydisperse suspensions in a good approximation frequency independent.

Then, the so-called intrinsic loss parameter independent of the frequency and amplitude of the AC magnetic field defined as

$$ILP = \frac{SAR}{H^2 v}$$

(17.18)

was recently proposed and introduced[17].

17.4.2 Heating Effects of Selected Materials

Before the consecutive discussion, however, let us mention that calorimetric measurements are often carried out in entirely different apparatuses and the corrections on the thermal losses, which can seriously influence the results, are very rare. Therefore, it is sometimes difficult to compare the results originating from various sources.

Another unfavourable effect that adversely affects the heating efficiency is a tendency of the particles to agglomerate. It is common for various systems and should be suppressed by a suitable coating.

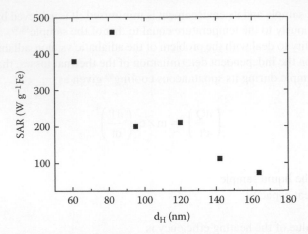

FIGURE 17.13 Evolution of the heating efficiency of magnetite cores of d = 14 nm with increasing hydrodynamic volume d_H, AC field parameters: 10 kA m^{-1} and 400 kHz. (Data from Dutz, S. et al., *J. Magn. Magn. Mater.*, 321, 1501, 2009.)

This behaviour is documented in Figure 17.13, where a rapid decrease of the heating efficiency with increasing hydrodynamic volume can be ascribed to the formation of agglomerates.

A similar effect observed on $La_{0.75}Sr_{0.25}MnO_3$ nanoparticles of d_{XRD} = 20 nm is reported[62].

17.4.2.1 Magnetite

As mentioned in Section 17.3.1.1, it is a magnetically soft material with magnetic properties sensitively dependent on the particle size and its distribution, particle shape and method of preparation. From the numerous results, one may conclude that the highest heating efficiency is obtained using AC hysteretic losses above the blocking temperature of the nanoparticles and thus their sizes of the order of 30 nm.

The situation at the transition from the superparamagnetic to ferrimagnetic state is illustrated in the schematic Figure 17.14 showing the hysteresis losses and losses due to Néel relaxation in dependence on the magnetite particle radius. The graph is based on the calculations for ν = 2 MHz, H = 6.5 kA m^{-1} and 300 K[18]. The relatively high Néel relaxation losses are connected to the assumed narrow distribution of the particle sizes, which is difficult to achieve in the preparation. For a more realistic case of broader distribution of particle sizes, the maximum of $P_{Néel}$ would be lower and

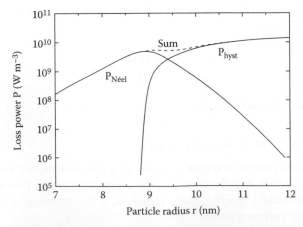

FIGURE 17.14 Loss power due to hysteresis and Néel relaxation calculated for ν = 2 MHz, H = 6.5 kA m^{-1} and 300 K. (From Hergt, R.A., d'Ambly, W., Hilger, C.G., Kaiser, I., Richter, W.A., Schmidt, U. et al., Physical limits of hyperthermia using magnetite fine particles, *IEEE Trans. Magn.*, 34, 3745–3754. Copyright 1998 IEEE.)

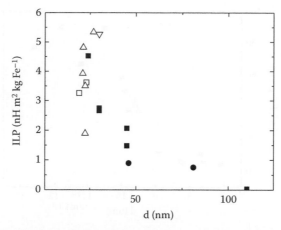

FIGURE 17.15 Dependence of the intrinsic loss power of ferrimagnetic magnetite nanoparticles on the mean particle diameter, calorimetric data evaluated from the initial slope at ~22°C–25°C stabilised suspensions: ■ [64], □ [65], △ [23], ▽ [66]; freshly dispersed powders: ● [67]. (Data from Gonzalez-Fernandez, M.A. et al., *J. Solid State Chem.*, 182, 2779, 2009; Hergt, R. et al., *J. Magn. Magn. Mater.*, 280, 358, 2004; Dutz, S. Nanopartikel in der Medizin, Verlag Dr. Kováč, Hamburg, Germany, 2008; Vergés, M.A. et al., *J. Phys. D Appl. Phys.*, 41, 134003 (10pp), 2008; Ma, M. et al., *J. Magn. Magn. Mater.*, 268, 33, 2004.)

broader and the hysteresis losses would prevail. Nevertheless, the rather sharp decrease of the hysteresis losses at the critical particle size (or alternatively blocking temperature) would be blurred.

This model calculation was more recently confirmed experimentally on the parent maghemite nanoparticles. Measured dependence of SAR on the size of particles in the range of 5–50 nm (27 kA m^{-1}, 700 kHz) showed the maximum for 16 nm diameter of the cores and diminution of the heating efficiency with decrease as well as increase of the size of particles[63].

The data on the intrinsic loss parameter for magnetite particles calculated from several papers are collected in Figure 17.15. It is clear that in the ferrimagnetic range of particle sizes, ILP increases with decreasing sizes down to ~20 nm.

The situation in the superparamagnetic state is complicated by the fact that susceptibility is no longer independent of frequency and the simple recipe how to confront results obtained at various frequencies and amplitudes of the AC field, that is, by comparing ILP instead of SAR, is only applicable with some caution.

Several commercially available magnetic fluid samples with particle sizes around 10 nm were calorimetrically measured at 900 kHz with a field amplitude of 5.66 kA m^{-1} [17]. The dependence of their heating power, expressed by means of SAR, is plotted in Figure 17.16 against their mean diameter derived from fitting of the DC magnetisation curves. In contrast to the foregoing case, the heating power rises with increasing particle size and for d = 11.5 ± 1 nm reaches considerable magnitude of about 80 W g^{-1}.

A similar tendency is seen in Figure 17.17 for particle systems obtained by magnetic fractionation of magnetic fluid based on iron oxide nanoparticles consisting of a mixture of magnetite and maghemite coated by dextran. Here, the particles are slightly smaller, down to 4 nm, and presumably possess narrower size distribution[68]. All the presented results indicate that the highest heating power of magnetite nanoparticles is reached in the vicinity of their transition to the superparamagnetic state and is mainly due to hysteresis losses.

17.4.2.2 Cobalt Ferrites

This is a typical hard magnetic material due to a rather high effective magnetic anisotropy, and thus, the critical particle size for the superparamagnetic–ferrimagnetic transition (static field, 300 K) is substantially shifted down to ~6 nm in a comparison with the magnetically soft magnetite.

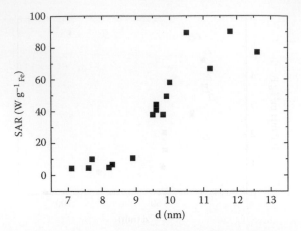

FIGURE 17.16 Evolution of SAR with size of magnetite cores (sizes calculated by fitting Langevin functions to measured superparamagnetic magnetisation curves). AC parameters: 5.66 kA m^{-1} 900 kHz. Commercial products. (Data from Kallumadil, M. et al., *J. Magn. Magn. Mater.*, 321, 1509, 2009.)

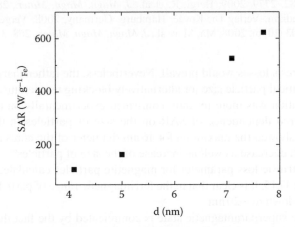

FIGURE 17.17 Increase of the specific absorption rate SAR with size of particles separated by magnetic fractionation from the magnetic fluids. AC field parameters: 12.5 kA m^{-1}, 500 kHz. (Data from Jordan, A. et al., *J. Nanoparticle Res.*, 5, 597, 2003.)

Therefore, the range of particle dimensions where the heating can be induced exclusively by Néel losses is very narrow. Heating of the suspensions of the nanoparticles of a size d < 6 nm, stabilised in water, depicted in Figure 17.18 can serve as an example.

The continuous evolution of the temperature terminates by a steady state achieved when the thermal losses are in equilibrium with the heating power. The values of SAR at 25°C derived from the initial nearly linear part of the measured dependences are 51 W $g_{(Co+Fe)}^{-1}$ and 11 W $g_{(Co+Fe)}^{-1}$, respectively.

A slight increase of the mean particle size to ~9 nm leads to a transition to essentially blocked single-domain particles but the behaviour may be still influenced by superparamagnetic fluctuations.

The direct comparison with Figure 17.18 concerned with the absolute magnitudes of heating power is, however, difficult as the frequency and field amplitude are different.

Table 17.1 shows a wide single-domain region of particle sizes and among them we devoted attention to the mean size of 18 nm possessing static coercivity of 19 kA m^{-1} at 300 K.

The calorimetric measurements, supported by the AC hysteresis loops, show an unexpected behaviour, see Figures 17.19 and 17.20, which is in a sharp contrast with the usual evolution of temperature with time similar to the dependences given, for example, in Figure 17.18.

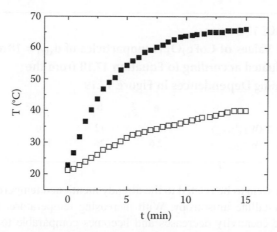

FIGURE 17.18 Magnetic heating of the superparamagnetic $CoFe_2O_4$, d ~ 5.6 nm, aqueous suspension. AC field parameters: ■—H_{max} = 50 kA m^{-1}, ν—266 kHz; □—H_{max} = 30 kA m^{-1}, ν—266 kHz. (Data from Kim, D.H. et al., *J. Magn. Magn. Mater.*, 320, 2390, 2008.)

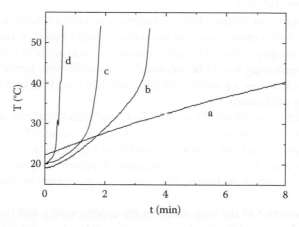

FIGURE 17.19 Magnetic heating of the aqueous suspension of the 18 nm particles. AC field parameters: H_{max} = 34 kA m^{-1} (a), 41 kA m^{-1} (b), 52 kA m^{-1} (c) and 64 kA m^{-1} (d), ν = 108 kHz. (Data from Veverka, M. et al., *Nanotechnology*, 18, 345704 (7pp), 2007.)

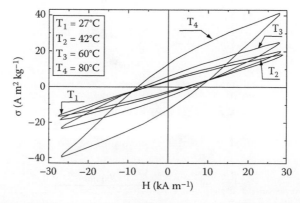

FIGURE 17.20 Evolution of AC hysteresis loops with temperature, measured on particles of the mean size d(XRD) = 18 nm at ν = 50 kHz. (Data from Veverka, M. et al., *Nanotechnology*, 18, 345704 (7pp), 2007.)

TABLE 17.3

SAR Values of $CoFe_2O_4$ Nanoparticles of $d_{XRD} = 18\,nm$ Evaluated according to Equation 17.18 from the Heating Dependences in Figure 17.19

	a	b	c	d
$SAR^{25°C}$ (W $g^{-1}_{(Co+2Fe)}$)	3	12	50	118
$SAR^{37°C}$ (W $g^{-1}_{(Co+2Fe)}$)	2.6	26	100	168

The observed effects seem to be related to the already mentioned temperature dependence of the constant of magnetocrystalline anisotropy. With increasing temperature, the magnetocrystalline anisotropy and thus the coercivity decreases and becomes comparable to the used amplitude of the AC field. As a result, hysteresis power losses dramatically increase, see Table 17.3, and overwhelm the increase of the thermal losses, which usually results in the saturation of the temperature increase[34].

17.4.2.3   Cobalt–Zinc Ferrites

From the foregoing discussion follows that it is necessary to have a reliable knowledge of the temperature dependence of the heating power of the particles and adjust it in two respects: to reach a high heating power in the range close to body temperatures and to adapt its temperature dependence in such a way that overheating would be inhibited. The cobalt–zinc ferrite with composition of $Co_{0.4}Zn_{0.6}Fe_2O_4$ was selected as a possible suitable candidate for MFH application exhibiting this self-controlled heating mechanism.

The used sample was a suspension of $Co_{0.4}Zn_{0.6}Fe_2O_4$ particles (mean size $d_{XRD} = 14\,nm$, concentration of the magnetically active species $c_{(Co+Fe)} = 2.5\,mg\ mL^{-1}$) coated by silica, see Figure 17.21, ensuring long-term stability and suppressing a tendency to agglomerate.

The measurements were carried out in the apparatus depicted in Figure 17.12, and some of the heating experiments are shown in Figure 17.22 and the corresponding data are summarised in Table 17.4.

Magnetic energy supplied to the suspension in the experiments a and b is sufficiently high to eliminate the thermal losses, and the steady state is achieved by approaching the Curie temperature where magnetisation gradually decreases. On the other hand, the input of the magnetic energy in the experiments c and d is insufficient to compensate the thermal losses. This effect is further evident from a decrease of the intrinsic losses power given in the Table 17.4.

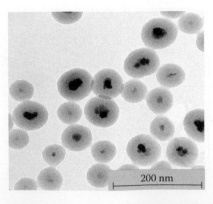

200 nm

FIGURE 17.21   Transmission electron micrograph of cobalt–zinc ferrite particles $Co_{0.4}Zn_{0.6}Fe_2O_4$ coated by silica.

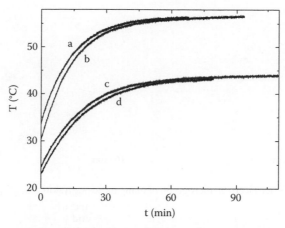

FIGURE 17.22 Magnetic heating of the $Co_{0.4}Zn_{0.6}Fe_2O_4$ and SiO_2 aqueous suspension, a, b, c, d, see Table 17.4.

TABLE 17.4

Magnetic Heating of the $Co_{0.4}Zn_{0.6}Fe_2O_4$ & SiO_2 Aqueous Suspension

	H_{max} (kA m⁻¹)	ν (kHz)	T_{max} (°C)	SAR (W $g_{(Co+Fe)}^{-1}$)	ILP (nHm² $kg_{(Co+Fe)}^{-1}$)
a	8.9	481	56	41	1.08
b	5.9	960	56	38	1.14
c	6	480	44	10	0.58
d	4	959	43	9	0.59

SAR, specific absorption rate (Equation 17.13); ILP, intrinsic loss power (Equation 17.18).

17.4.2.4 Ferromagnetic Perovskites $La_{1-x}Sr_xMnO_3$

As it follows from the basic magnetic properties, depicted in Figure 17.7a and b, the particles of the composition $La_{0.75}Sr_{0.25}MnO_3$ and size of ~20–25 nm seem to be a promising material. The required behaviour, reasonably high heating efficiency at 37°C, that is, the body temperature and simultaneously the self-controlled heating mechanism in the range of ~40°C–60°C by the proper modification of the Curie temperature, can be achieved.

The study was carried out in analogous conditions to those used for cobalt–zinc ferrite, that is, on the same apparatus and on the water suspensions of cores stabilised by silica coating. An example of the well-dispersed $La_{0.75}Sr_{0.25}MnO_3$ and SiO_2 particles quite comparable to the coated cobalt–zinc ferrite shown earlier is given in Figure 17.23.

A tendency to agglomerate, which could lead to a decrease of the heating efficiency, is suppressed[62]. Let us note that the effect was also reported for $CoFe_2O_4$[35] and seems to be valid in general.

Typical dependences of the temperature on the time of exposure to the AC field are illustrated in Figure 17.24. A rapid increase of the temperature for the applied frequency of 960 kHz saturates at the maximum achieved temperature of $T_{max} = 64°C$ in a rough correspondence with the determined Curie temperature of $T_C = 62°C$. There are, however, less steep dependences for lower frequencies, particularly of 107 kHz. The reason is in the decrease of the magnetic hysteresis losses and the equilibrium with the thermal losses thus occurs at lower temperatures.

A nearly linear dependence of SAR = $f(\nu H^2)$ measured in the range of 100 kHz–1 MHz on LSMO(0.25)@SiO₂ suspension of a polydisperse character is shown in Figure 17.25. It confirms, in agreement with Refs. [16,17], an independence of the imaginary component of the susceptibility χ'' in the relation (17.6) of frequency in this frequency range.

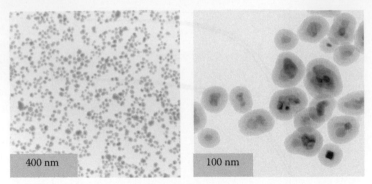

FIGURE 17.23 Transmission electron micrograph of manganese perovskite La$_{0.75}$Sr$_{0.25}$MnO$_3$@SiO$_2$ particles.

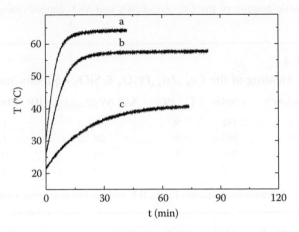

FIGURE 17.24 Magnetic heating of the LSMO(0.25)@SiO$_2$ aqueous suspension, AC field parameters: H$_{max}$ = 6.05 kA m^{-1}, ν = 960 kHz (a), 480 kHz (b) and 107 kHz (c). (From Pollert, E., Kaman, O., Veverka, P., Veverka, M., Maryško, M., Závěta, K. et al., Hybrid La$_{1-x}$Sr$_x$MnO$_3$ nanoparticles as colloidal mediators for magnetic fluid hyperthermia, *Phil. Trans. R. Soc. A*, 368, 4389–4405. Copyright 2010 The Royal Society.)

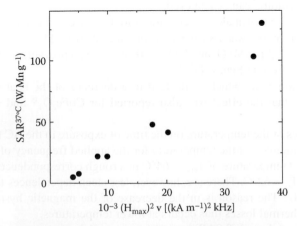

FIGURE 17.25 Dependence of SAR on the parameters of the applied AC magnetic field. (From Pollert, E., Kaman, O., Veverka, P., Veverka, M., Maryško, M., Závěta, K. et al., Hybrid La$_{1-x}$Sr$_x$MnO$_3$ nanoparticles as colloidal mediators for magnetic fluid hyperthermia, *Phil. Trans. R. Soc. A*, 368, 4389–4405. Copyright 2010 The Royal Society.)

17.5 SUMMARY

Nanoparticles for MFH usually consist of two components, and each of them contributes in a specific way to the resulting behaviour. While the inner cores act as magnetic heaters, the shells mediate suitable interaction between particles and surrounding liquid media. It is evident that the matter covers a large scale of tasks. Among them, we paid now attention to the various mechanisms of energy transfer *via* selected types of magnetic cores possessing specific features, which may differ from their bulk analogues.

Magnetic heating is realised by the conversion of magnetic energy to heat mediated by loss processes arising due to the changes of the MNP magnetisation under an alternating magnetic field, and three different mechanisms can be distinguished:

- Hysteresis power losses, originating in the irreversibility of the magnetisation process
- Néel relaxation, conditioned by the spontaneous rotation of the magnetic moments of the particles
- Friction losses, due to the Brownian rotation of the magnetic particles as a whole

From the point of view of achieving the highest possible heating power, it is thus necessary to adjust the parameters of the AC magnetic field to the characteristics of the system of particles or vice versa to tailor their properties to the available or admissible parameters of the AC field facility.

Further, regardless of the mechanism, the magnetic heating in the vicinity of T_C gradually decreases, and it is switched off when the transition temperature is achieved. Therefore, if the Curie temperature is suitably adjusted, this behaviour can be used as a mechanism of self-control, and local overheating of the healthy tissue can be prevented.

Two main aspects for the choice of suitable types of the materials thus should be pursued:

- Optimisation of the conversion of the energy from the AC magnetic field
- Effort to reach the required control of the temperature of their transition to paramagnetic behaviour by suitable parameters

These intentions are discussed on appropriately selected materials, namely,

- Magnetically soft magnetite
- Magnetically hard cobalt ferrites
- Cobalt–zinc ferrites
- Lanthanum–strontium manganese perovskites
- Primarily with respect to their magnetic properties, secondarily with respect to the specific use in magnetic heating

While the simple iron oxides, magnetite and maghemite, are already used in the clinical applications, particularly in the therapy of prostate, breast and glyoblastoma[69,70], the newly developed materials like cobalt–zinc ferrites and lanthanum–strontium manganese perovskites are now under biological investigations. Particular attention is paid to the manganese perovskites where the toxicity and *in vitro* tests are currently carried out[48,71]. Further experiments, *in vivo*, are on the way.

REFERENCES

1. Van der Zee, J. 2002. Heating the patient: A promising approach? *Ann Oncol* 13: 1173–1184.
2. Zaffaroni, N., Fliorentini, G., De Giorgi, U. 2001. Hyperthermia and hypoxia: New developments in anticancer chemotherapy. *Eur J Surg Oncol* 27: 340–342.
3. Moroz, P., Jones, S.K., Gray, B.N. 2002. Tumor response to arterial embolization hyperthermia and direct injection hyperthermia in a rabbit liver tumor model. *J Surg Oncol* 80: 149–156.

4. Néel, L. 1949. Théorie du traînage magnétique des ferromagnétiques en grains fins avec application aux terres cuites. *Ann Géophys* 5: 99–136.
5. Brown, W.F., Jr. 1963. Thermal fluctuation of a single-domain particle. *Phys Rev* 130: 1677–1686.
6. Stoner, E.C., Wohlfarth, E.P. 1948. A mechanism of magnetic hysteresis in heterogeneous alloys. *Philos Trans R Soc Lond Ser A* 240: 599–642.
7. Coey, J.M.D. 1971. Noncollinear spin arrangement in ultrafine ferrimagnetic crystallites. *Phys Rev Lett* 27: 1140–1142.
8. Haneda, K., Morrish, A.H. 1988. Noncollinear magnetic structure of $CoFe_2O_4$ small particles. *J Appl Phys* 63: 4258–4260.
9. Kodama, R.H., Berkowitz, A.E. 1996. Surface spin disorder in $NiFe_2O_4$ nanoparticles. *Phys Rev Lett* 77: 394–397.
10. Li, Z.W., Chen, L., Ong, C., Yang, Z. 2005. Static and dynamic properties of Co_2Z barium ferrite nanoparticles composite. *J Mater Sci* 40: 719–723.
11. Vasseur, S., Duguet, E., Portier, J., Goglio, G., Mornet, S., Hadová, E. et al. 2006. Lanthanum manganese perovskite nanoparticles as possible in vivo mediators for magnetic hyperthermia. *J Magn Magn Mater* 302: 315–320.
12. Atkinson, W.J., Brezovich, I.A., Chakraborty, D.P. 1984. Usable frequencies in hyperthermia with thermal seeds. *IEEE Trans Biomed Eng* 31: 70–75.
13. Hilger, I., Hergt, R., Kaiser, W.A. 2005. Towards breast cancer treatment by magnetic materials. *J Magn Magn Mater* 293: 314–319.
14. Reilly, J.P. 1992. Principles of nerve and heart excitation by time-varying magnetic fields. *Ann N Y Acad Sci* 649: 96–117.
15. Landau, L.D., Lifshitz, E.M. 1960. *Electrodynamics of continuous media*, Pergamon Press, London, U.K.
16. Rosensweig, R.E. 2002. Heating magnetic fluid with alternating magnetic field. *J Magn Magn Mater* 201: 370–374.
17. Kallumadil, M., Tada, M., Nakagawa, T., Abe, M., Southern, P., Pankhurst, Q.A. 2009. Suitability of commercial colloids for magnetic fluid hyperthermia. *J Magn Magn Mater* 321: 1509–1513.
18. Hergt, R., Andrä, W., d'Ambly, C.G., Hilger, I., Kaiser, W.A., Richter, U. et al. 1998. Physical limits of hyperthermia using magnetite fine particles. *IEEE Trans Magn* 34: 3745–3754.
19. Hergt, R., Dutz, S., Zeisberger, M. 2010. Validity limits of the Néel relaxation model of magnetic nanoparticles for hyperthermia. *Nanotechnology* 21: 015706 (5pp).
20. Pollert, E., Veverka, P., Veverka, M., Kaman, O., Závěta, K., Duguet, E. et al. 2009. Search of new core materials for magnetic fluid hyperthermia: Preliminary chemical and physical issues. *Progr Solid State Chem* 37: 1–14.
21. Krupička, S. 1973. *Physik der Ferrite und der verwandten magnetischen Oxide*, Academia, Prague.
22. Pollert, E. 1994. Structure and magnetic properties of ferrites. In *Materials Science Monograph 80, Structure and properties of ceramics*, ed. A. Koller, pp. 521–558. Elsevier, Amsterdam, the Netherlands.
23. Dutz, S. 2008. *Nanopartikel in der Medizin*, Verlag Dr. Kováč, Hamburg, Germany.
24. Műrbe, J., Rechtenbach, A., Töpfer, J. 2008. Synthesis and characterization of magnetite nanoparticles for biomedical applications. *Mater Chem Phys* 110: 426–433.
25. Sawatzky, G.A., Van der Moude, F., Morish, A.K. 1968. Cation distribution in octahedral and tetrahedral sites of ferrimagnetic spinel $CoFe_2O_4$. *J Appl Phys* 39: 1204–1205.
26. Pauthenet, R. 1952. Aimantation spontaneous des ferrites. *Ann Phys* 7: 710–747.
27. Bloembergen, N. 1956. Magnetic resonance in ferrites. *Proc IRE* 44: 1259–1269.
28. Tannenwald, P.E. 1955. Multiple resonances in cobalt ferrite. *Phys Rev* 99: 463–464.
29. Lee, S.W., Kim, C.S. 2006. Mössbauer studies on the superparamagnetic behavior of $CoFe_2O_4$ with a few nanometers. *J Magn Magn Mater* 303: e315–e317.
30. Baldi, G., Bonacchi, D., Innocenti, C., Lorenzi, G., Sagregorio, C. 2007. Cobalt ferrite nanoparticles. The control of the particle size and surface state and their effects on magnetic properties. *J Magn Magn Mater* 311: 10–16.
31. Berkowitz, A.E., Schuelle, W.J. 1959. Magnetic properties of some ferrite micropowders. *J Appl Phys* 30: 135S–136S.
32. Maaz, K., Mumtaz, A., Hasanain, S.K., Ceylan, A. 2007. Synthesis and magnetic properties of cobalt ferrite ($CoFe_2O_4$) nanoparticles prepared by wet chemical route. *J Magn Magn Mater* 308: 289–295.
33. El-Okr, M.M., Salem, M.A., Salim, M.S., El-Okr, R.M., Ashoush, M.A., Talaat, H.M. 2011. Synthesis of cobalt ferrite nano-particles and their magnetic characterization. *J Magn Magn Mater* 323: 920–926.
34. Veverka, M., Veverka, P., Kaman, O., Lančok, A., Závěta, K. 2007. Magnetic heating by cobalt ferrite nanoparticles. *Nanotechnology* 18: 345704 (7pp).

35. Kim, D.H., Nikles, D.E., Johnson, D.T., Brazel, C.S. 2008. Heat generation of aqueously dispersed $CoFe_2O_4$ nanoparticles as heating agents for magnetically activated drug delivery and hyperthermia. *J Magn Magn Mater* 320: 2390–2396.

36. Ammar, S., Helfen, A., Jouini, N., Flévet, F., Rosenman, I., Villain, F. et al. 2001. Magnetic properties of ultrafine cobalt ferrite particles synthesized by hydrolysis in a polyol medium. *J Mater Chem* 11 186–192.

37. Kim, Y.I., Kim, D., Lee, C.S. 2003. Synthesis and characterization of $CoFe_2O_4$ magnetic nanoparticles prepared by temperature – Controlled coprecipitation method. *Physica B* 337: 42–51.

38. Moumen, N., Veillet, P., Pileni M.P. 1995. Controlled preparation of nanosize cobalt ferrite magnetic particles. *J Magn Magn Mater* 149: 67–71.

39. Vaidyanathan, G., Sendhilnathan, S. 2008. Characterization of $Co_{1-x}Zn_xFe_2O_4$ synthesized by coprecipitation method. *Physica B* 403: 2157–2167.

40. Jeyadevan, B., Tohji, K., Nakatsuka, K. 1994. Structure analysis of coprecipitated $ZnFe_2O_4$ by extended X-ray absorption fine structure. *J Appl Phys* 76: 6325–6327.

41. Upadhyay, C., Verma, H.C., Sathe, V., Pimpale, A.V. 2007. Effects of size and synthesis route on the magnetic properties of chemically prepared nanosize $ZnFe_2O_4$. *J Magn Magn Mater* 312: 271–279.

42. Duong, G.V., Turtelli, R.S., Nunes, W.C., Schafler, E., Hahn, N., Grössinger, R. et al. 2007. Ultrafine $Co_{1-x}Zn_xFe_2O_4$ particles synthesized by hydrolysis: Effect of thermal treatment and its relationship with magnetic properties. *J Non-Crystal Sol* 353: 805–807.

43. Petitt, G.A., Forester, D.W. 1971. Mössbauer study of cobalt–zinc ferrites. *Phys Rev B* 4: 3912–3922.

44. Arulmurugan, R.G., Vaidyanathan, S., Sendhilnathan, B., Jeyadevan, B. 2006. Thermomagnetic properties of $Co_{1-x}Zn_xFe_2O_4$ (x = 0.1–0.5) nanoparticles. *J Magn Magn Mater* 303: 131–137.

45. Veverka, M., Veverka, P., Jirák, Z., Kaman, O., Knížek, K., Maryško, M. et al. 2010. Synthesis and magnetic properties of $Co_{1-x}Zn_xFe_2O_{4-\gamma}$ nanoparticles as materials for magnetic fluid hyperthermia. *J Magn Magn Mater* 322: 2386–2389.

46. Asamitsu, A., Morimoto, Y., Kumai, R., Tomioka, Y., Tokura, Y. 1996. Magnetostructural phase transitions in $La_{1-x}Sr_xMnO_3$ with controlled carrier density. *Phys Rev B* 54: 1716–1723.

47. Pollert, E., Knížek, K., Maryško, M., Kašpar, P., Vasseur, S., Duguet, E. 2007. New T_c-tuned magnetic nanoparticles for self-controlled hyperthermia. *J Magn Magn Mater* 316: 122–125.

48. Pollert, E., Kaman, O., Veverka, P., Veverka, M., Maryško, M., Závěta, K. et al. 2010. Hybrid $La_{1-x}Sr_x$ MnO_3 nanoparticles as colloidal mediators for magnetic fluid hyperthermia. *Philos Trans R Soc A* 368: 4389–4405.

49. Horyń, R., Zaleski, A.J., Sulkowski, C.E., Bukowska, E., Sikora, A. 2003. Magnetic features of $La_{1-2x}Na_xK_xMnO_{3+\delta}$ solid solution. *Physica C* 387: 280–283.

50. Pi, L., Hervieu, M., Maignan, A. 2003. Structural and magnetic phase diagram and room temperature CMR effect of $La_{1-x}Ag_xMnO_3$ solubility. *Solid State Commun* 126: 229–234.

51. Kuznetsov, O.A., Sorokina, O.N., Leontiev, V.G., Shlyakhtin, O.A., Kovarski, A.L., Kuznetsov, A.A. 2007. ESR study of thermal demagnetization processes in ferromagnetic nanoparticles with Curie temperatures between 40–60°C. *J Magn Magn Mater* 311 204–207.

52. Platil, A., Tomek, J., Kašpar, P. 2007. Characterization of ferromagnetic powders for magnetopneumography and other applications. *Sens Lett* 5: 311–314.

53. Mornet, S., Vasseur, S., Grasset, F., Duguet, E. 2004. Magnetic nanoparticle design for medical diagnosis and therapy. *J Mater Chem* 14: 2161–2175.

54. Glöckl, G., Hergt, R., Zeisberger, M., Dutz, S., Nagel, S., Weitschies, W. 2006. The effect of field parameters, nanoparticles properties and immobilization on the specific heating power in magnetic particle hyperthermia. *J Phys Condens Matter* 18: S2935–S2949.

55. Okawa, K., Sekine, M., Maeda, M., Tada, M., Abe, M. 2006. Heating ability of magnetite nanobeads with various sizes for magnetic hyperthermia at 120 kHz, a noninvasive frequency. *J Appl Phys* 99: 0H102-1–3.

56. Fortin, J.P., Wilhelm, C., Servais, J., Ménager, C., Bacri, J.C., Gazeau, F. 2007. Size-sorted anionic iron oxide nanomagnets as colloidal mediators for magnetic hyperthermia. *J Am Chem Soc* 129: 2628–2635.

57. Vasseur, S. 2007. Synthèse, caractérisation et échauffement par induction de nanoparticules magnétiqués hybrides à coeur $La_{0.75}Sr_{0.25}SrMnO_3$ pour des applications thermothérapeutiques en cancérologie, PhD thesis, Université Bordeaux 1.

58. Natividad, E., Castro, M., Mediano, A. 2008. Accurate measurement of the specific absorption rate using a suitable magnetothermal setup. *Appl Phys Lett* 92: 093116-1–3.

59. Natividad, E., Castro, M., Mediano, A. 2009. Adiabatic vs non-adiabatic determination of specific absorption rate of ferrofluids. *J Magn Magn Mater* 321: 1497–1500.

60. Kaman, O., Veverka, P., Jirák, Z., Maryško, M., Knížek, K., Veverka, M. et al. 2011. The magnetic and hyperthermia studies of bare and silica-coated $La_{0.75}Sr_{0.25}MnO_3$ nanoparticles. *J Nanopart Res* 13: 1237–1252.

61. Dutz, S., Clement, J.C., Eberbeck, D., Gelbrich, D., Hergt, R., Müller, R. et al. 2009. Ferrofluids of magnetic multicore nanoparticles for biomedical applications. *J Magn Magn Mater* 321: 1501–1504.
62. Kaman, O., Pollert, E., Veverka, P., Veverka, M., Hadová, E., Knížek, K. et al. 2009. Silica encapsulated manganese perovskite nanoparticles for magnetically induced hyperthermia without the risk of overheating. *Nanotechnology* 20: 275610 (7pp).
63. Levy, M., Wihelm, C., Servais, J., Menager, C., Bacri, J.C., Gazeau, F. 2008. Magnetically induced hyperthermia: Size dependent heating power of maghemite nanoparticles. *J Phys Condens Matter* 20: 204133 (5pp).
64. Gonzalez-Fernandez, M.A., Torres, T.E., Andrés-Vergés, M., Costo, R., de la Presa, P., Serna, C.J. et al. 2009. Magnetic nanoparticles for power absorption: Optimizing size, shape and magnetic properties. *J Solid State Chem* 182: 2779–2784.
65. Hergt, R., Hiergeist, R., Zeisberger, M., Glöckl, G., Weitschies, W., Ramirez, L.P. et al. 2004. Enhancement of AC-losses of magnetic nanoparticles for heating applications. *J Magn Magn Mater* 280: 358–368.
66. Vergés, M.A., Costo, R., Roca, A.G., Marco, J.F., Goya, G.F., Serna, C.J. 2008. Uniform and water stable magnetite nanoparticles with diameters around the monodomain–multidomain limit. *J Phys D Appl Phys* 41: 134003 (10pp).
67. Ma, M., Wu, Y., Zhou, J., Sun, Y., Gu, N. 2004. Size dependence of specific power absorption of Fe_3O_4 particles in AC magnetic field. *J Magn Magn Mater* 268: 33–39.
68. Jordan, A., Rheinländer, T., Waldöfner, N., Scholz, R. 2003. Increase of the specific absorption rate (SAR) by magnetic fractionation of magnetic fluids. *J Nanopart Res* 5: 597–600.
69. Johannsen, M., Gneveckow, U., Eckelt, L., Feussner, A., Waldöfner, N., Scholz, R. et al. 2005. Clinical hyperthermia of prostate cancer using magnetic nanoparticles: Presentation of a new interstitial technique. *Int J Hyperth* 21: 637–647.
70. Thiessen, B., Jordan, A. 2008. Clinical applications of magnetic nanoparticles for hyperthermia. *Int J Hyperth* 24: 467–474.
71. Kačenka, M., Kaman, O., Kotek, J., Falteisek, L., Černý, J., Jirák, D. 2011. Dual imaging probes for magnetic resonance imaging and fluorescence microscopy based on perovskite manganite nanoparticles. *J Mater Chem* 21: 157.

Emil Pollert graduated in 1961 from the Institute of Chemical Technology in Prague where later he became an associate professor, and he obtained the degree Doctor of Sciences from Charles University. He started his research work in the Institute of Solid State Physics, now Physical Institute of the Czech Academy of Sciences, Prague, in 1962. His main research activity includes problems of the synthesis and characterisation of magnetic oxides, like phase equilibria, growth of single crystals, crystal chemistry and magnetic properties. In the last years, he turned his interest to the MNPs for diagnostic and therapeutic applications in medicine. The orientation of his work has been influenced by the long-term and close cooperation with L'Institut de Chimie de la Matière Condensée de Bordeaux starting since 1973.

His leisure time activity has been devoted for several decades to whitewater canoeing on top international level.

Karel Závěta, born 1933, graduated in 1956 at the Charles University, Prague, and has been working at the Inst. of Technical Physics, finally renamed to Inst. of Physics of ASCR. In 1961/1962, he worked for 6 months at the Moscow State University. In 1966, he received the degree RNDr from the Charles University, Prague. In 1968/1969, he was a visiting professor at the Department of Physics and Astronomy, University of Maryland, College Park, Maryland. In 1976–1989, he collaborated with IFW AW DDR, Dresden, DDR, studying the magnetic properties and relaxation effects of metallic glasses. In 1993–2009, he headed the Joint Lab. of Mössbauer Spectroscopy at the Faculty of Mathematics and Physics of the Charles University, Prague. His main research activities concern magnetic properties of nanocrystalline systems, magnetisation processes and domain structures, study of magnetic nanomaterials by Mössbauer spectroscopy, for many years in collaboration with Inst. of Physics PAS, Warsaw. For several decades, he was a member of the International Advisory Board of the SMM Conferences and a member of the Editorial Board of *Czechoslovak Journal of Physics*. He was editor or joint editor of numerous proceedings of international conferences (e.g. as special issues of *Hyperfine Interactions*, *Low Temperature Physics*, *Czechosl. J. Phys.*).

18 Magnetic Liposomes and Hydrogels towards Cancer Therapy

Manashjit Gogoi, Manish K. Jaiswal,
Rinti Banerjee, and Dhirendra Bahadur

CONTENTS

18.1 INTRODUCTION TO HYPERTHERMIA AND DRUG DELIVERY

Cancer is one of the major challenges for modern medicine. It is a life-threatening disease that causes abnormal and uncontrolled growth of cells. Despite successful discovery of many potential anti-cancer drugs, every year cancer kills more than six million people and the number is growing. Success in cancer treatment is often limited by inadequate delivery of anti-cancer agents to the tumour and their severe side effects in normal tissues[1]. Liposomes and nanohydrogels have been drawing attention due to their promising applicability for drug delivery and magnetic field–assisted hyperthermia.

Liposomes are self-assembled vesicles composed of a lipid bilayer encapsulating an aqueous volume, known for their low toxicity, biocompatible and biodegradable nature[2,3], while polymeric hydrogels are water-rich 3D cross-linked network structures having capacity to carry magnetic nanoparticles (MNPs) and drugs together. For anti-cancer drug delivery, liposomes and nanohydrogels need to fulfil three basic requirements: (i) prolonging blood circulation, (ii) allowing sufficient accumulation in tumour and (iii) showing controlled triggered release at tumour site. Initially, conventional liposomes suffered from very quick clearance in blood by the reticuloendothelial system (RES), including liver, spleen and bone marrow[1]. Papahadjopoulos and Allen[4–6] discovered a way to extend the blood circulation from ~30 min to ~24–48 h by using polyethylene glycol (PEG)-derivatised lipids in the liposomes. The PEG coating increases blood circulation time by making the drug delivery system highly hydrated and protecting its surface from proteins and hence

479

from opsonisation. Both conventional and sterically stabilised drug delivery systems accumulate in tumours through leaky tumour vasculature and are retained there due to a poor lymphatic drainage system. This is called the enhanced permeability and retention (EPR) effect. Long-circulating liposomes were found to be accumulated more in tumour tissues[7]. Despite higher accumulation of doxorubicin (Dox)-loaded liposomes in tumour tissues, the therapeutic effect of many liposome formulations was not effective in treating cancer due to low bioavailability of drug within the tumour tissues. In order to treat tumour tissues effectively, drug must be released on demand in the tumour sites.

Hyperthermia is based on the fact that cancerous cells are more sensitive to heat than normal cells. Cancerous cells tend to die when heated to a temperature of 45°C, either by apoptosis or by mitotic death. The cells die due to protein denaturation. The heat at this temperature damages the cytoskeleton, membranes and causes changes in enzyme complexes responsible for DNA synthesis and repair. The effect of hyperthermia on cells varies in different cell cycles. Mitotic phase shows the highest sensitivity to heat[8].

Morphology of tumour vasculature is highly complex, which causes lower dissipation of heat from them in comparison to normal tissues. Unlike normal tissues, cancerous tissues do not expand in response to heat, and they prevent compensatory vasodilation due to their ill-formed morphology, which have no smooth muscle. Hence, the rate of heat dissipation is also low. This makes them susceptible to thermal damage. An increase in tissue temperature from 37°C to 45°C leads to a series of responses in blood flow in the normal and the tumour tissue. In the case of normal tissue, hyperthermia causes vasodilation and increased blood flow to the heated tissue[9,10] whereas tumour tissue becomes susceptible to occlusion and haemorrhage in the range of 43°C–45°C. Above 45°C, vascular damage occurs in both normal and tumour tissues, resulting in occlusion and haemorrhage. Still, sufficient proof is not there to confirm that heat damages the tumour vasculature in human[11].

Mild hyperthermia has been found to increase the tumour blood flow. Song[12] showed that tumour blood flow was increased by 1.5 ± 2 times during half an hour of mild hyperthermia at 41°C–42°C, compared to unheated control. Horsman and Overgaard[13] showed increase in tumour oxygenation during hyperthermia due to increase in tumour blood flow. Increase in tumour blood flow due to hyperthermia resulted in more extravasation of liposomes in tumour tissues than normal tissues[14]. Hyperthermia has been found to enhance the tumour microvasculature permeability. At normothermic conditions, the ill-formed and chaotic natured tumour microvasculatures are found to be leakier than normal blood vessels[15–17]. So, by applying mild hyperthermia, blood flow can be enhanced within the tumour.

Temperature-sensitive materials may be combined with MNPs for hyperthermia applications. Yatvin et al.[18,19] conceptualised the idea of using thermosensitive liposomes for heat-triggered release. Liposomes can be made thermosensitive by tuning the composition of lipid bilayer to undergo a temperature-dependent phase transition from gel to liquid phase in the temperature range of 42°C–44°C. This can be done by changing the type and molar ratio of lipids. Once the thermosensitive liposomes reach tumour sites, heat can be applied to trigger the release of liposomal contents. Thus, heat-triggered drug release from thermosensitive liposomes can provide targeted bioavailability at the target site and thereby increased the efficacy of treatment procedure[20]. Among the nanohydrogels, N-isopropylacrylamide (NIPA)-based hydrogels draw more attention due to its thermoresponsive property and low cytotoxicity[21,22]. Heat facilitates the release of drug and kills the tumour cells. Application of heat or hyperthermia is one of the most common strategies for triggered drug release at tumour sites.

18.2   MAGNETIC LIPOSOMES

Magnetic liposomes are liposomes that contain MNPs either in the core or in the lipid bilayer. Formation of phospholipid bilayers around lauric acid stabilised iron oxide nanoparticles was established with both theoretical and experimental data by Cuyper and Joniau[23]. Phospholipids from the inner monolayer were found to bind strongly with iron oxide nanoparticles by orienting the polar head groups towards the nanoparticles surface while the outer layer assembles through interacting with the exposed hydrocarbon chain of the inner layer. Moreover, they claimed to introduce high-gradient magnetophoresis as a highly efficient method to separate magnetoliposomes from non-magnetic liposomes of identical size and charge. The adsorption kinetics of phospholipids on the surface of lauric acid stabilised Fe_3O_4 nanoparticles was also studied by Cuyper and Joniau[24]. Results suggested that adsorption of inner monolayer of phospholipids is a first-order reaction, and generation of outer layer is retarded by back fluxes. Overall adsorption kinetics is independent of concentration of lipids and Fe_3O_4 nanoparticles, but dimethyl sulphoxide (DMSO) enhances the adsorption rate by weakening the lipid–lipid contact of donor vesicles and increasing the aqueous solubility of lipids. Magnetic liposomes can be prepared by thin-film hydration (TFH) and reverse-phase evaporation (RE) method or double emulsion (DE) method. In, former method, lipids are dissolved in organic solvent mixture, that is, either chloroform/methanol (2:1 v/v) or chloroform/isopropyl ether (1:1 v/v) along with hydrophobic drug or MNPs, and, solvent mixture is removed at 40°C using a rotary vacuum evaporator to form a thin film. Hydration of the thin film is done with aqueous solution at a temperature above the phase transition temperature of lipids for 0.5–1 h. Hydrophilic nanoparticles are usually added during hydration process. Then, unencapsulated MNPs and excess drug or therapeutic molecules are removed either by series of centrifugation steps or by size exclusion chromatography.

Thermosensitive liposomes are composed of lipids that undergo a phase transition from a gel to a liquid expanded state at a particular temperature called the phase transition temperature. The lipids are chosen such that this phase transition temperature matches with the hyperthermia temperatures of 42°C–45°C.

Heating ability of 1,2-dipalmitoyl-*sn*-glycero-3-phosphocholine (DPPC) magnetic liposomes containing monodispersed iron oxide nanoparticles of different sizes using TFH method was evaluated[25]. Results showed that the magnetic liposomes prepared with 11 nm size iron oxide nanoparticles were able to generate maximum heat within the biologically accepted range. The encapsulation efficiency of MNPs, as well as heating ability of magnetic liposomes prepared with TFH and DE method, was compared, and it was found higher for TFH method[26]. Maximum MNPs were encapsulated in egg PC/cholesterol (2:1 molar ratio) composition, and this magnetic liposome formulation had the highest heating ability, that is, specific absorption rate (SAR) value.

A modified RE method was reported to prepare magnetic liposomes with high-density encapsulation of Fe_3O_4 nanoparticles[27]. Transmission electron microscopy (TEM) images confirmed the formation of magnetic liposomes with densely packed Fe_3O_4 nanoparticles with mean diameter of about 225 nm. Sabaté et al.[28] prepared Fe_3O_4 nanoparticle-loaded magnetic liposomes from soy phosphatidylcholine by thin film hydration (TFH) method. Size of the extruded magnetic liposomes was 197 ± 9 nm. Encapsulation efficiency of MNPs varied between 18% and 99%, depending upon the initial concentration of MNPs used.

Teagafur-loaded thermosensitive magnetic liposomes were prepared using RE method[29]. TEM result showed that size of magnetic liposomes was in the range of 50–130 nm, and mean hydrodynamic diameter was around 126 nm. Size and zeta potential and magnetic moment of these magnetic liposomes remained almost same over a period of 6 months, suggesting that these particles were highly stable. *In vitro* release study showed that cumulative teagafur in rat serum was 25% over a period of 24 h. Pereira da Silva Gomes et al.[30] prepared highly stable polyelectrolyte coated γ-Fe_2O_3-loaded magnetic liposomes with TFH followed by layer-by-layer (LbL) method. The objective of this study was to prepare highly stable nanocapsules under physiological conditions. Results showed that average size of extruded liposomes was 164 nm and increased to about 200–400 nm after four layers of poly(allylamine hydrochloride)/poly(sodium 4-styrenesulphonate) (PAH/PSS) coating. These magnetic liposomes were very stable and were not disrupted by Triton TX-100 and phospholipase enzymes.

Both TFH and RE methods are used equally for preparation of magnetic liposomes. MNP encapsulation was reported to be more in the liposomes prepared by TFH than the liposomes prepared by RE method[26]. But, these results were not consistent with results reported by Zheng et al.[31] Further comparative study is required to evaluate MNPs encapsulation in liposomes prepared with these methods. Drug encapsulation efficiencies can be achieved as high as 60%–65% by RE method[32], whereas drug encapsulation efficiency was reported up to 90% in magnetic liposomes by TFH method[33]. Both small and large molecules can be encapsulated by RE method. But, organic solvents used in this method are not eco-friendly, and they can degrade biomolecules like protein and DNA. So, encapsulation of such biomolecules with RE method is a challenge[32].

18.3 MAGNETIC NANOHYDROGELS

Polymeric hydrogels are 3D hydrophilic, physically/chemically intermingled segments having capacity of absorbing fluids into its pores. Thermosensitive polymers represent class of hydrophilic hydrogels, which become hydrophobic upon heating by expelling out absorbed aqueous content. The behaviour of hydrophilic to hydrophobic transition can well be understood in terms of swelling of sub-micron-sized (~200–500 nm) hydrogels at low temperature to shrinking at higher temperature characterised by their lower critical solution temperature (LCST). Therefore, there occurs a reduction in size of these hydrogels as shown in schematic diagram (Figure 18.1). It may be noted that certain polymeric hydrogels turn again hydrophilic upon heating beyond a certain temperature defined as upper critical solution temperature (UCST).

These typical sub-micron-sized spherically shaped cross-linked porous hydrogels[35,36] provide easy platforms for entrapment of MNPs as well as drugs. The polymer segments swell in fluid and encapsulate the content dissolved into it and then deliver these by expelling out fluid upon heating beyond LCST. Furthermore, the particulate morphology not only helps in carrying drugs/molecules and circulating through narrow blood vessels, but also provides easy mechanism of tailoring the surface, which may further help target the vehicle to the specific site of interest.

There are numerous polymer-based hydrogels under exploration for drug delivery and hyperthermia purposes. Table 18.1 lists some of the popular thermoresponsive polymers and their respective transition temperatures. Among all, poly(N-isopropylacrylamide) (PNIPAAm) has attracted great attention due to its biocompatibility and close proximity of its LCST (which is ~32°C) to the body temperature and relatively insensitive to slight change in environmental conditions such as pH, concentration, external applied magnetic field, etc. (Figure 18.2). The underlying idea is to combine PNIPAAm with MNPs, which would provide heat under alternating current (AC) field to facilitate the shrinkage of hydrogels due to expulsion of aqueous media and thereby releasing out the encapsulated anti-cancer drug from its cross-linked pores. The combination of PNIPAAm-based hydrogels along with MNPs may take place either *via* encapsulating MNPs into hydrogels[34], by coating

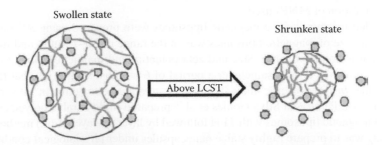

FIGURE 18.1 Polymer hydrogel going through swollen to shrunken state above its transition temperature by expelling out water content trapped inside its pores. (Reprinted from *Colloids Surf. B Biointerfaces*, 81, Jaiswal, M.K., Banerjee, R., Pradhan, P., and Bahadur, D., Thermal behavior of magnetically modalized poly(N-isopropylacrylamide)-chitosan based nanohydrogel, 185–194, Copyright 2011, with permission from Elsevier.)

TABLE 18.1
Thermoresponsive Polymers

	Thermosensitive Polymers	Transition Temperature (°C)
Natural polymers	Gelatine	~40
	Hydroxypropylcellulose	45–55
	Methylcellulose	~80
Synthetic polymers	Poly(N-isoprophylacrylamide), PNIPAAm	32–34
	Poly(N-isopropylmethacrylamide), PNIPMAAm	38–44
	Poly(methylvinylether), PMVE	~37
	Poly(vinylalcohol), PVA	~125
	Poly(vinylpyrrolidone), PVP	~160
	Poly(methacrylic acid), PMAc	~75
	Poly(N-vinylcaprolactam)	30–50
Composite copolymers	Poly(dimethylaminoethyl methacrylate), PDMAEMA	~50
	PNIPAAm–PEG	~30–39
	Poly(NIPAAm-co-DMAAm)	~38
	PNIPAAm–PLGA	34–50

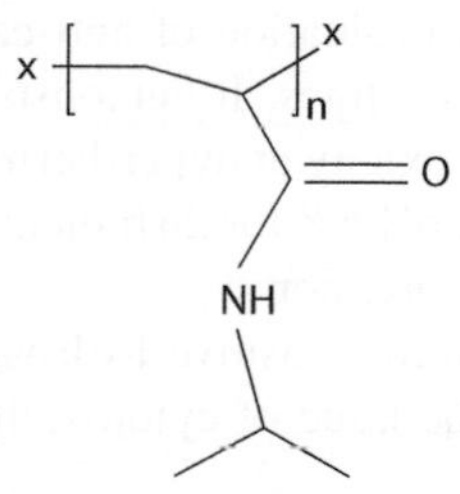

FIGURE 18.2 Chemical structure of PNIPAAm. (Reprinted from *Colloids Surf. B Biointerfaces*, 81, Jaiswal, M.K., Banerjee, R., Pradhan, P., and Bahadur, D., Thermal behavior of magnetically modalized poly(N-isopropylacrylamide)-chitosan based nanohydrogel, 185–194, Copyright 2011, with permission from Elsevier.)

MNPs onto the surface of growing hydrogels during synthesis[37], or *via* constructing a system of core–shell morphology wherein MNPs form a core and PNIPAAm forms a shell-like structure[38].

But the main challenge in the use of PNIPAAm is to raise its LCST (~32°C) above hyperthermia temperature, which is 42°C–43°C. Many researchers have paid considerable attention to enhance the LCST either by copolymerising it with another hydrophilic polymer[38–41] or by grafting it with another polymer in order to change the pore size[34,44] or by incorporating MNPs to it[34,37].

LCST of PNIPAAm was raised well above 42°C by copolymerising it with hydrophilic poly(acrylamide) (AAm)[39]. The synthesis was carried out using free radical polymerisation method in the presence of surfactant sodium dodecylsulphate (SDS). Mean diameter of the thus obtained sub-micron (50–400 nm) nanohydrogels was controlled by changing the feed ratio of surfactant during the synthesis.

The introduction of MNPs to the hydrogel systems can also cause upward shift in LCST, possibly due to magnetic interaction among nanoparticles, which prevents these structures from shrinking it easily. PNIPAAm-based hydrogels (size ~250 nm) covered with γ-Fe_2O_3 MNPs were synthesised using surfactant-free radical polymerisation method[37]. Increasing the amount of MNPs into the hydrogel system helped shift LCST of PNIPAAm monotonously. Quantitatively, 18 (w/w%) MNPs added to hydrogel system caused LCST to enhance above 40°C.

Apart from having LCST above 42°C, longer circulation time is also of utmost importance. PEGylated (polyethyleneglycol)-PNIPAAm-based sub-micron size hydrogel system having LCST well above 42°C was reported[40]. The introduction of PEG provides the system longer circulation time by not allowing the protein adsorption onto the particle surfaces. The study revealed that it reduced protein adsorption by nearly 50% in comparison with non-PEGlayted particles. It further reported the encapsulation and release studies of anti-cancer drug Dox triggered by temperature given externally with the help of water bath. The physical diffusion method employed for encapsulation gave nearly 85% entrapment of Dox into hydrogel system. The release profile showed a sharp jump in Dox release (from nearly 40% to 90%) when temperature of the medium changed from 37°C (pH ~ 7.4) to 40°C (pH ~ 5.5), demonstrating its applicability for tumour where pH of the medium is normally acidic compared to normal cells. Thus, the system provides a promising candidate for MNPs incorporated hydrogels wherein the heating can be obtained from external magnetic field to release the anti-cancer drugs.

The efficacy and potential of dual pH and thermoresponsive hydrogel systems were investigated, which may act as promising candidate for combined therapy of hyperthermia and chemotherapy to treat cancerous cells[41]. It reported the synthesis of sub-micron size PNIPAAm- and acrylic acid (AA)-based composite hydrogel (PNA) particles conjugated with anti-cancer drug Dox *via* acid-liable hydrazone linkage. The attempt was carried out to achieve the LCST well above body temperature by changing the pH of the media. The results are shown in Figure 18.3.

There occurred an upward shift in LCST of the PNIPAAm–AA system with increasing pH, which can suitably be tuned above hyperthermia temperature. The advantage of having such a dual temperature and pH-tuned system is that when MNPs are encapsulated into hydrogels, they can provide an excellent platform for site-specific release of drug in the presence of externally applied magnetic field at lower at cancerous sites.

This study further reported *in vitro* evaluation of anti-cancer drug Dox conjugated with PNA hydrogels done in carcinoma HepG2 cell lines. It demonstrated that lowering pH of the medium enhances the drug release and cell cytotoxicity at hyperthermia temperature. Addition of Dox to the system caused nearly 70% cell death at pH 6.8 for 24 h incubation period, much higher than that at pH 7.4, which caused only 20% cell suppression.

The application of pH and temperature-sensitive hydrogels as dual functionalities for *in vitro* hyperthermia was investigated[41], but the issue of cytotoxicity and biodegradability of poly(AAm)

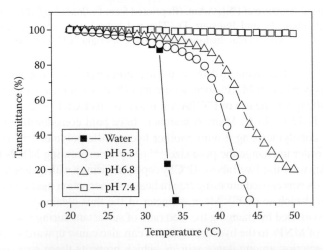

FIGURE 18.3 Temperature-sensitive phase transitions of Dox–PNA in phosphate-buffered saline at different pH values. (Reprinted from *Colloids Surf. B Biointerfaces*, 84, Xiong, W., Wang, W., Wang, Y., Zhao, Y., Chen, H., Xu, H., and Yang, X., Dual temperature/pH-sensitive drug delivery of poly(*N*-isopropylacrylamide-*co*-acrylic acid) nanogels conjugated with doxorubicin for potential application in tumor hyperthermia therapy, 447–453, Copyright 2011, with permission from Elsevier.)

was overlooked. To address the problem, PNIPAAm–chitosan (CS)-based iron oxide MNPs encapsulated sub-micron size hydrogels using free radical polymerisation method was synthesised[34]. CS being natural a polymer offers not only biodegradability and pH sensitivity to the system, but also helped achieve LCST well above 42°C. They reported the biocompatibility of temperature optimised system and assessed the potential for hyperthermia application under AC field. The time-dependent temperature rise analysis showed its capability to raise the temperature of the medium above 42°C in 10 min for 1 mL dose having 10 mg MNPs, making it a suitable candidate for cancer treatment. Representative SEM and atomic force microscopy (AFM) images of PNIPAAm and its composition along with CS named Nanohydrogel-1 (NHG-1) (weight proportion of NIPAAm and CS in 95:05) and iron oxide MNP named magnetic nanohydrogel (MNHG) (25 w/w% of the composition) are shown in Figure 18.4.

So far we have been discussing the hydrophilic hydrogel systems having MNPs encapsulating within it or covered with. Their functions were by and large less controlled once the temperature had reached the onset of LCST, and therefore, the release of drug cannot be stopped when triggered. Alternatively, researchers have paid their attention to explore core–shell hydrogel systems due to their unique properties of releasing the drug in a controlled fashion. Core–shell type spherical sub-micron-sized hydrogel contains a hydrophobic core and a hydrophilic thermosensitive shell, which exhibits characteristics of both core and shell.

Ma et al.[42] attempted to achieve core–shell hydrogel of P(*N,N*-diethylacrylamide-*co*-methacrylic acid) P(DEA-*co*-MAA) of size around 100 nm having LCST of 37°C. The TEM image revealed that PMAA formed shell over the periphery of P(N,N-diethylacrylamide-co-methacrylic acid) core/shell hydrogel (PDEA) particles. It reported that the size of hydrogels was largely dependent on the pH of the media. Increasing the pH from 1.2 to 7.4 caused a change in size (from nearly 100 to 250 nm) of the hydrogel microspheres possibly due to the formation of hydrogen bonds among carboxyl groups in the media. *In vitro* cytocompatibility of composites on human cervical cancer cell lines (HeLa) was also reported using 3-(4,5-dimethylthiazol-2-yl)-2,5-diphenyltetrazoliumbromide (MTT) assay. The chosen concentration of various compositions was within 20–140 mg/mL incubated for 30 h. The observation showed the samples were biocompatible and with no apparent toxicity. The advantage of

FIGURE 18.4 Representative (a) SEM and (b) AFM images of poly(NIPAAm), NHG-1 and MNHG, respectively, with their respective size distributions. (Reprinted from *Colloids Surf. B Biointerfaces*, 81, Jaiswal, M.K., Banerjee, R., Pradhan, P., and Bahadur, D., Thermal behavior of magnetically modalized poly(*N*-isopropylacrylamide)-chitosan based nanohydrogel, 185–194, Copyright 2011, with permission from Elsevier.)

such systems based on P(DEA-*co*-MAA) could be used for the development of site-specific carriers of MNPs as well as drugs in intestinal cancers where pH of the medium is relatively low.

Apart from PNIPAAm and PDEA, many other polymer-based hydrogels have also been under exploration for potential application in hyperthermia treatment. Recently, PEG and methyl ether methacrylate (PEGMMA)- and dimethacrylate (PEGDMA)-based magnetic hydrogels have been investigated[43] for their *in vitro* hyperthermia potential, and it demonstrated that the temperature of the hydrogels can be controlled by tuning the alternating magnetic field (AMF) strength so that the gels either reached hyperthermic (42°C–45°C) or thermoablative (60°C–63°C) temperatures.

In a typical experiment, the M059K glioblastoma cell lines were cultured for 24 h and were placed directly on the hydrogel for heat treatment after medium decantation. The cells were then exposed to the AMF for 5 min and then returned to the incubator for 2 h to allow time for a cellular response to the heat treatment and then assayed. The results are summarised in Figure 18.5 where cells lying near the centre of the field are destroyed almost completely (Figure 18.5a) due to heat generated by MNPs. As one moves farther away from the centre, the influence of the field is diminished (Figure 18.5d and g). The right column (Figure 18.5c, f and i) served as control groups for comparison. This study demonstrated the potential of chosen composite for further *in vivo* animal trials in order to kill cancer cells due to hyperthermia under AC field.

Apart from hyperthermia, thermoresponsive hydrogels find applications in targeting also. The thermal targeting behaviour of PNIPAAm–AAm-based temperature optimised hydrogels *in vivo*

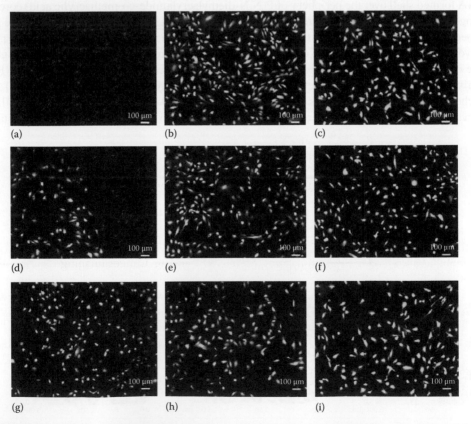

FIGURE 18.5 Fluorescent microscopy images of hyperthermia-treated M059K cells along with their control counterpart. The left column (a, d, g) represents the cells exposed to sample and AFM; middle column (b, e, h) represents the cells exposed to AFM only, while right column (c, f, i) represents cells neither exposed to sample nor AFM. (Reprinted from Meenach, S.A., Hilt, J.Z., and Anderson, K.W., *Acta Biomaterialia*, 6, 1039, 2010.)

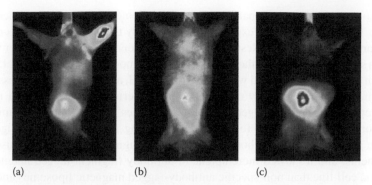

FIGURE 18.6 Fluorescence images of different healthy mice after 1 h of sample administration. (a) Mouse with hyperthermia treatment on the right leg and dye-loaded sample administration, (b) mouse with same hyperthermia treatment and pure dye injection and (c) mouse with same dye-loaded sample administration and without hyperthermia treatment. Bright fluorescence signals observed in all cases represent the accumulation of sample in liver, which is considerably low in (a). (Reprinted from *J. Control. Release*, 131, Zhang, J., Chen, H., Xu, L., and Gu, Y., The targeted behavior of thermally responsive nanohydrogel evaluated by NIR system in mouse model, 34–40, Copyright 2011, with permission from Elsevier.)

was demonstrated. A group of healthy mice were given heat using water bath to their leg up to 42°C and then near-infrared dye-12 (NIRD-12)-loaded sample was administered through the tail vein. The distribution analysis of sample inside the animal body was imaged through fluorescence microscopy. The observation (shown in Figure 18.6) after 1 h of sample administration clearly revealed its accumulation into the leg area, which is at higher temperature than rest of the animal body[39].

This result clearly provided the path to magnetic drug targeting with temperature-sensitive hydrogel-based carrier wherein incorporated MNPs generated heat and helped the carriers to reach the specific sites *in vivo*.

18.4 CANCER HYPERTHERMIA: SITE-SPECIFIC KILLING OF MALIGNANT CELLS

18.4.1 *In Vitro and In Vivo Hyperthermia*

Current cancer therapies are invasive, painful and have a lot of acute and chronic side effects[44]. Chemotherapy is most commonly employed for treatment of cancer patients but suffers from severe side effects. Novel treatment approaches like hyperthermia with gold nanoparticles[45,46], MNPs[47–51] and combination of hyperthermia and chemotherapy[33,52,53] are being evaluated for improving the patients' quality of life. Masuko et al.[50] studied anti-tumour activity of hyperthermia in AH60C (ascitic rat liver cancer) tumour-bearing Donryu rats using dextran magnetite-incorporated thermosensitive DPPC liposomes. Results showed that tumours were completely regressed in five out of six in once treated group and five out of five in twice treated group. Control animals did not show any regression in tumour. Histological analysis of tumour sections showed that magnetic liposomes remained in the tumour tissues immediately and 48 h after injection, and in the necrotic areas 2 weeks after injection.

The effectiveness of repeated doses of magnetic hyperthermia with cationic magnetic liposomes was investigated *in vivo*[51,54]. Results showed that complete tumour regression was observed in 20%, 60% and 87.5% of animals in group with single, double and triple treatments, respectively. Histological observation of haematoxylin–eosin-stained tumour sections showed that the necrotic areas spread in the tumours with repeated hyperthermia (RH) groups. Further, it was observed from microscopic observations of Prussian blue–stained organs that MNPs mostly accumulated in tumour and spleen.

In another study, magnetite nanoparticles–loaded liposomes were used for repeated rounds of hyperthermia[55]. The objective of this experiment was to study the therapeutic effects of single round of RH on tumours of different sizes. Results showed that tumours with 7 mm diameter were regressed completely with single round of hyperthermia, and 80% of tumours were regressed in 15 mm tumours. However, 15 mm tumours were treated with two rounds of RH, and the tumours regressed completely.

In order to increase the efficiency of magnetic hyperthermia, antibody-tagged magnetic liposomes were prepared. Antibody-tagged magnetic liposomes were found to accumulate more in tumour than normal liposomes due to their higher binding affinity towards antigens present in tumour. Human glioma–specific antibody fragment with magnetic liposomes was used *in vitro* and *in vivo* experiments, and it showed more uptake of antibody fragment-tagged magnetic liposomes in human glioma cell line than non-specific antibody-tagged magnetic liposomes and normal magnetic liposomes[56]. *In vivo* hyperthermia with antibody-tagged liposomes showed that growth of tumours was completely arrested over 2 weeks.

The targeting ability and cytotoxic effects of anti-human epidermal growth factor raceptor 2 (HER2)-tagged immunoliposomes (HML) containing Fe_3O_4 nanoparticles in tumour-bearing nude mice were studied[57]. Results showed that accumulation of HML was significantly higher in high HER2-expressing (BT474) tumours than low HER2-expressing (SKOV3) tumours. Tumours of mice treated under AC magnetic field were completely regressed in 10 weeks.

The various *in vivo* studies on different tumour models suggested that repeated magnetic hyperthermia is found to be effective in treating solid tumours or primary cancers. In most of the aforementioned experiments, surface temperature was maintained at 45°C, and hence, inside temperature was more than 45°C. At a temperature above 45°C, most of tumour cells might have died due to necrosis, but this is harmful for normal cells adjacent to the tumour also. Accumulation of immuno-magnetic liposome formulations was seen more in tumour than other vital organs of the body, suggesting application of immuno-magnetic liposomes did not heat up other organs except tumour and the effect of hyperthermia was observed only in tumour region. So, immuno-magnetic liposomes can play an important role in magnetic hyperthermia treatment.

18.5 DRUG DELIVERY VEHICLES

18.5.1 MAGNETIC DRUG TARGETING

To date, cancer chemotherapeutic drugs are neither specific nor are they targeted exclusively to cancer cells. So, they kill normal cells as well as cancerous cells. During the chemotherapy, patients suffer severely due to a lack of tumour-specific uptake resulting in damage of healthy tissues. Due to the severe side effects of drugs, the dose of drugs has to be reduced. This makes treatment of cancer very difficult[58]. So, focused strategies to deliver maximum amount of drug to the tumour sites while reducing the uptake by normal healthy tissues are required for the success of chemotherapy[59]. Guided transport of therapeutic agents to tumour sites allows optimal utilisation of chemotherapeutic drug with minimum dose[60,61]. In magnetic targeting, therapeutic agents are tagged with magnetic carriers and administered into the patient through systemic route, and these therapeutic agents are collected at the diseased site by applying a powerful magnetic field at the targeted site. Once the therapeutic agents reach the targeted site, they are released from the magnetic carriers to achieve the therapeutic effect[62]. Due to its non-invasive nature and high targeting efficiency, many researchers are pursuing magnetic targeting[63,64].

Dandamudi and Campbell[65] showed that the difference in accumulation of magnetoliposomes in murine melanoma cell line B16-F10 and endothelial cells human microvascular endothelial cells-1 (HMEC-1) with and without a permanent magnet was not significant. But, accumulation of neutral magnetic liposomes in an *in vitro* vascular model for B16-F10 cells made of capillary network was increased significantly. Similar results were reported *in vivo* in comparison to control group. The effect of magnetic field gradient on the cellular uptake of magnetic liposomes in human prostatic adenocarcinoma (PC3) cell line was studied, and the uptake of magnetic liposomes increases with the increase in magnetic field gradient[66].

Charlotte et al.[67] investigated the movements of maghemite containing rhodamine-labelled magnetic liposomes under the influence of a cylindrical magnet through a cranial window in C57BL/6 mice. The study revealed that the rhodamine-labelled magnetic liposomes were accumulated in the microvasculature under the magnet. The uptake of folic acid-tagged Dox-loaded magnetoliposomes in KB and HeLa cell lines under the influence of magnetic field was found to be higher due to combined effect of magnetic and folate targeting[33].

Nobuto et al.[68] evaluated the effect of systemic chemotherapy with magnetic liposomes containing Dox (magnetic Dox liposomes) under the influence of an externally applied magnetic force. They used Dox-loaded magnetic liposomes for treating primary tumour and lung metastasis model. In comparison to the control, growth of primary tumours was suppressed significantly by several Dox formulations. Magnetic Dox liposomes in combination with external magnetic field almost regressed the primary tumour in 2 week observation period. Magnetic Dox liposomes targeted with external magnetic field were found to be effective in reducing the tumour colonies in metastasis model. The pharmacokinetics of negatively charged paclitaxel-loaded magnetic liposomes was studied, and *in vivo* results suggested that accumulation of paclitaxel-loaded lyophilised magnetic liposomes in EMT-6 breast cancers in mice was higher than that of the lyophilised conventional liposomes following subcutaneous or intravenous injection under the influence of external magnetic field. In short, lyophilised paclitaxel-loaded magnetoliposomes were found to be more effective in treating breast cancer than other formulations administered *via* s.c. and i.p. route[108].

The results of the *in vivo* studies suggest that magnetic drug targeting can be a useful strategy to enhance the bioavailability of drugs at tumour sites and hence can improve the efficacy of chemotherapy.

18.5.2 TRIGGERED RELEASE OF ANTI-CANCER AGENTS *VIA* BOTH pH AND TEMPERATURE STIMULI

Conventional thermosensitive liposomes are composed of dipalmitoylphosphatidylcholine (DPPC), dipalmitoylphosphatidylglycerol (DPPG) or distearoyl phosphatidylcholine (DSPC)[18,19,69–71]. Phase transition temperatures of DPPC and DPPG are 41°C, while DSPC has the phase transition temperature of 58°C. Thermosensitive liposomes with a phase transition temperature of 42°C–44°C can be made by altering both type and molar ratio of lipids present in the bilayer of the liposomes[72]. Apart from what was discussed earlier, the lipids, cholesterol and PEG are often used in thermosensitive liposomes. Cholesterol is a small biomolecule. It interacts with the hydrophobic tail of the phospholipids and hence stabilises the liposomal membrane[73]. Addition of cholesterol lowers the phase transition temperature of liposomal membrane, but at the same time it broadens the phase transition temperature[74,75] and thereby reduces the temperature sensitivity of liposomes. So, optimum amount of cholesterol is necessary to prepare thermosensitive liposomes. Merlin[76] prepared a highly stable thermosensitive liposomal formulation with DPPC, DSPC and cholesterol. Addition of PEG does not affect the thermosensitivity of liposomes[77], but it extends circulation time of liposomes, which is necessary for a better drug delivery system.

Besides addition of cholesterol, there are a number of strategies used to prepare thermosensitive liposomes, including addition of lysolipids along with other saturated lipids, grafting polymers with lipids, encapsulation of thermosensitive block copolymers, etc.[78–82] During mild hyperthermia, lysolipid-containing liposomes undergo major morphological changes like formation of open liposomes, bilayer disc and pore-like defects[83–86]. Lysolipid-containing liposomes completely release Dox within 10–12 s at mild hyperthermia temperature, that is, 40°C–42°C[87–98]. During hyperthermia, encapsulated block copolymers disrupt the liposomal bilayer from inside and help in releasing liposomal contents quickly[90,91]. The mechanism of drug release from thermosensitive magnetic liposomes is schematically shown in Figure 18.7.

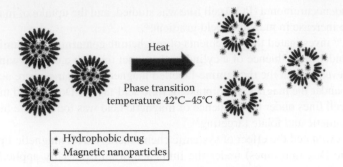

Hydrophobic drug
* Magnetic nanoparticles

FIGURE 18.7   Effect of heat on thermosensitive magnetic liposomes.

Viroonchatapan et al.[92] studied release behaviour of 5-fluorouracil from dextran stabilised iron oxide nanoparticle-loaded DPPC liposomes under an electromagnetic field in both *in vitro* and *in vivo* conditions. *In vivo* drug release study was done in B16-F10 melanoma tumour-bearing mice with a microdialysis apparatus. When tumour temperature reached 42°C, 50% of total encapsulated 5-fluorouracil was released in 20 min, and this amount of drug was sufficient to inhibit the growth of the tumours. In another report[93], Dox-loaded DPPC magnetic liposomes with Fe_3O_4 nanoparticles were prepared, and cumulative Dox release under AC magnetic field at 37°C and 42°C was found approximately 47% and 70%, respectively, within 6 min.

The release kinetics of carboxyfluorescein (CF) from cobalt ferrite ($CoFe_2O_4$) nanoparticle-loaded magnetic liposomes was reported at low-frequency (0.1–5 kHz) alternating magnetic field[94]. Results showed that CF release directly depended upon exposure time of magnetic field, frequency of magnetic field and amount of encapsulated MNPs. Results also proved that CF release from magnetoliposomes was more with negatively charged nanoparticles and with larger nanoparticles of mean diameter of 27 nm. Chen et al.[95] studied the release behaviour of bilayer-decorated DPPC magnetic liposome under alternating magnetic field. The presence of hydrophobic MNPs in the liposomal bilayer was found to stabilise the liposomes and reduce the spontaneous leakage of CF, and at the same time, it helps in releasing CF under the influence of electromagnetic field by partially rupturing the liposomal bilayer. Stability of liposome at normal condition and release of CF under AC magnetic field were found to be dependent directly on the amount of MNPs.

18.5.3   COMBINATION THERAPY: HYPERTHERMIA AND DRUG DELIVERY

Hyperthermia is minimally invasive and simple local treatment procedure, and it boosts the body's immune system[53,96]. Though hyperthermia is reported effective under *in vitro* conditions in killing cancer cell above 42°C, temperatures above 43°C cause irreversible damages to both tumour and surrounding normal cells or tissues *in vivo*. Since anti-tumour immunity is not enough, effective tumour regression needs RH. Under these circumstances, combined hyperthermia with chemotherapy can be promising in treating cancer[51].

Ito et al.[52] studied the efficacy of 4-S-cysteaminylphenol (4-S-CAP)-loaded magnetic liposomes in *in vitro* and *in vivo* experiments. *In vitro* results showed that combined effect of hyperthermia and chemotherapy on B16 melanoma cell line was additive. Results of *in vivo* experiment showed that application of magnetic hyperthermia in combination with chemotherapy was more effective than only hyperthermia and only chemotherapy. Yoshida et al.[53] investigated the combined effect of hyperthermia and chemotherapy with docetaxel-loaded cationic magnetic liposomes (DMLs) containing Fe_3O_4 nanoparticles. The results suggested that the minimum amount of docetaxel in magnetic liposomes required for complete regression of human MN 45 gastric tumour was 568 μg/mL without hyperthermia. However, dose of 60 μg/mL docetaxel in magnetic liposomes (almost 1/10th of required dose) was sufficient to regress the tumours with single dose of hyperthermia. Microscopic examination showed that DMLs gradually spread in the entire tumour. Many Berlin blue–stained

Fe_3O_4 nanoparticles were observed in the lymph nodes around inferior vena cava of DMLs injected mice 24 h post injection, and DMLs were found in the peripheral lymph nodes.

Pradhan et al.[33] investigated the combined biological and magnetic targeting abilities of Fe_3O_4 nanoparticle-loaded thermosensitive liposomes from DPPC/cholesterol/1,2-distearoyl-sn-glycero-3-phosphoethanolamine (DSPE)-PEG$_{2000}$/DSPE-PEG$_{2000}$-Folate *in vitro*. Results showed that cellular uptake of folate-tagged magnetic liposomes was increased significantly in the presence of a magnetic field. Cytotoxicity of this folate-tagged Dox-loaded magnetic liposome formulation was found to enhance synergistically due to hyperthermia.

18.6   CURRENT STATUS AND CLINICAL TRIALS

Magnetic liposome is found to be a successful system for cancer hyperthermia. The use of cationic magnetic liposomes in different animal models for cancer hyperthermia has been successfully demonstrated[50–57]. Solid tumour or primary cancer can be treated with magnetic hyperthermia, but not metastatic cancer. It is an invasive procedure, and for efficient treatment, magnetic liposomes should be distributed uniformly in the tumour in sufficient amount. Non-uniform distribution and insufficient amount of magnetic liposomes may affect the treatment efficiency. Moreover, pharmacokinetic and toxicity profile of drug encapsulated in magnetic liposomes is different from free drug, and hence, it is required to evaluate further before doing the clinical trial[97], for example, when Dox delivered in liposomes, cardio toxicity was virtually eliminated, but desquamation in skin folds was observed in patient[98]. These post-treatment complications due to long-circulating PEGylated liposomes may be much less morbid and toxic[99], but it is important to evaluate in advance. MNPs have been evaluated in clinical trial for magnetic hyperthermia application. Johannsen et al.[49] reported the first clinical trial done with magnetic fluid hyperthermia (MFH). They carried out a series of clinical trials to evaluate the feasibility of using MNPs[49] for cancer hyperthermia application and to evaluate patients' morbidity and quality of life and post-treatment process[49,100]. Results suggest that MFH can be used for the treatment of cancer, and desired hyperthermia temperature can be achieved by varying the strength of magnetic field. A magnetic field of 4–5 kA/m was found to be sufficient to achieve the temperature of the nanoparticles to hyperthermia range or even more. Nanoparticles were detected in the prostate even after 1 year of the thermotherapy, and no systemic toxicity was observed due to the presence of nanoparticles. Hauff et al.[101] did a phase I trial of intracranial thermotherapy using MNPs in combination with radiotherapy on 14 adult patients bearing glioblastoma multiforme. Treatment procedure was found to be within the tolerable limit of the patients, and no side effects were observed due to the treatment procedure. Finally, it may be concluded that MNPs could be used for hyperthermia treatment of glioblastoma multiforme patient with tolerable sufferings. Recently, Landeghem et al.[102] reported the neuropathological findings of three patients from an ongoing phase II clinical trial against glioblastoma multiforme. Post-mortem studies suggested that nanoparticles were present in the tumour in dispersed and aggregated state. This study reported practically no adverse effects related to MFH treatment; however, the MNPs were preferentially taken up by the macrophases than the tumour.

Lübbe et al.[103] conducted a clinical trial using magnetic microspheres (MMS) for the treatment of advanced solid cancer in 14 patients. However, 50% of 4′-epidoxorubicin-loaded MMS were found in liver due to their small particle size and low magnetic susceptibility. A phase I clinical trial had been reported with carbon-coated iron oxide microparticles of size in the range of 0.5–5 μm by a start up company named FeRx in San Diego[104]. Thirty-two patients had been treated with Dox-loaded microsphere without any treatment-related toxicity. They started a phase I/II clinical trial in China, Korea and USA to treat hepatocellular carcinoma with magnetic targeting[105]. Magnetic targeting is also used to deliver radioisotopes at tumour site. In this case, unlike chemotherapy, the isotopes were not released[106].

Clinical trials of magnetic targeting and MNPs-based hyperthermia suggest their efficacy and low toxicity. However, no trials have been reported on the combination of thermosensitive liposomes or hydrogels and MNPs for combined hyperthermia and temperature-triggered drug delivery. Application of magnetic liposomes for magnetic drug targeting was demonstrated successfully in *in vitro* and *in vivo* experiments[33,65–67,107]. The strategies of MNPs-mediated hyperthermia, magnetic drug targeting and combined drug delivery and hyperthermia appear promising. It is expected that these strategies will be translated into clinical practice as adjuvant therapies in near future.

18.7   CONCLUSIONS

Hyperthermia-mediated drug delivery with liposomes or nanohydrogel may be effective in treating cancer. Thermosensitive magnetic liposomes and nanohydrogels are particularly suited for this application. Improvement in liposomes and nanohydrogel system enhances the blood circulation time and accumulation in tumour tissues. Hyperthermia improves the treatment efficacy by increasing the bioavailability of drug at tumour sites with triggered release. Moreover, it improves the vascular perfusion and increases extravasation of nanoparticles in tumour vasculature. Magnetic drug targeting improves the bioavailability of drug at tumour sites and hence increases the effectiveness of treatment procedures. Results of *in vivo* magnetic drug targeting as well as magnetic hyperthermia are promising. Magnetic liposomes and nanohydrogels have the potential for successful use in cancer treatment in the future.

ACKNOWLEDGEMENT

Financial support from Nano Mission, DST, Government of India and DIT, Government of India are gratefully acknowledged.

REFERENCES

1. Andresen, T. L., Jensen, S. S. and Jørgensen, K. 2005. Advanced strategies in liposomal cancer therapy: Problems and prospects of active and tumor specific drug release. *Progress in Lipid Research*, 44, 68–97.
2. Klibanov, A. L., Maruyama, K., Torchilin, V. P. and Huang, L. 1990. Amphipathic polyethyleneglycols effectively prolong the circulation time of liposomes. *FEBS Letters*, 268, 235–237.
3. Bangham, A. D., Standish, M. M. and Watkins, J. C. 1965. Diffusion of univalent ions across the lamellae of swollen phospholipids. *Journal of Molecular Biology*, 13, 238–252.
4. Sharpe, M., Easthope, S. E., Keating, G. M. and Lamb, H. M. 2002. Polyethylene glycol-liposomal doxorubicin: A review of its use in the management of solid and haematological malignancies and AIDS-related Kaposi's sarcoma. *Drugs*, 62, 2089–2126.
5. Papahadjopoulos, D. and Gabizon, A. 1990. Liposomes designed to avoid the reticuloendothelial system. *Progress in Clinical Biology Research*, 343, 85–93.
6. Allen, T. M., Hansen, C. B. and Demenezes, D. E. L. 1995. Pharmacokinetics of long-circulating liposomes. *Advanced Drug Delivery Review*, 16, 267–284.
7. Wu, N. Z., Da, D., Rudoll, T. L., Needham, D., Whorton, A. R. and Dewhirst, M. W. 1993. Increased microvascular permeability contributes to preferential accumulation of Stealth liposomes in tumor tissue. *Cancer Research*, 53, 3765–3770.
8. Streffer, C. In Seegenschimicdt, M. H., Fessenden, P., Vernon, C. C. (Eds.), Molecular and cellular mechanisms of hyperthermia. *Thermoradiotherapy and Thermochemotherapy* 1995; 1, 47–74; Berlin; Springer Verlag.
9. Emami, B. and Song, C. W. 1984. Physiological mechanisms in hyperthermia: A review. *International Journal of Radiation Oncology, Biology, Physics*, 10, 289–295.
10. Dudar, T. E. and Jain, R. K. 1984. Differential response of normal and tumor microcirculation to hyperthermia. *Cancer Research*, 44, 605–612.
11. Waterman, F. M., Tupchong, L., Nerlinger, R. E. and Matthews, J. 1991. Blood flow in human tumors during local hyperthermia. *International Journal of Radiation Oncology Biology Physics*, 20, 1255–1262.
12. Song, C. 1984. Effects of local hyperthermia on blood flow and microenvironment: A review. *Cancer Research*, 44, 4721S–4730S.
13. Horsman, M. R. and Overgaard, J. 1997. Can mild hyperthermia improve tumour oxygenation? *International Journal of Hyperthermia*, 13, 141–147.
14. Kong, G., Braun, R. D. and Dewhirst, M. W. 2001. Characterization of the effect of hyperthermia on nanoparticle extravasation from tumor vasculature. *Cancer Research*, 61, 3027–3032.
15. Jain, R. K. 1987. Transport of molecules across tumor vasculature. *Cancer Metastasis Review*, 6, 559–593.

16. Wu, N. Z., Klitzman, B., Rosner, G., Needham, D. and Dewhirst, M. W. 1993. Measurement of material extravasation in microvascular networks using fluorescence videomicroscopy. *Microvascular Research*, 46, 231–253.

17. Yuan, F., Dellian, M., Fukumura, D., Leunig, M., Berk, D. A., Torchilin, V. P. and Jain, R. K. 1995. Vascular permeability in a human tumor xenograft: Molecular size dependence and cutoff size. *Cancer Research*, 55, 3752–3756.

18. Yatvin, M. B., Weinstein, J. N., Dennis, W. H. and Blumenthal, R. 1978. Design of liposomes for enhanced local release of drugs by hyperthermia. *Science*, 202, 1290–1293.

19. Yatvin, M. B., Muhlensiepen, H., Porschen, W., Weinstein, J. N. and Feinendegen, L. E. 1981. Selective delivery of liposome-associated *cis*-dichlorodiammineplatinum (II) by heat and its influence on tumor drug uptake and growth. *Cancer Research*, 41, 1602–1607.

20. Ponce, A. M., Vujaskovic, Z., Yuan, F., Needham, D. and Dewhirst, M. W. 2009. Hyperthermia mediated liposomal drug delivery. *International Journal of Hyperthermia*, 22, 205–213.

21. Chen, H., Zhang, J., Qian, Z., Liu, F., Chen, X., Hu, Y. and Gu, Y. 2008. *In vivo* non-invasive optical imaging of temperature-sensitive co-polymeric nanohydrogel. *Nanotechnology*, 19,185707.

22. Chearúil, F. N. and Corrigan, O. I. 2009. Thermosensitivity and release from poly *N*-isopropylacrylamide–polylactide copolymers. *International Journal of Pharmaceutics*, 366, 21–30.

23. Cuyper, M. D. and Joniau, M. 1988. Magnetoliposomes formation and structural characterization. *European Biophysics Journal*, 15, 311–319.

24. Cuyper, M. D. and Joniau, M. 1991. Mechanistic aspects of the adsorption of phospholipids on to lauric acid stabilized Fe_3O_4 nanocolloids. *Langmuir*, 7, 647–652.

25. Gonzales, M. and Krishnan, K. M. 2005. Synthesis of magnetoliposomes with monodisperse iron oxide nanocrystal cores for hyperthermia. *Journal of Magnetism and Magnetic Materials*, 293, 265–270.

26. Pradhan, P., Giri, J., Banerjee, R., Bellare, J. and Bahadur, D. 2007. Preparation and characterization of manganese ferrite-based magnetic liposomes for hyperthermia treatment of cancer. *Journal of Magnetism and Magnetic Materials*, 311, 208–215.

27. Wijaya, A. and Hamad-Schifferli, K. 2007. High-density encapsulation of Fe_3O_4 nanoparticles in lipid vesicles. *Langmuir*, 23, 9546–9550.

28. Sabaté, R., Barnadas-Rodriguez, R., Callejas-Fernández, J., Hidalgo-Álvarez, R. and Estelrich, J. 2008. Preparation and characterization of extruded magnetoliposomes. *International Journal of Pharmaceutics*, 347,156–162.

29. Zhaowu, Z., Xiaoli, W., Yangde, Z., Xingyan, L., Weihua, Z. and Nianfeng, L. 2009. Preparation and characterization of tegafur magnetic thermosensitive liposomes. *Pharmaceutical Development and Technology*, 14, 350–357.

30. Pereira da Silva Gomes, J. F., Rank, A., Kronenberger, A., Fritz, J., Winterhalter, M. and Ramaye, Y. 2009. Polyelectrolyte-coated unilamellar nanometer-sized magnetic liposomes. *Langmuir*, 25, 6793–6799.

31. Zheng, S., Zheng, Y., Beissinger, R. L. and Fresco, R. 1994. Microencapsulation of hemoglobin in liposomes using a double emulsion, film dehydration/rehydration approach. *Biochimica et Biophysica Acta (BBA) – Biomembranes*, 1196, 123–130.

32. Vemuri, S. and Rhodes, C. T. 1995. Preparation and characterization of liposomes as therapeutic delivery systems: A review. *Pharmaceutica Acta Helvetiae*, 70, 95–111.

33. Pradhan, P., Giri, J., Rieken, F., Koch, C., Mykhaylyk, O., Döblinger, M., Banerjee, R., Bahadur, D. and Plank, C. 2010. Targeted temperature sensitive magnetic liposomes for thermo-chemotherapy. *Journal of Controlled Release*, 142, 108–121.

34. Jaiswal, M. K., Banerjee, R., Pradhan, P. and Bahadur, D. 2010. Thermal behavior of magnetically modalized poly(*N*-isopropylacrylamide)–chitosan based nanohydrogel. *Colloids and Surfaces B: Biointerfaces*, 81,185–194.

35. Deng, Y., Wang, C., Shen, X., Yang, W., Jin, L., Gao, H. and Fu, S. 2005. Preparation, characterization, and application of multistimuli-responsive microspheres with fluorescence-labeled magnetic cores and thermoresponsive shells. *Chemistry A European Journal*, 11, 6006–6013.

36. Hu, S. H., Liu, T. Y., Liu, D. M. and Chen, S. Y. 2007. Controlled pulsatile drug release from a ferrogel by a high-frequency magnetic field. *Macromolecules*, 40, 6786–6788.

37. Retama, J., Zaferipoulos, C. E., Serafinelli, C., Reyna, R., Voit, B., Cabarcos, E. L. and Stamm, M. 2007. Synthesis and characterization of thermosensitive PNIPAM microgels covered with superparamagnetic γ-Fe_2O_3 nanoparticles. *Langmuir*, 23, 10280–10285.

38. Yuan, Q., Subramanian, V., Hein, S. and Misra, R. D. K. 2008. A stimulus-responsive magnetic nanoparticles drug carrier: Magnetite encapsulated by chitosan-grafted-copolymer. *Acta Biomaterialia*, 4, 1024–1037.

39. Zhang, J., Chen, H., Xu, L. and Gu, Y. 2008. The targeted behavior of thermally responsive nanohydrogel evaluated by NIR system in mouse model. *Journal of Controlled Release*, 131, 34–40.

40. Gulati, N., Rastogi, R., Dinda, A. K., Saxena, R. and Koul, V. 2010. Characterization and cell material interactions of PEGylated PNIPAAM nanoparticles. *Colloids and Surfaces B: Biointerfaces*, 79, 164–173.

41. Xiong, W., Wang, W., Wang, Y., Zhao, Y., Chen, H., Xu, H. and Yang, X. 2011. Dual temperature/pH-sensitive drug delivery of poly(*N*-isopropylacrylamide-*co*-acrylic acid) nanogels conjugated with doxorubicin for potential application in tumor hyperthermia therapy. *Colloids and Surfaces B: Biointerfaces*, 84, 447–453.

42. Ma, L., Liu, M., Liu, H., Chen, J. and Cui, D. 2010. *In vitro* cytotoxicity and drug release properties of pH- and temperature-sensitive core–shell hydrogel microspheres. *International Journal of Pharmaceutics*, 385, 86–91.

43. Meenach, S. A., Hilt, J. Z. and Anderson, K. W. 2010. Poly(ethylene glycol)-based magnetic hydrogel nanocomposites for hyperthermia cancer therapy. *Acta Biomaterialia*, 6, 1039–1046.

44. Martin, S. B., Choi, Y., Cano, R. M. S., Cosgrove, T., Vincent, B. and Barbero, F. A. 2005. Microscopic signature of a volume phase transition. *Macromolecules*, 38, 10782–10787.

45. Cherukuri, P., Glazer, E. S. and Curley, S. A. 2010. Targeted hyperthermia using metal nanoparticles. *Advanced Drug Delivery Reviews*, 62, 339–345.

46. Gobin, A. M., Lee, M. H., Halas, N. J., James, W. D., Drezek, R. A. and West, J. L. 2007. Near-infrared resonant nanoshells for combined optical imaging and photothermal cancer therapy. *Nano Letters*, 7, 1929–1934.

47. Gannon, C. J., Patra, C. R., Bhattacharya, R., Mukherjee, P. and Curley, S. A. 2008. Intracellular gold nanoparticles enhance non-invasive radiofrequency thermal destruction of human gastrointestinal cancer cells. *Journal of Nanobiotechnology*, 6 (2). doi:10.1186/1477-3155-6-2.

48. Jordan, A., Scholz, R., Wust, P., Fahling, H., Krause, J., Wlodarczyk, W., Sander, B., Vogl, T. and Felix, R. 1997. Effects of magnetic fluid hyperthermia (MFH) on C3H mammary carcinoma in vivo. *International Journal of Hyperthermia*, 13 (6), 587–605.

49. Johannsen, M., Gneveckow, U., Eckelt, L., Feussner, A., Waldofner, N., Scholz, R., Deger, S., Wust, P., Loening, S. A. and Jordan, A. 2005. Clinical hyperthermia of prostate cancer using magnetic nanoparticles: Presentation of a new interstitial technique. *International Journal of Hyperthermia*, 21, 637–647.

50. Masuko, Y., Tazawa, K., Sato, H., Viroonchatapan, E., Takemori, S., Shimizu, T., Ohkami, H., Nagae, H., Fujimaki, M., Horikoshi, I. and Weinstein, J. N. 1997. Antitumor activity of selective hyperthermia in tumor-bearing rats using thermosensitive magnetoliposomes as a new hyperthermic material. *Drug Delivery*, 4 (1), 37–42.

51. Yanase, M., Shinkai, M., Honda, H., Wakabayashi, T., Yoshida, J. and Kobayashi, T. 1998. Intracellular hyperthermia for cancer using magnetite cationic liposomes: An in vivo study. *Japan Journal of Cancer Research*, 89, 463–470.

52. Ito, A., Fujioka, M., Yoshida, T., Wakamatsu, K., Ito, S., Yamashita, T., Jimbow, K. and Honda, H. 2007. 4-S-cysteaminylphenol-loaded magnetite cationic liposomes for combination therapy of hyperthermia with chemotherapy against malignant melanoma. *Cancer Science*, 98, 424–430.

53. Yoshida, M., Watanabe, Y., Sato, M., Maehara, T., Aono, H., Naohara, T., Hirazawa, H., Horiuchi, A., Yukumi, S., Sato, K., Nakagawa, H., Yamamoto, Y., Sugishita, H. and Kawachi, K. 2010. Feasibility of chemohyperthermia with docetaxel-embedded magnetoliposomes as minimally invasive local treatment for cancer. *International Journal of Cancer*, 126, 1955–1965.

54. Shinkai, M., Yanase, M., Suzuki, M., Honda, H., Wakabayashi, T., Yoshida, J. and Kobayashi, T. 1999. Intracellular hyperthermia for cancer using magnetite cationic liposomes. *Journal of Magnetism and Magnetic Materials*, 194, 176–184.

55. Ito, A., Tanaka, K., Honda, H., Abe, S., Yamaguchi, H. and Kobayashi, T. 2003. Complete regression of mouse mammary carcinoma with a size greater than 15 mm by frequent repeated hyperthermia using magnetite nanoparticles. *Journal of Bioscience and Bioengineering*, 96 (4), 364–369.

56. Le, B., Shinkai, M., Kitade, T., Honda, H., Yoshida, J., Wakabayashi, T. and Kobayashi, T. 2001. Preparation of tumour specific magnetoliposomes and their application for hyperthermia. *Journal of Chemical Engineering Japan*, 34, 66–72.

57. Kikumori, T., Kobayashi, T., Sawaki, M. and Imai, T. 2009. Anti-cancer effect of hyperthermia on breast cancer by magnetite nanoparticle-loaded anti-HER2 immunoliposomes. *Breast Cancer Research and Treatment*, 113, 435–441.

58. Arnold, I. F. and Mayhew, E. 1986. Targeted drug delivery. *Cancer*, 58, 573–583.

59. Alexiou, C., Schmid, A., Klein, R., Hulin, P., Bergemann, C. and Arnold, W. 2002. Magnetic drug targeting: Biodistribution and dependency on magnetic field strength. *Journal of Magnetism and Magnetic Materials*, 252, 363–366.

60. Kuznetsov, A. A., Harutyunyan, A. R., Dobrinsky, E. K. et al. In Hafeli et al. (Eds), Ferrocarbon particles preparation and clinical applications. *Scientific and Clinical Applications of Magnetic Carriers* 1997; 379; New York; Plenum Press.

61. Kuznetsov, A. A., Filippov, V. I., Kuznetsov, O. A., Gerlivanov, V. G., Dobrinsky, E. K. and Malashin, S. I. 1999. New ferro-carbon adsorbents for magnetically guided transport of anti-cancer drugs. *Journal of Magnetism and Magnetic Materials*, 194, 22–30.

62. Hafeli, U. O. 2004. Magnetically modulated therapeutic systems. *International Journal of Pharmaceutics*, 277, 19–24.

63. Senyei, A., Widder, K. and Czerlinski, G. 1978. Magnetic guidance of drug-carrying microspheres. *Journal of Applied Physics*, 49, 3578–3583.

64. Gupta, P. K. and Hung, C. T. 1989. Magnetically controlled targeted micro-carrier systems. *Life Sciences*, 44, 175–186.

65. Dandamudi, S. and Campbell, R. B. 2007. Development and characterization of magnetic cationic liposomes for targeting tumor microvasculature. *Biochimica et Biophysica Acta (BBA) – Biomembranes*, 1768, 427–438.

66. Martina, M. S., Wilhelm, C. and Lesieur, S. 2008. The effect of magnetic targeting on the uptake of magnetic-fluid-loaded liposomes by human prostatic adenocarcinoma cells. *Biomaterials*, 29, 4137–4145.

67. Charlotte, R., Martina, M. S., Yutaka, T., Claire, W., Alexy, T., Christine, M., Elisabeth, P., Sylviane, L., Florence, G. and Jacques, S. 2007. Magnetic targeting of nanometric magnetic fluid-loaded liposomes to specific brain intravascular areas: A dynamic imaging study in mice. *Radiology*, 244, 439–448.

68. Nobuto, H., Sugita, T., Kubo, T., Shimose, S., Yasunaga, Y., Murakami, T. and Ochi, M. 2004. Evaluation of systemic chemotherapy with magnetic liposomal doxorubicin and a dipole external electromagnet. *International Journal of Cancer*, 109, 627–635.

69. Weinstein, J. N., Magin, R. L., Yatvin, M. B. and Zaharko, D. S. 1979. Liposomes and local hyperthermia: Selective delivery of methotrexate to heated tumors. *Science*, 204, 188–191.

70. Weinstein, J. N., Magin, R. L., Cysyk, R. L. and Zaharko, D. S. 1980. Treatment of solid L1210 murine tumors with local hyperthermia and temperature-sensitive liposomes containing methotrexate. *Cancer Research*, 40, 1388–1395.

71. Weinstein, J. N., Klausner, R. D., Innerarity, T., Ralston, E. and Blumenthal, R. 1981. Phase transition release, a new approach to the interaction of proteins with lipid vesicles application to lipoproteins. *Biochimica et Biophysica Acta (BBA)-Biomembranes*, 647, 270–284.

72. Ben-Yashar, V. and Barenholz, Y. 1989. The interaction of cholesterol and cholest-4-en-3-one with dipalmitoylphosphatidylcholine. Comparison based on the use of three fluorophores. *Biochimica et Biophysica Acta (BBA) – Biomembranes*, 985, 271–278.

73. Kirby, C., Clarke, J. and Gregoriadis, G. 1980. Effect of cholesterol content of small unilamellar liposomes on their stability in vivo and in vitro. *Biochemical Journal*, 186, 591–598.

74. Magin, R. L. and Niesman, M. R. 1984. Temperature-dependent drug release from large unilamellar liposomes. *Cancer Drug Delivery*, 1, 109–117.

75. Magin, R. L. and Niesman, M. R. 1984. Temperature-dependent permeability of large unilamellar liposomes. *Chemistry and Physics of Lipids*, 34, 245–256.

76. Merlin, J. L. 1991. Encapsulation of doxorubicin in thermosensitive small unilamellar vesicle liposomes. *European Journal of Cancer*, 27, 1026–1030.

77. Gaber, M. H., Hong, K., Huang, S. K. and Papahadjopoulos, D. 1995. Thermosensitive sterically stabilized liposomes: Formulation and in vitro studies on mechanism of doxorubicin release by bovine serum and human plasma. *Pharmaceutical Research*, 12, 1407–1416.

78. Lindner, L. H., Hossann, M., Vogeser, M., Teichert, N., Wachholz, K., Eibl, H., Hiddemann, W. and Issels, R. D. 2008. Dual role of hexadecylphosphocholine (miltefosine) in thermosensitive liposomes: Active ingredient and mediator of drug release. *Journal of Controlled Release*, 125, 112–120.

79. Lindner, L. H., Hossann, M. and Teichert, N. 2005. Alkylphosphocholines enhance the drug release rate of thermosensitive liposomes (Abstract). *Society for Thermal Medicine Annual Meeting*, Bethesda, MD.

80. Kono, K., Henmi, A., Yamashita, H., Hayashi, H. and Takagishi, T. 1999. Improvement of temperature-sensitivity of poly(N-isopropylacrylamide)-modified liposomes. *Journal of Controlled Release*, 59, 63–75.

81. Kono, K. and Takagishi, T. 2004. Temperature-sensitive liposomes. *Methods of Enzymology*, 387, 73–82.

82. Hayashi, H., Kono, K. and Takagishi, T. 1999. Temperature sensitization of liposomes using copolymers of N-isopropylacrylamide. *Bioconjugation Chemistry*, 10, 412–418.

83. Sandström, M. C., Ickenstein, L. M., Mayer, L. D. and Edwards, K. 2005. Effects of lipid segregation and lysolipid dissociation on drug release from thermosensitive liposomes. *Journal of Controlled Release*, 107, 131–142.

84. Mills, J. K. and Needham, D. 2005. Lysolipid incorporation in dipalmitoylphosphatidylcholine bilayer membranes enhances the ion permeability and drug release rates at the membrane phase transition. *Biochimica et Biophysica Acta (BBA) – Biomembranes*, 1716, 77–96.

85. Needham, D., Stoicheva, N. and Zhelev, D. V. 1997. Exchange of monooleoylphosphatidylcholine as mono-mer and micelle with membranes containing poly(ethylene glycol)-lipid. *Biophysical Journal*, 73, 2615–2629.

86. Zhelev, D. V. 1998. Material property characteristics for lipid bilayers containing lysolipid. *Biophysical Journal*, 75, 321–330.

87. Needham, D. and Dewhirst, M. W. 2001. The development and testing of a new temperature-sensitive drug delivery system for the treatment of solid tumors. *Advanced Drug Delivery Review*, 53, 285–305.

88. Needham, D., Anyarambhatla, G., Kong, G. and Dewhirst, M. W. 2000. A new temperature-sensitive liposome for use with mild hyperthermia: Characterization and testing in a human tumor xenograft model. *Cancer Research*, 60, 1197–1201.

89. Anyarambhatla, G. R. and Needham, D. 1999. Enhancement of the phase transition permeability of DPPC liposomes by incorporation of MPPC: A new temperature-sensitive liposome for use with mild hyperthermia. *Journal of Liposome Research*, 9, 499–514.

90. Wells, J., Sen, A. and Hui, S. W. 2003. Localized delivery to CT-26 tumors in mice using thermosensitive liposomes. *International Journal of Pharmaceutics*, 261, 105–114.

91. Chandaroy, P., Sen, A. and Hui, S. W. 2001. Temperature-controlled content release from liposomes encapsulating Pluronic F127. *Journal of Controlled Release*, 76, 27–37.

92. Viroonchatapan, E., Sato, H., Ueno, M., Adachi, I., Murata, J., Saikf, I., Tazawad, K. and Horikoshi, I. 1998. Microdialysis assessment of 5-fluorouracil release from thermosensitive magnetoliposomes induced by an electromagnetic field in tumor-bearing mice. *Journal of Drug Targeting*, 5, 379–390.

93. Babincová, M., Ćićmanec, P., Altanerová, V., Altaner, Ć. and Babinec, P. 2002. AC-magnetic field controlled drug release from magnetoliposomes: Design of a method for site-specific chemotherapy. *Bioelectrochemistry*, 55, 17–19.

94. Nappini, S., Bombelli, F. B., Bonini, M., Norden, B. and Baglioni, P. 2010. Magnetoliposomes for con-trolled drug release in the presence of low-frequency magnetic field. *Soft Materials*, 154, 154–162.

95. Chen, Y., Bose, A. and Bothun, G. D. 2010. Controlled release from bilayer decorated magnetoliposomes via electromagnetic heating. *ACS Nano*, 4, 3215–3221.

96. Yanase, M., Shinkai, M., Honda, H., Wakabayashi, T., Yoshida, J. and Kobayashi, T. 1998. Antitumor immunity induction by intracellular hyperthermia using magnetite cationic liposomes. *Japan Journal Cancer Research*, 89, 775–782.

97. Kong, G. and Dewhirst, M. W. 1999. Hyperthermia and liposomes. *International Journal of Hyperthermia*, 15, 345–370.

98. Alberts, D. and Garcia, D. 1997. Safety aspects of pegylated liposomal doxorubicin in patients with cancer. *Drugs*, 54, 30–35.

99. Muggia, F. 1997. Clinical efficacy and prospects for use of pegylated liposomal doxorubicin in the treat-ment of ovarian and breast cancers. *Drugs*, 54, 22–29.

100. Johannsen, M., Gneveckow, U., Taymoorian, K., Thiesen, B., Waldofner, N. and Scholz, R. 2007. Morbidity and quality of life during thermotherapy using magnetic nanoparticles in locally recurrent pros-tate cancer: Results of a prospective phase I trial. *International Journal of Hyperthermia*, 23, 315–323.

101. Hauff, M. K., Rothe, R., Scholz, R., Gneveckow, U., Wust, P. and Thiesen, B. 2007. Intracranial thermo-therapy using magnetic nanoparticles combined with external beam radiotherapy: Results of a feasibility study on patients with glioblastoma multiforme. *Journal of Neurooncology*, 81, 53–60.

102. Landeghem, F. K. H., Hauff, K. M., Jordan, A., Hoffmann, K. T., Gneveckow, U., Scholz, R., Thiesen, B., Brück, W. and Deimling, A. 2009. Post-mortem studies in glioblastoma patients treated with thermo-therapy using magnetic nanoparticles. *Biomaterials*, 30, 52–57.

103. Lübbe, A. S., Bergemann, C. and Riess, H. 1996. Clinical experiences with magnetic drug targeting: A phase I study with 4'-epidoxorubicin in 14 patients with advanced solid tumors. *Cancer Research*, 56, 4686–4693.

104. Lübbe, A. S., Alexiou, C. and Bergemann, C. 2001. Clinical applications of magnetic drug targeting. *Journal Surgical Research*, 95, 200–206.

105. Goodwin, S. 2000. Magnetic targeted carriers offer site-specific drug delivery. *Oncology News International*, 9, 22.

106. Johnson, J., Kent, T., Koda, J., Peterson, C., Rudge, S. and Tapolsky, G. 2002. The MTC technology: A platform technology for the site-specific delivery of pharmaceutical agents. *European Cells and Materials*, 3, 12–15.

107. Kuznetsov, A. A., Filippov, V. I., Alyautdin, R. N., Torshina, N. L. and Kuznetsov, O. A. 2001. Application of magnetic liposomes for magnetically guided transport of muscle relaxants and anti-cancer photody-namic drugs. *Journal of Magnetism and Magnetic Materials*, 225, 95–100.

108. Zhang, J. Q., Zhang, Z. R., Yang, H., Tan, Q. Y., Qin, S. R. and Qiu, X. L. 2005. Lyophilized paclitaxel magnetoliposomes as a potential drug delivery system for breast carcinoma via parenteral administration: In vitro and in vivo studies. *Pharmaceutical Research*, 22, 573–583.

Dr. Dhirendra Bahadur is a professor at the Department of Metallurgical Engineering and Materials Science and the Centre for Research in Nanotechnology and Science, Indian Institute of Technology, Bombay, India. He did his research work for PhD at IIT, Kanpur between 1973 and 1977. He served at various capacities at IIT, Kanpur and IIT, Bombay between 1978 and present. He was a visiting fellow of the Royal Society London-INSA exchange programme and worked at Cavendish Laboratory, 1985–1986. Professor Bahadur's main areas of research interest are nanomaterials, nanostructured magnetic materials, electronic ceramics, magnetic nanoparticulates and their biomedical applications in particular hyperthermia and targeted drug delivery.

Professor Bahadur has several honours and awards to his credit. Some of these are ICSC–MRSI award (2011), Medal of the Materials Research Society of India (1996), fellowship Royal Society London-INSA exchange programme (1985). He has served on several committees of national and international importance. He was a member of the Indo-French joint committee for the Indo-French Laboratory of solid-state chemistry. He served on the editorial advisory boards of Transactions Indian Institute of Metals 2004–2005 and of *Bulletin of Materials Science* since 2004–2010. He is coauthor/author of more than 230 publications in international journals and one book and three patents.

Dr. Rinti Banerjee is a professor of biomedical engineering at the Department of Biosciences and Bioengineering, and the Centre for Research in Nanotechnology and Science, Indian Institute of Technology, Bombay, India. She has an MBBS from BJ Medical College Pune, and a PhD in biomedical engineering from IIT Bombay. She did her postdoctoral at the University of California,

San Francisco, before joining IIT Bombay as a faculty in 2001. Her research interests include nano-medicine, drug delivery systems and pulmonary surfactant replacements. Dr. Banerjee has published over 90 papers in these areas, including invited editorials and books. She has received several awards in recognition of her work, including the Indo-US Frontiers of Engineering in 2010, Annual Felicitation Award from the Society for Cancer Research and Communications in 2008 and several young scientist awards from the International Foundation of Science, Sweden, DAE, India, DST, India, AICTE Career Award for Young Teachers. She is a member of the editorial board of international journals like *Drug Delivery and Translational Research, Biomaterials and Biodevices*, and is an associate editor of the *Journal of Biomaterials and Tissue Engineering*. She has many patented technologies to her credit, some of which are being transferred to industries for further development.

19 Magnetic Microbubbles

Eleanor Stride, Helen Mulvana, Robert Eckersley,
Meng-Xing Tang, and Quentin Pankhurst

CONTENTS

19.1 INTRODUCTION

Gas microbubbles stabilised by a surfactant or polymer coating have now been in clinical use as contrast agents for ultrasound imaging for several decades. More recently, their use in therapeutic applications, in particular drug delivery and gene therapy, has also become a highly active area of research. There remain, however, some significant challenges that must be overcome in order to fully realise the potential of microbubbles in these applications. In particular, the difficulty in controlling the concentration of microbubbles at a given site and in ensuring sufficient proximity between bubbles and target cells has frequently led to disappointing results from *in vivo* studies. Recent work has indicated that incorporating magnetic nanoparticles into the microbubble coating may provide an effective strategy for overcoming these challenges, by both enabling the bubbles to be localised using an externally applied magnetic field and enhancing the efficiency of cell uptake. This chapter will briefly describe the uses of microbubble agents in medical imaging and therapy and then review the recent studies on the development of magnetic microbubbles and their future applications.

19.2 BACKGROUND

19.2.1 MICROBUBBLE AGENTS

19.2.1.1 Microbubbles for Imaging Applications

The use of microbubbles as contrast agents for ultrasound imaging dates back to the accidental discovery in the late 1960s that the injection of a dye during an ultrasound examination of cardiac motion produced a temporary enhancement in the signal intensity[1]. Subsequent investigation demonstrated that it was not the dye but rather the formation of bubbles at the catheter tip that was responsible for this effect[2]. The further discovery that more prolonged contrast enhancement could be achieved using saline containing a small amount of a patient's blood[3] introduced the concept of microbubbles stabilised by a thin layer of adsorbed material, in this case serum albumin. The effect of the coating is both to reduce surface tension and inhibit gas diffusion, thus preventing the bubbles from dissolving too rapidly or agglomerating. A variety of different coating materials are now utilised in commercially available ultrasound contrast agents, including serum albumin, phospholipids and polymers such as cyanoacrylates and polycaprolactone[4]. The majority also contain gases such as perfluorocarbons with higher molecular weights and hence lower diffusivities than air to further improve bubble stability. The bubbles are supplied either in suspension or as a freeze dried powder, which is mixed with saline immediately before injection to provide *in vivo* concentrations of ~10^4 bubbles/mL. The mean bubble diameter is normally between 2 and 4 μm (Figure 19.1a and b), which enables them to safely traverse the pulmonary capillary bed without risk of embolisation.

The effectiveness of microbubbles as ultrasound contrast agents is due to their being filled with gas and their resulting compressibility. In ultrasound imaging, the amplitude of the backscattered signal, or 'echo', received normally depends on the difference in acoustic impedance (z) between different structures[5]. This in turn depends on the difference in the density (ρ) and speed of sound (c) of the materials from which the structures consist, since $z = \rho c$. Thus, despite the fact that micro-bubbles are similar in size to red blood cells, they provide a significant increase in acoustic impedance contrast when injected into the circulation. This is because the acoustic impedance of a gas is much lower than that of a liquid. The enhancement in the backscattered signal is, however, much larger than would be produced by a suspension of rigid spheres of the same acoustic impedance. This is due to the fact that the microbubbles are highly compressible and so absorb and reradiate the incident field, undergoing volumetric oscillations, rather than just passively scattering it. There is, moreover, a fortuitous coincidence between the frequencies used in diagnostic ultrasound imaging (2–15 MHz) and the range over which microbubbles with diameters between 2 and 4 μm undergo resonant oscillations. The most important property of microbubble behaviour in terms of contrast

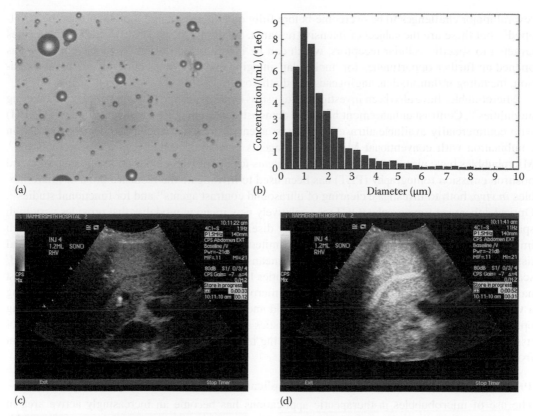

FIGURE 19.1 Microbubble ultrasound contrast agents: (a) micrograph showing microbubbles following suspension in saline, (b) microbubble size distribution, (c) and (d) ultrasound image before and after contrast enhancement.

enhancement is, however, the fact that at moderate ultrasound intensities, the microbubble oscillations become non-linear. The range of frequencies in the backscattered signal is therefore different from that of the incident field, containing an increasing number of whole and fractional harmonics with increasing ultrasound intensity. These non-linear components enable the signals from the bubbles to be easily distinguished from those from the surrounding tissue and several image processing techniques have been developed to exploit this phenomenon[6] for imaging blood flow. An example of contrast enhancement in an ultrasound image of the liver following intravenous injection of microbubbles is shown in Figure 19.1c and d.

The most widespread application of microbubbles currently is in echocardiography, specifically for ventricular opacification and delineation of endocardial borders[7,8]. The range of applications is continually increasing, however, and includes other cardiovascular applications such as detection and assessment of atherosclerosis[9], as well as non-vascular applications such as assessment of fallopian tube patency[10] and detection of ureteric reflux[11]. The development of non-linear imaging techniques, in particular in combination with 3D imaging, has led to significant advances in mapping of the microcirculation using microbubbles in the characterisation of tumour vascularity[12] in liver, breast and prostate and also in the brain, in particular, for the assessment of stroke patients[13]. Further details of the clinical applications of microbubbles may be found in Refs. [14–16].

A popular use of microbubbles initially in ultrasound imaging was a means of signal enhancement in Doppler studies[17], but bubbles have shown considerable potential in other types of quantitative imaging for the measurement of parameters such as relative vascular volume, flow velocity and perfusion rate. This has been demonstrated for a number of applications, including identification and classification of liver tumours[18,19] and assessment of myocardial function[20–22]. There remain several major challenges to be overcome before fully quantitative imaging protocols can be developed[23], but these are the subject of intensive research. Similarly, the ability to synthesise molecules targeted to specific cellular receptors, which can be incorporated into microbubble coatings, has opened up further opportunities for 'molecular' imaging and conditions currently under investigation, including inflammation, angiogenesis and atherosclerosis[24,25].

Microbubbles have also been investigated for use as contrast agents for a number of other imaging modalities[26]. Contrast enhancement has been demonstrated in magnetic resonance imaging (MRI) with commercially available ultrasound contrast agents[27], with hyperpolarised noble gases and in combination with conventional MRI contrast agents containing gadolinium and iron oxides[28,29]. Microbubbles have also been used as contrast agents for diffraction-enhanced x-ray imaging[30]; and positron emission tomography (PET) has been used to investigate the biodistribution of microbubbles *in vivo*, both to investigate clearing of ultrasound contrast agents[31] and for functional studies[32]. Optical coherence tomography (OCT) is a relatively new imaging modality that provides very high spatial resolution, particularly for investigating diseases of the eye and skin. The depth at which images can be acquired is, however, severely limited by the poor penetration of light in biological tissue. Microbubbles have been proposed as a means of increasing the amplitude of the reflected/scattered optical signal and providing opportunities for targeted imaging. Improved image contrast has been demonstrated in animal models with microbubbles coated with protein shells embedding a variety of nanoparticles including carbon, gold and iron oxide[33]. Imaging techniques combining optical and acoustic methods such as photoacoustics and acousto-optics are also limited in terms of tissue depth, and microbubbles are similarly being investigated as a means of signal enhancement from the region of interest[34–36].

19.2.1.2 Microbubbles for Therapeutic Applications

The use of microbubbles in therapeutic applications has become an increasingly active area of research in recent years. In particular, microbubbles have shown great potential as traceable vehicles for localised drug delivery and gene therapy[37–39]. They can be loaded with a range of therapeutic agents, traced to a target site under ultrasound imaging at low intensities and then destroyed with a high-intensity burst to release the encapsulated material locally, thus avoiding systemic administration, for example, of toxic chemotherapy drugs. There is, moreover, considerable evidence that motion of the microbubbles reversibly increases the permeability of both individual cell membranes and the endothelium[40,41], including temporary opening of the blood–brain barrier[42]. An example of this phenomenon, variously referred to as 'sonoporation' and 'sonophoresis', is illustrated in Figure 19.2 in an *in vitro* gene delivery experiment. In the absence of the microbubbles, very low levels of transfection are observed following ultrasound exposure. With microbubbles present, however, gene expression is very clear in the region corresponding to the ultrasound beam. The mechanisms underlying this uptake enhancement are not fully understood but are thought to relate to the stimulation of normal membrane transport processes[43], although there is also evidence of mechanical pore formation in the cell membrane at higher ultrasound intensities[44], and it is possible that both mechanisms may operate depending on the ultrasound exposure conditions.

Microbubbles have also been utilised in therapeutic applications involving much higher ultrasound intensities. In thrombolysis, they have been shown to markedly increase the effectiveness of tissue plasminogen activator (TPA) and the rate of clot lysis both *in vitro* and *in vivo*[45,46]. Similarly, in high-intensity-focused ultrasound (HIFU) surgery, microbubbles have been used as a means of promoting cavitation in the target region in order to increase both the rate of heat deposition and volume of tissue treated[47,48].

19.2.1.3 Current Challenges

Microbubble-mediated drug delivery offers several significant advantages: the bubbles can be administered intravenously and the treatment is thus minimally invasive. The fact that ultrasound can be focused into very small volumes (mm^3) provides a means of localising the area over which

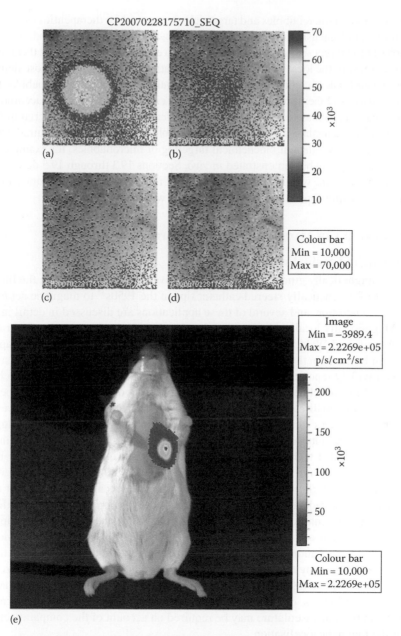

FIGURE 19.2 Microbubble-mediated gene therapy: Chinese hamster ovary cells expressing luciferase following exposure to (a) ultrasound and microbubbles, (b) ultrasound only, (c) microbubbles only, (d) sham exposure, and (e) *in vivo* expression in a mouse model following exposure to ultrasound and microbubbles. Details of the experiments are given in Sections 19.4 and 19.5.

the therapeutic effect is delivered. Microbubbles are inherently imageable under a range of modalities. A wide variety of therapeutics can be delivered. It is not necessary for the delivered molecule to be conjugated to the microbubbles, although several studies indicate that higher efficiencies are achieved with conjugation than co-delivery[49]. The unwanted side effects of viral delivery are avoided, and much higher rates of cell viability are maintained than in, for example, electroporation[50].

Despite the considerable potential of microbubbles for therapeutic applications, however, the results from *in vivo* studies have been disappointing, particularly for drug delivery and gene therapy[51]. There are a number of possible reasons that may be identified. First, it has been shown *in vitro* that close proximity between microbubbles and target cells is required for therapeutic effects to be realised at moderate ultrasound intensities and hence without unwanted damage[52]. Second, the concentration of microbubbles at the target site needs to be controlled to ensure that the required therapeutic effect is achieved, but increasing the local microbubble concentration by increasing the dose delivered is limited by the associated risk of embolism. Third, it is normally desirable for microbubbles to be administered intravenously, and their presence throughout the circulation can prevent accurate focusing of the ultrasound beam as well as resulting in unwanted release of therapeutic material in the overlying tissue. Thus, an effective method for localising microbubbles at a target site is required. Unfortunately, despite significant efforts, efficient biochemical targeting of microbubbles, for example, with monoclonal antibodies, has yet to be demonstrated *in vivo*. Sections 19.3 through 19.5 describe an alternative strategy that exploits the well-known phenomenon of remote manipulation using a magnetic field. First, the relevant existing applications of this technique will be briefly reviewed.

19.2.2 Applications of Magnetic Fields in Therapeutic Delivery

19.2.2.1 Magnetic Localisation

The concept of magnetically guided treatments is relatively well established in the literature, from the development of magnetically steered catheter tips in the 1950s[53] to magnetic actuation of stem cells for tissue regeneration, and several of these applications are discussed in detail in Chapters 13 through 16. Magnetically localisable drug carriers became a popular topic of research in the late 1970s. Senyei et al. and Widder et al.[54,55] demonstrated enhanced uptake of an encapsulated chemotherapeutic in tumours in the presence of an externally applied magnet field, and similar results have since been widely reported in a range of organs[56]. Magnetic carriers have also been utilised for the delivery of radionuclides in order to reduce patient exposure[57] and for chemoembolisation, where the blood supply to diseased tissue is deliberately restricted by occluding the vessels[58].

The carriers themselves range from polymeric microspheres to magnetically tagged proteins and cells; for example, chitosan particles have been used for delivery to tumours in the brain[59], while erythrocytes loaded with magnetic particles and anti-thrombotic drugs have been used to reduce the risk of clotting after surgery[60]. Once again, there are a number of challenges: as for any drug delivery application, the carrier must be biocompatible, encapsulate sufficient quantities of therapeutic material, degrade at the desired rate and be of an appropriate size to safely penetrate the target tissue. It must contain sufficient magnetic material to enable it to be manipulated at the required tissue depth using a magnetic field that can be practically generated. Magnetic guidance at depths of 10 cm has been demonstrated using moderate magnetic field strengths[61], but this remains a significant challenge for many applications. If the carrier is injected into the blood stream, then the target tissue must be sufficiently vascularised for material to reach it. Some additional 'targeting', for example, using antibodies, or reliance on physiological effects such enhanced permeability and retention (EPR) in tumour vasculature may be required on account of the comparatively low spatial selectivity of the magnetic localisation.

19.2.2.2 Magnetofection

A more recent development in magnetic therapy is the discovery that the manipulation of particles using a magnetic field can be beneficial not only in terms of bulk localisation but also for directly enhancing therapeutic uptake at the cellular level. This phenomenon has been termed magnetofection[62]. In this process, the molecules to be delivered (to date, these have been predominantly nucleic acids) are conjugated with magnetic nanoparticles through electrostatic attraction. A magnetic field is then used to concentrate the complexes on the target cells, thus maximising contact. Transport across the cell membrane occurs through normal mechanisms (endocytosis and pinocytosis), and both viral and non-viral transfections have been successfully demonstrated. Magnetofection has been demonstrated *in vitro* and *in vivo* with a range of nucleic acids (DNA, siRNA etc.) and cell lines and offers high efficiency with very low loss of cell viability.

19.3 MAGNETIC MICROBUBBLES

19.3.1 CONCEPT

As indicated in the previous section, both microbubble-mediated and magnetically guided therapeutic delivery are faced with a number of challenges. Combining the two approaches in the form
of magnetic microbubbles provides a means of addressing at least some of these. First, facilitating
manipulation of the microbubbles using an externally applied magnetic field enables them to be
localised to a target region, providing the required concentration for therapeutic effectiveness and
minimising the risk of premature release. Second, magnetic localisation can be used in conjunction with biochemical targeting to achieve greater specificity and good contact between bubbles
and cells, since by increasing the 'dwell' time in the target region, many of the issues with poor
attachment can be overcome[63]. Third, real-time treatment monitoring can be performed under ultrasound. Fourth, focusing the ultrasound beam provides much higher spatial resolution than can be
achieved with a magnetic field alone. Fifth, the enhanced uptake associated with both sonoporation and magnetofection may be exploited, although there are some questions with regard to the
latter as to whether the required geometry can be achieved *in vivo*. The requirement for the target tissue to have a good blood supply is not completely overcome unless the microbubbles are
administered by direct injection, but ultrasound has been shown to improve tissue penetration, for
example, into tumours[64], although this may require higher intensities. Similarly, the magnetic field
strength required to manipulate microbubbles at large tissue depths may still be quite high, although
microbubbles represent excellent vehicles for magnetically guided drug delivery in this sense. This
section will review the existing work on magnetic guidance of microbubbles and then describe the
development of a new magnetic microbubble agent.

19.3.2 PREVIOUS WORK

19.3.2.1 Magnetically Tagged Microbubbles

Soetanto and Watarai[65] reported the development of a new type of ultrasound contrast agent consisting of surfactant-coated microbubbles to which either 300 nm magnetite particles or 1 μm carbon-
coated iron particles were attached *via* electrostatic coupling. The magnetic particles were coated
with an anionic surfactant (sodium laurate) and mixed into a solution of a second anionic surfactant (sodium stearate) that had been agitated to form bubbles. Calcium chloride solution was then
added to achieve the required coupling between the bubbles and magnetic particles (Figure 19.3). That
the resulting complexes were magnetically responsive was confirmed macroscopically by holding a magnet against the vial containing the suspension, but a microscopic investigation was not

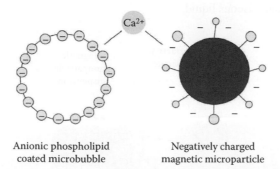

Anionic phospholipid   Negatively charged
coated microbubble   magnetic microparticle

FIGURE 19.3 Schematic representation of magnetically tagged microbubbles. Magnetic microparticles are
tagged to microbubbles through electrostatic coupling. (Reproduced with permission from Soetanto, K. and
Watarai, H., *Jpn. J. Appl. Phys. Part 1 Reg. Pap. Short Notes Rev. Pap.*, 39, 3230, 2002.)

presented. Acoustic measurements were carried out on the suspension to determine the attenuation coefficient at 3.5 MHz, which was not found to differ significantly from that of microbubbles prepared under the same conditions without magnetic particles. Varying the calcium ion concentration was found to have some effect on the microbubble resonance frequency, indicating that the number of magnetic particles per bubble was varying. Further studies of this formulation were, however, not reported.

19.3.2.2 Magnetic Lipospheres

Vlaskou et al.[66,67] reported *in vitro* and *in vivo* gene delivery with microparticles consisting of magnetic nanoparticles encapsulated with a nucleic acid (plasmid DNA or synthetic siRNA) and soybean oil using a cationic phospholipid. The particles were referred to as lipospheres but were prepared in a similar manner to microbubbles, *via* agitation in the presence of a gas (perfluoropropane, C_3F_8), and were found to be acoustically responsive. Exposure to an ultrasound field was found to produce no enhancement in gene delivery *in vitro* compared with exposure to a magnetic field alone. Similarly, there was found to be no statistically significant difference in delivery *in vivo* in a mouse model following tail vein injection. Plasmid deposition in a mouse skinfold chamber model was, however, found to increase upon combined exposure to a magnetic field and ultrasound following a carotid injection. These results will be discussed further in Section 19.4.

19.3.2.3 Magnetically Loaded Microbubbles

Stride et al.[68] conducted studies similar to those subsequently reported by Vlaskou et al. using gas-filled microbubbles with a phospholipid coating and an additional liquid layer containing magnetic nanoparticles (Figure 19.4). The assumption of a layered structure was based on the relative hydrophobicity of the different layers and results from electron microscopy. As will be discussed subsequently, the preparation was similar to that reported by Vlaskou et al. but with certain key differences. At the time of writing, these were the only studies available on magnetic microbubbles in the literature.

19.3.3 DESIGN AND CHARACTERISATION

19.3.3.1 Theoretical Description

In developing magnetic microbubbles, design calculations to determine the optimal composition must take into account the requirements for performance *in vivo*, namely magnetic localisation and acoustic response (for imaging and therapeutic effect). The bubble can be modelled as a gas-filled sphere of radius R_1, surrounded by a layer of viscous, hydrophobic liquid containing a fixed volume fraction, α, of solid spherical nanoparticles and onto which a phospholipid monolayer of negligible thickness is adsorbed. The bubble is considered to be suspended in isolation in an infinite volume of incompressible Newtonian viscous liquid.

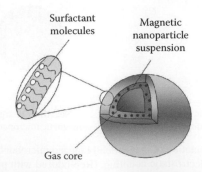

FIGURE 19.4   Schematic representation of the structure of a magnetic microbubble.

From Archimedes' principle, the force due to buoyancy (F_b) acting on the microbubble is given by

$$F_b = \frac{4}{3}\pi g \rho_L R_2^3 \tag{19.1}$$

where
  R_2 is the outer radius of the microbubble
  g is the acceleration due to gravity
  ρ_L is the density of the surrounding liquid

The weight (F_w) of the microbubble is given by

$$F_w = -\frac{4}{3}\pi g[\rho_G R_1^3 + (R_2^3 - R_1^3)((1-\alpha)\rho_o + \alpha\rho_{np})] \tag{19.2}$$

where
  ρ_G is the density of the gas in the gas core
  ρ_o is the density of the liquid in which the nanoparticles are suspended
  ρ_{np} is the density of the magnetic nanoparticles

The magnetic force (F_m) experienced by a microbubble upon exposure to a magnetic field of strength B is given by

$$F_m = -\frac{4\pi\chi(B\nabla)B\alpha(R_2^3 - R_1^3)}{3\mu_0} \tag{19.3}$$

where
  χ is the effective volumetric susceptibility of the magnetic nanoparticles suspended in the shell
  μ_0 is the permeability of free space

Once injected, microbubbles will also be subjected to a range of other forces. First, in order to determine the magnetic force required to localise microbubbles *in vivo*, the hydrodynamic forces due to blood flow need to be considered. The drag force on a sphere of radius R may be approximated as

$$F_D = 4\pi\gamma\mu_L R u \tag{19.4}$$

where
  u is the relative velocity between the liquid and the sphere
  γ is a constant depending on the flow Reynolds number
  μ_L is the dynamic viscosity of the flowing liquid[69]

Upon exposure to ultrasound, the microbubbles will experience an additional force due to the propagation of the incident field. This acoustic radiation force, F_A, may be determined as

$$F_A = -\langle V_B \nabla p \rangle \tag{19.5}$$

where
  V_B is the bubble volume
  p is the pressure in the liquid (which varies temporally and spatially due to the ultrasound field)
  $\langle ... \rangle$ denotes averaging with respect to time

In reality, a microbubble is not a rigid sphere and the validity of Equation 19.4 will depend largely on the local conditions. For example, in the capillary network around a tumour, the Reynolds number ($\rho_L u_L 2R_v/\mu_L$) will be relatively low (where R_v is the vessel radius, although in very narrow blood vessels, deformation of the bubble due to the presence of the blood vessel wall may need to be considered). The range of physiologically relevant flow rates varies widely, however, and in locations where the Reynolds number is very high, it may be more appropriate to use an alternative drag equation. Equation 19.4 also neglects the effects of other bodies, such as blood cells in the flow, and the net force on the microbubble will of course depend on the relative directions of the magnetic field gradient, ultrasound propagation and blood flow. Thus, detailed calculations must be performed for a given application.

The effect of the magnetic layer on the acoustic response of the microbubble can be determined by deriving the equation of motion for the bubble having the structure shown in Figure 19.4. Details of this treatment are beyond the scope of this chapter, but the derivation for a bubble surrounded by a liquid layer is given in Ref. [70]. The acoustic pressure radiated by the bubble can be found as

$$p_{rad}(r,t) = \rho_L \left(\frac{1}{r} \left(R_2^2 \ddot{R}_2 + 2R_2 \dot{R}_2^2 \right) - \frac{R_2^4 \dot{R}_2^2}{2r^4} \right) \tag{19.6}$$

where superscripted periods represent differentiation with respect to time. Figure 19.5 shows the effect of varying layer thickness on the amplitude of bubble oscillations and the corresponding frequency spectrum of the radiated pressure. As can be seen, the harmonic content of the radiated pressure is significantly reduced with increasing oil layer thickness (it should be noted that the bubble resonance frequency changes as the oil layer thickness changes, and this accounts for the change in amplitude of the radial oscillations in Figure 19.5a).

19.3.3.2 Composition and Preparation

A variety of methods are available for generating microbubbles[4]. The most commonly used and the most efficient in terms of large scale preparation are methods based on mechanical agitation, and it was this type of method that was adopted by both Stride et al. and Vlaskou et al. The quantities of the different constituents used were based on calculations using the equations in the previous section. Since a bulk preparation method was employed, this analysis could only provide approximate ratios of the components required. The materials were deliberately selected to provide as simple a bubble composition as possible in order to minimise the sources of variability in the subsequent experiments. There have been numerous studies, however, on the composition of non-magnetic microbubbles demonstrating the effect of different phospholipids on microbubble stability, acoustic response, conjugation with therapeutic components and the addition of components such as polyethyleneglycol for improving biocompatibility[71]. The results of these studies may be readily utilised in developing improved formulations. Vlaskou et al. employed a similar preparation method but used a mechanical shaker, different phospholipids and magnetic nanoparticles and added soybean oil to their formulation. They also incorporated DNA directly into their liposphere for the therapeutic studies. The significance of these differences in the context of the experimental observations will be discussed in later sections.

19.3.3.3 Characterisation

There are a number of characteristics that need to be determined for a magnetic microbubble suspension. First, as for any microbubble agent, the mean size and size distribution of the bubbles need to be measured to ensure that the maximum bubble size does not exceed ~10 μm (to avoid risk of embolism *in vivo*), to estimate the bubble resonance frequency and also to assess the degree of uniformity in the suspension. Bubble size can be determined most directly using optical microscopy, and an example of a magnetic microbubble suspension and the corresponding size distribution are

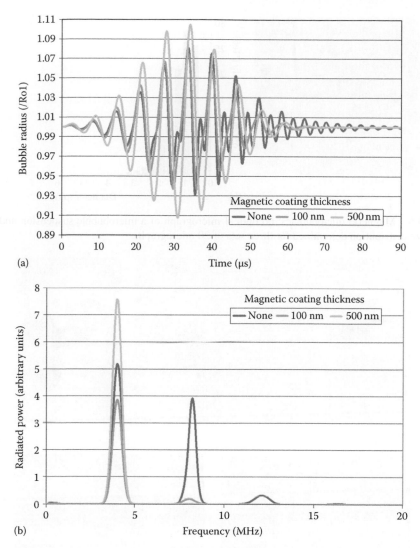

FIGURE 19.5 Response of a magnetic microbubble (initial diameter 2 μm) to ultrasound excitation: (a) variation of microbubble radius with time in response to an ultrasound pulse (4 MHz centre frequency, 100 kPa peak negative pressure) for magnetic coating layers of different thicknesses and (b) the corresponding frequency spectra for the radiated pressure from the microbubble showing the reduction in non-linear content with increasing layer thickness.

shown in Figure 19.6. The size distribution can be modified by varying the duration and acoustic intensity/repetition rate of the sonication and agitation stages during bubble preparation. It is usually desirable, however, to obtain a mean size similar to that found in commercial contrast agents (2–3 μm diameter) and as small a standard deviation as possible. The latter can be improved by filtering and/or centrifuging the bubble suspension[72] or by using an alternative preparation technique such as a microfluidic method, whereby variations in bubble diameter of less than 1% can be achieved[73].

The response of the bubbles to a magnetic field can be determined macroscopically (Figure 19.7a and b) and/or microscopically (Figure 19.7c and d) by simple application of a known magnetic field to ensure that sufficient magnetic material has been incorporated into the bubbles to allow them to be manipulated. Quantification of the magnetic material content of individual bubbles may be determined by observing the velocity of microbubble motion in the presence of uniform magnetic

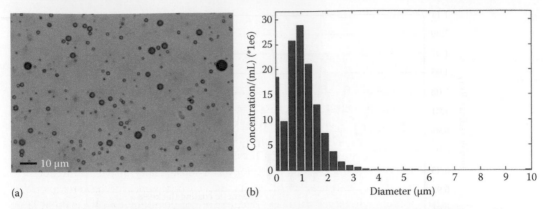

FIGURE 19.6 Magnetic microbubbles: (a) optical micrograph of a microbubble suspension and (b) corresponding size distribution determined for three sets of 20 micrographs.

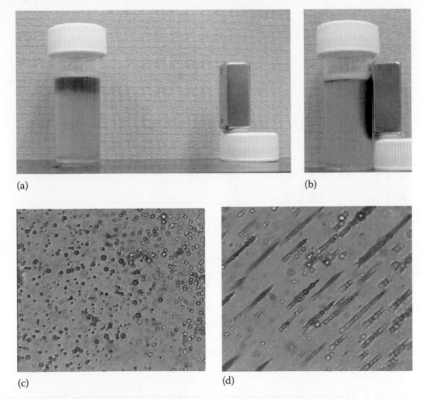

FIGURE 19.7 Response of magnetic microbubbles to an imposed magnetic field (a) and (b) macroscopically (c) and (d) microscopically. (Reprinted from Stride, E. et al., *Ultrasound Med. Biol.*, 35, 861, 2009.)

field gradient. The extent to which a bubble suspension may be manipulated may also be assessed using a flow phantom in which the liquid flow rate can be controlled[74].

The acoustic response of the microbubbles can similarly be determined both for the bubble suspensions and for individual bubbles. Figure 19.8 shows the ratio of scattering to attenuation of an ultrasound signal propagated through suspensions of magnetic and non-magnetic microbubbles. Full details of the experiment may be found in Refs. [75,76]. The scattering to attenuation ratio (STAR) provides a means of assessing the efficiency of contrast agents[77], since it is only the scattered signal that contributes to image formation. Also shown in Figure 19.8 is the corresponding ratio for the non-linear component of the signal (nSTAR), since it is this component which is normally of most

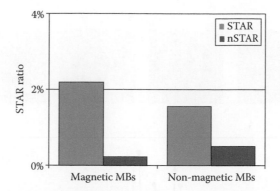

FIGURE 19.8   Linear and non-linear STAR of 3.5 MHz ultrasound pulses for magnetic and non-magnetic microbubble suspensions.

relevance for contrast-enhanced ultrasound imaging*. As may be seen, the STAR values are similar for both the magnetic and non-magnetic bubbles with nSTAR being slightly lower for the magnetic microbubbles in agreement with the theoretical results shown in Figure 19.5.

Figure 19.9a and b correspond to a series of frames from high-speed camera footage of a magnetic microbubble undergoing volumetric oscillations in response to an ultrasound pulse. Again, full details of the experiment may be found in Refs. [75,76]. Figure 19.9c shows results derived from high-speed camera videos of a series of individual bubbles demonstrating how the maximum amplitude of radial oscillation varies with initial bubble size. In agreement with the results shown in Figure 19.9, the response of the magnetic microbubbles is reasonably similar to that of the non-magnetic bubbles (SonoVue®). The resonant size is slightly different, which is likely to be due to the fact that this is largely governed by the radius of the air core, which will be different for magnetic and non-magnetic bubbles due to the layer of magnetic material in the former. Interestingly, the oscillations of the magnetic microbubbles did not appear to be significantly affected by the presence of a magnet used to pull the bubbles to the side of the cellulose tube in which the bubbles were enclosed. Also shown are results for magnetic micelles, which contained the same magnetic nanoparticle suspension coated with phospholipid (see Section 19.4.2.1) but negligible quantities of air. They are thus unresponsive to ultrasound excitation.

19.4   *IN VITRO* STUDIES

19.4.1   INTRODUCTION

Both Stride et al.[68] and Vlaskou et al.[66] have reported *in vitro* cell transfection experiments in which transfection rates were compared in the presence of magnetic microbubbles with exposure to ultrasound and/or a magnetic field. The former study also compared transfection rates achieved with non-magnetic microbubbles and magnetic particles which were not acoustically active. The experiments conducted in this study will be reviewed here and the results and key differences of both studies are discussed.

19.4.2   PROTOCOL

19.4.2.1   Microbubble Preparation

The aim in this initial study of magnetic microbubbles was to compare gene transfection rates to those obtained in previous studies of commercial (non-magnetic) ultrasound contrast agents. The protocol developed by Rahim et al.[52] was therefore followed, but, as mentioned previously, three different types of particle were prepared: A, B and C as indicated in Table 19.1.

---

* It is not currently known which are the most relevant parameters for therapeutic applications.

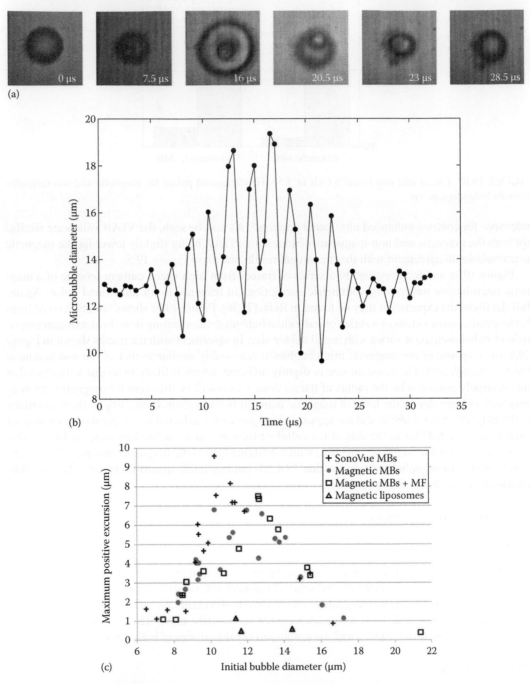

FIGURE 19.9 Microbubble response to ultrasound excitation: (a) single frames from high-speed camera video microscopy of a magnetic microbubble confined in a cellulose tube (200 μm inner diameter) exposed to a 0.5 MHz centre frequency 12-cycle Gaussian windowed ultrasound pulse peak negative pressure ~100 kPa. Images were captured at 2.5 million frames per second and the final image resolution was 20 pixels/μm. (b) Radius time curve for a magnetic microbubble plotted from high-speed camera data. (c) Variation in the amplitude of radial expansion with initial bubble diameter for different types of microbubbles as measured from the high-speed camera footage.

TABLE 19.1
Microbubble Preparations Used in the Experiments

Type	Gas	Coating	Magnetic Content	Diameter, Range (Mean)/μm
A	Air	L-α-Phosphatidylcholine	None	5–20 (6)
B	None	L-α-Phosphatidylcholine	Hydrocarbon 10% suspension of 10 nm spherical magnetite	1–2 (1.5)
C	Air	L-α-Phosphatidylcholine	Hydrocarbon 10% suspension of 10 nm spherical magnetite	1–20 (2)
SonoVue	Sulphur hexafluoride	Phospholipid (proprietary)	None	1–20 (2.25)

As described in Section 19.3.3.3, each suspension was examined *via* optical microscopy to determine the average size and concentration of microbubbles/micelles present. It was shown also under optical microscopy that the bubbles could be successfully manipulated using a magnetic field (see Figure 19.8) and that the minimum coating thickness required to overcome bubble buoyancy had been exceeded. This was essential for the experiments as described in the following. The scattering and attenuation coefficients for the bubble suspensions were also measured as described previously to confirm that the bubbles also contained a sufficient volume of air to be acoustically responsive.

19.4.2.2 Cell Culture

Chinese hamster ovary (CHO) cells were grown in commercial cell-culture devices (Opticell®) whose membranes are both acoustically and optically transparent. Prior to use, the medium in each Opticell was replaced with medium containing 50 μg pLuc plasmid DNA, following which the cells were returned to the incubator until required.

19.4.2.3 Apparatus

The apparatus used by Rahim et al. was modified to enable the cells to be exposed simultaneously to ultrasound and an applied magnetic field (Figure 19.10). The magnetic force was provided

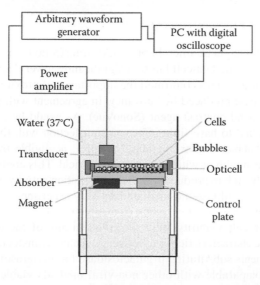

FIGURE 19.10 Apparatus used for the *in vitro* gene delivery experiments. (Reprinted from Stride, E. et al., *Ultrasound Med. Biol.*, 35, 861, 2009.)

by rectangular block N52 grade NdFeB permanent magnets, five of which were arranged in an aluminium frame in a Halbach array configuration with their transversal magnetisations (1.5 T) orientated at angles of 90° from one to the next. The Opticell was placed above the array with a 3 mm gap between the magnet surface and the lower Opticell membrane. This gap was filled with an acoustically absorbing material to reduce the amplitude of reflections from the magnet. In addition, in order to avoid any variations in exposure, an aluminium bar was substituted for the array during experiments in the absence of the magnet. Ultrasound exposure was provided from above by a single-element 1 MHz transducer driven at a pulse repetition frequency of 1 kHz with 40 cycle 1 MHz sinusoidal pulses. The peak-to-peak acoustic pressure was 1 MPa and the exposure time was 10 s. The transducer was in direct contact with the Opticell membrane and the bubbles/cells were therefore in the near field of the transducer.

19.4.2.4 Exposure Conditions

Immediately prior to each test, 0.4 mL of the relevant bubble/particle suspension (Table 19.1) was added to the Opticell, which was then immersed in the water bath and allowed 1 min (in the presence of the magnetic field, where used) for settling and temperature equilibration. The cells were deliberately grown on only one membrane of the Opticell, so that depending on which way up the Opticell was placed in the bath, non-magnetic bubbles would either rise towards or away from the cells under buoyancy. The same would be true for magnetic bubbles in the absence of the magnetic field, but with the magnetic field, they would be drawn down onto the cells. A minimum of three repetitions for each set of conditions were performed (for most $n = 6$). Control experiments in which the cells were exposed to DNA in the absence of microbubbles, to the magnetic field alone, to ultrasound alone and to ultrasound and the magnetic field in the presence of the nanoparticle suspension (as purchased, without phospholipid) were also carried out.

19.4.2.5 Cell Imaging

Thirty minutes after ultrasound and/or magnetic field exposure, the medium in each Opticell was replaced with 10 mL fresh growth medium. The cells were then incubated for 24 h before being examined. Luciferase expression was evaluated by bioluminescence imaging 2 h after replacement of the medium with a solution containing 15 µg/mL luciferin.

19.4.3 RESULTS

In the presence of the non-magnetic microbubbles (A), transfection was observed when the cells were on the upper surface of the Opticell but to a significantly lower degree (~10 times) when the cells were on the lower surface. This confirmed the importance of proximity between the cells and the microbubbles, in this case produced by buoyancy, in agreement with previous results obtained with a commercial ultrasound contrast agent (SonoVue). As would be expected, the presence of the magnetic field was found to have little effect on transfection with the non-magnetic bubbles. In the absence of microbubbles and/or magnetic micelles, negligible transfection was observed upon exposure to ultrasound with or without the magnetic field. This confirmed that the ultrasound intensity was insufficiently high to produce significant cavitation activity, which has been identified as contributory factor in other ultrasound-mediated delivery experiments[44]. Similarly, there was no transfection with the sham exposure or exposure to the magnet alone.

No adverse effects on cell viability were observed in any of the experiments, and this is clearly a highly desirable characteristic for a delivery system. Transfection efficiency (estimated by repeating the experiments substituting β-galactosidase for luciferase) was, however, relatively low (~10%), although comparable with other non-viral methods viable for use *in vivo*. Studies with non-magnetic microbubbles have indicated improved efficiency at larger peak negative pressures, the benefits of which may outweigh the associated damage to cells[78]. Increasing the ultrasound peak negative pressure to 1 and then to 2 MPa with the apparatus, described previously, in fact increased the observed transfection rates without any loss of cell viability, although the increase became proportionally smaller with increasing pressure. It should also be noted that these experiments were carried out without conjugating the DNA to the microbubbles. Again, studies with non-magnetic bubbles have shown that this can be used to significantly improve efficiency.

19.4.4 DISCUSSION

In the experiments described previously, the magnetic microbubbles were found to be much more effective for transfection of CHO cells with naked plasmid DNA under simultaneous exposure to ultrasound and a magnetic field than either magnetic micelles (by a factor of ~4) or non-magnetic microbubbles (which produced no observable transfection). Compared with exposure to either a magnetic field or ultrasound alone, transfection was enhanced by a factor of ~3. In the experiments reported by Vlaskou et al. with magnetic lipospheres, there was no significant difference observed between transfection achieved with simultaneous exposure to ultrasound and the magnetic field and the magnetic field alone. The most likely explanation is the difference in the ultrasound exposure conditions. Vlaskou et al. used a commercial device (Sonitron 2000D) with a smaller probe (3 mm diameter compared with 20 mm). The centre frequency was the same (1 MHz), but the intensity was slightly lower (2 W/cm^2 corresponding to ~200 kPa peak negative pressure) and the exposure time longer (30 s). The duty cycle was 20%. As discussed previously, the peak negative pressure and hence intensity of the ultrasound has been shown to have a significant effect on transfection enhancement.

There were a number of other differences in the experiments which should also be noted, however. First, different types of cells were transfected, although Vlaskou et al. did report similar results for mouse fibroblasts and human lung epithelial cells with DNA and HeLa cells with siRNA. The spatial distribution of the magnetic field was different and the strength somewhat lower in the latter study, which could have affected the concentration of bubbles in the region of interest. Moreover, the definition of the region of interest itself may have affected the results. Stride et al. designed the magnetic array to produce a region of maximum magnetic force coincident with area exposed to the ultrasound beam, and this was the area subsequently analysed. This was not the case in the study conduced by Vlaskou et al. The size of the microbubbles in the latter study was slightly larger, although it is not clear why this would have made them less acoustically responsive (at 1 MHz, larger bubbles would be expected to be closer to resonance, assuming the outer diameter was proportional to the volume of gas inside). Unfortunately, it is not possible to directly compare the concentrations of bubbles used in both studies as this was not observed during the experiments after exposure to the magnetic field. The composition of the microbubbles used by Vlaskou et al. was also different as mentioned earlier. The material to be transfected was incorporated into the bubbles (which probably explains why transfection efficiency was higher). The core gas was different (perfluoropropane rather than air), which should actually have increased their stability. The phospholipids used for the coating were different, and this can affect the amplitude of oscillation[71], although probably not to a very great extent at the pressures used. The magnetic nanoparticles were surfactant coated rather than suspended in a liquid carrier, which could potentially have produced a stiffer coating. Also, soybean oil was added during the microbubble preparation, which might have affected the volume of gas and hence the acoustic responsiveness, depending on how much was incorporated into the bubbles. Unfortunately, although electron microscopic analysis was carried out, it is not possible to determine the exact composition of individual bubbles and to predict whether or not they have a core–shell (Figure 19.4) or multilamellar structure such as that discussed in Ref. [79].

19.5 *IN VIVO* STUDIES

19.5.1 INTRODUCTION

The same research groups undertook corresponding *in vivo* studies to investigate the effectiveness of magnetic microbubbles for gene transfection in a mouse model[66,75]. As mentioned previously, an overview of the procedure in the former study will be described, and the results of both studies are then compared and discussed.

19.5.2 EXPERIMENTAL PROCEDURE

The same plasmid was used in all the *in vivo* experiments encoding the *Photinus pyralis* luciferase gene, *luc2*. Six- to eight-week-old, female, CD1 mice were anaesthetised, and a region of the chest of each mouse was shaved. Organ location and imaging were conducted using low-intensity (~220 kPa) high-frequency (14 MHz) ultrasound with the acoustic focus coincident with the heart. A total of 150 μL of the magnetic microbubble suspension and 50 μL of the plasmid preparation were administered simultaneously *via* tail vein injection. This was immediately followed by exposure of the region of interest to ultrasound and/or a magnetic field for 2 min. The ultrasound scanner settings were adjusted to give an exposure of 1.54 MPa peak negative pressure at 6 MHz. Following exposure, the probe and/or magnet were removed and the mouse was allowed to recover in a hot box. Three days after transfection, mice were anaesthetised as mentioned previously and injected (intraperitoneally) with luciferin. After 10 min, mice were euthanised and the organs recovered for bioluminescence imaging.

19.5.3 RESULTS AND DISCUSSION

Figure 19.11 shows the mean heart and lung transfection rates obtained from the bioluminescence images for 20 mice that were exposed to both ultrasound and the magnetic field. Also shown are the results for mice exposed to either ultrasound or the magnetic field alone. Unfortunately, due to the constraints of the study, the subject numbers were smaller for these groups. The data show a clear trend, with animals treated with both ultrasound and the magnetic field showing much greater transfection than with either alone. In addition, for the mice undergoing simultaneous exposure, transfection was consistently higher in the right lung than in other organs. This is compared with the results from previous studies[80], using the same exposure conditions but non-magnetic bubbles, in which transfection was consistently localised to the heart. Thus, it may be concluded that the application of the magnetic field did enable the transfection site to be successfully translated.

Again there were found to be discrepancies between the results obtained by Mulvana et al. and Vlaskou et al. When Vlaskou et al. administered their magnetic liposphere to mice *via* tail injection, they observed no increase in transfection rates with simultaneous exposure to ultrasound and a magnetic field, in agreement with their *in vitro* results. Interestingly, however, they did observe enhancement when the microbubbles were injected directly into a dorsal skin flap chamber model. As with the *in vitro* studies, there were some differences in the exposure conditions: first, the ultrasound frequency used by Vlaskou et al. in both sets of *in vivo* experiments was 1 MHz. In the tail vein injection study, the ultrasound intensity was 4 W/cm² and in the skin flap study it was 2 W/cm², corresponding to much lower pressures (200–300 kPa) than used by Mulvana et al. The corresponding duty cycles were 50% and 100%. The exposure times were considerably longer, 10 min in the tail vein study and 30 and 5 min, respectively, for the magnetic field and ultrasound in the skin flap study. A larger probe diameter was used than in the *in vitro* experiments (6 mm diameter), but this was considerably smaller than the scanner probe used by Mulvana et al., although this would be compensated to some extent by the fact that the latter was focused and the centre frequency was higher.

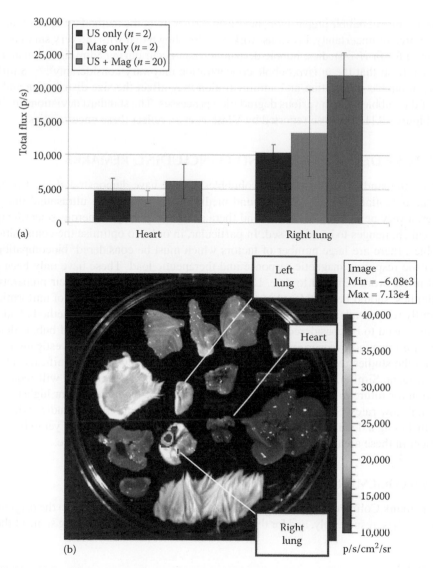

FIGURE 19.11 (a) Luciferase expression promoted *in vivo* in a mouse model by non-magnetic bubbles with exposure to ultrasound and/or a magnetic field, the sample size is indicated in the figure, and (b) example of *ex vivo* tissue from which measurements were obtained.

The fact that the mean bubble size was larger in the Vlaskou et al. study may have resulted in a number of bubbles being immobilised in the lung capillary bed, and this was noted by the authors. That the species of mice were different is probably of secondary importance but cannot be ignored completely. The use of different anaesthetics should be considered on account of the potential effect on the dissolution rates of the bubbles, likewise the fact that the bubbles contained different gases (sulphur hexafluoride and perfluoropropane). The volume of microbubble suspension administered was similar in both studies, but unfortunately the concentration of bubbles was not accurately known in either case. The magnetic field strengths appear to have been similar, although this will depend to some extent on the orientation of the magnets used, which again was not known precisely. It was also not possible to reliably assess the uniformity of the microbubbles in terms of their stability, magnetic or acoustic response in either study. The high-speed camera studies (see Section 19.3.3.3) did provide some indication, however, and showed that there could be considerable variation in the behaviour of bubbles which appeared identical under optical microscopy. Every effort was made to ensure consistency in bubble preparation, but given the nature of the method used, this must be identified as a source of uncertainty. Previous work has also shown that with the very small needles and catheters used for *in vivo* studies, bubble destruction can occur during administration and is sensitive to flow rate so that the *in vivo* bubble concentration may vary considerably[81,82]. Similarly, the time between bubble preparation and administration may affect the size distribution and acoustic response of the bubbles due to various degradation processes. The standard deviations in the results shown in Figure 19.11 and those reported by Vlaskou et al. reflect these uncertainties.

19.6   FUTURE DEVELOPMENTS AND CONCLUDING REMARKS

In summary, the studies of magnetic microbubbles to date have demonstrated great potential for realising the potential of localised ultrasound-mediated therapy. Unlike ultrasound imaging, this is, however, a very new area of research, and there are consequently numerous scientific questions and practical challenges to be addressed. In particular, in order to optimise the composition of the microbubbles, there are large number of factors which must be considered: biocompatibility, stability, acoustic response, magnetic response and therapeutic load. These have only been partially addressed in the work reported to date. Likewise, alternative methods for their preparation, such as microfluidics, should be considered in order to achieve the required degree of uniformity, which will hopefully remove a large contribution to experimental uncertainty. The method of administration may also need to be considered, or at least studied to assess the degree of bubble destruction. The optimal ultrasound exposure parameters require considerable further investigation, since the results from the studies reviewed in this chapter indicate that they have a significant influence on transfection rates. Similarly, the strength and direction of the magnetic field will require careful consideration for future *in vivo* studies, especially at large tissue depths and for higher physiologically relevant flow rates. Subsequently, more extensive *in vitro* and *in vivo* studies will clearly be required, including a detailed analysis of toxicity and elimination, which have yet to be conducted. Development in these areas will undoubtedly be forthcoming in the near future.

ACKNOWLEDGEMENTS

The authors thank Colin Porter and Richard Browning for their contributions to the original *in vitro* and *in vivo* studies, respectively, and for their subsequent assistance in preparing some of the figures.

REFERENCES

1. Feigenbaum, H., Stone, J. M., Lee, D. A., Nasser, W. K., and Chang, S. 1970. Identification of ultrasound echoes from left ventricle by use of intracardiac injections of indocyanine green. *Circulation*, 41, 615–621.
2. Kremkau, F. W., Gramiak, R., Carstensen, E. L., Shah, P. M., and Kramer, D. H. 1970. Ultrasonic detection of cavitation at catheter tips. *American Journal of Roentgenology, Radium Therapy and Nuclear Medicine*, 110, 177–183.
3. Feinstein, S. B., Tencate, F. J., Zwehl, W., Ong, K., Maurer, G., Tei, C., Shah, P. M., Meerbaum, S., and Corday, E. 1984. Two-dimensional contrast echocardiography. 1. In vitro development and quantitative-analysis of echo contrast agents. *Journal of the American College of Cardiology*, 3, 14–20.
4. Stride, E. and Edirisinghe, M. 2008. Novel microbubble preparation technologies. *Soft Matter*, 4, 2350–2359.
5. Hill, C. R., Bamber, J., and Ter Haar, G. 2004. *Physical Principles of Medical Ultrasonics*, John Wiley & Sons Ltd., Chichester, UK.
6. Harvey, C. J., Pilcher, J. M., Eckersley, R. J., Blomley, M. J. K., and Cosgrove, D. O. 2002. Advances in ultrasound. *Clinical Radiology*, 57, 157–177.
7. Al-Mansour, H. A., Mulvagh, S. L., Pumper, G. M., Klarich, K. W., and Foley, D. A. 2000. Usefulness of harmonic imaging for left ventricular opacification and endocardial border delineation by optison. *American Journal of Cardiology*, 85, 795–799.

8. Strauss, A. L. and Beller, K. D. 1999. Persistent opacification of the left ventricle and myocardium with a new echo contrast agent. *Ultrasound in Medicine & Biology*, 25, 763–769.

9. Lee, S. C., Carr, C. L., Davidson, B. P., Ellegala, D., Xie, A., Ammi, A., Belcik, T., and Lindner, J. R. 2010. Temporal characterization of the functional density of the vasa vasorum by contrast-enhanced ultrasonography maximum intensity projection imaging. *Journal of American College of Cardiology: Cardiovascular Imaging*, 3, 1265–1272.

10. Prefumo, F., Serafini, G., Martinoli, C., Gandolfo, N., Gandolfo, N. G., and Derchi, L. E. 2002. The sonographic evaluation of tubal patency with stimulated acoustic emission imaging. *Ultrasound in Obstetrics and Gynecology*, 20, 386–389.

11. Darge, K., Moeller, R. T., Trusen, A., Butter, F., Gordjani, N., and Riedmiller, H. 2005. Diagnosis of vesicoureteric reflux with low-dose contrast-enhanced harmonic ultrasound imaging. *Pediatric Radiology*, 35, 73–78.

12. Huang, S. F., Chang, R. F., Moon, W. K., Lee, Y. H., Chen, D. R., and Suri, J. S. 2008. Analysis of tumor vascularity using three-dimensional power Doppler ultrasound images. *IEEE Transactions on Medical Imaging*, 27, 320–330.

13. Meairs, S. 2008. Contrast-enhanced ultrasound perfusion imaging in acute stroke patients. *European Neurology*, 59, 17–26.

14. Becher, H. and Burns, P. 2008. *Handbook of Contrast Echocardiography*, Berlin: Springer–Verlag.

15. Cosgrove, D. and Lassau, N. 2010. Imaging of perfusion using ultrasound. *European Journal of Nuclear Medicine and Molecular Imaging*, 37, S65–S85.

16. Quaia, E. 2005. Contrast media in ultrasonography – Basic principles and clinical applications, New York: Springer–Verlag.

17. Bleeker, H., Shung, K., and Barnhart, J. 1990. On the application of ultrasonic contrast agents for blood flowmetry and assessment of cardiac perfusion. *Journal of Ultrasound in Medicine*, 9, 461–471.

18. Leen, E., Ceccotti, P., Kalogeropoulou, C., Angerson, W. J., Moug, S. J., and Horgan, P. G. 2006. Prospective multicenter trial evaluating a novel method of characterizing focal liver lesions using contrast-enhanced sonography. *American Journal of Roentgenology*, 186, 1551–1559.

19. Leen, E., Ceccotti, P., Moug, S. J., Glen, P., MacQuarrie, J., Angerson, W. J., Albrecht, T., Hohmann, J., Oldenburg, A., Ritz, J. P., and Horgan, P. G. 2006. Potential value of contrast-enhanced intraoperative ultrasonography during partial hepatectomy for metastases – An essential investigation before resection? *Annals of Surgery*, 243, 236–240.

20. Lindner, J. R., Wei, K., and Kaul, S. 1999. Imaging of myocardial perfusion with SonoVue (TM) in patients with a prior myocardial infarction. *Echocardiography: A Journal of Cardiovascular Ultrasound and Allied Techniques*, 16, 753–760.

21. Lindner, J. R., Villanueva, F. S., Dent, J. M., Wei, K., Sklenar, J., and Kaul, S. 2000. Assessment of resting perfusion with myocardial contrast echocardiography: Theoretical and practical considerations. *American Heart Journal*, 139, 231–240.

22. Wei, K., Jayaweera, A. R., Firoozan, S., Linka, A., Skyba, D. M., and Kaul, S. 1998. Quantification of myocardial blood flow with ultrasound-induced destruction of microbubbles administered as a constant venous infusion. *Circulation*, 97, 473–483.

23. Stride, E., Tang, M., and Eckersley, R. J. 2008. Physical phenomena affecting quantitative imaging of ultrasound contrast agents. *Applied Acoustics*, 70, 1352–1362.

24. Kaufmann, B. A. and Lindner, J. R. 2007. Molecular imaging with targeted contrast ultrasound. *Current Opinion in Biotechnology*, 18, 11–16.

25. Schneider, M. 2008. Molecular imaging and ultrasound-assisted drug delivery. *Journal of Endourology*, 22, 795–801.

26. Kogan, P., Gessner, R. C., and Dayton, P. A. 2010. Microbubbles in imaging: Applications beyond ultrasound. *Bubble Science, Engineering & Technology*, 2, 3–8.

27. Cheung, J. S., Chow, A. M., Guo, H., and Wu, E. X. 2009. Microbubbles as a novel contrast agent for brain MRI. *NeuroImage*, 46, 658–664.

28. Anderson, D., Anchan, R., Johnston, M., Duryee, M., Xie, F., Thiele, G., Klassen, L., Boska, M., and Porter, T. 2008. Ultrasound and magnetic resonance imaging: The role of microbubbles and gadolinium-labeled microbubbles in imaging after balloon angioplasty. *Circulation*, 118, S997–S997.

29. Yang, F., Li, Y. X., Chen, Z. P., Zhang, Y., Wu, J. R., and Gu, N. 2009. Superparamagnetic iron oxide nanoparticle-embedded encapsulated microbubbles as dual contrast agents of magnetic resonance and ultrasound imaging. *Biomaterials*, 30, 3882–3890.

30. Arfelli, F., Rigon, L., Menk, R. H., and Besch, H. J. 2003. On the possibility of utilizing scattering-based contrast agents in combination with diffraction enhanced imaging. *Medical Imaging 2003: Physics of Medical Imaging, Parts 1 and 2*, SPIE, 5030, 274–283.

31. Walday, P., Tolleshaug, H., Gjoen, T., Kindberg, G. M., Berg, T., Skotland, T., and Holtz, E. 1994. Biodistributions of air-filled albumin microspheres in rats and pigs. *Biochemical Journal*, 299, 437–443.

32. Willmann, J. K., Cheng, Z., Davis, C., Lutz, A. M., Schipper, M. L., Nielsen, C. H., and Gambhir, S. S. 2008. Targeted microbubbles for imaging tumor angiogenesis: Assessment of whole-body biodistribution with dynamic micro-PET in mice. *Radiology*, 249, 212–219.

33. Suslick, K. S. 1990. Sonochemistry. *Science*, 247, 1439–1445.

34. Hall, D. J., Hsu, M. J., Esener, S., and Mattrey, R. F. 2009. Detection of ultrasound-modulated photons and enhancement with ultrasound microbubbles. *Photons Plus Ultrasound: Imaging and Sensing 2009*, SPIE, 7177.

35. Honeysett, J., Stride, E., and Leung, T. 2010. Monte Carlo simulations of acousto-optics with microbubbles. *Photons Plus Ultrasound: Imaging and Sensing 2010*, SPIE, 7564.

36. Kim, C., Qin, R. G., Xu, J. S., Wang, L. V., and Xu, R. 2010. Multifunctional microbubbles and nanobubbles for photoacoustic and ultrasound imaging. *Journal of Biomedical Optics*, 15, 010510.

37. Bull, J. L. 2007. The application of microbubbles for targeted drug delivery. *Expert Opinion on Drug Delivery*, 4, 475–493.

38. Pichon, C., Kaddur, K., Midoux, P., Tranquart, F., and Bouakaz, A. 2008. Recent advances in gene delivery with ultrasound and microbubbles. *Journal of Experimental Nanoscience*, 3, 17–40.

39. Tartis, M. S., McCallan, J., Lum, A. F. H., LaBell, R., Stieger, S. M., Matsunaga, T. O., and Ferrara, K. W. 2006. Therapeutic effects of paclitaxel-containing ultrasound contrast agents. *Ultrasound in Medicine & Biology*, 32, 1771–1780.

40. Iwanaga, K., Tominaga, K., Yamamoto, K., Habu, M., Maeda, H., Akifusa, S., Tsujisawa, T., Okinaga, T., Fukuda, J., and Nishihara, T. 2007. Local delivery system of cytotoxic agents to tumors by focused sonoporation. *Cancer Gene Therapy*, 14, 354–363.

41. van Wamel, A., Kooiman, K., Harteveld, M., Emmer, M., ten Cate, F. J., Versluis, M., and de Jong, N. 2006. Vibrating microbubbles poking individual cells: Drug transfer into cells via sonoporation. *Journal of Controlled Release*, 112, 149–155.

42. Meairs, S. and Alonso, A. 2007. Ultrasound, microbubbles and the blood–brain barrier. *Progress in Biophysics & Molecular Biology*, 93, 354–362.

43. Juffermans, L. J. M., Kamp, O., Dijkmans, P. A., Visser, C. A., and Musters, R. J. P. 2008. Low-intensity ultrasound-exposed microbubbles provoke local hyperpolarization of the cell membrane via activation of BKCa channels. *Ultrasound in Medicine & Biology*, 34, 502–508.

44. Hallow, D. M., Mahajan, A. D., McCutchen, T. E., and Prausnitz, M. R. 2006. Measurement and correlation of acoustic cavitation with cellular bioeffects. *Ultrasound in Medicine & Biology*, 32, 1111–1122.

45. Culp, W. C., Porter, T. R., Lowery, J., Xie, F., Roberson, P. K., and Marky, L. 2004. Intracranial clot lysis with intravenous microbubbles and transcranial ultrasound in swine. *Stroke*, 35, 2407–2411.

46. Molina, C. A., Ribo, M., Rubiera, M., Montaner, J., Santamarina, E., Delgado-Mederos, R., Arenillas, J. F., Huertas, R., Purroy, F., Delgado, P., and Alvarez-Sabin, J. 2006. Microbubble administration accelerates clot lysis during continuous 2-MHz ultrasound monitoring in stroke patients treated with intravenous tissue plasminogen activator. *Stroke*, 37, 425–429.

47. Kaneko, Y., Maruyama, T., Takegami, K., Watanabe, T., Mitsui, H., Hanajiri, K., Nagawa, H. A., and Matsumoto, Y. 2005. Use of a microbubble agent to increase the effects of high intensity focused ultrasound on liver tissue. *European Radiology*, 15, 1415–1420.

48. Luo, W., Zhou, X. D., Tian, X., Ren, X. L., Zheng, M. J., Gu, K. J., and He, G. B. 2006. Enhancement of ultrasound contrast agent in high-intensity focused ultrasound ablation. *Advances in Therapy*, 23, 861–868.

49. Christiansen, J. P., French, B. A., Klibanov, A. L., Kaul, S., and Lindner, J. R. 2003. Targeted tissue transfection with ultrasound destruction of plasmid-bearing cationic microbubbles. *Ultrasound in Medicine & Biology*, 29, 1759–1767.

50. Pepe, J., Rincon, M., and Wu, J. 2004. Experimental comparison of sonoporation and electroporation in cell transfection applications. *Acoustics Research Letters Online*, 5, 62–67.

51. Kodama, T., Aoi, A., Watanabe, Y., Horie, S., Kodama, M., Li, L., Chen, R., Teramoto, N., Morikawa, H., Mori, S., and Fukumoto, M. 2010. Evaluation of transfection efficiency in skeletal muscle using nano/microbubbles and ultrasound. *Ultrasound in Medicine & Biology*, 36, 1196–1205.

52. Rahim, A., Taylor, S. L., Bush, N. L., Ter Haar, G. R., Bamber, J. C., and Porter, C. D. 2006. Physical parameters affecting ultrasound/microbubble-mediated gene delivery efficiency in vitro. *Ultrasound in Medicine & Biology*, 32, 1269–1279.

53. Tillander, H. 1951. Magnetic guidance of a catheter with articulated steel tip. *Acta Radiologica*, 35, 62–64.

54. Senyei, A., Widder, K., and Czerlinski, G. 1978. Magnetic guidance of drug-carrying microspheres. *Journal of Applied Physics*, 49, 3578–3583.

55. Widder, K., Flouret, G., and Senyei, A. 1979. Magnetic microspheres – Synthesis of a novel parenteral drug carrier. *Journal of Pharmaceutical Sciences*, 68, 79–82.

56. Schutt, W., Gruttner, C., Hafeli, U., Zborowski, M., Teller, J., Putzar, H., and Schumichen, C. 1997. Applications of magnetic targeting in diagnosis and therapy – Possibilities and limitations: A mini-review. *Hybridoma*, 16, 109–117.

57. Schutt, W., Gruttner, C., Teller, J., Westphal, F., Hafeli, U., Paulke, B., Goetz, P., and Finck, W. 1999. Biocompatible magnetic polymer carriers for in vivo radionuclide delivery. *Artificial Organs*, 23, 98–103.

58. Pouponneau, P., Leroux, J. C., and Martel, S. 2009. Magnetic nanoparticles encapsulated into biodegradable microparticles steered with an upgraded magnetic resonance imaging system for tumor chemoembolization. *Biomaterials*, 30, 6327–6332.

59. Devineni, D., Kleinszanto, A., and Gallo, J. M. 1995. Tissue distribution of methotrexate following administration as a solution and as a magnetic microsphere conjugate in rats bearing brain-tumors. *Journal of Neuro-Oncology*, 24, 143–152.

60. Orekhova, N. M., Akchurin, R. S., Belyaev, A. A., Smirnov, M. D., Ragimov, S. E., and Orekhov, A. N. 1990. Local prevention of thrombosis in animal arteries by means of magnetic targeting of aspirin-loaded red-cells. *Thrombosis Research*, 57, 611–616.

61. Allen, L. M., Kent, T., Wolfe, C., Ficco, C., and Johnson, J. 1997. MTC(TM) – A magnetically targetable drug carrier for paclitaxel. *Scientific and Clinical Applications of Magnetic Carriers*, 481–494.

62. Plank, C., Schillinger, U., Scherer, F., Bergemann, C., Remy, J. S., Krotz, F., Anton, M., Lausier, J., and Rosenecker, J. 2003. The magnetofection method: Using magnetic force to enhance gene delivery. *Biological Chemistry*, 384, 737–747.

63. Lindner, J. R. 2004. Molecular imaging with contrast ultrasound and targeted microbubbles. *Journal of Nuclear Cardiology*, 11, 215–221.

64. Dromi, S., Frenkel, V., Luk, A., Traughber, B., Angstadt, M., Bur, M., Poff, J., Xie, J. W., Libutti, S. K., Li, K. C. P., and Wood, B. J. 2007. Pulsed-high intensity focused ultrasound and low temperature sensitive liposomes for enhanced targeted drug delivery and antitumor effect. *Clinical Cancer Research*, 13, 2722–2727.

65. Soetanto, K. and Watarai, H. 2000. Development of magnetic microbubbles for drug delivery system (DDS). *Japanese Journal of Applied Physics, Part 1: Regular Papers, Short Notes & Review Papers*, 39, 3230–3232.

66. Vlaskou, D., Mykhaylyk, O., Krotz, F., Hellwig, N., Renner, R., Schillinger, U., Gleich, B., Heidsieck, A., Schmitz, G., Hensel, K., and Plank, C. 2010. Magnetic and acoustically active liposphere for magnetically targeted nucleic acid delivery. *Advanced Functional Materials*, 20, 3881–3894.

67. Vlaskou, D., Mykhaylyk, O., Pradhan, P., Bergemann, C., Klibanov, A. L., Hensel, K., Schmitz, G., and Plank, C. 2010. Magnetic microbubbles as mediators of gene delivery. *Human Gene Therapy*, 21, 1429–1430.

68. Stride, E., Porter, C., Prieto, A. G., and Pankhurst, Q. 2009. Enhancement of microbubble mediated gene delivery by simultaneous exposure to ultrasonic and magnetic fields. *Ultrasound in Medicine & Biology*, 35, 861–868.

69. Prosperetti, A. 1984. Acoustic cavitation series. 3. Bubble phenomena in sound fields. 2. *Ultrasonics*, 22, 115–124.

70. Stride, E. 2008. The influence of surface adsorption on microbubble dynamics. *Philosophical Transactions of the Royal Society A: Mathematical, Physical and Engineering Sciences*, 366, 2103–2115.

71. Borden, M. A., Dayton, P., Zhao, S. K., and Ferrara, K. W. 2004. Physico-chemical properties of the microbubble lipid shell – Composition, microstructure and properties of targeted ultrasound contrast agents. *2004 IEEE Ultrasonics Symposium Proceedings*, Vols. 1–3, 20–23.

72. Feshitan, J. A., Chen, C. C., Kwan, J. J., and Borden, M. A. 2009. Microbubble size isolation by differential centrifugation. *Journal of Colloid and Interface Science*, 329, 316–324.

73. Talu, E., Hettiarachchi, K., Nguyen, H., Lee, A. P., Powell, R. L., Longo, M. L., and Dayton, P. A. 2006. Lipid-stabilized monodisperse microbubbles produced by flow focusing for use as ultrasound contrast agents. *2006 IEEE Ultrasonics Symposium Proceedings*, Vols. 1–5, 1568–1571.

74. Mahue, V., Mari, J. M., Eckersley, R. J., and Tang, M. X. 2011. Comparison of pulse subtraction Doppler and pulse inversion Doppler. *IEEE Transactions on Ultrasonics Ferroelectrics and Frequency Control*, 58, 73–81.

75. Mulvana, H., Eckersley, R. J., Browning, R., Tang, M., Pankhurst, Q., Wells, D. J., and Stride, E. 2010. Enhanced gene transfection in vivo using magnetic localisation of ultrasound contrast agents. *Proceedings of the IEEE International Ultrasonics Symposium*.

76. Mulvana, H., Stride, E., Hajnal, J. V., and Eckersley, R. J. 2010. Temperature dependent behavior of ultrasound contrast agents. *Ultrasound in Medicine & Biology*, 36, 925–934.

77. Bouakaz, A., de Jong, N., and Cachard, C. 1998. Standard properties of ultrasound contrast agents. *Ultrasound in Medicine & Biology*, 24, 469–472.

78. Sonoda, S., Tachibana, K., Uchino, E., Okubo, A., Yamamoto, M., Sakoda, K., Hisatomi, T., Sonoda, K. H., Negishi, Y., Izumi, Y., Takao, S., and Sakamoto, T. 2006. Gene transfer to corneal epithelium and keratocytes mediated by ultrasound with microbubbles. *Investigative Ophthalmology & Visual Science*, 47, 558–564.

79. Ferrara, K., Pollard, R., and Borden, M. 2007. Ultrasound microbubble contrast agents: Fundamentals and application to gene and drug delivery. *Annual Review of Biomedical Engineering*, 9, 415–447.

80. Alter, J., Sennoga, C. A., Lopes, D. M., Eckersley, R. J., and Wells, D. J. 2009. Microbubble stability is a major determinant of the efficiency of ultrasound and microbubble mediated in vivo gene transfer. *Ultrasound in Medicine & Biology*, 35, 976–984.

81. Barrack, T. and Stride, E. 2009. Microbubble destruction during intravenous administration: A preliminary study. *Ultrasound in Medicine & Biology*, 35, 515–522.

82. Talu, E., Powell, R. L., Longo, M. L., and Dayton, P. A. 2008. Needle size and injection rate impact microbubble contrast agent population. *Ultrasound in Medicine & Biology*, 34, 1182–1185.

Dr. Eleanor Stride is a reader in biomedical engineering in the Department of Mechanical Engineering at UCL. Her main research interests are advanced encapsulation technologies and biomedical ultrasonics, in particular the use of microbubble agents in ultrasound imaging and drug delivery and the biophysical effects of ultrasound exposure. Dr. Helen Mulvana, Dr. Robert Eckersley and Dr. Menxging Tang specialise in the physics of ultrasound microbubble contrast agents in the Imaging Sciences and Bioengineering Departments of Imperial College London, the United Kingdom. Professor Quentin Pankhurst is Director of the Davy–Faraday Research Laboratory and Wolfson Professor of Natural Philosophy at the Royal Institution of Great Britain, London.

20 Magnetic Particle Imaging for Angiography, Stem Cell Tracking, Cancer Imaging and Inflammation Imaging

Patrick Goodwill, Kannan M. Krishnan, and Steven M. Conolly

CONTENTS

20.1 INTRODUCTION

Vascular contrast studies, also known as angiograms, are crucial for diagnosing obstructions of the cardiovascular system. Vascular contrast imaging studies are typically used to diagnose high-morbidity and high-mortality conditions, including coronary artery disease, atherosclerosis and stroke. Vascular contrast agent studies are also used to diagnose and stage cancer, relying on the fact that abnormal neovasculature of tumours are more permeable to small-molecular-weight contrast agents. The most common contrast agents in use today are iodine for x-ray and computerised tomography (CT) and gadolinium for magnetic resonance imaging (MRI). The principal challenges with current contrast agent studies include ionising radiation (x-ray and CT), weak contrast with a venous injection, invasive catheterised delivery (x-ray) and the risk of damaging kidneys[1–3], especially for patients with poor renal function.

Magnetic particle imaging (MPI) angiography shows great promise as a non-invasive, nonionising replacement for the millions of contrast studies performed. MPI is an entirely new imaging modality with ideal image contrast for angiography[5]. MPI uses novel hardware and has imaging properties completely distinct from x-ray, CT, nuclear medicine or MRI. MPI works by imaging the *in vivo* distribution of a magnetic ferumoxide contrast agent called ultrasmall superparamagnetic iron oxide (USPIO) nanoparticles[4–7]. MPI's contrast already exceeds that of x-ray, CT and MRI because only the contrast agent is detected; background tissue is transparent and gives off no signal. The MPI process is safe, involving no ionising radiation and no dangerous magnetic fields. USPIOs have been found to be extremely safe, and several have been approved as MRI contrast agents in humans. Moreover, there is good evidence that ferumoxides are safer for chronic kidney disease (CKD) patients than the two alternative contrast agents, iodine and gadolinium. Indeed, one ferumoxide agent is now approved for *treating* anaemia[8] in late-stage CKD patients.

MPI experiments already demonstrate high speed, great sensitivity and excellent contrast. Spatial resolution is currently the only weak specification for MPI, and this is improving quickly with the advent of MPI-tailored USPIOs. We expect that human MPI scanners will soon become feasible after collaborative innovations in MPI hardware, MPI methodology and MPI-tailored magnetic nanoparticles.

20.2 OVERVIEW OF CURRENT ANGIOGRAPHY METHODS AND CONTRAST AGENTS

The Center for Disease Control ranks the three leading causes of death in the United States as follows: cardiovascular disease (25%), cancer (23%) and stroke (5.6%)[9]. The diagnosis and treatment planning for each of these diseases relies on angiographic imaging methods, including x-ray catheterised angiography, CT angiography (CTA) and MR angiography (MRA). In 2003, over 80 million intravascular contrast media injections were administered[2,10], dominated by x-ray angiography and CTA with iodinated contrast.

20.2.1 CATHETERISED X-RAY ANGIOGRAPHY

X-ray angiography has excellent in-plane spatial resolution. It uses an invasive catheterised arterial injection to boost the contrast bolus concentration about 50-fold higher than what would be provided by a safer venous contrast injection. This translates to an *enormous* gain in x-ray image contrast, since the attenuation scales *exponentially* with local contrast concentration. Unfortunately, x-ray angiography exposes both patient and medical staff to significant levels of ionising radiation. Despite these risks, x-ray angiography remains the standard for coronary angiography and is considered crucial for guiding interventional procedures in the x-ray fluoroscopy suite.

20.2.2 CT ANGIOGRAPHY

3D CT offers excellent through plane resolution (about 300 μm). The contrast is also excellent despite the intravenous iodine injection, because the 3D imaging format allows for separating blood vessels from background tissue. CTA poses no catheter risk, but it does subject the patient to a significant x-ray dose: 1 mSv in the head and up to 20 mSv for an abdominal study. CTA can be obscured by bone or calcium deposits, which mimic the iodine contrast agent.

20.2.3 Contrast MRA

3D MRI with intravenous gadolinium (MRA) can achieve 1 mm 3D isotropic spatial resolution. MRI is non-invasive, but it has poor contrast and sensitivity suffers from motion and fat suppression artefacts. The primary contrast agents, gadolinium chelates, are considered safe except for patients with weak kidney function.

20.2.4 Toxicity of Current Contrast Agents for Chronic Kidney Disease Patients

Iodine and gadolinium contrast agents significantly improve the conspicuity of blood vessels and tumours, but they are not without their own safety challenges. Physicians are especially concerned about the effects of iodine and gadolinium contrast agents on CKD patients. There are five stages of CKD, as measured by glomerular filtration rate (GFR)[11]; stage 5 patients need dialysis or transplantation to survive. CKD is especially prevalent in the elderly: 47% of American patients over 70 have CKD[11]. Because of the older age of CKD patients and co-morbidity with diabetes and heart disease, *25% of patients who present for an x-ray angiogram now have CKD*[12,13].

Unfortunately, both iodinated and gadolinium contrast agents are risky for CKD patients. Late-stage CKD patients have a 2%–8% risk of requiring dialysis[14] after a single iodine injection. The risk of contrast-induced nephropathy is considered increased and clinically important for stage 3 CKD or above, which includes many millions of patients[2]. Once nephropathy is induced by a contrast study, many patients are then permanently managed by kidney dialysis[15]. In patients requiring dialysis after x-ray angiography, studies have shown the 1 year mortality rate to be worse than 55%[16]. Gadolinium MRA is now contraindicated for late-stage CKD patients due to the risk of nephrogenic systemic fibrosis (NSF)[1,3,17,18]. Clearly, gadolinium and iodinated contrast should be replaced for late-stage CKD patients to prevent kidney failure in this patient population.

The MRI ferumoxide (SPIOs and USPIOs) contrast agents are considered much safer than gadolinium or iodine for CKD patients[19]. Three USPIO or SPIO contrast agents (AMI-121, OMP and AMI-25) are approved for human clinical studies[20]. These iron oxide nanoparticles are metabolised directly by the liver, rather than by the kidneys[21,22]. Consider also the evidence of USPIO safety in the CKD patient population from the recent Food and Drug Administration (FDA) approval of ferumoxytol (trade name Feraheme, AMAG Pharmaceuticals). Ferumoxytol is a USPIO agent approved by the FDA to *treat* a common complication of CKD patients, iron deficiency anaemia[8]. The FDA safety trial found intravenous ferumoxytol injection to be more effective for alleviating anaemia than oral iron, and, critically, that it was safe for CKD patients. The safety trial comprised N = 605 CKD patients (mostly stages 3 or 4, some in stage 5). The FDA found no loss of kidney function after treatment with the USPIO ferumoxytol at 510 mg dose (10 mmol Fe) injected twice daily, nearly 20 times higher concentration than normally used in MRI angiography. The recommended dose in a conventional MRI-USPIO angiogram is 0.5 mmol Fe, which translates to a bolus ranging from 3 mM to 100 µM in a typical human, with about 5 L of blood. By contrast, each injection in this anaemia trial was administered twice daily at 20 times higher dose: 10 mmol starts out at roughly 60 mM and settles to 2 mM after complete mixing in the 5 L of patient blood. Despite the higher dose, the adverse reactions to Feraheme were mild: nausea (3.1%), dizziness (2.6%) and hypotension (2.5%)[8]. Hence, there is exciting evidence that USPIOs are much safer than iodine or gadolinium for late-stage CKD patients.

In an MRI scan, USPIOs reduce both T_2^* and T_1, so this agent can produce either positive or negative contrast, depending on the concentration. T_1 reduction dominates at intermediate concentrations, leading to bright blood ('positive contrast'). However, at higher USPIO concentrations, T_2^* shortening dominates and the net effect is 'black blood' or negative contrast. For this and other reasons, USPIOs/SPIOs are not as popular as the dominant MRI contrast agent, gadolinium.

20.3   MAGNETIC PARTICLE IMAGING

MPI is a new medical imaging modality first published in 2005[4] that has tremendous promise as a vascular imaging modality. The first MPI conference was held in 2010[23], with just over 100 attendees. The key research groups involved in MPI include the Philips group in Hamburg, the University Luebeck, Dartmouth University Radiology, University of Washington, Case Western Reserve, Johns Hopkins, the University College of London and the University of California at Berkeley. Many MPI references can be found in the bionanomagnetics review article by Pankhurst et al.[24] For more information on subtopics in MPI, please see the following:

- Original MPI papers[4,25,26]
- Image reconstruction[27,28]
- Projection imaging field free line (FFL)[29–31]
- Temperature monitoring[32,33]
- Single-sided MPI[34]
- Relaxation theory[35–37]
- Narrowband MPI[7]
- Analysis of MPI and x-space MPI[5,6,38]
- Spectrometer[39,40]
- Nanoparticle optimisation[35,39,41–45]

20.3.1   MPI METHOD

USPIOs obey Langevin physics (see Figure 20.1), which means that all the electronic domains are completely aligned ('saturated') whenever the applied magnetic field magnitude is stronger than the saturation field. For the USPIO nanoparticles borrowed from the MRI field, the saturation field is weak (about 3200 A/m, equivalent to about 4.5 mT)[35,36], whereas the saturation magnetisation is very intense (0.6 T). Because MPI is performed within a blood vessel, the USPIOs are able to rotate in blood. Hence, the nanoscale magnetic moments align with the local magnetic field, following a short (tens of microseconds) magneto-viscous relaxation process. The intense saturation magnetisation means that MPI can detect even low-frequency motions of nanoscale magnetic moments at low concentration using simple inductive (Faraday) detection. Indeed, we have shown experimentally that it is possible to detect the motion of 65 ng of USPIOs moving at kilohertz rotation speeds with a simple litz wire detector coil.

The key imaging innovation from Gleich and Weizenecker[4] was to add a very strong magnetic field gradient to localise the magnetic signature from the USPIOs (shown in Figure 20.2). The 'field-free point' (FFP) region is where the field *magnitude* is weaker than the saturation field. All USPIOs

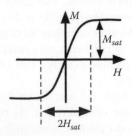

FIGURE 20.1 The contrast agent for MPI is a ferumoxide nanoparticle tracer, called USPIO agents. Ferumoxides obey Langevin physics magnetisation curve, as shown earlier. Typical values for MPI nanoparticles are as follows: $\mu_0 H_{sat} = 4.5\,\text{mT}$ and $\mu_0 M_{sat} = 0.6\,\text{T}$.

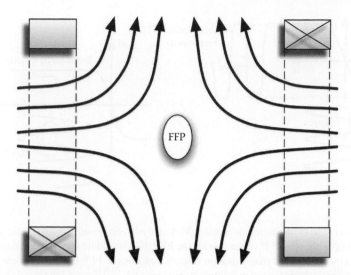

FIGURE 20.2 Gradient field. MPI relies on the axisymmetric field generated by a Maxwell gradient pair. Typically, in the range of 6 T/m, these gradients are often generated by PMs (NdFeB). The FFP is an elliptic region where the field magnitude is weaker than the saturation field of the nanoparticle.

outside the FFP are saturated, and all points inside the FFP are unsaturated. One can show that[5] the magnetic field created by a Maxwell gradient is

$$\vec{H}(x,y,z) = G_z\left(-\frac{1}{2}x\vec{i} - \frac{1}{2}y\vec{j} + z\vec{k}\right),$$

where
   x, y and z are the spatial coordinates
   $(\vec{i},\vec{j},\vec{k})$ represent the unit vectors in these coordinate directions

Hence, the magnitude of the gradient field is

$$|H(x,y,z)| = |G_z|\left(\frac{x^2}{4} + \frac{y^2}{4} + z^2\right)^{1/2}.$$

That is, the FFP is a 3D ellipse, and it is twice as broad in transverse (x and y) dimensions than in the axial (z) dimension.

20.3.2 MPI WITH HARMONIC DECOMPOSITION AND SYSTEM MATRIX RECONSTRUCTION

In Gleich and Weizenecker[4], the authors noted that the non-linear Langevin curve implies that a very low frequency (VLF) sinusoidal uniform field would generate harmonics. The key innovation was to localise those harmonics by adding an intense gradient field. This is illustrated in Figure 20.3.

One challenge with MPI is that direct feed through interference from the VLF transmit coil couples directly into the receiver coil. The direct feed through interference from the VLF transmit coil is several orders of magnitude stronger than the signal from USPIOs, but it is primarily isolated to the fundamental frequency of the VLF transmit tone. This enormous interference renders all lower frequency (fundamental and sometimes second harmonic) information contaminated. Fortunately, we can use frequency separation to remove the direct feed through interference, but this has the unwanted side effect of removing the low temporal frequencies of the USPIO signal. The non-linear

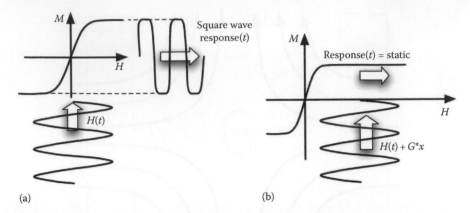

FIGURE 20.3 Illustration of the harmonic domain MPI analysis[5], contrasting the response to a sinusoidal $H(t)$ excitation waveform (a) within the FFP versus outside the FFP (b). The square wave response from within the FFP is rich in harmonic content. (b) The magnetisation response outside the FFP is static because the static gradient field offset keeps the nanoparticle frozen in saturation. Since the RF coil detects the rate of change of magnetisation, we receive zero voltage from nanoparticles outside the FFP and a rich spectrum of harmonics from within the FFP. Hence, the strong gradient localises the non-linear response of magnetic nanoparticles to the FFP.

response of the USPIOs creates a rich high-frequency signal, so the majority of MPI signal is well above the transmit frequency and is uncorrupted.

To reconstruct an image from MPI signal, a linear algebraic method was devised[27]. Here, the Philips group recommends a pre-scan calibration of impulse response images over the entire imaging field of view (FOV). They take the Fourier transform of the calibration signal at each location and for each impulse location to create a large dimensional harmonic 'system domain matrix'[27]. Then, subsequent imaging data are multiplied by the inverse of this system domain matrix to create an image.

There are some challenges with this image reconstruction scheme. First, the computation of the matrix inverse can be intractable for large matrix sizes. Moreover, the calibration scan must model the *in vivo* blood viscosity, but this varies between arterial and capillary blood. Hence, the *in vivo* viscosity is unlikely to be identical to the viscosity within the calibration phantom. This could introduce significant artefacts into the reconstructed images. Finally, the system matrix is often used to 'deconvolve' the blur due to the Langevin function. However, deconvolution is known to increase noise and has never been deemed reliable enough to improve spatial resolution of another clinical imaging modality, like positron emission tomography (PET). Hence, it would be preferred to avoid deconvolution by improving resolution with tailored nanoparticles and stronger imaging gradients.

20.3.3 x-Space Analysis of MPI

We have been pioneering an alternative method of analysing the MPI process called *x*-space analysis of MPI. Here, instead of analysing the Fourier harmonics of the MPI signal, we instead analyse the MPI time domain signal. We have shown that the instantaneous signal directly maps out the MPI native image. MPI occurs when we dynamically translate the FFP across the sample, either mechanically or electrically. Mechanical translation requires three axes of mechanical stages, and we simply translate the sample through the FFP or move the gradient across the sample. To electronically shift the FFP, we require three axes of homogeneous fields to create a 3D image. In *x*-space analysis, we consider the VLF transmit sinusoid to be just another field that shifts the instantaneous FFP. To see this, consider the sum of the gradient field and a three-axis homogeneous spatial shifting field:

$$\vec{H}(x,y,z,t) = \left(\left[-\frac{G_z}{2}x + H_x(t) \right]\vec{i} + \left[-\frac{G_z}{2}y + H_y(t) \right]\vec{j} + [G_z z + H_z(t)]\vec{k} \right).$$

Clearly, the instantaneous FFP is defined by the field magnitude to be at the location ($x_s(t)$, $y_s(t)$, $z_s(t)$):

$$|H(x,y,z)| = |G_z| \left(\frac{(x - x_s(t))^2}{4} + \frac{(y - y_s(t))^2}{4} + (z - z_s(t))^2 \right)^{1/2},$$

where we have defined the following:

$$x_s(t) \equiv \frac{2}{G_z} H_x(t),$$

$$y_s(t) \equiv \frac{2}{G_z} H_y(t),$$

$$z_s(t) \equiv \frac{-1}{G_z} H_z(t).$$

Note that this FFP translation is unique only within the region where the gradient field is linear and where the homogeneous fields are uniform.

As shown earlier, we can scan the FFP by applying a time-varying homogeneous shifting field or by mechanical translation. Note the field direction changes sign across the FFP in Figure 20.3. Scanning the gradient field across a pixel causes a 180° flip in the applied field direction at the FFP, as illustrated in Figure 20.4. The USPIO magnetisation remains locked to the applied field direction so it flips 180° as the FFP is scanned. This is detected as an electromagnetic force (EMF) in the receiver coil. Since we know the precise location of the FFP, it is simple to reconstruct the image by gridding the raw-induced MPI signal to the current FFP location in our reconstructed image. This is the conceptual basis for x-space MPI[5,6]. We have shown both theoretically and experimentally[5,6] that x-space reconstruction methods achieve precisely the *native* spatial resolution (the 'undeconvolved spatial resolution' from Rahmer et al.[27]).

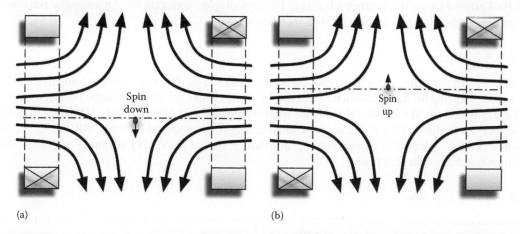

(a)   (b)

FIGURE 20.4   (a) Illustration of x-space MPI concept. Here, we have mechanically translated the FFP downwards from (a) to (b). The spin flips direction from 'down' in (a) to 'up' in (b) to remain locked to the local field direction. This 180° rotation is picked up as an EMF in the receiver coil. The voltage is directly proportional to the mass of USPIOs at the FFP location. To reconstruct an x-space image, we simply grid the instantaneous MPI signal to the current FFP location; this is far simpler and more robust than the harmonic domain system matrix reconstruction methods.

One advantage of the *x*-space analysis method is that it provides the first computationally trac-table image reconstruction method for large image matrices. There is also no calibration scan, no *a priori* assumptions about viscosity and no noise amplification due to deconvolution. Hence, *x*-space reconstruction provides a much more robust reconstruction method. We have recently shown that it is feasible to recover the lost low-frequency information[5] by enforcing boundary continuity when reconstructing overlapping partial field of view scans.

20.4   FUNDAMENTAL SENSITIVITY, CONTRAST, PENETRATION AND RESOLUTION

Here, we provide the theoretical limits to the fundamental imaging specifications for MPI.

20.4.1   Sensitivity Comparison to MRI

The USPIO moments saturate at 0.6 T magnetisation intensity. By comparison, the MRI nuclear paramagnetism of water is just 34 nT at 7.0 T[46]. Thus, MPI detects a magnetisation about 2–20 million times stronger than that of MRI. Also, the MPI relaxation time (about 20 μs) is much shorter than the T_1 recovery time of MRI (about 400 ms), which boosts signal to noise ratio (SNR) efficiency by the factor $\sqrt{400\,\text{ms}/20\,\text{microseconds}} \approx 141$. However, MPI loses three orders of magnitude sensitivity relative to MRI due to the loss of *k*-space averaging *x*-space. MPI also loses SNR by two additional factors of about 4 due to low-frequency detection and also due to the loss of the first harmonic signal. MPI nets a theoretical 10,000-fold gain in SNR efficiency over MRI. Since human tissue has about 50 M water and MRI has a high-resolution detection limit of about 500 mM water *in vivo*, one can predict that MPI scanners will ultimately be able to detect on the order of 50 μM USPIOs *in vivo*.

20.4.2   Contrast/Penetration

Living tissues are diamagnetic, and these tissues do not saturate during an MPI scan[46]. Hence, background tissue (even iron-rich blood) produces no MPI signal. Tissue is non-conductive and non-attenuating below 5 MHz. Hence, living tissue is transparent in MPI. The near-perfect suppression of background tissue is a distinct advantage for angiographic applications. An anatomic reference scan may be helpful, say, for cancer MPI applications.

20.4.3   Artefacts

Since we are scanning in *x*-space, motion artefacts should be far more benign compared to the coherent ghosting artefacts common to MRI and CT, where data are acquired in the Fourier domain. There may be several methods available for speeding up data acquisition. Unlike CT and MRI, MPI can image a partial FOV (e.g. the heart) without concerns about spatial aliasing. MRI and CT must sample k-space adequately to prevent spatial aliasing. Partial FOV MPI scanning will allow for dramatic reduction in scan time.

20.4.4   Resolution

As described in both *x*-space MPI and the original harmonic MPI method[6,27], the native (or 'undeconvolved') spatial resolution of MPI is

$$\Delta x \approx \frac{24 k_B T}{\pi G M_{sat} d^3}.$$

Here

d is the core diameter
M_{sat} is the saturation magnetisation in A/m
T is the absolute temperature
G is the gradient in A/m^2
k_B is Boltzmann's constant

Currently, with Resovist and 6 T/m gradients, we can expect about 1.5 mm spatial resolution[6,27]. While this is certainly adequate for human applications of MPI, it is not ideal for small animal imaging applications. We cannot improve the spatial resolution of MPI by temperature reduction (for obvious safety reasons) or by increasing the gradient strength, since 6 T/m is close to the physical limit of mouse-sized permanent magnet (PM) gradient coils. But we can improve spatial resolution eightfold in each spatial dimension simply by doubling the diameter, d, of the (single core) magnetic domain. In theory, 300 μm spatial resolution should be enabled by 30 nm particles[42,45].

20.5 HARDWARE: THE 3D BERKELEY MPI MOUSE SCANNER

In 2009, we constructed the 3D MPI mouse scanner shown in Figure 20.5[23]. The Berkeley 3D MPI mouse scanner is optimised for high resolution and high sensitivity. For high resolution, we built a 6.5 T/m PM gradient, as shown in Figure 20.5b. The system uses three 15 kW Copley audio amplifiers to electronically shift Helmholtz coils that scan the FFP with 80 mT fields. We detect the MPI signal with a noise-matched radio frequency (RF) receiver coil.

20.6 DATA FROM THE BERKELEY MPI SCANNERS

20.6.1 SENSITIVITY

Our Berkeley MPI scanner can now detect a miniscule sample of USPIOs (65 ng) with non-optimised electronics. This is equivalent to 1 mM USPIOs after complete mixing in blood. The Feraheme CKD anaemia treatment ranges from 60 mM during the initial first pass bolus and settles out at 2 mM. Therefore, we expect no problem with contrast to noise in MPI.

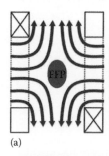

(a)　　　　　　(b)　　　　　　　　　(c)

FIGURE 20.5 Berkeley prototype MPI mouse scanner photos. (a) Magnetic field pattern created by an MPI gradient. USPIOs are saturated everywhere outside the FFP oval. We detect USPIOs within the FFP and scan the FFP to create an image. (b) 6.5 T/m NdFeB mouse gradient (9 cm free bore); (c) Completed Berkeley 3D mouse MPI scanner.

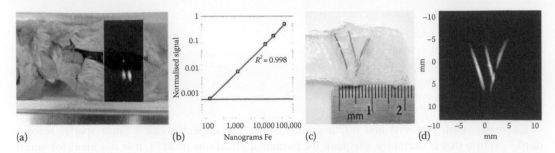

(a)   (b)   (c)   (d)

FIGURE 20.6 Experimental data from the Berkeley prototype MPI mouse scanner. Images from our first MPI scanner. (a) Seven minute MPI scan of 100 µg Fe injection into *ex vivo* mouse with photo overlay. (b) Demonstration of MPI signal linearity with iron mass. Flat line is the system noise floor (at 70 ng). (c) Photo of a vascular phantom with vessel diameter ID = 0.4 mm containing USPIO tracer. (d) Berkeley MPI angiogram of the vascular phantom showing 1.6 mm resolution (deconvolved) over 2.5 cm FOV. All magnets had modest 5% field homogeneity.

20.6.2 LINEARITY

Our preliminary data in Figure 20.6b show that the MPI signal varies linearly with the mass of USPIOs ($R^2 = 0.998$). This will be critical for quantitative MPI angiography. The sensitivity threshold in a mouse-sized RF coil is 65 ng now.

20.6.3 PHANTOM IMAGES

Consider also the preliminary *in vitro* Berkeley MPI scan in Figure 20.6c and d. Background suppression and spatial resolution (1.5 mm after deconvolution) are very encouraging. Also, magnetic fields were designed with modest 5% tolerances, so MPI will be far more robust to magnetic field disturbances (e.g. air, metal implants) than MRI.

20.6.4 PENETRATION AND CONTRAST

Figure 20.7 shows photographs of approximately 1 million cells labelled with USPIOs. While the labelled cells are no longer optically visible when buried in tissue, the MPI signal remains constant. This experimentally confirms that tissue is magnetically transparent and that it produces no MPI signal.

Figure 20.8 shows an *x*-space MPI scan complex phantom image of a 'UC' phantom. This image was acquired in just 18 s. The *x*-space MPI scan clearly shows that tubing had an air bubble (arrow); this is also seen in the digital photograph.

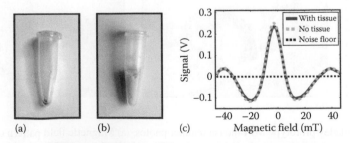

(a)   (b)   (c)

FIGURE 20.7 (a) Photograph of approximately 1 million cells labelled with USPIOs. (b) Photograph of the labelled cells after being mixed thoroughly with tissue. Note that the labelled cells are no longer visible since tissue is optically opaque. (c) The MPI signal is about 250 mV regardless of the presence of the covering tissue, experimentally confirming that tissue is magnetically transparent and that it produces no MPI signal.

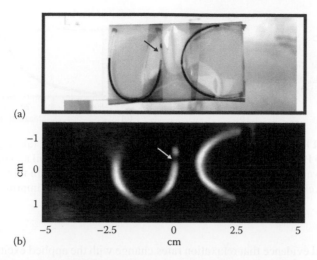

(a)

(b)

FIGURE 20.8 Preliminary x-space MPI scan of a complex phantom. (a) Photograph of a complex 2D phantom spelling 'UC'. (b) MPI of the phantom. We see no background signal, but just a little noise. This is ideal for angiography. Total acquisition time: 18 s, not including robot movement.

20.7 TAILORED MPI CONTRAST AGENTS, RELAXATION AND CHARACTERISATION

The best commercially available MPI particle is a tracer approved for human use in the European Union (Resovist), which has an estimated mean nanoparticle diameter of 16 nm. The current synthesis methods produce monodisperse nanoparticles for USPIOs under 20 nm diameter. But larger USPIOs may require new methods of synthesis; this is a crucial area of research for MPI. In theory, spatial resolution in MPI improves cubically in each dimension with increasing USPIO core diameter. See the simulated predictions in Figure 20.9.

A significant challenge with larger USPIOs is longer relaxation times. The important relaxation mechanisms are Brownian, Neel and magneto-viscous. These are governed by the core diameter and hydrodynamic diameter[37,41,43]. Slower relaxation rates blur the MPI image if care is not taken to mitigate their effects by tailoring scanning speed to the relaxation time.

MPI spectrometers measure the number of detectable harmonics[39] induced by a nanoparticle while being excited by 25 kHz magnetic field. While this may be suitable for harmonic space MPI, it is not adequate for x-space MPI. First, the spectrometers do not directly measure relaxation times, and 25 kHz may be simply too fast for the longer relaxation rates of larger nanoparticles.

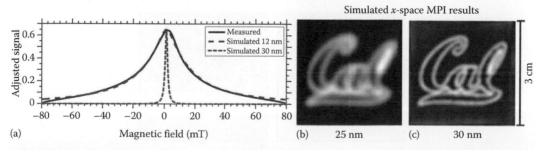

FIGURE 20.9 Resolution improvement with larger USPIO. (a) Comparison of measured and simulated PSF for a 12 nm particle. Note that theory matches the measured PSF and the finer resolution of 30 nm USPIO. (b and c) Simulated MPI scans of (b) 25 nm versus (c) 30 nm MNPs over a 3 cm FOV. This supports the prediction that 30 nm particles should give twice as fine resolution (300 mm) as the 25 nm MNPs.

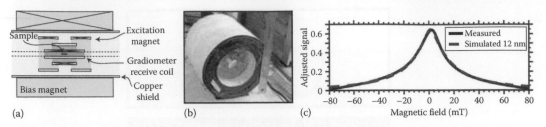

(a)   (b)   (c)

FIGURE 20.10 MPI Relaxometer measures PSF and relaxation of USPIOs. (a) Diagram of the Berkeley MPI Relaxometer. (b) Photograph of MPI Relaxometer showing outer bias coil and inner transmit/receive coil. (c) Representative PSF measured on our Relaxometer and compared with ferrofluid Langevin simulation. Notice the excellent fit. This analytic instrument will be critical for improving spatial resolution in MPI.

Finally, there is good evidence that relaxation rates change with the applied excitation field[37]. Hence, we constructed the Berkeley MPI Relaxometer, shown in Figure 20.10, to simultaneously measure both the 1D spatial resolution and relaxation times of USPIOs in the time domain. The MPI Relaxometer is essentially an x-space MPI scanner without a field gradient. The system has an excitation electromagnet (150 mT) and a variable constant bias field (80 mT). We measure the 1D point spread function (PSF) by scanning at 4.75 or 1 kHz. We estimate the relaxation time at both frequencies with standard data fitting methods. Figure 20.10 compares Relaxometer measurements with simulated PSFs for commercial USPIOs.

Ultimately, we hope to achieve 300 μm spatial resolution with a 6.0 T/m gradient by using larger (30 nm diameter) USPIOs in conjunction with innovations in scanning methods and hardware.

Our first experience with MPI-tailored nanoparticles is shown in Figure 20.11[41,43]. See also the very promising spatial resolution in Figure 20.12 of the University of Washington 20 nm USPIOs[41,43]. Here, the 8 mT magnetic field resolution translates to about 1.2 mm spatial resolution in a 6.5 T/m gradient field. While this is the finest native resolution we have seen in MPI, it falls a little short of our cubic scaling law for resolution with particle diameter. We suspect that relaxation effects are already a challenge at 20 nm because of the marked asymmetry of the measured PSF in Figure 20.12. Our preliminary results (Figure 20.12) indicate that we can expect a 20 μs relaxation time for 20 nm particle diameters and excitation strengths of 30 mT. A major future area of research is to tailor MPI scanning ('pulse sequences') to optimally trade-off spatial and temporal resolution to prevent relaxation blurring. We plan to mitigate relaxation blurring by scanning slower than 20 μs for each pixel in the image. This implies that a 192 × 256 image would require a 1 s frame rate.

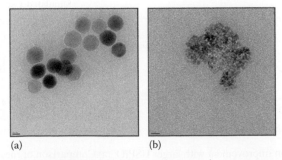

(a)   (b)

FIGURE 20.11 TEM images of USPIO nanoparticles. (a) TEM scans of monodisperse 20 nm particles synthesised for MPI by our University of Washington (UW) collaborators (Prof. Kannan N. Krishnan). (b) USPIO agent (Resovist) with diameter of 16 nm, which is the best commercially available agent now for MPI.

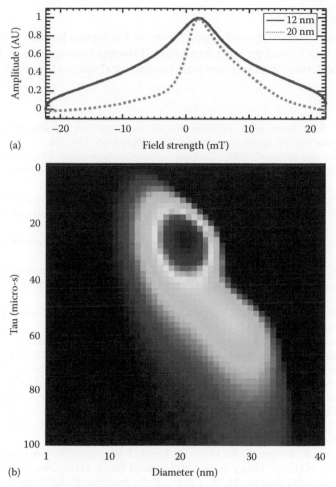

FIGURE 20.12 Preliminary Berkeley magnetic nanoparticle characterisation data. (a) Experimental comparison of the PSF for 12 and 20 nm USPIOs from Prof. Kannan N. Krishnan's lab at UW. The 20 nm particle has finer resolution. It also displays relaxation asymmetry. We expect larger USPIOs will obtain 300 micron native resolution, a breakthrough in MPI. (b) 2D relaxogram for the 20 nm USPIO from UW. The relaxogram measures MPI particles diameter and relaxation time constants.

20.8 CONCLUSION

MPI shows enormous potential for a new angiography method with greatly reduced risk of CKD patients developing life-threatening nephropathy from iodine from diagnostic x-ray and CT examinations. With unprecedented contrast to noise, sensitivity, safety and robustness, the only remaining technical challenge is to improve the spatial resolution of MPI and translate it to human scale scanners. We believe that a concerted effort in tailored nanoparticle design and imaging methodology will improve spatial resolution to 300 µm resolution MPI at 2–4 frames/s *in vivo*. Human scale MPI may require a superconducting 5 T/m gradient, but costs should be comparable to a low-field whole-body MRI scanner.

Once MPI hardware and nanoparticles are optimised, we believe that applications that require linearity and high contrast, including stem cell tracking, cancer imaging and inflammation imaging, will become critical new avenues of research.

MPI angiography requires innovations in MPI scanner hardware, data acquisition methods, tailored nanoparticle design and reconstruction algorithms. Hence, collaborative research between magnetic nanoparticle researchers, imaging scientists and physicians will be essential to achieve high-quality angiography with near-zero risk of contrast-induced nephropathy for CKD patients.

ACKNOWLEDGEMENTS

We thankfully acknowledge the research support from the Californian Institute of Regenerative Medicine (CIRM) Tools and Technology grant programme (Principal Investigator S.M. Conolly) and from the University of California (UC) Discovery grant programme (Principal Investigator S.M. Conolly).

REFERENCES

1. Broome, D. R. 2008. Nephrogenic systemic fibrosis associated with gadolinium based contrast agents: A summary of the medical literature reporting. *Eur J Radiol*, 66, 230–234.
2. Katzberg, R. W. and Haller, C. 2006. Contrast-induced nephrotoxicity: Clinical landscape. *Kidney Int Suppl*, 69, S3–S7. doi:10.1038/sj.ki.500366. http://www.nature.com/ki/journal/V69/n100s/full/5000366a.html
3. Mccullough, P. A. 2008. Contrast-induced acute kidney injury. *J Am Coll Cardiol*, 51, 1419–1428.
4. Gleich, B. and Weizenecker, J. 2005. Tomographic imaging using the nonlinear response of magnetic particles. *Nature*, 435, 1214–1217.
5. Goodwill, P. and Conolly, S. 2011. Multi-dimensional x-space magnetic particle imaging. *IEEE Trans Med Imaging*, 30, 1581–1590.
6. Goodwill, P. W. and Conolly, S. M. 2010. The X-space formulation of the magnetic particle imaging process: 1-D signal, resolution, bandwidth, SNR, SAR, and magnetostimulation. *IEEE Trans Med Imaging*, 29, 1851–1859.
7. Goodwill, P. W., Scott, G. C., Stang, P. P. and Conolly, S. M. 2009. Narrowband magnetic particle imaging. *IEEE Trans Med Imaging*, 28, 1231–1237.
8. Lu, M., Cohen, M. H., Rieves, D. and Pazdur, R. 2010. FDA report: Ferumoxytol for intravenous iron therapy in adult patients with chronic kidney disease. *Am J Hematol*, 85, 315–319.
9. Xu, J., Kochanek, K., Murphy, S. and Tejada-Vera, B. 2010. Deaths: Final data for 2007. *CDC: National vital statistics reports*, 58. U.S. Department of Health and Human Services, Centers for Disease Control and Prevention National Center for Health Statistics, National Vital Statistics System. Available online at: http://www.cdc.gov/nchs/data/nvss/nvss59/nvss59_02.pdf
10. Rudnick, M. R., Goldfarb, S. and Tumlin, J. 2008. Contrast-induced nephropathy: Is the picture any clearer? *Clin J Am Soc Nephrol*, 3, 261–262.
11. Coresh, J., Selvin, E., Stevens, L. A., Manzi, J., Kusek, J. W., Eggers, P., Van Lente, F. and Levey, A. S. 2007. Prevalence of chronic kidney disease in the United States. *JAMA*, 298, 2038–2047.
12. Ix, J. H., Mercado, N., Shlipak, M. G., Lemos, P. A., Boersma, E., Lindeboom, W., O'neill, W. W., Wijns, W. and Serruys, P. W. 2005. Association of chronic kidney disease with clinical outcomes after coronary revascularization: The Arterial Revascularization Therapies Study (ARTS). *Am Heart J*, 149, 512–519.
13. Reddan, D. N., Szczech, L. A., Tuttle, R. H., Shaw, L. K., Jones, R. H., Schwab, S. J., Smith, M. S., Califf, R. M., Mark, D. B. and Owen, W. F., Jr. 2003. Chronic kidney disease, mortality, and treatment strategies among patients with clinically significant coronary artery disease. *J Am Soc Nephrol*, 14, 2373–2380.
14. Goldfarb, S., Mccullough, P. A., Mcdermott, J. and Gay, S. B. 2009. Contrast-induced acute kidney injury: Specialty-specific protocols for interventional radiology, diagnostic computed tomography radiology, and interventional cardiology. *Mayo Clin Proc*, 84, 170–179.
15. Mccullough, P. A., Wolyn, R., Rocher, L. L., Levin, R. N. and O'neill, W. W. 1997. Acute renal failure after coronary intervention: Incidence, risk factors, and relationship to mortality. *Am J Med*, 103, 368–375.
16. Mccullough, P. 2006. Outcomes of contrast-induced nephropathy: Experience in patients undergoing cardiovascular intervention. *Catheter Cardiovasc Interv*, 67, 335–343.
17. Collidge, T. A., Thomson, P. C., Mark, P. B., Traynor, J. P., Jardine, A. G., Morris, S. T. W., Simpson, K. and Roditi, G. H. 2007. Gadolinium-enhanced MR imaging and nephrogenic systemic fibrosis: Retrospective study of a renal replacement therapy cohort. *Radiology*, 245, 168–175.
18. Sadowski, E. A., Bennett, L. K., Chan, M. R., Wentland, A. L., Garrett, A. L., Garrett, R. W. and Djamali, A. 2007. Nephrogenic systemic fibrosis: Risk factors and incidence estimation. *Radiology*, 243, 148–157.
19. Neuwelt, E. A., Hamilton, B. E., Varallyay, C. G., Rooney, W. R., Edelman, R. D., Jacobs, P. M. and Watnick, S. G. 2009. Ultrasmall superparamagnetic iron oxides (USPIOs): A future alternative magnetic resonance (MR) contrast agent for patients at risk for nephrogenic systemic fibrosis (NSF)? *Kidney Int*, 75, 465–474.

20. Laconte, L., Nitin, N. and Bao, G. 2005. Magnetic nanoparticle probes. *Mater Today*, 8, 32–38.
21. Ferrucci, J. T. and Stark, D. D. 1990. Iron oxide-enhanced MR imaging of the liver and spleen: Review of the first 5 years. *AJR Am J Roentgenol*, 155, 943–950.
22. Weissleder, R., Stark, D. D., Engelstad, B. L., Bacon, B. R., Compton, C. C., White, D. L., Jacobs, P. and Lewis, J. 1989. Superparamagnetic iron oxide: Pharmacokinetics and toxicity. *AJR Am J Roentgenol*, 152, 167–173.
23. Goodwill, P. W., Scott, G. C., Stang, P. P. and Conolly, S. M. 2010. Narrowband magnetic particle imaging in a mouse. *World Scientific Publishers: Proceedings of the First International Workshop on Magnetic Particle Imaging*, Singapore.
24. Pankhurst, Q. A., Thanh, N. K. T., Jones, S. K. and Dobson, J. 2009. Progress in applications of magnetic nanoparticles in biomedicine. *J Phys D: Appl Phys*, 42, 224001.
25. Gleich, B., Weizenecker, J. and Borgert, J. 2008. Experimental results on fast 2D-encoded magnetic particle imaging. *Phys Med Biol*, 53, N81–N84.
26. Markov, D. E., Boeve, H., Gleich, B., Borgert, J., Antonelli, A., Sfara, C. and Magnani, M. 2010. Human erythrocytes as nanoparticle carriers for magnetic particle imaging. *Phys Med Biol*, 55, 6461–6473.
27. Rahmer, J. R., Weizenecker, J. R., Gleich, B. and Borgert, J. R. 2009. Signal encoding in magnetic particle imaging: Properties of the system function. *BMC Med Imaging*, 9, 4.
28. Weizenecker, J., Borgert, J. and Gleich, B. 2007. A simulation study on the resolution and sensitivity of magnetic particle imaging. *Phys Med Biol*, 52, 6363–6374.
29. Knopp, T., Erbe, M., Biederer, S., Sattel, T. F. and Buzug, T. M. 2010. Efficient generation of a magnetic field-free line. *Med Phys*, 37, 3538–3540.
30. Knopp, T., Sattel, T. F., Biederer, S. and Buzug, T. M. 2010. Field-free line formation in a magnetic field. *J Phys A Math Theor*, 43, 012002.
31. Weizenecker, J., Gleich, B. and Borgert, J. 2008. Magnetic particle imaging using a field free line. *J Phys D Appl Phys*, 41, 105009.
32. Weaver, J. B., Rauwerdink, A. M. and Hansen, E. W. 2009. Magnetic nanoparticle temperature estimation. *Med Phys*, 36, 1822–1829.
33. Weaver, J. B., Rauwerdink, A. M., Sullivan, C. R. and Baker, I. 2008. Frequency distribution of the nanoparticle magnetization in the presence of a static as well as a harmonic magnetic field. *Med Phys*, 35, 1988–1994.
34. Timo, F. S., Tobias, K., Sven, B., Bernhard, G., Juergen, W., Joern, B. and Thorsten, B. 2009. Single-sided device for magnetic particle imaging. *J Phys D: Appl Phys*, 42, 022001.
35. Fannin, P. C. and Charles, S. W. 1989. The study of a ferrofluid exhibiting both Brownian and Neel relaxation. *J Phys D: Appl Phys*, 22, 187–191.
36. Kotitz, R., Fannin, P. C. and Trahms, L. 1995. Time-domain study of Brownian and Neel relaxation in ferrofluids. *J Magn Magn Mater*, 149, 42–46.
37. Shliomis, M. I. 1974. Magnetic liquids. *Uspekhi Fizicheskikh Nauk*, 112, 427–458.
38. Knopp, T., Biederer, S., Sattel, T., Weizenecker, J., Gleich, B., Borgert, J. and Buzug, T. M. 2009. Trajectory analysis for magnetic particle imaging. *Phys Med Biol*, 54, 385–397.
39. Biederer, S., Knopp, T., Sattel, T. F., Lüdtke-Buzug, K., Gleich, B., Weizenecker, J., Borgert, J. and Buzug, T. M. 2009. Magnetization response spectroscopy of superparamagnetic nanoparticles for magnetic particle imaging. *J Phys D: Appl Phys*, 42, 205007.
40. Biederer, S., Sattel, T., Knopp, T., Gleich, B., Weizenecker, J., Borgert, J. and Buzug, T. M. 2009. A spectrometer for magnetic particle imaging. *Imaging*, 22, 2313–2316.
41. Ferguson, R. M., Minard, K. R., Khandhar, A. P. and Krishnan, K. M. 2011. Optimizing magnetite nanoparticles for mass sensitivity in magnetic particle imaging. *Med Phys*, 38, 1619–1626.
42. Ferguson, R. M., Minard, K. R. and Krishnan, K. M. 2009. Optimization of nanoparticle core size for magnetic particle imaging. *J Magn Magn Mater*, 321, 1548–1551.
43. Krishnan, K. M. 2010. Biomedical nanomagnetics: A spin through possibilities in imaging, diagnostics, and therapy. *IEEE Trans Magn*, 46, 2523–2558.
44. Sānchez, J. H. and Rinaldi, C. 2009. Rotational Brownian dynamics simulations of non-interacting magnetized ellipsoidal particles in dc and ac magnetic fields. *J Magn Magn Mater*, 321, 2985–2991.
45. Yu, W. W., Falkner, J. C., Yavuz, C. T. and Colvin, V. L. 2004. Synthesis of monodisperse iron oxide nanocrystals by thermal decomposition of iron carboxylate salts. *Chem Commun*, 2306–2307.
46. Schenck, J. F. 1996. The role of magnetic susceptibility in magnetic resonance imaging: MRI magnetic compatibility of the first and second kinds. *Med Phys*, 23, 815–850.

Dr. Patrick Goodwill is a research associate and MPI project leader at the UC Berkeley Imaging Systems Laboratory. He works on MPI theory, imager design and imager construction. For his work on MPI, he was awarded the Young Investigator Award at the Society for Molecular Imaging in 2009. He received his PhD at UC Berkeley in Bioengineering under the mentorship of Prof. Steven Conolly and holds a BS and MS in Electrical Engineering from Stanford University.

Professor Kannan M. Krishnan works at the bio-nano-medical and magneto-electronic interfaces. He studies structural, magnetic, optical and transport properties of colloidal inorganic nanostructures, hybrid materials, thin films and lithographically patterned heterostructures. Prof. Krishnan was educated at IIT, Kanpur (B. Tech, 1978); SUNY, Stony Brook (MS, 1980); and UC, Berkeley (PhD, 1984). After holding various scientific and teaching positions at Lawrence Berkeley National Laboratory and UC, Berkeley, he joined the University of Washington, in 2001, as the Campbell chair professor of Materials Science and adjunct professor of Physics. Recently, along with two graduate students, he has founded a company, LodeSpin Labs, involved in the development of tailored magnetic carriers for a range of biomedical applications, including MPI.

Professor Steven M. Conolly focuses on the hardware, nanoparticle physics, systems theory and reconstruction algorithms for a brand new medical imaging modality called MPI, which has significant promise as a less invasive alternative for x-ray and CTA. He has overseen the construction of three MRI scanners at Stanford and four MPI scanners at UC Berkeley. He is also developing a novel susceptibility matching agent to improve fat suppression in conventional MRI. He has more than 25 U.S. patents in various stages of approval. He is a full professor in engineering at UC Berkeley, with appointments in both Bioengineering and Electrical Engineering and Computer Sciences.

21 Surgical Magnetic Systems and Tracers for Cancer Staging

Eric Mayes, Michael Douek, and Quentin Pankhurst

CONTENTS

While the previous chapters on *in vivo* applications of magnetic nanoparticles (MNPs) have focussed primarily on imaging and drug delivery, this chapter presents an application that is neither diagnostic nor medicinal. Despite being comparatively simple, magnetically marking internal anatomical sites or structures for external localisation is a valuable procedure that can improve the availability and economics of cancer staging.

21.1 SENTINEL LYMPH NODE BIOPSY IN CANCER STAGING

21.1.1 BACKGROUND

Cancer is a leading cause of death worldwide, accounting for around 13% of all deaths (7.6 million people) in 2008[1]. Deaths from cancer worldwide are projected to rise to over 11 million by 2030. While medicine has made significant strides in the detection and treatment of cancer, continued efforts to detect, treat and prevent cancer are essential. In view of the rising number of patients involved, decreasing costs and increasing the availability of care are additional important objectives. One important element of this care is the process of cancer staging. The stage of a cancer indicates its spread, from local tumour confinement to adjacent cells, tissues or lymph nodes, or to multiple tumours at distant sites. Cancer staging is not only a key predictor of survival, but also critical for establishing the sequence of additional treatment required.

Some important solid tumours, such as breast cancer, spread predominantly *via* the lymphatic system. When the cancer spreads, its cells migrate from the tumour and are carried away by the interstitial fluid in the lymphatic system[2]. In most cases, these cancer cells become trapped in one or more of the 500–600 or so lymph nodes that are distributed around the lymphatic system throughout the body. Lymph nodes are bean-shaped organs, ranging from a few millimetres to 1–2 cm in size, which act as filters to clean the lymph as it passes through.

The concept of a 'sentinel node' – a lymph node that receives the first lymphatic drainage from a tumour – dates to 1977, when Cabanas proposed that the lymphatic system was sequentially involved with cancer spreading from a primary tumour while studying patients with penile carcinoma[3]. Cabanas observed across a series of patients that a specific lymph node location was the preferential drainage area for squamous carcinoma cells in the penis. Moreover, he observed that this specific node was the first point of metastasis localisation, but that metastases continued their passage to subsequent nodes along the lymphatic drainage network. He proposed that the status of the entire lymphatic system could be determined by the localisation, dissection and histological examination of this 'sentinel lymph node' (SLN) – a procedure that became known as sentinel lymph node biopsy (SLNB).

While SLNB has been investigated as a technique for staging a variety of cancers, its implementation in the staging and treatment of breast cancer has been the most successful to date. In 1994, Giuliano et al.[4] published the first report on studies using a blue dye injected near a tumour to map the lymphatic drainage pathway to the SLN (Figure 21.1). This was extended shortly thereafter by Krag et al.[5], who used a radiolabelled colloid in conjunction with a handheld gamma-ray detector to identify the SLN pre-incision.

Prior to the development of SLNB, axillary lymph node dissection (ALND) was the only method for staging breast cancer[6]. In ALND, all the lymph nodes (around 30) in the regional basin in the axilla are removed for microscopic histology to determine whether cancer has spread. While this procedure is effective in determining whether cancer has spread to the lymphatic system, it has a number of unwanted side effects. In particular, the most common side effect is lymphoedema, or the permanent swelling of the upper limb due to its inability to process and drain lymph once the local lymph nodes are removed. Other long-term side effects include the potential for nerve damage or reduced mobility in the shoulder or arm.

Intra-operative SLNB addresses these issues and has the additional advantage of reducing hospital stay and improving the speed of recovery. By removing the sentinel node(s), through a smaller incision, clinicians can readily determine whether cancer has spread and whether further surgery (such as ALND), post-operative chemotherapy or radiotherapy is required. This minimally invasive

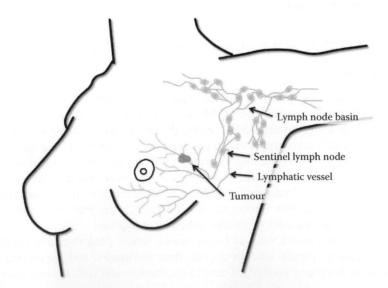

FIGURE 21.1 Lymphatic system draining the breast, indicating tumour and sentinel lymph node.

procedure retains the bulk of the local lymphatic system and so avoids the additional side effects of ALND in patients who are found to be sentinel node negative.

Between 1999 and 2004, Krag et al. enrolled 5611 patients for a randomised control trial across 80 sites in Canada and the United States. Of these, 3989 women had pathologically negative SLNs. The study published in 2010 confirmed that the disease-free and overall survival rates were equivalent between ALND and SLNB patient groups, confirming SLNB as the standard of care for breast cancer patients with clinically negative lymph nodes[7]. More recently, Giuliano et al.[8] published the results of the American College of Surgeons Oncology Group (ACOSOG) Z0011 randomised trial across 115 sites with 891 sentinel node positive patients. The trial did not reach the planned recruitment of 1900 patients since it was closed early (mortality rate was lower than expected). The authors concluded that among patients with limited sentinel node metastatic involvement, who also received chemotherapy, SLNB alone compared to ALND did not result in an inferior survival. This result suggests that axillary surgery is mainly a diagnostic procedure with a more limited therapeutic benefit.

Globally, 1.4 million new cases of breast cancer are diagnosed each year, making it the most frequent cancer among women[1]. Despite favourable trends in survival rates due to incremental improvements in systemic treatment and the introduction of mammographic screening in the West, it is still the most frequent cause of cancer death in both developed and developing regions. Increasing the availability of the standard of care of SLNB is an important element in addressing this issue.

21.1.2 CURRENT PRACTICE

The SLNB protocol outlined in Figure 21.2 generally involves injecting a technetium–sulphur colloid ([^{99m}Tc]-TSC) alongside isosulfan blue (N-[4-[[4-(diethylamino)phenyl] (2,5-disulphophenyl) methylene]-2,5-cyclohexadien-1-ylidene]-N-ethylethanaminium hydroxide), the so-called combined technique, interstitially into the breast around the tumour[9]. In the United Kingdom and some other countries, Patent Blue (Guerbet, Paris) is used instead of isosulfan blue. After allowing for the materials to localise in the lymphatic system, the clinician uses a handheld scintillation counter (a 'gamma probe') to locate the node or nodes receiving primary drainage from the tumour *via* lymphatic vessels. The blue dye assists in localisation post-incision, and lymph nodes that are radioactive, blue, or both, are judged to be 'SLNs'. The SLNs are then carefully removed and sent for microscopic histological examination. Most often, the pathologist finds that cancer has not spread to the lymph nodes, in which case no further axillary procedure is required. In broad clinical practice, the detection rate of the technique is generally accepted to be over 90%–95%, where 5%–10% of the patients have false-negative SLNB[10,11].

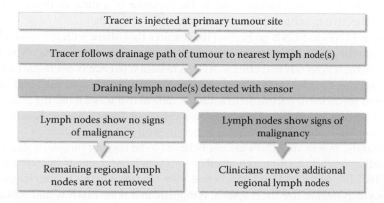

FIGURE 21.2 Procedural sequence for sentinel lymph node biopsy using a generic tracer (radioisotope or MNP).

While effective, this procedure has a number of limitations. [99m]Tc has a 6 h half-life, which limits its availability to proximal centres that handle its parent isotope of molybdenum-99. [99]Mo decays so quickly that it must be supplied to hospital nuclear medicine departments every other week, and it is made in just a few nuclear reactors worldwide. [99]Mo is a by-product of nuclear fission and, as such, has been subject to interruptions during refurbishments of the older nuclear reactors that supply it. Moreover, the global perception of nuclear reactors, sensitised by the Fukushima nuclear accident of March 2011, may further impact the future supply chain.

For hospitals without ready access to radioisotopes, their use presents administrators with an expensive logistical burden, requiring time and resource to be devoted to nuclear medicine in what is otherwise a routine operation. In addition, mandatory handling and waste disposal regulations add to the overhead, as does the training and licensing of operating theatre staff in the handling of radioactive materials. However, for hospitals with already established nuclear medicine facilities, these issues are less problematic as radioisotopes serve other procedures. In both cases, the 6 h half-life of the isotope presents challenges and limitations for theatre scheduling, especially since the injection is generally undertaken in the nuclear medicine department and not by surgeons. Finally the patients themselves express reluctance in relation to the use for radiation, particularly for diagnostic use.

These factors are a significant barrier to the widespread adoption of SLNB, and in hospitals and clinics without ready access to radioisotopes, it is sometimes simply not performed at all[12]. Consequently, the majority of women with breast cancer will not be offered SLNB treatment based on this method. While the incidence of cancer is growing, the SLNB procedure has plateaued with only around 60% of estimated 500,000 patients in the West having access to the procedure[13]. This figure drops to 5% in China, and is minimal in most of the rest of the world[14].

21.1.3 BENEFITS OF MNPs

MNPs, in particular iron oxide nanoparticles, exhibit the potential to replace the 'combined technique' in SLNB, as they share some features: (a) they can be externally detected in tissue, pre-incision; (b) they are of similar dimensions to the radiotracer colloid and (c) their brown-black appearance, at concentration, acts as a visual stain.

MNPs have been under clinical investigation and development as diagnostic and imaging agents for almost 40 years[15]. Recent clinical focus is on iron oxides such as maghemite (γFe_2O_3) and magnetite (Fe_3O_4) coated in a biocompatible molecule such as dextran. Below approximately 30 nm in diameter, these nanoparticles exhibit superparamagnetic behaviour, characterised by a response to an external magnetic field while retaining no magnetic remanence in its absence. This behaviour makes them ideal for SLNB, as they are prevented from agglomerating while being transported *via* lymph in the absence of an external field. In the presence of a static or dynamic external field, however, their collective moment can be sensed external to the body.

The benefits of MNPs in SLNB go beyond the features they share with the combined technique, however. In particular, most MNPs have a shelf life of several years, allowing them to be shipped globally to the furthest hospitals and clinics. Also, a radiation licence is not required to perform the injection, so surgeons have the freedom to perform SLNB on their schedule, and independent of that of the nuclear medicine department. In addition, radioactive waste handling issues are obviated. The scheduling and waste handling benefits have a corresponding economic benefit to the hospital or clinic, and the patients benefit by avoiding concerns with additional radiation.

Most importantly, iron oxide MNPs are not believed to be toxic or otherwise dangerous in clinical use. Depending on their coating and whether injection is intravenous or interstitial, iron oxide MNPs have a half-life between 1 and 36 h before they are taken up by macrophages in the mononuclear phagocyte system of the liver, spleen, lymphatic and bone marrow and broken down to be distributed across iron stores in the body[15]. The MNPs in clinical use have been employed traditionally as magnetic resonance imaging (MRI) contrast agents, with dosages of iron oxide around

25–100 mg. Such a dosage is roughly equivalent to 2–5 days of normal dietary intake of iron, which results in transient changes in serum iron, ferritin and iron-biding capacity, but does not raise a risk of iron overload. Studies on the toxic effects from iron oxide MNPs at increased dosages confirmed no acute or subacute toxic effects in rats or beagles that received 150 times standard dosage for MRI of the liver[16].

Although the MNP coating could also be a source of potential toxicity, most iron oxide MNPs in clinical use today are coated with dextran. Some individuals express anti-dextran antibodies, indicating that administration of such particles is a potential risk. However, no anaphylactic reactions have been reported for these materials[15]. Other coatings, such as carboxydextran, have been shown similarly to be safe and well tolerated at clinical doses[17].

By addressing the barriers to widespread adoption raised by the 'combined technique', MNPs offer a compelling SLNB alternative to radioisotopes and could change the face of cancer staging.

21.2   DEVICES FOR LYMPHATIC MAPPING WITH *IN VIVO* MNPs

21.2.1   LYMPHATIC SYSTEM

Lymph is the fluid that is formed when interstitial fluid enters the initial vessels of the lymphatic system. The lymphatic system is a part of the immune system that consists of a network of vessels that drain lymph into lymph nodes by either intrinsic contractions of the lymphatic passages or by extrinsic compression of the lymphatic vessels *via* external tissue forces (e.g. the contractions of skeletal muscles). The lymph nodes act as a type of filter removing foreign agents from the lymph, such as cancer cells or particles, but nodes also produce specialised white blood cells to fight agents such as bacteria or viruses. Foreign agents in the lymph enter the lymph node *via* the afferent lymphatic vessels, draining into each node through its convex surface as indicated in Figure 21.3.

The lymph node is surrounded by a fibrous capsule that partitions the interior into cortical lobes. Within each cortex is a meshwork of reticular fibres that supports the adhesion of macrophages, lymphocytes and dendritic cells in the germinal centre. Within this centre, a response to immunogens from bacteria or viruses leads to the production of specialised white blood cells that fight and process these foreign agents. Foreign agents without an immunological response pass through unhindered. After lymph flows from the afferent lymphatic vessels into the cortex to be processed

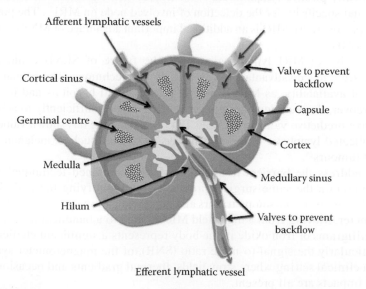

FIGURE 21.3   Simplified schematic anatomy of a lymph node (less blood vessels).

by the white blood cells, it then flows through cortical sinuses into medullary sinuses where the processed lymph drains alongside specialised white blood cells *via* the efferent lymphatic vessel emerging from the hilum. Larger particles will generally become trapped within the sinuses where macrophage pseudopods create a meshwork filter across the space.

21.2.2 LYMPHATIC MAPPING WITH MNPs

Interstitially injected particles exhibit differing behaviour in the lymphatic system depending on their size. For particles smaller than a few nanometres in diameter, they are generally exchanged through blood capillaries. At tens of nanometres in diameter, particles are absorbed into lymph capillaries and ultimately trapped in lymph node sinuses. At a few hundreds of nanometres or larger, particles become trapped in interstitial space or do not migrate at all[18]. Therefore, an ideal size for an SLNB tracer is with a diameter in the tens of nanometres.

The typical hydrodynamic diameter for interstitially injected ^{99m}Tc colloids is a few tens of nanometres, though there is variability in their production between different nuclear medicine departments[18]. Most clinically available MNPs meet this size criterion, allowing them to be used as markers for SLN imaging and localisation. Indeed, such MNPs are currently being explored for the imaging and diagnosis of lymph node metastases using MRI[19–26]. As opposed to SLNB, the concept of lymphangiography is to confirm nodal status without an operational intervention or subsequent histology.

MNPs have been traditionally used as contrast agents for MRI, as they alter magnetic field gradients therefore altering proton relaxation in the imaged tissue. Conventionally, iron oxide MNPs used as contrast agents are injected intravenously, which allows them to access nodes by moving into the medullary sinuses of the node *via* direct transcapillary passage, as well as directly accessing nodes *via* lymphatic flow as with interstitial injection. MNPs transported to the SLN act as 'negative' imaging contrast agents when using T_2- and T_2*-weighted pulse sequences, and have been shown to identify metastases independently of lymph node size[19]. As MNPs are filtered and taken up by macrophages, healthy lymph node tissue produces a dark signal using a T_2/T_2* sequence. If the lymph node is metastatic, the node will show up as bright (or partially bright), as the unhealthy node lacks macrophage activity. This technique allows for a non-invasive method of imaging the complete lymphatic drainage pathway from the tumour, and determining the involvement and location of any clinically positive lymph nodes[20–26]. MNPs, when injected intravenously, provide the best sensitivity and specificity for the detection of involved node on MRI[27]. The potential of being able to perform pre-operative MRI is an additional important advantage of MNPs over conventional techniques for SLNB.

With the promise of MRI lymphangiography, the future of SLNB could be questioned. However, there are a few disadvantages presented by this technique. The primary disadvantage returns to that of availability, as MRI is not yet present in all hospitals and it is operationally expensive. Moreover, the technique has not yet been developed sufficiently to achieve the specificity and positive predictive value of SLNB[24]. But this technique has shown value in elucidating the more complicated lymphatic drainage pathways in cancers such as melanoma, prostate and gynaecological tumours[25].

In order to address the barriers to adoption of the 'combined technique' in SLNB, it is preferable to maintain the same surgical protocol while modifying the underlying detection principle – replacing the radioisotope tracers and handheld gamma probe with MNPs and handheld magnetometer. Moving from a high-field MRI system to a handheld magnetometer for the detection of milligrams of iron oxide in the body represents a significant challenge. The sensitivity, and particularly the signal-to-noise ratio (SNR), of the magnetometer system is critical when used in a clinical setting where stray fields, thermal gradients and occasional mechanical vibrations and impacts are all present.

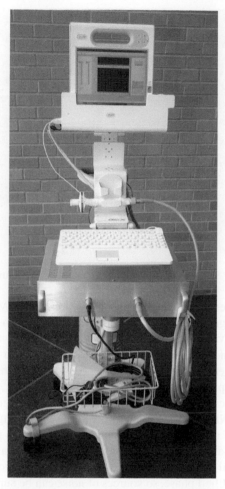

FIGURE 21.4 SQUID-based SentiMag – a prototype device, based on a cryogenically cooled superconducting magnetometer. (Courtesy of Prof. Audrius Brazdeikis, Texas Center for Superconductivity, University of Houston, Houston, TX.)

The first generation of magnetometers developed for SLNB employed cryogenically cooled superconducting quantum interference devices (SQUIDs)[28–31]. Tanaka et al.[28] performed a benchtop demonstration of the potential to use a high-T_c SQUID magnetometer to detect 1.6 μg of iron oxide MNPs at a distance of 40 mm. Joshi et al. developed an alternative SQUID design that demonstrated similar sensitivity, and used it clinically with patients that had been injected with the MNP, Endorem™ (Guerbet, Paris)[29–31]. This system was used to deliver the clinical results presented in Section 21.3 and was the prototype for a device under commercial development called the 'SentiMag', as shown in Figure 21.4. In both systems, a modulated alternating magnetic field is applied to the MNPs, while the SQUID sensor demodulates the induced magnetisation in the sample by using a lock-in amplifier.

While the SQUID-based systems demonstrated adequate sensitivity and SNR, their reliance on cryogenic liquids limited their utility and robustness. In moving to room-temperature sensors, however, a greater burden is placed on optimising SNR. Minamiya et al. are developing a magnetometer system that uses an uncooled Hall or magnetoresistive sensor, and they have successfully demonstrated its use in SLNB mapping of lung cancer[32–36]. However, the current system sensitivity is not comparable to that of a gamma probe, and they have needed to surgically bare the lymph

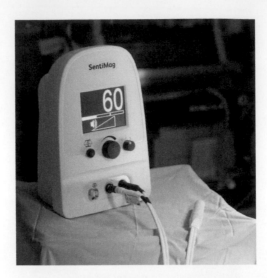

FIGURE 21.5 Non-cryogenic SentiMag™, commercially available in CE-mark accepted countries.

node of interest in order to get close enough to detect the presence of MNPs[34]. Ruhland et al.[37] have developed an alternative magnetic particle imaging (MPI) probe with greater sensitivity for use in breast SLN detection, but have yet to demonstrate it clinically.

Endomagnetics Ltd. (London, United Kingdom) has developed a system based on the earlier SQUID prototype of Joshi et al., but using a room-temperature sensing coil in a field-isolating geometry to significantly improve SNR (Figure 21.5).

The SentiMag™ comprises a probe that applies an alternating magnetic field *via* excitation coils to an injected dosage of MNPs. Sensing coils in the probe are arranged as a first-order gradiometer, enabling localisation of magnetised lymph nodes by moving the probe across the axilla while receiving both visual and audio queues for signal strength. Using an alternating magnetic field at a selected frequency, the signal returning from the induced magnetisation of the MNPs has the same frequency, with slightly modified phase, so sources of noise can be suppressed.

21.3 CLINICAL RESULTS OF INTRA-OPERATIVE SLNB FOR BREAST CANCER USING MNPs

21.3.1 METHODOLOGY

Clinical trials using the SQUID-based device of Joshi et al. started following Medicines and Healthcare Products Regulatory Agency (MHRA) advice and Ethics Committee approval for an initial pilot study at University College Hospital, London, in 2007. The pilot study protocol was designed to make a quantitative comparison between the combined technique and MNP-based detection. Ten patients with newly diagnosed breast cancer and scheduled for sentinel node biopsy were recruited prior to surgery.

These patients received a radioisotope injection (^{99m}Tc) and underwent lymphoscintigraphy a day prior to surgery. On the morning of surgery, they received a subcutaneous injection of 2 mL of Endorem in the MRI Department. Under general anaesthetic in the operating theatre, patients subsequently received an intradermal injection of patent blue dye (Guerbet, Paris). The sentinel nodes were localised using both a gamma probe and the prototype SentiMag. Skin localisation of the sentinel nodes with the prototype SentiMag and the gamma probe was identical.

Following the initial pilot study, the principal investigator (M. Douek) relocated to Guy's Hospital at Kings College, London, and a further 43 patients were recruited into an extended trial in 2009–2010. Consecutive patients with newly diagnosed breast cancer underwent SLNB following injection of patent blue dye (Guerbet, Paris), radioisotope (^{99m}Tc) and 2 mL of Endorem. Endorem was injected interstitially at the circumareolar aspect of the upper outer quadrant in the affected breast. Intra-operatively, the SLN was identified using a combination of blue staining, MNP staining and the gamma probe. Nodes were considered to be 'sentinel' if their count exceeded one-tenth of the highest *ex vivo* gamma count for any lymph node in that patient, or if they were blue[38]. Any other excised nodes were designated non-SLNs. The injection time, time of excision and node colour were recorded for each node. *Ex vivo*, gamma and magnetometer counts were compared.

21.3.2   RESULTS

In the initial 10-patient trial in the University College Hospital, a total of 19 sentinel nodes were resected from 9 patients. Intra-operative localisation using the combined technique was successful in detecting 19/19 (100%) nodes and using the SentiMag prototype alone in 19/19 (100%). Once found by the surgeon, most sentinel nodes were easily identified as black from MNP inclusion[39].

In the extended trial, the overall *ex vivo* SLN detection rate per patient was 100% for the combined technique but dropped to 87% for the MNP technique (Figure 21.6). However, the *ex vivo* SLN detection rate was higher in patients who received Endorem more than 1 h prior to surgery (93%), suggesting that a higher volume or concentration of MNP is required.

The optimal cut-off value with the prototype SentiMag for SLN detection is 20 counts. This corresponds to the signal expected from a mass of approximately 5 μg of injected MNP. Since 2 mL of Endorem contains 22.4 mg of iron oxide, the identified cut-off value is equivalent to 0.02% of the injected MNP.

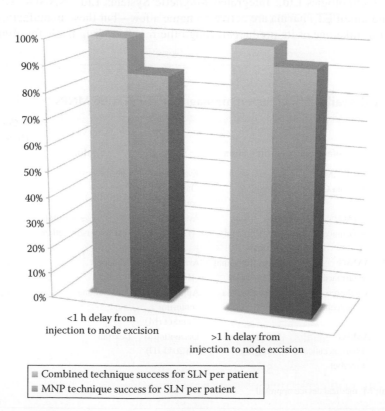

FIGURE 21.6   *Ex vivo* SLN detection rate per patient of the MNP compared to the combined technique.

These clinical results indicate promise for MNPs in SLNB, though further work is needed to be a compelling alternative to radioisotopes. The 87% *ex vivo* node detection rate for sub-1 h injection time could also be due to the fact that Endorem has a particle diameter of ~150 nm. Future work will focus on using an MNP with a size similar to the ^{99m}Tc colloid, as the dynamics of the particles in the lymphatic system should be commensurate.

21.4 CHALLENGES AND NEXT STEPS

21.4.1 AVAILABILITY CHALLENGES OF MNPs

While MNPs have been under clinical development as diagnostic and imaging agents for decades, the number of clinically available iron oxide MNPs has been in steady decline. This is due in part to competition from Gadolinium-based agents that, with their smaller size, are more general purpose and can be used as blood-pool contrast agents. Another possible reason for the decline is that the MRI contrast enhancement agent market is not substantial enough to support a variety of competing materials. For the future development of clinical MNP-based technologies and procedures, this decline represents a significant challenge.

The clinical results presented in Section 21.3 were based on Endorem, however, this material was discontinued and all remaining material in the market expired at the end of March, 2011. Reviewing the current clinically available MNPs paints a concerning picture (Table 21.1). In particular, there is no single globally approved material and, outside of the Japan-only Resovist®, the only small-diameter MNP available (Feraheme®, AMAG Pharmaceuticals) is not approved for imaging.

This is not to say that there are no additional manufacturers of MNPs – BioPAL, Innovative Bio-Medical Technologies Ltd., Integrated Magnetic Systems Ltd., NANOGAP Poligono de Milladorio and nanoPET Pharma are active, to name a few – but these manufacturers generally cannot afford to take one of its products through the requisite drug trials for clinical approval

---

TABLE 21.1

Current Clinical Availability of Superparamagnetic Iron Oxide MNPs

Trade Name	Manufacturer	Indication	Status	Coating	Core Diameter (nm)	Particle Diameter (nm)
Resovist®	FUJIFILM RI Pharma Co.	Liver/spleen imaging	Available, Japan only (D)	Carboxydextran	~4.5	45–65
Feraheme®	AMAG Pharmaceuticals	Iron replacement therapy	Available (D)	Polyglucose sorbitol carboxymethylether	~5	~30
GastroMARK™	AMAG Pharmaceuticals	Gastrointestinal imaging	Available (D)	Silicone	~10	~400
NanoTherm®	MagForce Nanotechnologies	Hyperthermia	Approved, entering market (M)	Aminosilane	~12	~16
Feridex I.V.®/ Endorem™	AMAG Pharmaceuticals/ Guerbet	Liver/spleen imaging	Discontinued in 2011 (D)	Dextran	~5	~150

D, Drug approval; M, medical device approval.

in a given indication, imaging or otherwise. If researchers developing the indication cannot afford the clinical trials either, then a funding stalemate could see the loss of many novel MNP applications.

However, it is promising to note that MagForce Nanotechnologies received medical device approval for its hyperthermia system, including the NanoTherm® MNP. Possibly its regulatory strategy represents a less expensive pathway for bringing new MNP applications to the market.

21.4.2 REGULATORY CHALLENGES OF MNPs IN TISSUE MARKING

The biggest regulatory challenge facing MNPs is that all imaging contrast agents are regulated as drugs, and all MNPs potentially could be used as imaging contrast agents. Developers are likely to face substantial time and expense with multiphase trials either having an existing or new MNP approved as a drug for their application. As potential patients, we all appreciate the high hurdles for safety and efficacy that are regulated on our behalf, but some technologies fall on a borderline between drugs and medical devices in a way that could limit innovation without any gain in safety or patient benefit. If an existing MNP with a known safety profile could be approved as a device, the time to market and associated expenses would be significantly reduced.

In Europe, medical device guidance documents (MEDDEVs) are issued to promote a common approach by manufacturers and Notified Bodies to assessing the conformity of any medical device. While MEDDEVs have no legal force, these documents facilitate common positions throughout the European Union. MEDDEV 2.1.3, rev. 2, July 2001 on Demarcation Between Directive 90/385/EEC on Active Implantable Medical Devices, Directive 93/42/EEC on Medical Devices and Directive 65/65/EEC Relating to Medicinal Products and Related Directives was published in response to discussions on the borderlines between these directives. The MEDDEV recommends that the following questions be considered to categorise a product as a medical device or medicinal product: (a) what is the intended purpose, including an assessment of the way the product is presented; (b) what is the method by which the primary mode of action is achieved, substantiated by manufacturer's labelling, claims and scientific data on mode of action?

If the method by which the primary mode of action is achieved is not pharmacological, immunological or metabolic, and the product is, for example, injected straight into a tumour for the purpose of hyperthermia ablation, then an MNP fitting this description would likely be regulated as a device rather than a drug.

It is clear that nanomaterials challenge this borderline and that definitions will necessarily be tightened globally to ensure patient safety and market clarity. Developing an understanding for, and maintaining an awareness of, this borderline is essential for anyone hoping to develop clinical applications of MNPs.

21.4.3 FUTURE DEVELOPMENTS

A multi-site trial using a sub-100 nm diameter MNPs will be initiated in 2011, co-ordinated by King's College London. A target of about 150 patients is planned to support a paired equivalence trial, where patients will receive injections of radioisotope, patent blue dye and MNP tracer. The percentage concordance for the detection of involved nodes will be compared (per patient and per node). Since each patient will have undertaken both localisation techniques, they will act as their own control. The trial will also evaluate procedure-related efficacy and morbidity. The primary end point will be the proportion of sentinel nodes detected (detection rate) with either the standard (blue dye and radioisotope) or the new technique (MNP and handheld magnetometer).

Once clinical equivalence for breast cancer is established, MNPs can also be used for staging cancers where SLNB appears suited such as colorectal, stomach, prostate, cervical, oesophageal and melanoma. While these areas are currently under clinical investigation using SLNB, the value to medicine is clear – excluding breast cancer, these represent 4.3 million cases per annum globally[1].

In the longer term, the development of multifunctional MNPs may contribute even further to the management of cancer. For example, Lee et al. studied in 2007 whether MNPs conjugated with a cancer-targeting antibody, trastuzumab, could be used to detect cancer *via* MRI. Trastuzumab is a monoclonal antibody that interferes with the HER2/neu receptor, a marker that is overexpressed in up to one-fifth of breast and ovarian cancer cases. *In vivo* results indicated that the high magnetic resonance (MR) sensitivity of MNP–trastuzumab conjugates enabled the MR detection of very small tumours (about 50 mg)[40]. Combining MNPs with such targeting moieties could allow for particles to bind preferentially to metastatic tissue, supporting cancer and tumour margin localisation as well as hyperthermia applications discussed in earlier chapters.

REFERENCES

1. Ferlay, J., Shin, H. R., Bray, F. et al. 2010. Estimates of worldwide burden of cancer in 2008: GLOBOCAN 2008, *International Journal of Cancer*, 127, 2893–2917.
2. Swartz, M. A. 2001. The physiology of the lymphatic system, *Advanced Drug Delivery Reviews*, 50, 3–20.
3. Cabanas, R. M. 1977. An approach for the treatment of penile carcinoma, *Cancer*, 39, 456–466.
4. Giuliano, A. E., Kirgan, D. M., Guenther, J. M. et al. 1994. Lymphatic mapping and sentinel lymphadenectomy for breast cancer, *Annals of Surgery*, 220, 391–398.
5. Krag, D. N., Weaver, D. L., Alex, J. C. et al. 1993. Surgical resection and radiolocalization of the sentinel lymph node in breast cancer using a gamma probe, *Surgical Oncology*, 2, 335–339.
6. Zavagno, G., Belardinelli, V., Marconato, R. et al. 2007. Sentinel lymph node metastasis from mammary ductal carcinoma in situ with microinvasion, *The Breast*, 16, 146–151.
7. Krag, D. N., Anderson, S. J., Julian, T. B. et al. 2010. Sentinel-lymph-node resection compared with conventional axillary-lymph-node dissection in clinically node-negative patients with breast cancer: Overall survival findings from the NSABP B-32 randomised phase 3 trial, *The Lancet Oncology*, 11, 927–933.
8. Giuliano, A. E., Hunt, K. K., Ballman, K. V. et al. 2011. Axillary dissection vs no axillary dissection in women with invasive breast cancer and sentinel node metastasis: A randomized clinical trial, *JAMA*, 305, 569–575.
9. Blessing, W. D., Stolier, A. J., Teng, S. C. et al. 2002. A comparison of methylene blue and lymphazurin in breast cancer sentinel node mapping, *American Journal of Surgery*, 184, 341–345.
10. Lyman, G. H., Giuliano, A. E., Somerfield, M. R. et al. 2005. American society of clinical oncology guideline recommendations for sentinel lymph node biopsy in early-stage breast cancer, *Journal of Clinical Oncology*, 23, 7703–7720.
11. Quan, M. L., Wells, B. J., McCready, D. et al. 2010. Beyond the false negative rate: Development of quality indicators for sentinel lymph node biopsy in breast cancer, *Annals of Surgical Oncology*, 17, 579–591.
12. Quan, M. L., Hodgson, N., Lovrics, P. et al. 2008. National adoption of sentinel node biopsy for breast cancer: Lessons learned from the Canadian experience, *The Breast Journal*, 15, 421–427.
13. Rescigno, J., Zampell, J. C., and Axelrod, D. 2009. Patterns of axillary surgical care for breast cancer in the era of sentinel lymph node biopsy, *Annals of Surgical Oncology*, 16, 687–696.
14. Leong, S. P. L., Shen, Z. Z., Jiu, T. J. et al. 2010. Is breast cancer the same disease in Asian and Western countries? *World Journal of Surgery*, 34, 2308–2324.
15. Wang, A. Z., Gu, F. X., and Farokhzad, O. C. 2009. Nanoparticles for cancer diagnosis and therapy. In *Safety of Nanoparticles: From Manufacturing to Medical Applications*, ed. Webster, T. J., pp. 209–236. New York: Springer.
16. Weissleder, R., Stark, D. D., Engelstad, B. L. et al. 1989. Superparamagnetic iron oxide: Pharmacokinetics and toxicity, *American Journal of Roentgenology*, 152, 167–173.
17. Reimer, P., and Balzer, T. 2003. Ferucarbotran (Resovist): A new clinically approved RES-specific contrast agent for contrast-enhanced MRI of the liver: Properties, clinical development, and applications, *European Radiology*, 13, 1266–1276.

18. Keshtgar, M. R. S., and Ell, P. J. 1999. Sentinel lymph node detection and imaging, *European Journal of Nuclear Medicine and Molecular Imaging*, 26, 57–67.

19. Jain, R., Dandekar, P., and Patravale, V. 2009. Diagnostic nanocarriers for sentinel lymph node imaging, *Journal of Controlled Release*, 138, 90–102.

20. Jung, C. W., Rogers, J. M., and Groman, E. V. 1999. Lymphatic mapping and sentinel node location with magnetite nanoparticles, *Journal of Magnetism and Magnetic Materials*, 194, 210–216.

21. Torchia, M. G., Nason, R., Danzinger, R. et al. 2001. Interstitial MR lymphangiography for the detection of sentinel lymph nodes, *Journal of Surgical Oncology*, 78, 151–156.

22. Gretschel, S., Moesta, K. T., Hunerbein, M. et al. 2004. New concepts of staging in gastrointestinal tumors as a basis of diagnosis and multimodal therapy, *Onkologie*, 27, 23–30.

23. Maza, S., Taupitz, M., Wegner, T. et al. 2005. Precise localisation of a sentinel lymph node in a rare drainage region with SPECT/MRI using interstitial injection of ^{99m}Tc-nanocolloid and superparamagnetic iron oxide, *European Journal of Nuclear Medicine and Molecular Imaging*, 32, 250.

24. Stadnik, T. W., Everaert, H., Makkat, S. et al. 2006. Breast imaging. Preoperative breast cancer staging: Comparison of USPIO-enhanced MR imaging and 18F-fluorodeoxyglucose (FDC) positron emission tomography (PET) imaging for axillary lymph node staging – Initial findings, *European Radiology*, 16, 2153–2160.

25. Maza, S., Taupitz, M., Taymoorian, K. et al. 2007. Multimodal fusion imaging ensemble for targeted sentinel lymph node management: Initial results of an innovative promising approach for anatomically difficult lymphatic drainage in different tumour entities, *European Journal of Nuclear Medicine and Molecular Imaging*, 34, 378–383.

26. Kimura, K., Tanigawa, N., Matsuki, M. et al. 2010. High-resolution MR lymphography using ultrasmall superparamagnetic iron oxide (USPIO) in the evaluation of axillary lymph nodes in patients with early stage breast cancer: Preliminary results, *Breast Cancer*, 17, 241–246.

27. Cooper, K. L., Meng, Y., Harnan, S. et al. 2011. Positron emission tomography (PET) and magnetic resonance imaging (MRI) for the assessment of axillary lymph node metastases in early breast cancer: Systematic review and economic evaluation, *Health Technology Assessment*, 15, 1–134.

28. Tanaka, S., Hirata, A., Saito, Y. et al. 2001. Application of high TcSQUID magnetometer for sentinel-lymph node biopsy, *IEEE Transactions on Applied Superconductivity*, 11, 665–668.

29. Joshi, T., Pankhurst, Q. A., Hattersley, S. et al. 2007. Magnetic nanoparticles for detecting sentinel lymph nodes, *European Journal of Surgical Oncology*, 33, 1135.

30. Joshi, T., Pankhurst, Q. A., Hattersley, S. et al. 2007. Magnetic nanoparticles for detecting cancer spread, *Breast Cancer Research and Treatment*, 106, S129.

31. Gunasekera, U. A., Pankhurst, Q. A., and Douek, M., 2009. Imaging applications of nanotechnology in cancer, *Targeted Oncology*, 4, 169–181.

32. Nakagawa, T., Minamiya, Y., Katayose, Y. et al. 2003. A novel method for sentinel lymph node mapping using magnetite in patients with non-small cell lung cancer, *Journal of Thoracic and Cardiovascular Surgery*, 126, 563–567.

33. Minamiya, Y., and Ogawa, J. I. 2003. A novel method for sentinel lymph node mapping using magnetite, *Nippon Geka Gakkai Zasshi*, 104, 759–761.

34. Minamiya, Y., Ito, M., Katayose, Y. et al. 2006. Intraoperative sentinel lymph node mapping using a new sterilizable magnetometer in patients with nonsmall cell lung cancer, *Annals of Thoracic Surgery*, 81, 327–330.

35. Minamiya, Y., Ito, M., Hosono, Y. et al. 2007. Subpleural injection of tracer improves detection of mediastinal sentinel lymph nodes in non-small cell lung cancer, *European Journal of Cardio-Thoracic Surgery*, 32, 770–775.

36. Minamiya, Y., Katayose, Y., Motoyama, S. et al. 2008. Sentinel lymph node mapping using MRI contrast medium and magnetic force, *Annals of Surgical Oncology*, 15, 19.

37. Ruhland, B., Baumann, K., Knopp, T. et al. 2009. Magnetic particle imaging with superparamagnetic nanoparticles for sentinel lymph node detection in breast cancer, *Geburtshilfe und Frauenheilkunde*, 69, 758.

38. Johnson, L., and Douek, M. 2010. How low should we go for gamma readings in sentinel lymph node biopsy of the breast? *European Journal of Surgical Oncology (EJSO)*, 36, 1116.

39. Johnson, L., and Douek, M. 2010. Nanoparticles in sentinel lymph node assessment in breast cancer, *Cancers*, 2, 1884–1894.

40. Lee, J. H., Huh, Y. M., Jun, Y. W. et al. 2007. Artificially engineered magnetic nanoparticles for ultrasensitive molecular imaging, *Nature Medicine*, 13, 95–99.

Eric Mayes is the CEO of Endomagnetics and has more than 10 years of experience in advanced materials and devices businesses. Dr. Mayes has a history in leading early-stage ventures, including his founding and leading NanoMagnetics Ltd. For his role in NanoMagnetics, he was named the Royal Society of Chemistry's 'Entrepreneur of the Year 2003'. He has a BSc in physics from Arkansas State University and a PhD in chemistry from the University of Bath.

Michael Douek is a Reader in Surgery at King's College London and consultant surgeon at Guy's and St Thomas' Hospital. Mr. Douek is the chief investigator of the SentiMag trial of sentinel node biopsy and principal investigator for the international randomised controlled trial of intra-operative radiotherapy (TARGIT trial), at Guy's Hospital. He graduated in medicine from the University of Dundee (Scotland) and trained in surgery in London, Oxford and Cambridge. He obtained his MD from the University of London in 2000.

Quentin Pankhurst is a professor of physics at University College London (UCL) (a world top-10 rated institution) and director of the Davy–Faraday Research Laboratories at the Royal Institution of Great Britain. He runs internationally leading research programmes aimed at making practical advances in the use of magnetics in healthcare and works closely with industry on such matters. Born in New Zealand, hc came to the United Kingdom in 1983 to study solid state physics at the University of Liverpool, where he stayed until joining UCL in 1994.

22 Safety Considerations for Magnetic Nanoparticles

Taher A. Salah, Hazem M. Saleh,
and Mahmoud H. Abdel Kader

CONTENTS

22.1 INTRODUCTION

This chapter will focus on major points concerning the safety of magnetic nanoparticles (MNPs) from a toxicological point of view and will provide key references to help the reader to delve further into the subject. It is important to know that toxicology as a discipline has been relatively new, since *Archiv fur Toxicologie*, the first journal expressly dedicated to experimental toxicology, started publication in Europe in 1930[1]. The Society of Toxicology, the pre-eminent toxicology organisation, was not founded until 1961[2]. The reader is referred to more reviews on the history of toxicology[3] and regulatory toxicology[4]. Assessing the potential hazards of nanomaterials is an emerging area in toxicology and health risk assessment. Nanotoxicology is a subspecialty of particle toxicology. It addresses the toxicology of nanoparticles (<100 nm diameter), which appear to have toxic effects that are unusual and not seen with larger particles. Because of the quantum size effects and large surface area to volume ratio, nanomaterials have unique properties compared with their larger counterparts. Nanotoxicological studies are intended to determine whether and to what extent these properties may pose a threat to the environment and to human beings[5].

The development of toxicity data sets and exposure assessments for various nanoparticles and nanomaterials is ongoing as new particles and materials are developed. A related issue in toxicology and risk assessment is the extent to which nanoparticles toxicity can be extrapolated from the existing toxicology databases for particles and fibres. From the toxicological point of view, it is very important to consider nanomaterials as new materials that must be tested with complete toxicological evaluation, even if their bulk forms have known toxicological data. Other information that need to be addressed are the environmental and biological fate, the transport and transformation of manufactured nanoparticles and the recyclability and over all sustainability of manufactured nanomaterials. Calls for tight regulation on nanotechnology have arisen alongside a growing debate related to human health and safety risks associated with nanotechnology. We will summarise and analyse selected studies on toxicity and risk assessment of nanomaterials with emphasis on MNPs. However, it is difficult to ascertain or deny the toxicity of a specific nanomaterial because its toxicity can also depend on dose, exposure and route of administration. In addition, the results from nanotoxicological studies on animals or on cultured cells cannot be perfectly extrapolated to human beings.

22.2   TOXICITY TESTING AND RISK ASSESSMENT OF NANOMATERIALS

Any nanomaterial should be considered a new material not related to its bulk form. Hence, the process of safety evaluation dedicated to new substances should be applied. The fine details will depend on the intended use. For most nanomaterials, safety evaluation starts with an overview of the physical and chemical properties (e.g. solubility, chemical reactivity and flammability) to address the safe handling and storage of such materials during industrial use. This information can also be used to study the effect of disposal on public health. For example, water soluble nanomaterials may raise concerns about contamination of water supplies and volatile ones about air quality[6]. Nanomaterials are also examined for their biological effects. Tests should study acute and chronic toxicity on whole organs and whole organisms. Mutagenicity should be tested (mostly *in vitro* using animal cells or bacteria), as well as ecotoxicity to assess the risk on the environment (e.g. acute toxicity to fish, invertebrates and algae). The methodology for these biological and chemical tests is annexed to legislations (e.g. The European Economic Community (EEC) 1992 for the European version of these tests)[7]. An international consensus exists also on the test methods (International Organization for Standardization (ISO) 1996)[8]. Nanomaterials intended to be used clinically, such as MNPs, are subjected, after passing the preliminary *in vitro* tests, to additional and gradually sophisticated series of non-clinical and clinical trials. These provide more in-depth investigations on human toxicity and metabolism that often include specific end points related to the intended use[9].

In assessing nanomaterials, there is a chance that these novel substances could have novel effects with no corresponding tests or suitable end points. The traditional toxicity tests are based on fundamental assumptions in the dose – response relationship. One of these assumptions is that the concentration of the toxin at the target (e.g. receptors on cell membranes of the test organism) is related to dose. Nanoparticles may aggregate at high ionic strength (e.g. in salty solutions) and may adsorb onto surfaces[10]. This suggests that nanomaterials will be trapped in the mucous layer on epithelial surfaces rather than being absorbed into the cells in a predictable dose-dependent manner[11]. Questions about how we should interpret data from current toxicity tests while testing nanoparticles should be raised. It may be necessary to modify the test methods to account for the chemistry or to add an extra uncertainty factor to risk calculations. Alternatively, we may need to suggest novel tests for nanomaterials in order to satisfy the fundamentals of the dose – response relationship. Risk characterisation for various routes of exposure for humans is described in Part I of the Technical Guidance Document on Risk Assessment published by the European Joint Research Centre[12]. This document does not mention nanoparticles but indicates that particle size distribution should be measured where relevant to respiratory exposure. The general principle for risk characterisation is to compare exposure concentrations with the non-observed adverse effect levels (NOAELs) or lowest observed adverse effect levels (LOAELs). If the environmental concentration of the new substance is greater than the NOAEL, then it is a chemical of 'concern' that requires some precautions (risk management). Information is therefore needed on exposure concentration for nanomaterials, if this approach is to be applied. Clearly, it is unethical to expose humans to nanomaterials to obtain no effect data, and NOAEL from animal studies can be used instead.

We do not have enough measurements of engineered nanomaterials in the environment to estimate reference doses for the human population, and there are no data giving NOAEL for humans incidentally exposed to engineered nanoparticles. To the best of our knowledge, nanotoxicologists are still in the data collection phase and are not able to make generic risk assessments in the context of public health.

22.3 SPECIAL CONSIDERATIONS IN NANOTOXICOLOGY

In nanotoxicology, the size of nanomaterials and the used animal model were found to be the most important factors affecting the results of toxicological study and deserve separate considerations.

22.3.1 NANOPARTICLE SIZE

As the particle size decreases, a greater proportion of atoms become orientated on the surface rather than within the interior of the material, hence allowing adjacent atoms and substances to interact more readily. The surface-to-volume ratio determines the potential number of reactive groups; the intrinsic properties of materials at the nanosized level are emphasised compared to their larger bulk counterparts. The enhanced activities could be either beneficial (e.g. antioxidation, carrier capacity for drugs, increased uptake and interaction with biological tissues) or disadvantageous (e.g. toxicity, instability and induction of oxidative stress) depending on the intended use[13,14]. One of the earliest observations was that nanoparticles (<100 nm) showed greater toxicity than microparticles (<2.5 μm) of the same material on a mass basis. This has been observed with different types of materials, including titanium dioxide, aluminium trioxide, carbon black, cobalt and nickel[15]. Oberdorster et al.[16] found that 21 nm titanium dioxide particles produced 43-fold more inflammation (as measured by the influx of polymorphonuclear leucocytes into the lung) than 250 nm particles based on the same mass instilled into animal lungs. Though multiple studies have shown that nanosized particles may be more toxic than micron-sized particles. However, this is not always the case. Intrinsic surface reactivity may also be as important as surface area. Warheit[17] found that the toxicity for cytotoxic crystalline quartz did not relate to particle size, but did relate to surface reactivity as measured by the haemoglobin release from cells *in vitro*.

The nanoparticles can easily penetrate the cell membrane and can diffuse from blood vessels due to their ultrafine size relative to human cells, where nanoparticles (<100 nm) are generally similar in size to proteins in the body. They are considerably smaller than many cells in the body.

Gornati et al.[18] illustrated the massive entry of MNPs into the cytoplasm using transmission electron microscope (TEM) imaging. When endothelial-like cells (ECV)-304 and HepG2 cells were exposed to Co_3O_4-NPs (nanoparticles), a massive internalisation of particles was observed. Thin section TEM images showed that Co_3O_4-NPs were taken up very readily. After 30 min of incubation, agglomerates of Co_3O_4-NPs were already inside the cells, while others were about to enter. The total amount of NPs within the cells increased with time. Although the mechanism of uptake was not investigated, it is more likely that the MNPs were internalised in an agglomerated form rather than taken up as individual particles. Objects with high electron density were mostly stored in vesicular structures, which may also contain amorphous cellular material. Moreover, although less frequently, Co_3O_4-NPs were also found very readily within cell nuclei, interestingly, some of these nuclear agglomerates consisted of MNPs lined up as a chain (Figure 22.1)*.

22.3.2 ANIMAL MODEL

The animal model plays an important role in nanotoxicology studies, where some models are more sensitive than others resulting in variations in the resultant adverse effects. For example, the rat

---

* Reprinted from *Toxicol. Lett.*, 189, Gornati, R., Papis, E., Rossi, F., Raspanti, M., Dalle-Donne, I., Colombo, G., Milzani, A., and Bernardini, G., Engineered cobalt oxide nanoparticles readily enter cells, 253–259, Copyright 2009, with permission from Elsevier B.V.

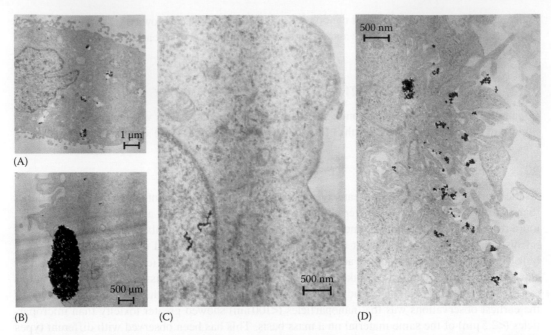

FIGURE 22.1   TEM images of ECV-304 (A–C) and HepG2 (D) cells exposed to Co_3O_4-NPs for 24 h (A) and 30 min (B–D). Interestingly, Co_3O_4-NPs are taken up very readily and NPs are present inside cytoplasmic vesicles and cell nuclei. (From Gornati, R. et al., *Toxicol. Lett.*, 189, 253, 2009.)

model, from which most of the nano- versus micro-size comparisons have been reported, is known to be an extremely sensitive species for developing adverse lung responses to particles, particularly to overload concentrations. As a consequence, long-term (2 years), high-dose inhalation studies in rats with poorly soluble, low-toxicity dusts can ultimately produce pulmonary fibrosis and lung tumours *via* an 'overload' mechanism. The tumour-related effects are unique to rats and have not been reported in other particle-exposed rodent species such as mice or hamsters, under similar chronic conditions. For the mechanistic connection, it has been postulated that the particle-overload effects in rats result in the development of 'exaggerated' lung responses, characterised by increased and persistent levels of pulmonary inflammation, cellular proliferation and inflammatory-derived mutagenesis, and this ultimately results in the development of lung tumours following high-dose, long-term exposures to a variety of particle types. In contrast to the response in rats, numerous studies demonstrate that particle-exposed mice and hamsters do not develop sustained inflammation, mesenchymal cell alterations and consequent lung tumours following high-dose, long-term exposures to low-toxicity dusts. Therefore, species differences in lung responses to inhaled particles are important considerations for assessing the health risks to nanoparticles. To complicate further the perceptions of nanoparticles toxicity, some recent evidences suggest that, on a mass basis, not all nanoparticles are more toxic than fine-sized particles of similar chemical composition[19,20].

22.4   TOXICOLOGICAL EVALUATION OF MNPs

The fabrication of nanomaterials with high magnetic moment density has attracted increasing research interests due to their potential biomedical applications such as bio-separation, bio-sensing, magnetic imaging, drug delivery and magnetic fluid hyperthermia[21]. Many medical innovations are under assessment now and claim benefits from MNPs. As all chemicals, MNPs are now subject to stringent government requirements for safety testing before they can be clinically applied. In addition to quantum size effects and large surface area to volume ratio, MNPs have unique properties compared to other nanomaterials due to the effect of their magnetic properties either with or without external applied magnetic field.

Knowledge on toxicity of MNPs is formulated on four levels: first, studying the physico-chemical properties including particle size, charges of capping materials and magnetic characteristics; second, specifying the intended clinical use, for instance, the toxicology protocol for MNPs used for magnetic resonance imaging (MRI) (applied once or repeated at long intervals) will differ from that used as a therapeutic drug (applied many times at short intervals); third, defining the mode of administration of MNPs either orally (where the effect on the gastrointestinal tract, gastric enzymes and liver must be deeply investigated), intramuscularly (where the effect on muscular absorptivity and MNPs agglomeration between the muscle fibres must be taken into consideration) or intravenously (where the effect on the cells, enzymes, plasma proteins and hormones circulating in the bloodstream in addition to the blood vessels must be an evaluation priority) and fourth, evaluating the biological effects in general by studies using cells (human, animal and plant), by studies using experimental animals or by data deduced from clinical trials or from accidental exposures of humans. In this section, examples of toxicological studies on MNPs are given to illustrate the principles of toxicological tests. Before any biomedical application of MNPs, thorough toxicological studies are often carried out and readers will find them abundant in the literature.

22.4.1 *In Vitro* MNPs Toxicological Studies

In vitro test systems do not employ intact higher organisms as models and include a large set of alternative models. These *in vitro* test systems have the benefit of being of low cost and have well-known mechanisms of action. They can be used for assessing or predicting the toxic effects of MNPs and for elucidating their mechanisms of action. The systems include the use of cell or tissue cultures, isolated cells, tissue slices, subcellular fractions, transgenic cell cultures and cells from transgenic organisms. The systems also include *in silico* modelling. Additional *in vitro* test systems are being developed for the use in high-throughput toxicology and pharmacology for the understanding of mechanisms of toxic action and for genomics, transcriptomics and proteomics applications[22,23]. The mostly used *in vitro* test systems are 3-(4,5-dimethylthiazol-2yl)-2,5-diphenyltetrazolium bromide (MTT) cell viability assay and Ames mutagenicity assay. Various *in vitro* test systems are described in the literature[22–25].

In vitro toxicity tests were considered the rapid way to answer necessary questions concerning MNPs cytotoxicity, such as what are the cytotoxic effects of MNPs compared to their ionic form, does the used model cell line affect the toxicological results, can the coating of MNPs affect its toxicity, what is the form of cytotoxicity: an effect on cellular viability or an induction of genetic modifications or an alteration of cellular metabolic processes and can the shape of MNPs affect the cellular toxicity? We will summarise answers to such questions from available published data.

Cobalt NPs were suggested as a contrast agent in MRI, alternative to iron NPs, due to their greater effects on proton relaxation[26]. Gornati et al.[18] evaluated the cytotoxicity of Co_3O_4-NP compared to cobalt ions, $CoCl_2$, incubated for 24, 48 and 72 h with endothelial-like cells (ECV-304) and hepatoma cells (HepG2) by measuring the cellular adenosine triphosphate (ATP) content. A dose and time-dependent reduction in cell viability was observed for both cobalt forms, although $CoCl_2$ showed a higher cytotoxic effect than the NPs. Moreover, hepatoma cells were less sensitive to cobalt exposure than ECV-304 cells. In the same work, Co_3O_4-NPs were investigated for their ability to induce reactive oxygen species (ROS) formation in ECV-304 and HepG2 cells. The oxidation of 2,7-dichloro-dihydrofluorescein diacetate (DCFH-DA) was used to estimate the production of ROS. Hydrogen peroxide (10 mM) was used as a positive control for the assay. One-hour exposure of DCFH-DA-pre-incubated cells to Co_3O_4-NPs resulted in a dose-dependent increase in ROS production in both cell lines, although hepatoma cells were less sensitive. ROS levels in treated samples were up to one order of magnitude higher than in controls. In the case of $CoCl_2$, no statistically significant increase in ROS formation could be observed as compared to untreated cultures, at any tested concentration. This can be explained by the fact that the cell membrane offers an excellent barrier for most ions, and it has been shown that cobalt, in ionic form, is taken up by the cells with lower efficiency[27] in contrast to the rapid and massive uptake of NPs by the cells. Co_3O_4-NPs being readily taken up by cells, this could be the basis of their *in vitro* toxicity, which therefore could represent a disadvantage for some applications. However, toxicity can probably be modulated by means of suitable coating materials with tailored characteristics. Conversely, NP toxicity can be exploited for other applications, such as cancer treatment.

In another valuable work, Häfeli et al.[28] studied the effect of coating on cell uptake and *in vitro* toxicity of superparamagnetic magnetite nanoparticles suspension coated with a polyethylenoxide (PEO)-polyurethane-PEO triblock copolymer for future use in ocular drug targeting. The prepared MNP suspensions were of narrow size distribution, superparamagnetic and showed no evidence of aggregation. Since the polymer coatings had not previously been tested for cytotoxic effects, they were evaluated in an MTT cell viability assay with relevant cell types. MNPs coated with longer chain triblock copolymers (5 and 15 kDa) were found to be non-toxic to human umbilical vein endothelial cells, human retinal pigment epithelial (HRPE) and prostate cancer (PC3 and C4-2) cells. Unlike MNPs coated with long-chain copolymers, the toxicity tests of the MNPs coated with the 2K-3-2K*, and even more so, the 0.75K-3-0.75K* copolymer, clearly showed dose-dependent toxic effects. This was strikingly obvious in the confocal microscopy studies (Figure 22.2) where many of the cells (all types) developed apoptotic signs including blebbing of the nucleus and condensing of the chromatin[29]. The toxic effects did not correlate with MNP uptake, as all cells pinocytosed or phagocytosed the MNPs with different coatings at similar rates, showing increasing numbers of MNPs inside their cytoplasm at later time points. Short-chain coating copolymers were found to be much more cytotoxic than long-chain copolymers of the same type.

Zange et al. saw similar toxic effects with poly(lactic-co-glycolic acid) (PLGA)-PEO-PLGA copolymers. They reported highly toxic effects in L929 fibroblast cells when they used a copolymer with 1 kDa PEO unit, whereas essentially no toxicity (MTT viability of 80%–100%) was seen in 4 and 10 kDa PEO unit copolymers[30]. Furthermore, Park et al. reported that oleic acid is non-toxic up to 8% of the blood volume, while the shorter chain decanoic and non-anoic acids are already toxic at 0.5%. All these fatty acids are used in coating MNPs. The fact that shorter polymer tails are more toxic than longer ones does not entail the same in the *in vivo* world as the mechanism of toxicity is not currently known[31]. One might expect that even the longer polymer coatings could convert from less toxic to more toxic molecules if they become degraded in the target tissue by enzymatic or hydrolytic action. For the currently tested copolymers, however, it is expected that such degradation processes will be minimal since the ether groups of PEO are rather stable, and the ester and amide bonds on the surface of the magnetite nanoparticles are protected from enzymatic attacks.

The content of magnetite within the MNPs did not seem to have a significant impact on cell viability since MTT assay for MNP 2K-3-2K (30%) and for MNP 2K-3-2K (50%) were not statistically different. Both MNPs possess identical PEO end blocks but differ in their iron oxide content. This suggests that the main factor in toxicity is the polymer coating and not the magnetite core. Successful magnetic targeting requires a large magnetic moment, and the finding that an increased amount of the magnetic component does not increase toxicity is thus valuable. In conclusion, MNPs coated with triblock copolymers containing PEO chains of 5 and 15 kDa are biocompatible as determined by a cell viability MTT assay. Furthermore, these MNPs at concentrations of up to 5 mg/mL do not disturb the growth of endothelial, epithelial and tumour cells, as verified by confocal microscopy studies. Such MNPs can thus be tested *in vivo* for magnetic ocular drug targeting.

It is necessary to address also the effect of variations in MNPs shape on cytotoxicity. The cytotoxicity of Fe nanowires (Fe NWs), which have potential biologic applications, was evaluated by Bi et al.[32] with direct microscopic observation and MTT assay. Previous studies showed that the long fibre was more toxic than the short one[33]. It was not known whether the same paradigm is relevant to Fe NWs, and thus, two different Fe NWs, prepared with an average length of about 2 and 5 mm,

---

* The nomenclature that has been adopted to describe these copolymers is the number average molecular weight of PEO (M_n) in kDa, followed by the average number of carboxylic acid groups in the central segment, and then by the M_n of the second PEO molecule; for example, 2K-3-2K.

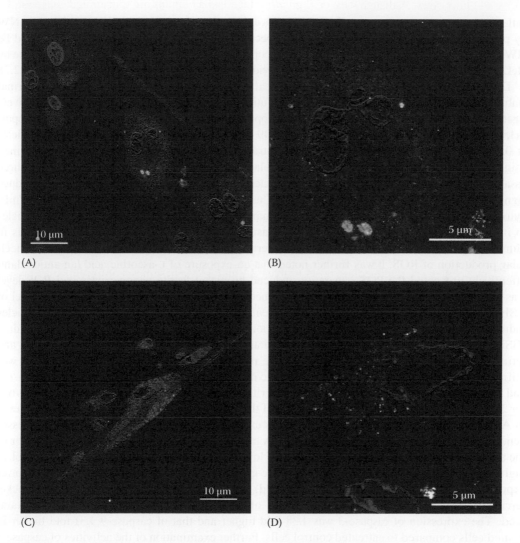

FIGURE 22.2 Confocal microscopic pictures showing MNPs 0.75K-3-0.75K with short PEO tails cytotoxicity to HRPE cells. Signs of toxicity and ongoing apoptosis are blebbing nuclei after 24 h of incubation (A and B) and condensed chromatin after 48 h of incubation (C, D). (Reprinted from Häfeli, U.O. et al., *Mol. Pharm.*, 6(5), 1417, 2009. Copyright 2009, with permission.)

were tested. At the same concentration, the short Fe NWs showed a higher toxicity than the long Fe NWs, as shown by a reduction in cell viability. This coincides with reports suggesting that smaller sized particles allow increased cellular interactions and therefore may possess enhanced intrinsic toxicity[34]. Cell viability was not significantly affected by the concentration of Fe NW since HeLa cells exposed to Fe NWs with increasing concentrations from 10:1 to 10,000:1 showed no reduction in viability. With changes in concentration, the cell viability remained at or above 100% after 24 h of incubation with Fe NWs. This study revealed that the toxicity of Fe NWs is affected mainly by the aspect ratio rather than by the concentration gradient.

In order to know whether the Fe NWs were taken up by the cells or whether their low toxic properties were principally mediated by their presence in the cell culture medium, uptake studies were performed. The interaction between HeLa cells and Fe NWs was investigated using phase contrast microscope. To observe the cell morphology obviously, HeLa cells incubated with Fe NWs were stained with Haematoxylin and Eosin (HE). Before fixation and staining, the cells were washed many times with phosphate buffered saline (PBS) to get rid of the free Fe NWs. It was found that an individual cell could take up more than one single NW. Since direct imaging of unlabelled, individual Fe NW by phase contrast microscopy was challenging because of the thin diameter, fluorescent marked Fe NWs imaging by confocal laser scanning microscopy (CLSM) was carried out to prove the uptake by HeLa cells. The results provided the first visual evidence for cellular uptake of Fe NWs (Figure 22.3).

Due to their interesting magnetic and electrical properties and their good chemical and thermal stabilities, nickel ferrite nanoparticles are being utilised in many applications. Akhtar et al.[35] investigated the cytotoxicity, oxidative stress and apoptosis induction by 26 nm nickel ferrite nanoparticles (in human lung epithelial, A549) cells. A549 cells were exposed to nickel ferrite nanoparticles at 0, 1, 2, 5, 10, 25, 50 and 100 μg/mL concentrations for 24 h and cytotoxicity was determined using MTT, neutral red uptake (NRU) assay and lactate dehydrogenase (LDH) assays. All the three assays have shown that nickel ferrite nanoparticles up to 10 μg/mL concentration did not produce significant cytotoxicity. As the concentration of nanoparticles increased to 25, 50 and 100 μg/mL, cytotoxicity was observed in a dose-dependent manner. The potential of nickel ferrite nanoparticles to induce oxidative stress was assessed by measuring the ROS and glutathione (GSH) levels in human lung epithelial A549 cells. The nickel ferrite nanoparticles significantly induced intracellular production of ROS. It was further noted that co-exposure of L-ascorbic acid (an antioxidant) effectively prevented the ROS generation induced by nickel ferrite nanoparticles and ROS level was reduced up to the control level in the presence of ascorbic acid. Also the intracellular level of GSH was significantly reduced. These results clearly demonstrated that nickel ferrite nanoparticles induced oxidative stress in A549 cells by induction of oxidants (ROS) and depletion of antioxidant (GSH). Since it was reported that ROS generation and oxidative stress lead to DNA damage and ultimately apoptotic cell death[36-38], quantitative real-time polymerase chain reaction (PCR) was utilised to analyse the mRNA levels of apoptotic markers (e.g. p53, survivin, bax, bcl-2, caspase-3 and caspase-9) in A549 cells exposed nickel ferrite nanoparticles at a concentration of 100 μg/mL for 24 h. Results showed that the mRNA levels of these apoptotic markers were significantly altered in A549 cells due to nickel ferrite nanoparticles exposure. The mRNA level of tumour suppressor gene p53 was 1.66-fold higher while the mRNA level of a well-known member of the inhibitor of the apoptosis protein (IAP) family was 1.34-fold lower in treated cells as compared to the control. A higher expression of mRNA of the pro-apoptotic gene bax (1.43-fold) was found as well as a lower expression of anti-apoptotic gene bcl-2 (1.41-fold) in exposed cells. Moreover, the effect of nickel ferrite nanoparticles on the mRNA expression of caspase-3 and caspase-9 enzymes was examined. The expression of caspase-3 was 1.38-fold higher and that of caspase-9 1.31-fold higher in treated cells compared to untreated control cells. Further examination of the activities of caspase-3 and caspase-9 enzymes at the concentrations of 25, 50 and 100 μg/mL showed that nickel ferrite nanoparticles induced the activities of both apoptotic enzymes in a dose-dependent manner.

From the nanotoxicological point of view, the MTT *in vitro* cell viability test, although easy to set-up and applicable to many cell types, has some drawbacks including relatively high standard deviations leading to high variability, short assay time (maximum 72 h) and non-specificity or sensitivity to mechanical (abrasive) particle effect which may promote cell loss during rinsing[39]. Preferable tests are ones that detect the apoptosis more specifically (e.g. caspase and terminal transferase dUTP nick end labelling (TUNEL) assays), tests that measure the expression of specific biomarkers (e.g. Ki67 and cytokines such as interleukin-8 in endothelial cells)[40] and tests that measure membrane integrity (e.g. LDH and live/dead assays)[41].

22.4.2 MNPs Translocation in the Body

Once in the animal body, some types of nanoparticles may have the ability to translocate and to be distributed to other organs, including the central nervous system. Cho et al.[42] evaluated the toxicity and tissue distribution of MNPs@SiO$_2$(RITC) in mice. After intraperitoneal administration for 4 weeks, MNPs were detected in diverse organs such as the brain, liver, lungs, kidneys, spleen, heart, testes and uterus (Figure 22.4). The particles were distributed in all organs, and the

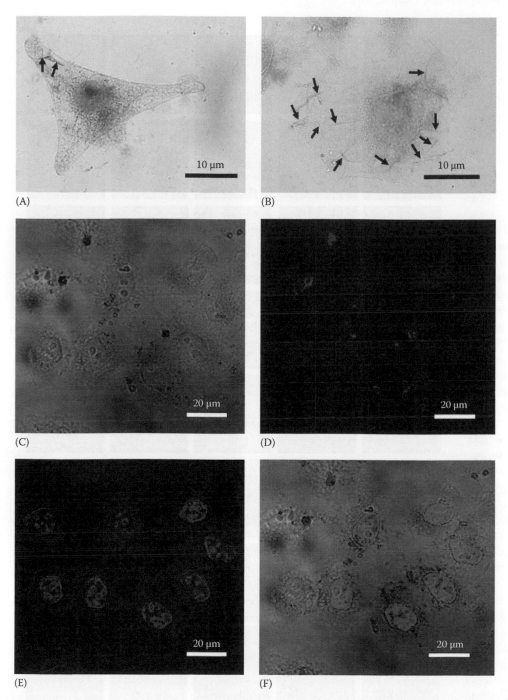

FIGURE 22.3 (A–B) Haematoxylin/Eosin-stained HeLa cells incubated with Fe NWs for 2 h; (C–F) CLSM images of T6219-stained Fe NWs uptaken by HeLa cells (stained with 4′,6-diamidino-2-phenylindole (DAPI)); (C) Phase contrast image; (D) The red rhodamine channel; (E) The blue DAPI channel; (F) Overlay of phase contrast image and fluorescence images of the blue and red channel. (A) showed some Fe NWs completely internalised by an individual HeLa cell and B showed an individual HeLa cell with some partially internalised Fe NWs. The black thin Fe NWs, marked with an arrow, could be distinguished from the HE-stained cell. It can be seen that an individual cell could uptake more than one single NW. In CLSM images shown in (C–F), the red represents the individual or bundles of Fe NWs or small Fe NWs aggregates, localised in the cytoplasm, around the blue nuclei. (Reprinted from Bi, H. et al., *Biomaterials*, 31, 1509, 2010. With permission.)

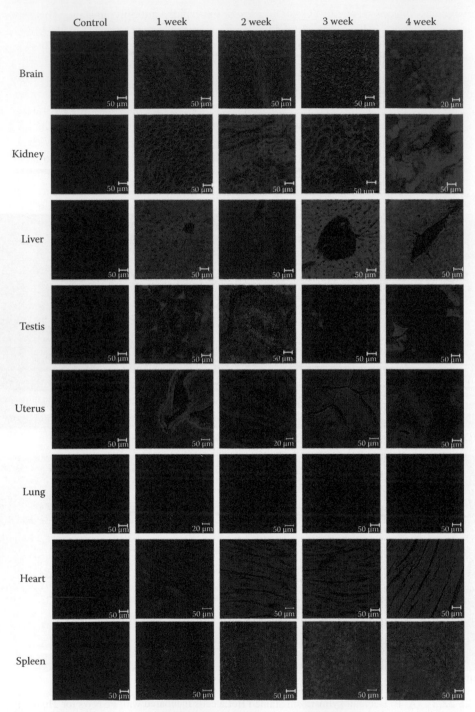

FIGURE 22.4 Tissue distribution of MNPs@SiO₂(RITC) in various organs of mice. The mice were treated with three different concentrations of MNPs@SiO₂(RITC) intraperitoneally for 4 weeks. The mice were sacrificed after designated time points (every week, thus, at four time points during the study). Each organ was removed from the mice and fixed in 4% paraformaldehyde solution. After fixation, the tissue samples were dehydrated and cryo sectioned at 10 mm thickness, then, observed under CLSM. Labels on the left side of the images indicate organs, and the times of observation are indicated above the top horizontal row. Please note that the tissue distribution pattern of female mice is identical to that of male mice except uterus; thus, data have not been shown (×200). (Reprinted from Cho, M.H. et al., *Toxicol. Sci.*, 89(1), 338, 2006. With permission.)

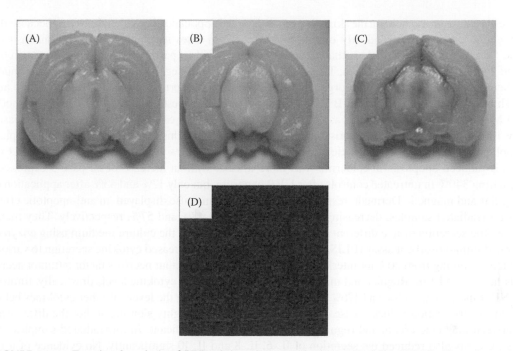

FIGURE 22.5 Functional analysis of BBB of the mice treated with MNPs@SiO$_2$(RITC). To evaluate the potential effects of MNPs@SiO$_2$(RITC) on BBB permeability, Evans blue (EB) dye was injected *via* the tail vein at 5 min prior to MNPs@SiO$_2$(RITC) administration and was allowed to remain in circulation for 30 min. Hyperosmolar mannitol was used as a positive control. A solution of 25% mannitol dissolved in 0.9% saline was injected into the mice at 10 min after intravenous injection of the EB dye. After 30 min, the mice were sacrificed and perfused and the brain was removed for further analysis. The samples taken from the brains were divided into three groups. (A) mice treated with saline solution; (B) mice treated with MNPs@SiO$_2$(RITC) (10 mg/kg); (C) mice treated with 25% mannitol; (D) fluorescence image of the mouse brain obtained by CLSM analysis (×800)[42]. (Reprinted from Cho, M.H. et al., *Toxicol. Sci.*, 89(1), 338, 2006. With permission.)

distribution pattern was time dependent. It was recorded that MNPs were detected in the brain, indicating that such nanosized materials can penetrate the blood–brain barrier (BBB) without disturbing its function or producing apparent toxicity (Figure 22.5). After 4 weeks of observation, MNPs@SiO$_2$(RITC) were still present in various organs without causing apparent toxicity. The ability of nanoparticles to move about the body may depend on their chemical reactivity, surface characteristics and ability to bind to body proteins. The main routes of nanomaterials transduction in the body are the dermal and inhalation exposures[43].

22.4.3 SKIN PENETRATION OF MNPs

The data published through the last few years leave no doubt about the ability of nanoparticles to penetrate through the skin. Particles in the micrometre range are generally thought to be unable to penetrate through the skin. From the histological and anatomical point of view, the outer skin consists of a 10 µm thick tough layer of dead keratinised cells (stratum corneum) that is difficult to pass for particles, ionic compounds and water soluble compounds. Milner et al.[44] dispersed a source of magnetic field (strontium hexaferrite nanomagnets and SHF nanomagnets) in a skincare preparation (Dermudt), in order to apply a magnetic field uniformly on the skin surface and to evaluate the biological effects of surface magnetism on human skin. They succeeded to produce a stable nanomagnetic cream, displaying a field of 1–2 gauss at 2 mm from its glass tube. This value largely exceeds the natural earth field (about 0.5 gauss), but remains below the pacemaker safety limit (5 gauss).

The actual magnetic field on the skin surface after topical application is weaker, however, because the magnetic particles are randomly oriented and the cream loses its global magnetisation. Instead, the 'operative' magnetic field is that created by each individual particle, which ranges between 0.5 and 2 gauss at a distance of 1 mm. They compared the action of regular and nanomagnetic Dermudt, when applied to human skin, in organ cultures for 60 h under normal conditions or 36 h before and 24 h after exposure to ultraviolet B (UVB). Cell toxicity was investigated by two different methods: an MTT assay and an enzymatic assay of late apoptosis effector caspase 3. MTT activity decreased by 10% after UVB exposure (statistically non-significant), and this trend was reversed after topical application of regular or magnetic cream, which induced a 30% decrease. The cream's effect on UVB-induced apoptosis in epidermis was more drastic, with an increase of caspase 3 expression reaching 340% in untreated controls after UVB exposure, but only 17% and 43% after application of regular and magnetic Dermudt, respectively. The two creams also displayed an anti-apoptotic effect in unirradiated samples, decreasing caspase 3 activity by 48% and 57%, respectively. They monitored the secretion of five different inflammatory cytokines into the culture medium using enzyme-linked immunosorbent assay (ELISA) tests. UVB irradiation increased cytokine secretion to various extents, ranging from 20% for interleukin (IL)-10 to 520% for tumour necrosis factor α (tumor necrosis factor (TNF)-α). Regular and magnetic Dermudt reduced all cytokine levels drastically, limiting TNF-α increase to 126% and 176%, respectively, and decreasing the levels of other cytokines below their pre-irradiation values. These changes were statistically highly significant, but the differences between magnetic cream and regular creams remained insignificant. In unirradiated samples, the two creams also reduced the secretion of IL-6, IL-8 and IL-10 significantly. No evidence of toxic effect or of altered anti-apoptotic properties associated with nanomagnets was found. Cream effects on UVB-induced cytokine secretion were also unchanged. It was found that MNPs can penetrate human skin irrespective to the hydrophobicity of the suspension media. The phenomenon is mainly size dependent. There are not enough available data concerning the transfer of MNPs from human skin to bloodstream nor how long its deposition underneath the epidermis is or what its metabolic fate inside the human body is.

22.4.4 INHALATION OF MNPs

The lack of information regarding the body distribution of inhaled nanoparticles may pose serious problems. In their study, Kwon et al.[45] addressed this question by studying the body distribution of inhaled nanoparticles in mice using approximately 50 nm fluorescent magnetic nanoparticles (FMNPs) as a model for inhalation of MNPs through a nose-only exposure chamber system. The body distribution of the inhaled FMNPs was examined by CLSM and MRI. FMNPs were distributed in various organs, including the liver, testis, spleen, lungs and brain. In contrast, only few FMNPs were distributed in the nasal cavity, heart, kidney and ovary. In the liver, the fluorescence intensity of FMNPs was strongest and distributed throughout the whole organ. In the spleen and testis, however, FMNPs were observed in specific regions. The brain tissue sections were inspected by MRI. In T_2-weighted spin-echo magnetic resonance (MR) images, it could be seen that FMNPs had penetrated the BBB. In fact, extensive size-dependent inhalation toxicities of various nanomaterials are under investigation. The differences in toxicity depending upon exposure routes (i.p. vs. inhalation) are quite interesting. Most likely, after i.p. administration, the first-pass effects by the portal circulation would account for the difference because the majority of i.p. administered FMNPs would be taken up by the liver *via* the first-pass effect and then be redistributed from the liver to other organs. However, inhalation exposure route would by-pass such liver-related first-pass effect. Kwon et al. studied the organs which are enriched with the reticuloendothelial system (RES), such as the liver, lung and spleen and non-RES organs such as the heart and kidney. Among the RES organs, the spleen was affected by inhalation of FMNPs, thereby suggesting that the body distribution of the FMNPs in the liver, lung and spleen might not be associated with the RES system. These results strongly suggest that other factors may be involved for tissue-specific distribution pattern of inhaled FMNPs. Among the unique properties of the endothelial cells forming the BBB is the presence of tight junctions between cells, where the gaps between the junctions are approximately 4 nm[46,47]. To gain entry to the brain, FMNPs would probably have to pass through the cell membrane of the endothelial cells of the brain, rather than between the endothelial cells. The penetration of any molecules into the brain is largely related to their lipid solubility and their ability to pass through the plasma membranes of cells forming the barrier[48]. However, FMNPs do not show this ability because they are highly water soluble. Moreover, the brain cells lack pinocytosis; therefore, FMNPs would have to gain access to the brain by some other routes. In the mature central nervous system, the spinal and autonomic ganglia as well as a small number of other sites within the brain, called the circumventricular organs, are not protected by the BBB[49]. The discontinuity of the barrier may allow the entry of FMNPs into the brain. Another possibility is that FMNPs may translocate along the olfactory nerve into the olfactory bulb. In fact, a recent report showed that translocation of inhaled nanoparticles along the neurons seemed to be more efficient pathway to the central nervous system (CNS) rather than *via* the BBB[50]. Such efficient translocation of FMNPs *via* the olfactory neuronal pathway in the nasal cavity could be responsible for the observed brain distribution of FMNPs. Moreover, FMNPs also penetrated the testis, which is protected by a blood–testis barrier (BTB). The production, differentiation and presence of male gametes represent inimitable challenges to the immune system as the existence of BTB protects the immune privilege of the testis[51]. In this regard, FMNPs may have penetrated BTB, thus disturbing the maintenance of the immune-privileged status of the testis. Since it is reported that inhaled FMNPs have penetrated the BBB and BTB, further studies are urgently needed to elucidate the precise mechanism by which FMNPs penetrate the BBB and the BTB and the potential outcome for the distribution of FMNPs in the brain and testis.

22.5 FUTURE PROSPECTIVE

As a result of their specific physico-chemical properties, it is expected that MNPs may interact with human matrixes, such as proteins, lipids, carbohydrates and nucleic acids. Innovative and interdisciplinary researches leading to novel risk assessment strategies for the MNPs should be encouraged, while maintaining or improving the current level of protection. Also, the development of validated testing methods would help to address specific nanotoxicological data gaps. It is recommended that MNPs must be toxicologically studied as new materials, and the toxicological protocols must match the proposed clinical application. Due to the rapid development of new MNPs with novel properties, it is very difficult to build a unique toxicity protocol convenient for all types of MNPs. In addition, future researches must focus on the effect of magnetic properties of MNPs beside their nanosize effects. It must be underlined that MNPs can be cleverly tailored to reduce or even to remove certain toxic effects, adding an effective coating is one of the possible solutions.

To step forward to clinical applications, researchers should be encouraged to design *in vivo* protocols using large animal models. *In vitro* researches, though easier to do, may not be sufficient to fulfil the needs to move for human trials.

In the coming years, it is anticipated that if their safety becomes well proven, MNPs would solve many life threatening conditions with low cost, minimally invasive as readily available remedies.

REFERENCES

1. Amdur, M.O., Doull, J., and Klaassen. 1991. *Casarett and Duoll's Toxicology: The Basic Science of Poisons*, 4th edn. Pergamon Press, New York.
2. Hays, H.W. 1986. *Society of Toxicology History*. Society of Toxicology, Washington, DC.
3. Gallo, M.A. and Duoll, J. 1991. History and scope of toxicology. In *Casarett and Doull's Toxicology*, 4th edn., Chapter 1. Amdur, Doull, Klaassen (eds.), Pergamon Press, New York, p15.
4. Gad, S.C. 2001. *Regulatory Toxicology*. Taylor & Francis, New York.
5. Borm, P.J. 2002. Particle toxicology: From coal mining to nanotechnology. *Inhalation Toxicology*, 14, 311–324.

6. Handy, R.D. and Shaw, B.J. 2007. Toxic effects of nanoparticles and nanomaterials: Implications for public health, risk assessment and the public perception of nanotechnology. *Health, Risk & Society*, 9(2), 125–144.

7. EEC Annex L 383A. 1992. Methods for the determination of physico-chemical properties, toxicity and ecotoxicity. *Official Journal of the European Community*, 35(L383A), A5–A233.

8. ISO 1996. Determination of the acute lethal toxicity of substances to a freshwater fish. [Brachydanio rerio Hamilton-Buchanan (Teleostei, Cyprinidae)]. Part 3: Flow-through method (ISO 7346-3:1996). International Standards Organisation, Geneva, Switzerland.

9. Verdier, F. 2002. Non-clinical vaccine safety assessment. *Toxicology*, 174, 37–43.

10. Lead, J.R. and Wilkinson, K.J. 2006. Aquatic colloids and nanoparticles: Current knowledge and future trends. *Environmental Chemistry*, 3, 159–171.

11. Handy, R.D. and Eddy, F.B. 2004. Transport of solutes across biological membranes in eukaryotes: An environmental perspective. In H.P. van Leeuwen and W. Köster (eds.) *Physicochemical Kinetics and Transport at Chemical–Biological Interphases*, IUPAC series. John Wiley, Chichester, U.K., pp. 337–356.

12. TGD-Part I, Technical guidance document on risk assessment. Part I. European Chemicals Bureau, European Commission Joint Research Centre, Ispra, Italy, 2003.

13. Oberdorster, G., Oberdorster, E., and Oberdorster, J. 2005. Nanotoxicology: An emerging discipline evolving from studies of ultrafine particles. *Environmental Health Perspectives*, 113, 823–839.

14. Nel, A., Xia, T., Madler, L., and Li, N. 2006. Toxic potential of materials at the nano level. *Science*, 311, 622–627.

15. Hallock, M., Greenley, P., DiBerardinis, L., and Kallin, D. 2009. Potential risks of nanomaterials and how to safely handle materials of uncertain toxicity. *Journal of Chemical Health and Safety*, 16, 16–23.

16. Oberdorster, G., Ferin, J., and Lehnert, B. 2004. Correlation between particle size, in vivo particle persistence, and lung injury. *Environmental Health Perspectives*, 102(Suppl. 5), 173.

17. Warheit, D.B., Webb, T.R., Reed, K.L., Frerichs, S., Christie, M., Sayes, C.M. 2007. Pulmonary toxicity study in rats with three forms of ultrafine-TiO_2 particles: Differential responses related to surface properties. *Toxicology*, 230, 90–104.

18. Gornati, R., Papis, E., Rossi, F., Raspanti, M., Dalle-Donne, I., Colombo, G., Milzani, A., and Bernardini, G. 2009. Engineered cobalt oxide nanoparticles readily enter cells. *Toxicology Letters*, 189, 253–259.

19. Lam, C.W., James, J.T., McCluskey, R., and Hunter, R.L. 2004. Pulmonary toxicity of single-wall carbon nanotubes in mice 7 and 90 days after intra tracheal instillation. *Toxicology Sciences*, 77, 126–134.

20. Maynard, A.D., Baron, P.A., and Foley, M. 2004. Exposure to carbon nanotube material: Aerosol release during the handling of unrefined single-walled carbon nanotube material. *Journal of Toxicology and Environmental Health, Part A: Current Issues*, 67, 87–107.

21. Pankhurst, Q.A., Connolly, J., Jones, S.K., and Dobson, J. 2003. Applications of magnetic nanoparticles in biomedicine. *Journal of Physics D: Applied Physics*, 36, R167.

22. Gad, S.C. and Kapis, M.B. 1993. Non-animal techniques in biomedical and behavioral research and testing. Lewis, Ann Arbor, MI.

23. Hakkinen, P.J. and Green, D.K. 2002. Alternatives to animal testing: Information resources via the internet and World Wide Web. *Toxicology*, 173, 3–11.

24. Huggins, J. 2003. Alternatives to animal testing: Research, trends, validation, regulatory acceptance. *ALTEX (Alternativen zu Tierexperimenten)*, 20(Suppl. 1), 3–61.

25. Meyer, O. 2003. Testing and assessment strategies, including alternative and new approaches. *Toxicology Letters*, 140–141, 21–30.

26. Parkes, L.M., Hodgson, R., Lu, L.T., Tung, L.D., Robinson, I., Fernig, D.G., and Thanh, N.T. 2008. Cobalt nanoparticles as a novel magnetic resonance contrast agent-relativities at 1, 5 and 3 tesla. *Contrast Media & Molecular Imaging*, 3, 150–156.

27. Colognato, R., Bonelli, A., Ponti, J., Farina, M., Bergamaschi, E., Sabbioni, E., and Migliore, L. 2008. Comparative genotoxicity of cobalt nanoparticles and ions on human peripheral leukocytes *in vitro*. *Mutagenesis*, 23, 377–382.

28. Häfeli, U.O., Riffle, J.S., Harris-Shekhawat, L., Carmichael-Baranauskas, A., Mark, F., Dailey, J.P., and Bardenstein, D. 2009. Cell uptake and *in vitro* toxicity of magnetic nanoparticles suitable for drug delivery. *Molecular Pharmaceutics*, 6(5), 1417–1428.

29. Wyllie, A.H., Kerr, J.F.R., and Currie, A.R. 1980. Cell death: The significance of apoptosis. *International Review of Cytology*, 68, 251–306.

30. Zange, R., Li, Y., and Kissel, T. 1998. Biocompatibility testing of ABA triblock copolymers consisting of poly (L-lactic-*co*-glycolic acid) blocks attached to a central poly (ethylene oxide) B block under in vitro conditions using different L929 mouse fibroblasts cell culture models. *Controlled Release*, 56, 249–258.

31. Park, S.I., Lim, J.H., Kim, J.H., Yun, H.I., Roh, J.S., Kim, C.G., and Kim, C.O. 2004. Effects of surfactant on properties of magnetic fluids for biomedical application. *Physica Status Solidi (b)*, 241, 1662–1664.

32. Bi, H., Song, M., Song, W., Wang, J., Wu, W., Sun, J., and Yu, M. 2010. Cytotoxicity and cellular uptake of iron nanowires. *Biomaterials*, 31, 1509–1517.

33. Donaldson, K. and Tran, C.L. 2004. An introduction to the short-term toxicology of respirable industrial fibres. *Mutation Research*, 53, 5–9.

34. Magrez, A., Kasas, S., Salicio, V., Pasquier, N., Seo, J.W., and Celio, M. 2006. Cellular toxicity of carbon-based nanomaterials. *Nano Letters*, 6, 1121–1125.

35. Akhtar, A., Siddiqui, M., Ahmad, J., Musarrat, J., Al-Khedhairy, A., AlSalhi, M., and Alrokayan, S. 2011. Oxidative stress mediated apoptosis induced by nickel ferrite nanoparticles in cultured A549 cells. *Toxicology*, 283, 101–108.

36. Paz-Elizur, T., Sevilya, Z., Leitner-Dagan, Y., Elinger, Y., Roisman, L.C., and Livneh, Z. 2008. DNA repair of oxidative DNA damage in human carcinogenesis: Potential application for cancer risk assessment and prevention. *Cancer Letter*, 266, 60–72.

37. Ahamed, M., Posgai, R., Gorey, T.J., Nielsen, M., Hussain, S., and Rowe, J. 2010. Silver nanoparticles induced heat shock protein 70, oxidative stress and apoptosis in *Drosophila melanogaster*. *Toxicology and Applied Pharmacology*, 242, 263–269.

38. Ahamed, M., Siddiqui, M.A., Akhtar, M.J., Ahmad, I., Pant, A.B., and Alhadlaq, H.A. 2010. Genotoxic potential of copper oxide nanoparticles in human lung epithelial cells. *Biochemical and Biophysical Research Communications*, 396, 578–583.

39. Häfeli, U.O. and Pauer, G.J. 1999. In vitro and in vivo toxicity of magnetic microspheres. *Journal of Magnetic Materials*, 194, 76–82.

40. Peters, K., Unger, R.E., Kirkpatrick, C.J., Gatti, A.M., and Monari, E. 2004. Effects of nano-scaled particles on endothelial cell function in vitro: Studies on viability, proliferation and inflammation. *Journal of Materials Science: Materials in Medicine*, 15, 321–325.

41. Häfeli, U.O., Aue, J., and Damani, J. 2007. The biocompatibility and toxicity. In Magnetic Cell Separation. M. Zborowski and J.J. Chalmers (eds.) Elsevier, Amsterdam, the Netherlands, pp. 163–223.

42. Cho, M.H., Kim, J.S., Yoon, T.J., Yu, K.N., Kim, B.G., Park, S.J., Kim, H.W., Lee, K.H., Park, S.B., and Lee, J.K. 2006. Toxicity and tissue distribution of magnetic nanoparticles in mice. *Toxicological Sciences*, 89(1), 338–347.

43. Semmler, M., Seitz, J., Erbe, F., Mayer, P., Heyder, J., Oberdörster, G., and Kreyling, W.G. 2004. Long-term clearance kinetics of inhaled ultrafine insoluble iridium particles from the rat lung, including transient translocation into secondary organs. *Inhalation Toxicology*, 16, 453–459.

44. Milner, Y., Zioni, T., Perkas, N., Wolfus, Y., Soroka, Y., Popov, I., Oron, M., Perelshtein, I., Bruckental, Y., Bregegere, F.M., Ma'or, Z., Gedanken, A., Yeshurun, Y., and Neuman, R. 2010. Strontium hexaferrite nanomagnets suspended in a cosmetic preparation: A convenient tool to evaluate the biological effects of surface magnetism on human skin. *Skin Research and Technology*, 16, 316–324.

45. Kwon, J.T., Hwan, S.K., Yoon, H.J., Mansoo, C., Byung-II, Y., Lee, J.K., Hua, J., Tae-Jong, Y., Kim, D.S., Duk-Young, H., Arash-Minai, T., Young-Woon, K., and Myung-Haing, C. 2008. Body distribution of inhaled fluorescent magnetic nanoparticles in the mice. *Journal of Occupational Health*, 50, 1–6.

46. Del Zoppo, G.J., Milner, R., Mabuchi, T., Hung, S., Wang, X., and Koziol, J.A. 2006. Vascular matrix adhesion and the blood–brain barrier. *Biochemistry Society Transactions*, 34, 1261–1266.

47. Stewart, P.A. 2000. Endothelial vesicles in the blood–brain barrier: Are they related to permeability? *Cell Molecular Neurobiology*, 20, 149–163.

48. Yonezawa, T., Ohtsuka, A., Yoshitaka, T., Hirano, S., Nomoto, H., Yamamoto, K., and Ninomiya, Y. 2003. A novel immunoglobulin superfamily protein localized to glia limitans formed by astrocyte end feet. *Glia*, 44, 190–204.

49. Oberdorster, G., Sharp, Z., Atudorei, V., Elder, A., Gelein, R., Kreyling, W., and Cox, C. 2004. Translocation of inhaled ultrafine particles to the brain. *Inhalation Toxicology*, 16, 437–445.

50. Elder, A., Gelein, R., Silva, V., Feikert, T., Opanashuk, L., Carter, J., Potter, R., Maynard, A., Ito, Y., Finkelstein, J., and Oberdorster, G. 2006. Translocation of inhaled ultrafine manganese oxide particles to the central nervous system. *Environmental Health Perspectives*, 114, 1172–1178.

51. Fijak, M. and Meinhardt, A. 2006. The testis in immune privilege. *Immunological Reviews*, 213, 66–81.

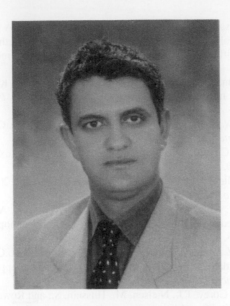

Taher A. Salah, PhD, is the head of Nanotechnology Centre and leads the bionanotechnology group in Agricultural Research Centre, Giza, Egypt. PhD in toxicology and safety of nanomaterials. His main research interests are the design and engineering of nanomaterials for bionanotechnology applications and their toxicological evaluations.

Hazem Mohamed Saleh, MD, DEA, is an associate professor of otorhinolaryngology, National Institute of Laser, Cairo University. Fellow in Head and Neck Oncologic Surgery, Gustave Roussy Institute, Paris, France. His main research interests are the use of safe nanomaterials in cancer diagnosis and therapy.

Mahmoud H. Abdel Kader, PhD, is the president of German University in Cairo. He leads the photodynamic therapy group in GUC and Cairo University, Cairo, Egypt. His main research areas of interests are photochemistry and photobiology. His current research interests are non-oncological application of nanomaterials and its environmental feedback.

Index